Soft Matter approaches to Structured Foods

Hof van Wageningen, the Netherlands
2–4 July 2012

FARADAY DISCUSSIONS

Volume 158, 2012

RSC Publishing

The Faraday Division of the Royal Society of Chemistry, previously the Faraday Society, founded in 1903 to promote the study of sciences lying between Chemistry, Physics and Biology.

EDITORIAL STAFF

Editor
Philip Earis

Deputy editor
Jane Hordern

Development editor
Heather Montgomery

Senior publishing editor
Anna Pendlebury

Publishing editors
Rachel Cooper, Helen Gray, Tegan Thomas

Publishing assistants
Aliya Anwar, Ella Mitchell, Claire Sissen

Publisher
Niamh O'Connor

Faraday Discussions (Print ISSN 1359-6640, Electronic ISSN 1364-5498) is published 6 times a year by the Royal Society of Chemistry, Thomas Graham House, Science Park, Milton Road, Cambridge, UK CB4 0WF. Volume 158 ISBN-13: 978-1-84973-449-3

2012 annual subscription price: print+electronic £709, US $1,322; electronic only £673, US $1,256. Customers in Canada will be subject to a surcharge to cover GST. Customers in the EU subscribing to the electronic version only will be charged VAT. All orders, with cheques made payable to the Royal Society of Chemistry, should be sent to RSC Distribution Services, c/o Portland Customer Services, Commerce Way, Colchester, Essex, UK CO2 8HP.
Tel +44 (0) 1206 226050;
E-mail sales@rscdistribution.org

If you take an institutional subscription to any RSC journal you are entitled to free, site-wide web access to that journal. You can arrange access *via* Internet Protocol (IP) address at www.rsc.org/ip. Customers should make payments by cheque in sterling payable on a UK clearing bank or in US dollars payable on a US clearing bank.

US Postmaster: send address changes to *Faraday Discussions*, c/o Mercury Airfreight International Ltd., 365 Blair Road, Avenel, NJ 07001. All despatches outside the UK by Consolidated Airfreight.

PRINTED IN THE UK

Faraday Discussions documents a long-established series of *Faraday Discussion* meetings which provide a unique international forum for the exchange of views and newly acquired results in developing areas of physical chemistry, biophysical chemistry and chemical physics.

Soft Matter Approaches to Structured Foods

Faraday Discussions

www.rsc.org/faraday_d

A General Discussion on Soft Matter Approaches to Structured Foods was held in Wageningen, The Netherlands on 2nd, 3rd and 4th July 2012.

RSC Publishing is a not-for-profit publisher and a division of the Royal Society of Chemistry. Any surplus made is used to support charitable activities aimed at advancing the chemical sciences. Full details are available from www.rsc.org

CONTENTS

ISSN 1359-6640; ISBN 978-1-84973-449-3

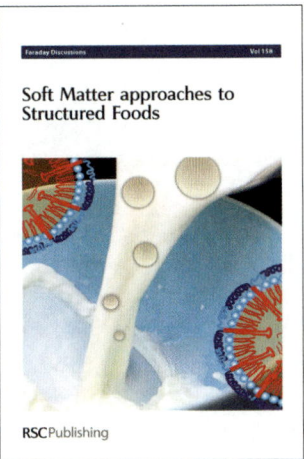

Soft Matter approaches to
Structured Foods

RSCPublishing

Cover
Soymilk oleosomes or oil bodies, a typical example of food microstructures and natural emulsions (photo by Natalie Russ, oleosome figure by David Baßler and Thomas A. Vilgis).

Image reproduced by permission of Mr Gustav Waschatko from *Faraday Discuss.*, 2012, **158**, 157.

INTRODUCTORY LECTURE

9   **Soft matter approaches to structured foods: from "cook-and-look" to rational food design?**
Job Ubbink

PAPERS AND DISCUSSIONS

37   **Designing colloidal structures for micro and macro nutrient content and release in foods**
David A. Garrec, Sarah Frasch-Melnik, John V. L. Henry, Fotis Spyropoulos and Ian T. Norton

51   **Protein cluster formation during enzymatic cross-linking of globular proteins**
Yunus Saricay, Surender Kumar Dhayal, Peter Alexander Wierenga and Renko de Vries

65   **Anomalies in moisture transport during broccoli drying monitored by MRI?**
Xin Jin, Antonius J. B. van Boxtel, Edo Gerkema, Frank J. Vergeldt, Henk Van As, Gerrit van Straten, Remko M. Boom and Ruud G. M. van der Sman

The Graduate School

VLAG

77   **Structural changes of deposited casein micelles induced by membrane filtration**
     R. Gebhardt, T. Steinhauer, P. Meyer, J. Sterr, J. Perlich and U. Kulozik

89   **Model for particle migration in bidisperse suspensions by use of effective temperature**
     H. M. Vollebregt, R. G. M. van der Sman and R. M. Boom

105 **General discussion**

125 **Stability of aqueous food grade fibrillar systems against pH change**
     Ardy Kroes-Nijboer, Hassan Sawalha, Paul Venema, Arjen Bot, Eckhard Flöter,
     Ruud den Adel, Wim G. Bouwman and Erik van der Linden

139 **Quinoa starch granules as stabilizing particles for production of Pickering emulsions**
     Marilyn Rayner, Malin Sjöö, Anna Timgren and Petr Dejmek

157 **Soy milk oleosome behaviour at the air–water interface**
     Gustav Waschatko, Ann Junghans and Thomas A. Vilgis

171 **Critical laminar shear-temperature effects on the nano- and mesoscale structure of a model fat
     and its relationship to oil binding and rheological properties**
     Nuria C. Acevedo, Jane M. Block and Alejandro G. Marangoni

195 **Surface shear rheology of hydrophobin adsorption layers: laws of viscoelastic behaviour with
     applications to long-term foam stability**
     Krassimir D. Danov, Gergana M. Radulova, Peter A. Kralchevsky,
     Konstantin Golemanov and Simeon D. Stoyanov

223 **Elucidation of density profile of self-assembled sitosterol + oryzanol tubules with small-angle
     neutron scattering**
     Arjen Bot, Elliot P. Gilbert, Wim G. Bouwman, Hassan Sawalha, Ruud den Adel,
     Vasil M. Garamus, Paul Venema, Erik van der Linden and Eckhard Flöter

239 **General discussion**

267 **New routes to food gels and glasses**
     Thomas Gibaud, Najet Mahmoudi, Julian Oberdisse, Peter Lindner, Jan Skov Pedersen,
     Cristiano L. P. Oliveira, Anna Stradner and Peter Schurtenberger

285 **Protein structure and interactions in the solid state studied by small-angle neutron scattering**
     Joseph E. Curtis, Arnold McAuley, Hirsh Nanda and Susan Krueger

301 **The role of quench rate in colloidal gels**
     C. Patrick Royall and Alex Malins

313 **Delayed solidification of soft glasses: new experiments, and a theoretical challenge**
     Yogesh M. Joshi, A. Shahin and Michael E. Cates

325 **Slow dynamics and structure in jammed milk protein suspensions**
     Peggy Thomar, Dominique Durand, Lazhar Benyahia and Taco Nicolai

341 **Arrested coalescence of viscoelastic droplets with internal microstructure**
     Amar B. Pawar, Marco Caggioni, Richard W. Hartel and Patrick T. Spicer

351 **General discussion**

371 **Viscoelastic phase separation in soft matter and foods**
     Hajime Tanaka

407 **Kinetic model for the mechanical response of suspensions of sponge-like particles**
     Markus Hütter, Timo J. Faber and Hans M. Wyss

425   **Nanoscale characteristics of triacylglycerol oils: phase separation and binding energies of two-component oils to crystalline nanoplatelets**
Colin J. MacDougall, M. Shajahan Razul, Erzsebet Papp-Szabo, Fernanda Peyronel, Charles B. Hanna, Alejandro G. Marangoni and David A. Pink

435   **Soft matter approaches as enablers for food macroscale simulation**
Ashim K. Datta, Ruud van der Sman, Tushar Gulati and Alexander Warning

461   **Numerical study of the effect of thiol–disulfide exchange in the cluster phase of β-lactoglobulin aggregation**
Rosanne N. W. Zeiler and Peter G. Bolhuis

479   **A multiscale approach to triglycerides simulations: from atomistic to coarse-grained models and back**
Antonio Brasiello, Silvestro Crescitelli and Giuseppe Milano

493   **General discussion**

CONCLUDING REMARKS

523   **Concluding remarks: the future of soft matter and food structure**
C. G. (Kees) de Kruif

ADDITIONAL INFORMATION

529   **Poster titles**
531   **List of participants**
533   **Index of contributors**

Soft matter approaches to structured foods: from "cook-and-look" to rational food design?

Job Ubbink[*ab]

Received 7th September 2012, Accepted 10th September 2012
DOI: 10.1039/c2fd20125a

Developments in soft matter physics are discussed within the context of food structuring. An overview is given of soft matter-based approaches used in food, and a relation is established between soft matter approaches and food technology, food creation, product development and nutrition. Advances in food complexity and food sustainability are discussed from a physical perspective, and the potential for future developments is highlighted.

Introduction

Over the last few decades, the application of physics, physical chemistry and materials science in food science and technology has experienced a rapid expansion. Foods, food ingredients and their transformations are increasingly being considered from the perspective of soft matter science and the awareness that various developments in fundamental physics are relevant for food development is on the rise. This includes aspects relating to food processing, storage and consumption and the physiology of digestion.[1-6]

While highlighting the emphasis in food development on structure building, the notion of "structured foods" is in fact a pleonasm. Not only are all foods structured, in a continuum of length scales from the molecular level up to the macroscopic scale of a food item, but the functional properties of a foodstuff also critically depend on at least one characteristic length scale.

The central contribution of soft matter science to food science and technology lies in the identification of relations between the critical elements of food structure and the associated physical phenomena on the one hand, and food functionality on the other hand. Potentially, soft matter approaches, including food materials science[7,8] and the physical chemistry of foods,[9] could thus strongly shape developments in the food field. For various reasons, however, the amalgamation of soft matter physics and food science and technology has not always been an easy process.

In the first place, in most foods, structure development by human intervention comes on top of a complex structure as laid down by nature. To have an impact on food development, soft matter science should therefore be able to deal with both nature- and man-induced structures and related physical properties. In this context, one should also not forget that, throughout human history, man has successfully manipulated the structure and properties of both plant and animal foods by systematic, biology-based approaches such as selective breeding.

In the second place, a central characteristic of foods is that they are complex. Foods display both a complex structure and a complex composition. Moreover, foods originate from a wide range of plant and animal sources, with major geographical

[a]Food Concept & Physical Design "The Mill", Mühleweg 10, CH-4112 Flüh, Switzerland. E-mail: job.ubbink@themill.ch; Tel: +41 61 271 12 51
[b]H.H. Wills Physics Laboratory, University of Bristol, Tyndall Avenue, Bristol BS8 1TL, United Kingdom

variations, and the foods as we currently know them are strongly shaped by long-standing empirical approaches. Furthermore, the appreciation of food is strongly determined by cultural preferences, which vary enormously over the world. The challenge to soft matter physicists is thus considerable, as the different complexity layers turn food into a rather unwieldy subject for fundamental physical investigation.

Finally, it does not make much sense to study a food as a static item; its transformation by cooking, processing or other preparation methods is an essential part of the game, as is the breakdown during consumption and digestion of the food product. Consequently, one cannot truly decouple the physics from the technology and craftmanship, from the physiology, or even from the behavioral sciences.

This introductory lecture is not intended as a general review of the field; its aim is rather to present a broad, personal sketch of the relation between physics and food. In this, I will often emphasize the food context rather than the physics, as I believe that increased awareness of elements relating to food preparation and processing, consumption, nutrition and sustainability may provide further incentive to the development of soft matter science of food and enhance its impact. Central to my discourse will be the following questions with a bearing on soft matter approaches in the food context:

1. How did – accidental or purposeful – efforts to food structuring impact what we eat, and how we eat?

2. What is the impact of soft matter science on the structuring of foods, prepared industrially as well as artisanally?

3. Does one need a dedicated physical science to deal with food-related issues? To which extent can principles, concepts and even quantitative results be transferred from more fundamental physical disciplines?

4. What future impact may we expect from soft-matter approaches towards food structuring? How will it impact food creation? Will it help in resolving societal challenges related to food, in particular those with a bearing on nutrition, health, and the sustainability of food production and consumption?

Even though in the following sections, I will regularly return to these questions, I will not be able to provide adequate answers but to a few loose elements. I, however, trust that in the remainder of this Discussion meeting, light is shed on some of the important issues in the soft matter science of food and nutrition, and that novel avenues are presented to advance our understanding of the physics of foods.

Soft matter approaches

Even if one limits oneself to the purely physical aspects of food, one will find that the field of soft matter physics of foods is very wide. This is because, in foods, most of the phases and states encountered in soft materials occur. Foods are particularly rich in examples of soft matter, comprising both phases exhibiting long-range order as well amorphous states. Central to the soft matter physics of food is that foods exhibit a complex free energy landscape (Fig. 1), which is characterized by many shallow free energy minima. Furthermore, it almost invariably turns out that in their most desirable form, foods are metastable: they either occupy a local free energy minimum, or are quenched into a state of fairly high free energy, but slow dynamics.

In foods, phase transitions thus readily occur, induced by thermal fluctuations or driven by external perturbations. The complex but fairly flat energy landscape characteristic of foods conveniently allows for a multitude of transformations; a fact that is thoroughly exploited in both food technology and in cooking. The metastability of many foods conversely renders them highly susceptible to undesired physical transformations, leading to a decay of food quality.

Foods do not only change their properties by phase transitions. Even in quenched or otherwise constrained systems, many physical parameters may more or less quickly relax to a state of local equilibrium when subjected to an external perturbation, such as a change in water content or relative humidity. In this sense, the soft matter physics of food is similar to other fields, such as the statistical mechanics of biopolymers (Table 1).

The rich phase behavior of foods results from a complexity in composition and a large variation in the physicochemical properties of food constituents, such as shape, (chain) stiffness, propensity to form hydrogen bonds and electric charge.[9] In turn, the phase behavior of foods and food ingredients may be used to create a multitude of mesoscopic and macroscopic structures in food.[1-6] In addition, food structures are often induced by templating. In particular in food ingredients structured by nature, various metabolism-driven biophysical mechanisms actively build highly complex food structures, such as plant cell walls and cellular assemblies, which by the thermodynamics of the constituents themselves would not materialize.

Table 1 Statistical mechanics of constrained systems

	DNA supercoiling	Water in carbohydrate glasses
Identify major physical phenomena	• Chain bending & torsion • Electrostatic interactions; • Chain entropy	• Osmotic elasticity of polymer glass • Hydrogen bonding • Mixing entropy
Coarse graining		water transfer $\tau_{CHO} \to \infty; \tau_{H_2O}$ finite $\quad \tau_{CHO} \approx \tau_{H_2O}$
Constraints	Topology: $\Delta Lk = \Delta Tw + Wr$	Physics: $T_g(\phi_w)$
Variables that attain equilibrium within constrained environment	superhelical radius r; linked variables p, a	water activity a_w
Minimal model	$F = F_{bending} + F_{torsion} + F_{pert}$ $F_{pert} = F_{et.stat} + F_{conf}$	• *Glass*: Sorption on heterogeneous sites • *Rubber*: Flory-Huggins solution theory
Check for consistency	• Limiting behavior towards tight supercoiling • Assumptions on parameter ranges	• Limiting behavior towards $a_w \to 0$; $a_w \to 1$
Validation	Comparison to experimental data and computer simulations	• Fitting to sorption data • Comparison to $T_g(a_w)$ by DSC • Global agreement with spectroscopic data
Refinements	Sequence-dependent bending	• Carbohydrate polydispersity • Aging of carbohydrate glass
Limits	• $\Delta Lk \to 0$ • Multivalent ions	• Prediction of dynamic phenomena • Prediction of details of molecular interaction

As mentioned above, it is not the aim of this introductory lecture to review the entire food-related soft condensed matter field; as far as I am aware, there are in fact no papers or books covering the whole extent of the field, although ref. 9 comes close, but it is intended as a textbook rather than a review of the scientific literature. Ref. 6 provides extensive coverage of research themes and researchers, but limits the discussion to generic aspects of soft condensed matter physics while stressing the potential for future advances. Specialized reviews and treatises, dealing for instance with food materials science,[8] emulsions and emulsifiers,[10] biopolymers,[11] phase transitions in foods[12,13] and the glassy state in foods,[13,14] are available, but do not aim to provide an integrated perspective on the whole of the field. Illustrative for the broad range of topics grouped under the umbrella of soft matter physics of food and the lack of internal coherence is that the available reviews rely, to a significant extent, on specific examples and case studies.[1-5] This introductory lecture, even more schematic in nature, will not be an exception in this respect.

The soft matter physics of food is faced with a dichotomy. One the one hand, in food science and technology, the focus is on a specific foodstuff. Soft matter physics on the other hand is departing from conjecture that the behavior of a soft matter system may be explained on the basis of generic physical mechanisms. As examples for the two approaches we may take bread as a specific food system (Fig. 1), and slow dynamics in relation to the glass transition as a generic physical phenomenon whose significance extends far beyond any material-specific system (Fig. 1).

For a specific food system such as bread and its intermediate state of dough, the physics should reflect as closely as possible the specific behavior of the product and its intermediates. The reductionist's approach is then to investigate whether the behavior of the whole system can be explained based on the understanding of the properties of the isolated components and their interactions. This means that dough and bread need to be considered as composite materials whose behavior depends on the physics of the main constituents, starch and gluten (Fig. 1). Interestingly, it is principally at the level of the individual components and their interactions that we encounter the physical concepts known from generic fields such as polymer physics (Fig. 1).

If we approach the food field from a physical perspective, one may attempt to find specific examples in which a generic physical model or concept may find application. For the case of the glass transition, a fundamental concept such as the energy landscape characteristic of supercooled fluids[15] may guide experimental and theoretical investigations of specific food-related model systems, such as jammed casein suspensions[16] as an intermediate step in the formation of casein gels, and the water-content dependent glass transition in amorphous carbohydrate matrices (Fig. 2).[7,12,13] Analyses at these more generic levels serve as a basis for the understanding of the behavior of food systems, such as milk powder (Fig. 1). One finds, however, that the complexities of the actual food system often require both a translation of key physical concepts, and a broadening of the scientific basis. In the case of milk powders, it is, for instance, useful to introduce the collapse temperature of the powder matrix as a glass transition-related parameter, and to extend the basis to incorporate chemical effects such as the Maillard reaction.[17]

Elements of the two approaches can be traced back, when looking at some of the important historical developments in the field. First, I will have a look at the scientific developments that led to the establishment of the so-called "stability map", as it has resulted in significant advances in the understanding of the relation between the thermodynamic state of amorphous food matrices and the kinetics of various physical, chemical and biological processes (Fig. 2). The original stability map, published in 1970, was conceived as a synthesis of the, at the time, available experimental results on the dependence on the water activity of the rate of various, mainly deteriorative, processes occurring in foods.[18] The stability map was intended as a general empirical guide, but subsequently inspired considerable amounts of research deepening the understanding of mechanisms behind the individual components of the map (different reactions, and conditions).[12,19-22]

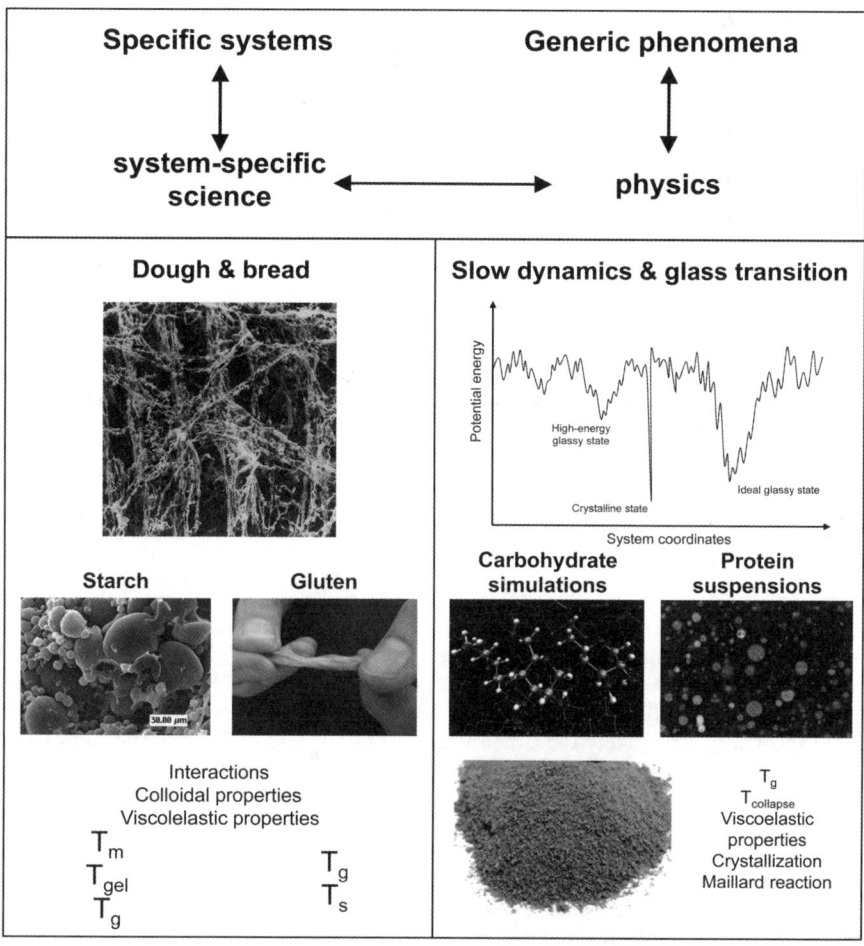

Fig. 1 Food science is traditionally concerned with the study of specific food systems, whereas physics is focused principally at the generic physical phenomena and mechanisms. Both disciplines, however, meet at the interface, as physical properties are needed to describe the detailed behavior of specific food systems. Left hand panel from top to bottom: micrograph of dough, polymer models for the two principal dough components (starch and gluten), physical properties of the combined system, and physical properties of starch (melting temperature T_m, gelatinization temperature T_{gel} and glass transition temperature T_g) and gluten (glass transition temperature T_g and thermosetting temperature T_s). Generic physical concepts, for the example energy landscape (right-hand panel), may serve as a basis to explore the physics of food-related model systems, such as concentrated carbohydrate systems and dense protein suspensions. Physical concepts derived from the two levels of description may be applied in the understanding of the behavior of actual food systems, such as amorphous food powders in the glassy state (collapse temperature $T_{collapse}$). The micrographs of dough and the protein suspension are reproduced with permission from ref. 148 and 16. The free energy diagram is reinterpreted from the original published in ref. 149. The images of the starch granules and the hydrated gluten sample, and the snapshot of the carbohydrate simulation are courtesy of M.-L. Dillmann, L. Forny and H.J. Limbach, respectively.

A major modification of the stability map occurred when, in the 1980s, it was realized that the water activity at the onset of the rapid increase in rate of the various physical, chemical and biological processes was linked to the water content-dependent glass transition of the food matrix (Fig. 2). At the time, the concept of the glass transition temperature had only recently been introduced into the food field.[7,13,14]

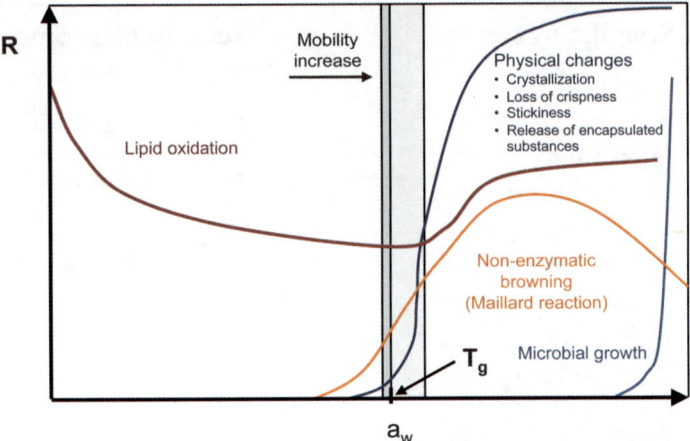

Fig. 2 Stability map for food, summarizing the dependence of the rate of various food deterioration processes on the water activity a_w. R denotes the relative rate at constant temperature, and T_g is the glass transition temperature of the food matrix. The figure is adapted from a figure by M. Karel (MIT).[150]

The late recognition of the importance of a general physical phenomenon such as the glass transition to foods might come as a surprise to physicists. However, as we will see later on, it is not the sole instance where essential physical concepts are adopted in the food field only after a significant delay. For the reasons of these lags we can only speculate; most likely the orientation of physics and of food science and technology are too different to easily bridge the gap between the two disciplines. On the one hand, food, with all its complexity, was for a long time neglected as a field for serious physical investigation. Food science, on the other hand, is a system-specific discipline, and for a long time generic concepts, mechanisms and principles were assumed to yield little in terms of concrete, applicable results.

As a second historical example I will consider the search for substitutes for butter, a typical case of ingredient replacement (Fig. 3). This example is of interest, not only as it has given rise to a major branch of the food industry, but also because physicochemical elements were used relatively early on in the developments. Margarine was invented in the second half of the 19th century as a low-cost alternative for butter.[23] As in butter, margarine consists of a high internal-phase dispersion of partially liquid fat droplets in a continuous aqueous medium, whereby a network of fat crystals covering the droplet surfaces and forming bridges between the fat droplets stabilizes the product matrix (Fig. 3).[24,25]

In its initial conception, the principal ingredients of margarine were lard and milk powder (itself an industrial invention). In the early 20th century, animal fats started to be replaced by (cheaper) vegetable fats. This replacement accelerated with the advent of chemical hardening processes,[23] as the mostly unsaturated fatty acids from vegetable oils are liquid at room temperature, and thus lack the solid lipid fraction needed to stabilize the fat droplets. By partial hydrogenation of the unsaturated vegetable oils, the melting profiles can be made to shift to higher temperatures, allowing a precise control over the solid lipid fraction at a defined temperature.[26]

The new avenue to structuring of lipid foods enabled by the industrial hardening of vegetable oils was for a long time considered to be the alpha and omega of fat structuring.[23] This was further supported from the 1950s onwards by evidence from nutritional and medical studies on comparative health benefits of vegetable fats and oils over fats from animal origin.[24] However, in the early 1990s, it was discovered that the *trans* isomers of the chemically hardened fats pose significant

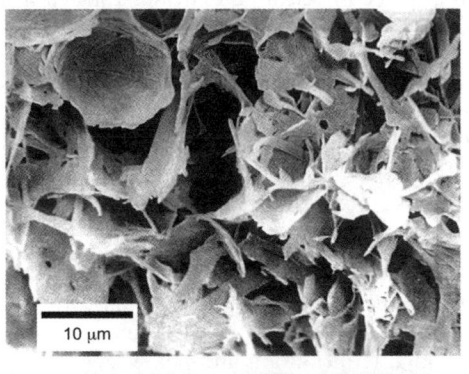

- **1869**: margarine invented as low-cost butter replacement; main ingredients lard and milk powder.
- **Early 20th century**: Hydrogenation (hardening) allows for flexible tuning of solid lipid fraction and use of novel (vegetable) lipid sources
- **1950s onwards**: margarine promoted as healthy substitute for butter
- **1980s**: Discovery of high content of trans fatty acids (TFAs) in hydrogenated fats
- **1990s-onwards**: Elimination of TFAs on going
- Current research on physically-inspired fat structuring to improve hardness

10 µm

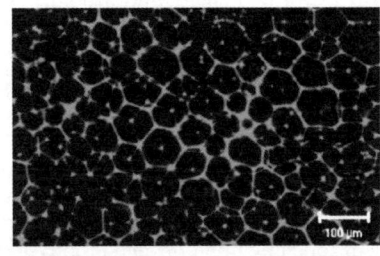

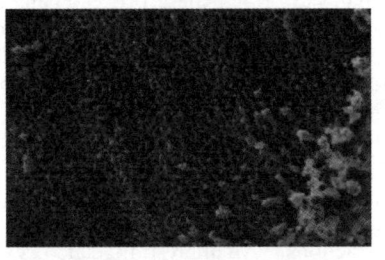

100 µm

Fig. 3 Top left panel: some of the key steps in the invention of margarine and in the development of hardened vegetable fats. Top right panel: Scanning electron micrograph of margarine structure, showing the fat crystal network encapsulating liquid oil in the matrix. Bottom left panel: protein network. Bottom right panel: self-assembled fibrillar organogels based on carnola oil. The micrographs of margarine, the oleogel, and the organogel are reproduced with permission from ref. 24, 31 and 151.

cardiovascular risks,[27] and efforts to eliminate *trans* fatty acids (TFAs) from food have gained momentum since the turn of the century.[24]

From a soft matter perspective, the most interesting avenue to avoid TFAs in lipid-based foods is to physically structure unsaturated vegetable oils (Fig. 3).[28] Even though the specific implementations vary widely, these efforts are based on the concept of structuring the oils using non-lipid additives, such as ethylcellulose[29,30] or proteins.[31]† Either by themselves, or in interaction with the fatty acids, these additives form a scaffold providing solid characteristics to the otherwise liquid oils. When successful, these developments signify an important transition from chemically- to physically-inspired food design.

From ingredients to foods

Whereas in the previous section I have emphasized developments in the food field from the soft matter perspective, in this section I will look at three specific foods ingredients and systems, and examine how and where soft matter approaches come into play. First, I will have a look into dairy-based products to highlight the

---

† An interesting example of the structuring of lipids in a different food system is formed by the use of sugar in chocolate. Whereas in conventional chocolate the sucrose crystals, cocoa particles, and optionally milk powder particles are dispersed in a continuous matrix of cocoa butter,[153] recent developments have attempted to provide a chocolate-like product with reduced content of cocoa butter and increased temperature stability by preparing a continuous phase of sintered sugar crystals trapping the cocoa butter.[154]

importance of scientific understanding in a field based on traditional know how. Second, the development of soluble coffee is highlighted as an example of an industrial food product, whose development was achieved in an, at times, uneasy interaction with physical chemistry and polymer science. Finally, I will look at complex coacervation, as it constitutes an example of a science-driven development to structure food ingredients.

Let us first have a global view on dairy structuring. In a highly simplistic way, I have organized these major dairy processes, as well as some of the principal products relating to these processes, in the categories "farm-based", "kitchen-based", "industry-based" and "science-based" (Table 2). These categories are definitely not absolute, but nevertheless serve as an important reminder on the importance of primordial discoveries and innovations for our today's diet. For instance, processes used in highly optimized form by the modern dairy industry, including curdling, fermentation and phase separation, are routinely used since times immemorial. Naturally, these processes have been transformed since, and are currently shaped by modern scientific and technological developments. The same holds true for most of the processes we associate with kitchen practice, such as whipping, and freezing, but which since have been transformed into highly optimized technologies for the food industry.

Dairy processes based on physical principles include homogenization of the milk fat in milk, implemented in industry in the first half of the 20th century,[32] and, more recently, the removal of bacteria out of raw milk by microfiltration as a soft process to increase the shelf life,[33,34] and the fractionation of butter fat in order to obtain more "hard" and "soft" fat fractions.[35] An interesting related development touching upon protein biophysics is the use of either thermal processes or enzymes to partially hydrolyze and denature casein in infant formula, in order to reduce potential allergenic effects.[36]

The current development most clearly shaped by soft matter science development is likely the self-assembly of β-lactoglobulin in various higher-order structures (Table 2, Fig. 4). The self-assembly of β-lactoglobulin is of particular interest, as β-lactoglobulin is the principal protein in whey, which currently is a waste product from cheese making, but which has significant nutritional qualities. Depending on the conditions, β-lactoglobulin can be made to self-assemble into structures as diverse as micelles (Fig. 4a),[37] wormlike fibers (Fig. 4b)[38] and spherulites (Fig. 4c).[39] Currently the object of scientific study, it will be interesting to see if these

Table 2 Major advances in dairy science and technology, grouped according to the environment in which the developments originally occurred. The overview is highly schematic; only a limited number of examples is shown. The vertical organization of the entries provides a rough chronology of the developments for the industry- and science-based examples

Farm-based	Kitchen-based	Industry-based	Science-based
Curdling (cheese)	Baking (butter in pastries)		
Fermentation (yoghurt)	Whipping (crème chantilly)	Sterilization	Pasteurization
Phase separation (butter; cream)	Freezing (ice cream)	Homogenization	
		Drying (milk powder)	
		Hydrolysis (proteins in infant nutrition)	
		Microfiltration (bacteria)	Isolation (milk oligos)
		Fractionation (milk fat)	Protein self-assembly (β-lactoglobulin)

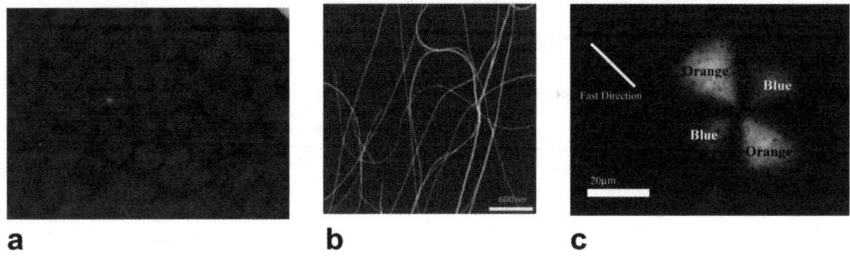

a   **b**   **c**

Fig. 4 Self-assembled structures of β-lactoglobulin: (a) micelles, (b) fibers and (c) spherulites. The micrographs of the micelles, fibers and spherulites are reproduced with permission from ref. 37–39, respectively.

structures will find application in food. A potential use of the micelles could be in the delivery of bioactives, or as a dispersed system to increase the whey content in various foods, while the β-lactoglobulin fibers could have potential in the production of meat replacers. The study of the formation of the spherulites has revealed similarities in formation mechanism with the protein spherulites thought to be responsible for amyloid diseases, and it could well be that the main impact of the understanding of spherulite formation will be in the area of medical biophysics.

An interesting example of a structured food is formed by the interplay of science and technology in the development of soluble coffee (Table 3).[40] For a long time, this product was intractable, as all efforts, dating back to at least the 19th century and aimed to prepare a dehydrated powder from a coffee extract, failed. This was because the resulting soluble coffee powder turned out to be highly unstable, quickly collapsing after manufacture and losing all aroma. Towards the end of the 1920s, efforts at the dairy-and-chocolate company Nestlé started towards the development of soluble coffee, apparently on the basis of a request from Brazil, a country that at the time was confronted with a significant surplus of coffee beans.

Table 3 The interface of science and technology: development of soluble coffee, and some key advances in the understanding of the glass-rubber transition and its application to foods. With hindsight, it is of interest to note that the development of soluble coffee has taken place without reference to the physics of the glassy state, even though part of the science already was in place

Technology	Science
19th–20th century – Numerous efforts to prepare powders from coffee extracts **1929–1932** – Unsuccessful project (supported by Staudinger)	**1920** – Staudinger proposes polymer concept
1935 – Importance of carbohydrates in "aroma fixation" realized **1936** – Stable spraydried powder from coffee extract and maltodextrins **1938** – Launch of Nescafé	**1930s** – T_g determinations for many compounds, incl. glycerol and glucose
1948 – Process modification elimination of added CHOs	**1948** – Kauzmann's paradox
1964 – Freezedried soluble coffee introduced	**1960s–1970s** – a_w approach in foods (Karel, Labuza)
1990s – First application of T_g in soluble coffee field	**1980s–1990s** – physical chemistry approach to foods (T_g based) (Levine and Slade)
2000s – Functionality: nutrition, consumer benefits	

Initial efforts turned out to be unsuccessful, for mainly the same reasons as listed above. However, persistent efforts by one scientist, in part pursued using improvised facilities at home, after some years led to an acceptable product with a sufficiently high physical and chemical stability.[40] A key step in the development was the realization that carbohydrates could help to "fix" the aromas in the product, and the soluble coffee powder, produced by spray-drying technology originally developed for milk powder, for the first years contained maltodextrins as an additive (Table 3).[40]

From a soft matter perspective, the development of soluble coffee is of interest, as it provides an example in which during the, ultimately highly successful, development of an industrially-prepared, structured food, physical considerations were often inappropriately used.[40] For instance, chemically relatively inert carbohydrates such maltodextrins do not bind or fixate aroma compounds; they rather form an amorphous shell encapsulating minute droplets of the aroma compounds in coffee oil.[41]

Moreover, it was not realized for more than half a century that the soluble coffee matrix constitutes, in effect, an amorphous system, the physics of which are governed by the glass transition (Table 3). In fact, it turns out that not only the basic product was developed using "cook-and-look" approaches, but also the later process modification leading to the elimination of the added carbohydrates took place on a purely empirical basis.[40] With hindsight, these process modifications can be interpreted as leading to a higher average molecular weight of the coffee extract, a concomitantly increased glass transition temperature and thus a higher physical stability of the resulting soluble coffee powder.

As a last ingredient-centred example I will have a look at coacervates, and at the process of complex coacervation (Fig. 5).[42] Coacervation is of interest for this discussion, as it is a prime example of a science-driven development (for a concise chronology, see Fig. 5a). In its use in food products, it has seen limited success, as

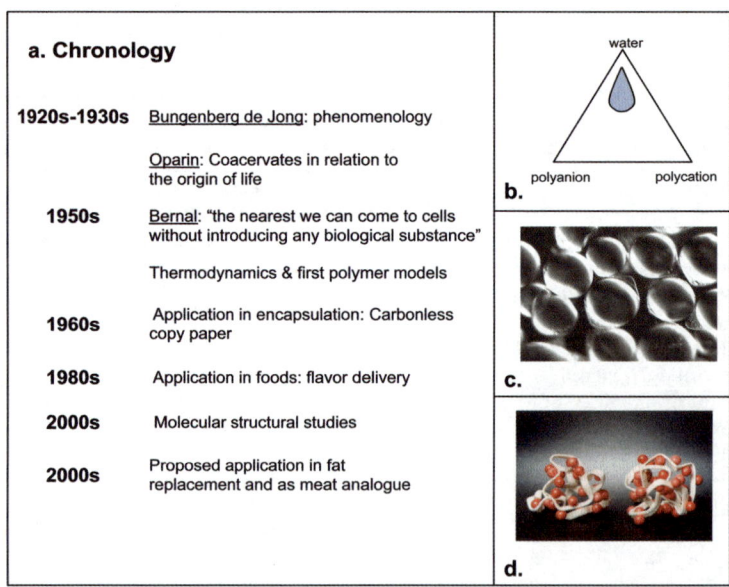

Fig. 5 Some key elements in the science and technology of complex coacervation: (a) chronology and (b) schematic phase diagram. The regime in which complex coacervates are formed is indicated in blue. (c) Flavor capsules based on complex coacervation. (d) Artist impression of the molecular structure of complex coacervates. Fig. 5d is reproduced with permission from ref. 42.

some food applications rely on coacervate-based structures, in particular in the encapsulation and controlled release of flavors (Fig. 5c).[41]

The phenomenon of complex coacervation was initially most extensively studied in biophysics;[43] the general definition of complex coacervation is the associative phase separation of two polymers, which are oppositely charged (Fig. 5b).[44] Consequently, in dilute systems, complex coacervation leads to the formation of small, globular polymer complexes, which are still highly hydrated.[42,44] The similarity in shape and in water content to cells has lead numerous scientists to argue that coacervation could have been one of the steps in the evolution of life, and that coacervates could be used as model systems in the study of complex living cells.[45,46]

In the food field, coacervation is mostly carried out using an anionic polysaccharide, such as gum arabic, and a protein, usually gelatin.[42,47] The amphiphilic nature of these compounds specifically allows the formation of thin shells around oil droplets dispersed in the aqueous medium; a fact that is exploited in the formation of capsules for the controlled release of actives (Fig. 5c).[41]

Over the last years, the complex coacervation of, in particular, dairy ingredients has been extensively explored for structure building in foods[47] and, in parallel, the molecular structure and properties of coacervate model systems have been elucidated in great detail (Fig. 5d).[42] However, as of today, very few food products are on the market in which the complex coacervation of food ingredients has been purposefully used to create defined food structures. In this, the development of coacervation as ingredient-structuring has suffered the typical fate of the "science-push" approach: while scientific approaches may have high potential for applications in food, it often takes longer than anticipated before the first major products materialize.

Interfaces

Food processing

As already testified by some of the examples discussed above, one of the principal fields for the application of soft matter physics in the food field is in food technology. However, as we have seen for the example of soluble coffee processing and for complex coacervation, the adoption of soft matter science in food technology is a slow process, as yet incomplete. Moreover, the interpretation and use of physical principles and concepts in the food field often differs subtly from the original physical definition and context. This is naturally impeding the communication between the food technology and soft matter communities. In addition, the focus on quick results and rapid product development in industry is not advantageous to the adoption of soft matter research by the food industry. Illustrative is the quote from an early review: "… despite their complexity, foods in no way differ in their behaviour from other systems, although the food industry does not necessarily think to turn to the physics literature to help them in their understanding".[1]

To some extent, the situation is similar to the state of the chemical industry in the early 1950s. At the time an industry still largely based on empirical knowledge, it shortly afterwards experienced a major transformation. From the 1950s to the 1970s, its fundamental basis was put on a solid theoretical foundation using approaches and results from statistical thermodynamics, the kinetic theory of gases, fluid mechanics, and reaction rate theory.[48,49]

However, for various reasons, the adoption of an unified fundamental physical basis underpinning the food industry will most likely not progress as swiftly and as far as in the chemical industry.

In the first place, the physics of the systems the food industry is confronted with is generally more complex. Although deviations from non-ideality are significant in the case of chemical processes, often it still makes sense to depart from the assumption of ideal behavior, and then amend. In food processing, this is often not possible: the systems of relevance are so far from ideality, and, moreover, so far from thermodynamic equilibrium, that one should be extremely cautious in the application of simple thermodynamic models or physical estimates.‡

In addition, the physics of relevance to a specific foodstuff is often more diverse than for many non-food materials. Essentially, all of the physical concepts required in the physics of foodstuff are used as well for chemicals, materials and for pharmaceutical products. Foods differ, however, from those classes of systems, as the structure of foodstuffs is generally more complex, and the relevant physics and chemistry is characterized by a spatial variation linked to the structural inhomogeneity of the product.

Milk powder, as an example of a still reasonably simple foodstuff, has the characteristics of an emulsion as well as a amorphous material.[17,50] It consists of a solid emulsion of droplets of milk fat embedded in a solid, glassy matrix largely made up of the disaccharide lactose, but containing whey protein as well. A third material dimension is added by the casein proteins: they form micelles, but are important in the stabilization of the fat droplets as well.[51-53] Recent developments in materials science, dealing with advanced, composite systems, may partially bridge the gap to food science and technology, as the complexity of these novel structured materials in some respects approaches food.

In the second place, from a chemical point of view, food ingredients are less pure than the raw materials used in chemical processing. Furthermore, whereas in the chemical industry further purification "up stream" usually leads to improved products with better specifications, this is generally not the case for foods. Nutritional studies increasingly show that the physiological impact of a diet is dependent on the interplay of macro- and micronutrients in food, and that the structure of the food or the ingredient plays a role as well.§ The choice of representative model systems to develop the fundamental basis of the field is thus more cumbersome for foods than for chemicals or pharmaceuticals. This is likely to be one of the reasons that many food science studies are of a descriptive nature.¶

In the third place, the specifications of a chemical or pharmaceutical product are usually more straightforward to define than those of a food product: simple chemical products should have a defined composition and purity, and have defined values for simple physical properties such as melting- and boiling points. This specifically holds true for pharmaceutical products. In case of more complex products, for example polymers, complex material properties, such as mechanical and rheological properties, may also be specified. For food products, the product specifications effectively include aspects relating to consumer liking and acceptance as well; these consumer-centric properties are, however, not easily translated into unambiguous physical specifications (see also the discussion below).

Notwithstanding these complications, over the last few decades, numerous interesting and successful efforts have demonstrated that food technology benefits when approached from a soft matter perspective. What are, then, the central prerequisites that have rendered these efforts successful?

---

‡ This is not to say that simple principles and concepts cannot be used: see *e.g.* Table 1 for a situation where a thermodynamic parameter (the water activity) achieves equilibrium on an experimentally acceptably short time scale, in a constrained environment of a food matrix (of possibly complex composition), which remains far from equilibrium.

§ For a discussion on the limits of the reductionist's approach in nutrition, see ref. 74. It is also becoming more and more evident that it is not the single nutrient that matters, but rather the whole food matrix. For a reappraisal of the role of dairy products in this context, see ref. 155 and 156.

¶ The food science literature abounds with papers on the composition and physical composition of fruits, meats and plant raw materials, specifying types, species and/or geographical origin. Differences between the various items are investigated, listed, but not usually interpreted in terms of a consistent physical or chemical framework.

From a global perspective, soft matter science has been successful in feeding into food technology by approaches that can be grouped into three axes:

1. Provision of analytical physical techniques.
2. Development of specific ingredients with defined functionality.
3. Establishment of a physical framework for the interpretation and guidance of food processing, and for the control of food properties.

Even though physical techniques have found widespread application in food technology,‖ it is mainly the contributions of the second and third axes that are of interest here.

A significant part of research in the food industry is concerned with the development of novel functional ingredients. These developments can be highly successful, as they were, for instance, with the introduction of essential nutrients such as vitamins (which I discuss below), or with the development of substitutes for butter (see the discussion above).

There are, however, risks associated with product development strategies based on novel ingredients. In the first place, it frequently turns out that the desired functionality of a food product cannot be realized by just adding or substituting one specific ingredient. For instance, it has turned out that the nutritional quality of a foodstuff depends on a sensitive balance between many different nutrients.§ In addition, the ingredient often needs to be formulated, structured, or processed as it does not itself exhibit the desired properties. Thirdly, the food industry is regulated by a strict regulatory framework: many of the novel ingredients are not food grade, and obtaining approval for food use often turns out to be not feasible, if only because of cost issues.

A most profound impact of soft matter science in food technology is on the understanding, control and modelling of food processing. For instance, emulsion technology is dominated by the framework set by colloid science: its concepts originate in colloid science, its processes are largely based on physical-chemical and hydrodynamic considerations, the schemes used to classify surfactants and emulsifiers are derived from soft matter science, and its formulation practices are founded on phase diagrams.[10]

An important field in food R&D, which is increasingly being based on soft matter science, is the control of food processing, stability and functionality by the use of phase and state diagrams.[12,13,54] As an example, let us consider the preparation of a dehydrated, glassy carbohydrate matrix starting from a carbohydrate solution (Fig. 6). In analyzing the physical state of the product matrix, and the three unit operations (freeze drying, concentration and spray drying), it is clear they are governed by a limited set of physical concepts. In addition, the use of state diagrams allows a quantitative prediction of many aspects of food processing, in particular when the state diagrams are supplemented with data on nucleation, crystallization, and viscosity.[13] Such state diagram-based approaches have successfully been used for numerous products, including milk powder,[17] bread,[55] cereals,[56] ice cream[57] and sugar confectionary.[58]

Not only quantitative results matter in the implementation of soft matter science in food technology and food development, but also qualitative insights and concepts. The use of physical concepts in food processing clarifies the often-diverse physical functionality of food ingredients, and thereby reduces the risk of inappropriate applications. As an illustration, let us consider the example of glycerol, a simple molecule widely used as a food additive. In foods, the application of glycerol

---

‖ Several techniques currently considered to be standard in food science and technology, but originally developed within the context of soft matter-related disciplines, include thermal analysis, rheology, NMR, and various diffraction and scattering techniques. Promising new developments are on-going in the application of advanced techniques such as neutron scattering,[157] X-ray tomography,[166,167] solid-state NMR,[165] rheology,[162,163] scattering in dense solutions and dispersions,[158,164] positron annihilation lifetime spectroscopy[161] and AFM.[159,160]

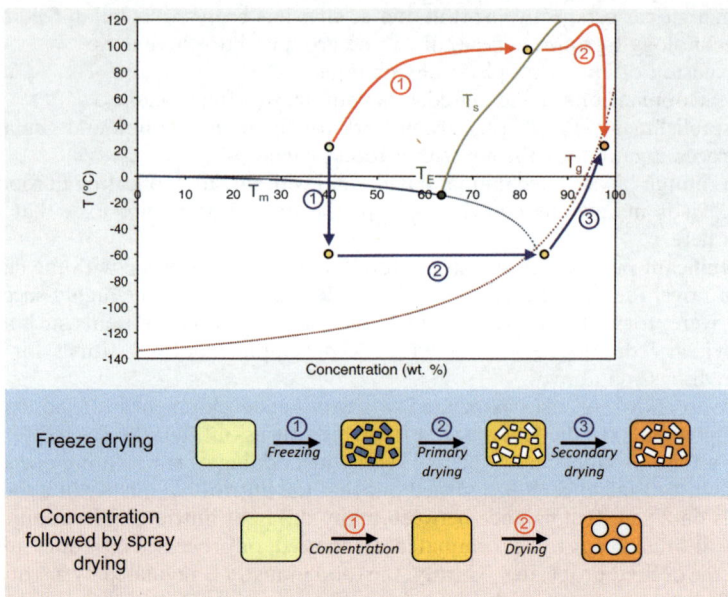

Fig. 6 Phase diagram of a typical food carbohydrate: sucrose. In addition to the equilibrium phase lines (solid lines), non-equilibrium transitions are shown (dashed lines). T_m: melting line; T_s: solubility line; T_g water content dependent glass transition. T_E is the eutectic temperature. Food structures can be created in a controlled way by following defined steps in the phase diagram; one of the final steps comprising the quenching of the food system into the glassy state. Two of the most important unit operations in food processing are plotted in the phase diagram: freeze drying and spray drying. In the embodiment shown here, the process at elevated temperatures is split into two distinct unit operations: a (slow) concentration step using *e.g.* a falling film concentrator, and a (quick) drying step using a spray-drying tower. The schematic structures of the carbohydrate matrices during processing are shown in the diagrams.

is often done on a heuristic basis; from a physical perspective, at least four distinct roles may, however, be discerned:

1. humectant
2. anti-plasticizer
3. plasticizer
4. anti-freezing agent

Glycerol is commonly used as a humectant to lower the water activity of foods of intermediate water content. The central concept here is the mixing entropy, which, when the food matrix is in the rubbery state, increases upon addition of glycerol. In the glassy state, however, the opposite behavior is often observed, and the water activity may increase upon the addition of glycerol. Here, the role as an anti-plasticizer must be invoked, and the addition of glycerol to a glassy food matrix leads to an increase in density, a reduction in molecular free volume[59,60] and a suppression of certain dynamic modes.[61] The anti-plasticization behavior of glycerol and other low molecular weight compounds such as mono- and disaccharides not only affects the water activity in the system, but also impacts the barrier properties and the biosta-bilization performance of glassy carbohydrates.[62] Glycerol also acts as a plasticizer: the glass transition of carbohydrate- and protein-based food matrices decreases upon the addition of glycerol.[12]

In frozen foods, the situation becomes even more complex, as in addition the freezing point reduction by glycerol should be considered. The downward shift in the melting line induced by glycerol leads to a higher fraction of unfrozen water, a phenomenon in the explanation of which, again, the colligative properties figure.**

For a food developer or a food technologist, the simple act of adding some glycerol to a food system may thus end up in a nightmare, as the food becomes excessively plasticized, or suddenly shows problems with the stability in the frozen state. The translation of fundamental physical findings and knowledge into directly applicable concepts should therefore be one of the principal goals of soft matter research in the food field.

Food creation

In this section, I will discuss the relation between soft matter science and diverse topics linked to food creation and food culture: cooking and culinary creation in an artisanal setting, and industrial food concept and product development.

A most promising interface of soft matter science and food is on cooking. Long seen as a purely artisanal field, pioneering efforts starting in the late 1960s have rendered cooking into a field whose approaches and methods are increasingly understood from a scientific perspective.[63–66]†† This is of importance not only to optimize the preparation of traditional dishes, but in particular to allow chefs to explore new ideas in a more rational way.

Commonly known as molecular gastronomy,[65,67] the scientific foundation of cooking is in fact a branch of food science.[67,68] It, however, differs in scope and approaches: it focuses on artisanally- rather than industrially-prepared foods, and is currently finding application in creative cooking rather than in industrial goals such as food preservation or the mimicking of existing tastes and textures. In this context it must, however, be stated that virtually all of the science that is finding application in science-based cooking (the term was coined in ref. 68), is directly derived from food chemistry and physics, as to date, little fundamental research has been carried out with cooking and culinary creation as the principal motivation.[67]

Even within a narrow field such as science-based cooking, several schools can be identified. On the one hand, one finds protagonists of the use of science-derived ingredients, such as industrial thickeners and gelling agents, and techniques such as freeze drying and industrial extraction. The use of these "novel" ingredients and techniques has enabled chefs to come up with a whole range of new and surprising structures and textures.‡‡ One the other hand, though, one finds scientists and chefs who are more interested in improving their understanding of the behavior of traditional foods and food ingredients. One example would be the science of dough and bread as discussed earlier, another the understanding of the cooking of meat as determined by the degree of unfolding of several key meat proteins.[64,69] From these examples it follows, however, that both schools base themselves strongly on soft-matter insights.

Because of differences in consumer orientation and food preparation scale, it is a major leap from creative cooking to concept development for the food industry. The two fields share an important challenge, however, and that is the translation of food concepts into actual products. Whereas in creative cooking, ideas for novel dishes are usually conceived by chefs, concepts for the food industry may be based on consumer insights and nutritional recommendations as well as innovative ideas.

In both cooking and industrial product development, the translation of food concepts into actual dishes or products is traditionally based on trial-and-error,

---

** Because of the coexistence of a cryo-concentrated phase and an ice phase, the water activity in frozen systems is defined solely by the temperature. The ice fraction and consequently the fraction of unfrozen water can vary considerably, dependent on, in particular, the molecular weight distribution of the food matrix.[168]

†† For accounts of science-based cooking, see also ref. 169–171.

‡‡ See e.g. ref. 171 for a review of the invention of "spherification": the development of gel beads for use in culinary creations.

or, in perhaps a more appropriate wording, "cook-and-look". Scientific insights, for instance from soft matter physics, invariably play a secondary role in such empirical endeavors, as those who are involved in the food development process are often unaware of the key physical principles governing their product applications. A typical example we have encountered above in the development of soluble coffee; such arduous product development trajectories are even now more common than one would wish to acknowledge.

In recent years, the challenge for food developers has in fact strongly increased, as food product development has shifted focus from raw materials and their transformation to the final product as characterized by defined consumer-centred[70,72] and health-oriented attributes.[73,74] This necessitates a reversed way of product development, which was termed the "retro-design approach"[75] or "reversal of the production chain".[76] The issue for food developers is that these defined product attributes are not usually formulated in physical, chemical or technological terms, which hampers rational product- and process development. As, in addition, the constraints on product safety, sustainability and the economics of processing are becoming more restrictive, food developers and food technologists are confronted with a playing field that is rapidly becoming more complex.[77]

Approaches based on soft matter science may facilitate food development in two principal ways. In the first place, as we have seen, soft matter approaches are instrumental in providing directions for the implementation of food manufacturing processes, and for the optimization of food shelf life and safety. In the second place, the increasing focus on consumer-centred food development and the retro-design of food products requires insights into various soft matter-related aspects of food and its physiological interactions. Specifically, it turns out that the understanding of the biophysics of food consumption, perception and digestion is important as it enables the rational translation of product attributes defined in the consumer- or health context into practical directions for product formulation and process development.

Biophysics of food, nutrition and perception

Food biophysics is, in itself, a wide and not very precisely demarcated field. One could for instance argue that the physical understanding of complex food ingredients, such as meat, milk and eggs, is part of food biophysics. For example, the physics of meat is closely linked to muscle biophysics, albeit with an emphasis on meat cooking and consumption, rather than muscle performance.[64,69,78,79] In a similar vein, the development of structure-property relationships for, for instance, milk and eggs, could be considered to be part of food biophysics rather than food materials science. In the context of the present paper, I will, however, limit the scope of food biophysics to those topics that are related to living organisms, in particular humans. Food biophysics is therefore centred on questions relating physical aspects of foods and food constituents and their interaction with the human body. A second area falling within the present definition of food biophysics is at the interface with biotechnology and microbiology: the biophysics of microorganisms in relation to food fermentation and human health.

Soft matter approaches concentrating on food and human physiology have recently seen a surge of interest, in particular in the context of the modelling of the human digestion, although efforts are also oriented towards the understanding of the physical aspects of perception and food consumption.

The modelling of the human digestion is mainly carried out experimentally using laboratory models of a varying degree of complexity (see ref. 80–83 and references contained therein). These digestions are developed with varying aims, such as to provide assays for nutrition bioavailability and bioaccessibility, and to help in understanding the role of the various parts of the digestive tract on the digestion process. From the perspective of soft matter science, the most interesting approaches are, however, centred on colloidal aspects of nutrient digestion and on the impact of food structure on the digestion and bioavailability. In particular, lipid and emulsion digestion[84,88] and the relation between the molecular organization and digestion of starch have been studied using *in vitro* models.[89,92] The colloidal basis of digestion modelling remains as yet incomplete, lacking in particular the formulation of appropriately formulated physicochemical hypotheses combined with quantitative experimentation. Moreover, there is a strong need to connect digestion modelling to advances in fundamental biophysics.[93–95]

The assessment of consumer liking and acceptance of food products is largely based on sensory evaluations. Sensory research has principally focused on developing statistically valid methods to discern, describe and quantify differences between various food formulations, and to enable identification of the food product with the highest chance of commercial success.[96,97] It, however, has turned out that the use of a limited set of standardized sensory methods limits the quality of the predictions. Whereas this can be partially compensated for by further developing the methods based on perception psychology,[98] biophysical insights in the mechanism of perception will further add to the quality of analysis and prediction of sensory properties and food perception.[99,100]

The important area of food perception and sensory evaluation has been covered extensively and important developments continue on flavor release and perception,[101–104] taste perception[105] and the assessment of food texture.[106,107] Two examples of texture perception may help to illustrate the opportunities for biophysical and soft matter research. First, it has been demonstrated that the perception of the mouthfeel of liquid products thickened with biopolymers, such as soups and sauces, is dependent on the solution behavior of the biopolymer thickeners. In particular, the critical overlap concentration of the biopolymer coils c^* appears to play a key role in determining dilution behavior and thereby the mouthfeel of the thickened product.[108,109] Second, numerous investigations have pointed out the importance of fracture behavior in assessing crispness of foodstuffs, and the physics of fracturing has been invoked in explaining the crispness experience during food consumption.[110–112] A true understanding of the mechanics of perception remains a distant perspective, however. For example, a recent investigation has demonstrated that the fundamental study of the mechanics of the mouth in relation to shape, size and physical properties of a food system may still lead to major surprises concerning human perception.[113]

Nutrient delivery

A major emphasis in the food industry is currently on the development of foods with nutritional benefits extending beyond those associated with common foods.[114,115] The development of these so-called functional foods heavily relies on insights from soft matter science.

When formulating a functional food, two strategies are basically possible:

1. Addition of bioactive ingredients conveying a specific physiological effect upon consumption of the food product.

2. Removal or reduction of certain ingredients considered undesirable and replacement by other ingredients that better comply with the desired nutritional profile.

In this context, a principal distinction is to be made between macro- and micronutrients. Whereas for micronutrients, the main emphasis is on the first strategy, the fortification of foods with defined micronutrients, the second strategy is relevant for both micro- and macronutrients (Fig. 7).

An example of a product reformulation concerned with a macronutrient is the development of margarine as a substitute for butter (Fig. 3), and the current efforts on the structuring of oils in the form of physical gels as a replacement for margarine-type products based on hardened fats (Fig. 3).[28]

Micronutrients	Macronutrients
Essential Vitamins Minerals and trace elements Essential amino acids **Non-essential** Carotenoids Flavonoids Polyphenols **Digestion & control of bioavailability** Probiotics Phytosterols → *Direct addition* → *Product reformulation* → *Encapsulation*	**Carbohydrates** → *Sugar replacement*: polysaccharides instead of mono- & disaccharides → Amorphous instead of crystalline sugars lead to issues on stability & water control **Fats/lipids** → *Physical structuring* to harden unsaturated lipids (monoglyceride, polysaccharide or protein structuring) → *Replacement of fats* by polysaccharides or emulsion-based systems to reduce fat content

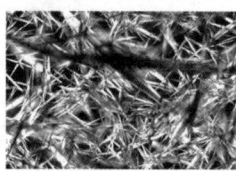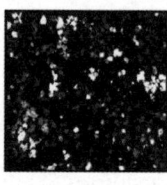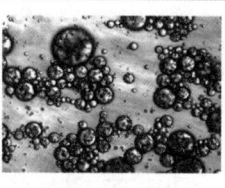

Fig. 7 Delivery of micro and macronutrients. Delivery strategies for various classes of micronutrients (left panel) and macronutrients (right panel) are listed. The images signify a non-formulated, crystalline bioactive (lycopene) with low bioavailability; a beverage mix with partially crystalline and partially amorphous carbohydrates and a W/O/W emulsion for the delivery of water-soluble bioactives. The left image is courtesy of M.-L. Dillmann; the right image is reproduced with permission from ref. 152.

The other class of macronutrients currently under consideration for reformulation are the carbohydrates. In order to lower the sugar load of numerous foodstuffs, and to lower their glycemic index, it is attempted to substitute a significant fraction of certain mono- and disaccharides, such as fructose, maltose and sucrose, by oligo- and polysaccharides, whereby the sweetness of the product is optionally retained by the addition of artificial sweeteners.

The substitution of simple sugars by complex, high-molecular weight carbohydrates strongly impacts the physical properties of the food matrix. In liquid, solution-type foods, the impact of polysaccharides on the rheological properties of the product is considerable, and the effects of concentration and the solution state of the added carbohydrate polymer on the viscosity should be taken into account in the product formulation. The major issues are, however, encountered in the reformulation of solid foods. Low molecular-weight sugars are often present in crystalline form in low water-content foods, whereas most polysaccharides are amorphous (Fig. 7). This leads to a dramatic change in the physical properties of the matrix, as instead of a melting point, it is now characterized by a glass transition. Moreover, the interaction with water changes completely: whereas a carbohydrate in its crystalline form has a defined hydration number, the water content of amorphous carbohydrates varies continually with the water activity. When not taken into account, this leads to issues with the stability and functionality of the food product.

The fortification of foods by micronutrient supplementation is well established in the case of vitamins and certain minerals (*e.g.* sodium iodide), micronutrients that are essential to human health (Fig. 8). More recently, the potential long-term health benefits of many classes of non-essential micronutrients are being exploited in a wide range of functional foods.[116]

In the most straightforward case, the novel micronutrient is simply added to the product matrix. Most often, the product reformulation turns out to be highly challenging, as the stability and functionality of the micronutrient are compromised by the food matrix. A micronutrient may, for instance, degrade by exposure to atmospheric oxygen, or it may react with a constituent of the food matrix and thereby lose its functionality. The bioavailability of a bioactive can also be negatively influenced by changes in its physical structure, or by aspects relating to its physical properties, such as the aqueous solubility. Further issues arise when the micronutrient influences the taste, texture or appearance of the food product. Many bioactives, for instance, have an undesirable, bitter taste, which renders the food product in which they are used unappealing.

A typical soft-matter approach to counteract issues arising from the fortification of foods, and to optimize the nutritional benefits of the added bioactive, is to use encapsulation.[75,117–121] The variety of encapsulation systems is, however, very wide, and highly diverse soft matter physics is underlying the properties and performance of the various systems, rendering a discussion of encapsulation systems beyond the context of the present introductory lecture. In the coacervate capsules discussed previously (Fig. 5c), as well as in the carbohydrates added to the glassy coffee matrix to enhance the glass transition temperature (Table 3), we have, however, already encountered a specific encapsulation system as well as an example of the physics controlling the performance of encapsulation systems.

An aspect unfortunately often underestimated in encapsulation is that one does not only need to formulate a capsule that can carry the bioactive, but that one should also tune the properties of the capsule in such a way that it enhances the stability or functionality of the bioactive.[75] In effect, one should attempt to match or compatibilize the bioactive and the food matrix by using the appropriate encapsulation technology. An effective and rational way to do so, and to systematically evaluate the options to successfully use a specific technology, is by using a so-called retro-design approach (Fig. 8).[75]

Central in this approach is that one starts by precisely defining the target specifications for the product, and from there works back to feasible strategies for product development. A key element in the retro-design approach is to acquire an

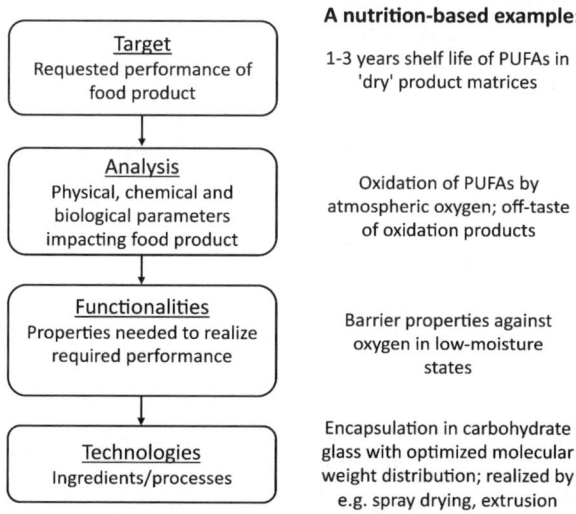

Fig. 8 retro-design approach for the delivery of bioactives in food, and outline of implementation for the stability by encapsulation of polyunsaturated fatty acids (PUFAs) for application in low-moisture foods.

appropriate understanding of the physics, chemistry and biology governing both the product application and the bioactive: apart from being indispensible in the development of an encapsulation system, soft matter insights are essential in the understanding of the state and performance of the bioactive compounds in the product matrix.

Even though the encapsulation and delivery of bioactives is one of the central areas where soft matter approaches have had an impact on food development, I would like to end this section with some words of caution. In the first place, one should realize that the formulation or reformulation of a product is only one of the many ways one could attempt to improve the human diet.[122,123] Factors relating to serving size, product labelling, the use of more healthy whole ingredients (such as vegetables, fruits and unrefined cereals) as well as the education of consumers are most likely having a larger impact than product reformulations. Second, food developers should approach clinical evidence of the nutritional benefits of specific bioactives very critically: "If the only real function 'of functional foods'... is to bolster profits, consumers and regulators will eventually see through."[124]

Sustainable foods

Food production will need to become more sustainable in the foreseeable future; environmental issues including pollution, global warming, soil erosion, biodiversity and loss of nature, and energy and water usage, are increasingly impacting the quality of life on the planet, and are expected to in the future also reduce the agricultural capacity of Earth.[125] The principal question of relevance to the present discussion is to what extent this has a bearing on food technology, and on the specific role of soft matter science therein.

Most of the factors mentioned previously do not relate directly to the food industry, but rather to agriculture, and, even though relevant, will be left out of the present discussion.[126,127] There are, however, important topics in food sustainability to which food-oriented soft matter science can provide guidance. These include the reduction of energy and water usage in food processing, which can be achieved in a more rational way when based on thermodynamic principles. In a similar vein, soft matter approaches can help to optimize food stability: foods that are more robust from a physical point of view (for instance with a sensitivity to heat shocks), will save in the energy costs of conditioned transport and storage,[128] and knowledge of food physics will allow for more soft processing.

Soft matter approaches may also help to shift the diet away from less sustainable foods: the food industry has, for instance, been reasonably successful in introducing plant-based meat analogues.[129] This will allow for a much more efficient use of food calories, as the caloric conversion rate of *e.g.* cereals into meat is typically only between 10 and 30%.[130]§§ These meat substitutes nonetheless still lack many of the desirable qualities of real meat, such as succulence, taste and tenderness. Substitutes based on soft-matter designs could possibly improve the situation.

A general problem with the sustainability of soft matter-based food design is that, as a reductionist physical discipline, it most comfortably works with systems that are relatively pure, whereas foods are, by definition, complex in composition. Past developments have, then, generally also shown that, when soft matter-approaches are employed in food, the composition of the food product tends to become more simple.

§§ The shift of the diet to include insect meat is increasingly promoted, as it is a good source of proteins while having a low ecological footprint.[173,174] Soft matter approaches could in principle assist to develop textures of insect containing foods that are appealing to the Western consumer. While the composition and nutritional qualities of insects have been investigated,[173,175] very little scientific work is done on the processing of insect-based foods.[176] It should be realized, however, that the appeal of insects as a food source is a cultural one: in many Asian, African and Latin-American cuisines, insects in recognizable form are traditionally a valued ingredient.[177]

From an ecological perspective, this, however, leads to a thermodynamic penalty: in refining of ingredients, considerable amounts of energy need to be expended. This energy is then lost as the refined ingredient is usually mixed into a complex food matrix. It is, therefore, imperative that soft matter-inspired food design develops methodologies that allow it to rationally operate using complex ingredients in their natural state.

A last remark concerns the basis of sustainability evaluations.[131] The judgement on the sustainability of a food depends on the entire food chain from farmer to consumer, and any changes may lead to unexpected consequences elsewhere in the food chain. If, for instance, the use of ecological practices by a farmer incites individual consumers to use their cars to individually pick up the produce,[132] the increased energy usage resulting from the last "food miles" may nullify any gain in sustainability at the farm, or worse.[133] Apart from changes directly affecting only a specific food production process, such as the reduction of water or energy usage, sustainability evaluations need to be founded on a quantitative basis, taking into account agricultural, environmental, and social-cultural factors.[134-136] Unfortunately, this basis still finds itself in its infancy and sustainability claims, therefore, face the risk of being derided as "greenwashing".[137]

Food complexity

Implicit in most research in the natural sciences, including soft matter science of foods, is the reductionist hypothesis: by investigating the properties of its parts, one can obtain a meaningful insight into the behavior of a complex higher-order system. While reductionist approaches have provided many conceptual breakthroughs, it has emerged that many important features of complex systems cannot be described by reducible properties at lower levels of complexity, but rather should be addressed at the system level: "The ability to reduce everything to simple fundamental laws does not imply the ability to start from those laws and reconstruct the universe. The constructionist hypothesis breaks down when confronted with the twin difficulties of scale and complexity. At each level of complexity entirely new properties appear. Psychology is not applied biology, nor is biology applied chemistry. We can now see that the whole becomes not merely more, but very different from the sum of its parts".[138]

Central to the notion of complexity is the phenomenon of emergence. A hotly debated concept, emergence is generally understood to relate to processes that lead to higher-level (structural) features that result from, but are not directly attributable to, processes at the underlying microlevel.[139-142] Emergence is often seen as being related to the phenomenon of self-organization, but is also considered to be different, as self-organization refers to the development of a stationary higher-order structure while emergent phenomena are understood to evolve in time.[142]

For the purpose of the present discussion, we may consider the characteristics of emergency to be defined following ref. 140:

1. Emergent phenomena are somehow constituted by, and generated from, underlying processes.

2. Emergent phenomena are somehow autonomous from underlying processes.

Emergent phenomena occur widely in systems showing non-linear behavior, including in physical systems, in mathematical and in physical networks and in social organizations. Important aspects of food, food consumption and preference, nutrition, and the societal aspects may thus be expected to show emergent behavior.

The utilization of the concept of emergence in the soft matter physics of food makes sense only if it adds new avenues for exploration and understanding. As an example, consider bread: bread exhibits clear structural characteristics, which cannot be predicted based solely on knowledge of the properties of gluten and starch. It is, however, likely that by using familiar concepts from soft matter physics relating to self-organization, one can derive a satisfactory framework for the analysis of bread structure and for the prediction of several relevant structural characteristics.¶¶ Emergent phenomena may, however, occur in situations where food structure has emerged in interaction with biological systems and organisms, such as in case of fermented foods.[143] In the case of bread, this means that emergent phenomena could become relevant if one is interested in the co-evolution of bread structure and yeast adaptation.

Emergent phenomena in food will most likely be identified, as well, in relation to food preference, food digestion and the societal aspects of food. For instance, the preference for specific foods and their tastes and textures has most likely evolved in a close relation with the availability of certain food ingredients, and the evolved skills at transforming these ingredients into foodstuffs. Food preference as a higher-order phenomenon is thus likely to emerge as a cultural phenomenon from underlying phenomena relating to food availability and processing. Rather than contributing to the elucidation of the complexities of food structure, I, therefore, expect that complexity theory may contribute towards forging a system connecting food structure and properties on the one hand and food preference on the other hand. This would fit well with the current focus on consumer-centric food design.

At an even higher level, it is not truly possible to predict societal consequences of changes in diet: changes in food policy or food development based on nutritional insights often lead to unexpected changes in food consumption.[144,145] An exploration of the network effects involved in positive and negative feedback loops could likely bring fresh insights in this intricate matter.

Given the fast development of complexity theory over the last years, it is surprising that as yet there are only few applications in the food field. Most of those studies are on the relation between food and its societal context, for instance on the relation between food prices and social unrest,[146] or on the global connections linking food, people and regions, and investigating the stability of the global food supply.[147]

Concluding remarks

In the introduction I posed four questions as to the nature, role, and present and future impact of soft matter approaches to food structuring. Since these questions have served as motivation for the discussion in the subsequent sections, it is appropriate to return to them in concluding this paper:

1. All foods are characterized by their structure, which is often of an intricate nature. It is this structure that, to a large degree, is responsible for the properties linked to the appreciation of the food. This includes factors such as mouthfeel, flavor release, tactile properties and appearance. Food structure also has an impact on the nutritional properties, as food digestion is, to a varying degree, directly dependent on the structure of the foodstuff, even though indirect structure-related differences in composition may have a higher impact on the digestion properties. Foods have changed over the ages, mostly on an evolutionary basis or by gradual improvements. There are, as yet, very few foods on the market of which the structure was purposefully designed. Therefore, one could conclude that the foods, which we know and which we are fond of, have evolved on a purely empirical basis.

2. Over the last decades, soft-matter approaches have started to make significant inroads in food development. These approaches are having an impact principally on industrial food process and product development. It is of interest to note that, even though concepts from physics and physical chemistry are used in these developments, advances in food science and technology have proceeded mostly independent of these basic scientific disciplines. The impact of soft matter science on currently available food products is still limited: it presently excludes essentially all fresh

¶¶ See *e.g.* the paper by Tanaka in this volume and the subsequent discussion.

and artisanally prepared foods, which still make up the bulk of all food consumed. Moreover, even in industrial food manufacturing, the number of foodstuffs that were developed using soft matter approaches is still rather restrictcd.

3. The notion of "Soft Matter approaches to structured foods" implies that the discipline is essentially an applied one, of which the theoretical framework and basic concepts are predominantly derived from more fundamental disciplines. This, however, does not exclude fundamental scientific developments taking place in a context primarily oriented towards food: the specific situations encountered in food provide a fertile ground for physical exploration, and food-inspired physical discoveries could have an impact beyond food-related sciences. In addition, the complexity of food, both composition-wise and structure-wise, necessitates approaches for its study, which can address these complexities. Problems of relevance to food are almost by definition of a multidisciplinary nature, and studies integrating soft matter approaches with food chemistry, biology and physiology should be strongly encouraged. Furthermore, physical approaches may deliver interesting results in conjunction with approaches emphasizing the psychological, behavioral and social aspects of foods.

4. Given the persistent drive towards rationalization in the food industry, and the increasing awareness that physical principles are underlying central aspects of food properties, including processing performance, shelf life and stability, and the behavior of food during consumption and digestion, it is highly probable that physical approaches increasingly will find application in the food industry. This trend is reinforced by a shift to food product development based on defined properties of the end product, often based on insights from consumer studies or on nutritional recommendations.

The future perspectives of soft matter approaches to structured foods are thus closely linked to developments in the fields of food science and food technology. However, the unique physically inspired perspective provided by soft matter science will also broaden the scope of food science and food technology, and will facilitate the interaction with other disciplines. In addition, soft matter approaches for food structuring will also help to formulate societal issues relating to food, such as health, environmental impact and food culture and diet in a language enabling a rational translation to explicit objectives for food development. Soft matter science will therefore definitely aid in the transition from a "cook-and-look" attitude towards food development to a more rational approach based on physical insights. Nevertheless, we should not forget that foods are intrinsically so complex, culturally interlinked and based on tradition, that physically-inspired approaches will also in the future constitute only one of the many elements needed in food design. For a viable and attractive food culture, human creativity, inventiveness and adaptability will remain indispensable ingredients as well.

Notes and references

1 A. M. Donald, *Rep. Prog. Phys.*, 1994, **57**, 1081.
2 A. M. Donald, *Nat. Mater.*, 2004, **3**, 579.
3 R. Mezzenga, P. Schurtenberger, A. Burbidge and M. Michel, *Nat. Mater.*, 2005, **4**, 729.
4 J. Ubbink, R. Mezzenga and A. Burbidge, *Soft Matter*, 2008, **4**, 1569.
5 R. G. M. van der Sman and A. J. van der Goot, *Soft Matter*, 2009, **5**, 501.
6 R. G. M. van der Sman, *Adv. Colloid Interface Sci.*, 2012, **176–177**, 18.
7 M. Karel, in: *Food Engineering 2000*, P. Fito, E. Ortega-Rodríguez and G. V. Barbosa-Cánovas, ed., Chapman and Hall, London, 1997.
8 J. M. Aguilera and P. J. Lillford, *Food Materials Science: Principles and Practice*, Springer, New York, 2008.
9 P. Walstra, *Physical Chemistry of Foods*, Marcel Dekker, Basel, 2003.
10 J. D. McClements, *Food Emulsions: Principles, Practices, and Techniques*, 2nd ed., CRC Press, Boca Raton, 2005.
11 S. Kasapis, I. Norton and J. Ubbink, ed., *Modern Biopolymer Science: Bridging the Divide between Fundamental Treatise and Industrial Application*, Academic Press, New York, 2009.

12 Y. Roos, *Phase Transitions in Foods*, Academic Press, New York, 1995.
13 M. Pilar Buera, Y. Roos, H. Levine, L. Slade, H. R. Corti, D. S. Reid, T. Auffret and C. A. Angell, *Pure Appl. Chem.*, 2011, **83**, 1567.
14 L. Slade and H. Levine, *Adv. Food Nutr. Res.*, 1995, **38**, 103.
15 P. G. Debenedetti and F. H. Stillinger, *Nature*, 2001, **410**, 259.
16 P. Thomar, D. Durand, L. Benyahia and T. Nicolai, *Faraday Disc.*, 2012, 158, DOI: 10.1039/C2FD20014G.
17 G. Vuataz, *Dairy Sci. Technol.*, 2002, **82**, 485.
18 T. P. Labuza, S. R. Tannenbaum and M. Karel, *Food Technol.*, 1970, **24**, 543.
19 T. P. Labuza, *Food Technol.*, 1979, **34**, 36.
20 Y. H. Roos, *J. Therm. Anal. Calorim.*, 2003, **71**, 197.
21 J. Chirife and M. del Pilar Buera, *Crit. Rev. Food Sci. Nutr.*, 1996, **36**, 465.
22 M. Karel, S. Angle, P. Buera, R. Karmas, G. Levi and Y. Roos, *Thermochim. Acta*, 1994, **246**, 249.
23 B. S. Ghotra, S. D. Dyal and S. S. Narine, *Food Res. Int.*, 2002, **35**, 1015.
24 O. Korver and M. B. Katan, *Nutr. Rev.*, 2006, **64**, 275.
25 S. Ghosh and D. Rousseau, *Curr. Opin. Colloid Interface Sci.*, 2011, **16**, 421.
26 A. J. Dijkstra, *Eur. J. Lipid Sci. Technol.*, 2009, **111**, 857.
27 D. Mozaffarian, M. B. Katan, A. Ascherio, M. J. Stampfer and W. C. Willett, *N. Engl. J. Med.*, 2006, **354**, 1601.
28 M. Pernetti, K. F. van Malssen, E. Flöter and A. Bot, *Curr. Opin. Colloid Interface Sci.*, 2007, **12**, 221.
29 M. A. Ruiz Martınez, M. Munoz de Benavides, M. E. Morales Hernandez and V. Gallardo Lara, *Farmaco*, 2003, **58**, 1289.
30 T. Laredo, S. Barbut and A. G. Marangoni, *Soft Matter*, 2011, **7**, 2734.
31 A. I. Romoscanu and R. Mezzenga, *Langmuir*, 2006, **22**, 7812.
32 http://www.raw-milk-facts.com/homogenization_T3.html. Accessed August 30, 2012.
33 L. V. Saboya and J.-L. Maubois, *Dairy Sci. Technol.*, 2000, **80**, 541.
34 Y. Pouliot, *Int. Dairy J.*, 2008, **18**, 735.
35 H. Kontkanen, S. Rokka, A. Kemppinen, H. Miettinen, J. Hellström, K. Kruus, P. Marnila, T. Alatossava and H. Korhonen, *Int. Dairy J.*, 2011, **21**, 3.
36 A. Clemente, *Trends Food Sci. Technol.*, 2000, **11**, 254.
37 C. Schmitt, C. Moitzi, C. Bovay, M. Rouvet, L. Bovetto, L. Donato, M. E. Leser, P. Schurtenberger and A. Stradner, *Soft Matter*, 2010, **6**, 4876.
38 J. Adamcik, J.-M. Jung, J. Flakowski, P. De Los Rios, G. Dietler and R. Mezzenga, *Nat. Nanotechnol.*, 2010, **5**, 423.
39 E. H. C. Bromley, M. R. H. Krebs and A. M. Donald, *Faraday Discuss.*, 2005, **128**, 13.
40 A. Pfiffner, "A real winner one day": die Entwicklung des "Nescafés" in den 1930er-Jahren. In: R. Rossfeld, editor, *Genuss und Nüchternheit.*, Hier + Jetzt, Baden, Switzerland, 2002, p. 123.
41 J. Ubbink and A. Schoonman, *Flavor Delivery Systems*, Kirk-Othmer Encyclopedia of Chemical Technology, 2003, DOI: 10.1002/0471238961.0612012221020209.a01.
42 C. G. de Kruif, F. Weinbreck and R. de Vries, *Curr. Opin. Colloid Interface Sci.*, 2004, **9**, 340.
43 H. G. Bungenberg de Jong, *Planta*, 1932, **15**, 110.
44 M. A. Cohen Stuart, *Colloid Polym. Sci.*, 2008, **286**, 855.
45 A. I. Oparin, *The Origin of Life*, Dover, Mineola NY, 1953.
46 A. E. Needham, *Q. Rev. Biol.*, 1959, **34**, 189.
47 C. Schmitt, C. Sanchez, S. Desobry-Banon and J. Hardy, *Crit. Rev. Food Sci. Nutr.*, 1998, **38**, 689.
48 R. Aris, *The mathematical theory of diffusion and reaction in permeable catalysts*, Oxford University Press, Oxford, 1975.
49 R. B. Bird, W. E. Stewart and E. N. Lightfoot, *Transport Phenomena*, 2nd ed., J. Wiley, New York, 1963.
50 M. E. C. Thomas, J. Scher, S. Desobry-Banon and S. Desobry, *Crit. Rev. Food Sci. Nutr.*, 2004, **44**, 297.
51 E. Dickinson, *Colloids Surf., B*, 2001, **20**, 197.
52 D. S. Horne, *Curr. Opin. Colloid Interface Sci.*, 2002, **7**, 456.
53 D. G. Dalgleish and M. Corredig, *Annu. Rev. Food Sci. Technol.*, 2012, **3**, 449.
54 M. S. Rahman, *Trends Food Sci. Technol.*, 2006, **17**, 129.
55 B. Cuq, J. Abecassis and S. Guilbert, *Int. J. Food Sci. Technol.*, 2003, **38**, 759.
56 P. M. Forssell, J. M. Mikkiläa, G. K. Moates and R. Parker, *Carbohydr. Polym.*, 1997, **34**, 275.
57 R. W. Hartel, *Trends Food Sci. Technol.*, 1996, **7**, 315.
58 R. W. Hartel and A. V. Shastry, *Crit. Rev. Food Sci. Nutr.*, 1991, **30**, 49.

59 M. Roussenova, M. Murith, A. Alam and J. Ubbink, *Biomacromolecules*, 2010, **11**, 3237.
60 M. Roussenova, J. Enrione, P. Diaz-Calderon, A. J. Taylor, J. Ubbink and M. A. Alam, *New J. Phys.*, 2012, **14**, 035016.
61 M. T. Cicerone and C. L. Soles, *Biophys. J.*, 2004, **86**, 3836.
62 J. Ubbink, Nanostructural Advances in the Understanding of Carbohydrate Glasses, in: *Modern Biopolymer Science: Bridging the Divide between Fundamental Treatise and Industrial Application*, S. Kasapis, J. Ubbink, I. Norton, ed., Academic Press, New York, 2009, p. 277.
63 N. Kurti, *Proc. Roy. Inst. Great Britain*, 1969, **42**, 451.
64 H. McGee, *On food and cooking*, 2nd ed., Scribner, New York, 2004.
65 H. This, *Nat. Mater.*, 2004, **4**, 5.
66 H. This, *Int. J. Pharm.*, 2007, **344**, 4.
67 P. Barham, L. H. Skibsted, W. L. P. Bredie, M. Bom Frøst, P. Møller, J. Risbo, P. Snitkjær and L. Mørch Mortensen, *Chem. Rev.*, 2010, **110**, 2313.
68 C. Vega and J. Ubbink, *Trends Food Sci. Technol.*, 2008, **19**, 372.
69 E. Tornberg, *Meat Sci.*, 2005, **70**, 493.
70 A. I. A. Costa, M. Dekker and W. M. F. Jongen, *Trends Food Sci. Technol.*, 2004, **15**, 403.
71 P. A. M. Oude Ophuis and H. C. M. Van Trijp, *Food Qual. Pref.*, 1995, **5**, 177.
72 K. G. Grunert, *Eur. Rev. Agric. Econ.*, 2005, **32**, 369.
73 J. B. German and H. J. Watzke, *Compr. Rev. Food Sci. Food Saf.*, 2004, **3**, 145.
74 I. Hoffmann, *Am. J. Clin. Nutr.*, 2003, **78**, 514S.
75 J. Ubbink and J. Krüger, *Trends Food Sci. Technol.*, 2006, **17**, 244.
76 A. R. Linneman, M. Benner, R. Verkerk and M. A. J. S. van Boekel, *Trends Food Sci. Technol.*, 2006, **17**, 184.
77 A. I. A. Costa and W. M. F. Jongen, *Trends Food Sci. Technol.*, 2006, **17**, 457.
78 P. García-Segovia, A. Andres-Bello and J. Martinez-Monzo, *J. Food Eng.*, 2007, **80**, 813.
79 V. Santé-Lhoutellier, T. Astruc, P. Marinova, E. Greve and P. Gatellier, *J. Agric. Food Chem.*, 2008, **56**, 1488.
80 S. Boisen and B. O. Eggum, *Nutr. Res. Rev.*, 1991, **4**, 141.
81 A. G. Oomen, A. Hack, M. Minekus, E. Zeijdner, C. Cornelis, G. Schoeters, W. Verstraete, T. Van De Wiele, J. Wragg, C. J. M. Rompelberg, A. J. A. M. Sips and J. H. Van Wijnen, *Environ. Sci. Technol.*, 2002, **36**, 3326.
82 S. J. Hur, B. Ou Lim, E. A. Decker and D. J. McClements, *Food Chem.*, 2011, **125**, 1.
83 Y. Zhu, *Food Chem.*, 2011, **128**, 820.
84 D. J. McClements and Y. Li, *Food Funct.*, 2010, **1**, 32.
85 A. Sarkar, K. K. T. Goh, R. P. Singh and H. Singh, *Food Hydrocolloids*, 2009, **23**, 1563.
86 H. Singh, A. Ye and D. Horne, *Prog. Lipid Res.*, 2009, **48**, 92.
87 P. J. Wilde and B. S. Chu, *Adv. Colloid Interface Sci.*, 2011, **165**, 14.
88 M. Golding and T. J. Wooster, *Curr. Opin. Colloid Interface Sci.*, 2010, **15**, 90.
89 J. Holm, I. Lundquist, I. Björck, A. C. Eliasson and N. G. Asp, *Am. J. Clin. Nutr.*, 1988, **47**, 1010–1016.
90 I. Björk, *Am. J. Clin. Nutr.*, 1994, **59**(suppl), 699S.
91 C. G. Oates, *Trends Food Sci. Technol.*, 1997, **8**, 375.
92 G. Zhang, M. Venkatachalam and B. R. Hamaker, *Biomacromolecules*, 2006, **7**, 3259.
93 M. C. Carey, D. M. Small and C. M. Bliss, *Annu. Rev. Physiol.*, 1983, **45**, 651.
94 A. Vila Verde and D. Frenkel, *Soft Matter*, 2010, **6**, 3815.
95 J. Maldonado-Valderrama, P. Wilde, A. Macierzanka and A. Mackie, *Adv. Colloid Interface Sci.*, 2011, **165**, 36.
96 H. L. Stone and J. L. Sidel, *Sensory Evaluation Practice*, 3rd ed., Elsevier Academic Press, San Diego, 2004.
97 M. A. Amerine, R. M. Pangborn and E. B. Roessler, *Principles of sensory evaluation of food*, Academic Press, New York, 1965.
98 E. P. Köster, *Food Qual. Preference*, 2003, **14**, 359.
99 T. van Vliet, G. A. van Aken, H. H. J. de Jongh and R. J. Hamer, *Adv. Colloid Interface Sci.*, 2008, **150**, 27.
100 E. T. Rolls, *Physiol. Behav.*, 2005, **85**, 45.
101 P. Overbosch, W. G. M. Afterof and P. G. M. Haring, *Food Rev. Int.*, 1991, **7**, 137.
102 K. B. de Roos, *ACS Symp. Ser.*, 2000, **763**, 126.
103 E. Guichard, *Food Rev. Int.*, 2002, **18**, 49.
104 M. Auvray and C. Spence, *Conscious. Cognit.*, 2008, **17**, 1016.
105 D. A. Yarmolinsky, C. S. Zuker and N. J. P. Ryba, *Cell*, 2009, **139**, 234.
106 A. S. Szczesniak, *Food Qual. Preference*, 2002, **13**, 215.
107 L. Engelen and A. van der Bilt, *J. Texture Studies*, 2008, **39**, 83.
108 Z. V. Baines and E. R. Morris, *Food Hydrocolloids*, 1987, **1**, 197.

109 A.-L. Ferry, J. Hort, J. R. Mitchell, D. J. Cook, S. Lagarrigue and B. Valles Pamies, *Food Hydrocolloids*, 2006, **20**, 855.
110 T. van Vliet, *Food Qual. Pref.*, 2002, **13**, 227.
111 L. Duizer, *Trends Food Sci. Technol.*, 2001, **12**, 17.
112 G. Roudaut, C. Dacremonta, B. Vallès Pàmies, B. Colasa and M. Le Meste, *Trends Food Sci. Technol.*, 2002, **13**, 217.
113 J. Strassburg, A. Burbidge, A. Delgado and C. Hartmann, *J. Biomech.*, 2007, **40**, 3533.
114 A. Weststrate, G. van Poppel and P. M. Verschuren, *Brit J. Nutr.*, 2002, **88**, S233.
115 C. M. Hasler and A. C. Brown, *J. Am. Dietectic Ass.*, 2009, **109**, 735.
116 C. M. Hasler, *J. Am. Coll. Nutr.*, 2000, **19**, 499S.
117 D. J. McClements, E. A. Decker, Y. Park and J. Weiss, *Crit. Rev. Food Sci. Nutr.*, 2009, **49**, 577.
118 L. Sagalowicz and M. E. Leser, *Curr. Opin. Colloid Interface Sci.*, 2010, **15**, 61.
119 E. Acosta, *Curr. Opin. Colloid Interface Sci.*, 2009, **14**, 3.
120 M. A. Augustin and Y. Hemar, *Chem Soc. Rev.*, 2009, **39**, 902.
121 J. Flanagan and H. Singh, *Crit. Rev. Food Sci. Nutr.*, 2006, **46**, 221.
122 C. M. Hasler, *J. Nutr.*, 2002, **132**, 3772.
123 A. Sibbel, *Social Sci. Med.*, 2007, **64**, 554.
124 The Economist, "The fad for functional foods: Artificial Success", September 24, 2009.
125 G. C. Daily and P. R. Ehrlich, *Ecolog. Appl.*, 1996, **6**, 991.
126 N. Schaller, *Agric. Ecosyst. Environ.*, 1994, **46**, 89.
127 H. C. J. Godfray, J. R. Beddington, I. R. Crute, L. Haddad, D. Lawrence, J. F. Muir, J. Pretty, S. Robinson, S. M. Thomas and C. Toulmin, *Science*, 2010, **327**, 812.
128 J. G. A. J. van der Vorst, S.-O. Tromp and D.-J. van der Zee, *Int. J. Prod. Res.*, 2009, **47**, 6611.
129 R. Egbert and C. Borders, *Food Technol.*, 2006, **60**, 29.
130 R. Goodland, *Ecolog. Econ.*, 1997, **23**, 189.
131 J. W. Hansen, *Agric. Syst.*, 1996, **50**, 117.
132 M. Pollan, *The Omnivor's Dilemma: A Natural History of Four Meals*, Penguin, 2007.
133 D. Coley, M. Howard and M. Winter, *Food Policy*, 2009, **34**, 150.
134 B. J. Brown, M. E. Hanson, D. M. Liverman and R. J. Merideth, *Environ. Manag.*, 1987, **11**, 713.
135 M. C. Heller and G. A. Keoleian, *Agric. Syst.*, 2003, **76**, 1007.
136 P. W. Gerbens-Leenes, H. C. Moll and A. J. M. Schoot Uiterkamp, *Ecolog. Econ.*, 2003, **46**, 231.
137 L. Sirieix, M. Delanchy, H. Remaud, L. Zepeda and P. Gurviez, *Int. J. Consumer Studies*, Published on line May 30, 2012, DOI: 10.1111/j.1470-6431.2012.01109.x.
138 P. W. Anderson, *Science*, 1972, **177**, 393.
139 J. Goldstein, *Emergence*, 1999, **1**, 49.
140 M. A. Bedau, *Philosoph. Perspec.*, 1997, **11**, 375.
141 J. P. Crutchfield, *Phys. D*, 1994, **75**, 11.
142 T. De Wolf and T. Holvoet, *LNAI*, 2005, **3464**, 1.
143 G. Campbell-Platt, *Food Res. Int.*, 2003, **27**, 253.
144 F. Kuchler, E. Golan, J. N. Variyam and S. R. Crutchfield, *Amber Waves*, 2005, **3**, http://webarchives.cdlib.org/sw1vh5dg3r/http://www.ers.usda.gov/AmberWaves/June05/Features/ObesityPolicy.htm. Accessed August 29, 2012.
145 J. E. Tilottson, *Annu. Rev. Nutr.*, 2004, **24**, 617.
146 M. Lagi, K. Z. Bertrand and Y. Bar-Yam, *arXiv*:1108.2455v1, 2011.
147 M. Ercsey-Ravasz, Z. Toroczkai, Z. Lakner and J. Baranyi, *PlosOne*, 2012, **7**, e37810.
148 R. C. Hoseney, *Cereal Foods World*, 1978, **23**, 362.
149 P. G. Debenedetti and F. H. Stillinger, *Nature*, 2001, **410**, 259.
150 M. Karel, personal communication, 2012.
151 M. A. Rogers, A. J. Wright and A. G. Marangoni, *Soft Matter*, 2008, **4**, 1483.
152 S. Frasch-Melnik, F. Spyropoulos and I. T. Norton, *J. Colloid Interface Sci.*, 2010, **350**, 178.
153 D. Rousseau and P. Smith, *Soft Matter*, 2008, **4**, 1706.
154 T. A. Stortz and A. J. Marangoni, *Trends Food Sci. Technol.*, 2011, **22**, 201.
155 J. B. German, R. A. Gibson, R. M. Krauss, P. Nestel, B. Lamarche, W. A. van Staveren, J. M. Steijns, L. C. P. G. M. de Groot, A. L. Lock and F. Destaillats, *Eur. J. Nutr.*, 2009, **48**, 191.
156 T. Tholstrup, *Curr. Opin. Lipidol.*, 2006, **17**, 1.
157 A. Lopez-Rubioa and E. P. Gilbert, *Trends Food Sci. Technol.*, 2009, **20**, 576.
158 D. L. Dalgleish and F. R. Hallett, *Food Res. Int.*, 1995, **28**, 181.
159 J. Ubbink and P. Schär-Zammaretti, *Micron*, 2005, **36**, 293–320.
160 A. Gunning, A. R. Mackie, P. Wilde and V. Morris, *Langmuir*, 2004, **20**, 116.

161 D. Kilburn, S. Townrow, V. Meunier, R. Richardson, A. Alam and J. Ubbink, *Nat. Mater.*, 2006, **5**, 632–635.
162 P. Fischer and E. J. Windhab, *Curr. Opin. Colloid Interface Sci.*, 2011, **36**, 16.
163 R. I. Tanner, Engineering rheology, Oxford University Press, New York.
164 C. Urban and P. Schurtenberger, *Phys. Chem. Chem. Phys.*, 1999, **1**, 3911.
165 M. J. Gidley, *Trends Food Sci. Technol.*, 1992, **3**, 232.
166 K. S. Lim and M. Barigou, *Food Res. Int.*, 2004, **37**, 1001.
167 B. R. Pinzer, A. Medebach, H. J. Limbach, C. Dubois, M. Stampanoni and M. Schneebeli, *Soft Matter*, 2012, **8**, 4584.
168 L. Slade, H. Levine and D. S. Reid, *Crit. Rev. Food Sci. Nutr.*, 1991, **30**, 115.
169 P. Barham, *The Science of Cooking*, Springer, Berlin, 2001.
170 N. Myrvold, C. Young and M. Bilet, *Modernist Cuisine: The Art and Science of Cooking*, The Cooking Lab, Bellevue WA, 2011.
171 C. Vega, J. Ubbink and E. van der Linden, *The Kitchen as Laboratory: Reflections on the Science of Food and Cooking*, Columbia University Press, New York, 2012.
172 C. Vega and P. Castells, in: *The Kitchen as Laboratory*, C. Vega, J. Ubbink and E. van der Linden, ed., Columbia University Press, New York, 2012. p. 25.
173 P. B. Durst, D. V. Johnson, R. N. Leslie and K. Shono, *Forest insects as food: humans bite back*, Food and Agriculture Organization of the United Nations, Bangkok, 2010.
174 Marcel Dicke, Why not eat insects? TED Talks, December 2010. http://www.ted.com/talks/marcel_dicke_why_not_eat_insects.html. Accessed on August 28, 2012.
175 J. Ramos-Elorduy Blásquez, J. M. Pino Moreno and V. H. Martínez Camacho, *Food Nutr. Sci.*, 2012, **3**, 164.
176 H. C. Klunder, J. Wolkers-Rooijackers, J. M. Korpela and M. J. R. Nout, *Food Control*, 2012, **26**, 628.
177 J. Ramos-Elorduy, *Entomol. Res.*, 2009, **39**, 271.

Designing colloidal structures for micro and macro nutrient content and release in foods

David A. Garrec, Sarah Frasch-Melnik, John V. L. Henry, Fotis Spyropoulos and Ian T. Norton

Received 16th February 2012, Accepted 20th March 2012
DOI: 10.1039/c2fd20024d

We report on how edible nano-emulsions can be designed and produced in order to remain stable on storage. Edible nano-emulsions can potentially be used to target and control delivery of micronutrients to the human gastrointestinal tract. A class of microstructures that offers enormous potential in foods is duplex (or double) emulsions. In this paper we report the ability to design and construct particle and low molecular weight emulsifier stabilised edible duplex emulsions; *i.e.* Pickering-in-Pickering emulsions. This novel design opens up routes for significant fat replacement in a way that is imperceptible to the consumer. Having demonstrated the ability to design novel emulsion structures for food applications, we finally present data on how fluid gel structures can be designed and used in foods to give fat-like lubrication properties in the absence of fat.

1. Introduction

Dietary related health problems are increasing throughout the world, most particularly associated with consumption of foods with high levels of calories and/or salt. These can increase the risk of developing numerous health conditions such as type II diabetes, strokes, increased blood pressure and cardiovascular diseases. Processed foods tend to contain high levels of salt, fat and sugar in order to meet consumers' demand for flavour and texture.[1] In recent years, increased pressure has been placed on the producers of processed foods to lower these levels. However, reduced fat, or salt, alternatives are often of inferior quality and are less favoured by consumers.[2] This is particularly problematic for the western world, where consumers tend not to comprise their short-term eating pleasure for their long term health benefits. The challenge for the future, therefore, is to develop healthier alternatives to processed foods without detriment to the consumers' experience. Structured foods are likely to play a large role in achieving this, where colloidal systems can be designed to provide health benefits in such a way that their bulk properties and in-mouth behaviour are not necessarily influenced.

The role of designer colloids in structured foods of the future will be discussed here in detail, where, for example, hydrophobic nutrients can be delivered 'discretely' within nano-scaled emulsion droplets invisible to the naked eye. To lower the fat content of emulsion based foods, double emulsions are structured where water is dispersed within the already dispersed oil phase, and salt release can be triggered within the mouth by the breakdown of the stabilising structure. Purely aqueous systems can also be devised whereby gelled particles, in the form of fluid gels, provide lubrication.

Department of Chemical Engineering, University of Birmingham, Edgbaston, Birmingham, B15 2TT

2. Materials and methods

2.1 Nano-emulsions

2.1.1 Preparation. Decane (Sigma Aldrich, purity > 99%), Medium Chain Triglyceride (MCT) (Myritol 318, Cognis Ltd, UK), soybean oil (soya bean oil, Sigma Aldrich) and deionised water were used to prepare oil-in-water emulsions with two emulsifiers: Tween 80 (polyoxyethylene sorbitan monooleate, Sigma Aldrich) and phosphatidylycerol (Cat. No. Lipoid S PG-3, Lot No. 264020-1, Lipoid GMBH, Germany).

Nano-emulsions were produced by first forming a pre-emulsion by adding oil (10 vol%) to an aqueous surfactant solution, then mixing for 60 s using a Silverson L4R Laboratory mixer with a fine emulsifier attachment. The pre-emulsions (10–30 μm) were then passed, once, through an M110-S Microfluidizer with a Microjet high-shear chamber at 1035 bar.

2.1.2 Droplet sizing. An ALV/CGS-8F S/N 052 laser Dynamic Light Scattering (DLS) goniometer was used as supplied by ALV (GmbH, D-63225 Siemensstrasse 4, Langen/Hessen Germany). Samples were measured immediately after diluting (10,000 times) with deionised water and were stored in a refrigerator (3 °C) between measurements. Droplet radii were determined *via* intensity autocorrelation using the cumulants method.[3]

2.2 Pickering-in-Pickering stabilised $W_1/O/W_2$ emulsions

2.2.1 Preparation. The primary emulsion (30 wt% water in sunflower oil) was stabilised by 0.5 wt% monoglyceride (an equal mixture of Dimodan P Pel/B and Dimodan HP, provided by Danisco) and 1% (wt/wt in the oil phase) tripalmitin (supplied by Sigma Aldrich), and was prepared by passing the homogenised mixture (at 80 °C) through a scraped surface heat exchanger and a pin stirrer (both cooled to 5 °C). The W_1 phase also contained 1.6% (wt/wt in W_1) KCl. The stability of such an emulsion has been shown to be >3 months when stored at 10 °C.[4]

To prepare the W_2 aqueous phase, 1% (wt/wt in W_2) silica particles (Aerosil 200, Degussa, UK) were dispersed at pH 2 in a 10% (wt/wt in W_2) glucose solution (D+-glucose, supplied by Sigma Aldrich), using a high intensity ultrasonic vibracell (Jencons-PLC). Finely dispersed silica particles (floc size ~100 nm) were obtained. 0.1% sodium azide (supplied by Sigma Aldrich) was added to this dispersion as a preservative.

As the particles carry no charge at pH 2, there is no electrostatic inter-particle repulsion and the interfacial layer is tightly packed with silica particles.[5] It was previously shown that vegetable oil-in-water emulsions containing 1% silica particles are stable without any additional emulsifiers,[5] even though the particles are hydrophilic. It is likely that impurities such as mono- and diglycerides in "natural" oils (*e.g.* rapeseed and sunflower) modify the wetting properties of these particles sufficiently for them to be attracted to the oil/water interface.

A 2% starch solution was prepared by dissolving 2% (wt/wt) OSA starch (n-creamer 46, National Starch, UK) in a 10% (wt/wt) glucose solution at room temperature. 0.1% (wt/wt) sodium azide (supplied by Sigma Aldrich) was added to this dispersion as a preservative.

To prepare the double emulsion, the primary emulsion was mixed with equal parts of sunflower oil in order to reduce the viscosity and to allow even dispersion in the aqueous W_2 phase. This mixture was then slowly added to the silica particle dispersion or OSA starch/glucose solution in a rotor/stator mixer (Silverson). The mixture was then sheared at 8,000 rpm for 3 min while cooling in an ice bath.

2.2.2 Particle sizing. The resulting emulsions were stored at 10 or 25 °C. Double emulsion droplet size was measured periodically using a Mastersizer 2000. Salt release from the W_1 phase was measured at regular intervals using a Mettler Toledo 7 Easy conductivity meter and probe (InLab 710). In order to minimise the impact of emulsion droplets on the conductivity measurements, samples were taken from the serum phase of the creamed emulsion.

2.3 Fluid gels

2.3.1 Preparation. Kappa-carrageenan (κC) (kindly provided by Unilever) and KCl (Sigma Aldrich) were used as supplied without further purification. To a stirred de-ionised KCl solution (0.3 wt%), κC was dispersed and stirred for 1 h at room temperature. This was then heated to $\sim$85 °C for 1 h before being fed into a jacketed pin-stirrer. The pin-stirrer was used at 1,438 rpm and the inlet and exit temperatures were $\sim$50 and 5 °C, respectively. The pin-stirrer thus provided sheared cooling of the gelling kappa-carrageenan to produce fluid gels.

2.3.2 Tribology. Friction coefficients are measured using a MTM2 (Mini Traction Machine) tribometer (PCS Instruments) at a rotating ball-on-disc contact mode. Balls and discs were made from PDMS (Sylgard 184; Dow Corning) which was set in moulds at 70 °C overnight. The MTM2 allows the user to control fluid entrainment speed (U), applied normal load (W) and relative sliding to rolling motions (slide-to-roll ratio; SRR). Friction coefficients (μ) are obtained ($\mu = F/W$) from the recorded tangential friction force, F. Stribeck curves are recorded at $W = 2$ N, $SRR = 50\%$ where U is ramped up from 1 to 1,000 mm s^{-1} and then ramped down; this procedure is repeated three times constituting one test. The tests are repeated three times for each sample where the ball and disc surfaces are used for one test only, and the resultant 18 Stribeck curves are averaged and plotted with error margins representing one standard deviation of error. The temperature is controlled and maintained at 10 °C throughout testing.

3. Results and discussion

3.1 Nano-emulsions

There has been a growing interest in the formation and use of nano-scaled emulsions for food formulation. Nano-emulsions have been defined as those where the dispersed phase is of the order of 10–100 nm in radius.[6] Their small size provides a large interface for the delivery of e.g. flavours or nutrients, and they can also be translucent provided that the droplet dimensions are smaller than the wavelength of light. This property is particularly beneficial for beverage formulation. Nano-emulsions can lead to enhanced stability against creaming/sedimentation, coalescence and flocculation as a consequence of the greater impact of Brownian motion as the size decreases; however, they are typically more susceptible to Ostwald ripening as a consequence of the increased Laplace pressure.

3.1.1 Edible nano-emulsions. The formation of nano-emulsions prepared with edible oils (medium chain triglyceride (MCT) and soybean oil) has been investigated using Tween 80. Nano sized emulsions were produced using a high pressure impinging jet device in which the impact is followed by elongational flow[7] which allows the low molecular weight emulsifier to enter and stabilise the interface before droplets collide. The data obtained here is compared with data for decane in terms of droplet size and stability against ripening.

Fig. 1 shows the dependence of Tween 80 concentration on the radius of oil-in-water emulsions using decane, MCT and soybean oil. Droplet radii decrease with Tween 80 concentration until a minimum, which, for all oils, was $\sim$15 mM. Since this is the concentration at which coalescence events are prevented,[7] emulsion droplet size in the presence of Tween 80 concentrations over 15 mM is determined by the process only. The limiting droplet size produced in this study is the same

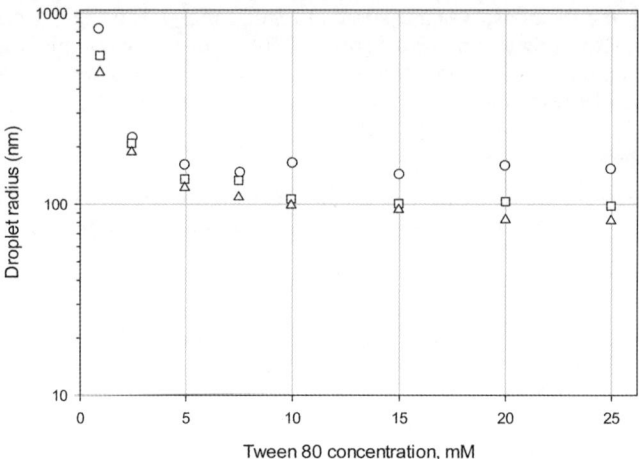

Fig. 1 Dependence of Tween 80 concentration on O/W droplet size with decane (○), soybean oil (□) and MCT (△).

as those reported earlier with decane as the oil phase. This demonstrates that the droplet size produced on impact is independent of the oil viscosity and that as long as droplet re-coalescence events are "removed", as a limiting step, edible nano emulsions can be produced.

Whilst the formulation of edible nano-emulsions seems promising from the droplet size data (Fig. 1), their stability must also be considered. To investigate mechanisms of instability, the ripening rates were assessed as a function of Tween 80 concentration. This is shown in Fig. 2 for decane, MCT and soybean oil, where the ripening rate was calculated over a 200 h period. The data for decane shows an "optimum" Tween 80 concentration (~2.5 mM) where the ripening rate is mini- mised, as previously discussed, whilst greater concentrations cause higher ripening rates *via* micelle mediated decane transport. However, the edible MCT and soybean oils do not suffer from such effects at any Tween 80 concentration; mechanistic reasons for this will be discussed later. After 2 weeks, the samples began to cream,

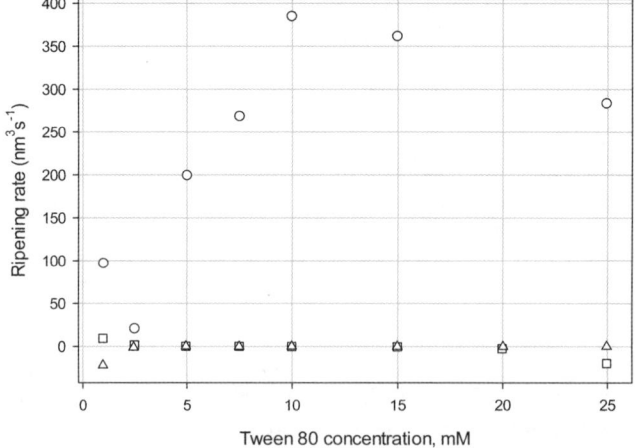

Fig. 2 Dependence of Tween 80 concentration on O/W ripening rate with decane (○), soybean oil (□) and MCT (△).

which could be prevented by addition of a "viscosity modifier", such as a hydrocolloid, to prevent gravitational separation. The radius cubed, a^3, is used to show volumetric changes to droplets (Fig. 3) as a function of time for decane, MCT and soybean oil, emulsified with Tween 80; data for decane emulsified with phosphatidylycerol are also shown. The effect of emulsifier can be seen for decane-in-water emulsions. Tween 80 mediates the ripening process, whereas phospholipids prevent it by forming aggregates at the decane–water interface that prevent/restrict phase transport.[8]

Edible O/W emulsions prepared from either MCT or soybean oil are stable against ripening at all Tween 80 concentrations (Fig. 2) and for up to 200 h (Fig. 3). This indicates that Tween 80 mediated oil transport does not occur, unlike the behaviour with decane. This dissimilarity is a result of the water solubilities of the oils used in this study; where decane is partially water-soluble and MCT and soybean oil are much less so. Thus transport through the aqueous phase, even with Tween 80 micelles present, is unfavourable for MCT and soybean oil.

3.2 Pickering stabilised double emulsions

Oil-in-water (O/W) emulsions (*e.g.* mayonnaise or salad dressings) are typically high in levels of fat, sugar and salt. One way to lower the fat content is to partially replace the oil content with water in the form of a water-in-oil-in-water emulsion, or a $W_1/O/W_2$ double emulsion. If these can be designed and constructed to osmotically separate the two aqueous phases, they can potentially be used to reduce salt and/or sugar in products. This is because it is the solute concentration that influences its perception within the mouth (rather than mass), and so by including significant amounts of the water into the fat droplets which contains no salt or sugar, the outer phase can be more highly concentrated and thus give the sensation required.

One of the main problems in $W_1/O/W_2$ double emulsions is that of retaining solutes within the W_1 encapsulated phase. The presence of two different types of emulsifiers in double emulsions is a requirement to provide stability to a system with oppositely curved interfaces. The close proximity of the two interfaces facilitates water and/or solutes transport from one aqueous phase to the other (depending on the direction of the osmotic pressure gradient) by formation of reverse micelles or "holes" in the oil layer.[9–11]

Previous work[12] has demonstrated that fat crystal-stabilised W/O emulsions can be produced with sintered crystal shells at the interface, effectively giving osmotic

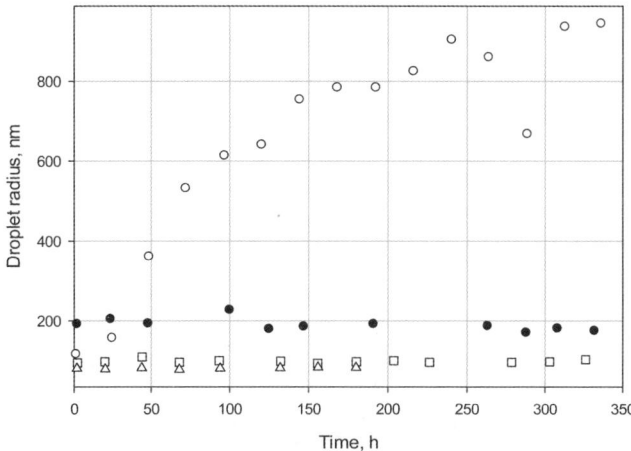

Fig. 3 Droplet size against time for Tween 80 stabilised O/W emulsions with decane (○), soybean oil (□), MCT (△), and decane stabilised with phosphatidylcylcerol (●).

isolation of the included water. When such emulsions are incorporated into a double structure water/solute transport can be reduced or even prevented, despite the presence of a hydrophilic surfactant, Na-caseinate, used for the stabilisation of the O/W_2 interface.[13] Although the double structure was preserved for several weeks storage, the use of Na-caseinate as secondary emulsifier (*i.e.* stabilising the interface between the primary W_1/O emulsion globules and the W_2 aqueous phase) did not prevent coalescence between double emulsion globules. It was proposed that coalescence between the double emulsion globules was caused by protruding fat crystals (stabilising the W_1/O interface) piercing the interfacial film of neighbouring globules.[13]

3.2.1 Pickering-in-Pickering stabilised $W_1/O/W_2$ emulsions.
Small particles adsorbed at droplet interfaces (Pickering emulsions) give a thicker and more rigid surface film than that of small molecular weight emulsifiers/surfactants and proteins.[14] In light of this, we have proposed that if particles are placed at the second interface in the duplex emulsion then this thicker and more rigid film should effectively protect oil globules from coalescence by physically preventing contact of the oil phase in each globule, despite the existence of fat crystals. In this investigation, two different types of Pickering materials were investigated as "emulsifiers" for the secondary interface: silica particles and colloidally solved OSA-starch.

Double emulsions stabilised by 1% silica particles in the W_2 aqueous phase and stored at 10 °C showed little variation in drop size distribution over time (Fig. 4). When the double emulsions were stored at 25 °C, a small shift in average globule size occurred during storage ($D_{3,2}$ after production ~20 ± 1 µm, after 8 weeks ~22 ± 1 µm). However, this shift was within the experimental error (see Fig. 4). These results show that it is possible to prevent coalescence between double emulsion globules containing fat crystal-stabilised water droplets when silica particles are used to stabilise the secondary interface.

Formulations containing 2% OSA starch were also assessed for their stability against coalescence when stored at 10 and 25 °C. Double globules were small, in the order of 10 ± 0.5 µm, due to the higher viscosity (600 mPa.s) of the starch containing W_2 phase compared to that in the silica experiments. Globule size distribution remained constant at a storage temperature of 10 °C, as shown in Fig. 5. Light microscopy confirmed that the double emulsion structure was retained during this time.

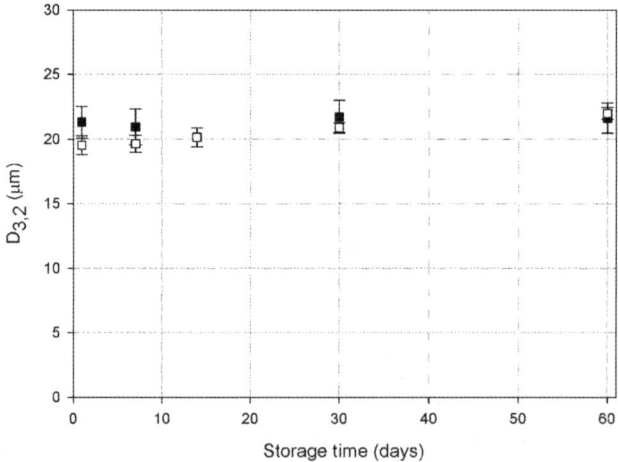

Fig. 4 Size distribution for double emulsions stabilised by fat crystals (primary W1/O interface) and 1% silica particles (secondary O/W2 interface). Storage temperature 10 (■) or 25 °C (□).

According to Tesch et al.,[15] OSA starch molecules are colloidally solubilised and sterically stabilise oil globules. Nilsson and Bergenståhl[16] have shown that the surface coverage of OSA-starch varies from 3 mg m^{-2}, to as high as 16 mg m^{-2}. A likely explanation for the stability against coalescence is therefore the surface loading of the starch grains on the double emulsion interface, giving a thick layer that prevents crystals from protruding from the oil globules and "damaging" neighbouring globules.

A higher storage temperature (25 °C) decreases the stability of starch-stabilised double emulsions against coalescence. Droplet size appeared to increase from 15 to 21 μm within the first month of storage and then decreased (Fig. 5). The reason for the decrease is the gradual formation of large aggregates of primary emulsion in the continuous aqueous phase. These are no longer part of the double structure and could not be accurately measured using the Mastersizer. The formation of such aggregates was reflected in the changing shape of the droplet size distribution curves (Fig. 6) and confirmed using light microscopy.

The instability observed by measuring droplet size was further checked using salt release as an indicator. The duplex emulsions stabilised with silica particles (1% at pH 2) or OSA starch (2%) were subjected to conductivity analysis and compared with the stability of double emulsions stabilised by Na-caseinate as the secondary emulsifier. The inner water phase (W$_1$) contained 1.6% KCl and the outer water phase (W$_2$) contained 10% glucose to match the osmotic pressure between the two aqueous phases. Conductivity measurements were performed on samples stored for several weeks at 10 or 25 °C.

Fig. 7 shows that double emulsions stabilised by 1% silica stored at 10 °C had released almost no salt to the continuous aqueous phase after 2 months of storage. When stored at 25 °C the same emulsion released around 25% of the encapsulated salt. The formulation containing 2% starch and 1% Na-caseinate had released less than 20% salt to the W$_2$ phase after two months of storage at 10 °C. The same samples released 30% salt when they were stored for the same time period at 25 °C. These results support the droplet size measurements that show that the secondary emulsifier influences stability and that Pickering-in-Pickering double emulsions can be constructed to give stability over extended periods of time.

The reason behind the observed instability for samples stored at room temperature is the presence of liquid, and therefore mobile, monoglyceride molecules within

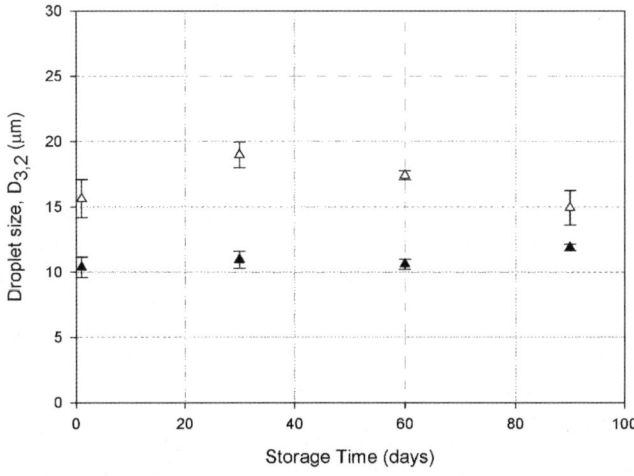

Fig. 5 Droplet sizes of double emulsions stored at 10 (▲) or 25 °C (△) and stabilised by 2% OSA starch in W2.

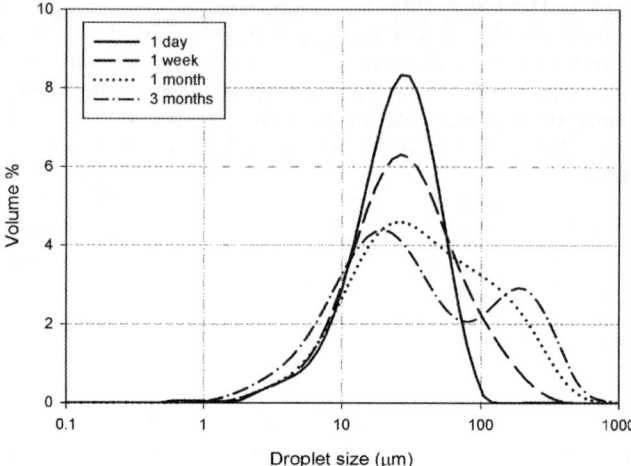

Fig. 6 Size distributions of double emulsions stored at 25 °C containing 2% OSA starch as secondary emulsifier.

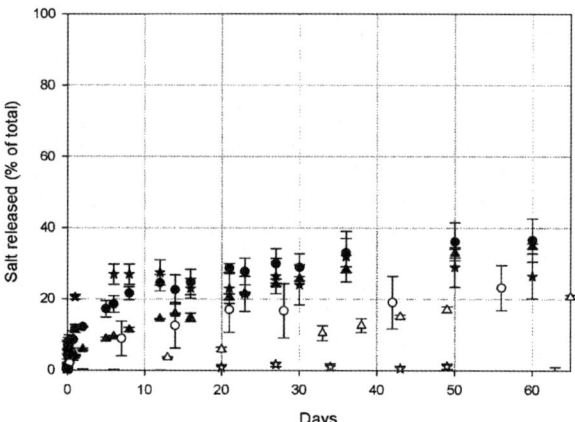

Fig. 7 Release of KCl from the encapsulated W1 phase to the continuous W2 phase, when 1% silica particles (★☆), 2% OSA-starch (▲ △) or 1% na-caseinate (◆ ◇) were used as secondary emulsifiers. The open symbols denote storage temperatures at 10 °C, and the closed symbols at 25 °C.

the oil phase. At 10 °C, monoglycerides are in their crystalline state and therefore incorporated into the "shell" structure surrounding the primary emulsion droplets. This prevents their migration between interfaces and subsequent destabilisation of the secondary interface.[13] However, monoglycerides become increasingly soluble in the oil phase at temperatures greater than 20 °C and this increases their mobility between the two interfaces. Mobile monoglycerides may decrease the stability of an OSA-starch stabilised interface because monoglycerides have a higher surface activity than OSA starch and could therefore displace it from the secondary interface. Interfacial tension measurements were performed to confirm this: when the water phase contained 2% OSA starch and the oil phase was sunflower oil, the interfacial tension, measured using the Wilhelmy plate method in a Krüss K100 Tensiometer at room temperature, was measured as 20 ± 1 mN. On the other hand, when the oil phase contained 0.5% monoglyceride and 1% tripalmitin, and the aqueous phase was distilled water, the interfacial tension was 3.3 ± 0.1 mN (again at room temperature).

A key part of the mechanism has been tested and it has been shown that chemical potential drives water/solute transport in double emulsions despite the absence of an osmotic pressure gradient between the two aqueous phases. However, the existence of a crystalline "shell" surrounding the primary emulsion water droplets minimises water transport. Some salt is nevertheless released through imperfections in these "shells". In order to be released to the continuous aqueous phase, this "leaked" salt must pass through the oil phase and the secondary interface. At 10 °C the availability of small-molecular weight surfactants is limited because most monoglycerides are incorporated into the fat crystal network. At higher temperatures (greater than 25 °C) this is no longer the case and the greater availability of "free" monoglycerides can compromise the double emulsion structure.

3.2.1.2 Triggered release. Assuming that the main stabilisation mechanism for the primary emulsion is the existence of sintered crystals at its interface, then it should be possible to take advantage of the melting characteristics of these shells, thus the removal of the Pickering particles from the primary emulsion interface, to trigger release of encapsulated material. This hypothesis was tested by heating samples in a water bath set at 50 °C with gentle stirring ensuring an even temperature distribution. Continuous temperature and conductivity measurements were taken until complete salt release was achieved. Fig. 8 shows that as the temperature rose beyond 32 °C, rapid and complete salt release occurred within 50 s in all formulations. Fig. 9 shows micrographs taken at 10 °C (before heating) and at 50 °C (after heating), of an emulsion containing 1% silica particles as the secondary emulsifier (although all other formulations show the same behaviour). These micrographs confirm that the emulsion had lost its double structure character during the heating process. The reason for the destabilisation of the primary emulsion at this temperature is that the stabilising fat crystal network surrounding the primary emulsion droplets melts.[4] The ensuing large-scale coalescence of the primary emulsion droplets leads to a rapid release of all water and encapsulated solutes to the continuous aqueous phase, and the emulsion loses its double character.

This work has shown direct controlled release of encapsulated substances in double emulsions using temperature as a trigger. Very stable Pickering-in-Pickering double emulsions have been created using fat crystals on one-interface and silica particles on the other. The secondary interface is also effectively stabilised when using OSA-starch instead of silica particles. Both substances form rigid interfacial

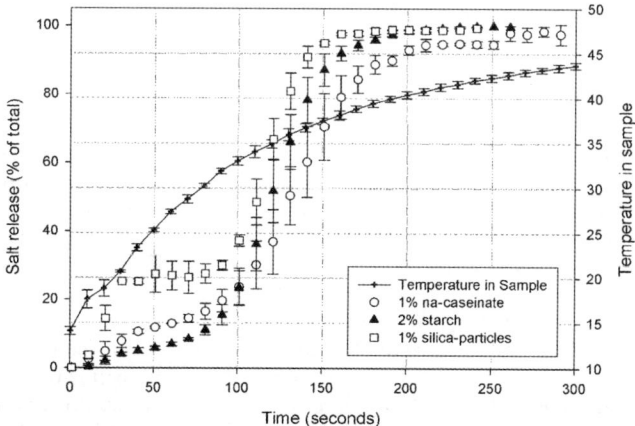

Fig. 8 Salt release of different formulations at elevated temperature: rapid complete salt release from W1 to W2 as the sample temperature rises above 30 °C.

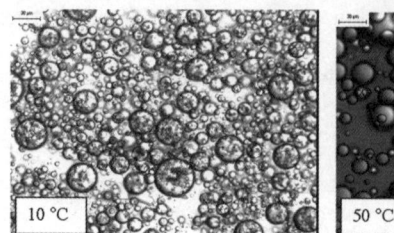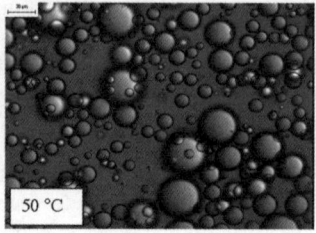

Fig. 9 Heating causes the emulsion microstructure to change from double to single O/W emulsion.

structures[14,15] that are resistant to damage by fat crystals on the included primary emulsion interface. Salt release can be triggered at temperatures that are higher than the melting point of the fat crystals.

3.3 Fluid gels

Fluid gels are a colloidal system as they are suspensions of solid gelled particles dispersed within a non-gelled matrix. They are formed by cooling a gelling biopolymer solution within a sheared environment. Particle formation is initiated *via* spinodal decomposition at the onset of gelation, and the applied shear maintains the particulate composition, provided it is sustained throughout the gelation process. Fluid gels can thus be produced solely from gelling biopolymer solutions *i.e.* with water contents >99%. Their rheological properties are highly tuneable through choice of biopolymer concentration (which influences particle elasticity) and volume fraction, either through dilution or centrifugation.

The shear required to produce fluid gels can be provided within a continuous process pin-stirrer. This technique has been used to yield average particle diameters as low as 100 μm with agarose[17] and 1 μm with kappa-carrageenan.[18] This biopolymer dependence on particle size is likely to arise from the rate of the coil to helix transition which is both biopolymer and, for salt mediated gelation, ionic concentration dependent. The kappa-carrageenan fluid gel particles are undetectable to the naked eye and upon touch/oral consumption.[18] Therefore, it can be argued that if the rheology and particle size are controlled, along with the interaction between the particles (*i.e.* bridging), fluid gels have the potential to be used in food formulations where they partially substitute the dispersed oil phase of emulsion based foods. We have tested this design rule using soft tribology and rheology.

3.3.1 Fluid gel tribology. The ability of fat-containing foods to lubricate motions made by the tongue and palate during consumption can influence fat-related textural attributes.[19,20] Thus solid-surface lubrication is an important factor to study in reduced fat food formulation. Soft-tribology is a well characterised technique[21] and can be used to correlate friction coefficients with sensory data.[22] Soft (PDMS) surfaces are used to replace the steel components that are typically used in tribological setups. This provides low contact pressures as expected for the cases of tongue-on-palate interactions during food consumption. Additionally, PDMS can be rendered hydrophilic by incorporating a surfactant into the rubber;[23] such surfaces mimic oral mucosa that is coated in salivary proteins.

In order to develop fluid gel products that have the perception of fat when little or none is present, we need to understand how the gelled particles behave in the mouth. The physical sensation of textural attributes, and indeed, fat perception, tends to be one of a mixture of tribological and rheological response, thus both should be probed. Work by Gabriele *et al.*[17] has shown Stribeck curves for 1 wt% agarose fluid gels (Fig. 10). The data shows two Stribeck curves (friction coefficient *versus* disc

This journal is © The Royal Society of Chemistry 2012

speed): one obtained by ramping the entrainment speed up from 1–100 mm s^{-1}, and the other by ramping down over the same range of speeds. There is a marked difference in these Stribeck curves. Ramping up shows friction coefficient initially decreasing until ~15 mm s^{-1}, where there is an increase in friction, which is maintained until ~30 mm s^{-1}, then friction continues to decrease with speed. Ramping down, however, shows friction continually increasing with speed. This hysteresis behaviour is not typical of fluid lubrication, but can be explained by understanding particle entrainment. At speeds below 15 mm s^{-1}, the film thickness between the ball and disc elements of the equipment is small and only the continuous phase is able to entrain the gap and provide lubrication. As the speed increases, however, there will be a film thickness/entrainment speed at which the particles begin to enter the contact zone, and at this stage the friction temporarily increases due to, effectively, a thinner layer of fluid between the surfaces of the ball–particle(s)–plate. Greater speeds (>30 mm s^{-1}) allow multiple layers of particles and bulk fluid to enter the gap and thus provide lubrication; i.e. a reduction in friction. The ramping down behaviour is different because particles reside in the contact from the onset of the measurement and presumably remain there even at the lowest speeds.

Fig. 11 shows Stribeck curves for the two kappa-carrageenan fluid gels (0.5 and 2 wt%) produced in this study and water. The data shows the average of 6 runs where the speed was ramped up then down. As can be seen, the data is in contrast to that for agarose fluid gels with no friction hysteresis on ramping up and down. At all entrainment speeds, the friction coefficient for κC fluid gel lubrication is lower than that of water (Fig. 11) and friction is lower for the fluid gel prepared with the higher κC concentration. This implies that the presence of the particles is beneficial for lubricity, where increasing their stiffness through biopolymer concentration lowers friction. The fact that friction is lower in the presence of particles, and with increasing their stiffness, together with the superimposable ramped up and down tests, suggests that particle exclusion does not occur during these tribological experiments. This can be explained by considering the dimensions of the particles in relation to the roughness of the surface. The PDMS ball and disc have root mean square (R_q) roughness values of 324 and 894 nm, respectively, and their profiles have numerous regions where the peak to trough value exceeds 1 μm. Thus it would seem that κC particles (~1μm)[18] are small enough to be entrained into the contact due to the surface roughness, even in boundary lubrication conditions where the two surfaces are effectively in full contact. On the contrary, agarose particles

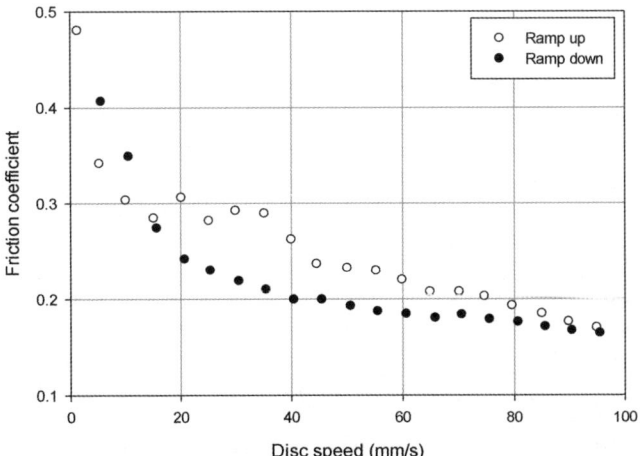

Fig. 10 Stribeck curves (a) for fluid gels prepared with 1 wt% agarose; ramping up (○) and down (●) adapted from Gabriele et al.[17]

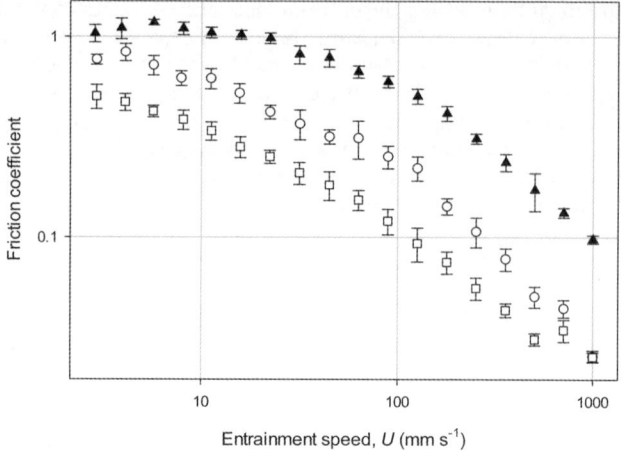

Fig. 11 Stribeck curves for deionised water (▲) and fluid gels prepared from 0.5 (○) and 2 (□) wt% κC.

($\sim$100 μm)[17] are much larger and thus they are excluded until the film thickness increases with rotation speed eventually allowing entrainment to occur.

4. Conclusions

Edible nano-emulsions (80–100 nm) can be produced which are stable on storage. This requires a process that induces elongational flow immediately after droplet breakup, thus allowing the emulsifier time to adsorb at the interface before droplets collide causing coalescence. In addition, the oil needs to have low solubility in water, thus suppressing Ostwald ripening.

Duplex emulsions designed and constructed with particles at both oil/water interfaces are stable and, if fat crystals are used to stabilise the inner interface, their melting can be used to trigger the release of addenda even of low molecular weight; *i.e.* salt or vitamins.

Fluid gels can be formed with particle diameters ($\sim$1 μm) small enough that they are not detected on consumption and provide lubrication to a sliding/rolling contact.

Acknowledgements

We are grateful to Cargill, Unilever and the EPSRC for sponsorship. We also wish to thank the Royal Commission for the Exhibition of 1851 for the Industrial Fellowship for David Garrec.

References

1 I. Norton, S. Moore and P. Fryer, *Obes. Rev.*, 2007, **8**, 83–88.
2 I. Norton, P. Fryer and S. Moore, *AIChE J.*, 2006, **52**, 1632–1640.
3 D. E. Koppel, *J. Chem. Phys.*, 1972, **57**, 4814.
4 S. Frasch-Melnik, I. T. Norton and F. Spyropoulos, *J. Food Eng.*, 2010, **98**, 437–442.
5 R. Pichot, F. Spyropoulos and I. Norton, *J. Colloid Interface Sci.*, 2009, **329**, 284–291.
6 D. J. McClements, *Soft Matter*, 2011, **7**, 2297–2316.
7 J. V. L. Henry, P. J. Fryer, W. J. Frith and I. T. Norton, *J. Colloid Interface Sci.*, 2009, **338**, 201–206.
8 J. V. L. Henry, P. J. Fryer, W. J. Frith and I. T. Norton, *Food Hydrocolloids*, 2010, **24**, 66–71.
9 J. Cheng, J. F. Chen, M. Zhao, Q. Luo, L. X. Wen and K. D. Papadopoulos, *J. Colloid Interface Sci.*, 2007, **305**, 175–182.

10 S. Matsumoto, T. Inoue, M. Kohda and K. Ikura, *J. Colloid Interface Sci.*, 1980, **77**, 555–563.
11 L. Wen and K. D. Papadopoulos, *J. Colloid Interface Sci.*, 2001, **235**, 398–404.
12 J. E. Norton, P. J Fryer, J. Parkinson and P. W Cox, *J. Food Eng.*, 2009, **95**, 172–178.
13 S. Frasch-Melnik, F. Spyropoulos and I. T. Norton, *J. Colloid Interface Sci.*, 2010, **350**, 178–185.
14 B. P. Binks, *Curr. Opin. Colloid Interface Sci.*, 2002, **7**, 21–41.
15 S. Tesch, C. Gerhards and H. Schubert, *J. Food Eng.*, 2002, **54**, 167–174.
16 L. Nilsson and B. Bergenståhl, *Langmuir*, 2006, **22**, 8770–8776.
17 A. Gabriele, F. Spyropoulos and I. T. Norton, *Soft Matter*, 2010, **6**, 4205–4213.
18 D. A. Garrec and I. T. Norton, *J. Food Eng.*, 2012, **112**, 175–182.
19 R. A. de Wijk and J. F. Prinz, *Food Qual. Preference*, 2005, **16**, 121–129.
20 R. A. de Wijk, L. J. van Gemert, M. E. J. Terpstra and C. L. Wilkinson, *Food Qual. Preference*, 2003, **14**, 305–317.
21 J. H. H. Bongaerts, K. Fourtouni and J. R. Stokes, *Tribol. Int.*, 2007, **40**, 1531–1542.
22 M. E. Malone, I. A. M. Appelqvist and I. T. Norton, *Food Hydrocolloids*, 2003, **17**, 763–773.
23 D. A. Garrec and I. T. Norton, *J. Colloid Interface Sci.*, 2012, **379**, 33–40.

Protein cluster formation during enzymatic cross-linking of globular proteins

Yunus Saricay,[a] Surender Kumar Dhayal,[b]
Peter Alexander Wierenga[b] and Renko de Vries[*a]

Received 22nd February 2012, Accepted 26th March 2012
DOI: 10.1039/c2fd20033c

Work on enzymatic cross-linking of globular food proteins has mainly focused on food functional effects such as improvements of gelation and enhanced stabilization of emulsions and foams, and on the detailed biochemical characterization of the cross-linking chemistry. What is still lacking is a physical characterization of cluster formation and gelation, as has been done for example, for cluster formation and gelation during heat-induced protein aggregation. Here we present preliminary results along these lines. We propose that enzymatic cross-linking of apo-α-lactalbumin is a good model system for studying the problem of cluster formation and gelation during enzymatic cross-linking of globular proteins. We present initial results on cluster sizes produced when cross-linking dilute solutions of apo-α-lactalbumin with a range of cross-linking enzymes: microbial transglutaminase, horseradish peroxidase, and mushroom tyrosinase. These results are used to highlight similarities and differences between different enzymes, when acting on the same substrate. Next we consider cluster growth and gelation in somewhat more detail for the specific case of cross-linking by horseradish peroxidase, under the periodic addition of H_2O_2. Upon increasing the initial concentration of apo-α-lactalbumin, at a fixed enzyme-to-substrate ratio and fixed reaction time, the size of the clusters at the end of the reaction increases rapidly, and above a critical concentration, gelation occurs. For the conditions that we have used, gelation occurred at very low initial apo-α-lactalbumin concentrations of 3–4% (w/v), indicating a very dilute cross-linked protein network, with a low average number of cross-links per protein. It is found that reactive protein monomers are first rapidly (1–2 h) incorporated into small covalent clusters. This is followed by a much slower phase (up to about 12 h) in which the small clusters are coupled together to form much larger covalent protein clusters. Consistent with this two-step mechanism, atomic force microscopy shows that the covalent protein clusters are very heterogeneous and seem to consist of smaller subclusters.

Introduction

Proteins play key roles in determining food structure and mechanics, both in bulk and at surfaces. In processed foods, various protein concentrates or isolates are being used to modify viscosity, as a gelling agent, or to stabilize foams and emulsions. These food-functional properties are coupled to both the internal structure and the aggregation state of the proteins. Controlling food-functional properties to a large extent means controlling the internal structure and aggregation state of

[a]Laboratory of Physical Chemistry and Colloid Science, Wageningen University, P.O.Box 8038, 6700 EKWageningen, the Netherlands. E-mail: Renko.deVries@wur.nl
[b]Laboratory of Food Chemistry, Wageningen University, P.O. Box 8129, 6700 EVWageningen

the proteins. Not surprisingly, heat-induced protein aggregation and protein interfacial behavior have been, and still are being investigated intensively, for a range of food proteins. In particular for the much investigated globular whey proteins α-lactalbumin and β-lactoglobulin from milk, there is a large body of work on heat-induced protein aggregation, and interfacial behavior. As for an understanding in the sense of being able to model the final protein structures and physical properties either in the bulk, or at interfaces, there have been many attempts, but only in a few cases some kind of universality has been clearly demonstrated, and models were shown to have some predictive power. One such case is that of the heat-induced aggregation of β-lactoglobulin at near-neutral pH values.[1-3] In this case, fractal aggregates are formed with fractal dimensions close to values predicted by simple aggregation models such as cluster-cluster aggregation.[4] Furthermore, it has been shown that the apparent reaction order of 1.5 at the initial stages of the reaction can be explained in terms of the chemistry of the disulphide exchange reaction that is known to be very important for the particular case of β-lactoglobulin.[5]

This case, however, is a well-investigated exception to the rule, and in general what is striking in particular heat-induced protein aggregation is an extreme sensitivity to environmental conditions, and an apparent lack of universality. In practical terms this means that it is hard to control protein aggregation as an industrial process, and that in food technology there is a continued need for new processes to control the aggregation/cluster state of proteins in a reliable way.

A route for controlling the aggregation/cluster state of proteins that is complementary to heat-induced aggregation is the use of certain enzymes that can induce covalent cross-links between proteins. Various enzymes are available, some of which are approved for use in the food industry. The best known of these is microbial transglutaminase, that catalyzes the formation of cross-links between lysine and to glutamine residues (schematically illustrated in Fig. 1a.). Another large class of cross-linking enzymes are oxidative cross-linking enzymes such as horseradish peroxidase, and mushroom tyrosinase. These catalyze the activation of a tyrosine residue that may then couple to either another tyrosine residue, or to other types of residues. Horseradish peroxidase, in the presence of H_2O_2, catalyzes the formation of radicals on the phenol groups of the tyrosine side chains. Two radicals may combine to form a dityrosine cross-link,[6] as illustrated in Fig. 1b. Mushroom tyrosinase, in the presence of dissolved O_2, catalyzes the conversion of the phenol group of tyrosine residues into a diquinone, that may subsequently react with a number of protein sidechains,[7] as illustrated in Fig. 1c.

Food functional effects of enzymatic protein cross-linking that have been found include, for example, improvements of gelation properties and emulsion stability,[8,9] and enhanced heat-stability[10] upon enzymatic cross-linking. Non-food applications of enzymatic protein cross-linking are also being considered, for example in tissue engineering, and for wool and leather.[11]

As mentioned, understanding the aggregation or cluster state is crucial to understanding and controlling the food-functional properties of proteins. For this reason, historically, many quite detailed studies of heat-induced protein aggregation for many types of proteins have been performed. It is therefore quite surprising that there is virtually no work on the architecture and physical properties of protein clusters formed during enzymatic cross-linking. Instead, studies either focus directly on functional effects, or on the biochemistry of the cross-linking reactions. Here we wish to present a first set of results on the physical characterization of covalent protein cluster formation during enzymatic cross-linking of globular proteins. In particular, we focus on the enzymatic cross-linking of the globular whey protein α-lactalbumin.

Whereas disordered, coil-like food proteins such as caseins and gelatins are excellent substrates for cross-linking enzymes due to the high accessibility of the reactive groups, they represent only a small subset of the total spectrum of food proteins.

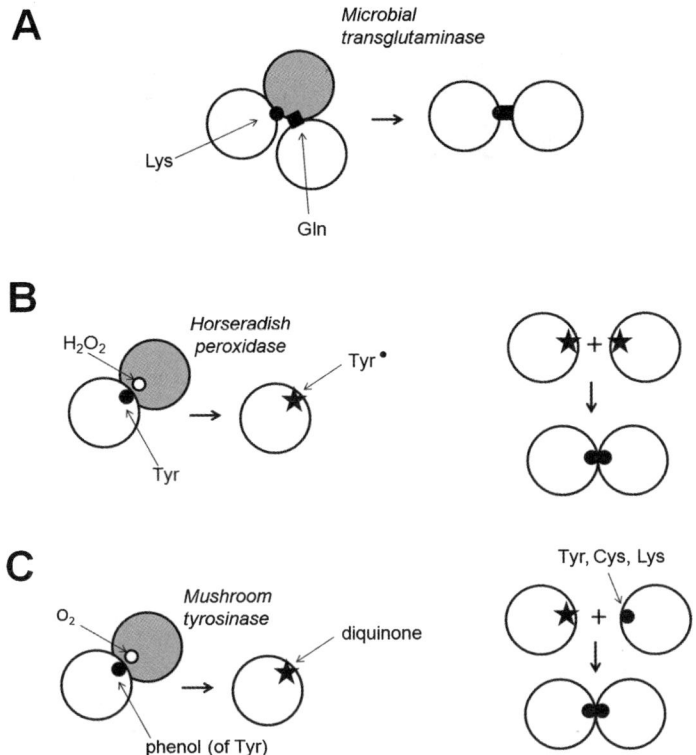

Fig. 1 Illustration of enzymatically catalyzed cross-linking of protein side-chains for a number of different enzymes. A) Microbial transglutaminase, B) horseradish peroxidase and C) mushroom tyrosinase.

Therefore, successful reactions on commercially important globular proteins, are potentially important. Generally, accessibility of the substrate reactive groups is a limiting factor when attempting to cross-link globular proteins with cross-linking enzymes. For this reason, enzymatic cross-linking globular proteins often requires some pretreatment that improves accessibility, such as adding denaturants, or a heat-treatment. A more subtle and controlled way of improving accessibility is possible for proteins that have intermediate folding states: the so-called molten-globule state.[12] For α-lactalbumin, a molten-globule state can be induced by taking away its Ca^{2+} ligand (apo-α-lactalbumin). Since α-lactalbumin is also a commercially important food protein, calcium free α-lactalbumin is an excellent model system for investigating the enzymatic cross-linking of globular food proteins.

A thorough chemical characterization of cross-linking of α-lactalbumin by microbial and mammalian transglutaminases has been performed by Matsumura and co-workers.[13–15] For the specific case of cross-linking in the absence of any dena-turants,[13] it was shown that hardly any cross-linking is observed for holo-α-lactal-bumin, but significant cross-linking was observed for the apo-α-lactalbumin. Fig. 2a shows a crystal structure of holo-α-lactalbumin in which the 12 lysine and 7 glutamine residues are highlighted. All of them are located on the surface, but nevertheless it was observed by Matsumura et al.[13] that only 3 lysines and only one glutamine (Gln54) of apo-α-lactalbumin were accessible for modification by microbial transglutaminase. A similar observation was made when DTT was used to improve accessibility.[15]

We have previously considered the initial stages of the oxidative cross-linking of α-lactalbumin catalyzed by horseradish peroxidase, under the periodic addition of

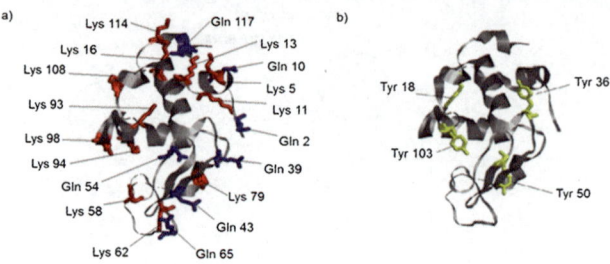

Fig. 2 a) Crystal structure of (holo) α-lactalbumin with lysines highlighted in red, and glutamines highlighted in blue. b) Crystal structure of (holo) α-lactalbumin with tyrosines highlighted in yellow.

peroxide.[6,16–18] As compared to other oxidative enzymes that use dissolved oxygen, the horseradish peroxidase system is a convenient model system since the peroxide concentration is tuned more easily than the concentration of dissolved oxygen. For eventual industrial applications other oxidative enzymes are probably better, since the peroxidase enzyme is inactivated rather quickly by excess peroxide. For this reason, in our work, to minimize inactivation we use periodic addition of small amounts of peroxide. As mentioned, under the influence of horseradish peroxidase and H_2O_2, the phenol sidechains of the 4 tyrosine residues of α-lactalbumin are converted into radicals that may combine to form a dityrosine cross-link. Fig. 2b shows the crystal structure of holo-α-lactalbumin with the 4 tyrosines highlighted. None of the phenol groups of the tyrosines are located completely at the surface, but they are not very far away from the surface either. Similar to the denaturant-free cross-linking of α-lactalbumin by microbial transglutaminase, cross-linking of holo-α-lactalbumin by horseradish peroxidase was observed to be less efficient than cross-linking of apo-α-lactalbumin.[6] Mass spectrometry data suggests[18] that initially, only dityrosine cross-links are formed between Tyr18 and Tyr50.

In the present paper we first present a brief comparison of the size distributions of covalent clusters generated after prolonged incubation of dilute of apo-α-lactalbumin (1–3%) with different cross-linking enzymes. In addition to the cases of microbial transglutaminase and horseradish peroxidase discussed above, we also consider the case of mushroom tyrosinase. Next, for the selected case of the cross-linking of apo-α-lactalbumin under the periodic addition of H_2O_2, catalyzed by horseradish peroxidase, we present initial data on covalent protein cluster formation: on-line light scattering for a series of concentrations, the rate of monomer consumption, and atomic force microscope images. The work that we present here may set the stage for further investigations of the architecture and physical properties of covalent clusters of globular proteins.

Materials and methods

Chemicals

Calcium-free α-lactalbumin was obtained from Davisco Foods International Inc., Le Sueur, MN, USA and was used as received. It is sold commercially under the name BioPURE α-Lactalbumin and is obtained from bovine milk. According to specifications of the manufacturer, the total protein content of the powder was 95% (w/w) of which 90% (w/w) was α-lactalbumin, and the calcium content of the powder was 0.055% (w/w). The latter implies that when dissolved, approximately 20% of the α-lactalbumin molecules could still be in the holo- rather than the apo-form. Horseradish peroxidase was obtained from Sigma and used as received. Tyrosinase from mushroom was obtained from Sigma (product code T3824) and used as received. Microbial transglutaminase was obtained from Ajinomoto. The crude enzyme preparation (Activa YG) was purified using cation exchange. Crude enzyme powder was dissolved in 10 mM ammonium acetate, pH $\simeq$ 6.5 to obtain around 150 ml at a concentration of 20% (w/w). This was mixed with ~300 ml of cation exchange beads (Streamline SP XL). Beads were washed around ten times with 10 mM ammonium acetate in a batch mode and the supernatant was discarded. Next the beads were washed with ~300 ml of 10 mM ammonium acetate containing 1 M NaCl. The resulting supernatant was kept frozen and was diafiltered and concentrated using 3 kDa centrifugal filters (Millipore, Amicon Ultra, Ultracel -3 K) with 100 mM ammonium acetate. All other chemicals were obtained from Sigma (analytical grade) and were used as received.

Cross-linking reactions

Reactions catalyzed with horseradish peroxidase were performed at 37 °C, in a 100 mM NH$_4$Ac buffer, pH 6.8. The enzyme to substrate ratio was 1 : 20, by weight. Reactions were carried out at 1%, 2%, 3% and 4% (w/v) of α-lactalbumin. For 1% of α-lactalbumin, H$_2$O$_2$ was titrated in at 10 min intervals. The amount of H$_2$O$_2$ added at each titration step corresponded to a change in the concentration of Δ[H$_2$O$_2$] = 0.1 mM. For the reaction at 2%, we have used Δ[H$_2$O$_2$] = 0.1 mM every 5 min, for 3% Δ[H$_2$O$_2$] = 0.1 mM every 3.3 min. For 4% of α-lactalbumin, we have used Δ[H$_2$O$_2$] = 0.2 mM every 5 min. Reactions were allowed to continue for 15 h. Reactions catalyzed by microbial transglutaminase were performed at 37 °C in 100 mM NH$_4$Ac, 100 mM NaCl, pH 6.8. The concentration of α-lactalbumin was 1.5% (w/w) and the enzyme to substrate ratio was 1 : 35, by weight. Reactions were allowed to continue for 24 h. Reactions catalyzed by mushroom tyrosinase were performed in 10 mM NH$_4$Ac pH 5.8. A 3% (w/v) solution of α-lactalbumin was first incubated for 1 h at 37 °C. Next, tyrosinase from a concentrated stock (1.5 mg ml^{-1} in demineralised water) was added to the protein solution at a protein-to-enyzme ratio of 1 : 20, by weight, and the reaction mixture was left at 37 °C for another 16 h.

Size-exclusion chromatography

Reaction products and a α-lactalbumin reference sample (50 µl) were applied to a Superose 6 10/300 GL column (GE Healthcare, Uppsala) connected to an Akta Purifier system at room temperature. The column was equilibrated and eluted with 0.1 M ammonium acetate buffer pH 6.8 at a flow rate of 0.5 mL min^{-1}. The eluate was monitored at 280 nm.

Light scattering

A cylindrical glass cell was used that allows for titration of H$_2$O$_2$, and stirring (using a small glass stirrer) combined with light scattering measurements at an angle of 90°. This cell was placed in an ALV light scattering instrument equipped with an ALV-5000/60X0 external digital correlator and a 300 mW solid state laser (Cobolt Samba-300 DPSS laser) operating at a wavelength of 532 nm. The angle of detection was 90°. A refractive index matching bath of filtered *cis*-decalin surrounded the cylindrical scattering cell, and the temperature was controlled at 37 $\pm$ 0.1 °C. Injections of H$_2$O$_2$ into the reaction solution were done using a Schott Gerate TA01 titration system. The measurement cycle consisted of a 30 s H$_2$O$_2$ injection, followed by 30 s of stirring and a series of 1 min light scattering measurements until the start of the next addition of H$_2$O$_2$.

Determination of monomer consumption using size-exclusion chromatography

Aliquots were taken at regular intervals from the 1% α-lactalbumin reaction catalyzed by horseradish peroxidase (see cross-linking reactions). A volume of 100 µL of the aliquots was applied to a Bio-Silect® SEC 400-5 column (BioRad), connected to a Biologic DuoFlow Chromatography system (Bio-Rad) and equilibrated with 0.1 M NH$_4$Ac (pH 6.8). Detection was done using UV absorbance at a wavelength of 280 nm, elution using a 0.1 M NH$_4$Ac buffer pH 6.8 at 1 mL min^{-1}. Peaks corresponding to the α-lactalbumin monomer were always well separated from cross-linked products, and the relative area of the monomer peak as compared to the area of the peak at the start of the reaction (= 100%) was used to quantify monomer consumption.

Atomic force microscope imaging

Reaction products of the 1% α-lactalbumin reaction catalyzed by horseradish peroxidase (see cross-linking reactions) were dialyzed against 0.1 M NH$_4$Ac at pH 6.8 at 4 °C with 300 kDa cellulose ester membrane (Spectra/Pro Biotech) to remove any leftover monomers and small oligomers. It was found that optimal imaging of the protein particles on mica required a pH of 5.5. Dialyzed protein particles were diluted to 10 μg mL^{-1} and the pH was adjusted to 5.5. Samples (20 μL) were incubated on a freshly cleaved mica surface for 1 min, rinsed with deionized water and dried with nitrogen. AFM imaging was performed using a Digital Instruments NanoScope V Multimode Scanning Probe Microscopy with an ultrasharp silicon cantilever (NT-MDT CSCS11), in the ScanAssyst imaging mode.

Results and discussion

Cluster formation from dilute solutions: comparison of different enzymes

Fig. 3 show chromatograms of the final reaction products for the cross-linking of dilute solutions of (calcium free) α-lactalbumin with the various enzymes (horseradish peroxidase, microbial transglutaminase, mushroom tyrosinase). The limit of separation of the particular column that was used corresponds to molar masses of order 10^6 g mol^{-1}, and anything eluting between 7 and 11 ml is too large to be separated by the column. Therefore, the shape of the chromatogram for elution volumes smaller than 11 ml does not reflect the true size distribution of the covalent protein clusters. Despite the fact that we cannot separate over the whole range of cluster sizes, there are marked quantitative and qualitative differences between the various reaction products.

The horseradish peroxidase mediated cross-linking (Fig. 3a) almost exclusively produces clusters with molar masses in excess of 10^6 g mol^{-1}. Reaction conditions for this case have already been tuned quite extensively in previous work[6,16–18] to achieve high conversion into high molecular weight products. A small fraction of the α-lactalbumin monomers is not converted. This fraction most likely corresponds to the Ca^{2+} containing fraction in the α-lactalbumin starting material, which is less reactive. As is shown by on-line light scattering (see next section), the growth of covalent protein clusters for this reaction stops after about 12 h of reaction time. The enzyme is quite easily inactivated by excess peroxide. Despite the fact that we add the peroxide stepwise, and in small amounts, in turns out that the reaction stops due to inactivation of the enzyme rather than due to a lack of reactive groups: when adding fresh enzyme at the end of a reaction (continuing the supply of peroxide), cluster growth proceeds (as detected by on-line light scattering).

For microbial transglutaminase, Matsumura has already demonstrated cross-linking of dilute apo-α-lactalbumin at neutral pH in a low ionic strength buffer.[13] Again at neutral pH, we have done a preliminary investigation of the role of ionic strength, and found that the conversion into high molecular weight products is somewhat larger at 100 mM of added NaCl than at 0 mM. A chromatogram of the reaction product at 100 mM is shown in Fig. 3b. The fraction of leftover monomeric α-lactalbumin is somewhat larger than for the horse radish peroxidase-mediated reaction, and may include both the holo-α-lactalbumin fraction that is

This journal is © The Royal Society of Chemistry 2012

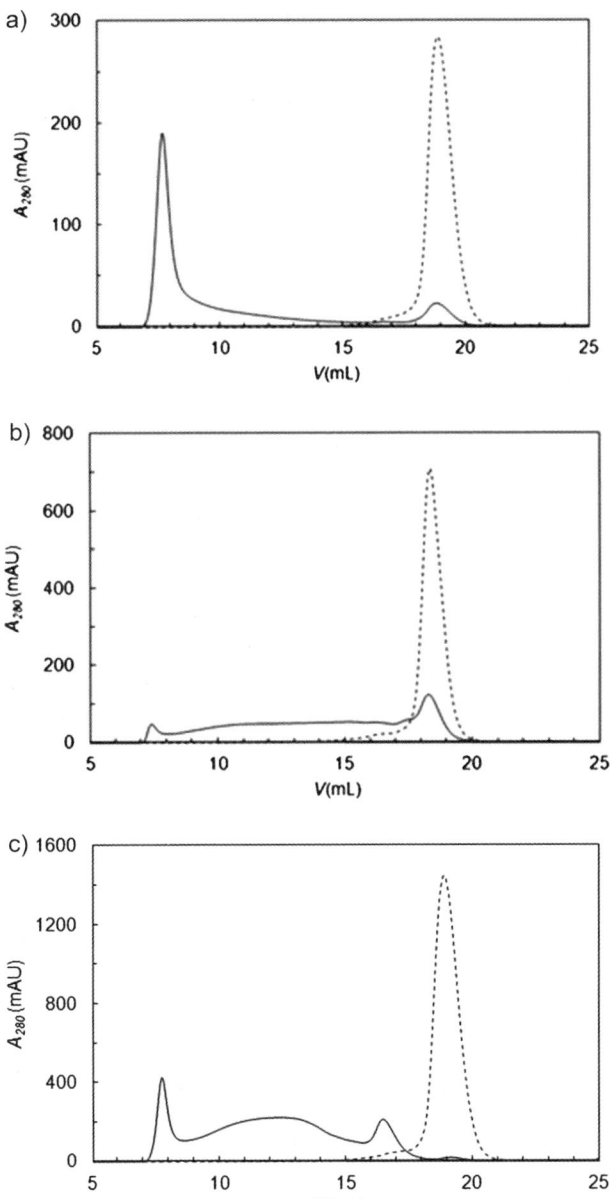

Fig. 3 Size exclusion chromatograms of reaction products of cross-linking dilute apo-α-lact-albumin for selected enzymes (absorbance at 280 nm *versus* elution volume). Dashed lines: uncross-linked apo-α-lactalbumin. Solid lines: dilute apo-α-lactalbumin. Cross-linked by a) horseradish peroxidase, b) microbial transglutaminase and c) mushroom tyrosinase.

essentially unreactive,[13] as well as unreacted apo-α-lactalbumin. Most of the trans-glutaminase cross-linked products elute before the limit of separation of approx. 10^6 g mol^{-1}; only a small peak elutes beyond the limit of separation. Comparing with the peroxidase system, microbial transglutaminase cross-linking clearly leads to much smaller protein clusters. But, a more extensive optimization with respect to solution conditions and enzyme/substrate ratios still has to be done, hence it may be that conditions will be found that lead to higher conversion into large clusters when cross-linking dilute apo-α-lactalbumin using microbial transglutaminase.

Cross-linking of α-lactalbumin by tyrosinase has previously been investigated by Thalmann.[19] Remarkably, this enzyme appears to be active even on holo α-lactalbumin. The enzyme requires dissolved oxygen and it is (tacitly) assumed that the concentration of dissolved oxygen is large enough not to be rate limiting. The chemistry of tyrosinase-induced cross-linking is different from that of horseradish peroxidase in that tyrosinase catalyzes the transformation of phenol group into a diquinone, that subsequently can attract and cross-link to a range of amino-acid side chains.[7] For α-lactalbumin, it was shown that a slightly acidic pH leads to the largest cross-linked products.[19] Additionally, we have performed a preliminary investigation of the ionic strength dependence at slightly acidic pH, and found that whereas a higher ionic strength leads to cross-linked products of higher molar mass, it also leads to precipitation of part of the reaction product. For this reason, we here only consider the case of low ionic strength. Fig. 3c shows the chromatogram for the reaction product of dilute apo-α-lactalbumin, cross-linked with tyrosinase. On-line light scattering (data not shown) indicates that, as compared with reactions catalyzed by horseradish peroxidase and microbial transglutaminase, this reaction is rather fast, with particle growth stopping completely in a few hours. The overall size distribution of the covalent protein-clusters generated by this reaction is intermediate between that of the two other reactions, with significant fractions eluting both inside and outside ($>10^6$ g mol^{-1}) the range of separation of the column. Consistent with the report of Thalmann $et\ al.$,[19] essentially all of the monomeric α-lactalbumin is incorporated into covalent clusters, including the fraction of holo-α-lactalbumin in our starting material. It remains to be established whether for tyrosinase there is still a reactivity difference between the holo and apo forms of α-lactalbumin, as was shown for the other enzymes. A final interesting finding is the peak at an elution volume of around 16.5 ml, possibly indicating some kind of preferred cluster size of around 90 kg mol^{-1} (estimated from a calibration with globular proteins).

These preliminary results serve to illustrate differences and similarities between enzymatic cross-linking reactions on dilute solutions of a single model substrate, apo-α-lactalbumin. Next we consider the particle growth, both as a function of reaction time, and as a function of the initial substrate concentration, for the selected case of cross-linking induced by horseradish peroxidase under the periodic addition of H_2O_2.

Peroxidase-mediated cross-linking: concentration-dependence

If protein clusters grow large enough, they may start to overlap, connect together and form a space-filling network or gel. Whether or not this happens depends in particular on the initial protein concentration. For the heat-induced aggregation of globular proteins, the sol–gel transition, and its dependence on the initial protein concentration has been investigated extensively. So far, hardly any work has been done on characterizing the sol–gel transition for enzymatically cross-linked globular proteins. In the previous section, the cross-linking of 1% apo-α-lactalbumin was considered, catalyzed by horseradish peroxidase, under the periodic addition of H_2O_2. Here we consider, for the same reaction, what happens upon increasing the initial concentration of α-lactalbumin, while keeping the enzyme to substrate and enzyme to H_2O_2 ratio constant. Fig. 4 shows the average hydrodynamic radius of the protein clusters during the reaction (determined from on-line dynamic light scattering) for initial α-lactalbumin concentrations of 1%, 2% and 3%. For the lowest concentrations, particle growth seems to saturate at around 12 h, whereas for the highest concentration, particle growth has not yet stopped after 15 h of reaction time. Final particle sizes (after 15 h of reaction time) increase very rapidly

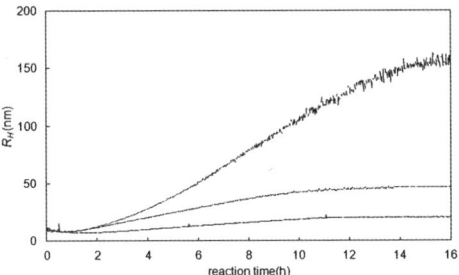

Fig. 4 On-line dynamic light scattering of cross-linking of apo-α-lactalbumin, catalyzed by horseradish peroxidase, under the periodic addition of H_2O_2. Average hydrodynamic radius of the reaction mixture *versus* reaction time. From bottom to top, initial concentration of apo-α-lactalbumin: 1%, 2%, 3% (w/v).

with initial protein concentration. Indeed, for a 4% initial concentration, the dynamic light scattering data shows that after about 7 h of reaction time, gelation occurs.

A critical protein concentration C_g somewhat below 4% (for gelation after 15 h of reaction time) is also consistent with the plot of the final average hydrodynamic cluster size *versus* initial concentration shown in Fig. 5. An estimate for the molar mass M_c of the clusters just prior to gelation is $M_c \approx N_{av} {}^4\!/_3 \pi R_H^3 C_g = O$ (10^9 g mol^{-1}) or about 10^5 monomer units, where we have used $C_g = 3.5\%$ and $R_H = 250$ nm. It still remains to be seen what kind of size-mass scaling applies for these covalent protein clusters. In any case, they must be very open, since the average internal protein concentration of the clusters is only a few percent close to the gelation threshold.

Peroxidase-mediated cross-linking: kinetics and final cluster architecture

Next we consider how the covalent clusters grow during enzymatic cross-linking, and how this connects to the final cluster architecture. To this end, for the reaction at a 1% initial concentration of apo-α-lactalbumin, we have determined the fraction of monomeric α-lactalbumin that is not (yet) incorporated into a cross-linked product, *versus* the reaction time. Results are shown in Fig. 6, together with a plot of the development of the average hydrodynamic radius of the clusters during cross-linking, from the on-line dynamic light scattering.

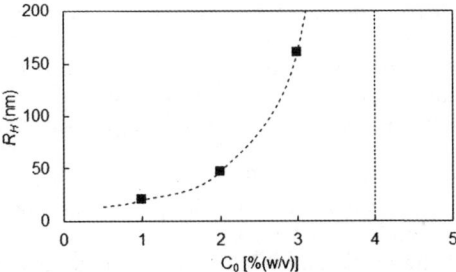

Fig. 5 Cross-linking of apo-α-lactalbumin, catalyzed by horseradish peroxidase, under the periodic addition of H_2O_2. Final average hydrodynamic radius of reaction mixture *versus* initial concentration of apo-α-lactalbumin at fixed enzyme/substrate ratio and fixed ratio of H_2O_2 addition rate to enzyme concentration. The dotted line at an initial concentration of 4% indicates that for this concentration the reaction mixture gels within the (fixed) reaction time.

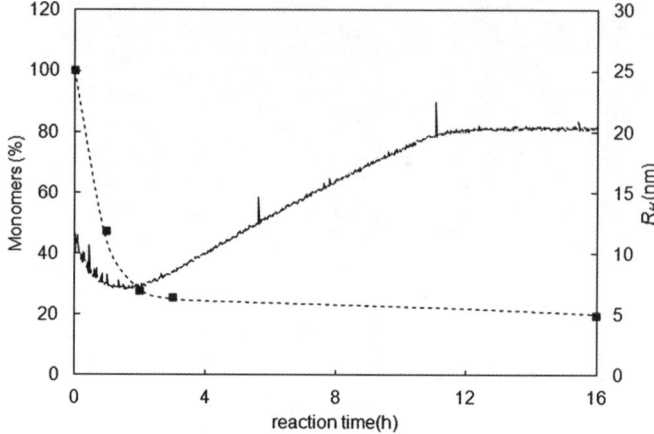

Fig. 6 Monomer consumption (filled squares connected by dashed line) and development of average hydrodynamic cluster size (solid line) during cross-linking of apo-α-lactalbumin (initial concentration 1% (w/v)), catalyzed by horseradish peroxidase, under the periodic addition of H_2O_2.

The initial average hydrodynamic radius is somewhat larger than the expected value for an α-lactalbumin monomer. A CONTIN fit of the dynamic light scattering data reveals that initially there are two distinct populations of particles: a dominant population with a hydrodynamic radius of a few nm, corresponding to the α-lactalbumin monomers, and a minority population with hydrodynamic radii O (100 nm). The latter most likely consists of a small amount of aggregates in the starting material that could not be removed using simple filtration and centrifugation. This also explains the initial decrease of the average hydrodynamic radius: as the covalent protein clusters start to increase in both number and size, their contribution to the scattering initially decreases the average size, until they completely dominate the scattering signal. For reaction times larger than about 2 h, the CONTIN analysis reveals a single dominant population, hence from that moment on, the average hydrodynamic radius properly reflects the growth of the α-lactalbumin clusters due to enzymatic cross-linking.

The main conclusion from the dynamic light scattering data is however that for the 1% reaction, cluster growth continues for a long time, until about 12 h of reaction time. In contrast, the fraction of monomers not (yet) incorporated into covalent clusters decreases much more rapidly: in about 2 h, almost all of the reactive apo-α-lactalbumin monomers have been incorporated into covalent clusters, leaving only the fraction of holo-α-lactalbumin in the starting material that is less reactive.

According to the dynamic light scattering, at the reaction time at which almost all reactive α-lactalbumin monomers have already been depleted, the average cluster size is still rather small, between 5 and 10 nm. In a subsequent, much slower phase of the reaction, these smaller covalent clusters are apparently linked together to form larger clusters.

This is consistent with preliminary AFM images of the reaction product of the 1% reaction, an example of which is shown in Fig. 7. After drying, the particles exhibit a wide distribution of in-plane sizes as well as heights. Although details of the internal structure of the particles are hard to determine, it is quite clear that the particles seem to consist of smaller sub-particles. Heights are mostly below 10 nm, corresponding to a few layers of proteins. This indicates that the clusters are flattened by drying, which is not entirely surprising given the fact that they seem to have a quite open structure with an internal protein content of only a few percent (at least close to the gelation transition).

7.5 nm

0.0   1: Height   1.0 μm

-7.5 nm

Fig. 7 Atomic force microscopy imaging (in air) of reaction product of cross-linking reaction of apo-α-lactalbumin (initial concentration 1% (w/v)), catalyzed by horseradish peroxidase, under the periodic addition of H_2O_2. Monomeric α-lactalbumin and small oligomers were removed prior to imaging by dialysis using a 300 kDa cut-off membrane.

Concluding remarks

A crucial quantity that certainly needs to be investigated in future work is the actual number of covalent bonds that are formed per protein, or in terms of the language of chemical networks, the functionality f of a cross-linking node, which in our case is a single α-lactalbumin monomer. The fact that the clusters produced by the horseradish peroxidase reactions seem to be very dilute suggest that for this case the average functionality f may be much lower than the theoretical maximum, which is $f = 4$ for the case of cross-linking of the 4 tyrosines of α-lactalbumin by horseradish peroxidase and H_2O_2. Our preliminary studies have already show that at early stages in these reactions there is a preference for a well defined single dityrosine bond to form first, namely between Tyr18 and Tyr50. This alone gives $f = 2$, corresponding to a linear topology. A true network cannot form if all proteins have $f = 1$ or $f = 2$, hence at later stages in the reaction, other cross-links must form too. A mixture of a large number of proteins with $f = 1$ and 2 plus a small number of proteins with $f = 3$ or 4, could indeed give rise to the kind of dilute network that we have found for the case of α-lactalbumin cross-linked by horse radish peroxidase and H_2O_2.

Similar considerations apply to the case of cross-linking by microbial transglutaminase, where, for the case of apo-α-lactalbumin, it has been shown that at least initially, and at low concentration, cross-linking occurs between only 1 out of 3 available glutamines and only 3 out of 12 available lysines.[13] For microbial transglutaminase, the average number of cross-links per protein can in principle be determined from the amount of ammonia released during cross-linking. Some work along these lines has already been done by Matsumura *et al.*,[13] but only for relatively short reactions at low initial concentrations of apo-α-lactalbumin, in which dimers, trimers and tetramers but no larger covalent clusters were formed.

For mushroom tyrosinase not much detailed biochemical characterization has been done yet on its activity in cross-linking apo-α-lactalbumin. It is quite remarkable though that this enzyme appears to be active even on holo-α-lactalbumin, and other globular proteins in their native state. Since the enzyme activates tyrosines that subsequently may react to a range of other protein side-chains, its cross-linking chemistry is quite complicated and heterogeneous. Hence, it is presumably less suitable as a model system for studying covalent protein cluster formation. On the other hand, since the enzyme appears to have access to its tyrosine substrate even for folded globular proteins, it could be a quite important enzyme for practical applications.

Future studies will also need to address the issue of the size-mass scaling for covalent protein clusters. The only data that is available so far is due to Matsumura,[14] for apo-α-lactalbumin in the presence of the denaturant DTT. When cross-linked by microbial transglutaminase, very large protein clusters are formed, with molar masses in the range 10^5–10^8, similar to what we have found for the cross-linking of apo-α-lactalbumin with horseradish peroxidase and H_2O_2. Using SEC-MALLS on the largest fraction of the protein clusters, a scaling of the gyration radius R_g with the cluster molar mass M_c was found of $R_g \propto M_c^\alpha$, with $\alpha \approx 0.31$. Unfortunately, no absolute values were given for the gyration radius and molar masses, such that it remains unclear whether these clusters were very dense or very open, which makes any interpretation of the exponent α at this stage premature.

The issues of the distribution of the cross-link functionalities f of the proteins at the end of the reaction, and the resulting open or dense network structures that are formed are crucially important for the potential applications of cross-linked proteins in food technology. Whereas dilute networks could have applications as protein-based thickeners, dense networks would be suitable for the creation of dense protein colloids to be used for example in high protein foods,[20] or as protein-based Pickering stabilizers of foams and emulsions.[21] Other important issues that have not yet been addressed here include the degree of secondary and tertiary structure left in enzymatically cross-linked globular proteins, and the resulting response of cross-linked protein to heating, and at interfaces. These, and other issues will be addressed in our future work.

References

1 R. Vreeker, L. L. Hoekstra, D. C. den Boer and W. G. M. Agterof, *Food Hydrocolloids*, 1992, **6**, 423.
2 J.-C. Gimel, D. Durand and T. Nicolai., *Macromolecules*, 1994, **27**, 583.
3 C. Le Bon, T. Nicolai and D. Durand, *Int. J. Food Sci. Technol.*, 1999, **34**, 451.
4 P. Meakin, *Phys. Scr.*, 1992, **46**, 295.
5 S. P. Roefs and K. G. de Kruif, *Eur. J. Biochem.*, 1994, **226**, 883.
6 G. Oudgenoeg, "*Peroxidase catalyzed conjugation of peptides, proteins and polysaccharides via endogenous and exogenous phenols*", PhD thesis, Wageningen University, 2004.
7 E. Monogioudi, N. Creusot, K. Kruus, H. Gruppen, J. Buchert and M.-L. Mattinen, *Food Hydrocolloids*, 2009, **23**, 2008.
8 E. Dickinson and Y. Yamamoto, *J. Agric. Food Chem.*, 1996, **44**, 1371.
9 E. Dickinson, *Trends Food Sci. Technol.*, 1997, **81**, 334.
10 C. H. Tang and C. Y. Ma, *Eur. Food Res. Technol.*, 2007, **225**, 649.
11 Y. Zhu and J. Tramper, *Trends Biotechnol.*, 2008, **26**, 559.
12 K. Kuwajima, *Proteins: Struct., Funct., Genet.*, 1989, **6**, 87.
13 Y. Matsumura, Y. Chanyongvorakul, Y. Kumazawa, T. Ohtsuka and T. Mori, *Biochim. Biophys. Acta*, 1996, **1292**, 69.
14 Y. Matsumura, D.-S. Lee and T. Mori, *Food Hydrocolloids*, 2000, **14**, 49.
15 D.-S. Lee, S. Matsumoto, Y. Matsumura and T. Mori, *J. Agric. Food Chem.*, 2002, **50**, 7412.
16 W. H. Heijnis, P. A. Wierenga, A. E. M. Janssen, W. J. H. van Berkel and H. Gruppen, *Chem. Eng. J.*, 2010, **157**, 189.
17 W. Heijnis, P. Wierenga, W. J. H. van Berkel and H. Gruppen, *J. Agric. Food Chem.*, 2010, **58**, 5692.

18 W. Heijnis, H. L. Dekker, L. J. de koning, P. A. Wierenga, A. Westphal, C. G. de Koster, H. Gruppen and W. J. H. van Berkel, *J. Agric. Food Chem.*, 2011, **59**, 444.
19 C. R. Thalmann and T. Lötzbeyer, *Eur. Food Res. Technol.*, 2002, **214**, 276.
20 D. Saglam, P. Venema, R. de Vries, L. M. C. Sagis and E. van der Linden, *Food Hydrocolloids*, 2011, **25**, 1139.
21 E. Dickinson, *Curr. Opin. Colloid Interface Sci.*, 2010, **15**, 40.

Anomalies in moisture transport during broccoli drying monitored by MRI?

Xin Jin,[*a] Antonius J. B. van Boxtel,[a] Edo Gerkema,[c]
Frank J. Vergeldt,[c] Henk Van As,[c] Gerrit van Straten,[a]
Remko M. Boom[b] and Ruud G. M. van der Sman[b]

Received 7th March 2012, Accepted 24th April 2012
DOI: 10.1039/c2fd20049j

Magnetic resonance imaging (MRI) offers unique opportunities to monitor
moisture transport during drying or heating of food, which can render
unexpected insights. Here, we report about MRI observations made during the
drying of broccoli stalks indicating anomalous drying behaviour. In fresh
broccoli samples the moisture content in the core of the sample increases during
drying, which conflicts with Fickian diffusion. We have put the hypothesis that
this increase of moisture is due to the stress diffusion induced by the elastic
impermeable skin. Pre-treatments that change skin and bulk elastic properties of
broccoli show that our hypothesis of stress-diffusion is plausible.

1 Introduction

Magnetic resonance imaging (MRI) is an established technique for studying the
internal states in biological systems. When coupled to in-situ thermal treatments
MRI can be viewed as a rather novel enabling technology that yields unprecedented
insights into the physical transport phenomena that occur during food processing.[1-6]

In this work, we apply MRI to investigate the changes in moisture distribution
with time observed during *in-situ* drying of broccoli, which apparently violates
Fick's law.

Convective drying of vegetables is mostly considered as a diffusion-controlled
process.[7] Traditionally, in food science, diffusion is described by Fick's law, in which
the mass flux is linear with the gradient in moisture content. A few papers in food
science report deviations of moisture transport from Fick's law. Johnson *et al.*
(1998)[8] reported an increased moisture content in the product centre during drying
of plantain but without further explanation. Arnaud and Fohr (1988)[9] observed that
the intra-kernel moisture content gradient increases during drying and decreases
during tempering. Courtois *et al.* (2001)[10] and Toyoda (1988)[11] have considered
the internal structure of products as a reason for deviation. In rice and corn there
are internal regions with different moisture transport properties. Other deviations
from Fick's law are reported for the cooking of starch-rich products, with moisture
transport against the moisture gradient.[12,13] The deviations are said to be caused by
gelatinization of starch during heating, which results in changes of local water
holding capacity and water activity in the heterogeneous product. Different degrees
of starch gelatinization result in different potential maximum moisture contents
(ceiling moisture content). Therefore, the moisture transport during rice cooking

[a]Systems and Control Group, Wageningen University, P.O. Box 17, 6700AA Wageningen, The
Netherlands. E-mail: xin.jin@wur.nl; Tel: +31(0)317482014
[b]Food Process Engineering Group, Wageningen University, The Netherlands
[c]Laboratory for Biophysics, Wageningen University, The Netherlands

is driven by the difference of local moisture content and the ceiling moisture content. Watanabe *et al.* (2001)[14] proposed the so called "water demand" model to describe this transport phenomenon, which is not captured by Fick's law.

Furthermore, Wählby *et al.* (2001)[15] observed during experiments an increased moisture content in the product centre during the cooking of beef without clear explanation. Transport against the gradient in moisture content is thermodynamically possible if gradients in swelling pressure arise. Van der Sman (2007)[16] modelled swelling pressure-driven moisture transport (based on the Landau expansion of the Flory-Rehner theory) caused by protein denaturation near the product surface during the cooking of meat. In that model, the elasticity properties of the product play an important role and result in moisture transport in directions opposite to the moisture gradient. Recently, it has been shown that the full Flory-Rehner theory, indeed, holds for cooked meat.[17]

Another example is from the field of polymer physics where during drying the formation of a skin at the polymer surface resulted in water transport against the gradient in the moisture content. Okuzono and Doi (2008)[18] have called this phenomenon stress diffusion, and they formulated a generalized Fick's equation with an elastic term to describe the stress diffusion. In view of the preceding cited analysis and observations of non-Fickian moisture transport, we pose that the observed non-Fickian behaviour during broccoli drying is due to elastic stresses.

To test this hypothesis, we applied pre-treatments (blanching, freezing and peeling), which change the product structure and thus its textural and elastic properties. Experiments have been done in a MRI device with continuous and controlled *in-situ* hot air supply; the acquired MRI images provide data about moisture transport and shrinkage during drying. The different pre-treatments have been compared in terms of drying rates, shrinkage and moisture content profiles, *via* which the validity of our hypothesis is analysed.

2 Materials and methods

2.1 Materials

For all measurements, parts of the broccoli stalk were used. The sizes of the samples were about 0.01 m in height, 0.01 m in radius. Fig. 1 gives an example of a fresh broccoli sample.

2.2 Pre-treatments

In total six different pre-treatments were applied. An overview of the samples and pre-treatments is given in Table 1. After all pre-treatments, the free water at the sample surfaces was removed at room temperature with tissue paper.

Fig. 1 Cross section of a broccoli stalk sample.

Table 1 Overview of pre-treatments and experiments

Pre-treatment	Procedure
Peeling	1.0 mm–1.5 mm of skin was removed
Blanching	90 °C water, 3 min
Freezing	−25 °C, 48 hours

Experiment	pre-treatment
1	Non treated
2	Peeled
3	Non-peeled, blanched
4	Peeled, blanched
5	Non-peeled, frozen
6	Peeled, frozen

2.3 Drying in the MRI device

The sample was fixed by a stick on a sample supporter and inserted into a drying chamber in the MRI measurement device. The size of the drying chamber was 0.032 m in diameter and 0.2 m in length. A continuous flow of temperature-controlled air was supplied. The air temperature was controlled at 30 °C or 50 °C, the air velocity at 1.0 m/s and the relative humidity at 10%.

Drying was continued until the moisture content of the samples was constant. Depending on the material properties, the experimental time for the fresh broccoli stalks ranged from 12 to 48 h. Initial and final product moisture contents (M_0, M_i) were determined by oven drying (105 °C, 24 h).

2.4 MRI imaging equipment

All measurements were performed on a 3 T (128 MHz for protons) MRI system (Bruker, Karlsruhe, Germany), consisting of an Avance console, a superconducting magnet with a 0.5 m vertical free bore (Magnex, Oxford, UK), a 1 T m^{-1} gradient coil, and a birdcage RF coil with an inner diameter of 0.04 m.

2.5 MRI imaging

3D images were obtained using a Turbo Spin Echo (TSE) MRI sequence,[19] a repetition time TR of 2 s, an effective spin echo time TE of 3.35 ms and a spectral bandwidth SW of 50 kHz. Only 16 echoes were acquired in the TSE train to avoid blurring due to T2-weighting. Odd and even echoes were separately phase-encoded forming two different images to avoid Nyquist ghost's artefacts, so the turbo factor was 8. Two acquisitions were averaged to improve image quality. The field-of-view (FOV) was 35 × 35 × 35 mm^3 with a matrix size of 64 × 64 × 64 resulting in a spatial resolution of 0.55 × 0.55 × 0.55 mm^3. The interval time between measurements was 34 min.

T2 mapping was done using a multi spin echo (MSE) imaging sequence,[20] a TR of 2 s, a TE of 3.59 ms and a SW of 50 kHz. Per echo train 128 echoes were acquired; 16 acquisitions were averaged to improve image quality. The FOV was 35 × 35 mm^2 with a matrix size of 64 × 64 resulting in an in-plane resolution of 0.55 × 0.55 mm^2. The slice thickness was 3 mm. The interval time between measurements was 34 min.

2.6 Numerical methods and data analysis

MRI-measurement data handling for graphical interpretation and analysis was performed with home-built software written in IDL (RSI, Boulder, CO). For shrinkage calculations, pixels with an intensity value above 0.75 (with the maximum value of 11.42 per pixel) were counted and for each pixel a volume of 0.16 mm^3 was assigned. By summing up the volume of all counted pixels the instantaneous volume (V_t) of the sample was calculated.

The degree of shrinkage is defined as the ratio of the reduced volumes ($V_0 - V_t$) to the initial volume (V_0):

$$S_V = \frac{V_0 - V_t}{V_0} \tag{1}$$

and the fraction of moisture removed from the initial product is:

$$S_m = \frac{M_0 - M_t}{M_0} \tag{2}$$

Where M_0, the initial moisture content, was measured by the oven method and M_t is the moisture content at different sampling time t.

2.7 MRI data calibration

For wet samples, the signal intensity is assumed to be linear with moisture content.[21] However, for nearly dry samples the intensity images are weak and deviate from the linear relationship. The minimum detectable liquid water concentration is about 20 kg m^{-3}.[22,23]

Hence, to establish the extent of linearity between MRI signal intensity and moisture content, we have plotted the shrinkage data as a function of the fraction of removed moisture. Assuming incompressibility of the solid and water present in broccoli, and the absence of air, the reduced volume must be attributed to the loss of moisture. In Fig. 2 we show this relation, which also indicates the accuracy of the interpretation of signal intensity to moisture content. It shows that below $S_m = 0.85$ shrinkage is more or less linear with the fraction of removed moisture, whereas in the last phase of drying ($S_m > 0.85$) there is a deviation from linearity. In this region, with product moisture content below 0.3 kg water kg^{-1} dry matter, the measured values might be lower than the actual values.

3 Results and discussions

3.1 Drying pattern of fresh broccoli stalks

Fig. 3 presents the MRI measurements for the central cross-section of fresh broccoli samples dried at 30 °C and 50 °C. Differences in brightness indicate the distribution of moisture throughout the sample; the brighter the colour, the higher the moisture content. The bar in the figures provides a relative scale for the moisture content. The gap at the bottom of each image indicates the hole made by the stick, which supports the sample. In the first image of both figures a slightly higher moisture content is observed at the edges. This indicates that there is still some free water at the surface, which is the consequence of cutting.

Fig. 3 shows an anomalous drying behaviour for both the 30 °C and 50 °C drying experiment for fresh and non-treated samples: after several hours of drying the moisture in the centre has been increased for the sample dried at 30 °C, the brightness at the centre of the images of row 2 is above that of row 1, and for 50 °C drying the brightness of the first image of row 2 is above that of the initial image.

Fig. 4 gives the intensity (proportional to moisture content) for the cross-section of the image at the start of drying and at 18.1 hours for drying at 30 °C, and at 6.8 hours for drying at 50 °C. We observe clearly that during drying, the moisture content in the centre rises far above the initial moisture content at any location of the sample. Compared to its initial value, the moisture content in the centre increases by 50–60%. This anomaly in drying behaviour evidently deviates from the standard

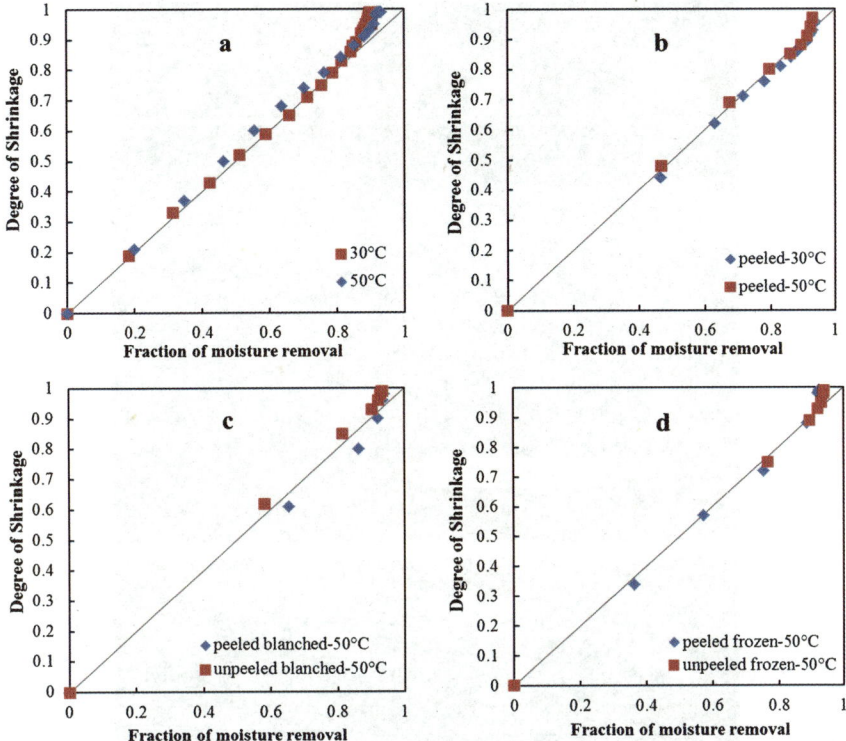

Fig. 2 Shrinkage as a function of the fraction of moisture removal. (a) unpeeled fresh samples at 30 and 50 °C drying, (b) peeled fresh samples at 30 and 50 °C drying, (c) peeled blanched samples and unpeeled blanched samples dried at 50 °C, and (d) peeled frozen samples and unpeeled frozen samples dried at 50 °C.

Fickian diffusion. According to the work of van der Sman (2007)[16] and Okuzono and Doi (2008),[18] shrinkage and deformation of the skin cause an internal pressure gradient, which results in a temporarily pressure-driven moisture transport towards the centre of the product.

Despite the increasing moisture content in the centre of the product, the drying curves for the full samples, which are given in Fig. 5, show monotonic decreasing moisture content.

Furthermore, the MRI images in Fig. 3 show the decreasing size of the samples due to shrinkage during drying. Fig. 2a shows the shrinkage quantitatively. The drawn lines in Fig. 2 correspond to the situation where shrinkage is equal to moisture removal. The data points for drying at 50 °C are above that for drying at 30 °C, which indicates that shrinkage at 50 °C is stronger than at 30 °C (Fig. 2a). This result can be explained by the relation of elasticity and moisture content. Krokida *et al.* (1998)[24] reported that in the high moisture content region, elasticity decreases with decreased moisture content, whereas in the low moisture content region, elasticity increases while moisture content decreases. During drying, the skin dries fast and due to the low moisture content the skin is more elastic and causes a centre-directed moisture transport.[16] Moreover, the skin forms a significant barrier for moisture transport and therefore moisture removal takes place in the longitudinal direction of the sample. It results in an early stage of drying in a "butterfly" shape.

The cross-sections in Fig. 4 also show that shrinkage differs for the height and width directions. For isotropic shrinkage of a cylindrical shape, the ratio (V_t/V_0) between the diameter at time t (d_t) and the initial diameter (d_0) is equal to the square

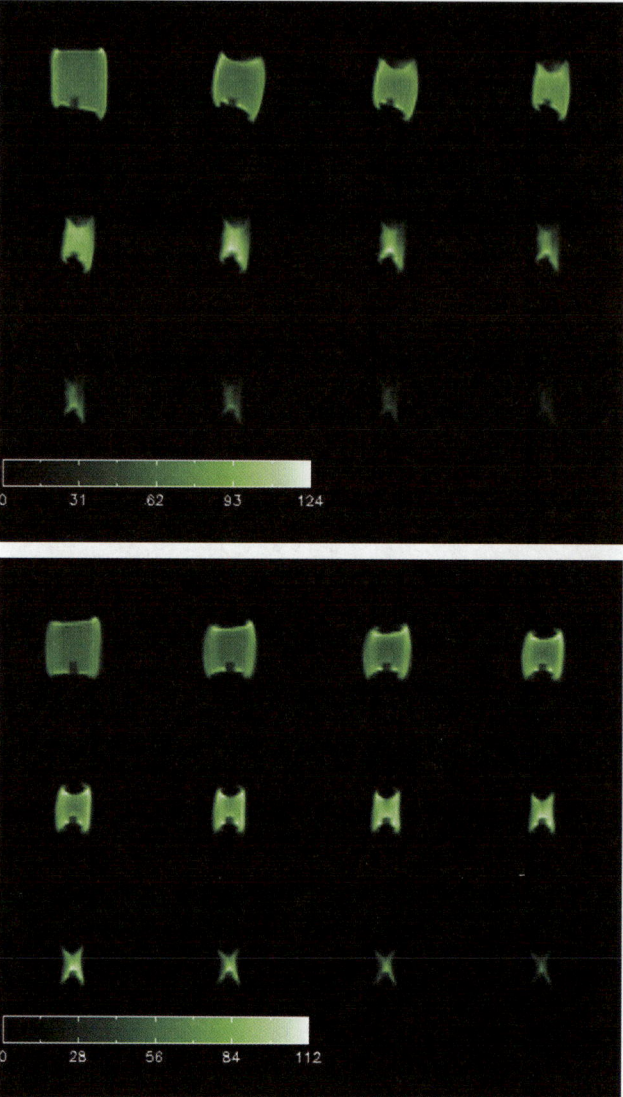

Fig. 3 Series of MRI intensity of the middle slice of fresh broccoli samples in time. Top: drying at 30 °C; time interval between samples 272 min, total time 50 h. Bottom: drying at 50 °C, time interval between samples 68 min, total time 12.5 h.

root of the volume ratio $(V_t/V_0)^{0.5}$ (with V_t the volume at time t, and V_0 the initial volume).

However, for the broccoli samples in Fig. 4, $(V_t/V_0) = 0.52$, while $(V_t/V_0)^{0.5} = 0.22$. The anisotropic shrinkage is caused by the impermeable and elastic structure of the skin and causes internal stress in the samples. From these results it is hypothesized that by applying pre-treatments that break down the wall structure the anomalies in drying behaviour could be reduced or even be cancelled.

3.2 Drying patterns after pre-treatments

To verify the hypothesis that the centre directed moisture transport is induced by the elastic properties of the product structure, product treatments were applied to break

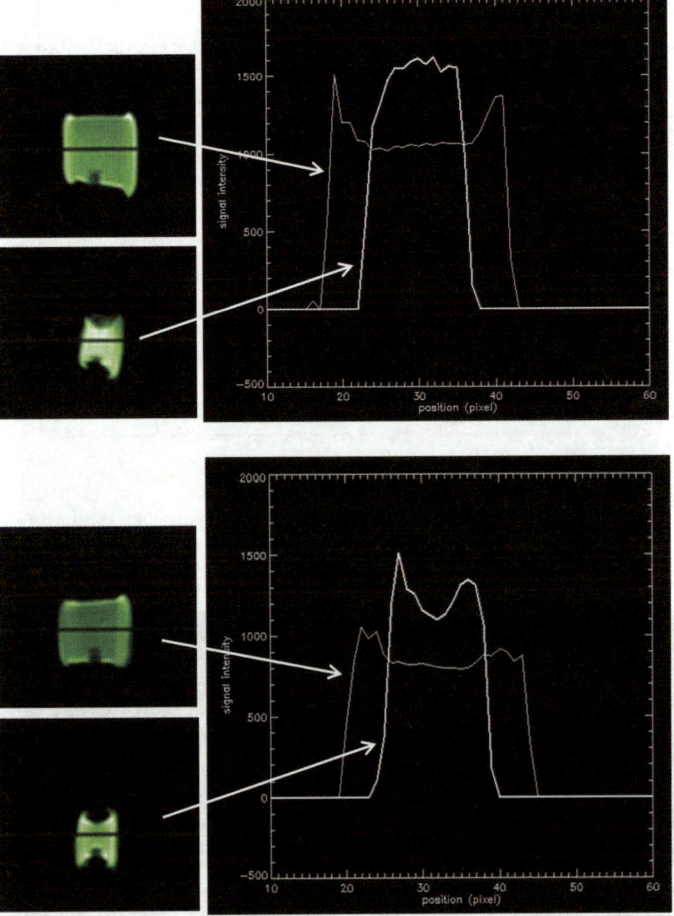

Fig. 4 MRI intensity values for a cross-section (given by the horizontal lines). Top: drying at 30 °C; initial sample and at 18.1 hours. Bottom: drying at 50 °C; initial sample and at 6.8 hours. In both cases, the moisture content in the centre of the samples surpasses the initial sample.

down the wall structure. The product treatments are (1) peeling: to remove the skin, (2) blanching: to soften the total tissue, both skin and core, thus to level out the elasticity differences between skin and center,[25] and (3) freezing and thawing: to break down the internal structure, and to change the elasticity of the internal structure.[5]

3.2.1 Peeling. The results for two drying temperatures are shown in the images of Fig. 6. Compared to Fig. 3, the increased moisture content hardly occurs, which confirms the role of elastic properties and the transport barrier of the skin. The first images also show increased moisture content at the surface, which is a result of moisture release at the surfaces where the skin was removed by cutting.

MRI images in Fig. 3 and Fig. 6 show different forms of shrinkage during drying. The peeled samples keep their original form for a long time, but towards the end of drying when the edge of the product approaches the glassy state, these samples also end with the "butterfly" shape. Fig. 2b presents the degree of shrinkage as a function of the fraction of removed moisture for the peeled samples at 30 and 50 °C. For a fraction of removed moisture below 0.85, shrinkage is linear to the fraction of removed moisture. For both temperatures the results coincide with the drawn line, which indicate the absence of an elasticity contribution to drying (see section 3.1).

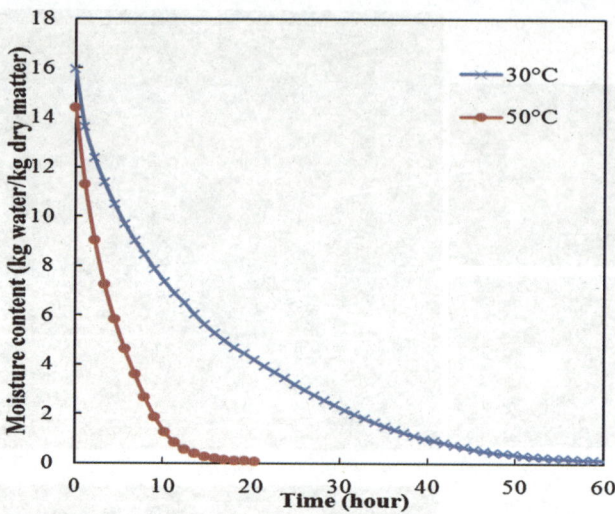

Fig. 5 Drying curves of fresh broccoli stalks at different drying temperatures (30 and 50 °C drying).

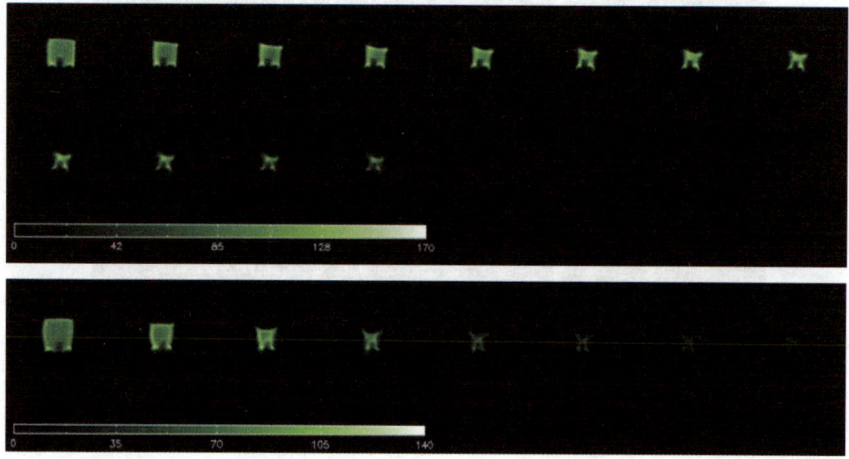

Fig. 6 MRI intensity of the middle slice of peeled broccoli stalks with respect to time. Top: drying at 30 °C; Bottom: drying at 50 °C. Total drying time, respectively, 12.5 and 9.0 h.

3.2.2 Blanching. Blanching results in tissue-softening and can level out the differences in mechanical properties between the core and the skin.[26] For the blanched samples only a small difference in drying behaviour between the peeled and unpeeled sample is found in Fig. 7. So, the barrier for mass transport by the skin is removed by only blanching. Not only the structure of the skin is softened, the internal matrix is also softened by blanching and, as a result, differences in elastic properties are levelled out, resulting in standard Fickian moisture transport. The drying time is reduced to about 4.5 hours. The drying rate of both samples is nearly equal, but the shrinkage of the peeled blanched sample is below that of the non-peeled sample (Fig. 2c).

3.2.3 Freezing. During freezing, ice crystals are formed in the tissue. Upon thawing, individual ice crystals merge into large complexes, which both break the structure and increase the internal pore size. With the destruction of internal structure,

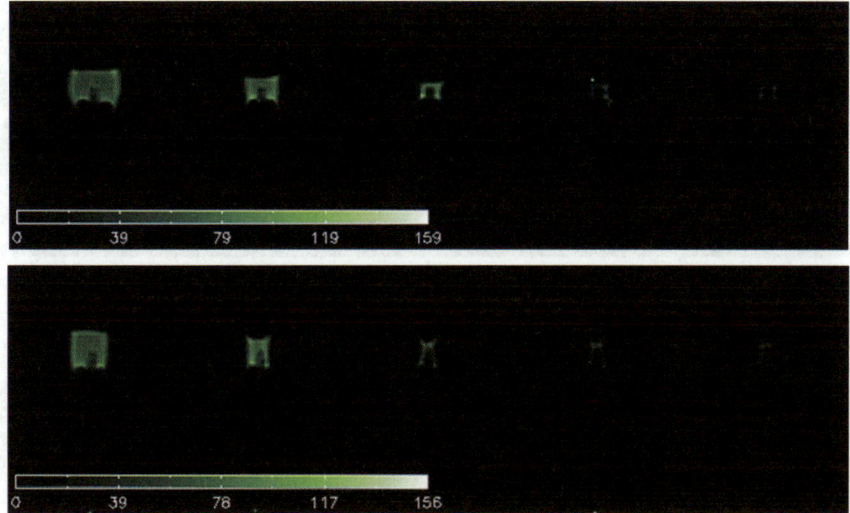

Fig. 7 MRI intensity of the middle slice of blanched broccoli stalks with respect to time. Samples are dried at 50 °C. Top: fresh unpeeled sample is blanched before drying. Bottom: sample first peeled then blanched before drying. Total drying time 4.5 hours.

elasticity is lowered. However, freezing is not effective enough to break down the barrier by skin. Results for peeled and non-peeled samples are given in Fig. 8. Due to the internal stress, the product shrinks easily during drying. The images in Fig. 8 show a strong "butterfly" form for the non-peeled frozen sample. The skin remains a mass transfer barrier. Drying occurs mainly along the longitudinal direction, and there is moisture accumulation just below the skin where the highest moisture content was detected.

The barrier for mass transport is absent for the peeled sample and therefore the peeled frozen sample does not have elastic stress-driven diffusion, and dries uniformly and the shape remains during drying and the moisture content in the centre decreases in time. For both samples, shrinkage was equal to the volume of lost moisture (see Fig. 2d). The change in the structure due to the freezing and thawing advances drying significantly and drying is completed within 4.5–5 hours.

4   Conclusions

From this investigation using in-situ MRI imaging of drying broccoli, and the above cited earlier reports, it is observed that moisture transport in foods can be due to moisture concentration gradients and gradients in elastic stress. In our case of pre-treated broccoli, the gradients in elastic stresses are probably induced by inhomogeneity in elastic properties of the skin compared to the core. During the drying of broccoli, the internal stress gradient can achieve a level that results in moisture transport against the gradients in moisture concentration. These observations are confirmed by pre-treatments that break down the internal structure and that of the product skin. Removing the skin of the broccoli sample by peeling results in a uniform product with drying behaviour close to Fickian diffusion. Blanching as a pre-treatment softens the skin and core of the product and creates uniform properties throughout the material to be dried. In this case drying can, indeed, be considered as a diffusion-driven process. Freezing and subsequently thawing as pre-treatment of fresh broccoli does not change the inhomogeneity of the elastic properties of the product during drying. In addition, we have observed that the pre-treatments peeling, blanching and freezing all enhance the drying rate significantly.

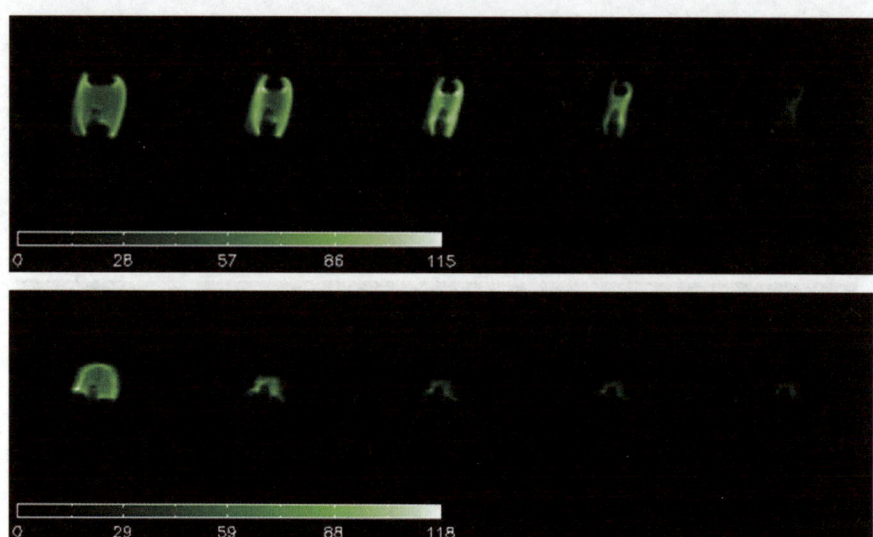

Fig. 8 MRI intensity of the middle slice of frozen broccoli stalks with respect to time. Samples are dried at 50 °C. Top: fresh sample frozen before drying. Bottom: sample first peeled then frozen before drying. Total drying time 4.5 hours.

It is also likely that the skin (cuticle) of fresh products forms a barrier for moisture transport, which amplifies the effects of the internal stress gradients. This leads to anisotropic shrinkage, and subsequent moisture transport towards the centre of product. Drying of fresh products no longer follows the standard Fick's law of diffusion. Drying models must be extended and the observed results show that stress-driven diffusion term must be included, similar to the work of Okozuno and Doi (2008).[18]

According to thermodynamics, it is even more proper to relate the moisture transport to gradients in chemical potential (or equivalently water activity or swelling pressure). If formulated in terms of the proper thermodynamic potential, there exists no anomaly. The formulation of transport in terms of gradients of thermodynamic potential is customary in the field of soft matter physics,[18,27] and food science can take advantage of this in adapting their framework.

Acknowledgements

This work is supported by the Energy Research Program EOS (EOS LT07043) of the Dutch Ministry of Economics.

References

1 K. P. Nott, L. D. Halla, J. R. Bows, M. Hale and M. L. Patrick, *Magn. Reson. Imaging*, 2000, **18**, 69–79.
2 A. LeBail, L. Boillereaux, A. Davenel, M. Hayert, T. Lucas and J. Y. Monteau, *Innovative Food Sci. Emerging Technol.*, 2003, **4**, 15–24.
3 X. Ye, R. Ruan, P. Chen and C. Doona, *LWT–Food Sci. Technol.*, 2004, **37**, 49–58.
4 K. Knoerzer, M. Regier, E. H. Hardy, H. P. Schuchmann and H. Schubert, *Innovative Food Sci. Emerging Technol.*, 2009, **10**, 537–544.
5 T. Lucas, D. Grenier, M. Bornert, S. Challois and S. Quellec, *Food Res. Int.*, 2010, **43**, 1041–1048.
6 M. Bouhrara, B. Lehallier, S. Clerjon, J.-L. Damez and J.-M. Bonny, *Magn. Reson. Imaging*, 2012, **30**, 422–430.
7 A. Mulet, N. Sanjuán, J. Bon and S. Simal, *Eur. Food Res. Technol.*, 1999, **210**, 80–83.

8 P. N. T. Johnson, J. G. Brennan and F. Y. Addo-Yobo, *J. Food Eng.*, 1998, **37**, 233–242.
9 G. Arnaud and J. P. Fohr, *Int. J. Heat Mass Transfer*, 1988, **31**, 2517–2526.
10 F. Courtois, M. Abud Archila, C. Bonazzi, J. M. Meot and G. Trystram, *J. Food Eng.*, 2001, **49**, 303–309.
11 A. G. F. Stapley, T. M. Hyde, L. F. Gladden and P. J. Fryer, *Int. J. Food Sci. Technol.*, 1997, **32**, 355–375.
12 N. C. Reis, R. F. Griffiths, M. D. Mantle and L. F. Gladden, *Int. J. Heat Mass Transfer*, 2003, **46**, 1279–1292.
13 T. W. J. Scheenen, D. van Dusschoten, P. A. de Jager and H. Van As, *J. Magn. Reson.*, 2000, **142**, 207–215.
14 H. Watanabe, M. Fukuoka, A. Tomiya and T. Mihori, *J. Food Eng.*, 2001, **49**, 1–6.
15 K. W. Waldron, M. L. Parker and A. C. Smith, *Compr. Rev. Food Sci. Food Saf.*, 2003, **2**, 128–146.
16 S. Takeuchi, M. Maeda, Y.-i. Gomi, M. Fukuoka and H. Watanabe, *J. Food Eng.*, 1997, **33**, 281–297.
17 U. Wählby and C. Skjöldebrand, *J. Food Eng.*, 2001, **47**, 303–312.
18 T. Okuzono and M. Doi, *Phys. Rev. E: Stat., Nonlinear, Soft Matter Phys.*, 2008, **77**, 030501.
19 R. G. M. Van der Sman, *Food Hydrocolloids*, 2012, **27**, 529–535.
20 H. T. Edzes, D. van Dusschoten and H. Van As, *Magn. Reson. Imaging*, 1998, **16**, 185–196.
21 M. J. McCarthy, E. Perez and M. Ozilgen, *Biotechnol. Prog.*, 1991, **7**, 540–543.
22 R. G. M. Van der Sman, *Meat Sci.*, 2007, **76**, 730–738.
23 X. D. Chen, *Drying Technol.*, 2006, **24**, 121–122.
24 M. K. Krokida, Z. B. Maroulis and D. Marinos-Kouris, *Drying Technol.*, 1998, **16**, 687–703.
25 V. Y. Martínez, A. B. Nieto, P. E. Viollaz and S. M. Alzamora, *J. Food Sci.*, 2005, **70**, E12–E18.
26 B. Hiranvarachat, S. Devahastin and N. Chiewchan, *Food Bioprod. Process.*, 2011, **89**, 116–127.
27 M. Doi and A. Onuki, *J. Phys. II France*, 1992, **2**, 1631–1656.

Structural changes of deposited casein micelles induced by membrane filtration

R. Gebhardt,[*a] T. Steinhauer,[a] P. Meyer,[a] J. Sterr,[b] J. Perlich[c] and U. Kulozik[a]

Received 15th February 2012, Accepted 5th April 2012

DOI: 10.1039/c2fd20022h

Casein micelles undergo shape changes when subjected to frontal filtration forces. Grazing incidence small angle X-ray scattering (GISAXS) and atomic force microscopy (AFM) allow a quantification of such structural changes on filtration cakes deposited on smooth silicon micro-sieves. A *trans*-membrane pressure of $\Delta p = 400$ mbar across the micro-sieve leads to an immediate film formation after deposition of casein solution. We observe significant changes in the GISAXS pattern depending on how many layers are stacked on top of each other. Compared to a deposit formed by one layer, GISAXS on a deposit formed by three layers of casein micelles leads to less scattering in the vertical and more scattering in the horizontal direction. Simulations show that the experimental results can be interpreted by a structural transformation from an originally spherical micelle shape to an ellipsoidal-deformed shape. The results are supported by AFM measurements showing a reduced lateral size of casein micelles deposited on top of a membrane pore. The observed shape changes could be due to filtration forces acting on densely packed deposits confining the micelles into ellipsoidal shapes.

1 Introduction

Microfiltration is a pressure driven separation process which enables the fractionation of complex mixtures of dissolved substances. During the microfiltration of milk for instance, casein micelles are retained by the porous membrane, while the smaller whey proteins should be able to pass through. Casein micelles deposited on the membrane however, interfere with the filtration process by forming, apart from the membrane, an additional mass transfer barrier or filtration resistance.[1,2] The major contribution to this filtration resistance originates from caseins forming surface-near deposits whose structure is experimentally unexplored. The molecular organization of casein micelle has been investigated at distances between 280 μm to 1 mm above the membrane using small angle X-ray scattering (SAXS) in transmission geometry.[3] As a result, a temporal variation of the casein concentration at different distances from the membrane could be deduced. Depending on a certain deposit height, the resistance of the deposit can be estimated according to the Kozeny–Carman equation.[4] For the computation, the equation considers global values such as porosity and tortuosity of the whole deposit as well as the surface area of individual particles. Variation in the surface area can be caused by a deformation

[a]Chair for Food Process Engineering and Dairy Technology Weihenstephaner Berg 1, 85354 Freising, Germany. E-mail: Ronald.Gebhardt@tum.de; Fax: +49 8161 71 4383; Tel: +49 8161 71 3536
[b]Lehrstuhl für Lebensmittelverpackungstechnik, Weihenstephaner Steig 22, 85350 Freising, Germany
[c]HASYLAB-DESY, Notkestr. 85, 22603 Hamburg, Germany

of the particles. The impact of deformable particles such as yeast, Ca-alginate and dextran-MnO$_2$ particles on the microfiltration process has been investigated as a function of the *trans*-membrane pressure during frontal filtration[5] and as a function of the shear stress in cross-flow microfiltration experiments.[6,7] The studies have shown that softer particles led to more compact deposits next to the membrane surface with higher filtration resistances. Studies on casein micelles, aiming to investigate the effect of filtration forces on their structure are lacking. Recent studies combining osmotic stress with SAXS have revealed that some internal parts of compressed micelles collapse while other parts resist deformation.[8,9] The key element for the explanation of such shape changes is the internal structural organization of the casein micelle.[10–12] Casein micelles are spherical protein aggregates[13] broadly lognormally distributed between 50–500 nm.[14] Preparation of size fractions with a reduced distribution width can be achieved by ultracentrifugation.[15,16] Main constituents of the casein micelle are four types of casein monomers (α_{S1}-, α_{S2}-, β- and κ-casein) and colloidal calcium phosphate. High charge densities of the phosphorylated serines and the high proportion of prolines, glutamine and asparagine residues in the primary structure of the caseins hinder the condensation into a globular state.[17] Instead, caseins adopt an open, flexible structure with rheomorphic behavior.[17–19] Caseins can be simply regarded as block copolymers with blocks consisting of amino acids with high levels of hydrophobic or hydrophilic residues.[20,21] Due to both main types of interactions the stability of the casein micelles strongly depends on the temperature, calcium content and pH of the medium.[22] The hydrophilic contacts between caseins and colloidal calcium phosphate and hydrophobic contacts among caseins are the energetic basis in both casein micelle models, the nanocluster[23] and the dual binding model.[24] Further developments of the interaction models including also water, which contributes with 3–4 g per 1 g protein, have been published recently.[9,10] Unevenly distributed regions filled with water are suggested to be inside the micellar structure, which could be involved in both the diffusive mass transfer between casein micelles and their environments and the structural deformation due to a different compressibility.

An outer layer of amphiphilic κ-casein acts as an interphase between the hydrophobic core of the micelle and the hydrophilic surrounding. Furthermore, the outer κ-casein layer provides steric and entropic stabilization and ensures colloidal integrity of the micelle.[25]

Here, we report for the first time on filtration-induced shape changes of casein micelles in close vicinity to the membrane. The findings were made possible by combining frontal filtration across silicon-micro-sieves with surface sensitive X-ray scattering (GISAXS). We have used GISAXS in a number of studies to investigate structural changes of casein micelles on surfaces in dependence of proteolytic treatment,[26] of the micellar size distribution as well as of pH, calcium and casein concentration.[27,28] The here reported experimental results provide new insights into the structural properties of casein micelle in membrane deposits under the influence of fluid forces and could have implications on the optimization of membrane separation processes.

2. Experimental

Materials

Casein micelles were prepared from commercial-grade skim milk by combined uniform *trans*-membrane pressure micro–filtration (mean pore diameter: 0.1 μm) and ultra–filtration. The dried casein powder was dissolved by thoroughly stirring for 5 h in 50 mM Tris-HCl and 50 mM CaCl$_2$, pH 7.5 and $T = 20\ °C$. A Beckmann centrifuge was used to prepare a size-fractionated sample with casein micelles sharply distributed around a mean size of $D = 150$ nm as described elsewhere.[29] The concentration of remaining whey proteins was close to zero.

Microfiltration cell

Micro-sieves (Nano-Filtertechnik-GmbH, München, Germany) with a defined pore size of 0.88 μm and an active area of 0.4–0.5% were used for the GISAXS and AFM measurements because of their low surface roughness. The micro-sieves have 75 membrane sections with a width of 100 μm.

Grazing incidence small angle-X-ray scattering - GISAXS

GISAXS measurements were performed at the BW4 beamline at HASYLAB/DESY.[30] The beam was generated by a wiggler from a positron (e$^+$) current inside the storage ring DORISIII. After passing a monochromator, which provided a wavelength of $\lambda = 0.138$ nm, the beam was collimated and moderately focused by beryllium compound refractive lenses. An ionization chamber in front of the sample was used to monitor the intensity of the beam which decreased gradually after injection due to the current decay in the storage ring. The frontal filtration cell of the micro-sieve was mounted on an $x/y/z$-translation stage and two motorized circle-segments to allow a proper GISAXS alignment.

The frontal filtration cell (C) for GISAXS measurements consisted of two parts, vacuum-sealed by a clamping ring. To place the micro-sieve (D), the upper part of the cell contained a rectangular rubber gasket around its central hole. The lower part was an adapter flange which was connected with a diaphragm pump (MZ2C, Vacuubrand) with a vacuum-controller 220 (E) *via* a vacuum-resistant tube. The diaphragm pump generated pressures up to $\Delta p = 600$ mbar with a precision of $\delta_P = 10$ mbar. During the deposition of the sample we used an acting *trans*-membrane pressure of $\Delta p = 400$ mbar. For sample deposition, a hydraulic pump (F) controlled outside the experimental hutch was available. GISAXS measurements and alignment were performed under normal pressure conditions *i.e.* without acting *trans*-membrane pressure. Within the GISAXS alignment the sample was tilted by $\alpha_i = 0.4149°$ by the circle segment relative to the incoming X-ray beam. This caused specular reflection from the sample at an outgoing angle $\alpha_f = \alpha_i$. The sample-to-detector distance was 2190 mm measured by a silver behenate standard. The size of the X-ray beam was 24 μm × 32 μm (vert. × hor.) which led to a sample area on the micro-sieve which was illuminated by the X-ray beam of 3.4

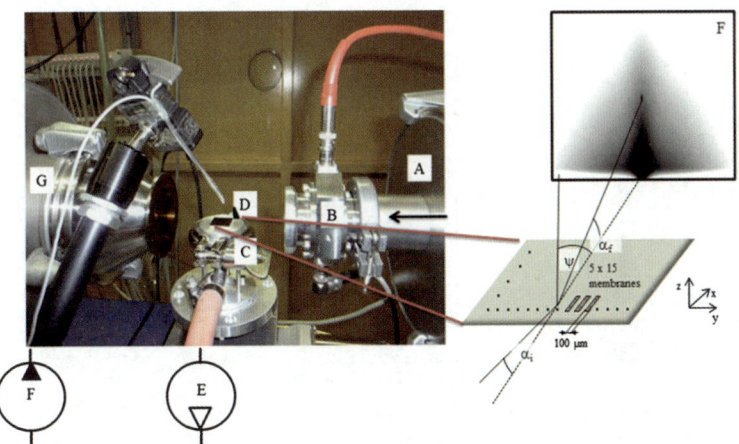

Fig. 1 Synchrotron radiation microbeam GISAXS setup: (A) guard slit system; (B) ionization chamber; (C) frontal filtration cell; (D) micro-sieve consisting of 75 membrane sections and GISAXS scattering geometry (see enlarged view on the right); (E) vacuum diaphragm pump; (F) hydraulic pump for droplet deposition; (G) evacuated flight tube; (F) typical GISAXS intensity distribution of the MAR165 CCD detector.

mm × 32 μm. Fig. 1 shows an enlarged view of the micro-sieve and the GISAXS scattering geometry.

Data analysis

We used the Fit2D software package for data reduction.[31] GISAXS patterns for the ellipsoidal form factor were simulated within the Born approximation, called the BA-LAYER using the IsGISAXS program.[32] Surface areas were calculated according to:

$$A = 2\pi \left(a^2 + c^2 \frac{i}{\tan(i)} \right) ; \tag{1}$$

with $i = \arccos\left(\frac{a}{b}\right)$; a: the half width and b: the half height of the prolate ellipsoids.

Atomic force microscopy (AFM)

An atomic force microscopy system (WITec-alpha500) was used for intermittent-contact measurements (AC mode) in air and under room conditions ($T = 23\ °C$, relative humidity: 30%). The cantilevers used for the measurements had a resonance frequency of 285 kHz and a spring constant of 42 N m^{-1}. Images were recorded with 300 Point/Line and a scan rate of 0.66 Hz.

Results

GISAXS pattern of casein deposits and micro-sieves

Fig. 2A shows a typical GISAXS pattern of casein micelles on a micro-sieve, which was subjected to frontal filtration. Beside the specular peak (shielded by a beam-stop) at $\alpha_f = \alpha_i = 0.41°$, the two-dimensional intensity distribution contains scattering contribution of the casein layer and its solid support, *i.e.* the micro-sieve. Fig. 2C shows in an enlarged view the intensity distributions of a casein droplet deposited on the micro-sieve (D). The intensity enhancement at $\alpha_f = 0.22°$ originates

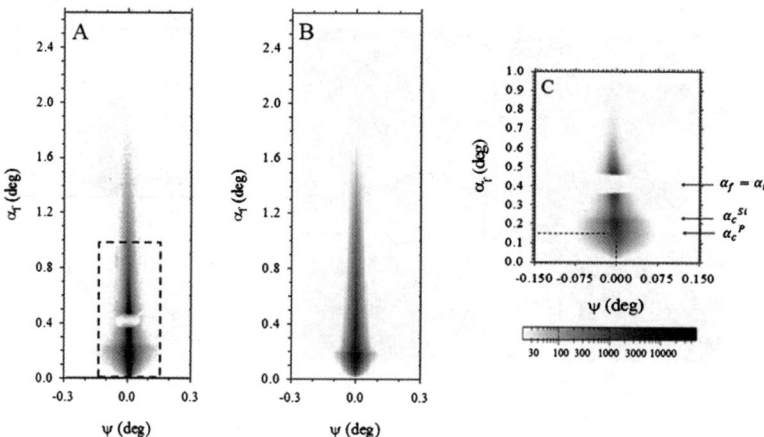

Fig. 2 Two-dimensional GISAXS intensity distribution as a function of the out-of plane angle ψ and the exit angle α_f, measured (A) and simulated by an elastic scattering approach (B). GISAXS-pattern of a casein deposit on top of the micro-sieve is displayed in an enlarged view (C). Intensity enhancements occur at $\alpha_f = \alpha_i$ (specular peak, shielded by the secondary beam-stop), at $\alpha_f = \alpha_C^{Si}$ (critical angle of silicon) and at $\alpha_f = \alpha_C^P$ (critical angle of protein). The positions of the out-of-plane cut at $\alpha_f = \alpha_C^P$ and the detector cut at $\psi = 0$ are indicated as dashed lines.

from the critical angle of silicon. The additional intensity enhancement at $\alpha_f = 0.15°$ is due to the critical angle of the adsorbed protein on the micro-sieve.

GISAXS scan across casein droplet profiles

Casein deposits were formed after deposition of casein droplets on the micro-sieve. We performed three GISAXS-scans with a positional resolution of 100 μm across one and the same profile of deposited casein micelles. After each scan a further casein droplet was stacked on top of the previous one, which led to three casein samples consisting of one, two and three casein micelle layers. A low-pressure of $p = 600$ mbar was generated below the micro-sieve using a vacuum diaphragm pump. The casein droplet deposition led to an immediate film formation. Prior to the GISAXS measurement the low-pressure was raised to normal pressure, since the acting *trans*-membrane pressure of $\Delta p = 400$ mbar led to a slide bending of the micro-sieve. Under normal pressure conditions, however, the initial GISAXS alignment was completely recovered. The GISAXS scan, performed across the deposited layer of casein micelles, is shown in Fig. 3 and 4. Fig. 3 depicts the positional variation of the detector cuts from the measured GISAXS intensity distributions of the scan across the profiles of one (A), two (B) and three (C) layers. The position of the primary beam ($\alpha_f = -0.41°$) and the specular beam ($\alpha_f = 0.41°$) remains constant with the positional coordinate. Small angle X-ray scattering at $\alpha_f = -0.3°$ (left) around the primary beam arises at positional coordinates $x = 1$ mm and $x = 5$ mm *i.e.* at the boundary of the droplet profile. The contribution of the micro-sieve to the scattering signal is indicated in Fig. 3. Outside the layer

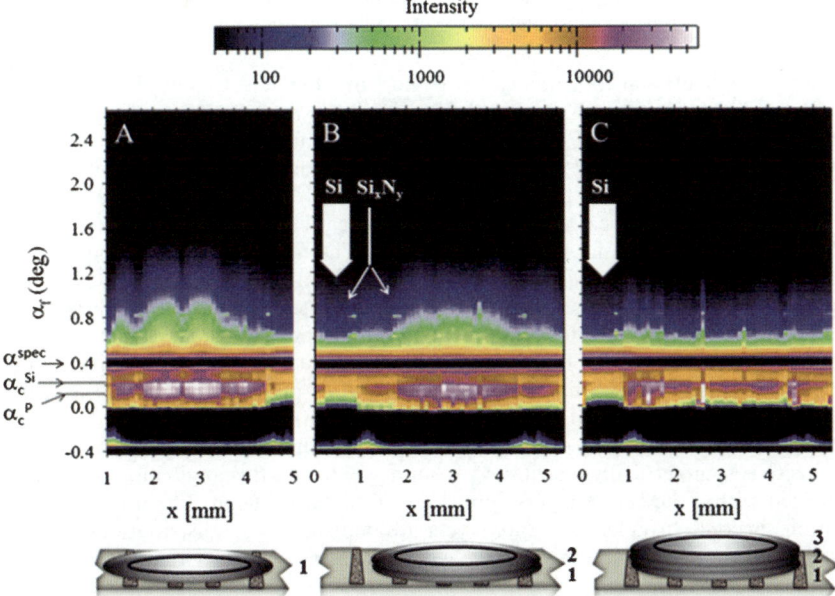

Fig. 3 Detector-cuts as a function of the positional coordinate of a GISAXS-scan across one (A), two stacked (B) and three stacked (C) casein droplets on a micro-sieve. Shielded primary beam at $\alpha_f = -0.4°$ and small angle X-ray scattering arising from the primary beam from the droplet rim; GISAXS-horizon at $\alpha_f = 0°$; Intensity enhancement at $\alpha_f = 0.15°$ and $\alpha_f = 0.22°$ due to the critical angles of the protein and the silicon micro-sieve; shielded specular beam at $\alpha_f = 0.4°$. Contributions of the micro-sieve to the GISAXS signal of the casein droplet (thick white arrows: silicon spacer; thin white arrows: silicon nitride membranes) are indicated. A sketch below illustrates the corresponding situation on the micro-sieve with the membrane areas indicated as dark patterned bars.

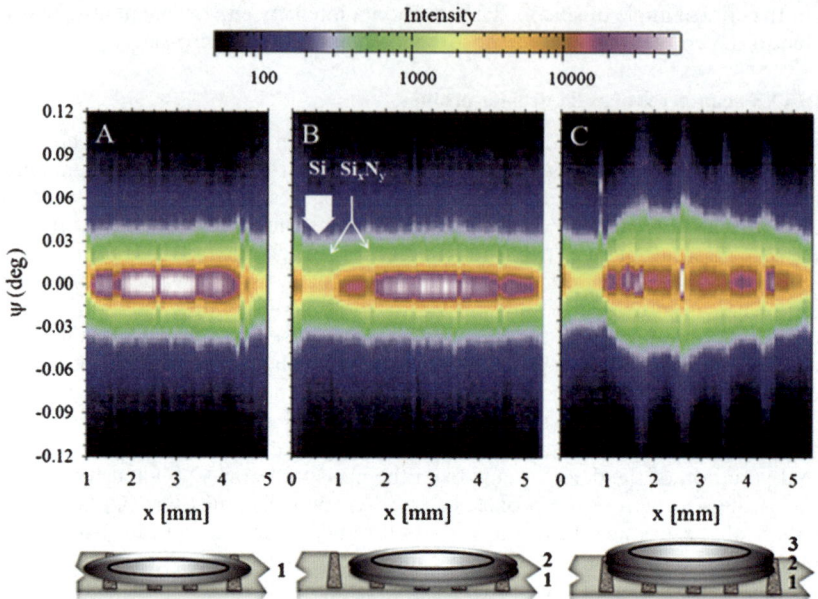

Fig. 4 Out-of-plane cuts taken on the critical angle of the protein ($\alpha_f = 0.15°$) as a function of the positional coordinate of a GISAXS-scan across one (A), two stacked (B) and three stacked (C) casein droplets. Out-of-plane cuts of the clean silicon micro-sieve (Si) and the membrane region (Si_xN_y) are indicated by white arrows. Diffuse out-of plane scattering increases from (A) to (C). A sketch below illustrates the corresponding situation on the micro-sieve.

between the positional coordinates $x = 0$ and 1 mm the detector cut of bare silicon is detected. The intensity is higher at the critical angle $\alpha_f = 0.22°$ of silicon. Periodic intensity spots (compare Fig. 2A) appear every 1 mm over a distance of 100 μm due to the arrangement of the porous membranes on the micro-sieve. Within the casein layer between $x = 1$ and 5 mm, the protein causes an additional intensity enhancement at its critical angle of $\alpha_f = 0.15°$. Scattering from the protein appears above the specular beam at angles $\alpha_f > 0.41°$. The more layers are deposited, the smaller the collected intensity in that angular range of the detector. The scan across one and two layers shows that the intensity increases from the boundary to the middle of the droplet profile. Interestingly, at the positions of the membrane on the micro-sieve (arrows in Fig. 3 marked by Si_xN_y), the scattered intensity drops.

Fig. 4 shows the positional variation of the out-of-plane cuts taken at the critical angle of the protein ($\alpha_f = 0.15°$) of the measured GISAXS intensity distributions of the scan across the profiles of one (A), two (B) and three (C) droplets. The scattered intensity is symmetrically distributed around $\psi = 0°$. At this angle more intensity is detected within the casein layers compared to the bare silicon. The intensity distributions become broader the more layers are deposited. The distributions broaden further on the sites of the micro-sieve where membranes are located (Si_xN_y).

Simulation of the GISAXS pattern

We performed simulations shown in Fig. 2B to analyze the representative GISAXS pattern in Fig. 2A. An ellipsoidal form-factor reproduced best the two-dimensional intensity distribution with the striking central rod at $\psi = 0°$. No interference effect between neighboring scattering objects *e.g.* structure factor was considered. The calculations were based on the Effective Layer Born Approximation in order to simulate the intensity enhancement near the critical angle at $\alpha_f = \alpha_C$. Hence, beside the scattering geometry the model considered only the form factor and the optical

This journal is © The Royal Society of Chemistry 2012

properties of the scattering objects. The model was used to simulate the in-plane and out-of-plan scattering shown in Fig. 3–4. The dimensions of the width W and height H of the ellipsoids were varied in such a way that the volume of the original spherical micelle (diameter of $D = 160$ nm) was kept constant. The dimensions of the ellipsoid were assumed to be log-normally distributed. We fixed the width of the log-normal distribution for both parameters $\sigma W/W$ and $\sigma H/H$ to 2.2. Three ellipsoidal geometries which differed in width and height (Tab. 1) were simulated in order to analyze the experimental results in Fig. 3 and 4. In order to analyse the experimental results, we simulated three ellipsoidal form factors which differed in width and height (Tab. 1). Compared to bare silicon, the protein contributes significantly between $\psi = 0.03$–$0.1°$ (Fig. 4) and between $\alpha_f = 0.6$–$1.2°$ (Fig. 3) to the scattering signal. We used both angular ranges to roughly fit the ellipsoidal form factor to the data (Fig. 5).

The lowest out-of-plane angle $\psi = 0.03°$ corresponds to a smallest scattering vector $Q_Y = 0.024$ nm^{-1} allowing resolution of distances up to 250 nm. In contrast, in-plane analysis is restricted to smaller distances due to the strong influence of the specular peak at $\alpha_f = 0.4°$ (intensity shielded by a beam-stop), which is not considered by our model. As a comparison of the model with the data shows, differences in the out-of plane and in-plane scattering between both samples can be well explained by an ellipsoidal deformation, by which the entire volume of the ellipsoid remains unchanged. Both data and model show a stronger scattering at large ψ and a weaker scattering at large α_f for the casein micelles in droplet 3. This indicates that micelles in droplet 3 become stronger compressed laterally and stretched in the vertical direction. That information can be extracted well by selecting a form factor of a simple ellipsoidal shape for the casein micelles. A more complex model considering further structural details of the micelle (substructure, surface-inhomogeneity *etc.*) increases further the number of fitting parameters and would be rather speculative. Fig. 6 depicts simulations for the detector-cut (upper row) and the out-of-plane-scattering in the bottom row for comparison with the measurements presented in Fig. 4 and 3. The assumed ellipsoidal micellar shape is shown as an insert.

The first ellipsoid simulated (index 1) had a width of $W = 113$ nm and a height of $H = 320$ nm. The corresponding simulated GISAXS (Fig. 6A$_1$, 6B$_1$) most closely resembled the GISAXS signal measured on the spacer region in the middle of the first droplet deposited (Fig. 3A and 4A). According to the used color the measured and simulated intensity transition from green to blue proceeds at $\alpha_f = 1°$ and $\psi = 0.04°$. If the width of the ellipsoid decreases while its height increases, the transition shifts towards smaller α_f and larger ψ angles. The effect can be seen in the GISAXS measured on membrane regions (Si$_x$N$_y$) or with increasing droplet number *i.e.* with increasing concentration of micelles on the micro-sieve. The simulated scattering signal of the second ellipsoid ($W = 90$ nm, $H = 500$ nm, Fig. 6 A$_2$,B$_2$) fits well both the scattering recorded on the membrane region in the middle of droplet 1 and the scattering from the silicon spacer region in the middle of two-stacked droplets (Fig. 3B and 4B). The GISAXS pattern of the third ellipsoid ($W = 82$ nm, $H = 600$ nm) was simulated to reproduce the scattering from three stacked droplets. The

Table 1 Specifications of three ellipsoidal form factors

Index	Width, W [nm]	Height, H [nm]	Surface Area [nm^2][a]	Increase in surface area relative to a sphere of equal volume [%]
1	113	320	9.3×10^4	16
2	90	500	1.1×10^5	42
3	82	600	1.2×10^5	51

[a] calculated according to Formula 1.

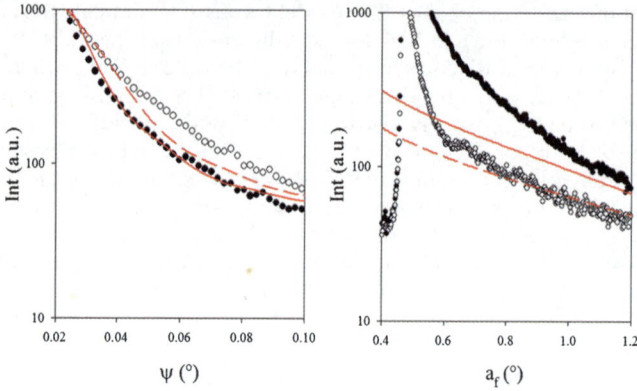

Fig. 5 Out-of-plane and detector cuts from measurements on droplet 1 (filled circle) and 3 (open circle) together with ellipsoidal form factor model 1 (solid line) and 3 (dashed line).

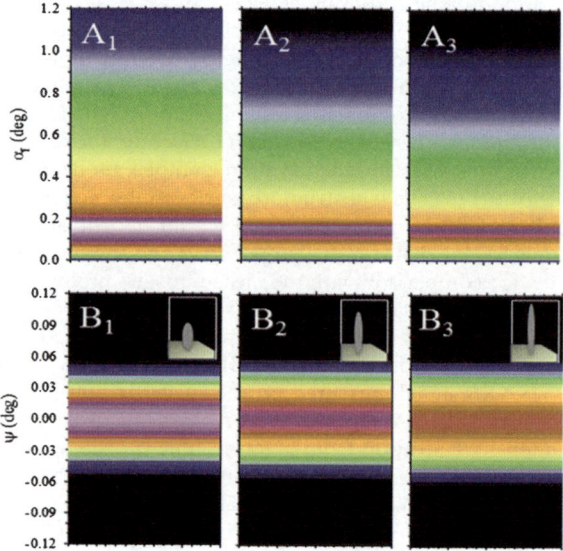

Fig. 6 Simulated GISAXS detector (A) and out-of plane (B) cuts for three different ellipsoidal form factors (A_1 and B_1 at $x = 2.2$ mm in Fig. 3 and 4A; A_2 and B_2 at $x = 3$ mm in Fig. 3 and 4B; and A_3 and B_3 at $x = 2$ mm in Fig. 3 and 4C). Insets in B_1–B_3 show the corresponding ellipsoidal objects in real space.

green-blue color transition of the intensity distribution occurs at $\alpha_f = 0.6°$ (Fig. 6A_3) and $\psi = 0.05°$ (Fig. 6B_3) in accordance with the experimental data (compare Fig. 3C, 4C).

The specifications of three ellipsoidal form factors are listed in Tab. 1. All deposited casein micelles are elongated perpendicularly to the membrane. The degree of elongation depends on both the deposit position on the micro-sieve and on how many deposited layers were stacked on top of each other. Micelles deposited above the membrane region are always more elongated than those above the spacer regions. The same applies for micelles which became deposited later. Micelles from layers (*e.g.* droplet 2) covering existing casein layers (droplet 1) are more strongly affected. We performed AFM measurements in order to validate the GISAXS results in real space.

This journal is © The Royal Society of Chemistry 2012

AFM-measurements

We used the same setup and *trans*-membrane pressure ($\Delta p = 400$ mbar) for the sample preparation as for the GISAXS measurements. We recorded AFM images on the membrane areas of the silicon micro-sieves. Fig. 7A shows a representative micrograph of deposited casein micelles around a membrane pore. The dimensions of the pore (radius: 400 nm) are marked by a dashed circle. In order to increase the quality of the images, we used a dilute sample, resulting in a surface which is only partly covered by casein micelles. Deposited casein micelles on top of the pores are more closely spaced and considerably reduced in lateral size compared to those outside the pore. Fig. 7B shows the 1D height profile resulting from a cut (black line in Fig. 7A) across the pore. A size estimate of the peak profiles (Fig. 7B) revealed a lateral size of $d \sim 90$ nm on top of and $d \sim 170$ nm outside the pore. The pore blocking by smaller particle sizes can take place either by pore narrowing/constriction due to the adsorption of casein micelle layers[33] or by hydrodynamic bridging.[34]

Discussion

We have investigated shape changes of casein micelles in their deposits on top of micro-sieves during filtration. For the study we used grazing incidence small angle X-ray scattering in combination with a newly developed special filtration set-up. The set-up includes a sufficiently smooth membrane (silicon micro-sieve with a surface roughness below 4 nm, which is important to avoid too strong background scattering), a fully auto-controlled sample deposition and an adjustment of the *trans*-membrane pressure by a vacuum diaphragm pump. After deposit formation, structural states of casein micelles induced by filtration forces were frozen in a film matrix and remained unchanged until the end of the experiment. Since filtration and film formation proceeded within seconds, the experiments were sensitive to structural changes induced by filtration forces. In contrast, drying effects proceed on a much longer time scale. For that reason, casein micelle deformation as detected in compressed dried films[35] can be excluded. The small beam size (24 µm × 32 µm) allowed for position-sensitive X-ray scans,[36] or rather GISAXS-scans as demonstrated in previous studies.[37] The detected GISAXS pattern from the membrane differed significantly from the interjacent spacers region and allowed a precise localization of the measurement position on the micro-sieve. The obtained GISAXS patterns of the casein micelle deposits could be best described by a form factor model of an ellipsoid (see Tab. 1). The results indicate deformability of the spherical casein

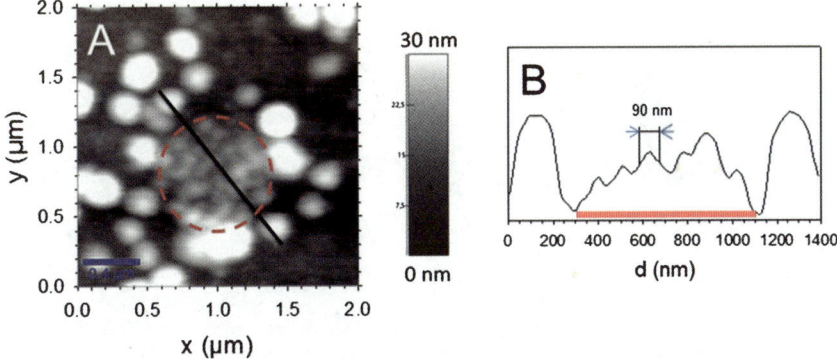

Fig. 7 A) Atomic force micrograph of casein micelle deposits subjected to frontal filtration ($\Delta p = 400$ mbar). A pore of the micro-sieve is indicated by a dashed circle. B) Cut across the AFM image (black line in Fig. 7A). The characteristic lateral dimension of a casein micelle above a pore and the dimension of the pore itself (red bar) are indicated.

micelles as reported from small angle X-ray scattering experiments under osmotic stress.[8,9] In contrast, for the data presented here we do not have any indication that the degree of hydration of casein micelles changes, since their volume remains constant after ellipsoidal deformation. A changed water binding behaviour of biopolymers under stress can be quantified by either applying Flory–Rehner theory in simulation studies[38] or by fitting Carnahan-Starling type models to osmotic pressure data.[8] Information about the structure of water cannot be obtained from small angle X-ray experiments because of their low resolution. However, if this would be the case, loss of water can be approximated from the decrease in the radius of gyration, as demonstrated recently in GISAXS[26] and SAXS[9] experiments.

Our experimental results indicate that packing effects and fluid forces are involved in the deformation mechanism. Densely packed micelles on top of the pores were reduced in lateral sizes as AFM images showed (Fig. 7) while isolated micelles in the periphery were less affected. Furthermore, the degree of deformation was enhanced the more casein micelle layers were stacked for deposit formation. During addition of layers, the concentration of casein micelles increases on top of the microsieve and the packing of the deposit becomes denser. How does the deformation process proceed? Since packing effects are involved, casein micelles could be confined into an ellipsoidal shape by the elongational filtration tension acting perpendicularly to the membrane and by the compression from their surroundings. The mechanism, which is used in microfluidics to generate particles with a nonspherical shape,[39–41] is illustrated in Fig. 8A. Elongated particles are formed when the diameter of the used drops exceeds the diameter of the outlet channel w. For a casein micelle the surrounding micelles constitute the lateral confinement. We have recently shown that size-fractionated casein micelles tend to order hexagonally when cast on surfaces due to their sharp size distribution.[29] The resulting surrounding of a casein micelle in such a scenario is depicted in Fig. 8B. During the filtration all micelles become equally elongated which results in a vertical elongation and a lateral compression of the deposit (Fig. 8C). The flexible, rheomorphic structure of casein monomers[17,18] could facilitate such large shape changes. For that reason our experimental results cannot be generally transferred to the majority of other native proteins, which are more rigid due to α-helical and β-sheet structures.[19] A protein system comparable with casein micelles is fibroin and its higher aggregated forms. Similar to caseins, silk fibroin also possesses a hydrophilic–hydrophobic copolymer structure. In case of the silk protein, assembling into fibers has been demonstrated in a microfluidic device.[42,43] As a molecular basis, elongation and alignments

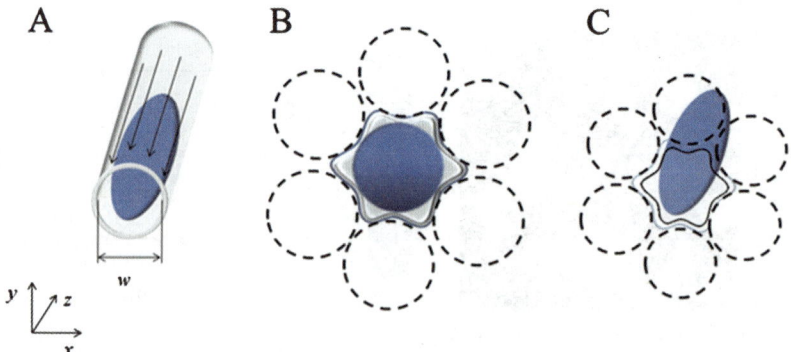

Fig. 8 Mechanism of the ellipsoidal shape change. A) Shape change under confinement by microfluidic synthesis; B) assumed hexagonal confinement of a casein micelle surrounded by six spheroidal neighbors; C) shrinking confinement leading to smaller lateral sizes due to ellipsoidal shape changes and *vice versa* when subjected to elongational filtration forces acting in the z-direction.

of fibroin micelles into fibrillar structures by physical shear has been suggested.[44–46] An increase in the ellipsoidal deformation causes an increase in the surface area. Related to a sphere with an equal volume, the increase in surface area is 16% for an ellipsoid with a height/width ratio of $R = 2.8$ and 51% for $R = 7.3$ (cf. Tab. 1). As a result, the filtrate flow is more strongly affected since the friction surface between the fluid and the deposit increases.

If closely packed, several systems such as cells or emulsion deform into a hexagonal shape. Yeast cells for instance deform by flattening in regions of cell–cell contact.[47] Such facets between highly ordered particles give rise to intensity streaks normal to that face in the GISAXS patterns, which we did not observe in our experimental data. However, small flattening effects as observed recently in dried casein films by high resolution AFM[35] could also occur in filtered deposits. If so, our GISAXS model would treat such effects by adjusting the width distribution σW/W.

Conclusions

The results demonstrate the ability of spherical casein micelles to undergo a transformation towards an ellipsoidal shape when subjected to a filtration flow. In contrast to previous studies,[5,6] we showed that the longitudinal axis of the deformed particles is oriented perpendicularly and not parallel to the membrane surface. The structural changes require the embedding of the micelles in the densely packed deposit on the membrane by ensuring at the same time the integrity of a single micelle. Therefore, stabilizing colloidal interactions might have an important impact on the structure formation processes and should be investigated in detail in future work. Furthermore, the influence of the filtration force could be investigated by varying the *trans*-membrane pressure. An important fact for the shape change is also the intrinsic structure of the casein micelles showing rheomorphic behaviour, which could be compared with those from more rigid, structured systems. The shape changes lead to a noticeable increase in the surface area which might have implications on the filtration resistance of the deposited casein micelles and on filtration models for casein micelles in general.

Acknowledgements

We thank C. Ederer and E. Schneider for manufacturing the frontal filtration module for the GISAXS setup and P. Müller-Buschbaum, S.V. Roth and M. Heinrich for helpful discussions. Beamline access and financial support was kindly provided from DESY/HASYLAB (Hamburg) within the long-term project II-20100114.

References

1 W. Kühnl, A. Piry, V. Kaufmann, T. Grein, S. Ripperger and U. Kulozik, *J. Membr. Sci.*, 2011, **352**, 107–115.
2 A. Piry, W. Kühnl, T. Grein, A. Tolkach, S. Ripperger and U. Kulozik, *J. Membr. Sci.*, 2008, **325**, 887–894.
3 C. David, F. Pignon, T. Narayanan, M. Sztucki, G. Gesan-Guiziou and A. Magnin, *Langmuir*, 2008, **24**, 4523–4529.
4 Y. Endo, D.-R. Chen and D. Y. H. Pui, *Powder Technol.*, 2002, **124**, 119–126.
5 K.-J. Hwang, Y.-T. Wang, E. Iritani and N. Katagiri, *J. Membr. Sci.*, 2009, **341**(1–2), 286–293.
6 W.-M. Lu, K.-L. Tung, C.-H. Pan and K.-J. Hwang, *J. Membr. Sci.*, 2002, **198**(2), 225–243.
7 K.-J. Hwang, Y.-T. Wang, E. Iritani and N. Katagiri, *J. Membr. Sci.*, 2010, **365**(1–2), 130–137.
8 A. Bouchoux, P.-E. Cayemitte, J. Jardin, G. Gésan-Guiziou and B. Cabane, *Biophys. J.*, 2009, **96**, 693–706.
9 A. Bouchoux, G. Gésan-Guiziou, J. Pérez and B. Cabane, *Biophys. J.*, 2010, **99**, 3754–3762.
10 D. G. Dalgleish, *Soft Matter*, 2011, **7**, 2265–2272.

11 D. S. Horne, *Curr. Opin. Colloid Interface Sci.*, 2006, **11**, 148–153.
12 J. H. M. Farrell, E. L. Malin, E. M. Brown and P. X. Qi, *Curr. Opin. Colloid Interface Sci.*, 2006, **11**, 135–147.
13 G. A. Morris, T. J. Foster and S. E. Harding, *Biomacromolecules*, 2000, **1**, 764–767.
14 P. F. Fox and A. Brodkorb, *Int. Dairy J.*, 2008, **18**, 677–684.
15 S. Marchin, J.-L. Putaux, F. Pignon and J. Leonil, *J. Chem. Phys.*, 2007, **126**, 045101.
16 M. Heinrich and U. Kulozik, *International Dairy Journal*, 2011, **21**, 664–669.
17 C. Holt and L. Sawyer, *J. Chem. Soc., Faraday Trans.*, 1993, **89**, 2683–2692.
18 E. Smyth, C. D. Syme, E. W. Blanch, L. Hecht, M. Vašák and L. D. Barron, *Biopolymers*, 2001, **58**, 138–151.
19 A. Gaspar, M. S. Appavou, S. Busch, T. Unruh and W. Doster, *Eur. Biophys. J.*, 2008, **37**, 573–582.
20 D. S. Horne, *Curr. Opin. Colloid Interface Sci.*, 2002, **7**, 456–461.
21 S. R. Euston and D. S. Horne, *Food Hydrocolloids*, 2005, **19**, 379–386.
22 R. Gebhardt, N. Takeda, U. Kulozik and W. Doster, *J. Phys. Chem. B*, 2011, **115**, 2349–2359.
23 C. Holt, C. G. de Kruif, R. Tuinier and P. A. Timmins, *Colloids Surf., A*, 2003, **213**, 275–284.
24 D. S. Horne, *Int. Dairy J.*, 1998, **8**, 171–177.
25 R. Tuinier and C. G. d. Kruif, *J. Chem. Phys.*, 2002, **117**, 1290–1295.
26 R. Gebhardt, S. V. Roth, M. Burghammer, C. Riekel, A. Tolkach, U. Kulozik and P. Müller-Buschbaum, *Int. Dairy J.*, 2010, **20**, 203–211.
27 P. Müller-Buschbaum, R. Gebhardt, E. Maurer, E. Bauer, R. Gehrke and W. Doster, *Biomacromolecules*, 2006, **7**, 1773–1780.
28 P. Müller-Buschbaum, R. Gebhardt, S. V. Roth, E. Metwalli and W. Doster, *Biophys. J.*, 2007, **93**, 960–968.
29 R. Gebhardt, W. Holzmüller, Q. Zhong, P. Müller-Buschbaum and U. Kulozik, *Colloids Surf., B*, 2011, **88**, 240–245.
30 J. Perlich, J. Rubeck, S. Botta, R. Gehrke, S. V. Roth, M. A. Ruderer, S. M. Prams, M. Rawolle, Q. Zhong, V. Korstgens and P. Muller-Buschbaum, *Rev. Sci. Instrum.*, 2010, **81**, 105105.
31 A. Hammersley, Available at: http://www.esrf.fr/computing/scientific/FIT2D/.
32 R. Lazzari, *J. Appl. Crystallogr.*, 2002, **35**, 406–421.
33 A. Saxena, B. P. Tripathi, M. Kumar and V. K. Shahi, *Adv. Colloid Interface Sci.*, 2009, **145**, 1–22.
34 V. Ramachandran and H. S. Fogler, *J. Fluid Mech.*, 1999, **385**, 129–156.
35 R. Gebhardt, C. Vendrely and U. Kulozik, *J. Phys.: Condens. Matter*, 2011, **23**, 444201–444208.
36 C. Riekel, *et al.*, *IOP Conf. Ser.: Mater. Sci. Eng.*, 2010, **14**, 012013.
37 S. V. Roth, T. Autenrieth, G. Grubel, C. Riekel, M. Burghammer, R. Hengstler, L. Schulz and P. Müller-Buschbaum, *Appl. Phys. Lett.*, 2007, **91**, 091915.
38 R. G. M. van der Sman, *Food Hydrocolloids*, 2012, **27**, 529–535.
39 J. I. Park, A. Saffari, S. Kumar, A. Günther and E. Kumacheva, *Annu. Rev. Mater. Res.*, 2010, **40**, 415–443.
40 S. Xu, Z. Nie, M. Seo, P. Lewis, E. Kumacheva, H. A. Stone, P. Garstecki, D. B. Weibel, I. Gitlin and G. M. Whitesides, *Angew. Chem.*, 2005, **117**, 3865–3865.
41 D. K. Hwang, D. Dendukuri and P. S. Doyle, *Lab Chip*, 2008, **8**, 1640–1647.
42 S. Rammensee, U. Slotta, T. Scheibel and A. R. Bausch, *Proc. Natl. Acad. Sci. U. S. A.*, 2008, **105**, 6590–6595.
43 A. Martel, M. Burghammer, R. J. Davies, E. Di Cola, C. Vendrely and C. Riekel, *J. Am. Chem. Soc.*, 2008, **130**, 17070–17074.
44 H.-J. Jin and D. L. Kaplan, *Nature*, 2003, **424**, 1057–1061.
45 F. Hagn, L. Eisoldt, J. G. Hardy, C. Vendrely, M. Coles, T. Scheibel and H. Kessler, *Nature*, 2010, **465**, 239–242.
46 R. Gebhardt, C. Vendrely, M. Burghammer and C. Riekel, *Langmuir*, 2009, **25**, 6307–6311.
47 M. Meireles, C. Molle, M. J. Clifton and P. Aimar, *Chem. Eng. Sci.*, 2004, **59**, 5819–5829.

Model for particle migration in bidisperse suspensions by use of effective temperature

H. M. Vollebregt,[*a] R. G. M. van der Sman[a] and R. M. Boom[b]

Received 23rd February 2012, Accepted 17th April 2012
DOI: 10.1039/c2fd20035j

A model for the particle migration in a bidisperse flowing suspension is proposed and compared to experimental data. A mixture formulation, describing the suspension velocity and pressure and the concentrations of two solid fractions is derived from a multi-fluid model. In the multi-fluid model the liquid phase and both dispersed phases are interpenetrating phases. The closure relations are based on a mean field approach extending closure relations of a monodisperse suspension. The model is used to predict segregation based on particle size in channel flow where the particles are subjected to Brownian motion and shear-induced migration. The comparison of the model results with experimental data shows that particle migration is predicted well by the given formulation.

1 Introduction

Particle segregation and mixing play important roles in several applications, such as fluidised beds and filtration applications in different industries. A better understanding of the behaviour of the different fractions leads to improvements in design and operation of such applications.[7,29] In practice applications are not monodisperse, but polydisperse. For several applications the polydispersity can be approximated by a bidisperse system given that the particles can be divided in two groups with different sizes. We are specifically interested in applications with particles in the range of 0.1–10 μm. Examples of applications are beer clarification with the yeast particles as the large particle group and polyphenol–protein aggregates as the small particle group, or milk fractionation with fat globules as the larger sized fraction and casein micelles as the smaller sized fraction.

We are interested in flow through narrow geometries with relatively concentrated suspensions. The higher concentration leads to interactions between three or more particles resulting in so-called shear-induced migration (SIM).[16,30] This makes particles deviate from their streamlines, and migrate towards regions with lower shear rates. This chaotic behavior of particles in linear shear has been named shear-induced diffusion (SID).[31] Applications where SIM plays a key role, are crossflow microfiltration,[4] and particle suspension fractionation.[27,7,10] For micron-sized particles SIM is the dominant interaction mechanism resulting in cross streamline particle migration. However, depending on geometry, flow and concentration Brownian motion and lift are also important.[4]

Experiments with bi- and polydisperse suspensions have shown that the presence of different sized particles influences the movement of the particles, resulting in behaviour that differs from behaviour in a monodisperse suspension. For instance, the interactions between the different particles can lead to separation between the

[a]Food and Biobased Research, Wageningen University & Research Centre, P.O. Box 17, 6700 AA Wageningen, The Netherlands. E-mail: martijntje.vollebregt@wur.nl
[b]Food Process Engineering, Wageningen University, The Netherlands

fractions, to the formation of stratified layers or to complex pseudo-periodic segregation and mixing.[2,3,12,25,34,44,52]

Modelling the behaviour of bi- and polydisperse suspensions is done on different scales: from the scale of the particles itself (fully resolved) to the scale of the suspension viewed as a continuum (underresolved).[32,46] The ongoing developments in the fully resolved techniques have extended the possibilities beyond Stokes flow and to large numbers of particles, see for example,[1] and see[53] for a recent example on shear-induced migration in concentrated suspensions. Underresolved methods have been applied in sedimentation and gas fluidized beds.[8,51] An example of the use of underresolved methods to simulate the low Reynolds number pressure-driven flow of a bidisperse suspension subjected to buoyancy and shear-induced migration to obtain the steady state solution can be found in ref. 41.

One way to describe solid–fluid systems on a continuum scale is to make use of separate mass and momentum conservation balances for each fraction. The system then consists of two interpenetrating phases. This is a multi-fluid approach. In the field of gas-fluidised beds two-fluid models are a commonly used approach. It is less used in the field of suspensions flow. For a monodisperse solid–fluid system the two-fluid model with separate mass and momentum balances for the fluid and the dispersed phase has been used as the starting point to obtain a closed set of equations.[50] We follow again this route to obtain the mixture formulation for bidisperse suspensions. A mixture formulation is more natural for shear-induced migration as the diffusion happens at a much slower time scale than the hydrodynamics, which can be taken at a quasi-steady state.

Suspension flow of micron-sized particles can be regarded as driven soft matter, where the physics is dominated by hydrodynamic interactions. Such a system is an out-of-equilibrium system which cannot be described by classical thermodynamics. However, approximate thermodynamic descriptions can be used with use of an effective temperature. The concept of effective temperature was first used in driven granular media[33] and states that the kinetic energy of velocity differences between the phases scales linearly with an effective temperature. The presence of slow dynamics in suspension flow, the diffusion like behaviour of the particle migration, gives the temperature concept the necessary thermodynamic meaning.[42,5,48] Also from the field of granular media the linear relation between the particle pressure and the effective temperature is introduced,[33] including the relation with the equation of state, similar to the relation for a Brownian suspension. We showed that shear-induced migration can be reformulated in terms of an effective temperature,[50] this facilitates thermodynamic descriptions.[47] Adoption of the concept of effective temperature makes the use of many thermodynamic paradigms possible, for instance the relation between the free energy, particle pressure and chemical potential.[45]

Underresolved models require closure relations for the relevant fluid–particle interaction and particle behaviour. Closures for dilute systems are often not capable of capturing the behaviour of more concentrated systems. For monodisperse suspensions either a phenomenological formulation of the closures is proposed or more physical derivations are given. The work on closure relations for bi- and polydisperse closures is limited. We showed that monodisperse closures can be rewritten in terms of the ratio between the particle volume fraction ϕ and the random close packing ϕ_{max}, $\tilde{\phi} = \phi/\phi_{max}$. With use of the mean field theory these closures were extended for bidisperse suspensions for the viscosity, particle pressure and the friction factors. The mean field theory states that the free volume is correlated to the ratio between the total particle volume fraction and the random close packing for a bidisperse suspension, in analogy to the monodisperse formulation. The applicability and limitations of this approach were investigated by comparison with empirical closures and experimental data.[47] Similarly to the monodisperse case, the closures obtained for the mixture model can be translated back to be used in a multi-fluid model.[50] Furthermore, closures for properties of the individual solid components are not available and are difficult to measure experimentally. For instance, this holds for the particle stresses. However, the particle stress of the suspension as a whole (a mixture) is measurable.

In the mixture approach, the mass balances for the solid fractions are rewritten to convection–diffusion equations. This can be done under the assumption of negligible inertia, for which the momentum exchange between the different phases consists of buoyancy and drag force. The resulting diffusion contributions to particle migration are given by the gradients in chemical potentials. For a monodisperse suspension direct use of the Gibbs–Duhem relation and the closure for the particle pressure resulted in the chemical potential.[50] However, for a bidisperse configuration this cannot be used directly, because there are no closures available for the particle pressures of the different solid fractions. Instead, we make use of the known relation between particle pressure, Helmholtz free energy and chemical potentials to derive the chemical potentials from the closure for bidisperse particle pressure.[45]

In Section 2 the governing equations for bidisperse suspension flow in a channel are given. The required closure relations and the derivation of the chemical potentials are discussed in Section 3. The results from the simulations are presented in Section 4 and the conclusions are summarised in Section 5.

2   Governing equations

We describe the bidisperse particle suspension as three interpenetrating continua, a multi-fluid extension of the two-fluid approach.[22] Similar to the model for a monodisperse suspension the mixture model for the bidisperse suspension is derived from the equations for the separate suspension fractions.[50] As for the monodisperse case we limit ourselves to the Stokes flow regime. The interaction between the solid and liquid phases is *via* drag and buoyancy forces. The continuum and momentum equations for a bidisperse suspension are thus given by

$$\partial_t(1 - \phi) + \nabla \cdot (1 - \phi)\mathbf{u}_f = 0 \tag{1}$$

$$\partial_t\phi_i + \nabla \cdot \phi_i\mathbf{u}_i = 0 \tag{2}$$

$$0 = \nabla \cdot \sigma_f - \mathbf{f}_1 - \mathbf{f}_2 + (1 - \phi)\rho_f\mathbf{g} \tag{3}$$

$$0 = \nabla \cdot \sigma_i + \mathbf{f}_i + \phi_i\rho_i\mathbf{g} \tag{4}$$

with $\phi = \phi_1 + \phi_2$ the total particle volume fraction, $\mathbf{u}_f$ and $\mathbf{u}_i$ the velocity of the fluid and particle phase i, $i = 1,2$, σ_f the fluid stresses, σ_i the particle stress of particle phase i, $\mathbf{f}_i$ the momentum exchange between the fluid phase and particle phase i, ρ_f the density of the fluid and ρ_i the density of particle phase i. Particle–fluid interactions are incorporated in the momentum exchange between particle and fluid, $\mathbf{f}_i$, and particle–particle interactions in the particle stresses, σ_i.

Typical applications of interest are food streams with particles in the size range 0.1–10 μm. In this size range the relevant migration mechanisms are Brownian motion and shear-induced migration, and close to the upper limit of this limit lift as well.[4] Lift is not included in the current implementation. Lift is due to inertia effects, which conflicts with the assumption that the drag force is linear with the slip velocities. However, incorporation of the lift force in the particle flux enlarges the applicability of the model from Stokes flow to flows with finite Reynolds numbers[28] and to larger particle sizes.

To solve the resulting system of equations it is advantageous to analyse the different dynamics, or time scales, in the system. In case of large differences in the dynamics the fastest modes can be considered quasi-steady state and thus be removed. From scale analysis one can find that the model contains fast time scales related to the drag forces.[13,19] This time scale can be removed by reformulation in slow and fast variables. The slow variable is the average suspension velocity (or suspension momentum) and the fast variables are the slip velocities of the individual solid phases with the liquid phase. For the fast variable a quasi-steady state approach is valid, reducing the complexity of the model.

Summation of the individual continuum balances (1)–(2) gives the mixture continuum balance

$$\partial_t \bar{\rho} + \nabla \cdot \overline{\rho \mathbf{u}} = 0 \tag{5}$$

with $\bar{\rho} = (1 - \phi)\rho_f + \phi_1\rho_1 + \phi_2\rho_2$ the average density of the mixture and $\overline{\rho \mathbf{u}} = (1 - \phi)\rho_f \mathbf{u}_f + \phi_1\rho_1\mathbf{u}_1 + \phi_2\rho_2\mathbf{u}_2$ the mixture momentum. For a density matched mixture this reduces to $\nabla \cdot \bar{\mathbf{u}} = 0$.

Summation of the individual momentum balances (3)–(4) gives the mixture momentum balance

$$0 = \nabla \cdot \sigma_f + \nabla \cdot (\sigma_1 + \sigma_2) + \bar{\rho}\mathbf{g} \tag{6}$$

Momentum exchange consists of the buoyancy force and the drag force. Different formulations exist for the buoyancy force, these are equivalent to each other with a pressure gauge transformation, see for instance.[36,14,50] Here the following is used:

$$\mathbf{f}_i = \phi_i \nabla \cdot \sigma_f + \phi_i(\bar{\rho} - \rho_i)\mathbf{g} + \mathbf{f}_i^* \tag{7}$$

with $\mathbf{f}_i^*$ the momentum exchange due to the drag force between particle phase i and the fluid. The drag force is linear with the slip velocity $\mathbf{w}_i$

$$\mathbf{f}_i^* = \zeta_i(\mathbf{u}_i - \mathbf{u}_f) = \zeta_i\mathbf{w}_i \tag{8}$$

with ζ_i the drag coefficient for solid phase i.

The fluid and particle phase mass balances (1)–(2) can be rewritten in terms of the average mixture velocity and the slip velocities between the particles and the fluid, see for instance.[8] The mixture velocity and the slip velocities are given by

$$\bar{\mathbf{u}} = (1 - \phi)\mathbf{u}_f + \phi_1\mathbf{u}_1 + \phi_2\mathbf{u}_2 \tag{9}$$

$$\mathbf{w}_i = \mathbf{u}_i - \mathbf{u}_f \tag{10}$$

With (9) and (10) the solid phase velocities can be written as, with $i \neq j$

$$\mathbf{u}_i = \bar{\mathbf{u}} + (1 - \phi_i)\mathbf{w}_i - \phi_j\mathbf{w}_j \tag{11}$$

Substitution of (11) in (2) leads to, with $i \neq j$

$$\partial_t \phi_i + \nabla \cdot \phi_i\bar{\mathbf{u}} = -\nabla \cdot (\phi_i(1 - \phi_i)\mathbf{w}_i - \phi_i\phi_j\mathbf{w}_j) \tag{12}$$

Expressions for the slip velocities follow from the given relation between the slip velocity and the drag force (8) and the momentum balances of the individual suspension components. The fluid stress tensor is decomposed in a contribution due to pressure, p, and the viscous forces, $\sum_f$,

$$\sigma_f = -p\mathbf{I} + \sum_f \tag{13}$$

In general the Gibbs–Duhem relation applies to stress and chemical potentials μ_i

$$\sum_i \nabla \cdot \sigma_i = \nabla \cdot \sigma_{\text{total}} = \nabla \cdot \Pi = \sum_i \phi_i \nabla \mu_i \tag{14}$$

with $\sigma_{s,\text{total}} = \sigma_1 + \sigma_2$ and Π the osmotic pressure of the suspension. It is not clear how to divide the stresses over the individual fractions. We assume that the stress is divided over the different fractions according to volume fraction, therefore, the following can be used:

$$\nabla \cdot \sigma_i = \phi_i \nabla \mu_i \tag{15}$$

With use of the formulation of the momentum exchange (7), the decomposition of the fluid tensor (13) and the division of the stress (15) the momentum balances (3) and (4) can be written as

$$0 = -(1 - \phi)\nabla p + (1 - \phi)\nabla \cdot \Sigma_f - \mathbf{f}_1^* - \mathbf{f}_2^* - \phi_1(\bar{\rho} - \rho_1)\mathbf{g} - \phi_2(\bar{\rho} - \rho_2)\mathbf{g} + (1 - \phi)\rho_f \mathbf{g} \tag{16}$$

$$0 = -\phi_1 \nabla p + \phi_1 \nabla \mu_1 + \phi_1 \nabla \cdot \Sigma_f + \mathbf{f}_1^* + \phi_1 \bar{\rho} \mathbf{g} \tag{17}$$

$$0 = -\phi_2 \nabla p + \phi_2 \nabla \mu_2 + \phi_2 \nabla \cdot \Sigma_f + \mathbf{f}_2^* + \phi_2 \bar{\rho} \mathbf{g} \tag{18}$$

From these equations expressions for the momentum exchanges can be obtained. An expression for $\mathbf{f}_1^*$ is obtained by substitution of (18) as expression for $\mathbf{f}_2^*$ in (16) and elimination of the pressure gradient from the resulting expression and (17) by weighted subtraction. An identical procedure is performed for the fluid–particle momentum exchange of the second fraction. This results in the following expressions for the drag forces, with $i \neq j$:

$$\mathbf{f}_i^* = -(1 - \phi_i)\phi_i \nabla \mu_i + \phi_i \phi_j \nabla \mu_j = \zeta_i \mathbf{w}_i \tag{19}$$

Note, that the slip velocities are only functions of the gradients in the chemical potentials. This is a direct consequence of the chosen formulation for the momentum exchange (eqn (7)).

Use of the resulting expression for the slip velocities in (12) results in, with $i \neq j$

$$\partial_t \phi_i + \nabla \cdot \phi_i \bar{\mathbf{u}} = \nabla \cdot (M_{ii} \nabla \mu_i - M_{ij} \nabla \mu_j) \tag{20}$$

The different mobilities M are given by, with $i \neq j$

$$M_{ii} = \frac{\phi_i^2(1 - \phi_i)^2}{\zeta_i} + \frac{\phi_i^2 \phi_j^2}{\zeta_j} \tag{21}$$

$$M_{ij} = \phi_i \phi_j \left(\frac{\phi_i(1 - \phi_i)}{\zeta_i} + \frac{\phi_j(1 - \phi_j)}{\zeta_j} \right) \tag{22}$$

Note that with $M_{ij} = M_{ji}$ for $i \neq j$ these equations satisfy the Onsager relation.

Eqn (20) describes the evolution of the solid phases due to convection with the suspension velocity and due to diffusion due to gradients in the chemical potentials. The first term on the rhs of (20) is equal to the diffusive term present in the mixture formulation for a monodisperse suspension.[50] Compared to this the second part in the $\nabla\mu_i$ term represents the increase in the diffusive flux of particle phase i due to the presence of particle phase j. And the $\nabla\mu_j$ term is an additional (depletion) correction due to gradients in ϕ_i or shear rate of ϕ_j. The chemical potentials of the different fractions are functions of both fractions. Thus, also the gradients in the chemical potential of one fraction are affected by the presence of the other fraction.

The final governing equations, the mixture model, for flow of a bidisperse suspension are given by the mixture mass balance (5), the mixture momentum balance (6) and the mass balances for the solid phases (20). These equations determine the mixture velocity, the pressure and the partial volume fractions of the two dispersed phases. A closed set of equations is found by supplementing the governing equations with closure relations for the particle stress, viscosity, friction factors and chemical potentials given in the next section. The given formulation can be extended to more than two fractions by addition of mass balances and adjustment of the closure relations.

3 Closure relations

In this section the used bidisperse closures for the particle stress, friction factors and viscosity are given and the derivation of the chemical potentials is presented. A detailed discussion on the particle stress, friction factors and viscosity relations can be found in ref. 47.

3.1 Viscosity

The viscosity η_{eff} for a bidisperse suspension is obtained from application of the mean field theory on the monodisperse formulation.[47] It was shown that with combined use of the expressions of McGreary[37] and Shauly[44] the random close packing of a bidisperse suspension is well predicted. Given the particle sizes in the experimental dataset to which the model is compared, with $\lambda = a_1/a_2 \approx 2$, we use the following:

$$\eta_{\text{eff}} = \eta_f (1 - \tilde{\phi})^{-2.5\phi_{\text{max,bi}}} \tag{23}$$

$$\phi_{\text{max,bi}} = \phi_{\text{max}} \left(1 + \beta |b|^{3/2} \frac{\phi_1}{\phi}^{3/2} \frac{\phi_2}{\phi} \right) \tag{24}$$

with η_f the viscosity of the fluid, $\tilde{\phi} = \phi/\phi_{\text{max,bi}}$ the normalised total particle volume fraction, with $\phi = \phi_1 + \phi_2$, $\phi_{\text{max,bi}}$ the random packing fraction of the bidisperse suspension, $\beta = 1.75$ and $b = (\lambda - 1)/(\lambda + 1)$. Eqn (24) is the correlation as proposed by Shauley et al.[44] with adjusted β.

3.2 Friction factor

The mean field theory is also used to determine the friction factors, which are given by

$$\zeta_i = \gamma_i n_i = \frac{2\eta_{\text{eff}}}{9a_i^2} \phi_i \tag{25}$$

with $\gamma_i = 6\pi\eta_{\text{eff}}a_i$ the drag coefficient of a single particle in the suspension and $n_i = \phi_i/V_{pi}$ the number of particles, with V_{pi} the particle volume. The friction factors appear in the mobilities, see eqn (21)–(22). Fig. 1 shows the magnitude of the different mobility components for $a_1 = 1.5$ μm, $a_1/a_2 = 2$ and $a_1/a_2 = 5$. The figures show that depending on the ratio between the particle sizes and the particle volume fractions the contributions to the diffusion behaviour of the gradients in the different chemical potentials is different. For larger size ratios the mobility of the smallest

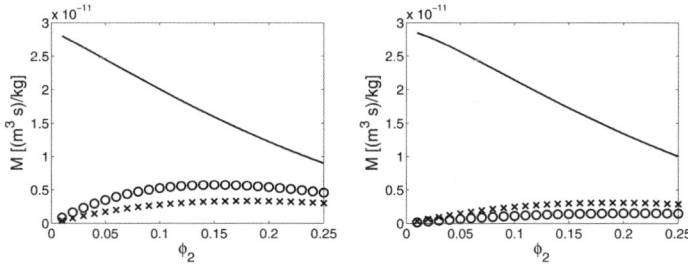

Fig. 1 Mobilities as function of ϕ_2 with $\phi_1 = 0.1$ for $a_1 = 1.5\mu m$, $a_1/a_2 = 2$ (left) and $a_1/a_2 = 5$ (right). Solid line: M_{11}; $\times$ symbols: $M_{12} = M_{21}$; $\bigcirc$ symbols: M_{22}.

fraction is smaller. The depleting contribution to migration of the largest fraction due to gradients in the chemical potential of the smallest fraction (and *vice versa*) is less influenced by the change in particle size ratio.

3.3 Osmotic pressure

The particle pressure is the result of the contact forces between the suspension phases. In sheared suspensions the lubrication forces are dominant, which scale linear with the shear rate. Below the monodisperse formulation with both Brownian motion and shear-induced migration relevant is given. After that the bidisperse extension is discussed.

The osmotic pressure Π scales with the product of an effective temperature T_{eff} and the compressibility Z

$$\Pi = (nkT)_{\text{eff}}Z \tag{26}$$

with $n = \phi/V_p$ the number of particles, with V_p the particle volume. The compressibility depends on ϕ, several formulations are given in Table 1. The effective temperature depends on the specific particle migration mechanisms. In case Brownian motion and shear-induced migration are relevant it can be written as

$$(nkT)_{\text{eff}} = n(k_B T + c\eta_{\text{eff}}\dot{\gamma}V_p) \tag{27}$$

with k_B the Boltzmann constant, T the temperature and c a constant. There are different formulations for the compressibility given in the literature. Under the assumption of the particles being hard spheres these should match with the Carnahan–Starling relation, with the adjustment that diverging behaviour close to the random close packing fraction is preferred. The closure relation by Morris & Boulay[40] performs well in predicting particle migration for suspensions with

Table 1 Different formulations of the osmotic pressure

Reference	$\phi Z(\phi)$
Carnahan Starling18	$\phi(1 + \phi + \phi^2 - \phi^3)/(1 - \phi)^3$
Enskog9	$\phi/(1 - \tilde{\phi}^{1/3})$
20	$3\phi/(1 - \tilde{\phi})$

Reference	$\Pi/(\eta_f\dot{\gamma})$
40	$0.75\tilde{\phi}^2/(1 - \tilde{\phi})^2$
39	$0.65\tilde{\phi}^{1/3}/(1 - \tilde{\phi})^2$
This work	$\Pi = 1/9\eta_{\text{eff}}\dot{\gamma}\phi Z(\phi)$

shear-induced migration as main migration mechanism.[20,38] Some researchers have derived ϕ-dependent expressions for the osmotic pressure without a decomposition in the compressibility and the effective temperature ($\sim\eta_{\text{eff}}\dot{\gamma}$). Comparison between different particle pressure formulations showed that $c = 1/9$.[47]

For the particle sizes of interest, 0.1–10 μm, both Brownian motion and shear-induced migration are relevant. The formulation given in eqn (27) is compared with another monodisperse model formulation based on several scaling arguments and results from direct particle simulations and experiments.[20] Fig. 2 shows the comparison for different Peclet numbers, with $\text{Pe} = \dfrac{H^2/D_\phi}{L/H}$, with $D_\phi = M\dfrac{\partial\mu}{\partial\phi}$, H the channel height and L the channel length. It shows that the formulations compare well with each other for $\phi \geq 0.3$, for smaller ϕ the formulations differ by a factor of 2.

Using the free volume/mean field theory as presented in[47] the bidiperse formulation of the particle pressure is obtained by using $\tilde{\phi} = \phi/\phi_{\text{max,bi}}$ with $\phi = \phi_1 + \phi_2$ and $\phi_{\text{max,bi}}$ the random close packing for a bidisperse suspension. We assume equi-partitioning of the kinetic energy over the different fractions, implying a single effective temperature depending on the local composition. We have proposed the following formulation for the bidisperse osmotic pressure:[47]

$$\Pi_{\text{bi}} = (nkT)_{\text{eff,bi}}Z_{\text{bi}} \tag{28}$$

with $Z_{\text{bi}} = Z_{\text{bi}}(\phi_1,\phi_2,a_1,a_2)$ a bidisperse formulation of the compressibility. Similar to the monodisperse case both Brownian and shear-induced migration can be included in the formulation of the effective temperature (27). Then $n = \phi/\bar{V}_p$ is the number density of the total particle phase and the viscosity equals the bidisperse formulation. The formulation of the particle volume of the entire suspension $\bar{V}_p$ requires attention, the following is used:

$$\bar{V}_p = \phi_1/\phi V_{p1} + \phi_2/\phi V_{p2} \tag{29}$$

The Carnahan–Starling equation of state has been extended for bidisperse compositions to the Boublik–Mansoori–Carnahan–Starling–Leland relation (BMCSL)[6,35]

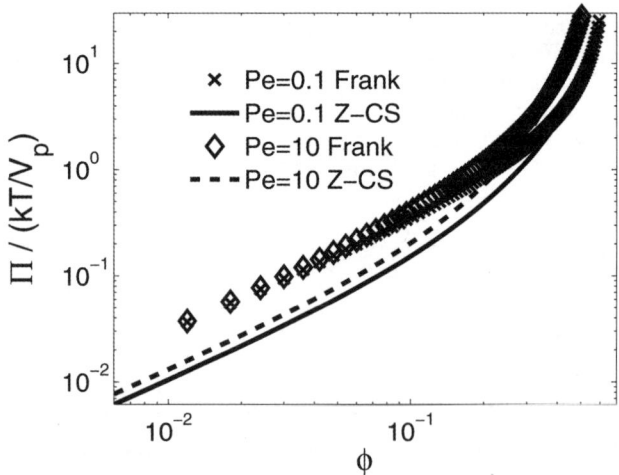

Fig. 2 Dimensionless osmotic pressure including both Brownian motion and shear-induced migration based on the particle stress formulation of Frank *et al.*[20] (Frank) or as product of an effective temperature and compressibility (Z-CS) for $a = 1$ μm for different Peclet numbers.

$$Z_{\text{BMCSL}} = \frac{1}{1-\phi} + \frac{3\phi}{(1-\phi)^2} \frac{\langle d \rangle \langle d^2 \rangle}{\langle d^3 \rangle} + \frac{\phi^2(3-\phi)}{(1-\phi)^3} \frac{\langle d^2 \rangle^3}{\langle d^3 \rangle^2} \tag{30}$$

with $\phi = \dfrac{\pi \langle d^3 \rangle n}{6}$ the total particle volume fraction, n the total number of particles and $\langle d^n \rangle = \sum_{i=1}^{N} \dfrac{n_i}{n_s} d_i^n$ the n-th moment of the particle diameters, with N the number of different particle fractions, n_i the number of particles of fraction i and d_i the particle diameter of fraction i. The BMCSL expression does not diverge at random close packing. Alternatively, following mean field arguments, one could use relations similar to the relations given in Table 1 with $\tilde{\phi}$ formulated for bidisperse suspensions. It is shown that these approaches agree well with each other for a large range of ϕ_1/ϕ_2 and a_1/a_2 ratios.[47]

3.4 Chemical potentials

Formulations for the chemical potentials can be derived from the osmotic pressure of a bidisperse suspension as described by Van Sint Annaland et al.[45] and references therein. The non-ideal parts of the pressure $P_s^{\text{ex}} = P_s - P_s^{\text{id}}$ and the chemical potentials $\mu_n^{\text{ex}} = \mu_i - \mu_i^{\text{id}}$ are related to the Helmholtz free energy A^{ex} as follows:

$$P_s^{\text{ex}} = n^2 \left(\frac{\partial A^{\text{ex}}}{\partial n} \right) \quad \mu_i^{\text{ex}} = \frac{\partial (nA^{\text{ex}})}{\partial n_i} \tag{31}$$

with n_i the number of particles of fraction i.

With use of the following dimensionless notation, with z^{ex} the compressibility and $\theta_s = kT$,

$$z^{\text{ex}} = \frac{P_s^{\text{ex}}}{\theta_s n}, \quad a^{\text{ex}} = \frac{A^{\text{ex}}}{\theta_s} \tag{32}$$

the non-ideal parts of the pressure and the chemical potentials can be rewritten as

$$a^{\text{ex}}(\phi, \theta_s) = \int_0^{\phi} \frac{z^{\text{ex}}(\phi', \theta_s)}{\phi'} d\phi', \quad \frac{\mu_i^{\text{ex}}}{\theta_s} = a^{\text{ex}} + n \frac{\partial a^{\text{ex}}}{\partial n_i} \tag{33}$$

Thus, given an equation of state, describing the pressure or the compressibility, eqn (33) can be used to obtain the chemical potentials of the different fractions. Alternatively, the procedure of Santos et al. can be followed, as described in ref. 45, by which the compressibility of a multi-component mixture is derived from the excess compressibility of a single component system z^{I} by use of

$$z^{\text{ex}} = f^{\text{I}} z^{\text{I}} + f^{\text{II}} z^{\text{II}} \tag{34}$$

with $z^{\text{II}} = \dfrac{\phi}{1-\phi}$, $f^{\text{I}} = 1/2(m_1 + m_2)$, $f^{\text{II}} = 1 + m_1 - 2m_2$, $m_1 = \dfrac{\langle d \rangle \langle d^2 \rangle}{\langle d^3 \rangle}$ and $m_2 = \dfrac{\langle d^2 \rangle^3}{\langle d^3 \rangle^2}$. Following the procedure given above the chemical potential of fraction i is given by

$$\frac{\mu_i^{\text{ex}}}{\theta_s} = y_i^3 \left(f^{\text{I}} z^{\text{I}} + f^{\text{II}} z^{\text{II}} \right) + g_i^{\text{I}} a^{\text{I}} + g_i^{\text{II}} a^{\text{II}} \tag{35}$$

with $y_i^n = \dfrac{d_i^n}{\langle d^n \rangle}$, $g_i^{\text{I}} = 1/2(m_1(y_i^1 + y_i^2 - y_i^3) + m_2(3y_i^2 - 2y_i^3))$ and $g_i^{\text{II}} = 1 + m_1(y_i^1 + y_i^2 - y_i^3) - 2m_2(3y_i^2 - 2y_i^3)$.

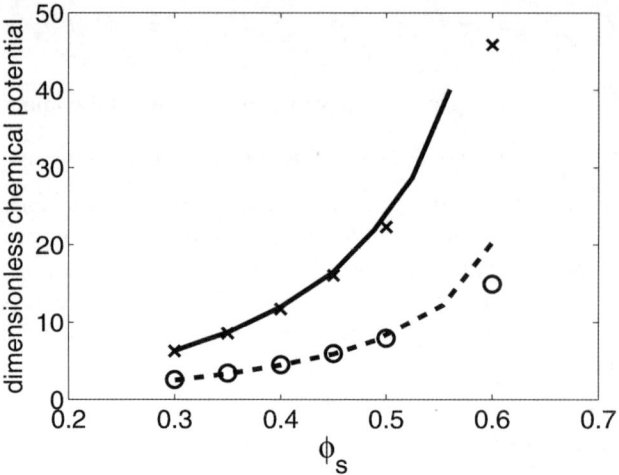

Fig. 3 Dimensionless chemical potentials in a bidisperse suspension with $n_1/n = n_2/n = 0.5$ and $a_1/a_2 = 1.67$. Lines: data from Van Sint Annaland (solid: largest fraction, dashed: smallest fraction). Symbols: resulting from use of Z_{BMCSL} ($\times$: largest fraction, $\bigcirc$: smallest fraction).

Fig. 3 shows the dimensionless chemical potential for the largest fraction of a bidisperse suspension derived from $z_{BMCSL} = Z_{BMCSL} - 1$ with (33). The data of Van Sint Annaland was obtained using the procedure of Santos on the Carnahan–Starling equation. The difference is due to the non-divergent formulation of Z_{BMCSL}.

The above described procedure results in analytical expressions for the chemical potentials in $\phi_1, \phi_2, a_1, a_2, \phi_{\max,bi}$ (the latter for diverging formulations of the compressibility). Applying the given procedure on the Boublik–Manssori–Carnahan–Straling–Leland equation of state the chemical potential of particle volume fraction i is derived to be equal to

$$\mu_i = \left(\frac{3d_i^2\langle d^2\rangle^2}{\langle d^3\rangle^2} - \frac{2d_i^3\langle d^2\rangle^3}{\langle d^3\rangle^3} - 1\right)\ln(1-\phi)$$

$$+\left(\frac{\pi d_i^3 n}{6} + \frac{3d_i}{\langle d^3\rangle}\phi\left(\langle d^2\rangle + d_i\langle d\rangle - \frac{d_i^2\langle d\rangle\langle d^2\rangle}{\langle d^3\rangle}\right)\right)\frac{1}{1-\phi}$$

$$+\left(\frac{3\langle d\rangle\langle d^2\rangle}{\langle d^3\rangle}\frac{\pi d_i^3 n}{6} + \phi\left(\frac{3d_i^2\langle d^2\rangle^2}{\langle d^3\rangle^2} - \frac{2d_i^3\langle d^2\rangle^3}{\langle d^3\rangle^3}\right)\right)\frac{1}{(1-\phi)^2}$$

$$+\phi(3-\phi)\frac{\langle d^2\rangle^3}{\langle d^3\rangle^2}\frac{\pi d_i^3 n}{6}\frac{1}{(1-\phi)^3} \tag{36}$$

with $d_i = 2a_i$.

Fig. 4 shows the magnitude of the dimensionless chemical potentials for a bidisperse suspension with $a_1 = 1.5$ μm and $a_2 = 0.7$ μm for different combinations of particle volume fractions. The differences in magnitude are due to changes in ϕ_1 or ϕ_2 resulting from differences in particle size.

4 Simulations of bidisperse suspension channel flow

4.1 Simulation method

The governing system of eqn (20), (5), and (6) for p, $\bar{u}$, ϕ_1 and ϕ_2 was solved. The equations are in closed form combined with the closures (23), (24), (25), (28), (29),

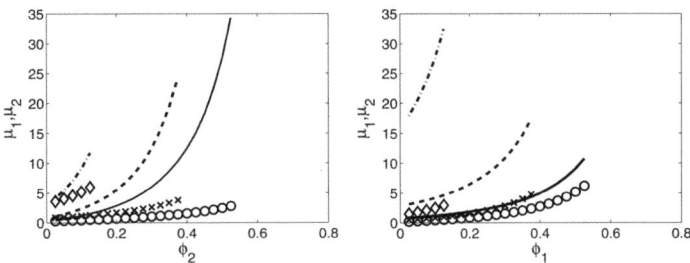

Fig. 4 Dimensionless chemical potential in a bidisperse suspension with $a_1 = 1.5$ μm and $a_2 = 0.7$ μm based on Z_{BMCSL}. Lines: μ_1 for three values of ϕ_1 (left) or ϕ_2 (right); symbols: μ_2 for three values of ϕ_1 (left) or ϕ_2 (right). Values of $\phi_{1,2}$: solid line and $\bigcirc$: 0.1; dashed line and $\times$: 0.25; dotted line and diamonds: 0.5. The curves are limited in length due to $\phi_1 + \phi_2 < \phi_{\text{max,bi}}$.

(30) and (36). The flow and mass balances were coupled *via* the dependency of the particle pressure on the shear rate and *via* the ϕ-dependent viscosity. Given the much faster dynamics of the flow the system was solved as weakly coupled.

The flow was solved with the Lattice Boltzmann method.[49] With this method the local shear rate was directly obtained during the calculations. The local shear rate is needed to calculate the particle pressure and the chemical potentials.[26] The coupling between the flow and mass balances is through the particle volume fractions' dependency of the viscosity and the particle pressure, and through the convection and the shear rate dependency of the chemical potentials. The mass balances for the concentrations were solved with the Finite Volume method given certain initial concentration profiles at the beginning of the channel. The convection was discretised with an upwinding scheme. The time integration was explicit, with simple Euler formulation. The particle fraction mass balances can also be solved with other methods, however, the simplicity of the Finite Volume implementation and the direct use of the formulation with gradients in the chemical potentials is preferred. The particle volume fractions are homogeneous at the inlet of the channel. At the end of the channel free outflow is assumed.

The model is applied to flow between two parallel plates. The applicability is not limited to this configuration, also other geometries can be studied, for instance a channel with a contraction.[38] The boundaries influence the resulting migration. The results show that there is additional physics that is not included in the given formulation. Furthermore, the geometry, specifically the ratio between the channel height and the particle size, determines the magnitude of the shear-induced migration.

4.2 Results

Fig. 5 shows a comparison between simulation results and data from Semwogerere[43] for a bidisperse suspension with $\phi_1 = 0.1$, $a_1 = 1.5$μm, $\phi_2 = 0.1$ and $a_2 = 0.7$ μm at different locations along the channel length during the development of the profile. The suspension flows with an average flow rate of 0.30 μl min^{-1} and $\text{Pe} = \dfrac{6\pi\eta_f\dot{\gamma}a_1^3}{kT} = 480$. The location x/H along the channel length is expressed in the channel height. The steady state profile corresponds to location $x/H = 1600$.

The results in Fig. 5 show that the model captures the general trend of the migration of the different fractions well. We note that the results are obtained without any parameter estimation. It appears that near the wall additional physics play a role, which are not yet captured in the presented model. We assume that these are depletion effects. One possible extension of the model is to incorporate this wall effect by adjustment of the maximum packing close to the wall. The data shows a smaller

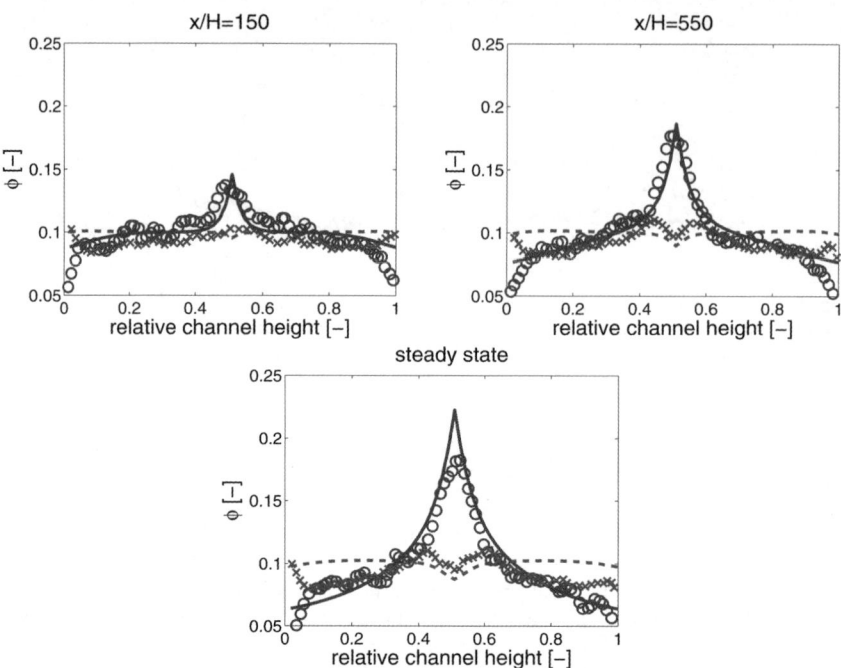

Fig. 5 Comparison of data (symbols) and simulation results (lines) for a bidisperse suspension with $\phi_1 = \phi_2 = 0.1$, $a_1 = 1.5$ μm, $a_2 = 0.7$ μm and 0.30 μl min^{-1} at three different locations along the channel length. The steady state location corresponds to $x/H = 1600$. Symbols: ×: small particles; ○: large particles. Lines: dashed: small particles; solid: large particles. Data from Semwogerere.[43]

peak in the center of the channel compared to the model prediction. This is probably due to the finite sizes of the particles and can be improved by incorporating finite particle size effects, such as by applying particle size based averaging, see for instance.[20,38]

Confinement itself also influences properties of the suspension, it affects the viscosity, mobility and the random close packing (see[47] and references therein). Given the small channel height to particle size ratio in the experiment (16.7 for the large particles and 71.4 for the small particles) it can be expected that this is relevant as well.

5 Conclusion and discussion

A mixture model for the particle migration in a bidisperse flowing suspension was derived from a multi-fluid approach and compared to experimental data. The model uses the assumptions of equipartitioning of stresses and of kinetic energy over the particle phases, resulting in a single effective temperature. The mixture model is used to predict particle segregation based on particle size in channel flow where the particles are subjected to Brownian motion and shear-induced migration. The comparison of the model results with experimental data shows that particle migration is predicted well by the given formulation. Since the comparison is made without parameter fitting, this can be regarded as a true prediction. Therefore, we have confidence that effective temperature is a powerful concept, and extends probably to the inertial regime.[17]

Use of the mean field approach to extend closures for a monodisperse suspension results in closure relations for the viscosity, friction factors and particle pressure for a bidisperse suspension.[47] The particle pressure is formulated as product of an effective temperature and compressibility.[50] The effective temperature contains the relevant particle interaction mechanisms. In bidisperse granular media two temperatures are assigned, different for each solid fraction.[21,24] Possibly this issue can be resolved *via* fully resolved simulations. The chemical potentials of the different fractions are derived based on relations between the particle pressure, free energy, and the chemical potentials.

The model can be used to design fractionation processes. Simulations can be performed for different suspension and geometry configurations. For instance, the effect of variations in particles sizes, volume fractions, and ratio between particle size and channel height can be analysed with the given model. The effects can, for instance, be expressed in development of the profiles or in the resulting concentration differences over the channel height between the fraction (selectivity). Based on these results the best location for an actual fractionation device, possibly an insert channel or non-selective membrane,[15] can be determined. However, the range of applicability of the model must be determined first. This is the topic of ongoing research.

Acknowledgements

The research is financed by ISPT, The Netherlands, FO-00-01, and by the European project CAFE, FP7-KBBE-2007-2-3-01.

References

1 M. Abbas, E. Climent, O. Simonin and M. R. Maxey, Dynamics of bidisperse suspensions under Stokes flows: linear shear flow and sedimentation, *Phys. Fluids*, 2006, **18**, 121504.

2 C. Barentin, E. Azanza and B. Pouligny, Flow and segregation in sheared granular slurries, *Europhys. Lett.*, 2004, **66**(1), 139–145.

3 G. K. Batchelor and R. W. J. van Rensburg, Structure formation in bidisperse sedimentation, *J. Fluid Mech.*, 1986, **166**, 379.

4 G. Belfort, R. H. Davis and A. L. Zydney, The behaviour of suspensions and macromolecular solutions in crossflow microfiltration, *J. Membr. Sci.*, 1994, **96**, 1–58.

5 L. Berthier and J. L. Barrat, Nonequilibrium dynamics and fluctuation-dissipation relation in sheared fluid, *J. Chem. Phys.*, 2002, **116**, 6228.

6 T. Boublík, Hard-sphere equation of state, *J. Chem. Phys.*, 1970, **53**, 471–472.

7 G. Brans., *Design of membrane systems for fractionation of particle suspensions*, Thesis Wageningen University, The Netherlands, 2006.

8 R. Bürger, K. H. Karlsen, E. M. Tory and W. L. Wendland, Model equations and instability regions for the sedimentation of polydisperse suspensions of spheres, *Z. Angew. Math. Mech.*, 2002, **82**, 699–722.

9 Y. A. Buyevich, Particle distribution in suspension shear flow, *Chem. Eng. Sci.*, 1996, **51**(4), 635–647.

10 C. Contado and M. Hoyos, SPLITT cell analytical separation of silica particles. Non-specific crossover effects: does shear-induced diffusion play a role?, *Chromatographia*, 2007, **65**, 453–462.

11 R. H. Davis and H. Gecol, Hindered settling function with no empirical parameters for polydisperse suspensions, *AIChE J.*, 1994, **40**, 570–575.

12 A. Deboeuf, G. Gauthier, J. Martin and D. Salin, Segregation and periodic mixing in a fluidized bidisperse suspension, *New J. Phys.*, 2011, **13**, 075005.

13 W. M. Deen, *Analysis of Transport Phenomena*, Oxford University Press, New York, 1998.

14 J. J. Derksen and S. Sundaresan, Direct numerical simulations of dense suspensions: wave instabilities in liquid-fluidized beds, *Journal of Fluid Mechanics*, 2007, **587**(1), 303–336.

15 A. M. C. Van Dinther, C. G. P. H. Schroën and R. M. Boom, High-flux membrane separation using fluid skimming dominated convective fluid flow, *J. Membr. Sci.*, 2011, **371**, 20–27.

16 E. C. Eckstein, D. G. Bailey and A. H. Shapiro, Self-diffusion of particles in shear flow of a suspension, *J. Fluid Mech.*, 1977, **79**(01), 191–208.

17 A. Fall, A. Lemaître, F. Bertrand, D. Boon and G. Overlaz, Shear thickening and migration in granular suspensions, *Phys. Rev. Lett.*, 2010, **105**(26), 268303.

18 N. F. Carnahan and K. E. Starling, Equation of state for noninteracting rigid spheres, *J. Chem. Phys.*, 1969, **51**, 635.

19 S. M. Fielding and P. D. Olmsted, Kinetics of the shear banding instability in startup flows, *Phys. Rev. E: Stat. Phys., Plasmas, Fluids, Relat. Interdiscip. Top.*, 2003, **68**(3), 036313.

20 M. Frank, D. Anderson, E. R. Weeks and J. F. Morris, Particle migration in pressure-driven flow of a Brownian suspension, *J. Fluid Mech.*, 2003, **493**, 363–378.

21 D. Gidaspow., *Multiphase Flow and Fluidization*, Academic Press, San Diego, 1994.

22 Recent progress towards hydrodynamic modelling of dense gas-particle flows, *Recent research developments in chemical engineering*, Transworld Research Network, India, pp. 273–292, 2000.

23 K. Höfler, S. Schwarzer, The structure of bidisperse suspensions at low Reynolds numbers. In: Sändig, A.-M., W. Schiehlen, W. L. Wendland (ed.), *Multifluid problems: State of the art*, Springer Verlag, Berlin, 42–49, 2000.

24 L. Huilin, D. Gidaspow and E. Manger, Kinetic theory of fluidized binary granular mixtures, *Phys. Rev. E: Stat. Phys., Plasmas, Fluids, Relat. Interdiscip. Top.*, 2001, **64**, 61301–61309.

25 D. M. Husband, L. A. Mondy, E. Granani and A. L. Graham, Direct measurements of shear-induced particle migration in suspensions of bimodal spheres, *Rheol. Acta*, 1994, **33**, 185–192.

26 J. Kromkamp, A. Bastiaanse, J. Swarts, G. Brans, R. G. M. van der Sman and R. M. Boom, A suspension flow model for hydrodynamics and concentration polarisation in crossflow microfiltration, *Journal of Membrane Science*, 2005, **253**(1–2).

27 J. Kromkamp, A. van der Padt, C. G. P. H. Schroen and R. M. Boom, Shear induced fractionation of particles, Patent EP 1 673 975 A1, European Patent Office, 2006.

28 P. M. Kalkarni, *Suspension mechanics at finite inertia*, thesis City University of New York, 2009.

29 T. Kulrattanarak, R. G. M. van der Sman and C. G. P. H. Schroën, Analysis of mixed motion on deterministic ratchets *via* experiment and particle simulation, *Microfluid. Nanofluid.*, 2011, **10**, 843–853.

30 D. Leighton and A. Acrivos, The shear-induced migration of particles in concentrated suspensions, *J. Fluid Mech.*, 1987, **181**, 415–439.

31 D. Leighton and A. Acrivos, Measurement of shear-induced self-diffusion in concentrated suspensions of spheres, *J. Fluid Mech.*, 1987, **177**, 109–31.

32 E. Loth, Numerical approaches for motion of dispersed particle, droplets, and bubbles, *Prog. Energy Combust. Sci.*, 2000, **26**, 161–223.

33 C. K. K. Lun, S. B. Savage, D. J. Jeffrey and N. Chepurniy, Kinetic theories for granular flow: inelastic particles in Couette flow and slightly inelastic particles in a general flowfield, *J. Fluid Mech.*, 1984, **140**, 223–256.

34 M. K. Lyon and L. G. Leal, An experimental study of the motion of concentrated suspensions in two-dimensional channel flow. Part 2. Bidisperse systems, *J. Fluid Mech.*, 1998, **363**, 57–77.

35 G. A. Mansoori, N. F. Carnahan, K. E. Starling and T. W. Leland, Equilibrium thermodynamic properties of the mixture of hard spheres, *J. Chem. Phys.*, 1971, **54**, 1523–1525.

36 M. Marchioro, M. Tanksley and A. Prosperetti, Mixture pressure and stress in disperse two-phase flow, *Int. J. Multiphase Flow*, 1999, **25**(67), 1395–1429.

37 R. K. McGreary, Mechanical packing of spherical particles, *J. Am. Ceram. Soc.*, 1961, **44**(10), 513–522.

38 R. M. Miller and J. F. Morris, Normal stress-driven migration and axial development in pressure-driven flow of concentrated suspensions, *J. Non-Newtonian Fluid Mech.*, 2006, **135**, 149–165.

39 P. Mills and P. Snabre, Apparent viscosity and particle pressure of a concentrated suspension of non-Brownian had spheres near the jamming transition, *Eur. Phys. J. E*, 2009, **30**, 309–316.

40 J. F. Morris and F. Boulay, Curvilinear flows of noncolloidal suspensions: the role of normal stresses, *J. Rheol.*, 1999, **43**(5), 1213–1237.

41 J. T. Norman, B. O. Oguntade and R. T. Bonnecaze, Particle-phase distribution of pressure-driven flow of bidisperse suspensions, *J. Fluid Mech.*, 2008, **594**, 1–28.

42 I. K. Ono, C. S. O'Hern, D. J. Durian, S. A. Langer, A. J. Liu and S. R. Nagel, Effective temperatures of a driven system near jamming, *Phys. Rev. Lett.*, 2002, **89**(9), 95703.

43 D. Semwogerere and E. R. Weeks, Shear-induced migration in binary colloidal suspensions, *Phys. Fluids*, 2008, **20**, 043306.

44 A. Shauly, A. Wachs and A. Nir, Shear-induced particle migration in a polydisperse concentrated suspension, *J. Rheol.*, 1998, **42**(6), 1329–1348.

45 M. Van Sint Annaland, G. A. Bokkers, M. J. V. Goldschmidt, O. O. Olaofe, M. A. van der Hoef and J. A. M. Kuipers, Development of amulti-fluid model for poly-disperse dense gas-solid beds, Part I: Model derivation and numerical implementation, *Chem. Eng. Sci.*, 2009, **64**, 4222–4236.

46 R. G. M. Van der Sman, Simulations of confined suspension flow at multiple length scales, *Soft Matter*, 2009, **5**, 4376–4387.

47 R. G. M. Van der Sman, H. M. Vollebregt, Effective temperature for sheared suspensions: a route towards closures for migration in bidisperse suspension, submitted, 2011.

48 T. Speck and U. Seifert, Extended fluctuation-dissipation theorem for soft matter in stationary flow, *Phys. Rev. E: Stat., Nonlinear, Soft Matter Phys.*, 2009, **79**(4), 40102.

49 S. Succi., *The lattice Boltzmann Equation for Fluid Dynamics and Beyond*, Oxford University Press, New York, 2001.

50 H. M. Vollebregt, R. G. M. van der Sman and R. M. Boom, Suspension flow modelling in particle migration and microfiltration, *Soft Matter*, 2010, **6**(24), 6052–6064.

51 B. G. M. Van Wachem, J. C. Schouten and C. M. van den Bleek, CFD modeling of gas-fluidized beds with a bimodal particle mixture, *AIChE J.*, 2001, **47**(6), 1292–1302.

52 R. H. Weiland, Y. P. Fessas and B. V. Ramarao, On stabilities arising during sedimentation of two-component mixtures of solids, *J. Fluid Mech.*, 1984, **142**, 383.

53 K. Yeo and M. R. Maxey, Numerical simulations of concentrated suspensions of monodisperse particles in a Poiseuille flow, *J. Fluid Mech.*, 2011, **682**, 491–518.

General discussion

Dr Ettelaie opened the discussion of the paper by Professor Norton: In the case of duplex emulsions, the monoglycerides from the primary emulsions will eventually equilibrate between the inner and outer interfaces. Does this have any effects on the stability of secondary emulsions?

Professor Norton answered: In our study this doesn't happen as we have a crystalline shell.

Dr van der Sman commented: How does the controlled release in the pickering stabilized double-emulsion happen? Do the fat crystals melt, and migrate towards to the outer interface? How do the particles on the outer phase become detached from the outer interface?

Professor Norton answered: Controlled release is *via* melting of the crystals at the inner interface. Once they have melted then there is no Pickering stabilisation of the inner interface. We think that as the inner droplets (without particles on the interface) approach the outer interface the particles at the outer interface can bridge between the two interfaces and cause coalescence.

Dr Bot commented: Pickering-in-Pickering emulsions are a very elegant solution to create storage-stable $w_1/o/w_2$ duplex emulsions because they circumvent the problem of diffusion of the emulsifier between the interfaces. A remaining source of instability could be the escape of the w_1 water droplets into the continuous w_2 phase. This process would lead to a reduction of the amount of dispersed w_1 phase and to mixing of the particles that stabilise the o/w_2 interface with particles that stabilise the w_1/o interface. Does such a process occur in your systems, or did you take any specific precautions to prevent this process from happening?

Professor Norton answered: The process you mention is potentially possible. However, it would require the removal of the particles from the inner water/oil interface. With our system we do not see this.

Prof. Nicolai opened discussion of the paper by Dr De Vries: One would not expect dense aggregates for charged globular proteins. Perhaps, if the electrostatic repulsion was decreased by lowering the pH, and thus the charge density of the proteins, one could favor the formation of denser aggregates.

Dr De Vries responded: This is a very good suggestion, this certainly needs to be examined. One potentially complicating factor is that pH also strongly affects the enzymatic activity and it may be difficult to distinguish between pH effects on substrate–substrate and substrate–enzyme interactions, *versus* the direct pH effect on enzymatic activity (*e.g.* due to the pH dependent dissociation of amino acids in the active site).

Professor Kulozik asked: Why was destabilized (apo)-α-lactalbumin used for the cross-linking experiments? (apo)-α-Lactalbumin is very unstable and quickly loses its native globular state to become an aggregate. Was the state of the molecule controlled? If it was the intention to work with a more accessible molecule, then Na-caseinate, which is commercially even more important than α-La, could have been used.

Dr De Vries responded: Since so many important food proteins are globular proteins, we have been interested in the enzymatic cross-linking of globular food proteins as a complementary route to create protein clusters. Hence, the issue of enzymatic accessibility becomes the key issue. We have chosen to use partially desta-bilized (apo)-α-lactalbumin as a model case, since the calcium-free destabilized state of α-lactalbumin is relatively well defined and well characterized, as compared to other destabilization methods such as the addition of denaturants, or heat treat-ments.

Professor Marangoni continued the discussion of the paper by Professor Norton: The size of the emulsions decreases exponentially to 100 nm as you increase Tween 80 concentration to 10 mN. Monolayer coverage for the 100 nm droplets considering a polar headgroup surface area of 1 nm^2 for Tween 80 and a MW of 1310 g mol^{-1} is estimated at a 10 mM surfactant concentration. This agrees quantitatively with your result that a minimum in emulsion droplet size is achieved at 10 mM Tween 80. Monolayer coverage for the 50 nm emulsions is thus estimated at 20 mM Tween 80. Why did you not see a reduction in droplet size to 50 mM even if you had enough surfactant to do so? Are microfluidizers incapable of making smaller droplets? If so, why?

Professor Norton replied: We were expecting a reduction as the Tween concentra-tion was increased. However, the 100 nm limit for the materials we used seems to be a consequence of the microfluidiser design.

Professor Marangoni remarked: The stability of the emulsions was related to the solubility of triacylglycerols and decane in water. You mentioned that decane was more soluble than triacylglycerols in water. What is your model of a decane molecule "solubilized" in water? Even though decane is a smaller molecule than a TAG, it is more apolar than a TAG. Could you elaborate on how mass transport of decane *vs.* TAGs affects emulsion stability?

If a decane molecule folds into a "ball" trying to exclude water from its core, then one could possibly have a smaller particle, stabilized by a clathrate structure that fits into the water structure more favorably than a TAG. One can envision this happening at very low concentrations, but at higher concentrations, one would expect phase separation. TAGs, by mere virtue of their size, cannot adopt such tightly packed conformations and would thus not be transported as efficiently in water. What would be the driving force for the decane to leave the emulsion droplet and migrate to water? The mechanism stated above would not be favored by ther-modynamics!

Dr Ettelaie asked: How would the radius of gyration of hydrocarbon chains alter, if at all, when they are part of the triglyceride as opposed to being a linear chain on their own? Can they present a much larger effective size?

Professor Dr Wilde responded to the discussion regarding the solubility of decane *vs.* triglyceride in water during ripening in emulsions: Probably the most important factor is the solubility in surfactant micelles rather than water, which facilitates the ripening process. It was observed that the ripening of decane emul-sions occurs in the presence of a micelle forming surfactant (Tween80), but did not occur in the presence of a phospholipid, which does not form micelles. Decane is more likely to reside in surfactant micelles (mw = 142g mol^{-1}) than a triglyceride molecule (mw ca 800g mol^{-1}).

Professor Dr Stoyanov commented: The flavor industry uses Gam Arabic (GA) or beta-pectins to produce submicron flavor emulsions and stabilize them against ripening. Why have you used Tween 80 in your nano-emulsion studies and not GA?

Pure Tweens are known to be poor stabilizers for systems prone to ripening, such as oils with high solubility in the water phase and/or foams.

Arguments such as "these are big molecules and they could not diffuse fast to the interface to be able to produce stable emulsions with very small droplets" are not very relevant, since the process of emulsification is highly turbulent and the adsorption process is driven by convection. Due to this, big does not always mean slow and secondly this can be overcome by increasing the bulk concentration (in typical flavor emulsions 10+wt% of GA is used). Of course, there can be an adsorption barrier that slows down the processes even if the convective transport to the sub-surface is fast, but the fact that emulsions using GA are routinely produced indicates that this is likely not the case. Another argument could be that GA is less understood and more complex then Tween, but I do not think that for the point of illustrating the ability to produce stable nano emulsions this is relevant. Most of the food ingredients and systems are inherently complex by nature and our role, as the soft matter community, is to understand the complexity.

Professor Norton replied: The paper reports a scientific study in which we are developing the understanding of colloids for use in food construction. A product development route would not add any value to the study. The range of papers we have published, including this one, give a good mechanistic understanding. If the molecules at the interface aggregate or cross link they can resist the ripening process. Whether gum arabic does this is an outstanding question. However, it would be better to know what the molecule is that has the inter-facial activity before speculating about it's mechanism.

Professor Dr Stoyanov opened the discussion of introductory lecture by Dr Ubbink: In your introductory talk you made a very clear demarcation between the emerging food soft matter physical approach and the more traditional food science and food engineering approaches. You gave a few examples where this new approach has been beneficial and can generate new results and understanding faster and more effectively. I would like to ask if you have some counter examples where using this approach has failed?

Dr Ubbink answered: Some reasons why soft-matter approaches may fail in industrial food development are: (1) Food systems may turn out to be too complex to be described by the relatively simple physics, which can be straightforwardly applied in food product and process development. For instance, problems may be ill defined as the controlling physical parameter turns out to be very sensitively dependent on small deviations of the actual food system from a simple, generic model. For example, some years ago, I encountered a food system in which a key structural parameter was, in principle, well defined, but only for a monodisperse system. For such monodisperse systems, control of this parameter would allow the precise tuning of the structure of the system and thereby its functional properties. However, in this case, it turned out that even a fairly narrow distribution of this parameter around its mean value obscured the relation to the physical model. Due to the polydispersity, the direct correspondence between the physical parameter and the structure of the system was severed. The way forward for this system was to renounce the use of a physical model, and instead rely on extensive imaging techniques combined with a pragmatic, sample-specific interpretation.

(2) It occasionally happens that the proper physics is not used to address the problem at hand. For instance, the concept of the glass transition temperature (T_g) and its dependence on the water content of a food system has found widespread application in food development, as it allows the rationalization of observations on food stability and behaviour during processing. This works very well for all situations in which the product matrix is brought (transiently) into the rubbery state (during processing, storage above T_g, uptake of water, etc.). It does not directly

bear on situations in which the product properties depend on the physics of the glassy state. However, I have regularly encountered cases in which correlations between the glassy state properties and T_g were searched for, as expected, in vain. In order to predict *e.g.* glassy-state barrier properties and properties related to bio-stabilization, one should make use of different physical concepts.

(3) Solutions that have been developed using physical approaches may turn out to be too expensive to implement, or it may turn out that for solutions that work perfectly well on small scale, no large-scale technologies are yet available. In addition, a specific risk for soft matter approaches in the food field is that solutions are found that make use of ingredients that are not food grade. For projects that heavily rely on the use of specific ingredients, it is therefore imperative to consider the specific chemistry of the ingredients from the start on.

(4) A major risk when companies adopt soft matter approaches in their product development is that they do not show enough perseverance to really benefit. It usually takes time and effort to implement a soft matter approach, as some experimental techniques and methods need to be setup and a basic body of data needs to be collected. Often, however, it turns out that the required efforts are not very significant, as experimental plans have already foreseen direct product development projects and can be extended to include the needs for a soft matter approach. This requires some good planning, though.

(5) The solutions for process and product development based on soft matter science may be slow to catch on, as, sometimes, they can be considered to be unconventional and not fitting to a specific business. I have experienced numerous situations where fully worked out concepts for new products or processes were initially shelved, as they would not smoothly fit into the existing capabilities of the organization. However, it turns out that important elements of these "sleeping beauties" will find application within a couple of years, provided that the key drivers keep pushing the novel technologies, and keep an open eye for opportunities to implement even small parts of the novel developments.

Dr van der Sman continued the discussion of the paper by Professor Norton: You have mentioned that the mouth perception (lubrication) of o/w emulsions relate more to physical properties than to the chemical identity of the colloids. Which physical properties are thus relevant to this perception? How does this hold for the fluid gel case? Is the expression of the fluid under compression from the fluid gel important for the lubrication?

Professor Norton answered: I think that the physics of what happens in the mouth is at an early stage of research. It seems that tribology, rheological aspects and fracture all have a part to play for all soft solid materials that are consumed.

Dr Rayner commented: In the section on Pickering-in-Pickering stabilized emulsions there is a comparison to other "non-Pickering" secondary emulsifiers. Fig. 6 in the paper shows the ageing of double emulsions during their storage of 3 months with 2% OSA starch as the secondary emulsifier. As the storage time increases the size distribution (d43) shifts to larger emulsion drop sizes with a shoulder or secondary peak forming at >100 microns. Do you believe this to be coalescence or aggregation? As you also write "...the gradual formation of large aggregates of primary emulsions in the continuous phase. These are no longer part of the double structure...", Do you mean that they no longer have an internal W1 phase? In Fig. 4 and 5 of the paper, the evolution of the surface mean diameter is plotted as a function of storage time for 1% silica particles and 2% OSA modified starch as secondary emulsifiers. The increase and then decrease in the d32 was explained by the change in proportion of small particles *versus* larger particles in the volume distribution shown in Fig. 6. Was this phenomena observed in the double emulsions stabilized by 1% silica particles as the secondary emulsifier? How do their particle size distributions

This journal is © The Royal Society of Chemistry 2012

look (uniformity, any free silica particles *etc.*) over 3 months? I would have liked to see a similar plot to Fig. 6 for the silica particles as well, for comparison, as they have constant d32 over 60 days and remarkable stability in general.

Professor De Kruif remarked: We have previously investigated the perceived creaminess of skimmed milk with various fat contents.[1] We found that up to 1% fat the milk was perceived as not creamy. Interestingly, precisely at that fat content the friction coefficient decreased exponentially. Also, we observed the buildup of a fat layer on the surface. This is very similar to what is observed in steel cutting.

In your research, the friction coefficient was measured for fluid polysaccharide gels. The suggestion is made that these gels can replace the perception of fat droplets. Therefore, the question is: how did the perceived mouthfeel depend on the fluid gel concentration and was it correlated with the friction coefficient?

1 A. Chojnicka-Paszun, H. H. J. de Jongh and C. G. de Kruif, Sensory perception and lubrication properties of milk: Influence of fat content, *International Dairy Journal*, 2012, **26**, 15–22.

Professor Norton responded: You will find most of the answers in the earlier works:

M. E. Malone, I. A. M. Appelqvist and I. T. Norton, Oral behaviour of food hydrocolloids and emulsions. Part 1. Lubrication and deposition considerations., *Food Hydrocolloids*, 2003, **17**, 763–773.
M. E. Malone, I. A. M. Appelqvist and I. T. Norton, Oral behaviour of food hydrocolloids and emulsions. Part 2. Taste and aroma release., *Food Hydrocolloids*, 17, 775-784.
The fluid gel lowers the friction and is concentration dependent as you suggest. For example please see:

A. Gabrielle, F. Spyropoulos and I. T. Norton, A conceptual model for fluid gel lubrication, *Soft Matter*, 2010, **6**, 4205–4213.

Dr Menut continued the discussion of the paper by Dr De Vries: In this paper, you do not comment about the possibility that the enzymes do not only generate inter-molecular cross-links (between proteins), but could also generate intra-molecular cross-links (inside a single protein). In this last case, the objects molecular weight shouldn't change, but it might affect their capacity to create further inter-molecular cross-links. Is it possible to associate the differences that you observed between enzymatic systems to differences in their ability to produce inter- *versus* intra-molecular cross-links?

Also, is it possible that the inter- to intra-molecular cross-linking ratio ("efficiency") depends on the level of aggregation, and could thus evolve upon aggregation, for example due to internal molecular rearrangements?

Dr De Vries answered: Dityrosine bonds have a characteristic signature in UV spectroscopy. By performing UV spectroscopy on chromatographic fractions of the reaction products and reaction intermediates, we have found that only dimers, trimers *etc.* show a dityrosine signal. Monomers never showed the dityrosine signal, suggesting that, at least for our reaction conditions, cross-linking is predominantly inter-molecular.

Dr van der Sman commented: Tyrosinase is known for enhancing browning in mushrooms and vegetables. Does this not hold for protein-based products too?

Dr De Vries replied: Indeed, after prolonged cross-linking with tyrosinase, the cross-linked protein products develop a dark, brownish color.

Dr van der Sman asked: For globular proteins some denaturation/melting is required for crosslinking/aggregation. Can shear also provide the impetus for the melting? Is this milder, or more reversible?

Dr De Vries answered: Most likely, shear is too weak to affect protein secondary structure. However, for larger clusters (>100nm), strong shear fields during cross-linking may certainly affect the overall architecture of the cross-linked protein.

Dr van der Sman enquired: What is the condition for stopping the enzymatic reaction ? Can you control this? Do you have some control over the final structure?

Dr De Vries responded: For the peroxidase system, simply stopping the supply of peroxide stops the reaction. For transglutaminase and tyrosinase, heat-inactivation may be a "food-grade" way of stopping the reaction, without resorting to inhibitors. For the peroxide-based cross-linking we have already shown that heating has only a minor influence on the structure of the cross-linked proteins. For transglutaminase and tyrosinase cross-linked α-lactalbumin, the heat-stability is still under investigation. In our current work we have let reactions continue until no further changes occurred. Indeed, stopping the reactions at earlier stages, *e.g.* by a short heat treatment, would give additional control over the reaction products.

Dr van der Sman remarked: At some stage of the process you have coexistence of aggregated state and a fluid phase containing monomers. Do you expect in this stage some partitioning of the enzymes? Partitioning of enzymes is known to happen in aqueous two-phase systems.

Dr De Vries replied: Partitioning could occur in more concentrated systems. So far we have only considered rather dilute systems (10%), but phase separation during gelation could indeed be an issue.

Dr van der Sman asked: Do you expect that *via* enzymes (directed assembly) one can have better control over protein structuring than *via* pH and temperature induced aggregation?

Dr De Vries answered: As we write in the paper, enzymatic cross-linking may especially give a larger degree of control over the size and architecture of the protein structures than can be obtained *via* heat- or pH-induced aggregation. Furthermore, since the structures are very different (in terms of architecture and chemistry) from protein aggregates, they may also be expected to have different food functional properties, although it is not immediately clear which ones, since so little is known yet about the cross-linked protein structures.

Dr Zanchetta asked: In Fig. 6 in the paper, you show that two time-scales apparently coexist; one for the depletion of monomers, around 2 h, and one for the further increase of aggregates sizes, around 12 h. In the text, you also report that large aggregates are present from the beginning, but stop contributing to the size estimate after around 2 h. Is this the case for all the concentrations investigated? Do the big, insoluble aggregates stop playing a role always at the same time as the monomers "disappear"? If this is the case, can you exclude the possibility that the already formed aggregates do contribute to the overall cluster formation?

I also have a second question: for the higher protein concentrations, you are running dynamic light scattering experiments at 90 degrees on rather big aggregates; in your size estimate, do you take into account the contribution of the rotational diffusion coefficient to light dephasing?[1,2] Such a contribution can lead to a quite significant underestimate of the hydrodynamic radius, larger for larger aggregates, which can lead to a distorted time evolution.

1 H. M. Lindsay, R. Klein, D. A. Weitz, M. Y. Lin and P. Meakin, Effect of rotational diffusion on quasielastic light scattering from fractal colloid aggregates, *Phys. Rev. A*, 1988, **38**,
2614–2626.
2 M. Lattuada, P. Sandkühler, H. Wu, J. Sefcik and M. Morbidelli, Aggregation kinetics of
polymer colloids in reaction limited regime: experiments and simulations, *Adv. Colloid Interface Sci.*, 2003, **103**, 33–56.

Dr De Vries responded: We have also performed some experiments on extensively
centrifuged and filtered α-lactalbumin that did not show the aggregate peak in the
starting material. These experiments showed exactly the same further development
of monomer depletion and particle size development, so we are confident that the
small amount of aggregates initially present do not strongly determine the course
of the reaction.

As to the scattering angle, indeed, the larger size values in Fig. 4 and 5 in the paper
should be considered as estimates only, and a proper characterization would involve
investigating the angular dependence of the scattering as well.

Dr Ubbink queried: Would you expect that by changing the reaction conditions
(T, ionic strength, pH) higher values of the functionality factor f could be obtained?

Dr De Vries answered: Increasing the functionality factor f requires increasing the
enzymatic accessibility of the reactive residues. Presumably, partial unfolding would
help here, so one would expect that, especially, heat treatments before cross-linking,
or elevated temperatures during cross-linking (provided the enzyme stability is good
enough) could increase the functionality factor f and give rise to more dense cross-
linked protein structures.

Professor Kulozik commented: On the second page of the paper it is said that there
is virtually no work on the architecture and physical properties of protein clusters
formed during enzymatic cross-linking. Depending what is meant by architecture
and physical properties, the authors may want to consider for citation the works outlined in ref. 1–5.

1 R. Sharma, P. C. Lorenzen and K. B. Qvist, Influence of transglutaminase treatment of skim
milk on the formation of ε-(γ-glutamyl)lysine and the susceptibility of individual proteins
towards crosslinking, *International Dairy Journal*, 2001, **11**(10), 785–793.
2 C. Gauche, J. T. C. Vieira, P. J. Ogliari and M. T. Bordignon-Luiz, Crosslinking of milk whey
proteins by transglutaminase, *Process Biochemistry*, 2008, **43**(7), 788–794.
3 R. Aboumahmoud and P. Savello, Crosslinking of Whey Protein by Transglutaminase, *Journal of Dairy Science*, 1990, **73**(2), 256–263.
4 M. P. Bönisch, A. Tolkach and U. Kulozik, Inactivation of an indigenous transglutaminase
inhibitor in milk serum by means of UHT-treatment and membrane separation techniques,
International Dairy Journal, 2006, **16**(6), 669–678.
5 S. Lauber, T. Henle and H. Klostermeyer, Relationship between the crosslinking of caseins by
transglutaminase and the gel strength of yoghurt, *Eur. Food Res. Technol.*, 2000, **210**, 305–
309.

Professor De Kruif opened the discussion of the paper by Ms Jin: I find it hard to
understand how drying on the outside leads to an increase of water at the centre. I
suppose the broccoli stalk is incompressible, *i.e.* does not contain empty pockets.
Suppose we put a rubber band around the stem, would that have the same effect?

Ms Jin answered: For fresh plant tissues, with well-organized cellular structures,
the drying mechanism is much more complex than we expected. For many plants,
they have thick bark, which has the role of protection, with low moisture content.
They can be dried faster and reach the glassy state after a relatively short time
and the rigid glassy structure forms a barrier to moisture transport, which has
been reported by several other papers, but they did not report the increased moisture

content phenomena.[1,2] In the case of broccoli, the role of the bark becomes significant, together with shrinkage; it increases the internal stress, which results in stress driven moisture transport towards the centre of the product. I think if we put a rubber band around the stem, we would increase the external stress; probably this could validate our hypothesis.

1 H.S.P. Kumar, K. Radhakrishna, P.K. Nagaraju and D.V. Rao, Effect of combination drying on the physic-chemical characteristics of carrot and pumkin, *Journal of Food Processing and Preservation*, 2001, **25**, 447–460.
2 N. Leeratanarak, S. Devahastin and N. Chiewchan, Drying kinetics and quality of potato chips undergoing different drying techniques, *Journal of Food Engineering*, 2006, **77**, 635–643.

Dr Ubbink queried: In the materials section of your paper, you state that for the measurements, parts of the broccoli stalks are used, which are of approximately similar sizes (radius approx. 1 cm, height approx. 1 cm). If this is true (please check and clarify this for me), and you have used slices from the broccoli stalks, I do not understand that the peel has such a large effect on the drying kinetics and on the water distribution during drying, as I would guess that water can easily escape from the stalks *via* capillaries oriented in the longitudinal direction. Why does this not happen? Would it be of interest to investigate the drying of broccoli slices with varying thickness, in order to get a better understanding of when and how the peel starts to affect the kinetics of water loss and of the water redistribution?

Ms Jin replied: For the samples used in all the measurements, they had similar sizes as shown in the paper, Fig. 1. We used part of the stalks in the edible part (relatively thin), so all the fresh samples had the thick peels. Actually if we look at the overall moisture profiles, although moisture content in the centre increases during drying, moisture indeed escapes from the stalks *via* the longitudinal direction, as the overall moisture content is decreasing.

Dr Ubbink further asked: To which extent would you expect the drying to induce irreversible changes in the structure of the stalks? Did you investigate if the differently treated broccoli samples did show different rehydration behaviour upon exposition to controlled relative humidity or when immersed in water?

Ms Jin replied: We did not do experiments on the rehydration of the dried broccoli. It might be interesting to investigate the rehydration behaviour, as this can give information on the influences of drying and pre-treatments on the plant tissue structures.

Dr Ettelaie opened the discussion of the paper by Dr Gebhardt: Can the degree of flexibility of micelles be a probe to the arrangement of their internal structure? That is to say, can it provide stronger support in favour of one or the other of the many different proposed casein micelle models?

Dr Gebhardt replied: Using this approach, this would first of all demand experimental tests on whether isolated sub-micelles are non-deformable under the described conditions. If so, we could indeed expect for a submicellar model that the mean sub-structure distances (shoulder at $q \sim 0.3$ nm^{-1} in the scattering functions) remain unchanged during deformation, while variations from that distance may support other models (nanocluster model).

Dr van Gruijthuijsen remarked: You made an effort to fractionate the caseins to obtain a highly monodisperse casein sample. I assume that this was done to enable the detailed modelling of the GISAXS data. To what extent is a well-defined size distribution required for data interpretation and modelling? Can you get qualitative or even quantitative information from the 2D patterns for polydisperse systems; notably for the shape change from spheres to ellipsoids?

Dr Gebhardt replied: Generally, GISAXS patterns resulting from casein micelles with a natural broad size distribution are dominated by diffuse scattering and the reflected contributions leading to the specular or Yoneda peaks are poorly visible. Under such conditions, experiments cannot be properly evaluated since the transformation of the GISAXS detector image to the q-scale demands a well-defined specular peak. The impact of a sharp casein micelle size distribution on the formation of a smooth and well defined film surface was discussed in our recent paper.[1] In general, qualitative and quantitative information of a polydisperse sample can be obtained by fitting the 2d GISAXS pattern by an elastic scattering approach using the IsGISAXS program as demonstrated in Gebhardt et al.[2] In this approach, all parameters (size in all dimensions and their distributions) are adjusted by the fit. In the work presented here, we show that casein micelles assume an ellipsoidal shape when subjected to filtration forces and that their elongation towards the membrane increases with increasing *trans*-membrane pressure applied. This effect could be modeled by increasing the height and decreasing the lateral dimension of the micelles and letting their volumes and polydispersities unchanged, as the comparison of the data with the simulations shows.

1 R. Gebhardt, C. Vendrely and U. Kulozik, *J. Phys.: Condens. Matter*, 2011, **23**, 444201.
E. Metwalli, J.-F. Moulin, R. Gebhardt, R. Cubitt, A. Tolkach, U. Kulozik and P. Müller-Bushbaum, *Langmuir*, 2009, **25**, 4124–4131.

Dr van Gruijthuijsen queried: I understood from your paper that the GISAXS measurements were performed directly after cessation of the shear/pressure. Did you remeasure the caseins after some more time (either with GISAXS or any other method) to see whether the elongation of the caseins is reversed and the caseins become spherical again?

Dr Gebhardt answered: We did not check the reversibility of the elongation. It was difficult to get enough material for a corresponding experiment because of the small quantity of deposited sample and the sensitivity of the micro-sieve surface. For that reason, a recovering of the sample by mechanical forces or by dissolution was not possible.

Dr Hütter asked: My question concerns the determination of the mechanical properties of casein micelles. I understand from your contribution that the micelles are deformed at (close to) constant volume. Nevertheless, your experiments could be used to assess the mechanical characteristics of the micelles, namely their shear modulus. You have already presented measurements of the micelle deformation from a spherical to an ellipsoidal shape. Is it possible to give an estimate of the stresses acting on the (initially) spherical micelles? If so, it would in turn be interesting to infer an estimate of the shear modulus that leads to the actually measured deformation, and to compare that value for the shear modulus with literature values.

Dr Gebhardt replied: In our opinion such an estimation of the shear modulus should be possible from the data presented in the paper and might be a matter of concern for a forthcoming paper.

Dr Hütter continued the discussion of the paper by Ms Jin: My question concerns the rationalization of the observed effects, particularly the effect of stress, *i.e.*, of pressure. If I understand the last paragraph in the conclusions well, the experimental findings can be explained by realizing that the driving force for the concentration change is not given in terms of the gradient of the concentration itself, but rather in terms of the gradient of the chemical potential. This is in line with classical irreversible thermodynamics. Under isothermal conditions, the chemical potential gradient can thus be written in the form $\nabla\mu = a(\nabla\Phi/\Phi) + b(\nabla p/p)$, with moisture content Φ. The coefficients are given by $a = \Phi \, \partial\mu/\partial\Phi$, $b = p \, \partial\mu/\partial p$.

It can be observed that the relative importance of the gradients in concentration and pressure is dictated by the ratio $a/b = (\Phi \, \partial\mu/\partial\Phi)/(p \, \partial\mu/\partial p)$. It consists purely of thermodynamic properties, and I would, hence, assume that this ratio can be measured on a sample under isotropic, homogeneous conditions (analogous to a PVT-experiment). It would be interesting to discuss what sort of experiments would have to be performed to assess this ratio. What can be said about an approximate estimate of the ratio $(\Phi \, \partial\mu/\partial\Phi)/(p \, \partial\mu/\partial p)$ with the current data available? Can this estimate be used in support of the findings of the drying experiments?

Dr van der Sman, as co-author, responded: We guess that Flory-Rehner theory applies to brocolli. In a current study we have found that Flory-Rehner applies to carrot and mushrooms. If the vegetable is relatively wet, the elastic contribution is of similar order as the osmotic contribution (which determines the water holding capacity of cell wall materials). Likewise, as we have done for carrot and mushroom, the elastic contribution to the total chemical potential can be obtained *via* centrifugation experiments, where the moisture is expelled *via* a porous bottom of the sample container. For carrots and mushrooms we have obtained an elastic modulus of about 2 MPa. In Flory-Rehner theory we do not include the elastic stress as a state-variable, but take it as a function of volume fraction and elastic modulus. Hence, a direct computation of a/b we can not perform.

Professor Dr Wilde commented: To me, the fact that the diffusion of water from plant tissue is non-Fickian is not surprising, as the internal cellular structures of the plant tissue are full of impermeable barriers such as cell walls, which are designed to hold on to as much water as possible. There are some structures within the plant tissue, such as lignified tissue, which, under normal circumstances, contains less water than a turgid cell, and is of a size that cannot be resolved by MRI. Therefore, can conventional optical microscopy be used to look at the structural changes that have occurred during drying, which may help explain the re-distribution of the water within cells and tissues?

Ms Jin replied: Based on our knowledge, for the drying of vegetables, there is not much work indicating non-fickian diffusion during drying. Most of the work still considers drying of fresh vegetables can be considered as a Fickian diffusion process. For all the fresh plant tissues, the cellular structures are often well organized, which makes the drying mechanism much more complex than we had expected. During drying or pre-treatments, the cellular structures can be changed. The aim of this work is to state the influence of the structure. With MRI the cellular structure cannot be detected, it might be helpful to use conventional optical microscopy to monitor the structural changes during drying, but it must be noted that there may be a loss in time resolution.

Professor Norton opened the discussion of the paper by ir. Vollebregt†: What does chemical potential mean in terms of your particles? Why are all 3 phases continuous?

Dr van der Sman replied: A thermodynamic quantity like chemical potential always arises after averaging over microscopic states, which are the particle positions and their velocities in our case. The chemical potential is computed from

---

† ir. Vollebregts paper was presented by her co-author, Dr van der Sman.

the osmotic pressure, which can be obtained from the microscopic states *via* the virial theorem. It involves averaging over the crossproduct of the vector connecting the particle positions and the force due to hydrodynamic interaction, for which we have taken the lubrication approximation (linear in the difference of particle velocities). Using a purely continuous description is a matter of choice, and length scales.[1] We are also interested in the development of the density profile along the length of the channel.[2] This entrance length can be quite long, and using a particulate description would be very demanding for computer simulations. A particulate description of confined suspensions requires fully resolved Lattice Boltzmann simulations. We have done that in the past,[3] however, 3D-parallel calculations of a simple periodic box can take months. We have initiated a new method using semi-resolved particles,[1,4] but this requires further development.

1 R. G. M. van der Sman, Simulations of confined suspension flow at multiple length scales, *Soft Matter*, 2009, **5**(22), 4376–4387.
2 J. Kromkamp, A. Bastiaanse, J. Swarts, G. Brans, R. G. M. van der Sman and R. M. Boom, A suspension flow model for hydrodynamics and concentration polarisation in crossflow microfiltration, *J. Membr. Sci.*, 2005, **253**, 67–79.
3 J. Kromkamp, D. van den Ende, D. Kandhai, R. G. M. van der Sman and R. Boom, Lattice Boltzmann simulation of 2D and 3D non-Brownian suspensions in Couette flow, *Chem. Eng. Sci.*, 2006, **61**(2), 858–873.
4 R. G. M. van der Sman and G. Brans, Subgrid particle method for porous media and suspension flow, *Advances in Fluid Mechanics*, 2008, DOI: 10.2495/AFM080291.

Professor Norton asked: What is the direction of the applied force? Are you causing jamming?

Dr van der Sman answered: In the application of a suspension flowing in a microchannel the driving force is the pressure difference across the channel. The direction of the driving force is perpendicular to the migration direction of the particles. In case of microfiltration there is an extra driving force along the migration direction due to the pressure difference over the membrane.[1] This driving force acts against the particle migration, and leads to the build-up of a cake layer. The thickness of a cake layer remains finite if the applied shear (due to crossflow) is sufficient. The shear of the crossflow induces migration, which is exactly balanced by the drag exerted by the flow induced by the transmembrane pressure. The cake layer can be regarded as jamming. In the application where we consider only the microchannel, jamming can arise at the entrance - a phenomenon also called self-filtration.[2] Also in constrictions and bends one can expect jamming.

1 J. Kromkamp, A. Bastiaanse, J. Swarts, G. Brans, R. G. M. van der Sman and R. M. Boom, A suspension flow model for hydrodynamics and concentration polarisation in crossflow microfiltration, *J. Membrane Sci.*, 2005, **253**, 67–79.
2 M. D. Haw, Jamming, Two-Fluid behaviour, and "Self-Filtration" in Concentrated Particulate Suspensions, *Phys. Rev. Lett.*, 2004, **92**(18), 185506.

Dr Ettelaie remarked: The closure equation, eqn (23) in your paper, gives an intrinsic viscosity that seems to be inversely proportional to the random close packing, $\Phi_{max,bi}$, and therefore varies with the fraction of small and large particles in the bidisperse system. However, as is well known for a distribution of hard spheres, even a polydisperse one, the intrinsic viscosity is given by Einstein's relation and should be a constant (*i.e.* 2.5). Are there any particular reasons as to why there is such a deviation in the system you have been considering?

Dr van der Sman answered: This is a typing mistake. We have proposed either a generalization of Krieger-Doghourty or Quemada. Krieger-Doghourty would indeed follow the Einstein relation. The exponent in eqn (23) in the paper should read $2.5\Phi_{max,bi}$. In the limit of dilute suspensions it follows the Einstein relation. For the Quemada relation the exponent is 2.0 - and it already deviates from the Einstein relation for the monodisperse case. However, the Quemada relation is propagated by Brady and Morris in their models of shear-induced migration. In our simulations we do not observe large differences between these closures.

Dr Royall queried: Firstly: have you considered that, in the case of zero flow, binary hard spheres can exhibit phase separation due to depletion effects? For the parameters listed in Fig. 5 of the paper, the system may be close to phase separation, if not in the two-phase region.

Secondly: this phase separation should lead to an inversion in the structure of the suspension in a channel. Under zero flow conditions, one expects the large particles to experience a strong (depletion) attraction to the wall, while under flow the small particles are favoured at the wall. Thus, there should be a non-equilibrium transition in the structure as a function of flow rate.

Dr van der Sman replied: In this paper we use the BMCSL equation of state, which does not reproduce the phase separation in zero flow. In a forthcoming paper,[1] we have considered the free volume expression for the compressibility factor $Z(\phi)$.[2] Hence, we write the particle pressure as: $\sigma = n(kT)_{eff}Z(\phi)$ $\qquad$ (1) with the effective temperature defined by $(kT)_{eff} \sim \eta_{eff}\dot{\gamma}V_p$, with $V_p = 4/3\pi a^3$, and

$$Z(\phi) = \frac{\tilde{\phi}^{\frac{1}{3}}}{1 - \tilde{\phi}^{\frac{1}{3}}} \qquad (2)$$

with $\tilde{\phi} = \phi/\phi_{II,max}$ being the normalized particle volume fraction. Note that $\phi = \phi_1 + \phi_2$ is the sum of volume fractions of both small and large particles. $\phi_{II,max}$ is the random close packing for a bidisperse suspension, which depends on the size ratio of the particles, and the relative amount of small particles. The original theory is developed for monodisperse suspensions,[2] but we have extended that by taking $\phi_{II,max}$ for a bidisperse suspension,[1] by using an empirical relation by Shauly.[3] For various instances, the free volume theory predicts similar results as the BMCSL equation of state. In regions of the phase separations the free volume expression deviates from BMCSL, and we expect it to reproduce phase separation - but this has to be tested yet. The use of the free volume expression is hindered by the fact that we can not obtain an analytical expression for the chemical potentials from the osmotic pressure, if we follow the procedure described in our paper.

In shear flow, the resulting profile is due to gradients in the volume fraction, and gradients in the shear rate (due to the parabolic flow profile). This results in the large particles migrating to the middle of the channel – in the realisation we have studied. It can happen that in suspensions with the smaller particles in the majority – the smaller particles migrate to the middle of the channel.

We do expect effects of the confining walls, which is not yet captured based on our current model, based on the BMCSL equation. If a free volume expression can be used, we can adapt $\phi_{II,max}$ in the neighbourhood of the wall, which is known to be lowered by the confining wall.[4]

1 R.G.M. van der Sman and H.M. Vollebregt, Effective temperature for sheared suspensions: a route towards closures for migration in bidisperse suspension, *Adv. Colloid Interface Sci.*, 2012, DOI: 101016/j.cis.2012.08.006.
2 R.D. Kamien and A.J. Liu, Why is random close packing reproducible? *Phys. Rev. Lett.*, 2007, **99**(15), 155501.
3 A. Shauly, A. Wachs and A. Nir, Shear-induced particle migration in a polydisperse concentrated suspension, *J. Rheol.*, 1998, **42**, 1329.
4 K.W. Desmond and E.R. Weeks, Random close packing of disks and spheres in conned geometries, *Phys. Rev. E*, 2009, **80**(5), 051305.

Dr Hütter said: I would like to comment on the effective temperature defined in eqn (27) of your paper. It is stated that the effective temperature depends in a linear fashion on the shear rate, particularly on the absolute value of the shear rate. I understood in the presentation that the linear dependence can be derived from more microscopic considerations, which is a good basis for eqn (27). However, this implies that due to the occurrence of the absolute value, the effective temperature has got a non-analytic dependence on shear rate around equilibrium. While I cannot pinpoint a clear conflict by doing so, I find the occurrence of the absolute value quite remarkable, particularly for a statement derived from a more microscopic picture.

Furthermore, I would like to ask about the further usage of the effective temperature. In the presented manuscript, the effective temperature appears only in the expression for the osmotic pressure, eqn (26). If this is indeed the only occurrence, why can one not interpret eqn (26) rather as two contributions (namely, elastic and viscous) to the osmotic pressure, since it involves a transport coefficient (the viscosity)? Or, alternatively, is it anticipated that the effective temperature defined in eqn (27) has also ramifications for quantities other than the osmotic pressure, e.g. friction, diffusion, or Brownian contributions?

Dr van der Sman responded: Actually, we compute the shear rate, $\dot{\gamma}$, from the square root of the second invariant of the viscous stress tensor, cf. the models by Brady and Morris.[1] The viscous stress tensor is readily available from the Lattice Boltzmann simulations.[2]

Our derivation of the osmotic pressure from microscopic details is as follows. The particle stress tensor σ has a contribution due to contact forces,[3] and follows from the Kirkwood–Irving virial theorem:

$$\sigma = \frac{1}{V} \sum_{i<j} F_{ij} \times r_{ij} \qquad (1)$$

where V is the volume of the system, F_{ij} is the lubrication force, and r_{ij} is the distance between particles. The magnitude of the lubrication force for monodisperse suspensions scales is:

$$F_{ij} = \eta_{\text{eff}} \dot{\gamma} \frac{a^2}{h_{ij}} \qquad (2)$$

where η_{eff} is the suspension viscosity, $\dot{\gamma}$ is the shear rate, a is the particle radius and h_{ij} the gap between particles, indexed with i and j. Here, we have followed the mean field argument of Snabre and Mills,[4] that in a concentrated suspension the lubrication force scales with the suspension viscosity, rather than the fluid viscosity. Snabre and Mills have also argued that the particle pressure (the trace of the particle stress) is linear with the average distance between particles, a result we will use below:

$$\sigma = n\eta_{\text{eff}} \dot{\gamma} a^2 < h/a > \qquad (3)$$

with n being the number density of the suspended particles, and $< h/a >$ is the average gap between particles, normalised against the radius a. Hence, from the above, it follows that the particle stress is linear with the viscous stress, which can be equated to an effective temperature.[5,6,7] Van den Brule and Jongschaap[8] and others[9] have argued that the average distance between particles is a function of the volume fraction, which is very similar to the free volume expression for the compressibility factor $Z(\phi)$.[10] In this microscopic derivation, we have assumed that $\dot{\gamma}$ is the absolute value (or rather the square root of the second invariant of the viscous stress tensor). Because also the earlier models of Brady and Morris work with the absolute value, we do not think there is an issue with the non-analytic dependence on shear rate around equilibrium. Furthermore, we do not expect the effective temperature to enter the friction coefficient, or Brownian diffusion coefficients. The hydrodynamic interactions give rise to velocity fluctuations independently of the thermal fluctuations – and, therefore, the effective temperature should not enter the transport coefficients related to Brownian motion. Of course, the osmotic pressure or rather particle stress can also be regarded as having two independent (elastic and viscous) contributions.

1 R. M. Miller and J. F. Morris, Normal stress-driven migration and axial development in pressure-driven flow of concentrated suspensions, *J. Non-Newton. fluid*, 2006, **135**(2), 149–165.
2 J. Kromkamp, A. Bastiaanse, J. Swarts, G. Brans, R. G. M. Van Der Sman and R. M. Boom. A suspension flow model for hydrodynamics and concentration polarisation in crossflow microfiltration, *J. Membr. Sci.*, 2005, **253**(1), 67–79.
3 A. Lemaitre, J. N. Roux and F. Chevoir, What do dry granular ows tell us about dense non-brownian suspension rheology? *Rheol. Acta*, 2009, **48**(8), 925–942.
4 P. Mills and P. Snabre, Apparent viscosity and particle pressure of a concentrated suspension of non-brownian hard spheres near the jamming transition, *Eur. Phys. J. E: Soft Matter and Biological Physics*, 2009, **30**(3), 309–316.
5 A. Negi and C. Osuji, Dynamics of a colloidal glass during stress-mediated structural arrest, *Europhys. Lett.*, 2010, **90**, 28003.
6 H. M. Vollebregt, R. G. M. van der Sman and R. M. Boom, Suspension flow modelling in particle migration and microfiltration, *Soft Matter*, 2010, **6**(24), 6052–6064.
7 C. Eisenmann, C. Kim, J. Mattsson and D.A. Weitz, Shear melting of a colloidal glass, *Phys. Rev. Lett.*, 2010, **104**(3), 35502.
8 B. Van den Brule and R. J. J. Jongschaap, Modeling of concentrated suspensions, *J. Stat. Phys.*, 1991, **62**(5), 1225–1237.
9 R. F. Probstein, M. Z. Sengun and T.C. Tseng, Bimodal model of concentrated suspension viscosity for distributed particle sizes, *J. Rheol.*, 1994, **38**, 811.
10 R. D. Kamien and A. J. Liu, Why is random close packing reproducible?, *Phys. Rev. Lett.*, 2007, **99**(15), 155501.

Professor Tanaka commented: The concept of the effective temperature seems to work well and reproduce the experimental results on a satisfactory level. However, the mechanical stress is a tensorial quantity, and shear flow, for example, has a compressional and an extensional axis, for which the effects on structuring are very different. This clearly indicates that, strictly speaking, effects of the mechanical stress cannot be represented by the effective temperature, which is a scalar quantity. The more natural way may be to treat the mechanical stress in the kinetic equations in an explicit manner (see, *e.g.*, eqn (1)–(4) in my paper, which will be discussed in the final session). In this scheme, the relative motion of the components of a mixture is driven not only by the osmotic force, but also by the mechanical force. This provides us with a very natural foundation for stress-diffusion coupling. In the simplest Newtonian liquid, we can express the mechanical stress as $\eta(\vec{\nabla}\vec{v} + (\vec{\nabla}\vec{v})^t)$, where η is the viscosity dependent upon the composition and t represents the transpose operator. This scheme has a firm physical basis, though phenomenological, and I expect that similar behaviour may be predicted with this scheme. The concept of the effective temperature is also often used for sheared glassy solids, but shear flow generally induces anisotropic fluctuations, as pointed out above, and, thus, this concept must break down in the strict sense.[1] Nevertheless, it is interesting to see the extent to which this scalar treatment practically can capture the basic effects of flow-induced mechanical stress, particularly after the reduction of dimensionality.

1 See, *e.g.*: A. Furukawa, K. Kim, S. Saito and H. Tanaka, *Phys. Rev. Lett.*, 2009, **102**, 016001.

Dr van der Sman responded: You are completely correct that the mechanical or particle stress is a tensorial quantity, while the effective temperature is a scalar quantity. The effective temperature only relates to the trace of the particle stress, which can be equated to an osmotic pressure.[1] In the general case, the particle stress of sheared suspensions is anisotropic, and can have normal stress differences.[2] In the planar microchannel, and in microfiltration applications, the migration is predominantly in one-direction and only the trace of the particle stress has a non-zero value.[3,4] A model taking a kinetic equation for the particle stress, using a tensorial order parameter, is presented by Philips and Powell.[5] This would be indeed a natural way to describe the particle migration for the general case - and which might account for wall effects. However, the kinetic equations are formulated phenomenologically. Further research is required. A more viable route might be via the Kirkwood–Irving virial theorem, where the particle stress is computed as a tensor. In the averaging procedure one has to insert the particle distribution function, $g(r)$, which is shown to be anisotropic in the case of sheared suspensions.[6,7] To apply the effective temperature concept to the general case would imply a generalization of the effective temperature concept from a scalar to a tensorial representation, as suggested in the reference given previously by Professor Tanaka[8] and in ref. 9 and 7. As in the sheared supercooled liquid, the shear-induced diffusivity is anisotropic, and this would imply comparable physics between sheared supercooled liquids and suspensions.

1 A. Deboeuf, G. Gauthier, J. Martin, Y. Yurkovetsky and J.F. Morris, Particle pressure in a sheared suspension: A bridge from osmosis to granular dilatancy, *Phys. Rev. Lett.*, 2009, **102**(10), 108301.
2 R. M. Miller and J. F. Morris, Normal stress-driven migration and axial development in pressure-driven flow of concentrated suspensions, *J. Non-Newton. fluid*, 2006, **135**(2), 149–165.
3 J. Kromkamp, A. Bastiaanse, J. Swarts, G. Brans, R. G. M. van der Sman and R. M. Boom, A suspension flow model for hydrodynamics and concentration polarisation in crossflow microfiltration, *J. Membrane Sci.*, 2005, **253**(1), 67–79.
4 H. M. Vollebregt, R. G. M. van der Sman and R. M. Boom, Suspension flow modelling in particle migration and microfiltration, *Soft Matter*, 2010, **6**(24), 6052–6064.
5 K. Yapici, R. L. Powell and R. J. Phillips, Particle migration and suspension structure in steady and oscillatory plane poiseuille flow, *Phys. Fluids*, 2009, 21:053302.
6 J. F. Morris, A review of microstructure in concentrated suspensions and its implications for rheology and bulk flow, *Rheol. Acta*, 2009, **48**(8), 909–923.
7 V. Chikkadi, S. Mandal, B. Nienhuis, D. Raabe, F. Varnik and P. Schall, Shear-induced anisotropic decay of correlations in hard-sphere colloidal glasses, 2012, *ArXiv preprint*, arXiv:1205.6070.
8 A. Furukawa, K. Kim, S. Saito and H. Tanaka, Anisotropic cooperative structural rearrangements in sheared supercooled liquids, *Phys. Rev. Lett.*, 2009, **102**, 016001.
9 H. Mizuno and R. Yamamoto, Mechanical responses and stress fluctuations of a supercooled liquid in a sheared non-equilibrium state, *Eur. Phys. J. E: Soft Matter and Biological Physics*, 2012, **35**(4), 1–14.

Professor Kulozik asked: The model presented is for regular pipe flow. How do you see the model evolving if flow towards the wall, *i.e.* in membrane filtration, is present?

Dr van der Sman answered: The presented model is valid for suspension flow in microchannels, with a slit-shaped cross-section. For suspension flows through channels with other cross-sections, like a circular tube, normal stress differences arise – and the particle stress tensor is not isotropic anymore.[1] Our model has to be modified to account for that. Anisotropic particle stresses are easily accounted for, but how to derive chemical potentials from that is still a question. Currently, we are thinking along the lines of (particle) stress division along the phases, which does not require the definition of a chemical potential. Hence, we have some questions for Professor Tanaka, which we will ask during the session discussing his paper.

In an earlier publication, we modelled the effect of a membrane incorporated in the wall.[2] That model was based on the model of Leighton and Acrivos. The later models of Brady and Morris build on that work. In our review, we have shown all these models are equivalent, only the correlation for the shear-induced diffusion by Leighton and Acrivos is not correct.[3]

Hence, we can extend our model along the lines of our earlier work.[1] The membrane is represented as a porous section in the wall, allowing fluid flow – but rejecting the particles. Below the porous wall, we apply a pressure lower than the pressure in the crossflow channel, creating a transmembrane pressure and subsequent flow, which draws particles to the membrane. A cake layer is formed if the particle volume fraction exceeds random packing. For the cake layer we use local rules similar to cellular automata. Applications with backpulsing can yet not be modelled, which require an adaption of the current CA rules. We require a rule for the fluidization the cake again. For bidisperse suspensions the random close packing would be a local value, depending on the local ratio of small to large particles.

1 R. M. Miller and J. F. Morris, Normal stress-driven migration and axial development in pressure-driven flow of concentrated suspensions, *J. Non-Newton. fluid*, 2006, **135**(2), 149–165.
2 J. Kromkamp, A. Bastiaanse, J. Swarts, G. Brans, R. G. M. van der Sman and R. M. Boom, A suspension flow model for hydrodynamics and concentration polarisation in crossflow microfiltration, *J. Membrane Sci.*, 2005, **253**(1), 67–79.
3 H. M. Vollebregt, R. G. M. van der Sman and R. M. Boom, Suspension flow modelling in particle migration and microfiltration, *Soft Matter*, 2010, **6**(24), 6052–6064.

Dr Ettelaie then continued the discussion of the paper by Ms Jin: The rate of change of moisture depends on the gradient of the diffusion flux, which not only is a function of moisture gradient but also the diffusion coefficient. Are there practical cases where the diffusion coefficient actually decreases with increasing moisture, and if so can this lead to an accumulation of moisture in regions close to the centre?

Ms Jin responded: Actually, in our other paper,[1] we predicted the diffusion coefficient first as increasing then as decreasing. Similar phenomena can be found in other cases in the literature as well.[2,3] However, if it is a concentration driven process, if there is no external stress, although the diffusion coefficient can be decreasing, I still expect moisture may move towards the surface.

1 X. Jin, R.G.M. van der Sman and A. J. B. van Boxtel, Evaluation of the Free Volume Theory to Predict Moisture Transport and Quality Changes During Broccoli Drying, *Dry. Technol.*, 2011, **29**, 1963–1971.
2 R. J. Aguerre and C. Suarez, Diffusion of bound water in starchy materials: application to drying, *J. Food Eng.*, 2004, **64**, 389–395.
3 V. T. Karathanos, G. Villalobos and G. D. Saravacos, Comparison of Two Methods of Estimation of the Effective Moisture Diffusivity from Drying Data, *J. Food Sci.*, 1990, **55**, 218–223.

Dr Cardinaels commented: You showed that you can speed up drying by removing the gradients in elastic properties by blanching or freezing the broccoli. This way, you make the material homogeneous and soft. However, I can imagine that consumers would rather like to receive crispy broccoli. Do you think it is possible to obtain a similar improved drying if you reduced the gradients in elasticity by making the inside of the broccoli more stiff? I suggest that this can be done by using specific thermal/pressure treatments or by Ca^{2+} addition, which stimulates crosslinking of the pectin in the cell walls. This way, there will be no damage of the cell walls and a more crispy material could be obtained after drying.

Ms Jin responded: Although we did not do sensory evaluation, the suggestion to use alternative pre-treatments might be helpful to preserve the internal structures. Besides, it also should be noticed that drying conditions can influence the internal structure as well. At low temperature drying, with a low drying rate, the shrinkage ratio is equal to moisture removal, resulting in a firmed structure, which one could not expect to be crisp.[1] Furthermore, for the processing of many vegetables, it is important to use blanching to inactivate some enzymes prior to further processing, which can give a better colour for the final product, as well as better flavour.[2]

1 N. Wang and J. G. Brennan, Changes in structure, density and porosity of potato during dehydration, *J. Food Eng.*, 1995, **24**, 61–76.
2 C. Severini, A. Baiano, T. De Pilli, B. F. Carbone and A. Derossi, Combined treatments of blanching and dehydration: study on potato cubes, *J. Food Eng.*, 2005, **68**, 289–296.

Dr Defraeye asked: Did you consider also other MRI parameters, such as the T1 or T2 value, apart from the MRI image intensity? Parameters such as the T2 value could provide information on the water mobility of the sample and are often also used in MRI data analysis.

Ms Jin responded: We did consider T1 or T2 measurements, but 3D T2 or T1 measurements are too time consuming to be practical. Since we would like to record the dynamic drying process, every second the moisture content may change, therefore, we chose 3D RARE measurements instead of 3D T2 measurements, to have the benefits of short time measurements.

Professor Tanaka addressed Ms Jin and Dr van der Sman, saying: If we can apply locally the mechanical stress to a part of broccoli and measure the temporal development of the spatial gradient of the moisture induced by the stress by MRI imaging, we might be able to directly measure the stress-induced diffusion process (see eqn (2) in my paper, which will be discussed in a later session). Is such an experiment plausible?

Dr van der Sman responded: I can imagine an experiment, which may be hard to realize though. For water holding capacity measurements we are developing a method based on confined compression.[1] Here, one drives a piston on a cylindrical sample, confined in a tube with a porous bottom. Due to the exerted pressure of the piston, the food sample will expel water from the tissue. We measure displacement and the exerted force. If this experimental setup is placed in a MRI scanner, one could image the moisture distribution. To realize this experiment the setup has to be made from non-metallic materials - which would be a challenge. Without the imaging, confined compression could already give sufficient information for stress-diffusion.

1 M. R. Drost, P. Willems, H. Snijders, J. M. Huyghe, J. D.Janssen and A. Huson, Confined compression of canine annulus fibrosus under chemical and mechanical loading, J. Biomech. Eng., 1995, **117**, 390.

Dr Ettelaie then continued the discussion of the paper by Dr Gebhardt: Have you tried carrying out similar experiments to those in your paper with partially dialysed casein micelles? Assuming that the sub-micelle model of casein micelles is correct, the removal of calcium phosphate will change the strength of the bonds between sub-micelles and, hence, should alter the rigidity of the casein micelles. Thus, monitoring the distortion of the shape of micelles for different degrees of dialysis can potentially provide useful information on the role of calcium phosphate in the formation of such bonds.

Dr Gebhardt replied: We did not perform experiments with dialysed samples as proposed. Such experiments could show to what extent the colloidal calcium phosphate influences the mechanical properties of the casein micelles. However, such experiments are most probably accompanied by a size or volume change, making the determination of relative shape changes very difficult. A verification of the different casein micelle models is in our opinion not possible since (1) the location of colloidal calcium phosphate in casein sub-micelle models is contradictory[1] and (2) calcium phosphate on a different location could also affect the elastic properties.

1 D. G. Dalgleish, *Soft Matter*, 2011, 7, 2265–2272, and ref. 47-50 therein.

Dr Rayner remarked: Microfiltration of milk is usually conducted in a closed system in a wet state. The challenge with measuring the semi dry filter cake by AFM is that it is not just a question of long term drying, as is discussed in the work, but also that there will be drying induced compression at short times because there is so little water to lose. Furthermore, there is the question of the interfacial forces at the wet casein deposit–air interface to complicate the picture. Have you considered measuring the microfiltration in a liquid cell?[1]

Another question is the reversibility of casein micelle deformation. Gésan-Guiziou *et. al.*[2] show that the casein film relaxes as soon as you release the pressure. Do you see an effect after releasing the pressure? Are the casein micelles in the cake trapped as they try to squeeze through the 800nm pore? This is a very different situation from real microfiltration where the micelles cannot enter. In the experimental setup used in the paper, could you monitor the permeability to see if the membrane pores became blocked and to what extent this is reversible?

1 For example: Helstad *et al.*, 2007, Liquid droplet-like behaviour of whole casein aggregates adsorbed on graphite studied by nano-indentation with AFM, *Food Hydrocolloid.*, 2007, **21**(5), 726–738.
G. Gésan-Guiziou, A Bouchoux, Dead-end filtration of sponge-like colloids: the case of casein micelle, *J. Membrane Sci.*, 2012, **417**, 10–19.

Professor Kulozik, as co-author on the paper, responded: It is correct that microfiltration is industrially conducted in a wet state. We did not state that the drying process was a longterm operation. In contrast, it was said that the film formation and removal of capillary water is a question of seconds. However, in order to be able to assess the chemico-physical properties as they exist in the compressed deposited casein layer, the approach was taken as described, namely to arrest the deposited layer in the compressed state in order to avoid structural changes upon the release of pressure. It is correct that the deposited casein micelle layers relax after the release of pressure when in wet state, which, however, was not the question under investigation in this study. We can say from our data derived from the GISAXS experiments that the volume of the casein micelle remains unchanged, *i.e.* an additional structural change due to drying can be excluded as a side effect, as far as measurement times were investigated. We analyzed and found that initially some casein micelles entered the pores of the microsieve used in the study. However, upon deposit formation the decisive retention took place at the surface of the deposit. Hence, we did the study in a deposit controlled condition, *i.e.* the structure was able to be measured comparable to real microfiltration. In the case of the GI-SAXS study using the microsieves in a frontal filtration mode, we did not investigate the aspect of reversible deposit formation due to the small scale of the technical set-up. However, in real crossflow microfiltration using a 0.1 μm membrane we found that the deposition on normal microfiltration membranes is mostly reversible.

Dr van der Sman asked: What sort of deformation do you expect for yeast cells during microfiltration? Is this similar to the casein micelles? The deformation of micelles seems to be counterintuitive.

Dr Gebhardt answered: As described in the paper, yeast cells deform into a hexagonal shape by flattening in regions of cell-cell contact. The measured ellipsoidal form factor for casein micelles does not support a flattening of the contact areas. Enthalpic or entropic repulsion terms act in the deposit and ensure the integrity of the micelles (as shown in a number of AFM studies on casein films). Some studies on particles in the µm-range showed that they become compressed in the direction towards the membrane and expand in the lateral direction. As our GISAXS and AFM data show, for the casein micelles the opposite is the case. We explained their elongation towards the membrane by shear forces acting parallel to the micelle surface in the direction of the fluid flow and their compression in the lateral direction by normal forces originating from the neighbouring micelles acting perpendicularly to the surface.

Professor De Kruif was invited by the session chair, Professor Dr Martien Cohen Stuart, to comment on the "hardness" of casein micelles: The colloidal behaviour of casein micelles can adequately be described using (adhesive) hard sphere models from the liquid state theory developed for small molecules or atoms. In fact, these models hold if the range of the potential is small compared to the radius of the particle (this will be discussed further in the concluding remarks). Thus, using these (adhesive) hard sphere models we could describe the interaction and rheology of (concentrated) casein micelle dispersions quite adequately.[1–3]

Whether casein micelles can be described as hard spheres in a mechanical sense depends on the value of the elastic constant of a casein micelle. Uricanu *et al.*[4] measured a reduced Young modulus $E^* = 450$ kPa. $E^* = E/(2(1-v2)) = 4G'$ where $v = 0.5$ is the Poisson ratio and G' is the shear modulus. So $E \approx 260$ kPa, which is a very small value compared to classical materials. For example, rubber would be 10 times "harder". A casein micelle contains about 25% caseins. The water phase of a young Gouda cheese contains 25% protein as well. The shear modulus varies between 100 to 1000 kPa.

A compressive deformation of a casein micelle by 1% would involve an energy of 260 kPa $\times$ 0.01 $\times$ 4/3π (100 nm)^3 $\approx 1 \times 10^{-17}$ Joule, equivalent to 2600 kT. The Peclet number measures the shear stress *versus* the thermal stress and is defined as Pe $= \gamma a^2/D$ where γ is the shear rate, a is particle radius and D the diffusion coefficient. Using the Stokes–Einstein equation, $D = $ kT/(6πηa), we find that for a shear rate of 200 s^{-1}; Pe = 1. In order to match a thermal energy of 2600 kT, one must increase the shear rate to about 0.5×10^6 s^{-1}. A value that is probably not even reached in a homogenizer.

Suppose we would squeeze the micelle sphere between two glass plates. The force needed is Fn $\approx$ 4/3E* $a^{1/2}$ $\delta^{3/2}$ where δ is the (half) compression. Assuming we compress the sphere by 2% we find that the work done is: Work $\approx$ 4/3 450 kPa $\times$ (100nm)^3 $\times$ 1 nm^2 = 6×10^{-20} Nm or 15 kT. Tuinier[3] calculated the interaction potential of casein micelles by summing the Van der Waals attraction, electrostatic repulsion and the steric repulsion. In these calculations, the interaction potential rises on contact very steeply over a distance of about 4 nm to a value of 10 kT.

In conclusion; casein micelles are a very soft material but can be treated in experiment as hard spheres for all extents and purposes.

1 C. G. de Kruif, Casein micelles: diffusivity as a function of renneting time, *Langmuir*, 1992, **8**, 2932–2937.
2 C. G. de Kruif, Casein micelle interactions, *Int. Dairy J.*, 1999, **9**, 183–188.
3 R. Tuinier and C. G. de Kruif, Stability of casein micelles in milk, *J. Chem. Phys.*, 2002, **117**(3), 1290–1295.
4 V. I. Uricanu, M. H. G. Duits and J. Mellema, Hierarchical networks of casein proteins: An elasticity study based on atomic force microscopy, *Langmuir*, 2004, **20**, 5079–5090.

Dr Rayner said: Although I am not an expert on casein, sometimes soft materials on small scales are dominated by interfacial forces and geometric effects and act like liquid droplets.[1]

K. Helstad, M. Rayner, *et al.*, Liquid droplet-like behaviour of whole casein aggregates adsorbed on graphite studied by nanoindentation with AFM, *Food Hydrocolloid.*, 2007, **21**(5–6), 726–738.

Professor De Kruif added: Caseins and casein gels are usually viscoelastic and therefore the value of the moduli depend on the time scale as well. On a long time scale the casein gels will flow. The calculations refer to a short time scale.

Professor van der Linden commented: I would like to add that it is important, when trying to bring various experiments in accordance with one another, to consider the boundary conditions applicable to the casein micelle. This could be zero deformation (probably the case for interaction at a distance, implying validity of a hard colloid picture), deformation under constant volume but changing surface area, deformation under constant area but changing volume, or deformation under changing surface area and changing volume.

Stability of aqueous food grade fibrillar systems against pH change

Ardy Kroes-Nijboer,[a] Hassan Sawalha,[†a] Paul Venema,[a] Arjen Bot,[b] Eckhard Flöter,[‡b] Ruud den Adel,[b] Wim G. Bouwman[c] and Erik van der Linden[*a]

Received 20th February 2012, Accepted 10th April 2012
DOI: 10.1039/c2fd20031g

We report that the stability of an aqueous food grade fibril system upon pH change is affected by the presence of peptides that are formed during the process of fibril formation. We discuss several other relationships between food relevant properties and nano-scale characteristics, and compare these relationships for aqueous fibril systems to those of oil based fibril systems. In such fibril systems, dynamics, self-organisation, and sensitivity to external conditions, play an important role. These aspects are common to complex systems in general and define the future challenge in relating functional properties of food to molecular scale properties of their ingredients.

1. Introduction

Foods can exhibit considerable variations in internal structure, with sizes, in one or more dimensions, ranging between nanometers to millimeters, and morphologies like fibrillar, plate-like, spherical, and topologically more complex bi-continuous structures.

In aqueous surroundings fibrillar structures may be *e.g.* polysaccharides like xanthan, triple helices of gelatin, carrageenan bundles, or thin protein strands.

The functionality of fibrillar structures ranges from increasing viscosity, gelling agent, to stabilisers of foams, emulsions and capsules.[1-4] The extent of each functionality strongly depends on the length distribution of the fibrils and their aspect ratio (length *versus* diameter). For example, their high aspect ratio leads to the fact that at quite low volume fractions they form a space filling network, making them weight efficient structurants.

For application purposes it is desirable to be able to control and influence length distribution. In addition it is desirable to have flexibility in the choice of raw materials. This implies one needs to understand which molecules under which circumstances form or are fibrillar structures, and how to control details in form and structure.

[a] Laboratory of Physics and Physical Chemistry of Foods, Department of Agrotechnology and Food Sciences, Wageningen University, Bomenweg 2, NL-6703 HD Wageningen, The Netherlands
[b] Unilever Research and Development Vlaardingen, Olivier van Noortlaan 120, NL-3133 AT Vlaardingen, The Netherlands
[c] Department of Radiation, Radionuclides & Reactors, Faculty of Applied Sciences, Delft University of Technology, Mekelweg 15, NL-2629 JB Delft, The Netherlands

† Current address: Chemical Engineering and Material Science, An-Najah National University, P.O. Box 7, Nablus, Palestine.
‡ Current address: Food Process Engineering, Department of Food Technology and Food Chemistry, Technical University Berlin, Königin Luise Strasse 22, D-14195 Berlin, Germany.

In an aqueous phase, fibrils can be formed by proteins or their peptides. Over the last few decades, several food-grade proteins have been used as a base for such fibrils. Examples are egg proteins,[5-9] soy proteins,[10,11] kidney bean proteins[12] and milk proteins.[13-30] To prepare these fibrils, different conditions apply for different proteins. For β-lactoglobulin (β-lg), fibrils are formed while heating at pH 2 and 80 °C for several hours.

Apart from food applications, fibrils are also relevant in the field of material sciences[31-33] and biomedical sciences.[34-37]

Because most foods have a pH between 4 and 7, it is important to also consider the stability of fibril systems upon pH change in more detail. While a pH change does not change the fibril itself, fibril solutions as a whole become turbid around pH 5, close to the iso-electric point of β-lg.[38,39] These results are suggesting that the fibrils aggregate around pH 5. However, the fibril solutions contain a mixture of fibrils and non-aggregated peptides. Only ∼40% of the peptides are being incorporated into the fibrils. These peptides were found to be the more hydrophobic peptides with a high ability to form β-sheets.[40] The rest of the material remains in the solution as non-aggregated peptides. It has not been previously addressed which fraction is causing the turbidity around pH 5. This is why we investigated the effect of pH on the stability of the aqueous fibril system.

2. Formation of fibrils

The fibrils, as obtained by heating a β-lactoglobulin sample at pH 2 to 80 °C, were recently reported by Akkermans *et al.* to be a composed of a certain group of peptides that result from the acidic hydrolysis of the bonds before or after aspartic acid residues in β-lg.[40] The important role of hydrolysis was also previously shown for hen egg white lysozyme.[41,42] The resulting peptides are subsequently incorporated into fibrils.

In Fig. 1 a schematic representation is given of the fibril assembly. The β-lg monomers are hydrolyzed into small peptides and some of these peptides are assembled into fibrils.

Once fibrils have been formed, they do not fall apart upon dilution (even over a period of weeks) or pH changes.[38,43,44]

3. Stability of aqueous fibril system as a function of pH

We set out to analyze the stability relative to pH changes for the different fractions that are present in the initial fibril solution. First the fibrils and non-aggregated material were separated to be able to analyze the behaviour of the pure fibrils, the non-aggregated peptides and the initial fibril solution separately. Three different methods were investigated on their ability to separate the two fractions. The first method uses centrifugal filtration and is based on the method that Bolder *et al.*

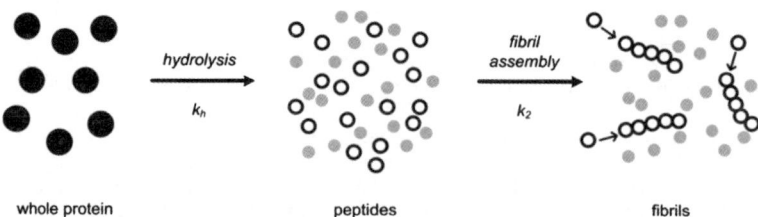

whole protein   peptides   fibrils

Fig. 1 Schematic representation of the fibril formation: first, the β-lg is hydrolysed into peptides with rate constant k_h, and second, some of these peptides assembly into fibrils with rate constant k_2. Reprinted from Kroes-Nijboer *et al.*, Langmuir, 27 (10), 5753–5761, (2011), with permission.

used to determine the conversion of protein into fibrils.[45] Next to this method also dialysis and ultracentrifugation were tested on their ability to separate the fibrils and the non-aggregated peptides. Subsequently, the behaviour of the initial fibril solution and both the pure fibrils and the non-aggregated material were analyzed within a pH range between 2 and 8.

3.1 Materials and methods

Fibril formation. β-Lg was obtained from Sigma (product no. L0130, lot. no. 095K7006). A stock solution (about 6 wt%) was made by dissolving the protein powder in HCl solution of pH 2. The pH of the protein solution was adjusted to pH 2 with 6 M HCl solution. Subsequently, this stock solution was filtered through a protein filter (FD 30/0.45 μm Ca-S from Schleicher & Schuell) to remove any traces of undissolved protein. The protein concentration of the stock solution was determined using an UV spectrophotometer (Cary 50 Bio, Varian) and a calibration curve of known β-lg concentrations at a wavelength of 278 nm. The stock solution was diluted to a protein concentration of 2 wt% with HCl solution of pH 2.

The β-lg solution was heated in small glass vials (20 ml) in a metal stirring and heating plate for 20 h at 80 °C. The protein solution was mildly stirred during heating.

Separation fibrils and non-aggregated peptides: centrifugal filtration. For the first method to separate the fibrils and the non-aggregated peptides, centrifugal filters were used. This separation method is based on the method developed by Bolder *et al.* to determine the conversion of protein into fibrils.[45] To separate the fibrils from the non-aggregated peptides, the fibril solution was diluted to a protein concentration of 0.2 wt% with HCl solution of pH 2. The solution was divided over 6 centrifugal filters (Amicon Ultra 100 K–15 Centrifugal Filters) and centrifuged at 3000 g and 15 °C for 30 min (Allegra X-22R Centrifuge, Beckman Coulter). After the centrifugation the filtrate was removed from the tubes and the retentate (containing the fibrils) was resuspended in ~10 ml of fresh pH 2 solution. Subsequently the samples were centrifuged again at 3000 g and 15 C for 30 min. In total 4 centrifugation steps were used. After the fourth centrifugation step the retentate was resuspended in 2 ml pH 2 solution instead of the original 15 ml.

Separation fibrils and non-aggregated peptides: dialysis. The second method used to separate the fibrils and the non-aggregated peptides was dialysis. To see whether it is an option to separate the fibrils and the non-aggregated peptides by dialysis, first it was investigated if the non-aggregated peptides would pass the dialysis tube. For this, the non-aggregated protein fraction (obtained by the ultracentrifugation method) was dialyzed against HCl solution of pH 2 overnight (MWCO 12–14 kDa). The protein concentration of the solution before and after dialysis was determined.

Separation fibrils and non-aggregated peptides: ultracentrifugation. The third method used to separate the fibrils and the non-aggregated protein was ultracentrifugation. To separate the fibrils from the non-aggregated protein, the fibril solution was diluted to a protein concentration of 1 wt% with HCl solution of pH 2. The solution was divided over 6 centrifuge tubes and centrifuged at 90 100 g and 15 C for 90 min. After the centrifugation the supernatant was immediately removed from the tubes and the pellets were resuspended in their original volume in a fresh pH 2 solution and stored overnight in a cold room (UCF 1). For 4 of the 6 tubes one extra washing step was performed by repeating the ultracentrifugation and resuspending step (UCF 2). For two of these 4 tubes, a third washing step was performed (UCF 3).

Protein concentration. The protein concentration of the various samples were determined using a UV spectrophotometer (Cary 50 Bio, Varian) using a calibration curve of known β-lg concentrations at a wavelength of 278 nm.

ThT fluorescence. To analyze the different fractions for fibrils after separation by ultracentrifugation, the various solutions were analyzed using a ThT assay. A ThT stock solution (3.0 mM) was made by dissolving 7.9 mg ThT in 8 ml phosphate buffer (10 mM phosphate, 150 mM NaCl at pH 7.0). This stock solution was filtered through a 0.2 μm filter (Schleicher & Schuell). The stock solution was diluted 50 times in a phosphate buffer (10 mM phosphate, 150 mM NaCl at pH 7.0) before use.

Aliquots of the solutions (48 μl) were mixed with 4 ml ThT solution and allowed to bind to the ThT for 1 min. The fluorescence of the samples was measured using a fluorescence spectrophotometer (Perkin Elmer LS 50 B). The excitation wavelength was set on 460 nm (slit width 4.0 nm) and the emission spectrum was recorded between 470 and 500 nm (slit width 2.5 nm) at a scanning speed of 200 nm min^{-1}. The fluorescence intensity peak was determined at 482 nm. The fluorescence intensity of the ThT solution itself was subtracted as a background. All samples were measured in duplicates.

Flow-induced birefringence. Flow-induced birefringence was used to analyze the length distribution of the fibrils that were present in the various fractions after ultracentrifugation. This method can be used to efficiently measure the length distribution of a large number of fibrils at the same time.[46-48] The decay curves of the flow-induced birefringence after the cessation of flow were measured with a strain-controlled ARES rheometer (Rheometrics Scientific) equipped with a modified optical analysis module.[49] From these decay curves the length distribution of the fibrils can be determined.[48]

Gel electrophoresis (SDS-PAGE). To check whether all non-aggregated peptides were removed from the fibril solution with the separation based on ultracentrifugation, SDS-PAGE was performed on the various fractions. For the gel electrophoresis a XCell *SureLock*™ Mini-Cell (Invitrogen Corporation, Carlsbad, California 92008, USA) was used. Samples were run on NuPAGE® Novex 4–12% Bis-Tris gels with NuPAGE® MES SDS running buffer (Invitrogen Corporation) under non-reducing conditions (no S-bonds are broken). The gels were stained with SimplyBlue™ SafeStain (Invitrogen Corporation).

Conversion of protein into fibril. The conversion of protein into fibrils was calculated using the various fractions obtained by centrifugal filtration. From the protein concentration of the various fractions and the weight of these fractions the conversion of protein into fibrils could be calculated as

$$C = \frac{1 - (F_1 + F_2 + F_3 + F_4)}{P} \times 100\%$$

where C is the conversion of protein in fibrils in wt%, F_1 is the weight of protein in filtrate 1 (g), F_2 is the weight of protein in filtrate 2 (g), *etc.*, and P is the weight of the initial amount of protein in the solution (g).

Visual appearance. For the visual observations, the initial fibril solution, the pure fibril solution and the solution containing the non-aggregated peptides were diluted to 0.1 wt% protein and brought to pH 5 and pH 8 using 0.1 M NaOH. The appearance of the various fractions at a protein concentration of 0.1 wt% and pH 2, 5 and 8 were analyzed visually and digitally recorded.

TEM. TEM pictures were taken from all fractions at pH 2, 5 and 8 (protein concentration 0.01 wt%). A droplet of the solution was put onto a carbon support film on a copper grid. The excess was removed after 15 s with a filter paper. Subsequently, a droplet of 2% uranyl acetate was put onto the grid and again removed after 15 s. Electron micrographs were taken using a JEOL electron microscope (JEM-1011, Tokyo, Japan) operating at 80 kV.

Mobility. For the electrophoretic mobility measurements, all fractions were diluted to 0.01 wt% protein with milliQ water. Samples at various pH were prepared by adding 0.1 M HCl or 0.1–0.01 M NaOH. The electrophoretic mobility of the different fractions at various pH was measured using a Zetasizer Nano (Malvern Instruments Ltd, Worcestershire, UK). Samples were measured in 5-fold.

Preventing aggregation. In analogy to the work of Jung *et al.* we investigated the addition of SDS to the various fractions to prevent clustering of the protein material in the solutions.[39] For this experiment the fibrils and the non-aggregated peptides were separated using ultracentrifugation. First a series of SDS solutions at pH 2 was made with SDS concentrations from 0 to 0.02 M. The initial fibril solution, the pure fibril solution and the solution containing the non-aggregated peptides were added so that the final protein concentration of the solutions was 0.005 wt%. The pH of all the samples were set at pH 3 using NaOH. The electrophoretic mobility of the samples were measured using a Zetasizer Nano (Malvern Instruments Ltd, Worcestershire, UK). Samples were measured in 5-fold.

Next, the solutions that contained enough SDS to prevent the fibrils and non-aggregated peptides from aggregation were adjusted to a range of pH values between 3 and 8 using NaOH. Also the electrophoretic mobility of these samples was measured.

3.2 Results and discussion

Separation of fibrils and non-aggregated peptides: centrifugal filtration. The first method that was used to separate the fibrils from the non-aggregated peptides was based on the method as introduced by Bolder *et al.* to determine the conversion of protein into fibrils.[46] In this method the fibrils and the non-aggregated peptides are separated using centrifugal filters. Four washing steps were performed to remove all the non-aggregated peptides from the fibril solution. No protein was found by UV spectroscopy measurements in the filtrate after the 4th washing step, indicating that the pure fibril solution was free from non-aggregated peptides after four washing steps. Besides, the fibrils do not pass the filters[40,46] and therefore no fibrils will be present in the filtrate.

Separation of fibrils and non-aggregated peptides: dialysis. The second method that was tested to separate the fibrils and the non-aggregated peptides was to dialyse the fibril solutions. To check whether the non-aggregated peptides would actually pass the dialysis tube, a prerequisite for this method, a solution containing the non-aggregated peptides (the supernatant obtained from the ultracentrifugation) was dialyzed against an HCl solution of pH 2. The protein concentration that was measured in the dialysis tube remained the same before and after dialysis, indicating that the non-aggregated peptides did not pass the dialysis tube and therefore making dialysis under the conditions used in this experiment unsuitable for the separation of the fibrils and the non-aggregated peptides. To be able to separate the non-aggregated peptides and the fibrils by dialysis, dialysis tubing with a higher MWCO than 12–14 kDa and longer dialysis times than ~12 h might be necessary. Jordens *et al.* dialysed a fibril solution against milliQ water of pH 2 for 7 days using dialysis tubing with a MWCO of 100 kDa to remove unreacted protein and low molecular weight residual peptides.[50] Since this method would become much more time consuming compared to the other separation methods, the separation based on dialysis was not further investigated.

Separation of fibrils and non-aggregated peptides: ultracentrifugation. The third method that was tested to separate the fibrils and the non-aggregated peptides was ultracentrifugation (~90 000 g for 90 min). After the first centrifugation step, the supernatant was removed and kept separate for further analysis. The transparent pellet that was obtained was resuspended in a fresh HCl solution of pH 2. These steps were repeated twice in an attempt to separate all the non-aggregated peptides from the fibrils. Protein concentration measurements showed that after three ultra-centrifugation steps, still protein was present in the supernatant. The protein in the supernatant can be present in the form non-aggregated peptides or fibrils that were not completely spun down. Therefore, a ThT fluorescence assay was performed to see in which fractions fibrils were present after the centrifugation steps (Fig. 2).

Fig. 2 shows that the supernatant after each centrifugation step still showed a fluorescence signal, indicating the presence of fibrils. It is estimated that in total about 60% of the fibrils remain in the supernatants. Flow-induced birefringence measurements confirmed that indeed fibrils were present in the supernatant (Fig. 3), showing that after three ultracentrifugation steps no solution could be obtained containing only the non-aggregated peptides.

Conversely, to check whether all the non-aggregated peptides were removed from the fibril solution, SDS PAGE was performed (Fig. 4) on the final resuspended pellet (P3). Fig. 4 shows that the final resuspended pellet (P3) indeed does not contain any detectable amount of small peptides, indicating that after three ultracentrifugation steps a pure fibril solution could be obtained.

From the ThT assay, the flow-induced birefringence and the SDS-PAGE it can be concluded that ultracentrifugation is a suitable method to remove all the non-aggregated material from the fibril solution to obtain a pure fibril solution. The drawback from this method is that it was not possible to obtain a solution that only contained non-aggregated peptides, since there were still fibrils present in the supernatant.

Comparing the separation using centrifugal filters with the other two methods, the separation using centrifugal filtration is more efficient in the sense that no fibrils are present in the non-aggregated peptide fraction and no non-aggregated peptides are present in the pure fibril solution. Therefore, the fractions obtained using centrifugal filtration were used to analyze the behaviour of the different fractions as a function of pH.

Behaviour of the various fractions as a function of pH: conversion of protein into fibrils. Based on the separation using the centrifugal filters three different solutions were now obtained:

(A) The initial fibril solution containing a mixture of fibrils and non-aggregated peptides.

(B) A solution containing only fibrils.

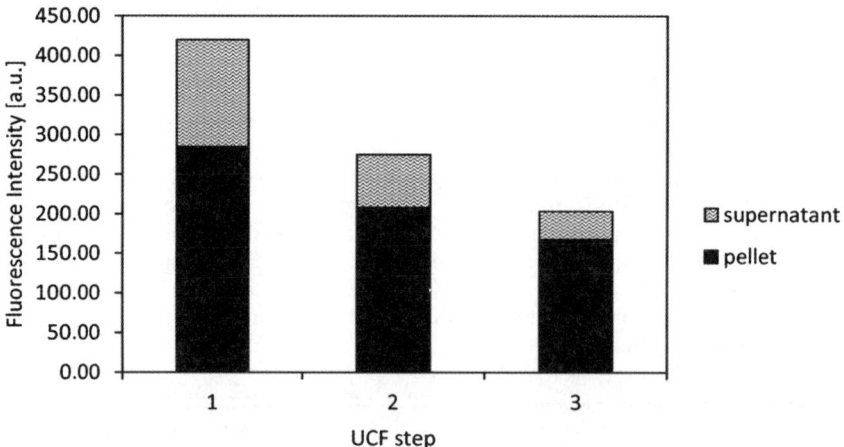

Fig. 2 ThT fluorescence intensities of the pellet and the supernatant after the 3 UCF steps.

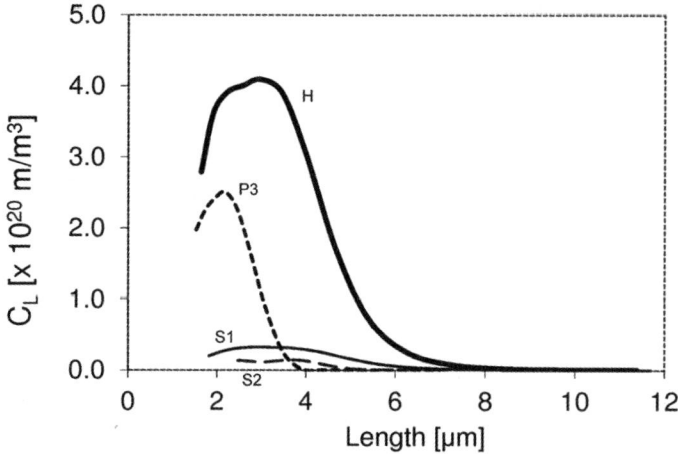

Fig. 3 Length distributions of the heated sample (H), pellet after 3 × UCF (P3), supernatant after 1 × UCF (S1), supernatant after 2 × UCF (S2).

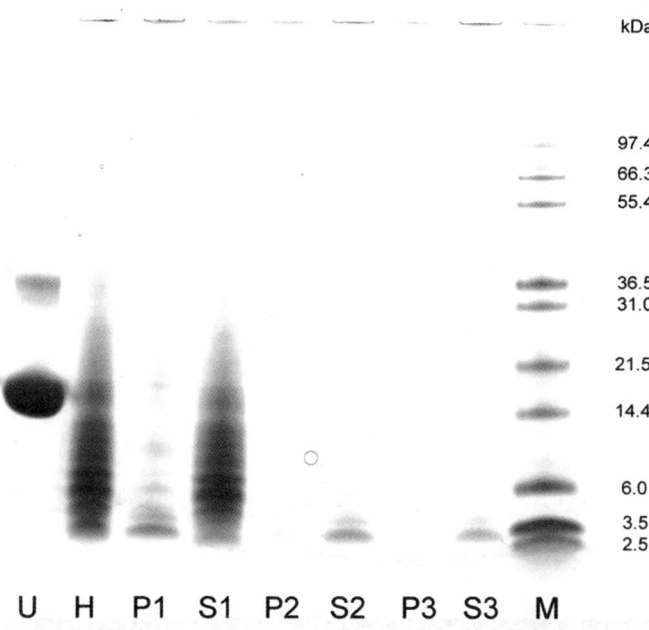

U   H   P1   S1   P2   S2   P3   S3   M

Fig. 4 SDS-PAGE. Unheated sample (U), heated sample (H), pellet after 1 × UCF (P1), supernatant after 1 × UCF (S1), pellet after 2 × UCF (P2), supernatant after 2 × UCF (S2), pellet after 3 × UCF (P3), supernatant after 3 × UCF (S3), marker (M).

(C) A solution containing only the non-aggregated peptides.

Using the weight and protein concentration that were obtained during the separation process, it was determined that about 46% (w/w) of the initial protein was present in the form of fibrils, whereas the rest of the protein material was present as small peptides. The behaviour of the three solutions A, B and C was analyzed as a function of pH. All solutions were transparent at pH 2.

Behaviour of the various fractions as a function of pH: visual appearance as a function of pH. The turbidity of the various fractions was visually inspected at pH 2, 5 and 8 and protein concentrations of 0.1 wt% (Fig. 5). It is found from Fig. 5 that solution A is transparent at pH 2, becomes turbid around pH 5, and becomes transparent again at pH 8. This behaviour was also observed by other researchers.[38,39] For solution C, a similar behaviour is seen, with the turbidity even more pronounced at pH 5. This more pronounced turbidity in solution C is possibly due to the fact that solution A and C have the same protein concentration, but in solution C all protein aqueous material exists of the non-aggregates peptides whereas in solution A also fibrils are present. In contrast to solution A and solution C, solution B stayed transparent at pH 5, only showing a slight haze. This is confirming that the peptides are the main cause for the turbidity around pH 5.

TEM as a function of pH. To analyze the samples on a microscopic scale in the solutions, TEM pictures were taken from the different fractions at pH 2, 5 and 8. The results are shown in Fig. 6. Linear fibrils are visible in the TEM pictures at pH 2 and pH 8 of solution A and solution B. At these pH values the fibrils are charged and the solutions were stable and transparent in line with the visual observations (Fig. 5). As expected in the TEM picture of solution C no fibrils were visible, confirming the efficiency of the separation.

In the TEM picture of solution A at pH 5, aggregates of fibrils were observed. Only spots of collapsed fibrils are visible and no elongated fibrils could be observed. In solution B, fibrils were visible at pH 5, but these fibrils looked smaller. In solution B, also aggregates of fibrils were visible at pH 5, however the aggregates look more open compared to the aggregates in solution A. These results are in agreement with the visual observations, where solution A was turbid, whereas solution B was still transparent at pH 5. The aggregates in solution B apparently are not optically dense enough to scatter light, resulting in a transparent solution. In these transparent pure fibril solutions at pH 5 some gel like particles were observed. The formation of these open aggregates might be caused by the separation method in which the fibrils have to be resuspended in the pH 2 solution after every washing step. However, it is more likely that the fibrils in solution B form aggregates because their net charge is close to zero at pH 5.

In solution C at pH 5 small aggregates are visible with a diameter of about 20 nm. At pH 8 also some small spherical aggregated were visible in this fraction, but these were much smaller than the ones at pH 5. This is in accordance with the visual observations that solution C is turbid at pH 5 and transparent at pH 8. The aggregates in solution C at pH 5 were large enough to scatter light in the visible wavelength, whereas the aggregates at pH 8 were too small to do this, resulting in a turbid solution at pH 5 and a transparent solution at pH 8.

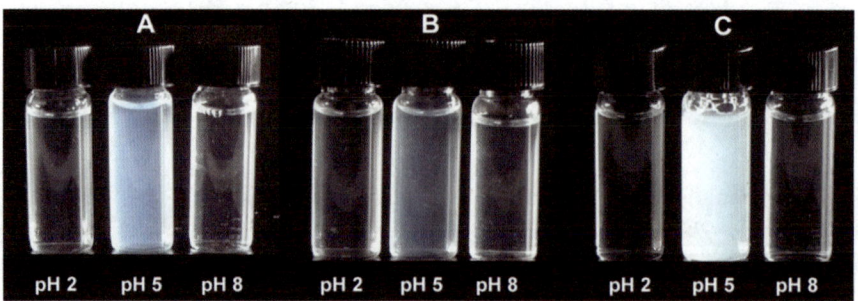

Fig. 5 Visual observations of solutions A, B and C at pH 2, 5 and 8, and 0.1 wt% protein.

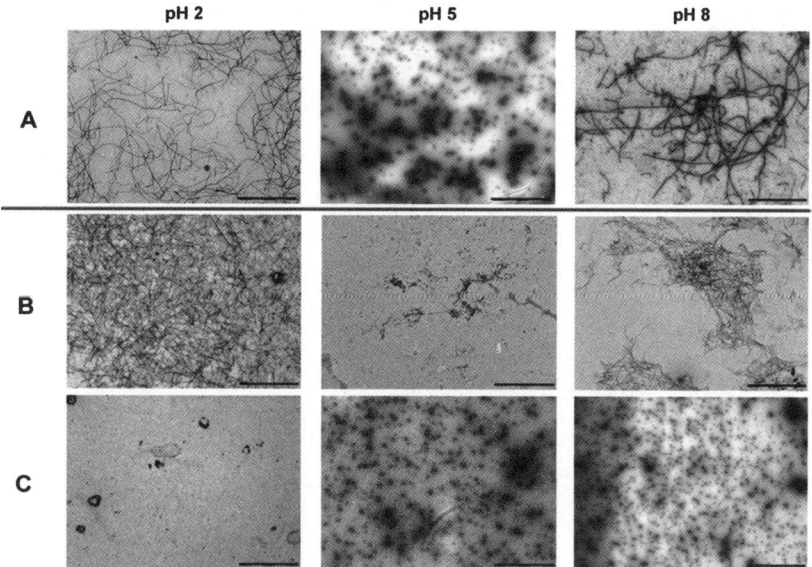

Fig. 6 TEM pictures of solutions A, B and C at pH 2, 5 and 8. Scale bars represent 2 μm.

Electrophoretic mobility as a function of pH. The electrophoretic mobility as a function of pH was measured for all solutions: A, B and C (Fig. 7).

Although the curves of the various fractions are not exactly the same over the whole pH range, the pH where the mobility of the samples is zero is for all fractions around pH 5, close to the iso-electric point of β-lg.[51] This is in agreement with the TEM pictures which showed that at pH 5 in all the fractions aggregates were present, indicating that both the non-aggregated peptides and the pure fibrils have no net charge at this pH. Furthermore, the curves of solution A and solution B are quite similar, compared to the curve of solution C. Note that more than 50% of the protein material in solution A is consisting of non-aggregated peptides.

Preventing of aggregation: coating with sodium dodecyl sulphate (SDS). It is shown that the turbidity is mainly coming from the non-aggregated peptides in the initial fibril solution and this can be avoided by removing the non-aggregated peptides. Yet it is also shown that the fibrils still form aggregates in the pure fibril solution around pH 5. One way to prevent the aggregation of the fibrils would be to coat the fibrils using anionic surfactants. This was investigated by Jung *et al.* who used sodium dodecyl sulfate (SDS) to coat the fibrils.[39] Two stages were distinguished in this study. In the first stage the SDS concentration is high enough to neutralize the charges of the protein, which is the case at SDS concentrations of 9×10^{-4} M, at 0.1 wt% protein. At this point a so-called SDS single layer is formed around the fibrils and precipitation of the coated fibrils is visible. At higher SDS concentrations, of 2 mM, a SDS double layer was formed around the fibrils and the coated fibrils could be completely redispersed in water.

In analogy to Jung *et al.*[39] we investigated various concentrations of SDS to our solutions A, B and C (Fig. 8). Note that the fibril solution used by Jung *et al.* can be compared with solution A in our research. By measuring the mobility as a function of the SDS concentration we observe a change in mobility for solution A, B and C from +1.5 μm cm V^{-1} s^{-1} at low SDS concentrations to a mobility of −2 μm cm V^{-1} s^{-1} at SDS concentrations of around 7 mM. No clear difference could be seen between the solutions A, B and C in their change in mobility as a function of SDS concentration. To investigate whether the coated fibrils and peptides would

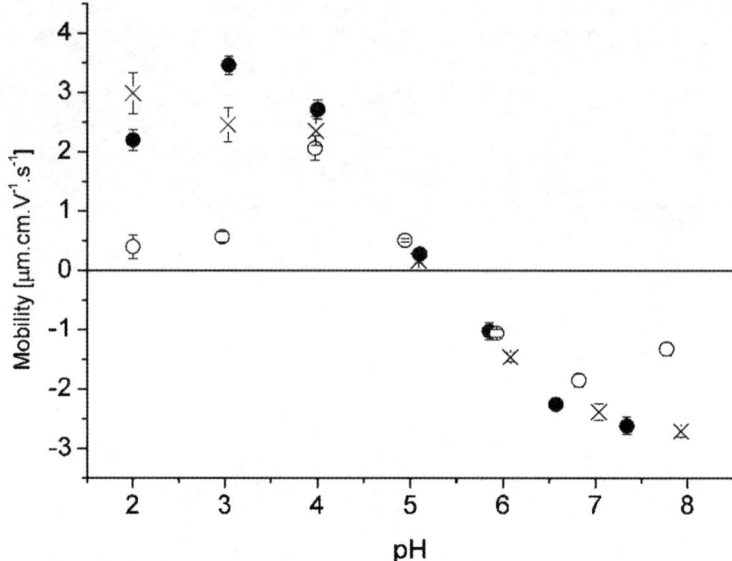

Fig. 7 Electrophoretic mobility of solution A (×), solution B (●) and solution C (○) as a function of pH. Protein concentrations are 0.01 wt%.

stay negatively charged over the whole pH range, solutions A, B and C containing 7 mM SDS were adjusted to a range of pH values between 3 and 8 using NaOH. SDS concentrations of 7 mM, only slightly below the critical micelle concentration of SDS (CMC ~8 mM)[39,52] could prevent the aggregation of both the pure fibrils and the non-aggregated peptides over the whole pH range at protein concentrations of 0.005 wt% (Fig. 9). All these samples were transparent over the whole pH range. With respect to food grade applications, alternative molecules should be tested that can coat and solubilize the fibrils and thereby prevent aggregation of the fibrils.

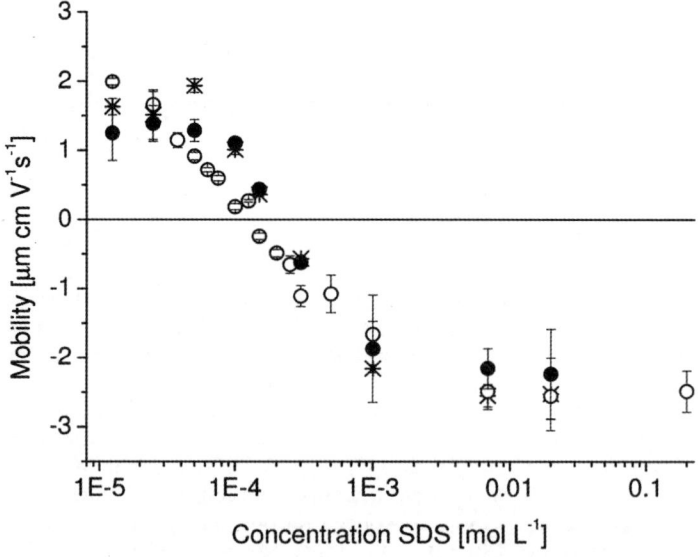

Fig. 8 Electrophoretic mobility of solution A (×), solution B (●) and solution C (○) as a function of SDS concentration at pH 3. Protein concentrations were 0.005 wt%.

We conclude for this section that the turbidity of a fibril solution at pH 5 is mainly caused by the aggregation of the non-aggregated peptides and can be prevented by removing the non-aggregated peptides from the fibril solution using centrifugal filters. The fibrils in the pure fibril solution still have the tendency to aggregate around pH 5, however these aggregates only scatter the light to a small extent. SDS is capable in solubilizing both the pure fibrils and the non-aggregated peptides.

4. Fibrillar structures in foods

The two main ingredients of foods are water and oil. Therefore it is interesting to investigate and compare the formation and properties of fibrillar structures in both solvents.

For the oil phase many organo-gelators are available to structure, but only a few of them are food-grade. As a result triglycerides rich in saturated fatty acids are commonly used to solidify the oil phase. However, saturated fatty acids are known to increase the risk of cardiovascular diseases by increasing the blood cholesterol levels. Several edible alternatives to structure oil have been proposed, amongst which mixtures of γ-oryzanol with β-sitosterol deserve special attention.[53–62] It was found that mixtures of γ-oryzanol with β-sitosterol are capable of forming firm, thermo-reversible organogels that are quite transparent up to high concentrations of struc-turants.[55] The transparency of these gels already hint in the direction of fibrillar aggregates that are too thin to scatter light. These aggregates were characterized by X-ray scattering and scanning electron microscopy (SEM) as hollow tubules that can be up to micrometers long, but have a thickness in the nanometer range.[62,63] For the rest of this section we will refer to these fibrils in oil as tubules.

Although the fibrils and tubules share a similar morphology, they differ in several aspects.

In regards to their assembly, in both cases a temperature dependent critical aggre-gation concentration can be determined below which no fibrils or tubules are formed. From these measurements it can be derived that the fibril assembly turns out to be entropy-driven and irreversible. This in contrast to the tubules where

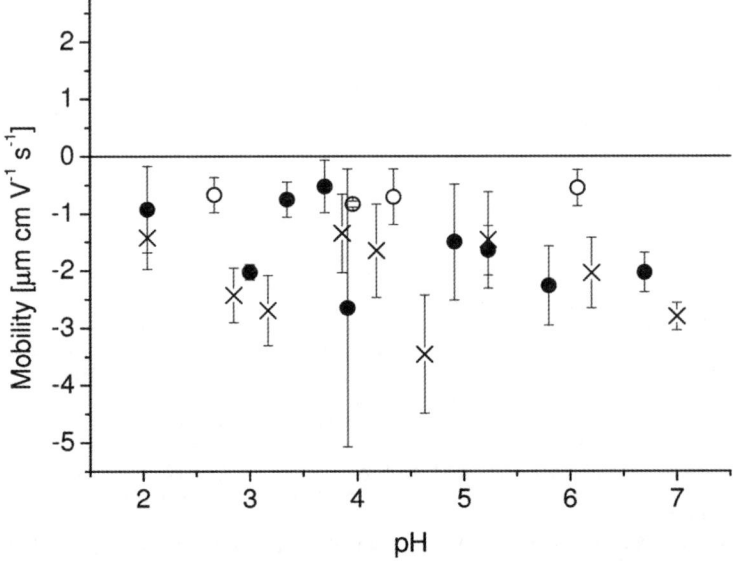

Fig. 9 Electrophoretic mobility of solution A (×), solution B (●) and solution C (○) as a function of pH, all containing 7 mM SDS. Protein concentrations were 0.005 wt%.

the assembly is found to be enthalpy-driven and reversible. The irreversibility and reversibility are referring to both dilution and temperature treatment. The typical concentrations (w/w) needed for fibril formation are typically an order of magnitude lower than for tubule formation. In both systems the aspect ratio is large, where the tubules are hollow cylinders with a diameter about $\sim$2–3 times larger than the diameter of the solid fibrils.[64]

Regarding the strength of gels, it is also interesting to note that, for the same length concentration, a gel formed by tubules is much stronger (G' $\sim$100 kPa)[55] than a gel formed by fibrils (G' $\sim$10 kPa).[30] This is possibly related to the difference in mechanical properties of tubules *versus* fibrils. From the theory of elasticity it follows that, for a sufficiently thick wall, the bending rigidity of a hollow cylinder is larger than the bending rigidity of a solid cylinder when the same amount of mass per unit length is used.

An important application field for foods are emulsions. When water is structured by fibrils the system is quite stable against the addition oil. In the case of oil structured by tubules, the addition of water quickly results in weakening or even breakup of the gel when a sufficient amount of water is added. This is caused because monohydrate β-sitosterol crystals form in the oil upon water addition, reducing the availability of β-sitosterol molecules for tubule formation. The kinetics of this process can be affected by affecting the transport of water from the water phase to the oil phase. This can be achieved by lowering the water activity using a solute (thermodynamic approach), or by reducing water solubility in the oil phase by lowering the polarity of that oil phase (kinetic approach).[63]

We conclude that formation and structure of the fibrils in both water and oil can be described well using generic principles of self-assembly. For the fibril formation it suffices to only consider the building blocks of the fibrils and neglect the non-incorporating peptides, while the presence of these peptides must be considered to understand the stability for the stability of the fibril system against pH change. From a short comparison between aqueous and oil based fibril systems it can be concluded that in such systems, dynamics, sensitivity to external conditions, and self-organisation play an important role. These aspects are common features to complex systems in general.

Acknowledgements

Financial support by the Dutch Microned consortium and by the Foundation Food and Nutrition Delta is gratefully acknowledged. In addition, the work of Mrs. E. Jansen in securely incorporating all the references and editing the entire text is sincerely appreciated.

References

1 N.-P. Humblet-Hua, L. M. C. Sagis, G. Scheltens, L. Yi and E. van der Linden, *Encapsulation systems based on proteins, polysaccharides, and protein-polysaccharide complexes*, Zurich, Switzerland, 2009.
2 L. M. C. Sagis, R. de Ruiter, F. J. R. Miranda, J. de Ruiter, K. Schroen, A. C. van Aelst, H. Kieft, R. M. Boom and E. van der Linden, *Langmuir*, 2008, **24**, 1608–1612.
3 F. J. Rossier-Miranda, K. Schroën and R. Boom, *Langmuir*, 2010, **26**, 19106–19113.
4 J.-M. Jung, D. Z. Gunes and R. Mezzenga, *Langmuir*, 2010, **26**, 15366–15375.
5 L. N. Arnaudov and R. de Vries, *Biophys. J.*, 2005, **88**, 515–526.
6 M. R. H. Krebs, D. K. Wilkins, E. W. Chung, M. C. Pitkeathly, A. K. Chamberlain, J. Zurdo, C. V. Robinson and C. M. Dobson, *J. Mol. Biol.*, 2000, **300**, 541–549.
7 C. Veerman, G. de Schiffart, L. M. C. Sagis and E. van der Linden, *Int. J. Biol. Macromol.*, 2003, **33**, 121–127.
8 M. Weijers, F. van de Velde, A. Stijnman, A. Pijpekamp and R. W. Visschers, *Food Hydrocolloids*, 2005, 1–14.
9 N.-P. Humblet-Hua, L. M. C. Sagis and E. van der Linden, *J. Agric. Food Chem.*, 2008, **56**, 11875–11882.

10 C. Akkermans, A. J. van der Goot, P. Venema, H. Gruppen, J. M. Vereijken, E. van der Linden and R. M. Boom, *J. Agric. Food Chem.*, 2007, **55**, 9877–9882.

11 C.-H. Tang and C.-S. Wang, *J. Agric. Food Chem.*, 2010, **58**, 11058–11066.

12 C.-H. Tang, Y.-H. Zhang, Q.-B. Wen and Q. Huang, *J. Agric. Food Chem.*, 2010, **58**, 8061–8068.

13 L. N. Arnoudov, R. de Vries, H. Ippel and C. P. M. van Mierlo, *Biomacromolecules*, 2003, **4**, 1614–1622.

14 P. Aymard, D. Durand and T. Nicolai, *Int. J. Biol. Macromol.*, 1996, **19**, 213–221.

15 P. Aymard, T. Nicolai and D. Durand, *Macromolecules*, 1999, **32**, 2542–2552.

16 D. Durand, J. Christophe Gimel and T. Nicolai, *Phys. A*, 2002, **304**, 253–265.

17 J. C. Gimel, D. Durand and T. Nicolai, *Macromolecules*, 1994, **27**, 583–589.

18 W. S. Gosal, A. H. Clark, P. D. A. Pudney and S. B. Ross-Murphy, *Langmuir*, 2002, **18**, 7174–7181.

19 W. S. Gosal and S. B. Ross-Murphy, *Curr. Opin. Colloid Interface Sci.*, 2000, **5**, 188–194.

20 S. Ikeda and V. J. Morris, *Biomacromolecules*, 2002, **3**, 382–389.

21 G. M. Kavanagh, A. H. Clark and S. B. Ross-Murphy, *Int. J. Biol. Macromol.*, 2000, **28**, 41–50.

22 M. Langton and A.-M. Hermansson, *Food Hydrocolloids*, 1992, **5**, 523–539.

23 C. Le Bon, T. Nicolai and D. Durand, *Macromolecules*, 1999, **32**, 6120–6127.

24 C. Le Bon, T. Nicolai and D. Durand, *Int. J. Food Sci. Technol.*, 1999, **34**, 451–465.

25 T. Lefevre and M. Subirade, *Biopolymers*, 2000, **54**, 578–586.

26 D. Renard and J. Lefebvre, *Int. J. Biol. Macromol.*, 1992, **14**, 287–291.

27 D. Renard, J. Lefebvre, M. C. A. Griffin and W. G. Griffin, *Int. J. Biol. Macromol.*, 1998, **22**, 41–49.

28 E. P. Schokker, *Int. Dairy J.*, 2000, **10**, 233.

29 C. Veerman, L. M. C. Sagis, J. Heck and E. van der Linden, *Int. J. Biol. Macromol.*, 2003, **31**, 139–146.

30 C. Veerman, H. Ruis, L. M. C. Sagis and E. van der Linden, *Biomacromolecules*, 2002, **3**, 869.

31 E. Gazit, *FEBS J.*, 2007, **274**, 317–322.

32 Y. D. Livney, *Curr. Opin. Colloid Interface Sci.*, 2010, **15**, 73–83.

33 S. Zhang, *Nat. Biotechnol.*, 2003, **21**(10), 1171–1178.

34 P. T. Lansbury and H. A. Lashuel, *Nature*, 2006, **443**, 774–779.

35 J. D. Sipe and A. S. Cohen, *J. Struct. Biol.*, 2000, **130**, 88–98.

36 M. Sunde, C. Blake, D. S. E. Frederic M. Richards and S. K. Peter, in *Advances in Protein Chemistry*, Academic Press, 1997, vol. 50, pp. 123–124, C111-C112, 125–159.

37 V. N. Uversky and A. L. Fink, *Biochim. Biophys. Acta, Proteins Proteomics*, 2004, **1698**, 131–153.

38 C. Akkermans, A. J. van der Goot, P. Venema, E. van der Linden and R. M. Boom, *Int. Dairy J.*, 2008, **18**, 1034–1042.

39 J.-M. Jung, G. Savin, M. Pouzot, C. Schmitt and R. Mezzenga, *Biomacromolecules*, 2008, **9**, 2477–2486.

40 C. Akkermans, P. Venema, A. J. van der Goot, H. Gruppen, E. J. Bakx, R. M. Boom and E. van der Linden, *Biomacromolecules*, 2008, **9**, 1474–1479.

41 E. Frare, P. Polverino de Laureto, J. Zurdo, C. M. Dobson and A. Fontana, *J. Mol. Biol.*, 2004, **340**, 1153–1165.

42 R. Mishra, K. Sorgjerd, S. Nystrom, A. Nordigarden, Y.-C. Yu and P. Hammarstrom, *J. Mol. Biol.*, 2007, **366**, 1029–1044.

43 C. Veerman, H. Baptist, L. M. C. Sagis and E. van der Linden, *J. Agric. Food Chem.*, 2003, **51**, 3880–3885.

44 S. G. Bolder, H. Hendrickx, L. M. C. Sagis and E. van der Linden, *Appl. Rheol.*, 2006, **16**, 258–264.

45 S. G. Bolder, *Int. Dairy J.*, 2007, **17**, 846.

46 S. G. Bolder, L. M. C. Sagis, P. Venema and E. van der Linden, *J. Agric. Food Chem.*, 2007, **55**, 5661–5669.

47 C. Akkermans, P. Venema, S. S. Rogers, A. J. van der Goot, R. M. Boom and E. van der Linden, *Food Biophys.*, 2006, 1.

48 S. S. Rogers, P. Venema, L. M. C. Sagis, E. van der Linden and A. M. Donald, *Macromolecules*, 2005, **38**, 2948–2958.

49 C. O. Klein, P. Venema, L. M. C. Sagis, D. V. Dusschoten, M. Wilhelm, H. W. Spiess, E. van der Linden, S. S. Rogers and A. M. Donald, *Appl. Rheol.*, 2007, **17**, 45210-45211–45210-45217.

50 S. Jordens, J. Adamcik, I. Amar-Yuli and R. Mezzenga, *Biomacromolecules*, 2011, **12**, 187–193.

51 M. Verheul, J. S. Pedersen, S. P. F. M. Roefs and K. G. de Kruif, *Biopolymers*, 1999, **49**, 11–20.

52 Y. Moroi, K. Motomura and R. Matuura, *J. Colloid Interface Sci.*, 1974, **46**, 111–117.
53 T. Laredo, S. Barbut and A. G. Marangoni, *Soft Matter*, 2011, **7**, 2734–2743.
54 M. Pernetti, K. F. van Malssen, E. Floter and A. Bot, *Curr. Opin. Colloid Interface Sci.*, 2007, **12**, 221–231.
55 A. Bot and W. G. M. Agterof, *J. Am. Oil Chem. Soc.*, 2006, **83**, 513–521.
56 J. Daniel and R. Rajasekharan, *J. Am. Oil Chem. Soc.*, 2003, **80**, 417–421.
57 L. Dassanayake, D. Kodali, S. Ueno and K. Sato, *J. Am. Oil Chem. Soc.*, 2009, **86**, 1163–1173.
58 M. A. Rogers, A. J. Wright and A. G. Marangoni, *Soft Matter*, 2009, **5**, 1594–1596.
59 H. M. Schaink, K. F. van Malssen, S. Morgado-Alves, D. J. E. Kalnin and E. van der Linden, *Food Res. Int.*, 2007, **40**, 1185–1193.
60 J. Toro-Vazquez, J. Morales-Rueda, E. Dibildox-Alvarado, M. Charó-Alonso, M. Alonzo-Macias and M. González-Chávez, *J. Am. Oil Chem. Soc.*, 2007, **84**, 989–1000.
61 H. Vaikousi, A. Lazaridou, C. G. Biliaderis and J. Zawistowski, *J. Agric. Food Chem.*, 2007, **55**, 1790–1798.
62 A. Bot, R. den Adel and E. Roijers, *J. Am. Oil Chem. Soc.*, 2008, **85**, 1127–1134.
63 H. Sawalha, R. den Adel, P. Venema, A. Bot, E. Flöter and E. van der Linden, *J. Agric. Food Chem.*, 2012, **60**(13), 3462–3470.
64 A. Bot, R. den Adel, E. Roijers and C. Regkos, *Food Biophys.*, 2009, **4**, 266–272.

Quinoa starch granules as stabilizing particles for production of Pickering emulsions

Marilyn Rayner,* Malin Sjöö, Anna Timgren and Petr Dejmek

Received 28th February 2012, Accepted 15th May 2012
DOI: 10.1039/c2fd20038d

Intact starch granules isolated from quinoa (*Chenopodium quinoa Willd.*) were used to stabilize emulsion drops in so-called Pickering emulsions. Miglyol 812 was used as dispersed phase and a phosphate buffer (pH7) with different salt (NaCl) concentrations was used as the continuous phase. The starch granules were hydrophobically modified to different degrees by octenyl succinic anhydride (OSA) or by dry heat treatment at 120 °C in order to study the effect on the resulting emulsion drop size. The degree of OSA-modification had a low to moderate impact on drop size. The highest level of modification (4.66%) showed the largest mean drop size, and lowest amount of free starch, which could be an effect of a higher degree of aggregation of the starch granules and, thereby, also the emulsion drops stabilized by them. The heat treated starch granules had a poor stabilizing ability and only the starch heated for the longest time (150 min at 120 °C) had a better emulsifying capacity than the un-modified native starch granules. The effect of salt concentration was rather limited. However, an increased concentration of salt slightly increased the mean drop size and the elastic modulus.

1 Introduction

Emulsions are heterogeneous mixtures of two immiscible phases where one phase is dispersed in the form of small drops into a continuous phase. Because the two phases are immiscible there exists an interfacial tension, which from a free energy perspective will drive the coalescence of drops to minimise their interfacial area. To prevent coalescence and maintain the stability of emulsions, amphiphilic molecules are usually included to lower the interfacial tension, to increase the steric hindrance or the electrostatic repulsion. However, the use of particles to stabilize emulsions has attracted increasing research interest over the past decade due to their distinctive characteristics and promising applications in a range of products.[1] Furthermore, recent advances in nano-technology have increased the availability and ability to design particles suitable for emulsion stabilization. However, the discovery of the ability of particles to stabilize emulsions, so called Pickering emulsions, is by no means new or limited to the field of nanotechnology. In the early 20th century Ramsden (1903)[2] and Pickering (1907)[3] independently observed that solid colloidal particles such as fine clays could stabilize the interface between two immiscible phases. Pickering emulsions are known to display long-term stability even without the addition of surfactant and are usually more stable against coalescence and Ostwald ripening compared to systems stabilized by surfactants.[1,4]

There are many types of particles reported in the literature used in generating Pickering type emulsions. They are often inorganic/synthetic particles such as silica,

Department of Food Technology, Engineering, and Nutrition, Lund University, P.O Box 124, SE 22100 Lund, Sweden. E-mail: marilyn.rayner@food.lth.se; Fax: +46 46 2224622; Tel: +46 46 222 0000

latex and clay, where food based stabilizing particles include fat crystals, globular proteins and aggregated hydrocolloids.[5] Recent works of particular interest in the food area include studies on insoluble flavonoid particles,[6] cellulose-ethyl cellulose complexes for stabilizing emulsions and foams,[7] freeze fractured starch granules and protein mixtures,[8,9] and chitin-nano crystals stabilized emulsions.[10] For comprehensive reviews on particle stabilized emulsions please refer to Binks (2002),[4] Aveyard *et al.* (2003),[1] Hunter *et al.* (2008),[11] and for food emulsion in particular see Dickinson (2006 and 2010).[12,13]

1.1 Particle stabilization

Food emulsions are generally stabilized by low-molecular weight emulsifiers such as surfactants, or higher molecular-weight bio polymers such as proteins and hydrocolloids.[14] In the case of surfactants, emulsion stability can be generally described by the DLVO theory, which is based on accounting for the long-range electrostatic repulsion and van der Waals attractive forces.[15] In systems stabilized by non-ionic surfactants and polymers, the film stability between two oil drops is usually explained by steric repulsion generated by the overlapping of hydrophilic heads of the surfactant/polymer molecules on the water side of the oil–water interface.[15] In the case of emulsions stabilized by solid particles, stabilization has been explained by a steric barrier created by the particles preventing contact between adjacent oil–water interfaces,[9] in addition to capillary forces, which appear as the menisci of the oil–water interface bend around the particles trapped in the emulsion films.[16]

The particle stabilization of emulsion droplets is possible due to partial dual wettability allowing for the spontaneous accumulation of particles at the oil–water interface. Finkel *et al.* in 1923 first described the correlation between the wettability of particles and their ability to stabilize emulsions.[17] More recent studies have also investigated particle wettability in relationship to emulsification performance.[18–23] Depending on the contact angle at the oil–particle–water interface, either an oil-in-water or water-in-oil droplet system will be favored, the point being that the side of the interface where the majority of the particle exists will likely be the continuous phase.

Emulsions stabilized by solid particles are generally more stable against coalescence compared to systems stabilized by surfactants.[1,4] The proposed explanation for this higher stability is that particles prevent interaction between the interfaces of neighboring drops by volume exclusion, and, once adsorbed, particles are strongly attached to the interface. Particles adsorbed at the oil–water interface create a physical barrier preventing contact between droplets. This is not fundamentally different from the steric barrier created by other emulsifiers such as proteins and hydrocolloids, however, in the case of Pickering emulsions, once particles absorb to the oil–water interface they are effectively trapped there, due to their large size and partial wettability.[1] This effect has been quantified by the energy of detachment, ΔG_{detach}, a function of interfacial tension, γ_{ow}, contact angle θ, and particle radius, R according to eqn (1) below[1]:

$$\Delta G_{detach} = \pi R^2 \gamma_{ow}(1 - |cos\theta|)^2 \tag{1}$$

If particles have favorable wetting conditions (*i.e.* not too close to zero or 180°) and are above a certain size (approximately 10 nm) their adsorption at the oil water interface is practically irreversible as the desorption energy per particle is several thousand kT, where k is the Boltzmann constant and T is the absolute temperature.[9] Under these conditions particles show irreversible adsorption, in contrast to low-molecular weight surfactant molecules, which exist in a dynamic equilibrium, rapidly adsorbing and desorbing from the interface on a short timescale.[24] The strong adsorption of particles at the interface can also explain their stability (even at large droplet sizes) over extended periods of time observed in particle stabilized emulsions.

1.2 Starch Pickering emulsions

Starch (including hydrophobically modified starch) is an accepted food ingredient and pharmaceutical excipient. Starch is used in many products as a food ingredient, where it contributes to the textural properties of these foods as a thickeners, colloidal stabilizers, as well as gelling, bulking, and water retention agents.[25] Starch granules are abundant, inexpensive, and can be isolated from a range of botanical sources. There is a large natural variation regarding size, shape, and composition of starch granules depending on the plant species they are obtained from. Native starch is not naturally hydrophobic, and, thereby, generally not particularly suitable to adsorb to the oil–water interface. The hydrophobicity of starch granules can be increased, for example, by chemical modification or by dry heating. A common way to chemically modify starches is by using octenyl succinic anhydride (OSA). OSA modified starch with a degree of modification less than 3% (E1450) is a well-established food ingredient with no specific limitations on its use. Dry heating can also be used to increase hydrophobicity by modifying the surface proteins of the starch granules to achieve a higher oil binding ability and affinity to the oil water interface.[26]

Previously, we have studied several varieties and types of starch granules,[27–29] however, in this study we have focused on different modifications of starch granules isolated from quinoa. Quinoa (*Chenopodium quinoa Willd.*) is a pseudo cereal native to the Andes, which grows in temperate climates and has been an important food crop in South America for the past 5000 years.[30] Quinoa starch granules were chosen as Pickering agents for several reasons; among them that quinoa does not contain gluten, and the granules are relatively small (0.5 to 3 μm in diameter) with a unimodal size distribution. Small size is of interest as it reduces the amount required (mg of starch per ml of oil) to stabilize a given emulsion droplet interface. The amount of starch required can be estimated by the theoretical maximum coverage, Γ_M [mg m^{-2}], using the following equation:

$$\Gamma_M = \rho_{sg}\frac{2}{3}d_{sg}\varphi \cdot 10^6 \qquad (2)$$

Where d_{sg} is the surface mean diameter of the starch granules, ρ_{sg} is the starch density (1550 kg m^{-3}) and φ is the packing density. The assumptions in this equation are that the starch particles are spherical and attached at the oil–water interface at a contact angle of 90° with an interfacial packing fraction $\varphi \approx 0.907$, *i.e.* hexagonal close packing.[9] By using the maximum surface coverage concept we can also roughly estimate the required starch mass per volume of oil to generate a starch granule stabilized oil-in-water emulsion of a given drop size.

$$C_{so} = 4\rho_{sg}\varphi\frac{d_{sg}}{d_{32}} \qquad (3)$$

Where C_{so} is the starch to oil ratio (mg ml^{-1}) and d_{32} is the surface mean diameter of the oil drops to be stabilized. The larger the granules the higher the required mass to cover an equal interfacial area. For example, barley starch granules with a surface mean diameter of 17 μm[29] need to have a starch to oil ratio almost 10 times higher than quinoa, which has granules with a surface mean diameter of 1.8 μm, to cover an equal surface of emulsion drops. Through the choice of the size of the stabilizing particles, the surface layer thickness can be manipulated, and the effective density of the emulsion droplets tailored to achieve positive, negative or close to neutral buoyancy.

In our previous studies, quinoa starch stabilized emulsions produced by high shear homogenization had droplet sizes of 9 to 70 μm, depending on the starch-to-oil ratio, which ranged from 36 to 3600 mg ml^{-1} of oil.[29] In Fig. 1 a typical droplet size distribution of a quinoa starch stabilized emulsion is shown, as well as images of a starch covered droplet and individual starch granules. In general, droplet size

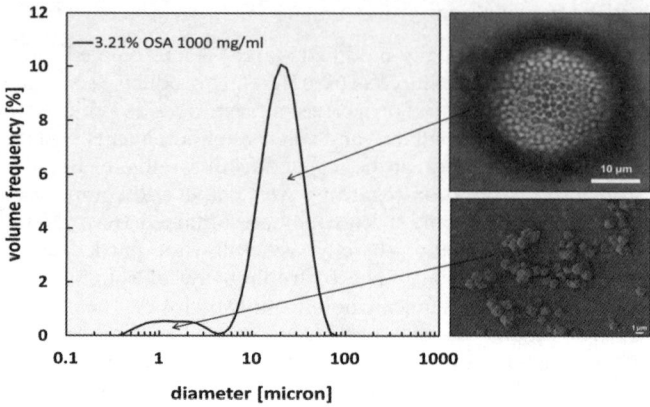

Fig. 1 Particle size distribution of quinoa starch stabilized Pickering emulsion (left), where the main peak represents the starch granule stabilized emulsion drops and the small peak free starch granules. Image of a starch granule covered drop (top right) and SEM image of quinoa starch granules (bottom right).

decreased with increasing starch to oil ratio, but was unaffected by the oil phase volume over a range of 5 to 33% oil (v/v).[29] During 8 weeks of storage, the emulsion drops were stable to coalescence with no measurable increase in droplet size and the emulsion index remained unchanged or even slightly increased.[27] The samples with a starch to oil ratio higher than 214 mg of starch per ml produced emulsions with a density higher than the continuous phase, and the starch covered droplets were sinking rather than creaming. As the starch concentration increased the resulting drop size decreased, and the amount of starch attached to the surface of the drops was higher in relation to the volume of the oil in the drop. In this way the overall drop could become heavier than the continuous phase. Thus, by choosing the granule size and amount of starch added we can control the density of the resulting emulsion drops.

The objectives of the present work were to study the effect of different levels and types of hydrophobic treatments on emulsion properties (droplet size distributions and rheological properties). Two types of hydrophobic treatment were considered; dry heating of quinoa starch granules at 120 °C for different lengths of time and chemical modification with OSA at different levels. The effects of electrolyte concentration and starch to oil ratio were also considered.

2 Material and methods

2.1 Isolation of quinoa starch granules

In this study, starch isolated from quinoa grains were used (Biofood AB, Stockholm, Sweden). Quinoa seeds were soaked in distilled water for 24 h in a cold storage room (4 °C) before the seeds were milled with distilled water in a blender (Philips HR 7625, The Netherlands) into a smooth pulp, which was filtered through a cheese cloth and rinsed with distilled water. The starch in the permeated liquid was allowed to settle and the supernatant was removed. The settled layer was re-dispersed in distilled water and centrifuged at 3000 × g for 10 min. The water and the grey top layer, consisting of proteins and seed-coat fragments, were removed. Fresh distilled water was added to the starch, which, after settling and removal of water, was dried in a vacuum-dryer at 20 °C for 4 d.

Proteins and any fibre residues in the dried starch were removed by washing the starch twice with 0.3% NaOH-solution, once with distilled water and once with citric acid (pH 4.5). Between each washing step the supernatant was removed by centrifugation at 3000 × g for 10 min. Finally, the starch was washed twice with distilled water and centrifuged. The starch was spread on stainless steel trays and dried at room temperature for at least 48 h. Before use the starch granules were dis-aggregated into a fine powder by grinding with mortar and pestle.

2.2 OSA-Modification of starch

The isolated starch granules were OSA-modified with n-octenyl succinic anhydride to three different degrees (1.95%, 3.21% and 4.66%). Starch powder equivalent to 50 g of dry weight was thoroughly suspended in water using a stainless-steel propeller and the pH was adjusted to 7.6 using 25% HCl and/or 1 M NaOH. OSA (3%, 6% or 10% based on the dry weight of the starch) was added in four equal portions at intervals of 15 min. During the modification the pH was maintained at 7.6 using automatic titration of 1 M NaOH. The modification was considered to be completed when the pH was constant for at least 15 min. The mixture of starch and water was centrifuged at 3000 × g for 10 min and the supernatant was removed. The starch was re-suspended and centrifuged, twice with distilled water and once with citric acid (pH 4.5). The OSA-modified and washed starch granules was spread on stainless steel trays and dried at room temperature for at least 48 h.

The degree of OSA substitution was determined using a titration method. The analyses were performed in duplicate for both the OSA-modified starch and the control starch, which was a non-modified sample of the same origin batch as the OSA-modified starch. Starch (2.5 g based on dry substance) was added to a 50 ml beaker and wetted with some drops of ethanol before 25 ml of 0.1 M HCL was added. The mixture was stirred with a magnetic stirrer for 30 min before centrifugation at 3000 × g for 10 min. The starch was washed once with 25 ml of ethanol and twice with distilled water and centrifuged between each washing step. The starch was added to a 500 ml beaker and mixed with 150 ml of distilled water and heated in a water bath at 95 °C for 10 min before being cooled to 25 °C. The gelatinized mixture was titrated with 0.1 M NaOH until the pH reached 8.3. The percentage of carboxyl groups from OSA on the granules was calculated by:

$$\% \, OSA = \frac{(V_{sample} - V_{control}) \cdot M \cdot 210}{W} \cdot 100\% \tag{4}$$

Where V is the volume (ml) of NaOH required for the sample and the control titration, respectively, M is the molarity of NaOH (0.1 M), W is the dry weight of the starch (2.5 mg) and 210 is the molecular weight of octenyl succinate group.

2.3 Thermal modification of starch

The isolated starch granules were also heat-treated in order to hydrophobically modify the surface proteins of the starch.[26] Dry starch was placed in an open petri dish in a 1–2 mm thick layer and heated at 120 °C for five different durations (30, 60, 90, 120, and 150 min). The heat-treated starch was cooled and left at room temperature for several hours before use.

2.4 Preparation of emulsions and starch granule dispersions

Varying amounts of quinoa starch granules, buffer solution, and oil were weighed according to Table 1 and put into test tubes, and stirred with a vortex mixer (VM20, Chiltern Scientific Instrumentation Ltd, UK) before being emulsified using a high shear Ystral mixer (D-79 282; Ystral Gmbh, Ballrechten-Dottingen, Germany) at 22000 rpm for 30 s. The dispersed phase used was the medium-chain triglyceride oil Miglyol 812 (Sasol AG, Witten, Germany, density 945 kg m^{-3}) and the continuous phase was a 5 mM phosphate buffer with pH 7, with varying levels of NaCl (0, 0.2, 0.4, 1 and 2 M). Number of replicates are indicated in Tables 2 and 3.

Table 1 Emulsions' composition

mg starch ml^{-1} oil	Miglyol (g)	Buffer (g)	Starch (g)
214	0.40	5.63	0.089
400	0.40	5.63	0.168
530[a]	2.33	3.50	1.17
800	0.40	5.63	0.336
1000	0.40	5.63	0.420
N/A[b]	0	5.63	0.089

[a] for rheology measurements. [b] starch granule dispersions.

2.5 Characterisation of emulsions

2.5.1 Light scattering. A laser diffraction particle size analyser (Mastersizer 2000 Ver.5.60, Malvern, Worcestershire UK) was used in order to determine the particle size distribution of the starch granules and the starch granule stabilized emulsion oil drops. The sample was added to the flow system containing milliQ-water and was pumped through the optical chamber at a pump velocity of 2000 rpm. The refractive index (RI) of the sample was set to 1.54 (starch),[31] the RI of the continuous phase was set to 1.33 (water) and the obscuration was between 10 and 20%.

2.5.2 Microscopy. The starch granule stabilized emulsions were diluted 5 times with the continuous phase and then samples were placed in a VitroCom 100 micron square channel (CMS Ltd., Ilkley, UK) using a 100× objective (Plan, Olympus, Tokyo, Japan) or on a microscopic slide without cover glass using lower magnifications (5×, 10×, 20× and 50×, LMPlanFL, Olympus, Tokyo, Japan). The light was transmitted using a polarization filter (U-ANT, Olympus, Tokyo, Japan) and a color tint plate (U-TP530, Olympus, Tokyo, Japan). Microscopy images of the emulsions were obtained using Olympus BX50 (Tokyo, Japan) and a digital camera (DFK 41AF02, The Imaging Source, Bremen Germany) 1 d after emulsification.

2.5.3 Rheology. Rheological measurements were performed on the starch granule stabilized emulsions with a rheometer (Malvern Kinexus, Worcestershire, UK) 24 h after emulsion preparation. The characteristics of the emulsions were analyzed at 25 ± 0.1 °C using a serrated plate–plate geometry (upper plate 40 mm diameter, lower plate 65 mm diameter, gap height 1.0 mm). All rheology experiments were completed in duplicate. Oscillatory measurements were performed in order to determine the linear viscoelastic region of the sample (amplitude sweep). The phase angles, shear viscosity (η, Pa s), storage (G', Pa) and loss (G'', Pa) moduli were investigated. The oscillatory test was performed in the shear stress range of 0.001–1000 Pa at a frequency of 1 Hz.

3 Results and discussion

3.1 Effect of hydrophobic modification on starch granules

Particle size distributions of native and hydrophobically modified quinoa starch granules are shown in Fig. 2. Before measurement by light scattering, the various starches were dispersed in buffer solution and sheared in the same manner as the emulsions. The mode (peak of the d_{43} distribution shown in Fig. 2 left) was very similar for the native and OSA modified starch granules (1.65 to 1.74 μm) but was slightly larger for the heat treated starch granules (2.22 μm). The main difference among the starch granules in the measured particle size distribution is seen in the magnified view of Fig. 2 (right). Here, we observe considerable differences in the

This journal is © The Royal Society of Chemistry 2012

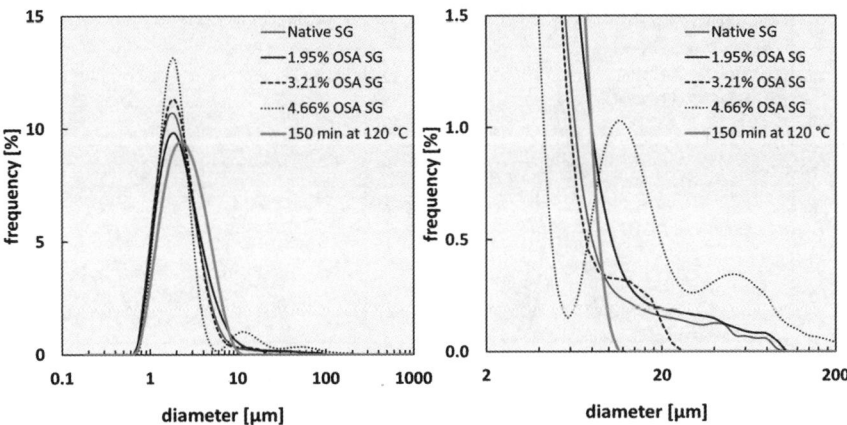

Fig. 2 Particle size distributions (d_{43}) of quinoa starch granules with varying degrees of OSA or heat treatment (left). The granules were dispersed in buffer solution and sheared in the same manner that emulsions were made with the exception that no oil was added. The magnified area of the same plot (right) shows differences in the degree of granule aggregation.

shoulder of the distribution. The particle size distribution of the 4.66% OSA modified granules have a secondary peak around 11 to 13 μm and a third one around 50 μm. This range in individual granule size is not observed in SEM images (for example Fig. 1, bottom right) and in our previous studies where we have imaged native, heat treated and OSA modified starch granules.[28] As the individual granules are not larger, the particle size distributions indicate a higher degree of aggregation of hydrophobically modified granules.

The tendency for granules to aggregate can be seen in the particle size data in Table 3. At 4.66% OSA modification the measured mean granule diameter (d_{43}) was significantly larger than for the 3.21% OSA, heat treated and the native starch granules. We suspect that here there is a higher degree of aggregation between starch granules and, in turn, the resulting emulsion droplets stabilized by them, as we increase the OSA level beyond what is necessary for adequate adsorption at the oil–water interface (discussed in section 3.2). If the particles are adsorbing more readily to the hydrophobic drop surface due to an increase in OSA% (or heating), then they will also be adsorbing at a higher extent to each other, resulting in a weakly aggregated structure.

3.2 Effect of hydrophobic modification on emulsion droplet size

To illustrate the varying effect of different hydrophobic treatments on the droplet size, the native quinoa starch, the 3 different OSA levels and 3 of the heat treatments are plotted at the concentration of 214 mg starch ml^{-1} oil in Fig. 3. OSA treatments were considerably more effective than heat treatments, and only after 150 min at 120 °C did the heat treated starch have an emulsifying capacity better than the native starch granules, which unexpectedly did show some ability to stabilize emulsions even without any modification. The overall particle size data of the starch granule stabilized emulsions is summarized in Table 2.

The measured drop size of the starch granule stabilized emulsions decreased with increased starch concentration for all OSA treatments, as seen in Fig. 4, which is in line with our earlier results. It has been previously shown that the average drop size of emulsions stabilized by solid particles decreases with increasing particle concentration (indicated in eqn (3)) as more particles are available to stabilize smaller drops, having a higher overall interfacial area in need of stabilization.[19,20,32] However, each system has probably a limiting drop size, which

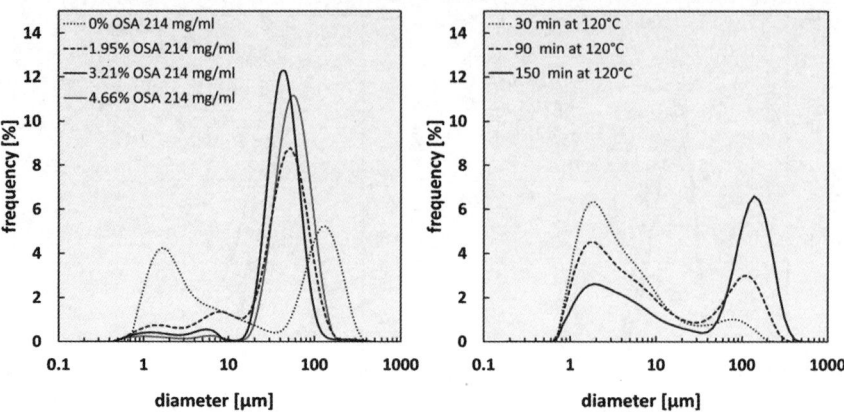

Fig. 3 Particle size distributions (d_{43}) of the quinoa starch granule stabilized Pickering emulsions for native starch (0% OSA) and 3 different levels of OSA modification (1.95%, 3.21%, and 4.66%, left) and 3 different heat treatment times (30, 90, and 150 min, right) at 120 °C. All are at the ratio of 214 mg starch ml^{-1} oil.

depends on the physical and mechanical properties of the system (*i.e.* the size of the particles and the emulsification method) and when this drop size is reached any excess of particles will be in the continuous phase.[33] We have previously observed this limit as a levelling out of the drop size above 1000 mg ml^{-1} oil[29] (and thus chose this to be the highest used in this study). We believe this is due to the limits of the emulsification method (high shear Ystral mixer), as we do not generate droplets smaller than approximately 10 μm, even at starch granule concentrations above 3000 mg ml^{-1} oil.[29] However, when performing a similar experiment using a low molecular weight emulsifier (Tween 20, 2 g L^{-1}) instead of starch granules, we obtain a somewhat smaller mean droplet size (d_{43}) of 8.7 μm. This can be explained by some of the features of the particles that are different from smaller emulsifiers, like proteins and surfactants. Specifically, they are much larger and have a non-negligible thickness at the interface in relation to the size of the emulsion drops they are stabilizing, they have slower adsorption kinetics (as there is a larger amount of material required to cover the interface), a potentially higher barrier to particle adsorption, and very high desorption energy.[34] A consequence of the slower adsorption kinetics is a shift in the balance between stabilization and re-coalescence during droplet–droplet and droplet–particle collisions during homogenisation. Even though the particles are essentially irreversibly adsorbed on the oil–water interface once there, the adsorption barrier has the effect of reducing the probability of particle attachment to the interface during homogenisation and particle adsorption layers may remain incomplete.[34] To overcome this barrier, higher hydrodynamic forces acting on the particle pushing it towards the oil–water interface are required. These hydrodynamic conditions are determined by the intensity of the homogenisation device (refer to Tcholakova *et al.* (2008)[34] for a more detailed discussion of hydrodynamics in relationship to particle adsorption kinetics).

In all the particle size distributions in Fig. 4 we can observe to varying degrees some non-adsorbed or free starch in the system as a second peak at around 1 to 2 μm in the particle size distributions. As expected, OSA treatments significantly improve the emulsifying capacity of the quinoa starch granules compared to the non-treated native (0% OSA). While the middle two OSA levels (1.95% and 3.21%) yielded smaller emulsion drops, *i.e.* volume mean diameter (d_{43}) and peak of the d_{43} distributions, than the highest OSA level (4.66%).

Table 2 Particle size data summary of starch granule stabilized emulsions with varying starch to oil ratios and hydrophobic treatments

Sample name	$d_{[4,3]}$ (μm)	St. dev.	Mode (μm)	St. dev.	$d_{[3,2]}$ (μm)	$d(0.1)$ (μm)	$d(0.5)$ (μm)	$d(0.9)$ (μm)
0% OSA 214 mg ml⁻¹ $n = 9$	59.6	9.7	121.4	11.4	3.7	1.2	12.3	165.5
0% OSA 400 mg ml⁻¹ $n = 6$	19.6	3.7	1.6	0.023	2.5	1.1	3.1	73.1
0% OSA 800 mg ml⁻¹ $n = 9$	36.4	5.2	1.6	0.015	2.9	1.1	4.1	124.0
0% OSA 1000 mg ml⁻¹ $n = 9$	29.1	3.9	1.6	0.007	2.7	1.1	3.5	106.0
1.95% OSA 214 mg ml⁻¹ $n = 9$	43.3	1.8	47.9	0.9	10.0	4.6	40.2	82.5
1.95% OSA 400 mg ml⁻¹ $n = 9$	27.4	0.57	30.0	1.0	8.3	4.0	25.3	51.1
1.95% OSA 800 mg ml⁻¹ $n = 9$	19.2	0.31	19.6	0.42	7.6	5.8	17.5	35.1
1.95% OSA 1000 mg ml⁻¹ $n = 9$	17.1	1.0	17.1	1.1	7.0	5.1	15.5	31.2
3.21% OSA 214 mg ml⁻¹ $n = 9$	42.0	3.6	40.1	1.5	13.7	19.0	38.8	67.5
3.21% OSA 400 mg ml⁻¹ $n = 9$	27.9	0.5	27.5	0.4	10.4	14.4	26.7	44.7
3.21% OSA 800 mg ml⁻¹ $n = 9$	24.7	1.0	24.0	0.8	9.4	10.9	22.9	41.9
3.21% OSA 1000 mg ml⁻¹ $n = 9$	19.8	0.6	19.4	0.5	8.2	7.7	18.2	34.6
4.66% OSA 214 mg ml⁻¹ $n = 9$	54.6	3.1	52.5	2.8	19.3	25.5	50.6	91.3
4.66% OSA 400mg ml⁻¹ $n = 9$	52.1	3.2	50.3	3.1	20.3	24.7	48.3	86.3
4.66% OSA 800 mg ml⁻¹ $n = 9$	37.0	3.1	36.2	3.0	14.9	17.4	34.6	61.4
4.66% OSA 1000 mg ml⁻¹ $n = 9$	27.1	2.5	26.8	2.4	11.6	12.4	25.5	45.2
Heat 30 min 214 mg ml⁻¹ $n = 3$	11.6	0.38	1.7	0.023	2.4	1.1	2.9	30.3
Heat 60 min 214 mg ml⁻¹ $n = 3$	19.6	0.90	1.6	0.011	2.5	1.1	3.1	73.4
Heat 90 min 214 mg ml⁻¹ $n = 3$	34.5	2.3	1.7	0.018	3.2	1.2	5.2	118.4
Heat 120 min 214 mg ml⁻¹ $n = 3$	37.3	6.3	1.7	0.025	3.1	1.2	4.6	127.6
Heat 150 min 214 mg ml⁻¹ $n = 3$	80.6	10.8	131.7	20.4	5.3	1.6	71.8	193.0

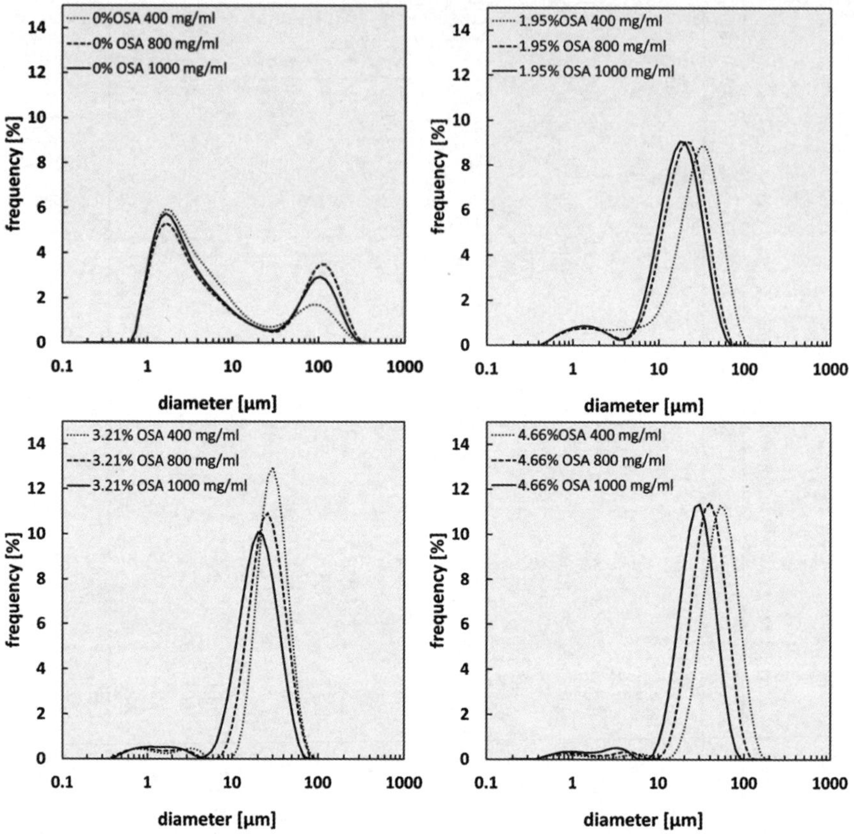

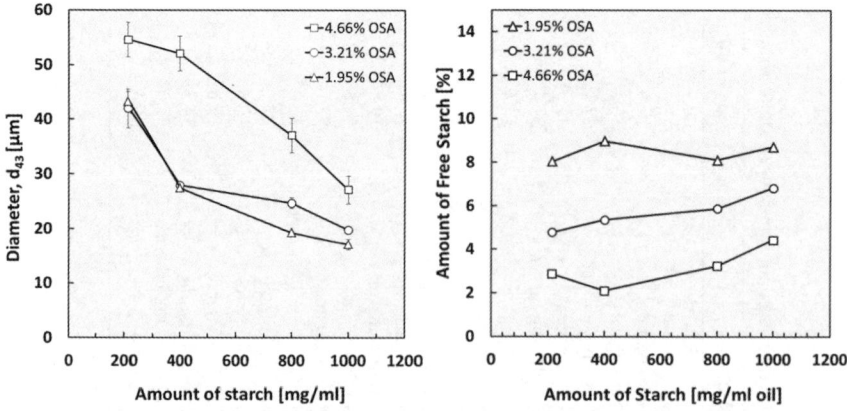

Fig. 4 Particle size distributions (d_{43}) of quinoa starch granule stabilized Pickering emulsions for native starch (0%, top left) and 3 different levels of OSA modification (1.95%, top right, 3.21%, bottom left, and 4.66%, bottom right) at 3 different starch concentrations (400, 800, and 1000 mg ml^{-1} oil), continuous phase 5mM phosphate buffer pH7.

Fig. 5 Volume mean diameter of quinoa starch granule stabilized Pickering emulsions plotted as a function of amount of starch expressed as mg starch ml^{-1} oil (left). Amount of free starch estimated from particle size distributions (right).

3.3 Effect of OSA level and starch concentration on fraction of non-adsorbed starch

The amount of non-adsorbed (or free starch) in the system was estimated from the cumulative particle size distributions. We set a size limit of twice the mean starch granule diameter, where any particle this size or smaller is interpreted to be starch only. The result of this analysis for the 1.95%, 3.21% and 4.66% OSA modifications at starch concentrations of 214–1000 mg ml^{-1} oil is plotted in Fig. 5, right and the corresponding drop radius of the emulsions in Fig. 5, left. The amount of free starch decreases with an increasing level of OSA modification. There is also a small trend towards increasing the amount of free starch at the highest concentration, which agrees with our previous work where we see that drop sizes do not decrease proportionally with starch concentrations higher than about 1000 mg ml^{-1}[29] and the majority of additional starch beyond this point seems to exist free in the emulsion. It was somewhat unexpected that the highest OSA concentration (4.66%) had both a larger droplet size and a lower fraction of free non-adsorbed starch. Intuitively, we would expect that if there are larger drops then there is less interfacial area to cover, and thus a lower potential area for starch to adsorb to. So, at the same starch to oil ratio, we expect more free starch, not less.

To consider this aspect further we have also estimated the apparent surface coverage of the emulsions based on the known dispersed phase composition (starch to oil ratio) and the measured droplet size distributions. The surface coverage is proportional to the interfacial packing fraction, which in eqn (2) and eqn (3) is assumed to be $\varphi \approx 0.9$, *i.e.* the highest packing ratio a single layer of spherical particles can achieve. However, in practice emulsions may have starch layers that are less tightly packed, or exist as a layer of more than one starch granule thick, or they may have clusters of starch adsorbed as satellites to the main drops. The first case would correspond to $\varphi < 0.9$, whereas in the second and third cases $\varphi > 0.9$. Sometimes φ is assumed to be equal to a set value such as that of hexagonal tight packing (*i.e.* 0.9), but in reality this is not always the case. Pickering emulsions have been shown to be effectively stabilized even by sparsely distributed particles. For example, in works by Vignati and Binks, emulsions were formed using silica (0.5–0.8 μm)[35] and spores particles ($\sim$25 μm)[36] despite the fact that the particles were highly unevenly distributed at the surface of the drops.

To make an estimation of the surface coverage, the particle size distributions were analysed where the relative amount of starch and oil in each bin size of the particle size distribution was considered. The volume frequency distribution in the results file from light scattering measurements (Mastersizer) is made up of a series of discrete bins, each representing a size interval that a fraction of the total volume of the particles measured falls in. If the bins were relatively small the error of taking the midpoint of the bin to represent the size of all the particles in the bin's size interval is acceptable, and we made the assumption that in any given bin the size of all the particles in that bin are the same size and $d_{\text{bin}} = d_{43} = d_{32} = d$. Using the definition of the specific surface area, S, of an emulsion, where Φ is the volume fraction of the disperse phase, and d_{32} is the surface mean diameter:

$$S \equiv \frac{6\Phi}{d_{32}} \quad (5)$$

In our case of starch stabilized Pickering emulsions we had both the surface area of the free starch granules, S_{fsg}, of diameter d_{sg}, and starch particle stabilized oil drops, S_{drops}, of diameter d_{drops}, contributing to the total surface area of the emulsion, S_{tot}.

$$S_{\text{tot}} = S_{\text{drops}} + S_{\text{fsg}} \quad (6)$$

$$\frac{\Phi_{tot}}{d_{32}^{tot}} = \frac{\Phi_{drops}}{d_{drops}} + \frac{\Phi_{fsg}}{d_{sg}} \qquad (7)$$

The dispersed phase volume fraction of the drops, Φ_{drops}, consists of oil, Φ_{oil}, and adsorbed starch granules, Φ_{asg}. The volume fraction of the free starch is Φ_{fsg} and is determined directly from the volume fraction in the bins smaller than $2 \times d_{sg}$, *i.e.* if the bin has a size smaller or equal to twice the diameter of the starch granule, then the total volume fraction of that bin is assigned to be just free starch. The specific surface area of the entire emulsion is the combined contribution of the surface of all bins.

$$\frac{\Phi_{tot}}{d_{32}^{tot}} = \sum_{bins>2d_{sg}} \frac{\Phi_{bin}}{d_{bin}} + \sum_{bins\leq2} \frac{\Phi_{fsg}}{d_{sg}} \qquad (8)$$

The amount of adsorbed starch and volume of oil in each bin larger than $2 \times d_{sg}$ was estimated by first calculating the oil volume of a starch covered drop with the diameter of the bin size, by subtracting the diameter of a starch mono layer.

$$V_{oil}^{bin} = \frac{\pi}{6}(d_{bin} - d_{sg})^3 \qquad (9)$$

The expected volume of starch covering an oil drop of this size was calculated *via* the surface area of the drop and the surface coverage (eqn (2)) expressed as volume. The volume of the starch represented in that bin is then equal to the surface coverage times the area of oil to be covered.

$$V_{starch}^{bin} = \frac{2\pi}{3} d_{sg}\varphi(d_{bin} - d_{sg})^2 \qquad (10)$$

Since we know the diameter of our starch granules and the bin size, the only unknown is the interfacial packing fraction φ.

$$\Phi_{asg}^{bin} = \Phi_{bin}\left(\frac{V_{starch}^{bin}}{V_{starch}^{bin} + V_{oil}^{bin}}\right) \qquad \Phi_{oil}^{bin} = \Phi_{bin}\left(\frac{V_{oil}^{bin}}{V_{starch}^{bin} + V_{oil}^{bin}}\right) \qquad (11)$$

By assuming conservation of mass in our system:

$$\Phi_{total} = \sum_{bin>2d_{sg}} \Phi_{oil}^{bin} + \sum_{bin>2d_{sg}} \Phi_{asg}^{bin} + \sum_{bin<2d_{sg}} \Phi_{fsg}^{bin} \qquad (12)$$

then all relevant information is known, *i.e.* the original starch to oil ratio (the emulsion recipe), the measured starch granule sizes, and the density of the starch, and thus can be combined to estimate the interfacial packing fraction φ. It was then possible to account for all the oil and starch in the system by adding up the contribution bin-wise in the volume frequency distribution (eqn (12)) and solving for the apparent interfacial packing fraction φ by iteration. This approach would satisfy the constraints of the known starch and oil volumes in the entire starch granule stabilized emulsion. The apparent interfacial packing fraction φ found by this analysis may not necessarily be the exact tightness of packing of a monolayer of particles, but rather an estimate of the amount of starch tightly associated with emulsion drops of a given size, particle size distribution, and known starch to oil ratio. Although here we are assuming that the surface coverage is the same for all drops in the emulsion, this procedure provides a more realistic estimation than just assuming that there is hexagonal tight packing at the interface. The result of this calculation is shown in Fig. 6, where the cumulative starch amount was plotted *versus* the bin size for emulsions with 1.95% and 4.66% OSA modified starch, respectively. The resulting apparent interfacial packing fraction φ was indicated as well as the total starch volume fraction, *i.e.* the total volume of dispersed phase that was

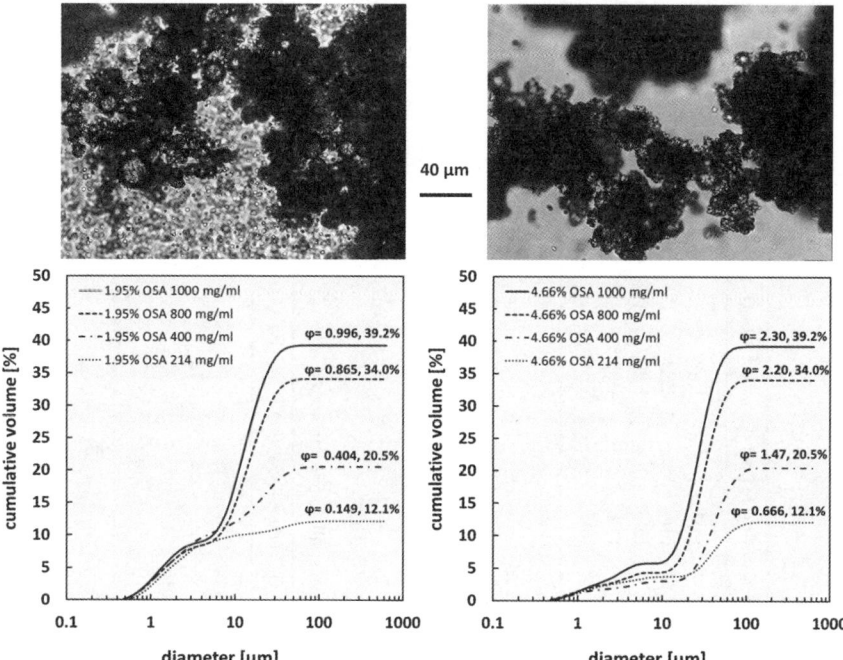

Fig. 6 Micrographs of quinoa starch granule stabilized Pickering emulsion (800 mg starch ml^{-1} oil in 5mM phosphate buffer with 0.2 M NaCl) 1.95% OSA (top left) and 4.66% OSA (top right). Estimated cumulative volume of starch plotted as a function of measured particle size for 1.95% OSA (bottom left) and 4.66% OSA (bottom right) for starch concentrations ranging from 214 to 1000 mg ml^{-1} oil. Starch volume fractions given as %.

starch, based on the mg starch ml^{-1} oil in the recipe. The measured mean starch granules sizes used in the calculations were for 1.95% OSA modified starch surface mean $d_{sg} = 1.84$ μm and d_{43} and for 4.66% OSA modified starch surface mean $d_{sg} = 1.76$ μm (see Table 3 and note the larger d_{43} of the high OSA modified starch granules also indicated some aggregation among the granules).

In previous work, we have observed that drops formed at a lower concentration of starch granules and were less covered by the granules as expected.[28] In contrast, in the present study higher levels of OSA modification were used (more than 3%), which, as discussed above, also made the starch granules more hydrophobic and more inclined to aggregation, creating a thicker multi-granule layer at the surface of drops. In the case of what was measured as larger drops in the 4.66% OSA starch granule stabilized emulsion, the apparent φ was also found to be significantly larger than 0.9 (much more than monolayer coverage). This was interpreted physically in two ways: there was a thicker layer of particles on drops and/or there were clusters of drops that caused a shift in the volume frequency distribution to larger diameters. A thicker layer and less free starch was observed under the microscope, see Fig. 6. Although the measured particle size distribution peak (mode of d_{43}) of the 4.66% OSA modified starch emulsion was 36.2 μm and almost twice that of the 1.95% OSA modified starch emulsion, 19.6 μm, the micrographs did not show drops of this larger size. From previous experience of this system, when measuring droplet size distributions by both counting using image analysis[26] and light scattering[27,28] under similar conditions, it was seen that both methods were in rather good agreement. However, until now only OSA modifications of starch up to 2.9% were used for the Pickering emulsions. The aggregation of measured starch granule stabilized emulsion drops and starch clusters at higher OSA% and more information on the

Table 3 Particle size data summary of starch granules with varying degrees of OSA of heat treatment

Sample name	$d_{[4,3]}$ (μm)	St. dev.	Mode (μm)	St. dev.	$d_{[3,2]}$ (μm)	$d(0.1)$ (μm)	$d(0.5)$ (μm)	$d(0.9)$ (μm)	Span (μm)
Native 0% OSA $n = 3$	2.51	0.157	1.69	0.0082	1.74	1.03	1.88	4.08	1.62
1.95% OSA $n = 3$	3.28	0.094	1.74	0.0151	1.84	1.06	1.99	4.77	1.86
3.21% OSA $n = 3$	2.31	0.044	1.73	0.0175	1.74	1.06	1.86	3.74	1.44
4.66% OSA $n = 3$	3.97	0.425	1.65	0.0012	1.76	1.06	1.80	7.84	3.77
heated 150 min at 120 °C $n = 6$	2.68	0.101	2.22	0.167	2.05	1.17	2.31	4.73	1.54

Table 4 Droplet size distributions and rheological data quinoa starch granule stabilized Pickering emulsions, all at 530 mg starch ml⁻¹ oil

Sample name	$d_{[4,3]}$ (μm)	St. dev. $n = 6$	$d_{[3,2]}$ (μm)	St. dev.	CV (%)	G_0 (Pa)	CV (%)	G''_0 (Pa)	CV (%)	η_0 (Pa s)	CV (%)	G' (45°)	CV (%)
1.8% OSA, 0.2 M NaCl	32.0	2.32	9.16	0.687	37	893	31	47	40	81	38	64	38
1.8% OSA, 2 M NaCl	36.7	6.91	10.5	0.422	17	1065	4	56	7	142	7	101	3
4.66% OSA, 0.2 M NaCl	42.9	2.64	16.5	0.902	44	2108	25	107	44	336	44	187	38
4.66% OSA, 2 M NaCl	54.1	3.15	22.2	1.18	28	2327	30	191	28	371	28	416	26

true droplet size of this apparently more aggregated system should be a topic of future work.

3.4 Effect of OSA level and salt concentration on droplet size and rheological properties

The effect of salt concentrations in the continuous phase on the measured starch granule stabilized emulsion drop sizes was rather limited. Drop size appeared to be mainly affected by the degree of OSA modification of the starch granules as there was no significant effect ($p < 0.05$) of the salt concentration for the three OSA modification degrees with the exception of 2 M salt at 4.66% OSA modification, which had a slightly larger measured droplet diameter than at the other salt concentrations. At 4.66% OSA modification there was also a trend towards larger diameters at higher salt concentrations, see Fig. 7.

Increasing the degree of OSA modification and salt concentration slightly increased the mean measured starch granule stabilized emulsion drop diameters of the particle size distributions, as well as the elastic modulus at which the gel structure in the emulsions was reduced to liquid like behaviour (G' at 45°), Fig. 8 and Table 4. The elastic and viscous modulus in the linear region was significantly affected only by the degree of OSA modification and not by the added salt.

At high dispersed phase fractions the more strongly interacting highly modified starch granules (4.66% OSA) gave rise to larger measured starch granule stabilized emulsion drops, interpreted as drop aggregates with more starch associated at the surface (high φ). Such granule aggregation could also explain the observed higher moduli of the coarser, more aggregated emulsions. The higher moduli would not be expected based on the higher deformability following from larger emulsion droplet size, but were in agreement with the aggregation effects that were also observed previously in emulsions stabilized with aggregating proteins.[37] Aggregation was indicated by the high apparent interfacial packing estimations and confirmed by micrographs showing substantial amounts of free starch at 1.95% OSA modification and extensive aggregation of starch and starch granule stabilized emulsion droplets,

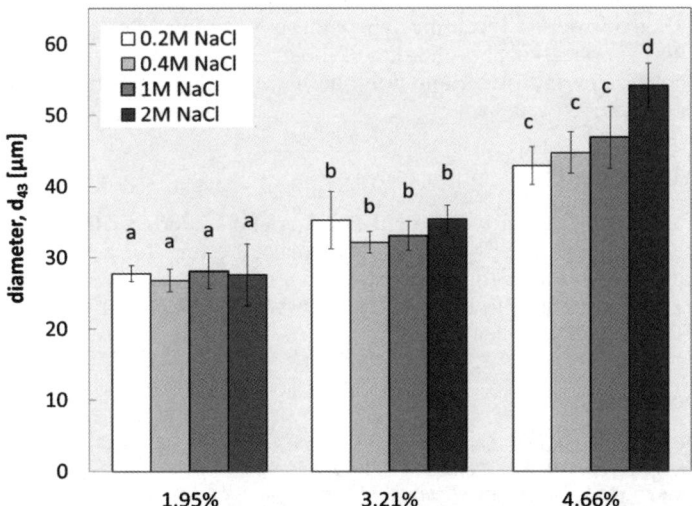

Fig. 7 Mean drop diameters (d_{43}) of quinoa starch granule stabilized Pickering emulsions for 3 different levels of OSA modification (1.95%, 3.21%, and 4.66%, right) and 4 different salt concentrations (0.2, 0.4, 1 and 2 M NaCl). All emulsions were made using 530 mg starch ml^{-1} oil. Columns labelled with different letters are significantly different (Mann-Whitney U-test $p < 0.05$).

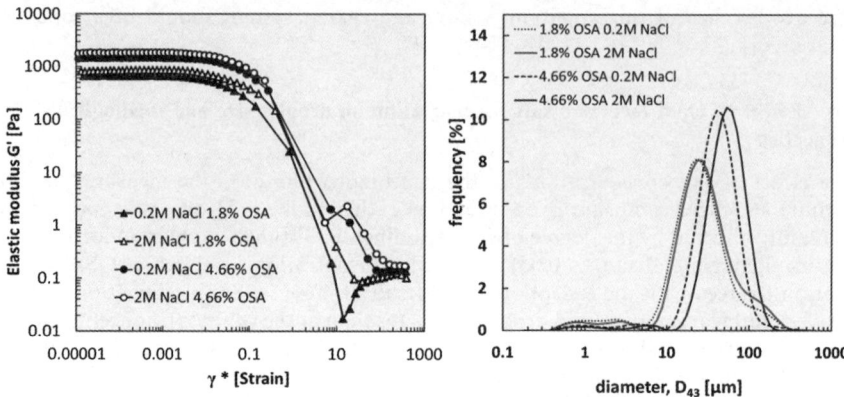

Fig. 8 Elastic modulus as a function of complex strain of quinoa starch granule stabilized emulsions at high and low salt concentration and OSA modification degree (left). Particle size distribution of emulsions tested in rheometer (right). All emulsions were made using 530 mg starch ml^{-1} oil, *i.e.* 50% total dispersed phase (starch and oil together) in 5mM phosphate buffer pH 7 with varying salt concentrations.

but almost no free starch granules at 4.66% OSA modification, as seen in Fig. 4 and Fig. 6 of the starch granule stabilized emulsion.

4 Conclusions

In what appeared to be an interesting coincidence, the level of 3% OSA modification of starch allowed by US and European authorities is close to optimal when using quinoa starch for emulsion formation and stabilization. A higher degree of modification caused aggregation of starch granules as well as of starch granule stabilized emulsion drops. OSA modification was substantially more efficient than heat treatment in providing hydrophobic character to starch granules. The emulsions were stable to all ionic strengths commonly encountered in foods and a range of other products. Starch granule Pickering type emulsion systems like the one presented here may have applications beyond that of food products, for example in the paper, cosmetic and paint industries and for pharmaceutical drug formulations where starch is an approved excipient.

Acknowledgements

The authors thank Gunnel Karlsson at Polymer and Materials Chemistry at Lund University for taking the SEM images, Nathanaële Leconte for her emulsion preparations and Diana Marku for her assistance with the Rheology measurements. The study was supported by the Lund University Antidiabetic Food Centre, which is a VINNOVA VINN Excellence Centre.

References

1 R. Aveyard, *et al.*, *Adv. Colloid Interface Sci.*, 2003, **100–102**, 503–546.
2 W. Ramsden, *Proc. R. Soc. London*, 1903, **72**, 156–164.
3 S. U. Pickering, *J. Chem. Soc. Trans.*, 1907, **91**, 2001.
4 B. P. Binks, *Curr. Opin. Colloid Interface Sci.*, 2002, **7**, 21–41.
5 E. Dickinson, *Curr. Opin. Colloid Interface Sci.*, 2010, **15**, 40–49.
6 Z. J. Luo, *et al.*, *J. Agric. Food Chem.*, 2011, **59**, 2636–2645.
7 B. S. Murray, *et al.*, *J. Agric. Food Chem.*, 2011, **59**, 13277–13288.
8 B. S. Murray, *et al.*, *Food Hydrocolloids*, 2011, **25**, 627–638.
9 A. Yusoff and B. S. Murray, *Food Hydrocolloids*, 2011, **25**, 42–55.

10 M. V. Tzoumaki, *et al.*, *Food Hydrocolloids*, 2011, **25**, 1521–1529.
11 T. N. Hunter, *et al.*, *Adv. Colloid Interface Sci.*, 2008, **137**, 57–81.
12 E. Dickinson, *Curr. Opin. Colloid Interface Sci.*, 2010, **15**, 40–49.
13 E. Dickinson, in *Colloidal Particles at Liquid Interfaces*, ed. B. P. Binks and T. S. Horozov, Cambridge University Press, Cambridge, 2006.
14 D. S. Horne, *Curr. Opin. Colloid Interface Sci.*, 1996, **1**, 752–758.
15 N. D. Denkov, *et al.*, *4th World Congress on Emulsions*, Lyon, France, 2006.
16 P. M. Kruglyakov and A. V. Nushtayeva, *Colloids Surf., A*, 2005, **263**, 330–335.
17 P. Finkle, *et al.*, *J. Am. Chem. Soc.*, 1923, **45**, 2780–2788.
18 I. Akartuna, *et al.*, *Langmuir*, 2008, **24**, 7161–7168.
19 B. P. Binks and S. O. Lumsdon, *Langmuir*, 2000, **16**, 2539–2547.
20 B. P. Binks and C. P. Whitby, *Langmuir*, 2004, **20**, 1130–1137.
21 B. P. Binks, *et al.*, *Phys. Chem. Chem. Phys.*, 2007, **9**, 6391–6397.
22 V. N. Paunov, *et al.*, *J. Colloid Interface Sci.*, 2007, **312**, 381–389.
23 L. G. Torres, *et al.*, *Colloids Surf., A*, 2007, **302**, 439–448.
24 B. P. Binks and R. Murakami, *Nat. Mater.*, 2006, **5**, 865–869.
25 J. Singh, *et al.*, *Food Hydrocolloids*, 2007, **21**, 1–22.
26 M. Seguchi, *Cereal Chem.*, 1984, **61**, 248–250.
27 A. Timgren, *et al.*, *Procedia Food Science*, 2011, **1**, 95–103.
28 A. Timgren, *et al.*, 2012, submitted.
29 M. Rayner, A. Timgren, M. Sjöö and P. Dejmek, *J. Sci. of Food Agric.*, 2012, **92**, 1841–1847.
30 L. E. Abugoch James, in *Advances in Food and Nutrition Research*, ed. L. T. Steve, Academic Press, 2009, vol. 58, pp. 1–31.
31 E. H. C Bromley and I. Hopkinson, *J. Colloid Interface Sci.*, 2002, **245**, 75–80.
32 B. R. Midmore, *J. Colloid Interface Sci.*, 1999, **213**, 352–359.
33 S. Tcholakova, *et al.*, *Langmuir*, 2003, **19**, 5640–5649.
34 S. Tcholakova, *et al.*, *Phys. Chem. Chem. Phys.*, 2008, **10**, 1608–1627.
35 E. Vignati, *et al.*, *Langmuir*, 2003, **19**, 6650–6656.
36 B. P. Binks, *et al.*, *Langmuir*, 2005, **21**, 8161–8167.
37 E. Dickinson, *Colloids Surf., B*, 2001, **20**, 197–210.

Soy milk oleosome behaviour at the air–water interface

Gustav Waschatko,[*ab] Ann Junghans[a] and Thomas A. Vilgis[a]

Received 24th February 2012, Accepted 24th April 2012
DOI: 10.1039/c2fd20036h

Soy milk is a highly stable emulsion mainly due to the presence of oleosomes, which are oil bodies and function as lipid storage organelles in plants, e.g., in seeds. Oleosomes are micelle-like structures with an outer phospholipid monolayer, an interior filled with triacylglycerides (TAGs), and oleosins anchored hairpin-like into the structure with their hydrophilic parts remaining outside the oleosomes, completely covering their surface (K. Hsieh and A. H. C. Huang, Plant Physiol., 2004, 136, 3427-3434). Oleosins are alkaline proteins of 15–26 kDa (K. Hsieh and A. H. C. Huang, Plant Physiol., 2004, 136, 3427–3434) which are expressed during seed development and maturation and play a major role in the stability of oil bodies. Additionally, the oil bodies of seeds seem to have the highest impact on coalescence, probably due to the required protection against environmental stress during dormancy and germination compared to, e.g., vertebrates' lipoproteins. Surface pressure investigations and Brewster angle microscopy of oleosomes purified from raw soy milk were executed to reveal their diffusion to the air–water interface, rupture, adsorption and structural modification over time at different subphase conditions. Destroying the surface portions of the oleosins by tryptic digestion induced coalescence of oleosomes (J. Tzen and A. Huang, J. Cell. Biol., 1992, 117, 327–335) and revealed severe changes in their adsorption kinetics. Such investigations will help to determine the effects behind oleosome stability and are necessary for a better understanding of the principal function of oleosins and their interactions with phospholipids.

Introduction

Evolution has developed different native structures to solubilise oil in water as small droplets: lipoproteins in vertebrates' circulatory systems (HDL, LDL, chylomicrons), fat globules in mammalian milk and oleosomes or oil bodies (OB) in plant seeds.[3,4] Their apparent similarity is based on the fact that nature has a limited amount of natural emulsifiers, i.e., phospholipids and proteins. However, on closer examination the lipid droplets produced in animals are totally different with regard to their surface proteins compared to plant oleosomes. One very high mass apolipoprotein, i.e., Hen Apolipoprotein B (UniProtKB: Q197X2) in egg yolk with its 4631 amino acids wound around the whole particle is located more or less closely to the surface made of phospholipids.[5,6] Thus it significantly constrains the dynamics of the phospholipids in the domains close to the proteins. The positive effect of the apolipoproteins is therefore obvious, they yield more stable droplets compared to "classical" emulsions where only phospholipids are used. Nevertheless, oleosomes appear

[a]Max Planck Institute for Polymer Research, Ackermannweg 10, 55128 Mainz, Germany.
E-mail: waschatko@mpip-mainz.mpg.de; Fax: +49 6131 379100; Tel: +496131 379548
[b]Institut für Pharmazie und Biochemie, Johannes Gutenberg-Universität Mainz, Johann-Joachim-Becher-Weg 30, 55128 Mainz, Germany

even more stable than LDL-particles, for example from egg-yolks. In accordance, soy lecithin gives more stable emulsions than egg lecithin.[7] The reason is a very different arrangement of stabilizing proteins, probably due to the required protection of the oil bodies of seeds against environmental stress during dormancy and germination. Even so, the usage of intact oil bodies in food products and food processing, even cooking (e.g., desserts, instant drinks, salad dressing) does not take place yet, especially not in the western world, whereas egg yolk, cream and other dairy products are very common in food emulsions. On the other hand, for the triacylglycerides (TAGs) and the phospholipids (soy lecithin), the two main components of the oil bodies, the situation is reversed: compared to animal fat sources, their usage and popularity is higher. More and more animal fat is replaced by vegetable oil and soy lecithin is the main emulsifier of processed food, for example in chocolate, bakery products, desserts, margarine or creams. Food manufacturers could easily profit from the pre-existing natural protection[8] of intact oleosomes to improve the stability of their products like dressings, sauces or desserts during storage and utilization. Neither emulsifiers nor homogenisation processes are needed, when oleosomes are used in those food systems.

Furthermore, the extraction process of oleosomes via flotation–centrifugation is of interest in two different research areas:

First, extracting oil from oil seeds with solvents like hexane is still the primary method in the food industry. However, safety and environmental requirements promote aqueous extraction processes, which can already be done on a pilot plant scale.[9] The use of enzymes like cellulase or pectinase to open the seeds cell wall[10] raises the yield of oleosomes. Moreover, disruption of oleosomes with proteases can improve the quantity and quality of the free oil.[11] Showing that a simple trypsin treatment is able to break open the otherwise extremely stable oil bodies is an important step in the direction of safer, more efficient and more environmentally friendly oil extraction. Secondly, oleosomes are attractive for genetic engineering and biotechnology, where they are used as plant expression systems for recombinant proteins or peptides, e.g., antibodies or vaccines. The protein–oleosin fusions are enriched in the seed oil bodies, which are easy to harvest via flotation–centrifugation, and after cleavage eukaryotic proteins without potential contamination of animal pathogens are obtained.[8] As the oil content of soybeans is around 20% there is a high amount of oleosomes and oleosins available from the soybean crops.

Simply put, oleosomes can be viewed as micelle-like structures with an outer phospholipid monolayer, an interior filled with TAGs, and associated proteins, the so called oleosins (Fig. 1).[4] Oleosins are alkaline proteins of 15–26 kDa which are expressed during seed development and maturation and are assumed to play a major role in the stability of oil bodies.[2] These oleosins are different to the apolipoproteins in LDL-particles, because they are not just located on the surface, but most likely anchored with their central hydrophobic stretch deep in the oil-phase. (Fig. 1) This location of the hydrophobic domain is commonly accepted as well as a proline knot forming a 180° turn of a hairpin-like sequence.[13] The secondary structure of this longest hydrophobic sequence (about 70 amino acids) known to date can either be antiparallel β-sheets[14] or α-helical.[15] The N- and C-terminal domains are more hydrophilic (e.g., amphipathic α-helix[2]), most probably remain umbrella-like outside the oleosome, and are less conserved among the plant species.

To investigate the impact of the individual compounds on the interfacial behaviour of the oleosomes further purification processes such as isolation of native oleosins are necessary, but currently not yet available. Regarding the oleosome proteins, the use of proteases provides analogous insight into their impact, such as the investigations of the low-density lipoproteins (LDL) in egg yolk by Dauphas et al.[16] As such enzymes completely destroy the functionality of the oleosins (e.g., phospholipid shielding, binding of phospholipids, water solubility, surface charge).

To better understand the nature of the oleosomes, systematic experiments have been performed. First, oleosomes are purified at pH 11 similar to a procedure

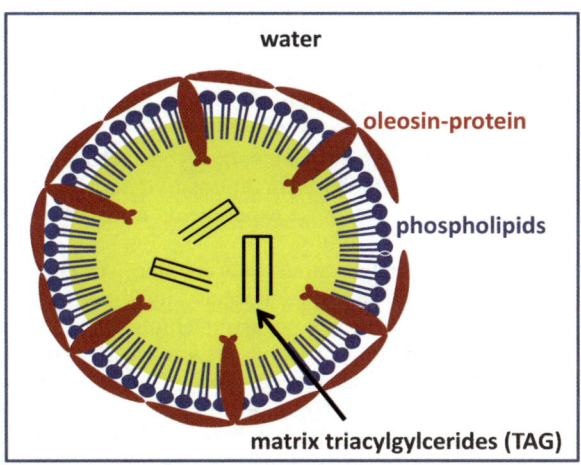

Fig. 1 2D-Model of a soybean oleosome in bulk water (depiction adapted from Huang[12]). The oleosins (red) are located at the oil–water interface of the intact oleosome. They penetrate the surface phospholipids (blue) into the oil matrix (yellow) with their hydrophobic hairpins forming a proline knot at the 180° turn. This central hydrophobic stretch is flanked by the different hydrophilic N- and C-terminal domains, interacting with the charged headgroups of the phospholipids and the surrounding water. Oleosomes are very stable against coalescence in bulk water, but break-up at interfaces, liberating their internal constituents.

introduced by Chen and Ono (2010).[17] Next, untreated and trypsin digested oleosomes are observed at the air–water interface of a film balance. This procedure allows the study of their stability and the behaviour of the three different constituents, *i.e.*, oleosins, phospholipids and TAGs. This method offers two basic observations. First, the kinetics, which show the instability and the "destruction" of the oleosomes at the air–water interface. The second issue is the behaviour of the oleosomes and their constituents under pressure (isotherms). Consequently, we can expect several physical scenarios: First of all, oil bodies immersed in the subphase will rise to the surface. Some oleosomes may stay intact and agglomerate at the interface. Others break up into a phase of triacylglycerides, phospholipids and oleosins. The oil spreads on the surface as film or droplets. The phospholipids separate from the oil droplets and either go to the air–water interface or form micelles in the subphase. The oleosins denature partially and form aggregates. Of course, these processes depend on many parameters, such as pH-value and ionic strength, especially when proteins are involved, since their behaviour depends strongly on the present charge defined by their amino acids exposed to the watery environment.

Thus, when the oleosomes are (partially) destroyed at the air–water interface four different components with competing interactions rule the physical picture at compression of the barriers of the film balance: TAGs, phospholipids, partially denatured or enzymatically cut oleosins, and still intact oil bodies, which will have different contributions to the pressure-surface dependence. Depending on their concentration, phospholipids will form different phases at the surface (see for example ref. 18–22), TAGs will partially wet the interface, oleosins may form micelles by clustering their hydrophobic parts (originally placed in the oil core) into collapsed, dense cores surrounded by the hydrophilic part (N-and C- terminal domains). These ideas and scenarios are depicted in Fig. 2 for intact and trypsin digested oleosomes on a pH 8 subphase.

Previous publications (Bonsegna *et al.*[23]) showed that Brewster angle microscopy (BAM) is the method of choice for visualising areas of different brightness due to different molecular density and/or refractive index in a thin interfacial layer and it has long been used to study for example lipids, proteins and mixed systems at the air–water interface.[24–27] This technique is indeed a very useful tool in the character- ization of the interfacial behaviour of oleosomes as marking them is not necessary for visualisation. Also, the surface pressure is recorded simultaneously to the BAM micrographs, so the changes in surface tension can be directly correlated with the surface active components seen by BAM.

In a previous paper,[28] oleosomes purified from soybeans have been studied at the air–water interface at different subphase conditions to determine the effects behind oleosome stability. Here, we repeat those experiments using trypsin digested oleo- somes.

Soybean oleosomes are highly robust micelle-like structures and provide extremely stable emulsions, preventing coalescence much better than egg lecithin. Before being able to use those naturally occurring emulsifiers in food applications, their compatibility with the human digestive system must be studied as proteins are the allergic components in food. Allergies occur when proteins or fragments of proteins are resistant to digestion, cannot be broken down in the digestive process and are tagged by immunoglobulin E, which triggers an immune response. To prevent food intolerances or allergic reactions the successful digestion and metabo- lism of all components needs to be guaranteed.

Additionally, such investigations are necessary for a better understanding of the principal function and structure of oleosins (*i.e.*, phospholipid binding sites on oleo- sins). They provide information on the mode of action of the digestive enzyme trypsin, and are of interest for food technique applications or carriers. The trypsin cleavage sites are located only on the hydrophilic part of oleosins, with the greater portion on the C-terminus.[29] The measurement depicted in Fig. 2 suggests a detailed scenario of the different physical processes occurring to the oleosome's components at the air–water interface. As already described, oleosomes contain three thermody- namically competing components: non-polar TAGs, phospholipids and oleosins. The oleosins are of special interest, since they consist of a long hydrophobic part and two mainly hydrophilic ends. Their secondary structure accounts for the high stability of the oleosomes by stretching its hydrophobic part into the oil droplets

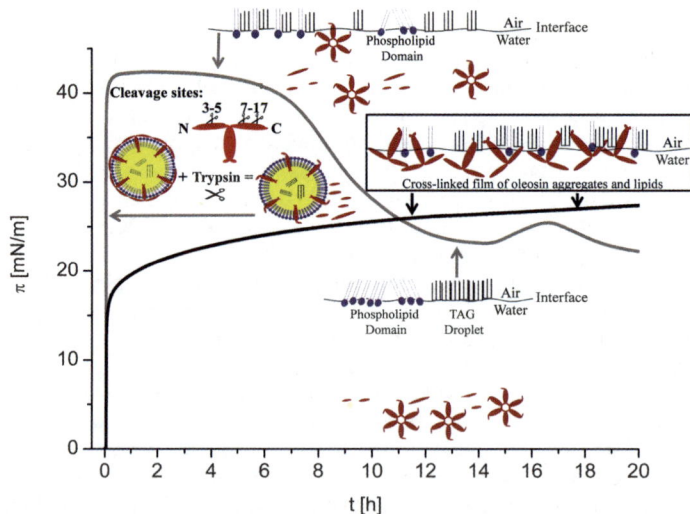

Fig. 2 Time evolution of the surface pressure and summary of the corresponding scenarios at the air–water interface for untreated (black) and trypsin digested (grey) oleosomes. The figure shows typical measurements of the time evolution of the pressure at open barriers at pH 8. Possible scenarios of the oil bodies when exposed to an air–water interface are sketched for illustration (oleosome depiction adapted from Huang[12]).

and arranging their hydrophilic ends into the continuous water phase. When the oleosins are partly digested by trypsin the hydrophilic part becomes cut in various places, leaving a small polar part attached to the hydrophobic sequence at the N-terminal domain. The remaining long hydrophobic parts of the oleosins turn to the surface, aggregate, collapse and form globules of mainly hydrophobic chains in the subphase, which are stabilized and held in the water phase by the short polar tails. The trypsin digested oleosins release most of the phospholipids which rise to the air–water interface quickly and increase the pressure. The TAGs form droplets and arrange themselves into domains of TAG droplets at the air–water interface. The fundamental difference between the digested and the non-digested case is shown in Fig. 2 by a significant difference in the development of the pressure with time. In the non-digestive case, a different arrangement of the components at the subphase and interface develops. The non-digested oleosins keep the oil bodies intact until they reach the air–water interface, then they burst and liberate oleosins, TAGs and phospholipids as indicated in Fig. 2.

Material and methods

For all preparation steps, ultra-pure water, filtered with a Millipore device (Billerica, MA/USA) was used and all experiments were carried out at room temperature.

Oleosome purification

Isolation of soybean oleosomes was performed by a modified aqueous flotation–centrifugation method proposed by Chen and Ono.[17] Dried soybeans from a local supermarket (Davert) were soaked in deionised water at 4 °C for at least 20 h. Then water was added to obtain a 10% soybean-to-water ratio and the mixture was ground in a Vorwerk Thermomix TM31 at a speed of 10 200 rpm for 90 s. The resulting slurry was filtered through two layers of Kimtech science precision wipes 21 × 11 cm (Kimberly Clark) to obtain raw soy milk. 25% sucrose (w/w) was added to the raw soy milk and the pH was adjusted to 11.0 with 1 mol l⁻¹ NaOH (AVS Titrinorm, Prolabo/VWR) solution. The solution was filled into six 50 ml centrifuge tubes (Roth), which were centrifuged in a Thermo Heraeus Multifuge X1R with 15 000 × g at 4 °C for at least 5 h. The resulting floating fractions (creamlayer, fat pat, oleosomes) were lifted with a small spoon and resuspended in 45 ml of 20% (w/w) sucrose in deionised water (pH 11) in a new centrifuge tube. This washing step (15 000 × g, 4 °C, 5 h) was performed twice. The resulting oleosomes were collected and dispersed in 20 ml of deionised water and dialyzed over night with Thermo Scientific Slide-A-Lyzer G2 Dialysis Cassettes (20 K MWCO).

Sodium dodecyl sulfate-polyacrylamide gel electrophoresis (SDS-PAGE)

The SDS-PAGE was performed with the invitrogen NuPAGE®-System. NuPAGE® MES-Running Buffer and a 10% NuPAGE® Novex® Bis-Tris Mini Gel were applied according to the manufacturer's instructions but without heating the samples, instead they were incubated overnight with the NuPage®LDS Sample Buffer and the NuPage®LDS Reducing Agent.

For the visualization of the protein bands Coomassie® G-250 SimplyBlue™ SafeStain (invitrogen) was used. After staining for 1 h, the polyacrylamide gel was destained twice with ultrapure water for 1 h and subsequently overnight.

Trypsin digestion

1 mg trypsin from porcine pancreas (Serva Electrophoresis GmbH, Tryptic activity: $\geq$ 50 U g⁻¹ (Ph. Eur.) or 4699 FIP-U g⁻¹) was added to 1 ml of dialysed

oleosomes (85% water content) and was incubated at 25 °C for 1 h in a HLC ThermoMixer MKR 13.

Buffer

For buffered subphases the following chemicals were used:
Phosphoric acid, ACS reagent, (Sigma-Aldrich, Munich, Germany, ≥85 wt. % in H_2O), acetic acid (Sigma-Aldrich, Munich, Germany, ≥99.8%) and monosodium phosphate (Fluka, Munich, Germany, ≥99%). For varying the ionic strength, the molarity was kept constant and adjusted by adding the required amount of sodium chloride (Prolabo/VWR, Darmstadt, Germany, min. 99.5%).

Film balance

The surface pressure was measured as a function of time (kinetics) and surface area (isotherms) on a Nima (Biolin Scientific, Västra Frölunda, Sweden) BAM Trough (available area at completely opened barriers = 715 cm²), with symmetric Delrin twin barriers. The trough was equipped with a surface pressure sensor that uses the Wilhelmy plate technique to determine the change in surface tension of the air–water interface in the presence of surfactant molecules. In general, measurements were repeated at least 3 times for any given set of parameters with a very good agreement between the different measurements.

Kinetics

The surface pressure was recorded as a function of time at a fixed surface area. If not denoted otherwise, 22 µl suspension of oleosomes in water were inserted into the subphase with a pipette (Eppendorf, Hamburg, Germany). Preliminary experiments have shown that no agitation was necessary to enable proper mixing of oleosomes in the subphase. The total volume of the trough was 500 ml.

Isotherms

Immediately after oleosome injection or subsequent after a kinetic, the surface pressure was recorded as a function of area, hence monitoring the changes in surface tension upon compressing. If not denoted otherwise, the barrier compression speed was 30 cm² min⁻¹.

Brewster angle microscopy

All BAM micrographs were taken on a EP³-BAM (Nanofilm Göttingen, Germany) with a lateral resolution of approx. 1 µm. The size of the micrographs was 600 × 500 µm and images were not processed in any way except for background correction carried out with the provided software. Simultaneously, the surface pressure was recorded.

Results and discussion

Oleosome purification, characterization and trypsin digestion

The dialyzed oleosomes used for all experiments had a water content of 85%. Their mean diameter of 300–350 nm was measured at room temperature and diluted in water by means of Dynamic Light Scattering (DLS) with a Nicomp particle sizer (model 380, PSS Santa Barbara, California) at a scattering angle of 90° (data not shown). The coomassie stained SDS-PAGE (Fig. 3) confirmed three proteins of 15–16, 17–18 and 23–24 kDa corresponding to the size of the four soybean oleosins known in protein databases (UniProtKB: Isoforms P29530 and P29531 23–2 kDa, C3VHQ8: 17–18 kDa, C6SZ13: 15–16 kDa).

This confirms that pH 11 extraction removed the unspecifically bound soybean storage proteins (glycinin and β-conglycinin), and potential allergenic proteins (such as Gly m Bd 30 K),[17] that exist in raw soy milk, from the surface of the oleosomes as shown in the SDS-PAGE.

Compared to other soybean oleosome purifications,[8] the oleosomes used here showed a very sharp aggregation behaviour between pH 4.4 and 5.7 (data not shown).

All four oleosins were digested during the trypsin treatment independently of the amount of protease used. The maximum remaining tryptic fragments were below approximately 8 kDa, including the hydrophobic hairpin and the proline knot motif.[2]

Film balance

Independent of the subphase condition, an increase in surface pressure from 0 mN m^{-1} to approx. 12 mN m^{-1} is observable immediately after oleosome injection together with the diffusion of round, very bright 3D particles in the size of the lateral resolution (approx. 1 μm) of the BAM to the air–water interface. With time, a further increase of surface pressure—whereby the gradient is dependent on subphase conditions such as pH, ionic strength *etc.*—can be monitored as well as a partial aggregation of the particles visible by BAM. When the area is decreased by compression, the shape of the subsequent isotherm depends on the subphase conditions as well as the time between injection and compression, but usually one to two transitions are observable: Liquid-expanded (LE) to coexistence of liquid-expanded/liquid-condensed (LE/LC) and/or LE/LC to liquid-condensed (LC). In accordance, the BAM micrographs show a two-phase system that is growing in intensity, and hence in optical density, with decreasing area and increasing surface pressure. In summary, intact oleosome behaviour at the air–water interface is driven by the charge of the oleosins.[28]

Trypsin digestion

Trypsin is a specific serine protease found in the digestive system of many vertebrates, where it hydrolyses proteins. It cleaves proteins at the carboxyl side of the basic amino acids lysine and arginine. This means the oleosins found in soybean oleosomes have 21–22 cleavage sites for the long and 10 for the short oleosins, with the main part located at the hydrophilic C-terminal domain (Fig. 2).[29]

When intact oleosomes were digested with trypsin, pronounced differences compared to native oleosomes were visible in the phase behaviour and BAM micrographs at the air–water interface.

After injection of the same amount of oleosome solution, trypsin digested oleosomes show a much higher increase in surface pressure (between 38 and 43 mN m^{-1}, depending on pH), whereas untreated, intact oleosomes only reach values between 24 and 27 mN m^{-1} after the same waiting time. (Fig. 4A)

This is a strong indication that trypsin cleaves the outer heads of the oleosins and hence destroys the stability of the oleosomes to a large extent, which leads to the breakage of the oil bodies. Their constituents become released and redistribute themselves in the subphase and air–water interface. As the surface pressure after injection is still considerably higher than the maximum observed for intact oleosomes after presumed breakage we assume that binding sites between the oleosins and phospholipids are destroyed during trypsin cleavage, leading to more free phospholipids.

Fig. 4A shows the pH-dependence of the surface pressure for intact as well as trypsin digested oleosomes. For intact oleosomes, lower surface pressure values are recorded at pH-values different from the isoelectric point (pI = pH 5), due to higher repulsion of the oleosins carrying a net charge. The pH-dependence of trypsin digested oleosomes is different. Here, the lowest surface pressure is recorded at pH 5.

As the charged terminal domains of the oleosin were cut by trypsin only a weak pH-dependency could be expected for such systems, because of the loss of almost all acidic and basic amino acids.[29] However, it is known from the literature[30] that the interfacial tension of the lipid membrane is dependent upon the hydrogen ion concentration of the surrounding solution. This connection explains the slightly lower surface pressure values at pH 5, as the interfacial tension of phosphatidyleth-anolamine, one of the main components of the present phospholipids,[31] peaks at pH 4, and shows lower values at pH 2 and 8.

This also explains the appearance of branched, starlike domains, that occur at pH 2 and pH 8 for trypsin digested oleosomes (Fig. 3B, 4, 5). Due to the high packing density, lipids cannot repel each other as much as necessary, which results in domain formation as this is energetically more favourable. The driving force for domain shape formation is the free energy $F = F_{el} + F_L$.[32] Here F_{el} is the free energy of the electrostatic repulsion and F_L the contribution from the line tension of the quasi two dimensional droplets perimeter. In the absence of charged headgroups (at the pI), the minimum energy domain shape is determined entirely by the short-range interaction line tension that favours circular structures. The isoelectric point of

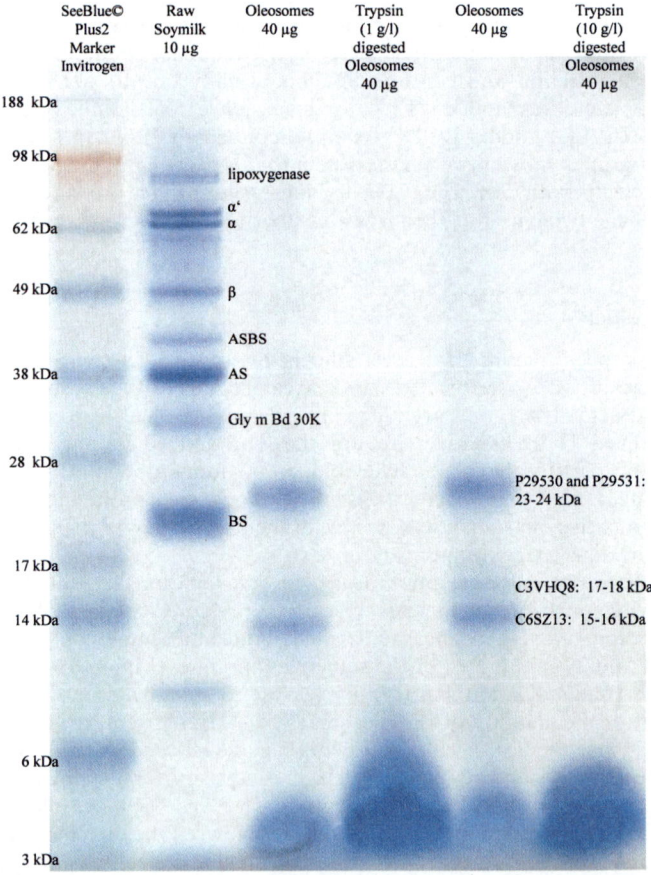

Fig. 3 Commassie stained SDS-PAGE gel of purified soybean oleosomes before and after digestion with trypsin. The raw soy milk shows the subunits of soybean storage proteins glyci-nin (AS: acidic subunit, BS: basic subunit) and β-conglycinin (α, α', β). Oleosomes were trypsin digested with 1 mg ml⁻¹ and 10 mg ml⁻¹ of protease at 25 °C for 1 h in a ThermoMixer. The tryptic fragments were below 8 kDa.

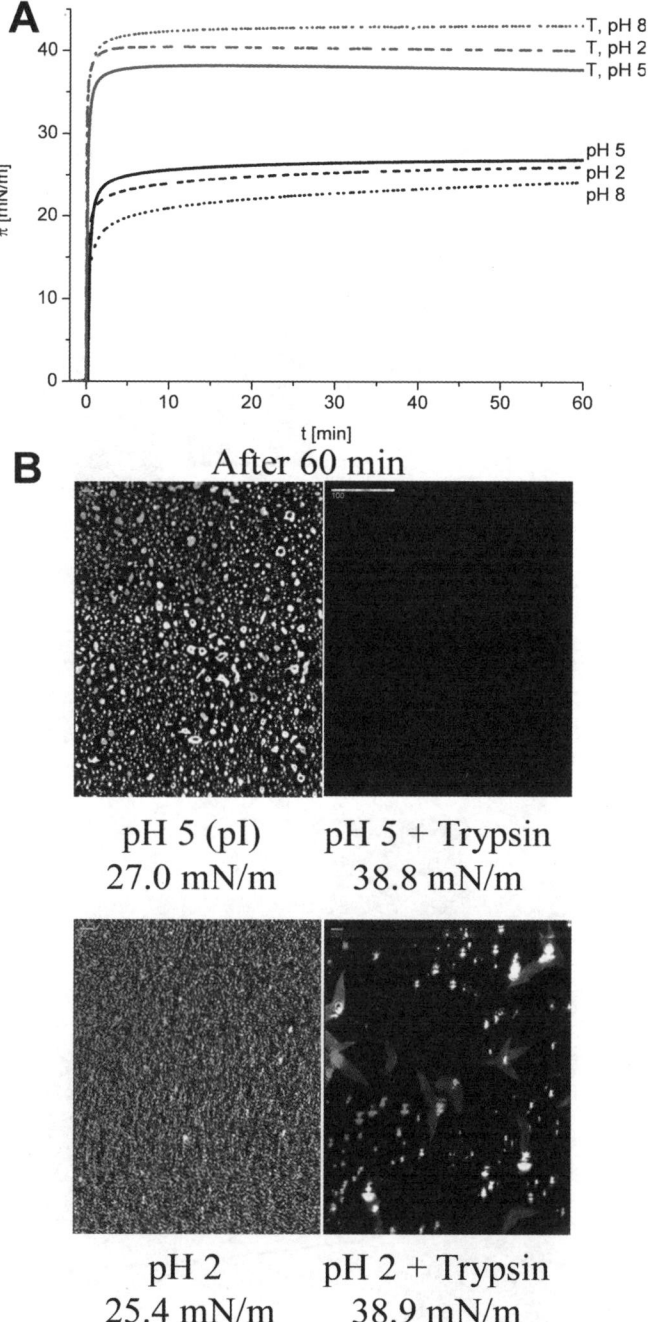

Fig. 4 (A) Measurement of surface pressure as a function of time at defined area at different pH (I = 5 mM). For intact oleosomes (black) at pH-values different from the isoelectric point (pH 5) lower surface pressure values are recorded due to higher repulsion of the oleosins. Trypsin digested oleosomes (grey) show higher surface pressure values and different pH dependencies. (B) BAM micrographs of soybean oleosomes at different pH after 60 min. At pH-values above and below the isoelectric point, repulsion between the charged oleosins occurs, leading to smaller aggregates for untreated oleosomes. Trypsin digested oleosomes show star-shaped phospholipid domains at pH 2 and small round lipid domains at pH 5. Conditions: 22 μl (6.6 mg l⁻¹) OB, pH 2: phosphoric acid, pH 5: acidic acid, pH 8: monosodium phosphate, all with 5 mM buffer concentration and I = 5 mM.

phosphatidylethanolamine is at pH 5.5[33] which leads to higher electrostatic repulsion at pH 2 and 8 between the lipids compared to pH 5, hence the long-range electrostatic dipolar repulsion F_{el} dominates, which leads to elongated, irregular structures.

The long-time development, shown in Fig. 5A, is also different for digested and intact oleosomes. For the latter, the surface pressure as well as the aggregate density

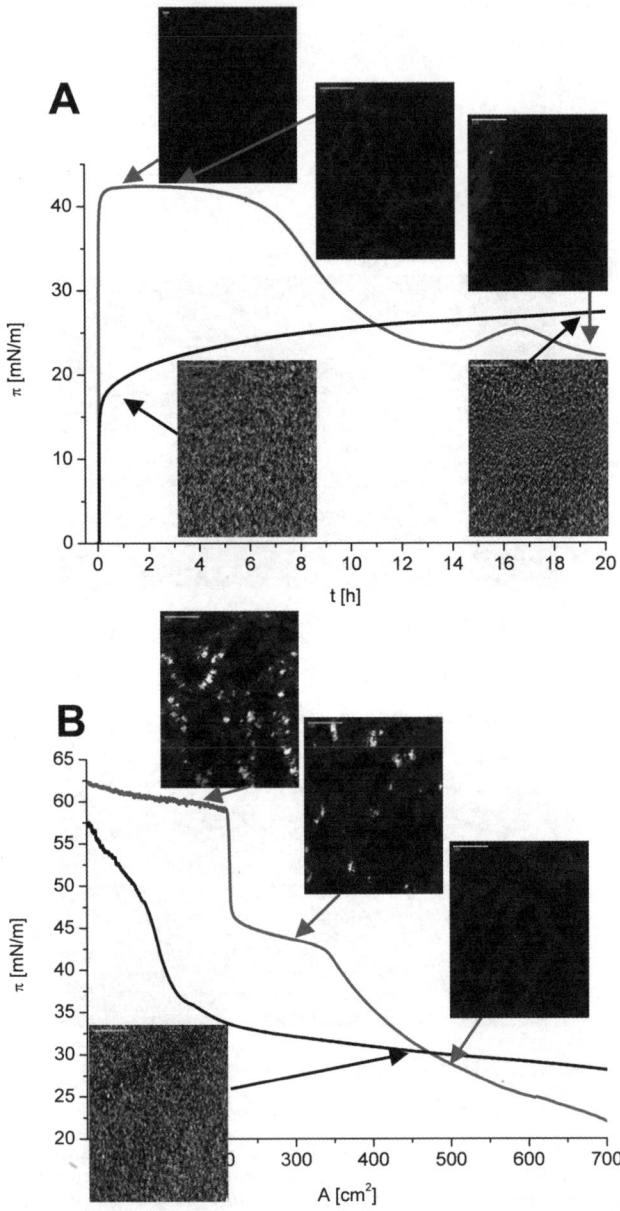

Fig. 5 (A) Measurement of surface pressure as a function of time at defined area at pH 8 ($I = 5$ mM) of intact (black) and trypsin digested oleosomes (grey). Trypsin digested oleosomes show star-shaped phospholipid domains and "drops of grease" which cause a decrease in surface pressure. (B) Compression ($v_B = 30$ cm^2 min^{-1}) isotherm of intact (black) and trypsin digested (grey) soybean oleosomes. Conditions: 22 μl OB (6.6 mg l^{-1}), 5 mM Tris, pH 8.

increase with time, presumably due to the breakup of the oleosomes. The situation is reversed when trypsin digested oleosomes are present. The surface pressure is then decreasing. These observations probably result from the faster aggregation of the longest tryptic fragments (<8 kDa, Fig. 3), including basically the hydrophobic hairpin of the oleosins. The remaining protein chains in water could be largely considered as unstructured random copolymers which tend to collapse. Several chains form dense, globular aggregates. Under these circumstances the size of the aggregates (*i.e.*, the number of chains involved) is determined by a balance of the surface tension of the aggregates, their bulk energy, and the interactions between the (polar) hydrophilic tails, which are distributed on the surface of the aggregate. The size of such aggregates diminishes with the polarity of the remaining short hydrophilic tail of the N-terminus. This also explains the occurrence of larger protein aggregates, *e.g.*, visible in Fig. 4B (pH 2) that are more likely to diffuse into the subphase due to their weight and less amphiphilic nature. Additionally, also here, a dependence on pH is noticeable: Close to the isoelectric point of phosphatidylethanolamine, where a presumably homogeneous film of phospholipids is present, no protein aggregates are visible. At pH 2, where domain formation occurs, large aggregates are observable in the first hours due to available interfacial space.

The setting of oleosins leaves only phospholipids and triglycerides at the air–water interface which continue to phase separate as the oleosins, which functioned as emulsifiers, start to sink. This can be seen in the growing of the phospholipid domains (Fig. 5A + 6A) and the appearance of circular TAG oil droplets (Fig. 6B). Those oil droplets coalesce with time (see flower-like structures in Fig. 5B), leading to bigger droplets with bigger area-to-perimeter ratio, which results in fewer binding sites and hence lower surface pressure.

These assumptions are further strengthened by the compression isotherms in Fig. 5B: In the isotherm of intact oleosomes only one distinct transition can be observed, whereas trypsin digested oleosomes show two transitions and a collapse at 60 mN m^{-1} that are very similar to phospholipid phase behaviour. Also the domains that are visible at high surface pressure (above 40 mN m^{-1}) reappear and increase in size during compression, confirming our hypothesis that these are indeed phospholipid domains that reform due to the decreasing area.

Trypsin digested Oleosomes at pH 8 after 20 h

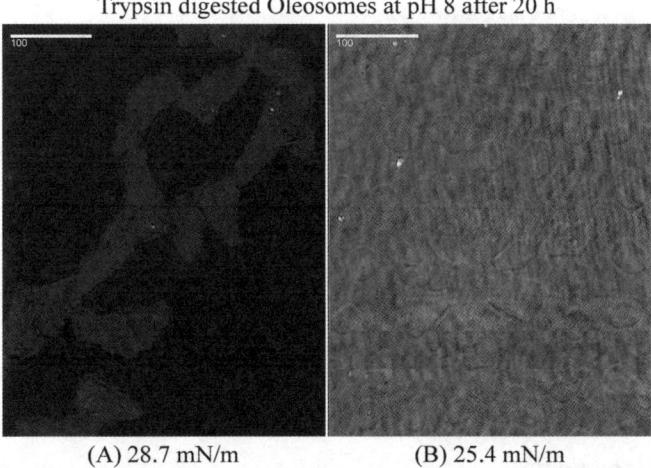

| (A) 28.7 mN/m | (B) 25.4 mN/m |

Fig. 6 BAM micrographs of trypsin digested soybean oleosomes at pH 8 after 20 h. (A) Left picture (100% brightness and contrast) shows irregular grown phospholipid domains. (B) The right picture (brightness and contrast +40%) shows TAG oil droplets and flower-like structures. Conditions: 22 μl (6.6 mg l^{-1}) OB, 5 mM Tris, pH 8, $I = 5$ mM.

Additionally, for the untreated oleosomes a highly viscous film is formed during the waiting time, which suggests a denaturation of oleosins and the formation of a protein-network.[28] Trypsin digested oleosins, due to the loss of the hydrophilic part, are no longer able to form this viscous detachable film after compression.

Conclusions

The combination of surface pressure measurements and simultaneous Brewster angle microscopy is used to reveal the behaviour at the air–water interface of trypsin digested oleosomes compared to untreated ones.

In principle, oleosomes are micelle-like structures with an outer phospholipid monolayer and an interior filled with triglycerides, but with oleosins sticking hairpin-like in the structure, with the hydrophilic parts remaining outside the oleosomes. After injection into the aqueous subphase, intact oleosomes diffuse immediately to the air–water interface due to flotation and their amphiphilic nature, which is visible in the steep increase in surface pressure and the bright, micrometer sized particles at the air–water interface during this diffusion. Rupture and subsequent coalescence take place when the packing of oleosomes exceed a crucial level and oil bodies come too close together. This rupture can be seen in an additional increase in surface pressure and increase in aggregate size visible by BAM.

Trypsin is a serine protease able to hydrolyse proteins. We assume that trypsin cleaves the N-and C-terminal domains of the oleosins, thereby destroying the stability of the oil bodies. This is supported by the increased surface activity compared to intact oleosomes, after injection into an aqueous subphase. Since the jump in surface pressure after injection of trypsin digested oleosomes into the subphase is much higher than the maximum measured for the same amount of untreated oleosomes we believe that the trypsin cleaved oleosins lead to different complexes with the phospholipids at the air–water interface. Binding sites between them are not available anymore, resulting in a large number of free phospholipids and a further decrease of the water surface tension. Furthermore, as inferred from the faster decrease in surface pressure, trypsin digested oleosomes presumably form larger protein aggregates and growing lipid domains. In general, the air–water interface after the injection of trypsin digested oleosomes is dominated by their phospholipids, showing star-shaped domains and collapse during compression, and their TAGs (circular oil droplets).

Soybean oleosomes are highly robust micelle-like structures and provide extremely stable emulsions but before their use in food applications their compatibility with the human digestive system has to be studied as proteins are the most common allergic component in food. Oleosomes consist of triacylglycerides and phospholipids, which can be emulsified and enzymatically digested, and oleosins. We could show that oleosins can be cleaved by trypsin, which is one indication that they can be safely used in food industries. Additionally, such investigations are requisites to better understand the principal function and structure of oleosins (i.e., phospholipid binding sites on oleosins), investigate their usability as plant expression system for recombinant proteins or peptides and provide information on the mode of action of the digestive enzyme trypsin.

Moreover, the use of proteases like trypsin yields a higher release of phospholipids (lecithin) and oil (TAGs) from the oleosomes during this safe and environmentally friendly aqueous oil extraction process.

Acknowledgements

The authors like to thank Markus Deserno, Johannes Franz, Sania Maurer, Christine Peter and Birgitta Schiedt for fruitful discussions and Sandra Ritz for the use of SDS-PAGE equipment.

References

1 K. Hsieh and A. H. C. Huang, *Plant Physiol.*, 2004, **136**, 3427–3434.
2 J. Tzen and A. Huang, *J. Cell Biol.*, 1992, **117**, 327–335.
3 V. Martinet, P. Saulnier, V. Beaumal, J.-L. Courthaudon and M. Anton, *Colloids Surf., B*, 2003, **31**, 185–194.
4 A. H. C. Huang, in *Annu. Rev. Plant Physiol. Plant Mol. Biol.*, 1992, vol. 43, pp. 177–200.
5 T. Hevonoja, M. O. Pentikäinen, M. T. Hyvönen, P. T. Kovanen and M. Ala-Korpela, *Biochim. Biophys. Acta, Mol. Cell Biol. Lipids*, 2000, **1488**, 189–210.
6 P. Jolivet, C. Boulard, V. Beaumal, T. Chardot and M. Anton, *J. Agric. Food Chem.*, 2006, **54**, 4424–4429.
7 L. Palacios and T. Wang, *J. Am. Oil Chem. Soc.*, 2005, **82**, 571–578.
8 D. Iwanaga, D. A. Gray, I. D. Fisk, E. A. Decker, J. Weiss and D. J. McClements, *J. Agric. Food Chem.*, 2007, **55**, 8711–8716.
9 L. Towa, V. Kapchie, C. Hauck, H. Wang and P. Murphy, *J. Am. Oil Chem. Soc.*, 2011, **88**, 733–741.
10 V. N. Kapchie, D. Wei, C. Hauck and P. A. Murphy, *J. Agric. Food Chem.*, 2008, **56**, 1766–1771.
11 L. Towa, V. Kapchie, G. Wang, C. Hauck, T. Wang and P. Murphy, *J. Am. Oil Chem. Soc.*, 2011, **88**, 1581–1591.
12 A. H. C. Huang, *Plant Physiol.*, 1996, **110**, 1055–1061.
13 F. Capuano, F. Beaudoin, J. A. Napier and P. R. Shewry, *Biotechnol. Adv.*, 2007, **25**, 203–206.
14 D. J. Murphy, J. N. Keen, J. N. O'Sullivan, D. M. Y. Au, E.-W. Edwards, P. J. Jackson, I. Cummins, T. Gibbons, C. H. Shaw and A. J. Ryan, *Biochim. Biophys. Acta, Gene Struct. Expression*, 1991, **1088**, 86–94.
15 L. Alexander, R. Sessions, A. Clarke, A. Tatham, P. Shewry and J. Napier, *Planta*, 2002, **214**, 546–551.
16 S. Dauphas, V. Beaumal, A. Riaublanc and M. Anton, *J. Agric. Food Chem.*, 2006, **54**, 3733–3737.
17 Y. Chen and T. Ono, *J. Agric. Food Chem.*, 2010, **58**, 7402–7407.
18 J. C. Watkins, *Biochim. Biophys. Acta, Lipids Lipid Metab.*, 1968, **152**, 293–306.
19 B. Alfred, *Biochim. Biophys. Acta, Biomembr.*, 1979, **557**, 32–44.
20 O. Albrecht, H. Gruler and E. Sackmann, *J. Colloid Interface Sci.*, 1981, **79**, 319–338.
21 V. von Tscharner and H. M. McConnell, *Biophys. J.*, 1981, **36**, 409–419.
22 H. Möhwald, *Annu. Rev. Phys. Chem.*, 1990, **41**, 441–476.
23 S. Bonsegna, S. Bettini, R. Pagano, A. Zacheo, V. Vergaro, G. Giovinazzo, G. Caminati, S. Leporatti, L. Valli and A. Santino, *Appl. Biochem. Biotechnol.*, 2011, **163**, 792–802.
24 D. Beaglehole, *Rev. Sci. Instrum.*, 1988, **59**, 2557–2559.
25 S. Henon and J. Meunier, *Rev. Sci. Instrum.*, 1991, **62**, 936–939.
26 D. Hoenig and D. Möbius, *J. Phys. Chem.*, 1991, **95**, 4590–4592.
27 D. Hönig and D. Möbius, *Thin Solid Films*, 1992, **210–211**(Part 1), 64–68.
28 G. Waschatko, A. Junghans and T. A. Vilgis, *J. Phys. Chem. B*, 2012, accepted.
29 E. Gasteiger, C. Hoogland, A. Gattiker, S. e. Duvaud, M. R. Wilkins, R. D. Appel and A. Bairoch, ed. J. M. Walker, Humana Press, 2005, pp. 571–607.
30 A. D. Petelska and Z. A. Figaszewski, *Biochim. Biophys. Acta, Biomembr.*, 2002, **1567**, 79–86.
31 J. M. Garcia, L. C. Quintero and M. Mancha, *Phytochemistry*, 1988, **27**, 3083–3087.
32 H. M. McConnell, *Annu. Rev. Phys. Chem.*, 1991, **42**, 171–195.
33 M. C. Phillips and D. Chapman, *Biochim. Biophys. Acta, Biomembr.*, 1968, **163**, 301–313.

Critical laminar shear-temperature effects on the nano- and mesoscale structure of a model fat and its relationship to oil binding and rheological properties

Nuria C. Acevedo,[a] Jane M. Block[b] and Alejandro G. Marangoni[*a]

Received 20th January 2012, Accepted 6th March 2012

DOI: 10.1039/c2fd20008b

This article reports on the effect of laminar shear on structural and mechanical properties of physical mixtures of fully hydrogenated soybean oil (FHSO) in soybean oil (SO). Blends were crystallized statically and under laminar shear rates of 30 and 240 s^{-1} at different wall temperatures (-10, 0, 20 °C). The micro- and nanocrystalline structures were characterized using Polarized Light Microscopy (PLM), and Cryogenic Transmission Electron Microscopy (Cryo-TEM). Rheological analysis was used to determine changes in mechanical properties. Oil-binding capacity was analyzed through the measurement of the oil lost from the fat samples (OL). Shearing greatly affected the structure at the nano- and mesoscale. At low shear rates, blends displayed the largest increase in crystal size with an increase in wall temperature at both the nano- and mesoscale. On the other hand, at shear rates of 240 s^{-1}, the effect of crystallization temperature was observed only at the nanoscale since no changes in meso-crystal sizes were observed at different temperatures. Crystallization under laminar shear promoted the growth of spherical crystalline particles at the mesoscale, called here "solid-lipid meso-particles". Crystallization under higher shear rates led to the formation of a weak network with low oil-binding capacity and promoted the asymmetric growth of nanoplatelets. In statically crystallized blends, nanoplatelets had an aspect ratio of ~2, while in sheared blends this value increased significantly. These results revealed the existence of critical shear rate values above which strong alterations in the structure of the solid crystalline network took place. Shearing also affected the material's strength. Laminar shear induced a decrease in elastic modulus and yield stress values which was more pronounced at higher shear rate–temperature combinations. Shear–temperature combinations were successfully used to structure fats at the nano and mesoscale.

1 Introduction

The successful replacement of unhealthy fats (containing *trans* and saturated fatty acids) with more healthy fats (containing mono- and polyunsaturated fatty acids) in industrial applications depends on being able to match their functionality and structure *via* judicious control of crystallization conditions under specific external

[a]Guelph-Waterloo Physics Institute, Centre for Food & Soft Materials Science, Department of Food Science, University of Guelph, 50 Stone Road East, Guelph, Ontario, Canada N1G 2W1. E-mail: amarango@uoguelph.ca; Fax: +1-519-824-6631; Tel: +1-519-824-4120 x 54340; nacevedo@uoguelph.ca; jmblock@cca.ufsc.br
[b]Department of Food Science and Technology, Santa Catarina Federal University, Rod. Admar Gonzaga, 1346, Itacorubi, Florianópolis, Santa Catarina, Postal Code: 88034-001, Brazil

temperatures and shear fields. Chemical composition and processing conditions influence the crystallization behavior of fats which in turn affects macroscopic mechanical properties, oil binding capacity and ultimately food product quality.[1,2] This crystallization behavior not only includes kinetics, but also polymorphism, stability and nano and mesoscale morphology along with size of the crystals created during the process. Several studies have shown that shear can have multiple effects during crystallization since it improves the rate of heat and mass transfer as well as enhances mixing and the formation of a homogeneous product. For instance, it has been recognized that shear can enhance nucleation and dramatically accelerate polymorphic transitions toward more stable phases.[3–6] Shearing during crystallization can also induce crystallite orientation.[4–7] Furthermore, it can strongly affect aggregation of fat crystals, either by preventing aggregation and/or breaking them down.[8] Additionally, shear forces can induce the rearrangement of the internal crystalline structure of crystal aggregates, promoting the formation of more close-packed structures.[9]

The capacity of a fat crystal network to trap oil is an important material property that directly influences oil migration or oil loss and will therefore affect the functionality and stability of food products.[10] The mechanisms responsible for oil migration through a fat matrix are not yet completely understood, however the capacity of fat crystals to bind and retain liquid oil within their crystal network has been reported to be dependent on molecular composition,[11] thermal properties,[12] intercrystalline interactions, wetting properties of fat crystals,[13] and crystal size.[14,15] Dibildox-Alvarado[14] et al. reported higher oil migration rates for materials containing larger crystals, where meso-crystal size was varied by crystallization of the fat at different cooling rates. They found that fats crystallized at slow cooling rates contained larger crystals which lead to a greater oil loss. However, to our knowledge; there are very few reports of the effects of crystallizing a fat under shear and its ability to retain oil.

Many reports in the scientific literature have specified that high levels of *trans*-fatty acids in the diet result in negative effects on both low-density lipoprotein cholesterol (LDL-C) and high-density lipoprotein cholesterol (HDL-C). In response to these reports, several health organizations have recommended the reduced intake of foods containing *trans*-fatty acids. Furthermore, many food manufacturers are considering means to reduce or eliminate *trans*-fats from their products. Candidate materials for this include fully hydrogenated oils. When liquid vegetable oil is fully hydrogenated almost no *trans* fatty acids remain in the material, however the fully hydrogenated fat is extremely hard and has a waxy texture. Even though the use of such materials would greatly decrease or completely remove *trans* fatty acids from the fat, it leads to an increase in saturated fatty acids. However, fully hydrogenated oils from oilseed crops contain mostly the saturated fatty acid stearic acid (18 : 0), which is transformed by the body to oleic acid (18 : 1).[16] Oleic acid has neutral effects on lipoprotein metabolism and therefore doesn't raise levels of bad cholesterol.[16] For this reason, fully hydrogenated fats (containing high proportions of stearic acid) are less harmful than partially hydrogenated fats containing high proportions of *trans* fatty acids. Moreover, fully hydrogenated oils are reasonably inexpensive and locally grown. These materials are thus considered a viable replacement for partially hydrogenated fats containing high proportions of *trans* fatty acids. However, mixtures of fully hydrogenated oil and liquid vegetable oil still tend to be very hard and waxy.

The purpose of this work was to determine the effects of external temperature and shear fields during crystallization on the nano- and mesoscale structure of a model mixture of fully hydrogenated soybean oil (FHSO) with liquid soybean oil (SO), and relate it to oil binding capacity (OBC) and mechanical properties. Our model fat included blends of fully hydrogenated soybean oil (FHSO) and soybean oil (SO). The ultimate objective of this work is to engineer the structure of these materials so as to predictably affect oil binding capacity and mechanical properties of the mixture. Such knowledge will lead to an improvement in the macroscopic functionality of such blends, which in turn will help enhance their use, leading to improvements in the nutritional quality of manufactured food products.

2   Experimental section

Materials

Fully hydrogenated soybean oil (FHSO) and soybean oil (SO) were generously provided by Bunge Canada (Toronto, Canada). All chemicals and organic solvents were purchased from Fisher Scientific and Sigma-Aldrich (ON, Canada).

Blend preparation

Blends of FHSO and SO were mixed in 40 : 60 and 45 : 55 (w/w) proportions respectively. The blends were melted and held at 80 °C for 30 min to erase crystal memory. Then, the samples were crystallized at 3 different wall temperatures (−10 °C, 0 °C and 20 °C), statically and under laminar shear rates of 30 and 240 s^{-1}. Sheared and non-sheared mixtures were crystallized for 30 min in a shear cell with a Searle configuration as previously described.[7] The shear cell consists of two concentric aluminum cylinders with a 2.5 mm gap in-between. The design of the cell includes a cooling jacket that allows a controlled cooling regime of the sample. A temperature-controlled bath filled with ethylene glycol was used to achieve the desired cooling temperatures. The cooling curves for each wall temperature were obtained using a copper-constant thermocouple (Omega Engineering, Inc., Stamford, Conn.); and the determined cooling rates were 9; 5 and 3 °C min^{-1} for wall temperatures of −10, 0 and 20 °C respectively. After crystallization, the mixtures were kept at 20 °C for 48 h to allow for material setting and subsequently stored at 4 °C until the moment of analysis.

Fatty acid composition

Lipids were extracted from fat sources by the method of Folch[17] *et al.* Briefly, 0.01 g of shortening or 10 µL of oil was added to 4 mL chloroform (Fisher, Cat#C298-4) : methanol (Fisher, Cat#A452-4) solution (2 : 1, v/v). Samples were vortexed for 1 min, flushed with nitrogen gas (Boc gases, Guelph, ON) and incubated at 4 °C overnight. On the following day, samples were centrifuged at 1000 rpm for 10 min (21 °C) to separate the phases. The lower chloroform layer was extracted and transferred to a fresh test tube and dried down with a gentle stream of nitrogen gas. The lipid was saponified in 0.5 M KOH in methanol and heated for 1 h at100 °C. Phospholipids were converted to fatty acid methyl esters with the addition of 14% boron trifluoride (Sigma, cat#B1252)/methanol and incubation at 100 °C for 1 h. Fatty acid methyl esters were quantified on an Agilent 6890N gas chromatograph equipped with flame ionization detection and separated on an Supelco SPTM-2560 fused-silica capillary column (100 m, 0.2 µm film thickness, 0.25 mm i.d.; Sigma, cat#24056). Samples were injected in splitless mode. The injector and detector ports were set at 250 °C. Fatty acid methyl esters were eluted using a temperature program set initially at 60 °C and held for 0.2 min, increased at 13 °C min^{-1} and held at 170 °C for 4 min, increased at 6.5 °C min^{-1} to 175 °C, increased at 2.6 °C min^{-1} to 185 °C, increased 1.3 °C min^{-1} to 190 °C and finally increased 13 °C min^{-1} to 240 °C and held for 13 min. The run time per sample is 37.77 min. The carrier gas was hydrogen, set to a 30 mL min^{-1} constant flow rate. Peaks were identified by retention times of fatty acid methyl ester standards (Nu-Chek-Prep, Elysian, MN) using EZchrom Elite version 3.2.1 software. Fatty acid concentrations were calculated as percent area.

Differential Scanning Calorimetry (DSC)

A differential scanning calorimeter (DSC; Q1000, TA Instruments, Mississauga, ON, Canada) was used in the thermal analysis of the different fat blends. The instrument heat capacity response was calibrated with sapphire, and the heat flow was calibrated with indium. Approximately 10 mg of the fat sample was placed in alodined pans and sealed hermetically (an empty pan served as reference). All measurements were performed at a heating rate of 5 °C/min. Thermograms were evaluated using TA Instruments Universal Analysis Software. The peak melting temperature (T_m) and the enthalpy of melting (ΔH_m) were determined. Four replicates of each sample were analyzed in this study.

Solid fat content determination

Crystallized samples were introduced into NMR glass tubes and stored for 24 h at 4 °C. Then, the tubes were incubated at the desired temperature for 30 min to allow a homogeneous distribution of temperature at the moment of measurement. Solid fat content (SFC) was measured by pulse nuclear magnetic resonance (p-NMR) using a Bruker Minispec spectrometer, (Bruker Optics Ltd., Milton, ON, Canada). The reported data correspond to the average of five individual measurements.

Polarized light microscopy and box counting fractal analysis

Polarized light microscopy (PLM) was used to observe fat microstructure. To obtain satisfactory reproducibility of slides and crystal appearance approximately 0.01 g of the crystallized sample plus 10 µl of soybean oil was weighed on a slide in order to maintain a 1 : 1 proportion; then the mixture was homogeneously spread in all directions and a cover glass was carefully laid over the fat to remove air and complete spreading the fat. Samples were imaged using a Leica DM RXA2 microscope with polarized light (Leica Microsystems, Richmond Hill, Canada) and equipped with a CCD camera (Q Imaging Retiga 1300, Burnaby, BC, Canada). All images were acquired using a 40X objective lens (Leica, Germany). The camera was set for autoexposure. Openlab 6.5.0 software (Improvision, Waltham, MA, USA) was used to acquire images. Focused images were stored as uncompressed 8-bit (256 grays) grayscale TIFF files with a 1280 × 1024 spatial resolution. Five images were captured from each of the five replicates prepared.

Microstructural analysis was carried out by image analysis employing the Adobe Photoshop CS 3 software (Adobe Systems Inc., San Jose, California, USA) and filters from the Fovea Pro 4.0 software (Reindeer Graphics, Inc., Asheville, NC, USA). A manual threshold was applied to all the pictures to convert the grayscale images to binary images, in order to discriminate between features and background and to measure the feature sizes. The microstructural elements were determined using the filter tools included in the Fovea Pro software.

Fat crystal network fractal dimensions (D_{box}) were determined by the box counting method.[18] Thresholded PLM images were processed using the software Benoit 1.3 (TruSoft Int'l Inc., St. Petersburg, FL) to calculate the 2D fractal dimension. A grid formed by boxes of decreasing sizes is placed over the binary images, and the number of occupied grids (N) is counted for a series of grid side length (L). Any box containing a number of particles greater than a threshold value is considered to be an occupied grid. The number of occupied boxes as a function of the size of the boxes is plotted, and the negative of the slope of the log–log plot is the box counting fractal dimension (D_{box}). An average of 25 replicates for each sample was determined.

Powder XRD analysis

XRD data were collected using a Rigaku Multiflex Powder X-ray Diffractometer (Rigakug, Japan). The copper lamp ($\lambda = 1.54$ Å for copper) was set to 40 kV and 44 mA. A 0.57 divergence slit, 0.57 scatter slit and 0.3 mm receiving slit were used. For the small angle X-ray diffraction analysis (SAXD) the samples were scanned from 0.9 to 8 degrees at 0.05°/min. The wide angle X-ray diffraction analysis (WAXD) was carried out for scanning the samples from 16 to 35 degrees at 0.5°/min. PeakFit software (Seasolve, Framingham, MA, USA) and MDI's Jade 6.5 software

(Rigaku, Japan) were used to analyze the obtained SAXD and WAXD patterns respectively.

From the SAXD patterns, the crystalline domain size (ξ) can be calculated by the well-known Scherrer formula which is limited to nano-scale particles and it is not applicable to sizes larger than about 100 nm:[19]

$$\xi = \frac{K\lambda}{\text{FWHM}\cos(\theta)} \tag{1}$$

where K is the shape factor, θ is the diffraction angle of the X-rays, FWHM is the full width at half of the maximum peak height in radians (usually from the first small angle reflection corresponding to the (001) plane) and λ is the wavelength of the X-ray. The dimensionless shape factor provides information about the "roundness" of the particle. For a spherical particle the shape factor is 1, for all other particles it is smaller than 1. A value of 0.9 is usually used for crystallites of unknown shape and the magnitude employed in this study.

Cryogenic transmission electron microscopy (Cryo-TEM)

In order to discard the oil fraction and isolate single crystals for observation, fat blends were treated as reported previously by Acevedo and Marangoni.[20] Samples were treated at 10 °C as follows. Fat blends were suspended in cold isobutanol at a ratio of approximately 1 : 50 using a glass stirring rod to obtain a uniform suspension. The fat plus isobutanol mixtures were homogenized at 30,000 rpm with a rotor-stator (Power Gen 125, Fisher Scientific) for 10 min. Then, the crystals were collected by vacuum filtration through a glass fiber filter of 1.0 μm pore size. After filtration, the recovered solid was re-suspended in cold isobutanol and re-homogenized for 10 min using the rotor-stator in order to obtain a suitable dispersion of crystals. Finally the mixtures were sonicated at 10 °C for 60 min using an ultrasonic processor (Bransonic 1210R-DTH, Branson Ultrasonic Corporation, Danburry, CT, USA) to complete the dispersion of the fat crystals. Five microliters of the obtained dispersion were placed on a copper grid with perforated carbon film (Canemco-Marivac, Quebec, Canada), and excess liquid was blotted automatically for 2 s using filter paper. A staining aqueous solution of 2% of uranyl acetate was used to enhance contrast. Subsequently, the sample was transferred to a cryo holder (Gatan Inc., Pleasanton, CA, USA) for direct observation at −176 °C in a FEI Tecnai G2 F20 Cryo-TEM operated at 200 kV in low dose mode (Eidhoven, The Netherlands). Images were taken using a Gatan 4k CCD camera. Micrographs were stored and analyzed using DigitalMicrograph™ software (USA). Image J 1.42q software (USA) was employed for a semiautomatic analysis procedure.

Oil loss determination

Oil loss studies were performed according with the technique described previously by Dibildox-Alvarado[14] et al. Once crystallized, fat blends were molded into discs of 22 mm diameter and 3.2 mm of thickness using polyvinyl chloride (PVC) molds and then transferred to filter papers (Whatman #5, 110-mm diameter). The amount of oil that each sample (prepared as discs) lost to filter papers was determined by the difference in weight of the filter papers before and after placing the fat disc on the paper for 24 h at 20 °C. A "blank" filter paper was included in all experiments to account for the effects of the treatments on the paper itself such as the influence of the humidity of storage environment. Filter papers must be large enough in order to avoid the paper saturation with oil during the period of measurement. An average and standard deviation of at least five replicates (five separate disks on individual filter papers) is reported. Oil loss (%) was calculated as:

$$OL(\%) = \frac{\text{wt. paper(24 h)-wt. paper(0 h)}}{\text{wt. paper(0 h)}} \times 100 \qquad (2)$$

Small deformation rheology

After crystallization samples were transferred to the wells of polyvinylchloride (PVC) disc molds of 3.2 mm thick and 20 mm in diameter. Rheological measurements were obtained using a TA Instrument AR2000 controlled stress dynamic Rheometer (TA Instruments, Mississauga, ON, Canada). A 20 mm diameter stainless steel flat plate was selected to carry out the experiments. The temperature of the sample was held at 20 °C during analysis achieved by a Peltier system located in the base of the measurement geometry.

The oscillatory stress sweep within the region of linear viscoelastic behavior (LVR) was performed from 0.1 to 1000 Pa (with a frequency of 0.1 Hz) for sheared samples and from 1 to 10000 Pa (with a frequency of 1 Hz) for statically crystallized samples. Compression was set to contact or to a normal force of 5 N for sheared and non-sheared blends, respectively. To prevent slippage, sandpaper (grade 60) was attached to the lower surface of the geometry and the upper surface of the Peltier base of the rheometer using Krazy glue. The yield stress (σ^*) values were determined from the stress sweep curves as the stress value (in Pa), above the LVR, at which the storage modulus (G') deviates 10% from a constant value. The reported data are the average of 6–10 individual replications.

Statistical analysis

Data were processed using GraphPad Prism 5 software (GraphPad Software, Inc., San Diego, CA, USA). Reported values correspond to means and standard errors of the determinations. Statistical analysis was performed by one-way ANOVA ($p < 0.001$) using Tukey's multiple comparisons as post-test ($p < 0.005$).

Table 1 Fatty acid composition of Fully Hydrogenated Soybean Oil (FHSO) and Soybean Oil (SO)

Fatty acid	SO	FHSO
16:00	11.18	12.81
17:00	0.14	0.34
18:00	4.58	83.91
18:1t6–8	0	0.07
18:1t9	0	0.11
18:1t10	0	0.12
18:1t11	0	0.31
18:1t12	0	0.12
18:1t13	0	0.20
18:1c6–10	21.73	0.50
18:1c11	1.47	0.11
18:1c12	0	0.16
18:2c9–t12	0.56	0
18:2t9c12	0.15	0
18:2n6	52.57	0.20
20:00	0.38	0.68
20:1c11	0.47	0
18:3n3	6.77	0
22:00	0	0.36

3 Results and discussion

The fatty acid compositions of FHSO and SO are shown in Table 1. FHSO contained 83.9% stearic acid (C18:0), and 12.8% of palmitic acid (C16:0) which represents 96.7% of the total fatty acids. The predominant fatty acids in SO were linoleic acid (C18:2) with 52.6% of the total; followed by oleic (C18:1), palmitic (C16:0) and stearic acid (C18:0) with 21.7, 11.2 and 4.6%, respectively. These results are in agreement with values reported previously for FHSO and SO.[21] Powder X-ray diffractions patterns showed short spacings at 3.6, 3.9 and 4.6 Å characteristic for the β polymorphic form[22] in all blends, independent of their FHSO proportion and crystallization conditions.

Furthermore, no significant differences were observed in the melting point of all the blends. The global average peak melting temperature was 63.38 ± 0.44 °C. On the other hand and as expected, processing did not affect significantly the enthalpies of melting of the fat mixtures. The obtained global enthalpies of melting were 86.58 ± 2.04 J/g and 98.08 ± 0.84 J/g for blends with 40 and 45% FHSO, respectively.

Polarized light microscopy (PLM) was used to study the effects of crystallization conditions on the mesostructure of the fat blends. Fig. 1 shows PLM micrographs of samples containing 45% FHSO crystallized at the three different wall temperatures (−10, 0 and 20 °C) and shear rates (0, 30 and 240 s⁻¹). Fat blends crystallized statically at any wall temperature displayed a spherulitic morphology showing slightly larger and more branched spherulites at the highest wall temperature used in this work (20 °C). The slowest cooling rate at which fat systems are subjected at 20 °C induced the lowest degree of supercooling leading to a crystallization regime that favoured the spherulitic growth. These observations are in agreement with previous work using comparable fat mixtures.[23]

Unexpectedly, laminar shear processing promoted the growth of individual spherical crystalline particles, which we took the liberty to call "spherical-lipid meso-particles (SLM)" in this work (Fig. 1D–I). These SLMs had a remarkably well-organized

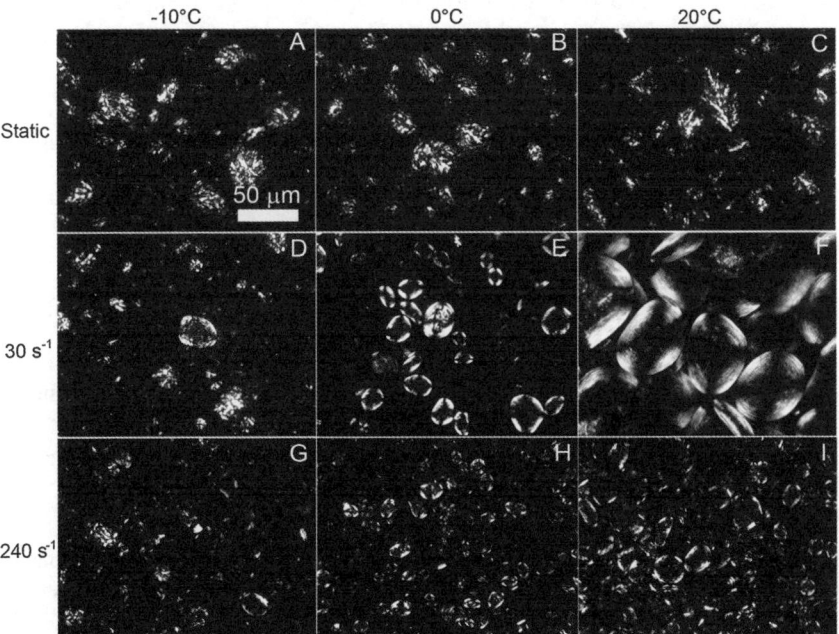

Fig. 1 PLM images of 45 : 55 FHSO : SO mixtures crystallized in static and under different wall temperature-laminar shear rate conditions.

internal structure. Furthermore, the growth of SLMs was not only dependent on the applied shear rate but also on the crystallization wall temperature. Fig. 1D–I show that an intermediate shear rate resulted in the largest SLMs in blends with 45% FHSO. Similar images and equivalent trends were obtained for mixtures containing 40% of FHSO (data not shown).

It is possible to explain the changes observed at the mesoscale by pointing out that crystallization under low shear or mild agitation promotes crystal aggregation.[24] At an intermediate shear rate of 30 s^{-1} there is a higher probability of crystals to coalesce and rearrange due to an increased collision frequency between crystals, resulting in the generation of large spherical particles. However, the application of a shear rate as high as 240 s^{-1} during crystallization can break the crystals resulting in more compact (smaller) aggregates,[9] in particular at 20 °C, where the extent of this effect is more pronounced. Although under these conditions SLMs still form, the shear forces generated are possibly high enough to exceed van der Waals attractive forces between crystals[6] causing breakage and/or preventing their further aggregation. Sonwai and Mackley[6] reported that the application of shear rate of 20 s^{-1} in cocoa butter appears to prevent the formation of spherulites to some extent, forcing some small crystallites to exist as individual crystal fragments. This is not surprising since the magnitude of the effect will depend on crystal size and the magnitude of inter-crystalline interactions within each system.

On another note, PLM microphotographs at −10 °C (Fig. 1 D, G) reveals that at this temperature, a granular texture combined with a small number of SLMs dominates the microstructure of the material, compared to the amount of spherical particles observed at other crystallization temperatures.

For static crystallization, no significant differences ($p < 0.0001$) were observed in meso-crystal size (equivalent diameter) of blends with both hardstock proportions at wall temperatures of −10 °C and 0 °C (Fig. 2A). Slightly larger ($p < 0.05$) meso-crystals were observed in blends crystallized at 20 °C, mainly due to the predominance of growth over nucleation at higher temperatures (and slower cooling rates). Nevertheless, it is important to mention that in this study the cooling rates during crystallization at −10 °C and 0 °C were not sufficiently different to induce significant changes between meso-crystal dimensions in samples crystallized under these conditions.

At an intermediate shear rate of 30 s^{-1} crystallization yielded larger meso-crystals, in particular at higher wall temperatures (Fig. 2B). These results are opposite to what was expected since shear enhances nucleation leading to the formation of smaller crystals. Garside and Davey[25] proposed that crystal collision in moderately sheared samples could provide a potential for surface damage, which favours a more rapid surface integration enhancing the growth rate, therefore leading to an increase in crystal size.

Under these intermediate conditions, it is worth noting that at lower wall temperatures (−10 °C and 0 °C) the effect of supersaturation on meso-crystal dimensions is still evident (Fig. 2B). Crystal sizes in blends with 40% of FHSO were between 14 to 26% larger than those found in mixtures with 45% FHSO. At −10 and 0 °C, the lower supersaturation of mixtures with a smaller amount of FHSO furnished a reduced viscosity suitable for crystal growth. However, at 20 °C, shear effects predominated over those of FHSO concentration and therefore no differences in crystal size were to be detected ($p < 0.05$) at both supersaturations.

The smallest meso-crystals were detected after crystallization at a shear rate of 240 s^{-1} (Fig. 2C), with sizes ranging from 1.1 to 1.3 µm. Moreover, the meso-crystal dimensions attained were not significantly different ($p < 0.05$) between all the blends, independent of supersaturation and wall temperature conditions. These results are not only a consequence of enhanced nucleation under these conditions, but also of the fact that high shear forces can lead to crystal breakage which results in the formation of a mesostructure composed of small crystals.[6] The increase and reduction in crystal size of fat mixtures at lower and higher shear rates, respectively were

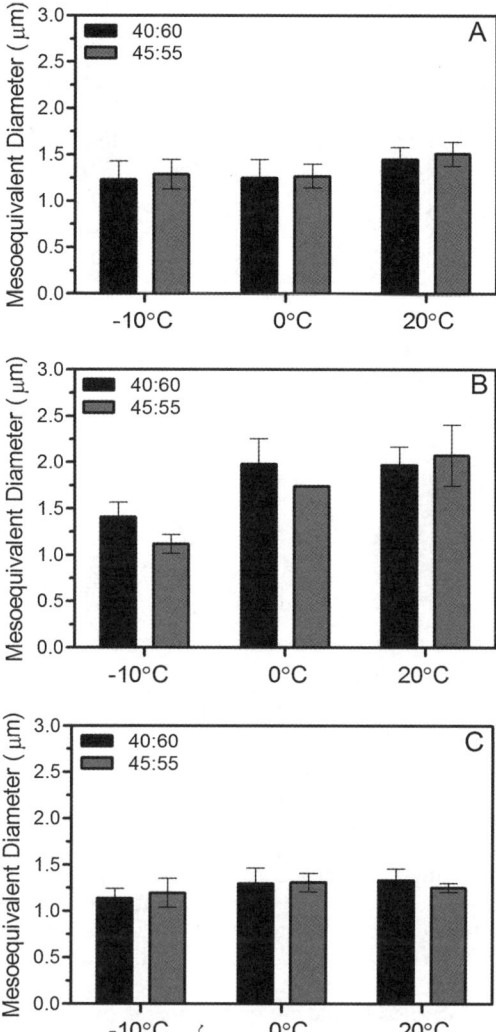

Fig. 2 Equivalent diameters of the mesostructural elements obtained from PLM image analysis of FHSO : SO blends. (A) Static crystallization; (B) crystallization at a laminar shear rate of 30 s^{-1}; (C) crystallization at a laminar shear rate of 240 s^{-1}. Error bars represent standard deviations.

consistent with the findings of Grall and Hartel[26] during butterfat crystallization, Herrera and Hartel[1] working with milk fat and Sonwai and Mackley[6] shearing cocoa butter samples.

As expected, shear effects, particularly at intermediate rates, on the microstructure were more pronounced at higher crystallization temperatures. Compared to statically crystallized mixtures, at a wall temperature of -10 °C, a crystal size increase of $\sim$15% and a reduction of $\sim$8% were observed when blends where subjected to shear rates of 30 and 240 s^{-1}, respectively. Instead, at 20 °C the meso-crystals grew $\sim$38% more after intermediate shearing and showed a $\sim$15% decrease when sheared at 240 s^{-1}.

In order to quantify the effects of shear on meso-crystal size, we determined the mean diameters of SLMs shown in Fig. 3. As observed in the PLM images (Fig. 1D–I), SLMs are considerably larger in blends crystallized at an intermediate

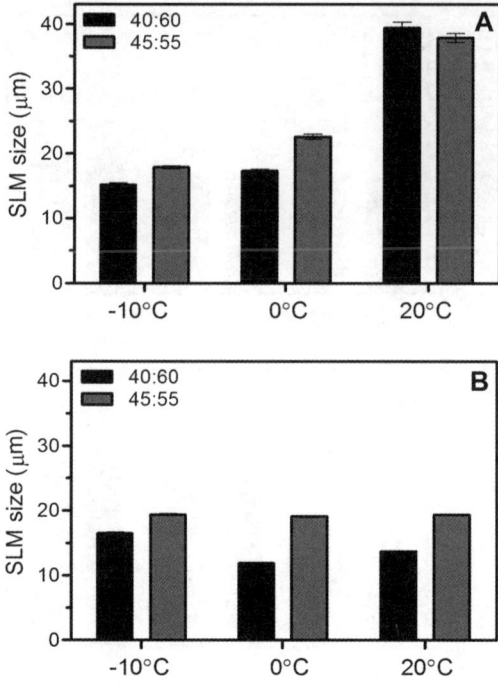

Fig. 3 Calculated mean diameters of SLMs observed in the PLM images of sheared blends. (A) 30 s^{-1}; (B) 240 s^{-1}. Error bars represent standard deviations.

shear rate (Fig. 3A). Additionally, in these fat blends the higher the wall temperature used in the crystallization process, the greater the average size attained.

Statistical analysis shows that shear processing at 240 s^{-1} resulted in a significant reduction in the size of the SLMs ($p < 0.001$). Not surprisingly, upon shearing at this rate, wall temperatures did not have an effect on SLM dimensions. These results seem to indicate that there is a critical shear rate above which SLM growth is no longer enhanced.

SLM size was also affected by the mixture's supersaturation (Fig. 3). Significantly larger particles ($p < 0.001$) can be noted in blends with a higher proportions of FHSO at both shear rates. These findings are contrary to what is expected since supersaturation is the driving force for both the initial nucleation step and the following crystal growth and therefore it governs the eventual crystal size. The degree of supersaturation controls whether the reaction is nucleation or growth dominated. At high supersaturations, crystallization is dominated by nucleation, which will generally lead to the formation of smaller crystals than for one that is dominated by growth.[25] However in this case we are dealing with a larger structural level governed to some extent by crystal aggregation phenomena, where SLM growth may be favoured by the greater concentration of saturated fat in the system. As demonstrated by Bremer and Smits,[27] aggregation is favoured at high solid concentrations and in systems where the collision between crystals is high, which characterizes this case.

Cryo-TEM was used to systematically study the effects of shear and crystallization temperatures on the nano-structure of FHSO : SO mixtures. In recent studies performed by Acevedo and Marangoni,[20] a new methodology based on a cold-solvent-extraction of solid lipids and the Cryo-TEM technique has been described for the characterization of fat nanostructure. Here, we used this technique to characterize the effects of shear on the nano-scale of a fat crystal network. Examples of

Cryo-TEM micrographs showing the extracted nanoplatelets from samples with 45% FHSO crystallized statically and under both shear rates at 3 different wall temperatures are presented in Fig. 4. Similar images and with analogous tendencies were observed for blends with 40% of FHSO (data not shown). As anticipated, platelet-like nano-crystals can be distinguished in Fig. 4, which constitute the primary crystals of the network.[20] Sheared blends (Fig. 4D–I) showed a predominance of larger nano-particles with higher aspect ratios (length to width ratio) than those in non-sheared blends. Again, as observed at the mesoscale, wall temperature seems to strongly influence the size of the nanoplatelets. Regardless of shearing conditions, as the wall temperature rises, so do nano-crystal dimensions.

Cryo-TEM images were analyzed by measuring the lengths and widths of each individual nanoplatelet. From the obtained dimensions, the frequency size distributions were plotted and the corresponding mean (median) dimensions determined. Fig. 5 illustrates the changes taking place in the average platelet lengths and widths with different shear rates and wall temperatures. In general, when samples were exposed to higher temperatures during crystallization, the platelets increased in size. This is understandable given the prevalence of crystal growth over nucleation at higher temperatures. Shearing caused a significant increase in nanoplatelet size, in particular at 0 and 20 °. These results show that crystallization at an intermediate shear rate yielded platelets with lengths and widths, respectively, up to 2.6 and 1.5 times larger than those in non-sheared samples. Meanwhile, at shear rates of 240 s^{-1} lengths and widths were only up to 1.8 and 1.3 times greater, respectively, than the values obtained from blends crystallized statically. Hence, low shear rates of 30 s^{-1} led to the largest nano-crystal sizes. These results are consistent with the

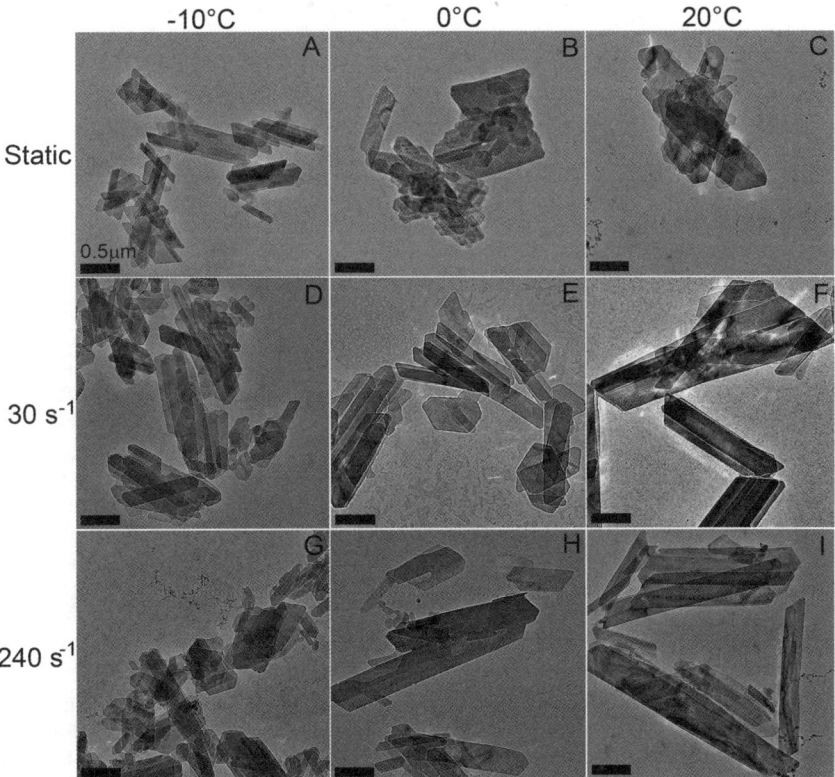

Fig. 4 Observation by Cryo-TEM of the nano-crystals formed during crystallization at different wall temperatures and under static and sheared conditions.

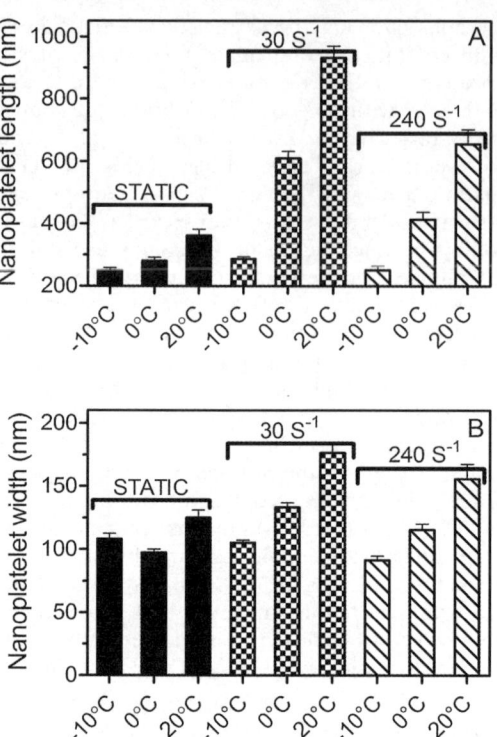

Fig. 5 Variations observed by Cryo-TEM in average nanoplatelet lengths (A) and widths (B) induced by changes in crystallization conditions.

visual observation of the Cryo-TEM images that indeed have shown greater platelet sizes in blends sheared at low rates.

Fig. 6 shows changes in nano-crystal thickness, obtained by the Scherrer analysis of the (001) plane reflection in the small angle region of a powder XRD spectrum. Results indicate that shear and crystallization temperature may have an effect on the thickness of the nanoplatelets. Even though the mean values are not significantly different ($p < 0.1$), it is possible to observe a decreasing trend in particle thickness with an increase in wall temperature at the intermediate laminar shear rate used. These results combined with those found for nanoparticle lengths and width suggest that under mild shear forces, length and width increase while thickness decreases.

Acevedo and Marangoni[28] demonstrated that cooling and shear rates markedly affected the nanoscale of crystalline FHCO networks. They found that both high cooling rates ($10\,°C\,min^{-1}$) and shearing at $300\,s^{-1}$ favoured the formation of smaller nano-crystals. Likewise, Maleky[7] *et al.* in a recent work applied shear rates of $340\,s^{-1}$ during the crystallization of cocoa butter at $20\,°C$ observing the growth of considerably smaller nano-crystals under this shearing condition. One might have thus predicted that a greater shear rate always leads to the formation of smaller nanocrystallites. However, our results suggest that this is the case above a critical shear rate only. Below this critical shear, the opposite trend is observed. Increases in shear from static conditions ($0\,s^{-1}$) to $240\,s^{-1}$ lead to an increase in crystallite size, while at shear rates greater than $300\,s^{-1}$, a decrease is observed. Unlike the mesoscale, at the nanoscale we do not have aggregation considerations. What might be the case is that above a certain shear rate, crystal fragments begin breaking off, inducing secondary nucleation and the creation of a greater number of smaller crystals. Thus, the eventual nano-crystal size is increased due to enhanced mass transfer

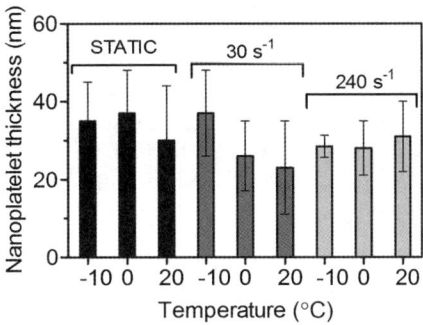

Fig. 6 Changes observed in the nanoplatelet thickness at different wall temperatures and laminar shear rates. Data were obtained by the Scherrer analysis of the SAXRD patters.

up to a critical shear rate (240 s^{-1} in our case), but any further increases in shear rate lead to nano-crystal fragmentation and an enhancement of secondary nucleation. This value of 240 s^{-1} is very rough and better estimates (narrower ranges) should be sought. From a practical point of view, this finding defines shear and temperature ranges in which crystal size and number can be affected dramatically by judiciously breaking them.

Da Pieve[29] *et al.* investigated the effect of shear rates between 0 and 2,000 s^{-1} during crystallization of monoglycerides in oil, and although these matrixes are not fats, the author's findings can be related to our results. Domain sizes, which represent nano-crystal thickness, were larger in sheared systems compared to non-sheared samples, however size decreases were observed with successive increases in the shear rate from values of 100 s^{-1} to 2,000 s^{-1} which may be indicating the presence of a critical shear rate between 0 and 100 s^{-1} for the monoglyceride system studied.

As can be inferred from the calculated nanoplatelets' aspect ratios shown in Table 2, there is an enhancement in the longitudinal growth of the nano-crystals as the crystallization temperature increases at the two shear rates applied during crystallization. There is reasonable agreement between this work and that of Acevedo and Marangoni[28] where a large increase (more than twice) in the nanoparticles' aspect ratios was reported in samples crystallized at a high cooling rate of 10 °C min^{-1} from the melt, even though sizes were considerably reduced. The results of this work reveal a substantial increase in nanoplatelet longitudinal growth at intermediate shear rates. However, the relative magnitude of this effect diminishes when the rate is increased from 30 s^{-1} to 240 s^{-1}. For example, at 20 °C, when compared to statically crystallized samples, the aspect ratios increased by 83% and 45% when rates of 30 and 240 s^{-1} were applied, respectively.

Earlier studies pointed out that shear induces the orientation of crystallites.[4,7] The orientation of particles suspended in a flowing system is determined by the interplay

Table 2 Nanoplatelet aspect ratios obtained from the analysis of the Cryo-TEM micrographs of non-sheared and sheared blends of Fully Hydrogenated Soybean Oil (FHSO) and Soybean Oil (SO)

	Non-sheared (s^{-1})	Sheared (s^{-1})	
Wall temperature (°C)	0	30	240
−10	2.3	2.7	2.7
0	2.9	4.6	3.6
20	2.9	5.3	4.2

between interparticle interactions, and the relative magnitude of shear and Brownian forces.[30] In the melt at low volume fractions of solids, interparticle forces do not come into play due to the large relative distance between crystals. Therefore, for instance, in an environment dominated by an ordering effect of shear forces, it is reasonable to assume that the suspended non-spherical particles can develop preferred orientations parallel to the direction of the flow (and rapid rotations fast perpendicular to the shear field). These phenomena induced by shear forces would allow a predominantly unidirectional growth of the nanoplatelets parallel to the direction of the shear flow. Additionally, it has been suggested that the Peclet number (Pe_r) at a given shear rate characterizes the transition from Brownian to shear regime; where the rotational Peclet number is defined by:

$$Pe_r = \frac{\dot{\gamma}}{D_r} \tag{3}$$

where $\dot{\gamma}$ is the shear rate in s^{-1} and D_r is the rotational diffusivity. The nanoplatelets can be assumed to have an ellipsoid-like shape whose rotary diffusivity is given by the Perrin equation:[31]

$$D_{r\text{-platelet}} = \frac{k_B T}{8\pi\eta a^3} \times \left\{ \frac{3}{2} \frac{\left(\frac{2-\left(\frac{1}{r^2}\right)}{\left(\sqrt{1-(1/r^2)}\right)}\right)\ln\left[r(1+\sqrt{1-(1/r^2)}\right]-1}{1-\left(\frac{1}{r^4}\right)} \right\} \tag{4}$$

In this equation T is the temperature, η the medium viscosity, k_B Boltzmann's constant, a and r are the magnitude of the semi-axis and the aspect ratio of the particles, respectively. The corresponding liquid fat viscosity and temperature were obtained before the nucleation occurs, under the external wall temperature/shear rate conditions.

Larson[30] proposed that at $Pe_r > 10$ orientation effects induced by the shear forces are active. The dependence of the rotational Peclet number on the particle equivalent diameter at both shear rates, 30 and 240 s^{-1} is shown in Fig. 7A. At all temperature/laminar shear rate combinations the Pe_r for the incipient nanoplatelets is higher than 10 which suggests particles are oriented even at the lowest shear rate used in these experiments. Our results are consistent with those reported by Mazzanti[4] et al. and Maleky[7] et al. who also found an average preferred orientation of the asymmetric crystals in the presence of laminar shear fields during crystallization.

The high Peclet numbers obtained for our shearing conditions could explain the enhancement of the platelets' longitudinal growth due to a predominant alignment of the crystallites parallel to the shear field. Furthermore, in order to better understand particle-orientation effects due to the applied shear for different particle geometries, we also plotted Peclet numbers as a function of crystal size considering nano-crystals as spheroid particles (Fig. 7C). In this case the rotary diffusivity, $D_{r\text{-sphere}}$ is equal to $k_B T/8\pi\eta r^3$, r being the sphere equivalent-radius. At 30 s^{-1} and the lowest wall temperature ($-10\,^{\circ}$C) Pe < 10; thus the shear induced perturbations are restored by Brownian forces. On the other hand, at higher wall temperatures, Pe > 10 indicating that hydrodynamic forces exceed the Brownian disordering forces. Therefore the results show that it is evident that 30 s^{-1} is the critical shear rate at which the nano-crystals' orientation is beginning to be affected by the applied external shear field.

Particular interest has been focused on the study of oil migration mechanisms in fat systems, including, among others, the effects of variations in crystal size and orientation with respect to the plane of oil migration. However, very little

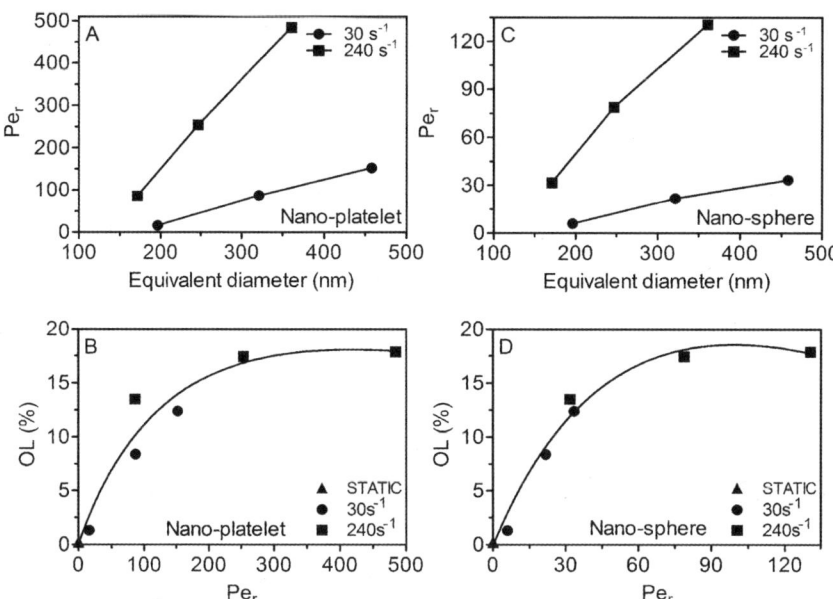

Fig. 7 Peclet numbers (Pe_r) as a function of the nano-crystal dimensions for sheared samples (A) shape of nano-platelet; (C) shape of rigid sphere. (B, D) Pe_r dependence with oil loss (OL) for all studied blends.

information on the influence of the different structural levels in the network on oil loss can be found. Oil binding capacity (OBC) is the physicochemical property of fat crystals to bind and trap liquid oil and is determined in this study through the measurement of the steady state (maximum) oil loss (OL, eqn (2)) which is the mass of oil lost from the fat sample under specific conditions.

It can be seen in Fig. 7B,D that OL increased in an asymptotic fashion as a function of Pe_r which in turn increased as the shear rate-temperature combinations were more pronounced. These results demonstrate that there is a strong relationship between the induced particle orientation and the rate of the oil movement within the network. It is manifested that there is an important and linear increase in the oil loss at the intermediate shear rate, where the alignment of the particles along the direction of the flow field begins to be evident. OL seems to reach a maximum value at a shear rate of 240 s^{-1} and 0 °C, and hardly changes when Pe_r values are higher than 250 or 78 for platelets or spheres respectively. Under these conditions the structural network disruption is such that the matrix is incapable of retaining liquid TAGs within the network, independently of the Pe_r.

OL values are summarized in Table 3. It can be seen that at both 40 and 45% FHSO, OL values increase significantly as the laminar shear rate increases, suggesting that shear affects the structure of the system in such a way that the oil loss from the crystal network is enhanced. Furthermore, at each shear rate, an increase in wall temperature leads in general to higher values of OL. However, the temperature influence on OL was not as noticeable as that of the shear rate, in particular under static conditions. In addition, it is interesting to note that the higher the wall temperature, the smaller the change in OL when changing the shear rate from 30 s^{-1} to 240 s^{-1}. Maleky and Marangoni[7] reported opposite results; they documented that laminar shear crystallization decreased oil migration in a cocoa butter matrix. Nevertheless, these researchers worked at a significantly higher shear rate (340 s^{-1}) which translates into a different crystalline network organization and therefore a dissimilar movement of oil through the sample. Also, the fact that they reported 'oil migration

Table 3 Steady state oil loss (%) of non-sheared and sheared blends of Fully Hydrogenated Soybean Oil (FHSO) and Soybean Oil (SO). Superscript letters represents statistically significant differences between the values ($P < 0.05$)

FHSO : SO	Wall temperature (°C)	Non-sheared (s⁻¹)	Sheared (s⁻¹)	
		0	30	240
40 : 60	−10	0.60 ± 0.70^a	2.30 ± 0.64 [c]	4.70 ± 0.83 [f]
40 : 60	0	0.50 ± 0.12^a	9.00 ± 0.96^d	21.70 ± 0.90^g
40 : 60	20	1.10 ± 0.15^b	23.80 ± 1.70^e	24.20 ± 0.70^e
45 : 55	−10	0.30 ± 0.06^a	1.33 ± 0.53^b	13.50 ± 0.53^h
45 : 55	0	0.32 ± 0.07^a	8.40 ± 0.71^d	17.46 ± 0.74^i
45 : 55	20	0.23 ± 0.11^a	12.40 ± 0.42^h	17.90 ± 0.75^i

rate' using the model proposed by Ziegleder *et al.*[32] instead of 'steady state oil loss', as in our case, may contribute to the contradictory results. Moreover, these authors worked on cocoa butter, which is a significantly more complex system than ours. Another important observation is that these researchers reported not only the growth of smaller nano-particles but also the decrease in their aspect ratios under laminar shear fields. Thus, it seems that both platelet characteristics, size and aspect ratio, play a fundamental role in the process of liquid TAG diffusion through a crystalline fat matrix.

We were interested in identifying the structural characteristics that were correlated most to oil binding capacity. In order to do so, we determined the equivalent diameters of crystals at the nano- and mesoscale, as well the equivalent diameter of the SLMs. Fig. 8A–C shows the results obtained for mixtures with 45 : 55 (w/w) FHSO : oil. As expected, particle diameters were significantly greater when wall temperatures were higher at both the nano- and mesoscales. Exceptions to this generalization were observed at the meso-crystal and SLM scales in highly sheared samples, where dimensions remained constant upon changing crystallization wall temperatures. Once again, samples crystallized at an intermediate shear rate displayed the largest particle sizes.

The relationship between OBC (represented by OL values) and particle diameters of physical blends composed of 45 : 55 (w/w) FHSO : SO are shown in Fig. 8D–F. What can be appreciated in this case is that, regardless of the structural level analyzed, the higher the shear rate used during crystallization, the greater the oil loss or oil migration in the blends. This suggests that shearing lead to a decrease in the oil binding capacity of the system. This behaviour can be attributed in part to the unusual microstructures created in each sample under laminar shear. We believe that the formation of SLMs largely contributes to the higher OL values. As a result of crystal breakage and/or reorganization stimulated by shear forces, SLMs begin to form as a consequence of the crystal tumbling, and oil appears to be excluded from these growing particles. Thus, a higher local amount of free oil becomes available in the network (outside the SLM) and consequently the driving force for oil loss is greater. When sheared at the same rate, the higher the wall temperature used, the larger the SLM and therefore the higher amount of oil expelled from the SLM. This allowed the liquid oil to migrate more easily out of the solid network compared to the denser and more homogenous network of non-sheared blends.

Additionally, and as anticipated, there is a good correlation between the increase in size of nano- and meso-crystals and the increase in OL, since permeability is strongly dependent on crystal size. The size changes occurred in the structural elements at the nano and mesoscale may contribute to the mechanism of oil loss

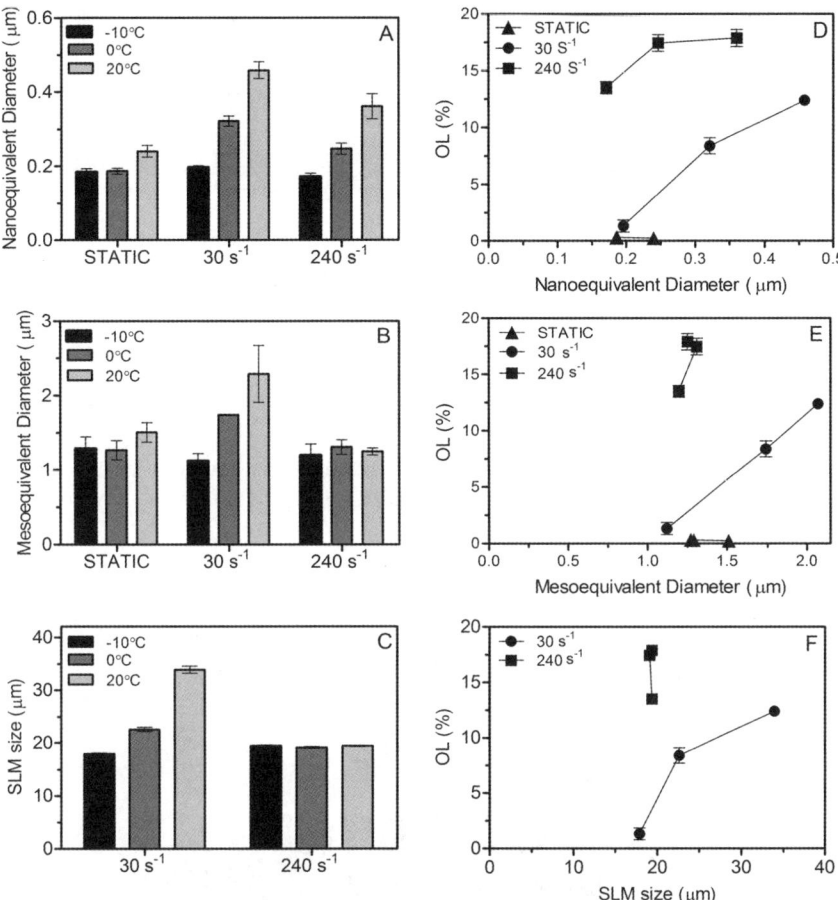

Fig. 8 Changes in particle diameters at the nano- (A), and micro-scale (B and C) induced by different processing conditions. Relationship between oil loss values (OL) and particle diameters [nano- (D); meso (E) and SLM particles (F)]. Error bars represent standard deviations.

exhibited in the blends. It is well known that solid crystalline networks structured by small crystals have an increased specific surface area, which improves contact with the oil, tend to be stronger and therefore have decreased OL.[23,33]

Normally, the capacity to bind oil in plastic fats increases as the total amount of crystalline network mass increases. Therefore, not surprisingly, significantly higher OL values were observed in mixtures containing 40% of FHSO (data not shown) relative to those with 45% of hardstock. Furthermore for both FHSO concentrations crystallization under shear seemed to induce an increase in the SFC, in particular at intermediate laminar shear rates, probably due to the high nucleation rate induced and the formation of highly developed crystals in relation to the formation of the crystal network under static and highly sheared conditions (Table 4).

The shear storage moduli G' for the different crystallization conditions are shown in Fig. 9A. The G' of these elastoplastic materials provides an indication of the elastic, or solid-like character of fats and is a strong function of the solid fat content.[34–36] Obviously, it was substantially higher ($p < 0.05$) in samples with higher % of FHSO, independent of processing conditions (Table 4).

We also determined the yield stress (σ^*) of all the blends since it has been previously shown that the yield stress of a fat correlates strongly to the macroscopic

Table 4 Solid Fat Content (SFC) at 20 °C of non-sheared and sheared blends of Fully Hydrogenated Soybean Oil (FHSO) and Soybean Oil (SO). Superscript letters represents statistically significant differences between the values ($p < 0.05$)

FHSO : SO	Wall temperature (°C)	Non-sheared (s⁻¹) 0	Sheared (s⁻¹) 30	240
40 : 60	−10	38.4 ± 0.2^a	39.3 ± 0.1^b	39.0 ± 0.1^b
40 : 60	0	38.2 ± 0.3^a	41.0 ± 0.3^c	41.0 ± 0.1^c
40 : 60	20	38.1 ± 0.1^a	41.9 ± 0.2^d	39.4 ± 0.3^b
45 : 55	−10	43.8 ± 0.3^e	43.2 ± 0.1^e	41.7 ± 0.2^d
45 : 55	0	42.7 ± 0.1^f	46.6 ± 0.2^g	46.1 ± 0.2^i
45 : 55	20	42.7 ± 0.1^f	45.5 ± 0.2^h	44.0 ± 0.2^i

functionality, such as hardness and spreadability.[37] Marangoni and Rogers[38] proposed a structural definition for the yield stress (σ^*) of a particle network:

$$\sigma^* \approx \frac{6\delta}{a}\Phi^{\frac{1}{d-D}} \tag{5}$$

where the yield stress is determined by the crystal-melt interfacial tension (δ), the primary particle diameter (a), the Euclidean dimension of the embedding space (d) and the fractal dimension (D).

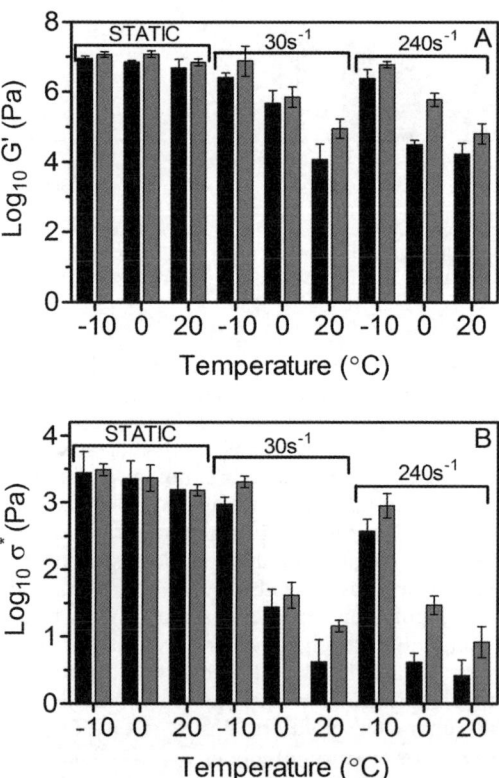

Fig. 9 Comparison between the LogG' and Log σ^* values obtained for samples with 40% (black bars) and 45% (grey bars) FHSO, processed at different crystallization conditions.

In general, our results agree with this model. Fig. 9B shows that when shear is applied during crystallization, σ^* is substantially lower and this decreasing trend is more pronounced when wall temperatures are higher. These results can be explained by the increase in primary crystal size induced by the external fields applied. As expected, G' exhibited a similar behaviour as σ^*.

An interesting observation that merits discussion is that yield stress seems to be more sensitive to processing conditions than elastic moduli. For instance, samples with 45% hardstock displayed an 80% decrease in yield stress when sheared at 30 s^{-1} and 20 °C relative to static crystallization, meanwhile the reduction in G' was only 40% for the same blends.

The shear modulus (G') is related to the volume fraction of solids (Φ) in a power law manner as a function of the network fractal dimension (D):[36]

$$G' = \lambda \Phi^{\frac{1}{3-D}} \tag{6}$$

The fractality of a fat crystal network lies in the aggregation of small crystallites (~300 nm) into larger clusters (~20–50 μm). Therefore, the solid-like macroscopic properties of these materials are highly dependent on their structure at the nanometer range (nanostructure), as well as micrometer range (microstructure). The fractal dimension is a parameter that describes the combined effects of morphology and spatial distribution patterns of the fat crystal clusters in the fat crystal networks.

In this work, the G' ratio between blends with 40 and 45% of FHSO was used to calculate the tridimensional fractal dimension for the arrangement of crystals within fractal flocs of crystals (D_r), namely:

$$D_r = 3 - \frac{\ln \frac{\Phi_1}{\Phi_2}}{\ln \frac{G'_1}{G'_2}} \tag{7}$$

Furthermore, the box-counting fractal dimension of the blends (D_b) was determined from image analysis of the PLM images and by adding "1" to the two-dimensional fractal dimension obtained to convert it to a 3-dimensional one.

Interesting trends were evident from this analysis (Fig. 10). Although D_r and D_b were not identical, their values showed similar trends for the different shear rate/wall temperature combinations. It is interesting to note that the rheological fractal analysis of samples crystallized statically yielded a fractal dimension D_r between 2.5 and 2.7, in line with a 3D diffusion-limited aggregation. These D_r values are in agreement with the dendritic pattern observed in the PLM micrographs (Fig. 1A–C), reminiscent of a fractal structure. However, under any of the two shear levels, the rheological fractal analysis yielded a D_r of ~2.9–3.0, suggesting a Euclidean dimensionality and which agree with the spherical and non-fractal looking clusters observed in Fig. 1D–I. Thus, under these specific shear conditions the regime has transformed from a fractal to near non-fractal. In general, an increase in D_b with temperature was observed for non-sheared samples and samples sheared at 30 s^{-1}. Meanwhile, no changes were observed at 240 s^{-1} (Fig. 10B). The decrease in G' observed at 240 s^{-1} as the temperature increases would have been expected to be the result of an increase in D_b, however the substantial structural changes induced may counterbalance the changes in D_b. Interestingly, the largest changes in D_b were observed at the intermediate shear rate studied, 30 s^{-1}, thus reflecting the correlation between network microstructure and small-deformation mechanical properties.

The relationship between the oil loss (OL) and the G' and σ^* of a fat are shown in Fig. 11. As expected, for all samples, OL is inversely correlated, to a greater or lesser extent, to both rheological parameters. Samples with lower elastic moduli or yield stresses, e.g. the sheared blends, displayed the greatest OL. The slopes for the linear

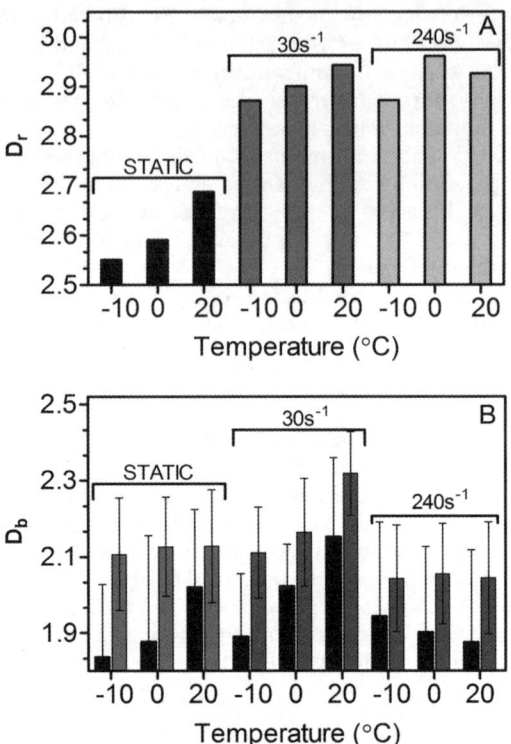

Fig. 10 Changes in the fractal dimension values obtained by box counting (D_b) and rheological data (D_r) of samples crystallized at different wall temperature/laminar shear rate conditions.

correlations in 45 : 55 FSHO:SO are ~0.3, ~5 and ~2 for samples crystallized under shear rates of 0, 30 and 240 s^{-1}, respectively. The significantly lower absolute values of the slopes at the highest shear rate are probably due to the greater structural damage inflicted upon the network by shearing.

In order to further investigate the relationship between structure and oil transport, we determined the permeability coefficient (B) of the oil through the crystal network *via* Darcy's law, as used by Bremer[39] *et al.*:

$$B = \frac{a^2}{\tau} \Phi^{\frac{2}{D_b - 2}}$$ (8)

In this equation a is the meso-particle size, τ is the tortuosity of the migration path (considered unity in this work due to the high volume fraction of solids in the system), Φ is the solids' volume fraction, and D_b is the fractal dimension obtained by microscopy with the box-counting method.

This model predicts that an increase in fractal dimension and particle size and a reduction in SFC lead to a higher permeability coefficient.

Not surprisingly the permeability coefficients (B) were higher in blends sheared at intermediate rates, at wall temperatures of 0 and 20 °C (Fig. 12A) indicating that these intermediate crystallization conditions may lead to maximum permeability. B values in highly sheared blends (240 s^{-1}) are similar to those observed for static crystallization. Under a shear rate of 30 s^{-1} the results show B values up to three times larger than those determined for static conditions and shear rates of 240 s^{-1}. Our results are in close agreement to data reported previously on similar

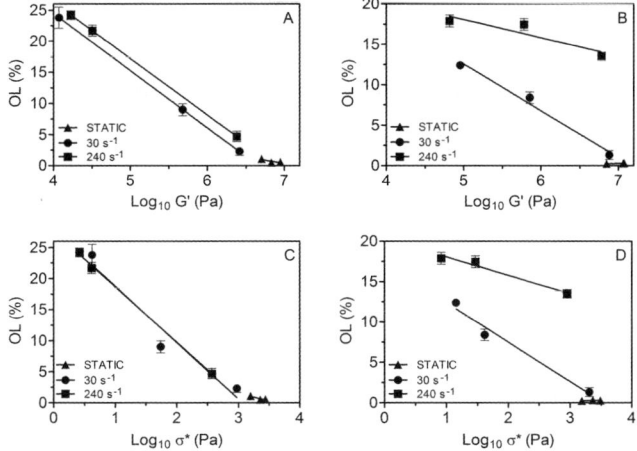

Fig. 11 OL (%) as a function of LogG' (A, B) and Logσ^* (C, D) for blends with 40% (A, C) and 45% (B, D) FHSO.

matrices.[7,14,40] As shown before in Fig. 8 OL values were higher in samples sheared at 240 s^{-1}, in contrast with B values which decreased at the highest shear rate after reaching their maximum at 30 s^{-1}.

It is worthwhile to note that there is only a linear correlation (with a high correlation coefficient, r^2 of 0.96) between OL and B at shear rates of 30 s^{-1} (Fig. 12B). In contrast to this trend, the permeability coefficient does not mirror OL trends for

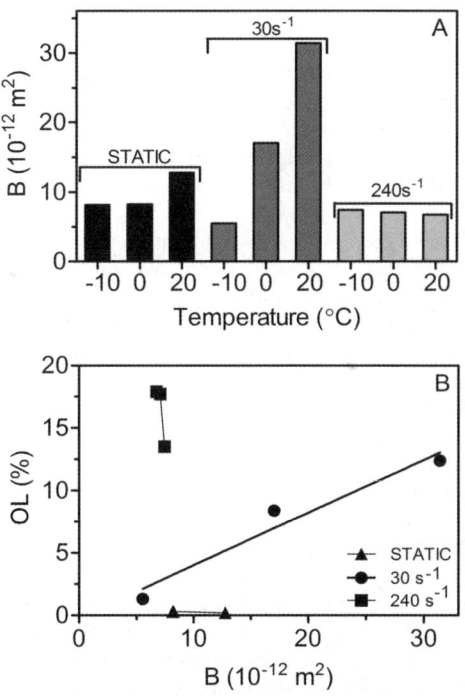

Fig. 12 Permeability coefficient values (B) obtained for blends crystallized statically, at shear rates of 30 s^{-1} and 240 s^{-1} (A). Relationship between OL and permeability coefficients (B) for all analyzed mixtures.

static conditions and the higher shear rates. A still unconsidered factor may play a significant role in explaining the different trends between oil loss and permeability coefficient for these matrices. Another possibility is that oil leakage from the network is a different phenomenon than a Darcy-type transport through a particle bed.

4 Conclusions

The present work describes the relationship between changes induced to the crystalline meso- and nano-structure and the oil binding capacity of FHSO : SO blends. Fat samples were subjected to different external shear and temperature fields with results revealing the existence of critical shear rate values above which strong alterations in the structure and permeability of the crystalline solid network took place.

It is worth mentioning that although the critical shear rate seems to be under 240 s^{-1} at both length scales, its effects on the structural element's sizes near this threshold is different. According to our results meso-crystal and SLM sizes did not change at different crystallization temperatures in highly sheared materials. Meanwhile nanoplatelets continuously increased in size under the same conditions. These findings would suggest that the nanostructure may be less sensitive to changes in processing conditions than the mesoscale. This is understandable since the mesoscale crystal sizes involve both crystal growth as well as aggregation phenomena of larger particles.

Additionally, as the rate applied during crystallization increases, effects on nanocrystal longitudinal growth are more predominant up to a certain value of shear. Once this threshold is exceeded, nucleation begins to dominate over growth. Furthermore, the breakage of the incipient crystals induced by the shear forces contributes to the formation of smaller particles. This explains the findings reported in previous works[7,28] where a decrease in the size of the particles under high shearing conditions was reported.

The uniqueness of the present study lies in the fact that we have been able to correlate effectively structural and mechanical properties of the system with oil loss. We believe that observed increases in oil leakage from laminar sheared samples is a combination of a generation of solid spherical crystalline particles (rather than porous spherulites), with a resulting exclusion of liquid oil from their structure, and an increase in the size of the meso-crystals, leading to a greater permeability of the system.

Furthermore we are aware that we have determined in this study values of steady state oil loss which are related to thermodynamic factors influencing oil binding capacity. However, oil binding capacity in fats is a complex concept that is influenced by both kinetic and thermodynamic factors. Thermodynamic factors would influence the total amount of oil that is retained by the network, while kinetic factors would influence the rate at which the oil is lost. A rapid oil loss does not necessarily mean that more oil will be lost in the end; therefore both aspects need to be determined, and reported, and they could be related to different structural factors. There is still much work to be performed and we are currently working on this area to determine the rate of oil loss by other methods, such as MRI, dye migration, FRAP or the slope in the early stages of the oil leakage behavior.

In addition, we found a correlation between oil loss and the permeability coefficient (Fig. 12B) only at a laminar shear rate of 30 s^{-1} and not under static conditions or after shearing at 240 s^{-1}. The results suggests that are factors other than crystal size, solid fat content, fractal dimension of the network and network tortuosity at play. For instance, the strength of intermolecular interactions is not considered in the models and these could have a significant effect on oil loss.

As a final observation and as expected, the analysis of OLs of the oil/fat mixtures showed that there is an inverse relationship between the capacity of the matrix to trap oil effectively and the mechanical properties of the fat, which in turn correlate inversely with meso- and nanocrystal sizes. Nevertheless, the fact that at the highest laminar shear rate used in this study, meso-crystals did not show further increase with increasing crystallization temperature (when OL and mechanical properties kept increasing) implies that the nanoscale is the length scale responsible for the mechanical strength and oil binding capacity rather than the mesoscale.

5 Acknowledgments

The authors acknowledge The Natural Sciences and Engineering Research Council of Canada and Advanced Foods and Materials network for the financial support. The Brazilian National Council of Technological and Scientific Development (CNPq – Conselho Nacional de Desenvolvimento Científico e Tecnológico) provided a postdoctoral fellowship for Dr Jane M. Block. We also wish to thank Professor David Ma (Department of Human Health and Nutritional Sciences, University of Guelph) for his analysis of the fatty acid composition of the oil mixtures.

References

1 M. L. Herrera and R. W. Hartel, *Ibid.*, 2000, **77**, 1177–1187.
2 R. Campos, S. S. Narine and A. G. Marangoni, *Food Res. Int.*, 2002, **35**, 971–981.
3 S. D. MacMillan, K. J. Roberts, A. Rossi, M. A. Wells and M. C. Polgreen, *Cryst. Growth Des.*, 2002, **2**, 221–226.
4 G. Mazzanti, S. E. Guthrie, E. B. Sirota, A. G. Marangoni and S. H. J. Idziak, *Cryst. Growth Des.*, 2003, **3**, 721–725.
5 G. Mazzanti, A. G. Marangoni and S. H. J. Idziak, *Phys. Rev. E: Stat., Nonlinear, Soft Matter Phys.*, 2005, **74**, 041607.
6 S. Sonwai and M. R. Mackley, *J. Am. Oil Chem. Soc.*, 2006, **83**, 583–596.
7 F. Maleky, A. Smith and A. G. Marangoni, *Cryst. Growth Des.*, 2011, **11**, 2335–2345.
8 W. Kloek, T. van Vliet and P. Walstra, *J. Texture Stud.*, 2005, **36**, 544–568.
9 P. Walstra, W. Kloek, T. van Vliet, in *Crystallization Processes in Fats and Lipid Systems*, ed. N. S. K. Garti. Marcel Dekker Inc., New York, USA, 2006, pp. 289–321.
10 J. M. Aguilera, M. Michel and G. Mayor, *J. Food Sci.*, 2004, **69**, 167–174.
11 P. Chawla and J. M. deMan, *J. Am. Oil Chem. Soc.*, 1990, **67**, 329–332.
12 L. deMan, J. M deMan and B. Blackman, *Fat Sci. Technol.*, 1995, **97**, 55–60.
13 D. Johansson and B. Bergenstahl, *Ibid.*, 1995, **72**, 205–211.
14 E. Dibildox-Alvarado, J. Neves Rodrigues, L. A. Gioielli, J. F. Toro-Vazquez and A. G. Marangoni, *Cryst. Growth Des.*, 2004, **4**, 731–736.
15 S. Marty, K. W. Baker and A. G. Marangoni, *Food Res. Int.*, 2009, **42**, 368–373.
16 R. P. Mensink, P. L. Zock, A. D. Kester and M. B. Katan, *Am. J. Clin. Nutr.*, 2003, **77**, 1146–55.
17 J. Folch, M. Lees and G. H. Sloane Stanley, *J. Biol. Chem.*, 1957, **226**, 497–509.
18 D. M. Tang and A. G. Marangoni, *J. Am. Oil Chem. Soc.*, 2006, **83**, 377–388.
19 A. R. West, In *Solid State Chemistry and Its Applications*, John Wiley & Sons, Chichester, West Sussex, England, 1984.
20 N. C. Acevedo and A. G. Marangoni, *Cryst. Growth Des.*, 2010, **10**, 3327–3333.
21 A. P. B. Ribeiro, R. Grimaldi, L. A. Gioielli and L. A. G. Gonçalves, *Food Res. Int.*, 2009, **42**, 401–410.
22 F. D. Gunstone and F. A. Norris, in *Lipids in foods: chemistry, biochemistry and technology.* Pergamon Press, Oxford, England, 1983.
23 T. S. Omonov, L. Bouzidi and S. S. Narine, *Chem. Phys. Lipids*, 2010, **163**, 728–740.
24 S. Chaiseri and P. S. Dimick, *J. Am. Oil Chem. Soc.*, 1995, **72**, 1497–1504.
25 J. Garside and R. J. Davey, *Chem. Eng. Commun.*, 1980, **4**, 393–424.
26 D. S. Grall and R. W. Hartel, *J. Am. Oil Chem. Soc.*, 1992, **69**, 741–747.
27 G. G. Bemer and G. Smits, *Proceedings of the 2nd World Congress of Chemical Engineering: Volume I*, 1981, 369–371.
28 N. C. Acevedo and A. G. Marangoni, *Cryst. Growth Des.*, 2010, **10**, 3334–3339.
29 S. Da Pieve, S. Calligaris, E. Co, M. C. Nicoli and A. G. Marangoni, *Food Biophys.*, 2010, **5**, 211–217.
30 R. G. Larson, in *The structure and rheology of complex fluids*, Oxford Univ. Press, Oxford, U.K., 1999.
31 F. Perrin, *J. Phys. Radium Ser.*, 1934, **5**, 499–511.

32 G. Ziegleder, C. Moser and J. Geier-Greguska, *Fett/Lipid*, 1996, **98**, 196–199.
33 A. Bot, Y. S. J. Veldhuizen, R. den Adel and E. C. Roijers, *Food Hydrocolloids*, 2009, **23**, 1184–1189.
34 D. Rousseau, K. Forestiere, A. R. Hill and A. G. Maragoni, *J. Am. Oil Chem. Soc.*, 1996, **73**, 963–972.
35 B. Liang, Y. Shi and R. W. Hartel, *J. Am. Oil Chem. Soc.*, 2008, **85**, 397–404.
36 S. S. Narine and A. G. Marangoni, *Food Res. Int.*, 1999, **32**, 227–248.
37 A. J. Haighton, *J. Am. Oil Chem. Soc.*, 1959, **36**, 345.
38 A. G. Marangoni and M. A. Rogers, *Appl. Phys. Lett.*, 2003, **82**, 3239–3241.
39 L. G. B. Bremer, T. vanVliet and P. Walstra, *J. Chem. Soc., Faraday Trans. 1*, 1989, **85**, 3359–3372.
40 S. Marty and A. G. Marangoni, *Cryst. Growth Des.*, 2009, **9**, 4415–4423.

Surface shear rheology of hydrophobin adsorption layers: laws of viscoelastic behaviour with applications to long-term foam stability

Krassimir D. Danov,[a] Gergana M. Radulova,[a] Peter A. Kralchevsky,[a] Konstantin Golemanov[ab] and Simeon D. Stoyanov[bc]

Received 9th February 2012, Accepted 19th March 2012
DOI: 10.1039/c2fd20017a

The long-term stabilization of foams by proteins for food applications is related to the ability of proteins to form dense and mechanically strong adsorption layers that cover the bubbles in the foams. The hydrophobins represent a class of proteins that form adsorption layers of extraordinary high shear elasticity and mechanical strength, much higher than that of the common milk and egg proteins. Our investigation of pure and mixed (with added β-casein) hydrophobin layers revealed that their rheological behavior obeys a compound rheological model, which represents a combination of the Maxwell and Herschel–Bulkley laws. It is remarkable that the combined law is obeyed not only in the simplest regime of constant shear rate (angle ramp), but also in the regime of oscillatory shear strain. The surface shear elasticity and viscosity, E_{sh} and η_{sh}, are determined as functions of the shear rate by processing the data for the storage and loss moduli, G' and G''. At greater strain amplitudes, the spectrum of the stress contains not only the first Fourier mode, but also the third one. The method is extended to this non-linear regime, where the rheological parameters are determined by theoretical fit of the experimental Lissajous plot. The addition of β-casein to the hydrophobin leads to softer adsorption layers, as indicated by their lower shear elasticity and viscosity. The developed approach to the rheological characterization of interfacial layers allows optimization and control of the performance of mixed protein adsorption layers with applications in food foams.

1. Introduction

The foams are thermodynamically unstable systems, which spontaneously decompose with time due to breakage of the liquid films intervening between the bubbles and due to the phenomenon Ostwald ripening (foam disproportionation).[1] These destructive processes, if not completely arrested, can be considerably slowed down, so that the foam could be considered stable in the time-scale of its practical applications. One of the strategies to stabilize foams, including foams in structured foods, is to "solidify" the bubble surfaces by covering them with a viscoelastic adsorption layer, which enhances the foam longevity at least in three aspects. First, the immobilization of the air/water interface slows down the film and foam drainage.[2–5] Second, the stability of the foam films is strongly enhanced, and their

[a]Department of Chemical Engineering, Faculty of Chemistry, Sofia University, 1164 Sofia, Bulgaria
[b]Unilever Research & Development, 3133AT Vlaardingen, The Netherlands
[c]Laboratory of Physical Chemistry and Colloid Science, Wageningen University, 6703 HB Wageningen, The Netherlands

permeability to gas transfer is considerably reduced. Third, the produced bubbles are considerably smaller and the foam viscoelasticity is markedly enhanced, which is important for the properties of many consumer products.[6-8] In the case of foams and emulsions in the food industry, the main substances that are used to enhance the interfacial rheology are proteins, which form dense and elastic layers at the surfaces of gas bubbles and emulsion drops. The interfacial rheology of protein adsorption layers has been intensively studied in relation to the properties of protein-stabilized foams,[9-11] emulsions,[12-16] and mixed systems such as proteins with lipids,[17,18] and proteins with surfactants.[19-22] Detailed information on the investigated systems, experimental techniques and theoretical models can be found in review articles on interfacial shear rheology.[22-27]

Recently, it was established that the hydrophobins are very promising stabilizers of foams[9-11] and emulsions.[28-30] They represent a class of proteins that are contained in filamentous fungi, including the common button mushroom.[31] The hydrophobins are relatively small proteins, whose structure is stabilized by several disulfide bridges, so that they do not undergo structural changes upon adsorption at interfaces.[31,32] The most remarkable property of the hydrophobins is that their adsorption layers at the air/water interface exhibit a considerable shear elasticity, which is higher than that of other investigated proteins.[11,33] The adsorption layers from hydrophobins solidify soon after their formation.[33,34] In applications, the hydrophobins can be used in mixtures with other proteins and/or surfactants,[35,36] which can modify their foam-stabilizing effect. For this reason, it is important to quantify and compare the surface rheological properties of adsorption layers of hydrophobins and their mixtures with other amphiphiles.

In the experiment, rotational surface rheometers in oscillatory regime are often used, which give the phenomenological surface storage and loss moduli, G' and G'' as functions of the frequency and time.[9-11] If the viscoelastic layer complies with the Kelvin law, then G' and G'' characterize, respectively, the elastic and viscous properties of the adsorption layer. The oscillatory regime does not give direct information on the applicability of the Kelvin or Maxwell model to a given system. However, an adsorption layer can be characterized with shear elasticity and viscosity, E_{sh} and η_{sh}, only in the framework of an adequate rheological model.

In a recent study,[36] we investigated the surface shear rheology of hydrophobin adsorption layers by a rotational rheometer in angle-ramp regime (increase of the rotation angle at a constant angular velocity). The results showed that the rheological behavior of the system complies with a combined Maxwell–Herschel–Bulkley law. From a physical viewpoint, if a rheological law is obeyed by a given continuous medium in a given kinetic regime, it should be obeyed by the same medium also in all other kinetic regimes. Our main goal in the present study is to investigate whether the combined Maxwell–Herschel–Bulkley model can also describe experimental data obtained in the oscillatory regime. This includes derivation of expressions for calculating the surface shear elasticity and viscosity, E_{sh} and η_{sh}, from the experimentally determined phenomenological moduli G' and G'', and verification of whether the same E_{sh} and η_{sh} are obtained in the angle-ramp and oscillatory regimes. In this way, the adequacy of the used model will be confirmed.

For this goal, we carried out experiments with hydrophobin and mixed, hydrophobin + β-casein adsorption layers, which are described in Sections 2 and 3. Further, the combined Maxwell–Herschel–Bulkley model was applied to interpret the data and to determine E_{sh} and η_{sh} as functions of the shear-rate amplitude (Section 4). This analysis was carried out in the regime of not-too-large amplitudes, at which the rheological response of the system is quasi-linear, i.e. the stress oscillated with the basic frequency (that of the applied sinusoidal strain), the higher-order Fourier modes in the strain being negligible. However, the Herschel–Bulkley law is nonlinear, and at greater amplitudes the higher Fourier modes give non-negligible contributions to the stress, so that we are dealing also with the case of a nonlinear rheological response. The latter case is considered in Section 5. At sufficiently large

This journal is © The Royal Society of Chemistry 2012

frequencies, the viscoelastic network of protein molecules at the interface can be broken as indicated by the Lissajous plots. The approach developed here for the investigation of hydrophobin adsorption layers is applicable also to any other visco-elastic interfacial layers.

2.   Materials and methods

The used protein HFBII represents a class II hydrophobin.[31,32] It was isolated from the fungus *Trichoderma reesei* following a procedure described elsewhere.[33] A stock solution of concentration 0.1 wt% was prepared. Before each experiment, this solution was sonicated in an ultrasound bath for 5 min to break-up the protein aggregates. Then, the necessary portion of the stock solution was diluted to 0.005 wt%.

The used β-casein from bovine milk, cat. No: C6905 by electrophoresis (assay >98%) was a product of Sigma. Mixed protein solutions were prepared with 0.005 wt% HFBII and 0.03 wt% β-casein. In view of the molecular masses of HFBII (7.2 kDa) and β-casein (24 kDa), the molar ratio β-casein/HFBII in the used solutions was 1.8.

Just before the rheological measurements, the working solution was sonicated again to break-up any newly formed protein aggregates. All solutions were prepared with deionized water of specific resistivity 18.2 MΩ·cm (Milli-Q purification system, Millipore, USA). In all experiments, the working temperature was 25 °C.

The surface shear rheology of the protein adsorption layers at the air/water interface was investigated by the rotational rheometer Bohlin Gemini, Malvern UK. This rheometer is equipped with a bi-conical tool for surface shear rheology measurements. The bi-conical tool is placed in a working cell, where the investigated solution is poured up to the edge of the tool. The outer radius of the bi-cone is $R_1 = 2.81$ cm; the inner radius of the wall of the cylindrical cell is $R_2 = 3.00$ cm, and the distance between them is $\Delta R = 0.19$ cm. The latter represents the width of the ring-shaped protein adsorption layer that is subjected to shear deformation.

Before each run, the solution in the experimental cell was replaced with a new portion. After loading the solution, we waited for 5 min before the start of the rheological measurements. This period of time is needed for the formation and consolidation of the adsorption layer. In general, the rheological properties of the protein adsorption layers vary with the surface age.[13,15,19,37,38] For this reason, in our experiments the aging time was the same, 5 min in all runs. A similar aging time has been used also by other authors.[15]

Experiments in two different regimes have been carried out. First, in the *angle-ramp regime* the bi-conical tool rotates with a fixed angular velocity, $\dot{\theta}$, and the increase of the torque, τ, is recorded as a function of time, t. Second, in the *oscillatory regime* the rotation angle θ oscillates and the corresponding periodic variations in the torque τ are registered. Each experiment was repeated at least six times to be sure that the results are reproducible. The direct measurements with pure water showed that at the used low angular velocities ($\dot{\theta} \leq 0.132$ rad s^{-1}), no effect of the viscosity of the bulk aqueous phase is registered. In other words, the measured torque is completely due to the viscoelastic adsorption layer.

The primary data for $\theta(t)$ and $\tau(t)$ recorded by the apparatus have been used for further processing and interpretation. The values of the surface shear stress, τ_{sh}, were calculated from the measured torque, τ, as follows:[15,39]

$$\tau_{sh} = g_f \tau, \qquad g_f \equiv \frac{1}{4\pi} \left(\frac{1}{R_1^2} - \frac{1}{R_2^2} \right) \tag{1}$$

Here, g_f is a geometric factor; with $R_1 = 2.81$ cm and $R_2 = 3.00$ cm, eqn (1) gives $g_f = 12.36$ rad m^{-2}. Eqn (1) is applicable to surface layers of arbitrary viscoelastic behavior in the case of narrow gap, *i.e.* $(R_2 - R_1)/R_1 \leq 0.1$. In our case the latter ratio is 0.0676.

The adsorption layers of β-casein alone (even at a relatively high concentration, 0.17 wt%) gives a very weak rheological response, which is below the sensitivity threshold of the used apparatus—the registered torque τ is practically zero. Nonzero τ has been measured only in the presence of hydrophobin.

3. Experimental results

3.1. Solutions of 0.005 wt% HFBII

As mentioned above, in the oscillatory regime the rotation angle θ oscillates and the corresponding periodic variations in the torque τ are registered. First, we verified whether the $\theta(t)$ dependence generated by the apparatus is perfectly harmonic. For this goal, the experimental dependence $\theta(t)$ was fitted with a sinusoid:

$$\theta(t) = \theta_a \sin(\omega t_e + \phi) \qquad (2)$$

see Fig. 1. Here, θ_a is the amplitude of oscillations; $\omega = 2\pi\nu$ is the angular frequency and ν is the conventional frequency; φ is a phase-shift angle, and t_e is the experimental time, as registered by the apparatus. The fits of the experimental data for $\theta(t)$ with eqn (2) showed that the oscillations of θ are sinusoidal with a very high precision (regression coefficient of at least 0.99995). For each experiment, θ_a and φ have been determined from the fit. For the needs of the subsequent theoretical analysis, it is convenient to introduce the theoretical time, t, and the strain, γ, as follows:

$$t = t_e + \phi/\omega, \ \gamma = \tan\theta \qquad (3)$$

Because the angle θ is small in our experiments, we have $\gamma \approx \theta$ and amplitudes $\gamma_a \approx \theta_a$ within an accuracy of at least six significant digits. For this reason, hereafter we will use the notation γ for both strain (measured in %) and rotation angle (measured in radians); 1 mrad = 0.1%. Likewise, $\gamma_a = \theta_a$ will denote the strain amplitude, and $\dot\gamma = \dot\theta$ will be the rate of strain, which represents also the angular velocity and the shear rate.

In view of eqn (3), eqn (2) acquires the form:

$$\gamma(t) = \gamma_a \sin(\omega t) \qquad (4)$$

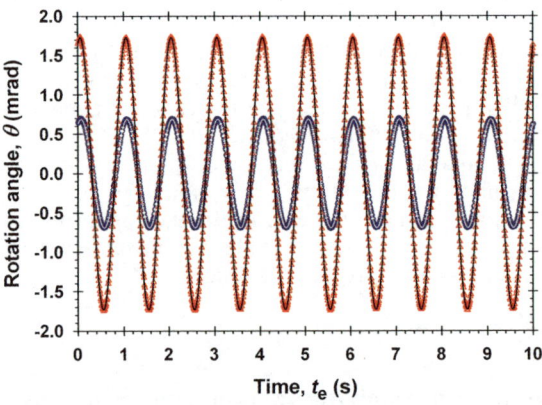

Fig. 1 Typical experimental dependence of the rotation angle, θ, on time, t_e, for 0.005 wt% HFBII at frequency $\nu = 1$ Hz, and at two different amplitudes: $\theta_a = 0.68$ and 1.74 mrad. The solid lines, which are fits by eqn (2), indicate that the strain is a perfect sinusoid.

For sufficiently small amplitudes, $\gamma_a \leq 5.235$ mrad, the registered torque τ (and shear stress $\tau_{sh} = g_f\tau$) also exhibits harmonic oscillations with phase shift, so that it can be expressed in the form:

$$\frac{\tau_{sh}}{\gamma_a} = G'\sin(\omega t) + G''\cos(\omega t) \tag{5}$$

see Fig. 2; G' and G'' are the storage and loss moduli. Eqn (5) means that at small amplitudes the adsorption layer exhibits a quasi-linear rheological response (for a true linear response, the determined surface shear elasticity and viscosity have to be independent of the shear rate, $\dot{\gamma}$, whereas for a quasi-linear response they may depend on $\dot{\gamma}$; see Section 4.2 for details). Fig. 2a is typical for all measurements carried out at frequencies $\nu \leq 2$Hz and amplitudes $\gamma_a \leq 5.235$ mrad. In contrast, Fig. 2b illustrates a special case at a higher frequency $\nu = 4$ Hz, at which the data for $\tau_{sh}(t)$ represent an experimental band, rather than a curve.

Our experiments show that at greater amplitudes, $\gamma_a \geq 10.5$ mrad, the periodical variations of τ_{sh} are not sinusoidal (Fig. 3), i.e. the adsorption layer exhibits

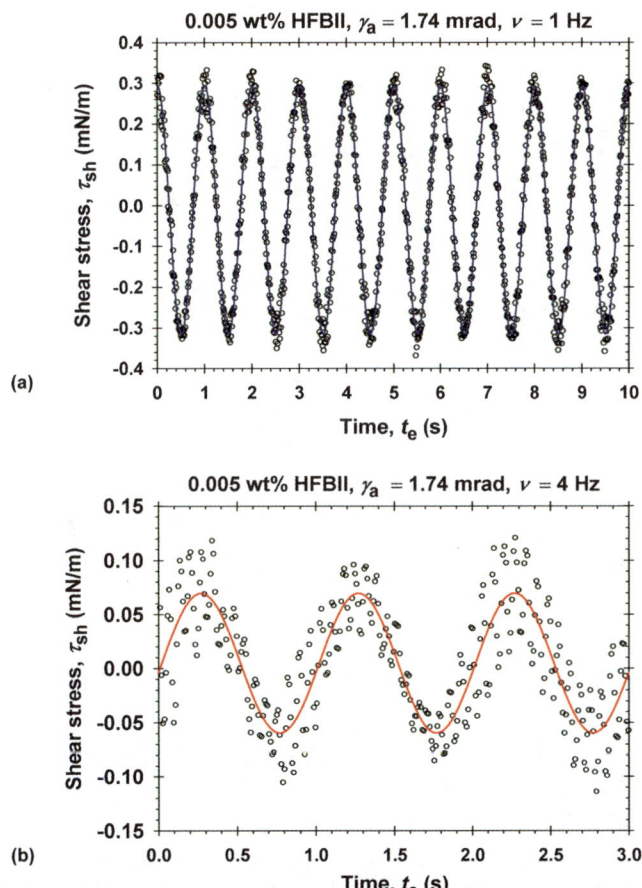

Fig. 2 Dependences of the measured stress, τ_{sh}, on time, t_e, for 0.005 HFBII, at strain amplitude $\gamma_a = 1.74$ mrad. The solid lines are the best fits with eqn (5). (a) The frequency is $\nu = 1$ Hz; this plot is typical for all measurements carried out at frequencies $\nu \leq 2$Hz and amplitudes $\gamma_a \leq 5.235$ mrad. (b) The frequency is $\nu = 4$ Hz; the data for $\tau_{sh}(t)$ represent an experimental band, rather than an experimental curve.

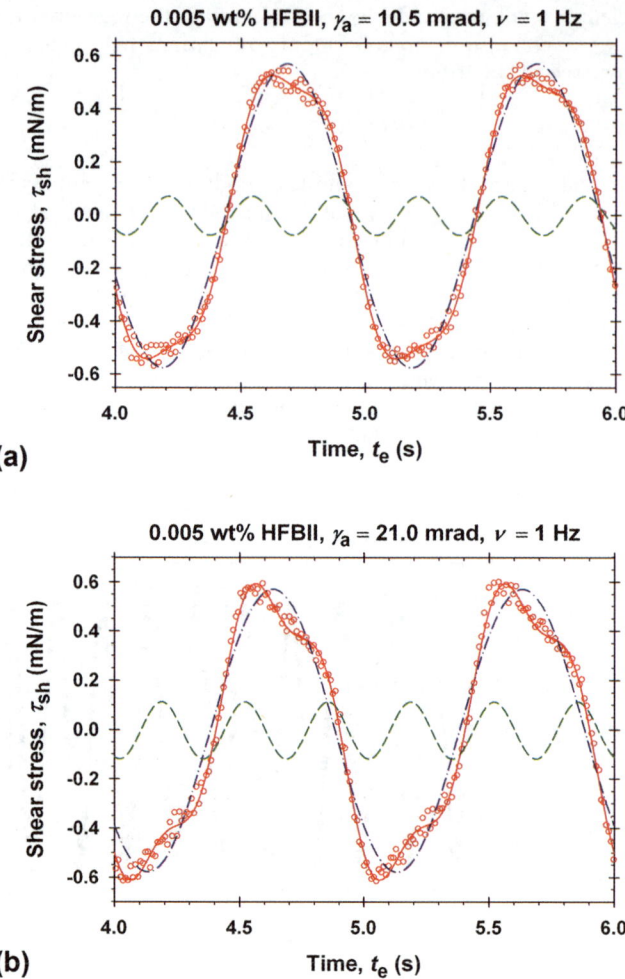

Fig. 3 Experimental dependences of the stress, τ_{sh}, *vs.* time, t_e, for 0.005 wt% HFBII in the case of nonlinear response; the frequency is $\nu = 1$ Hz; the strain amplitudes are (a) $\gamma_a = 10.5$ mrad; (b) $\gamma_a = 20.1$ mrad. The points are experimental data; the solid lines represent the best fit by eqn (6); the dash–dot and dashed lines are, respectively, the first and third Fourier modes.

a nonlinear rheological response. In such a case, instead of eqn (5), we have a Fourier series:

$$\frac{\tau_{sh}}{\gamma_a} = \sum_{k=1,3,5,\dots} [G'_k \sin(k\omega t) + G''_k \cos(k\omega t)] \qquad (6)$$

It can be proven theoretically[40] that the Fourier expansion, eqn (6), can contain only odd modes ($k = 1, 3, 5, \dots$). This follows from the circumstance that rotations clockwise and anticlockwise are mechanically equivalent. For the data in Fig. 3, only the first two odd modes ($k = 1, 3$) are significant. The solid line in Fig. 3 shows the fit with eqn (6), whereas the dashed lines show the separate modes with $k = 1$ and 3. The values of the coefficients in eqn (6) determined from the fits of the data in Fig. 3 are given in Table 1, where the errors reflect the accuracy of the measurements,

Table 1 The storage and loss moduli, G_k and G_k'', $k = 1,3$, determined from the data in Fig. 3

γ_a (mrad)	First Fourier mode		Third Fourier mode	
	G_1 (mN m^{-1})	G_1'' (mN m^{-1})	G_3 (mN m^{-1})	G_3'' (mN m^{-1})
10.50	33.0 ± 0.3	43.4 ± 0.4	-4.65 ± 0.06	5.20 ± 0.07
20.99	5.73 ± 0.06	26.8 ± 0.3	-5.52 ± 0.06	0.55 ± 0.07

rather than their reproducibility. In general, the experiment indicates that the storage and loss moduli in eqn (6) depend on both amplitude γ_a and frequency ν:

$$G_k = G_k(\gamma_a, \nu), \ G_k'' = G_k''(\gamma_a, \nu), \ k = 1, 3, 5, \dots \quad (7)$$

It is useful to plot the experimental data also in the form of Lissajous curves representing the stress $\tau_{sh}(t)$ vs. the strain $\gamma(t)$;[40] see Fig. 4. In the case of quasi-linear response ($\gamma_a \leq 5.23$ mrad), the Lissajous curves are ellipses, whereas in the case of nonlinear response ($\gamma_a = 10.5$ and 21.0 mrad), they look like curvilinear parallelograms of width that increases with the rise in γ_a. The Lissajous plots can be useful in several aspects.[40] First, by plotting the raw data for $\tau_{sh}(t)$ vs. $\gamma(t)$, the experimentalist can immediately verify whether the rheological response of the system is linear or nonlinear (ellipse vs. parallelogram). Second, if the Lissajous curves are wider than an ellipse (in nonlinear regime), as in Fig. 4, this indicates shear thinning, whereas a Lissajous curve that is concave with respect to an ellipse indicates shear thickening.[40] Last but not least, as demonstrated in Section 5 the parameters of the rheological model can be determined by a fit of the Lissajous curve without using a Fourier analysis.

In the oscillatory experiments with the rotational rheometer, there are two mechanical degrees of freedom: to vary the amplitude γ_a at a fixed frequency ν, and to vary ν at a fixed γ_a; see also eqn (7). Fig. 5a presents the dependences of G_1 and G_1'' on γ_a at $\nu = 1$ Hz. The data show that G_1 decreases with the rise of the amplitude γ_a, whereas G_1'' initially increases until reaching G_1, and then exhibits a tendency to decrease.

Fig. 5b presents the dependences of G_1 and G_1'' on the frequency ν at amplitude $\gamma_a = 1.74$ mrad. Initially, G_1 increases with the rise in ν, but at the highest

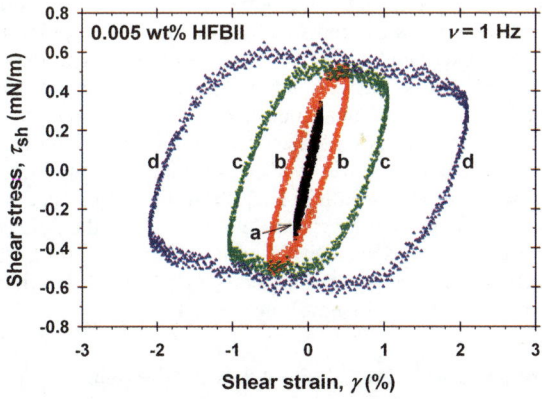

Fig. 4 Lissajous plots of the stress, $\tau_{sh}(t)$ vs. the strain, $\gamma(t)$, for 0.005 wt% HFBII at frequency $\nu = 1$ Hz, and at four different values of the strain amplitude, γ_a: (a) 1.74 mrad; (b) 5.23 mrad; (c) 10.5 mrad, and (d) 21.0 mrad.

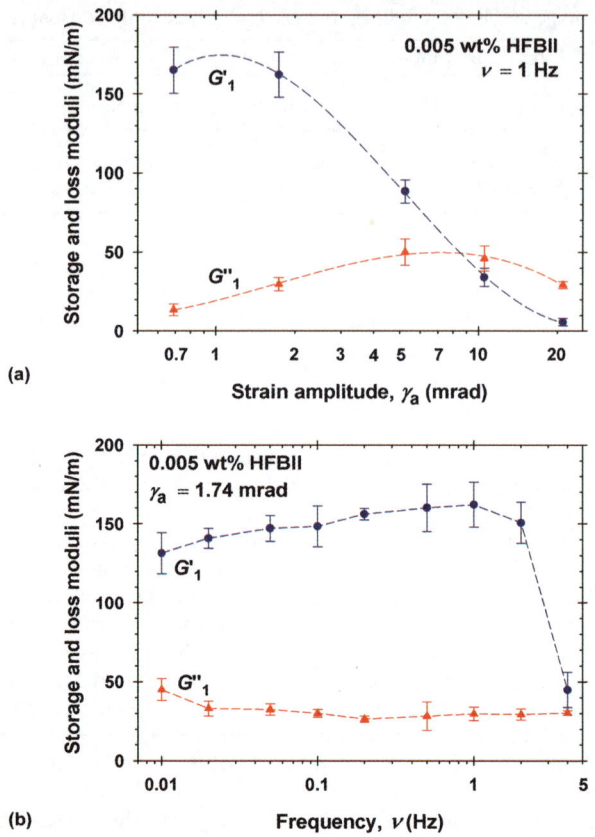

Fig. 5 Dependence of the storage and loss moduli, G'_1 and G''_1 (a) on the strain amplitude, γ_a, at fixed frequency, $\nu = 1$ Hz; (b) on the frequency, ν, at fixed strain amplitude, $\gamma_a = 1.74$ mrad. The data are for the adsorption layer at the surface of 0.005 wt% HFBII solution. Each point is the average value from 12 experiments. The dashed lines are guides to the eye.

investigated frequency it quickly decreases. In contract, G''_1 initially slightly decreases, and then it levels off at the higher values of ν. Note that the highest investigated frequency is $\nu = 4$ Hz, which corresponds to the graph in Fig. 2b.

In Fig. 5a and b, every experimental point is an average of twelve independent experiments. Each of them is carried out by pouring a new portion of the investigated solution in the experimental cell of the rheometer. The experimental errors shown in Fig. 5a and b (unlike those in Table 1) represent the reproducibility (rather than the accuracy) of the rheological measurements, characterized by the standard deviation. As a rule, the accuracy of these measurements is much better than their reproducibility. An exclusion is observed at the highest frequency in Fig. 5b, $\nu = 4$ Hz, which corresponds to the graph in Fig. 2b. In this special case (which probably represents a broken viscoelastic layer; see Section 5) the data for $\tau_{sh}(t)$ are rather scattered, but they can be fitted well with a sinusoid (Fig. 2b). Interestingly, this sinusoid is well reproducible (the experiment was repeated 24 times), which is reflected by the relatively small error bars in Fig. 5b at $\nu = 4$ Hz.

3.2. Solutions of 0.005 wt% HFBII and 0.03 wt% β-casein

As mentioned above, one of our goals is to give a self-consistent interpretation of data obtained in oscillatory and angle-ramp regimes by the same rheological model. Data in angle-ramp regime have been reported in our previous study,[36] for both

HFBII and mixed HFBII + β-casein solutions. In the measurements reported here, we used a new sample of β-casein, which turned out to give a slightly different rheological behavior of the mixed layers—a lower elasticity in comparison with the β-casein sample used in ref. 36. Here the data in both the angle-ramp and oscillatory regime have been obtained with the new sample of β-casein.

Fig. 6 shows experimental curves for the variation of the stress, τ_{sh}, with the increase of the rotation angle, γ, in the *angle-ramp* regime. Each curve corresponds to a fixed shear rate, $\dot{\gamma}$ = const. The dependence $\tau_{sh}(\gamma)$ exhibits a clear tendency to level off at the larger γ. In addition, τ_{sh} increases with the rise of the shear rate, $\dot{\gamma}$.

Fig. 7 presents data for the storage and loss moduli G'_1 and G''_1 as functions of the frequency, ν, in *oscillatory* regime at fixed amplitude $\gamma_a = 1.74$ mrad ($\gamma_a = 0.174\%$). At this amplitude, the rheological response of the adsorption layer is quasi-linear, *i.e.* the effect of the higher-order harmonics (G'_3 and G''_3) is negligible. The comparison of Fig. 5b and 7 indicates that the addition of β-casein leads to lowering of both G'_1 and G''_1. Moreover, the decrease of G'_1 with the rise of ν begins at lower frequencies.

4. Theoretical models *vs.* experimental results

As demonstrated below, the obtained experimental data comply with a compound rheological model, which combines the Maxwell model with a modified Herschel–Bulkley law. We begin with a brief overview and discussion on the applicability of simpler models.

4.1. Models with constant E_{sh} and η_{sh}

The Kelvin model, known also as Kelvin–Voigt model,[41,42] is characterized with a parallel connection of an elastic and a viscous element (Fig. 8a). The latter two elements are characterized by the conventional stress–strain relationships:

$$\tau_{sh,e} = E_{sh}\gamma, \; \tau_{sh,v} = \eta_{sh}\dot{\gamma} \tag{8}$$

where $\tau_{sh,e}$ and $\tau_{sh,v}$ are, respectively, the elastic and viscous stresses; E_{sh} and η_{sh} are the coefficients of surface shear elasticity and viscosity and (as usual) γ and $\dot{\gamma}$ are the shear strain and rate-of-strain. Note that in the conventional Kelvin model, E_{sh} and η_{sh} are constants independent of time, t.

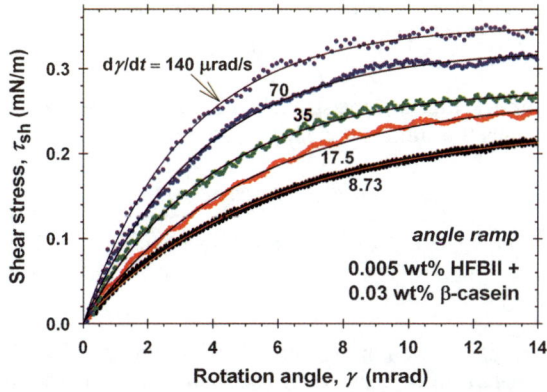

Fig. 6 Experimental data obtained in angle-ramp regime for adsorption layers at the surface of 0.005 wt% HFBII + 0.03 wt% β-casein solutions: Plots of the stress, τ_{sh}, *vs.* the rotation angle, γ, at three different fixed angular velocities: $\dot{\gamma}$ = 8.73, 17.5, 35, 70 and 140 μrad s⁻¹; the time is expressed as $t = \gamma/\dot{\gamma}$. The solid lines are fits by the Maxwell model, eqn (15).

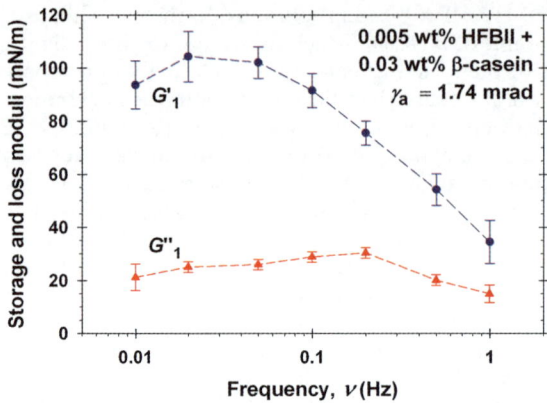

Fig. 7 Dependence of the storage and loss moduli, G'_1 and G''_1 on the frequency, ν, at fixed strain amplitude, $\gamma_a = 1.74$ mrad. The data are for the adsorption layer at the surface of 0.005 wt% HFBII + 0.03 wt% β-casein solution. Each point is the average value from 12 experiments. The dashed lines are guides to the eye.

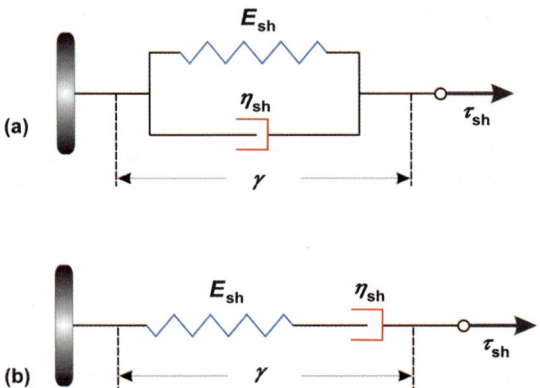

Fig. 8 Sketch of the two basic compound models of viscoelastic behavior composed of an elastic element (spring) of elasticity E_{sh}, and a viscous element (dash-pot) of viscosity η_{sh}. (a) The Kelvin model: parallel connection. (b) The Maxwell model: consecutive connection. τ_{sh} is the applied stress; γ is the total strain.

In the Kelvin model (Fig. 8a), the strain is the same for the two elements, whereas the total stress equals the sum of the elastic and viscous stresses:

$$\tau_{sh} = E_{sh}\gamma + \eta_{sh}\dot{\gamma} \tag{9}$$

Substituting γ and τ_{sh} from eqn (4) and (5) into eqn (9), we obtain:

$$G' = E_{sh}, \; G'' = \eta_{sh}\omega \tag{10}$$

In other words, if a given body obeys the Kelvin model, then the storage modulus G' is constant and equal to the elasticity, whereas the loss modulus G'' increases linearly with the frequency, $\omega = 2\pi\nu$, both of them being independent of the amplitude γ_a.

As seen in Fig. 5 and 7, our experimental system does not obey the Kelvin law (G' and G'' depend on γ_a, and G'' does not increase linearly with ν). Moreover, in the

This journal is © The Royal Society of Chemistry 2012

angle-ramp regime ($\dot{\gamma}$ = const.) eqn (9) gives $\tau_{sh} = E_{sh} \dot{\gamma}t$ + const., *i.e.* the stress has to linearly increase with time, which cannot explain the experimentally observed tendency of τ_{sh} to level off (Fig. 6).

The Maxwell model,[43] is characterized with a consecutive connection of an elastic and a viscous element (Fig. 8b). Hence, the stress, τ_{sh}, is the same for the two elements, whereas the total rate-of-strain equals the sum of the elastic and viscous rates-of-strain:

$$\frac{1}{E_{sh}}\frac{d\tau_{sh}}{dt} + \frac{\tau_{sh}}{\eta_{sh}} = \dot{\gamma} \tag{11}$$

Substituting γ and τ_{sh} from eqn (4) and (5) into eqn (11) and setting equal the coefficients before the sine and cosine, we obtain:

$$G' = \frac{E_{sh}(\eta_{sh}\omega)^2}{E_{sh}^2 + (\eta_{sh}\omega)^2}, \qquad G'' = \frac{E_{sh}^2\eta_{sh}\omega}{E_{sh}^2 + (\eta_{sh}\omega)^2} \tag{12}$$

The two relations in eqn (12) can be solved with respect to E_{sh} and η_{sh}:

$$E_{sh} = \frac{G'^2 + G''^2}{G'}, \qquad \eta_{sh} = \frac{G'^2 + G''^2}{G''\omega} \tag{12a}$$

Eqn (12) implies that in the conventional Maxwell model (at constant E_{sh} and η_{sh}), G' and G'' depend on the frequency ω, but are independent of the amplitude, γ_a. The latter is in contradiction with our data in Fig. 5a.

Note that the rheological response of the system is characterized with a specific frequency ν_{ch} (or characteristic time $t_{ch} = 1/\nu_{ch}$) defined as follows:

$$\nu_{ch} \equiv \frac{G''}{G'}\omega = \frac{E_{sh}}{\eta_{sh}} \tag{13}$$

where eqn (12) has been used at the last step.

4.2. Generalization of the Maxwell model

In angle-ramp regime, the shear rate is constant, *i.e.* $\dot{\gamma}$ = const. Then, integrating eqn (11) along with the initial condition $\tau_{sh}|_{t=0} = 0$, we obtain:

$$\tau_{sh} = \eta_{sh}\dot{\gamma}[1 - \exp(-\nu_{ch}t)] \tag{14}$$

where the definition (13) has been also used. Because in angle-ramp regime $t = \gamma/\dot{\gamma}$, eqn (14) can be expressed also in the form:

$$\tau_{sh} = \eta_{sh}\dot{\gamma}[1 - \exp(-\nu_{ch}\gamma/\dot{\gamma})] \tag{15}$$

At each fixed angular velocity $\dot{\gamma}$ (angle-ramp regime), the experimental τ–*vs.*–γ curve excellently agrees with eqn (15). This is illustrated in Fig. 6, for the mixed system HFBII + β-casein. Similar results have been obtained for pure HFBII.[36] From the fits, the parameters η_{sh} and ν_{ch} have been determined. The results for η_{sh} and $E_{sh} = \nu_{ch}\eta_{sh}$ are listed in Table 2 and plotted in Fig. 9. The values in Table 2 are average from six experiments, and the errors of these values reflect the reproducibility of the experiments. (The errors of E_{sh}, η_{sh} and ν_{ch} determined from the errors of the parameters of each separate fit in Fig. 6 are much smaller.)

The Herschel–Bulkley law[44] reads:

$$\tau_{sh} = K\dot{\gamma}^n = \eta_{sh}\dot{\gamma} \Rightarrow \eta_{sh} = K\dot{\gamma}^{n-1} \tag{16}$$

Table 2 Parameters determined form the fit of the experimental curves in Fig. 6 with eqn (15); 0.005 wt% HFBII + 0.03 wt% β-casein; angle-ramp regime

$\dot{\gamma}$ (μrad s^{-1})	E_{sh} (mN m^{-1})	η_{sh} (N.s/m)	ν_{ch} (10^{-3} Hz)
8.73	42 ± 5	26 ± 5	1.6 ± 0.3
17.5	54 ± 8	15 ± 3	3.5 ± 0.7
35.0	73 ± 8	8.0 ± 0.8	9.1 ± 0.8
70.0	91 ± 9	4.7 ± 0.1	19 ± 3
140.0	113 ± 5	2.6 ± 0.1	45 ± 4

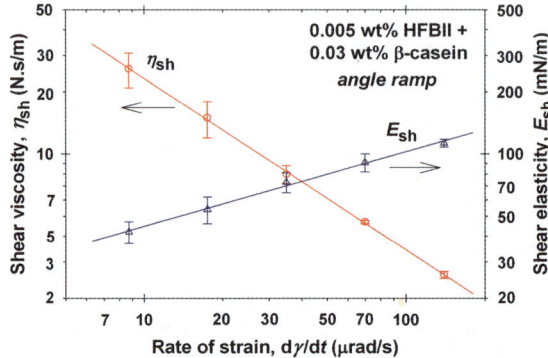

Fig. 9 Shear viscosity, η_{sh}, and elasticity, E_{sh}, vs. the rate of strain (the angular velocity) $\dot{\gamma}$: Plots of the data from Table 2 for 0.005 wt% HFBII + 0.03 wt% β-casein obtained in the angle-ramp regime in accordance with eqn (17). Each point is the average value from 6 experiments. The slopes of the linear regressions are given in Table 3.

where K is the consistency and n is the flow behavior index. At $n = 1$, the continuous medium behaves as a Newtonian fluid (η_{sh} = const.); for $n < 1$ and $n > 1$, the medium exhibits, respectively, shear thinning and thickening.

The data in Fig. 9 indicate that in the considered interval of shear rates η_{sh}, E_{sh} and ν_{ch} can be expressed as power functions:

$$\eta_{sh} = K|\dot{\gamma}|^{n-1}, \; E_{sh} = A|\dot{\gamma}|^{p} \tag{17}$$

$$\nu_{ch} = Q|\dot{\gamma}|^{m} \text{(modified Herschel–Bulkley law)} \tag{18}$$

where A, K, Q, m, n and p are constant parameters ($m = p - n + 1$), see eqn (13). The modulus of $\dot{\gamma}$ was inserted in view of subsequent generalization to oscillatory regime, for which $\dot{\gamma}$ can be both positive and negative. Note that the elasticity and viscosity are independent of whether the rotation is clockwise or anticlockwise, so in general they must depend on $|\dot{\gamma}|$:[40]

$$\eta_{sh} = \eta_{sh}(|\dot{\gamma}|), \; E_{sh} = E_{sh}(|\dot{\gamma}|) \tag{19}$$

In other words, η_{sh} and E_{sh} are even functions of $\dot{\gamma}$, which leads to the conclusion that the Fourier expansion, eqn (6) can contain only odd harmonics.[40]

Thus, the experiments in angle-ramp regime imply that the rheological behavior of protein adsorption layers can be described by using the basic equation of the

Maxwell model, eqn (11) with viscosity and elasticity, which depend on the shear rate $\dot{\gamma}$; see eqn (17)–(19).

4.3. Oscillatory regime with quasi-linear response

Here, it will be demonstrated that the combined Maxwell–Herschel–Bulkley model, based on eqn (11) and (18), describes also the rheological behavior of the investigated protein layers in oscillatory regime. In this regime, $\dot{\gamma} = \gamma_a \omega \cos(\omega t)$ is a periodic function of time (rather than constant as in the angle-ramp regime). In view of eqn (18) and (19), E_{sh}, η_{sh} and $\nu_{ch} = E_{sh}/\eta_{sh}$ are also periodic functions of time.

As discussed above, for small strain amplitudes (in our case, $\gamma_a \leq 5.235$ mrad; see Fig. 2) the investigated protein layers exhibit quasi-linear response, $i.e.$ the higher-order harmonics have negligible amplitudes, so that the stress obeys eqn (5). Then, substituting eqn (4) and (5) into eqn (11), we derive:

$$(\nu_{ch}G'' + G')\cos(\omega t) + (\nu_{ch}G' - G'')\sin(\omega t) = E_{sh}\omega\cos(\omega t) \qquad (20)$$

The multiplication of eqn (20) by $\sin(\omega t)$ and $\cos(\omega t)$, with a subsequent integration and some transformations, yields:

$$\langle \nu_{ch} \rangle \equiv \frac{G''}{G'}\omega = \frac{2}{\pi}\int_0^\pi \nu_{ch}\sin^2\xi\,d\xi \qquad (21)$$

$$G' + \frac{2G''}{\pi\omega}\int_0^\pi \nu_{ch}\cos^2\xi\,d\xi = \frac{2}{\pi}\int_0^\pi E_{sh}\cos^2\xi\,d\xi \qquad (22)$$

$\xi \equiv \omega t$ is an integration variable. In view of eqn (13), the left-hand side of eqn (21) can be considered as an average characteristic frequency, $\langle \nu_{ch} \rangle$. If ν_{ch} is independent of time (as in the angle-ramp regime), then eqn (21) acquires the form $\langle \nu_{ch} \rangle = \nu_{ch}$, as it should be expected. If ν_{ch} and E_{sh} depend on time (as in the oscillatory regime), the two algebraic expressions in eqn (12a) are not valid; instead of them we have eqn (21) and (22).

In the case of oscillatory regime, we substitute $\nu_{ch}(\dot{\gamma})$ from eqn (18) with $\dot{\gamma} = \gamma_a \omega \cos(\xi)$ in the right-hand side of eqn (21). The integral can be solved and the result can be presented in the form:

$$\langle \nu_{ch} \rangle = Q\langle\dot{\gamma}\rangle^m \qquad (23)$$

$$\langle\dot{\gamma}\rangle \equiv \mu\gamma_a\omega, \qquad \mu \equiv \left[\frac{\Gamma(m/2 + 0.5)}{\pi^{1/2}\Gamma(m/2 + 2)}\right]^{1/m} \qquad (24)$$

$\Gamma(x)$ is the gamma function. In analogy with eqn (18), eqn (23) expresses the mean characteristic frequency, $\langle \nu_{ch} \rangle$, as a power function of the mean shear rate, $\langle\dot{\gamma}\rangle$, defined by eqn (24).

Eqn (23) allows us to compare the experimental results obtained in angle-ramp and oscillatory regimes, see Fig. 10. In the $angle\text{-}ramp$ regime, the characteristic frequency is calculated from the expression $\nu_{ch} = E_{sh}/\eta_{sh}$, where E_{sh} and η_{sh} are determined as adjustable parameters from fits of the data like those in Fig. 6. In Fig. 10, ν_{ch} is plotted $vs.$ the shear rate $\dot{\gamma}$ in double logarithmic scale; m and Q are determined from the slope and intercept of the obtained linear dependence. The used data for 0.005 wt% HFBII + 0.03 wt% β-casein are these in Fig. 6 and Table 2, whereas the respective data for 0.005 wt% HFBII (without added β-casein) have been obtained in ref. 36 in the angle-ramp regime.

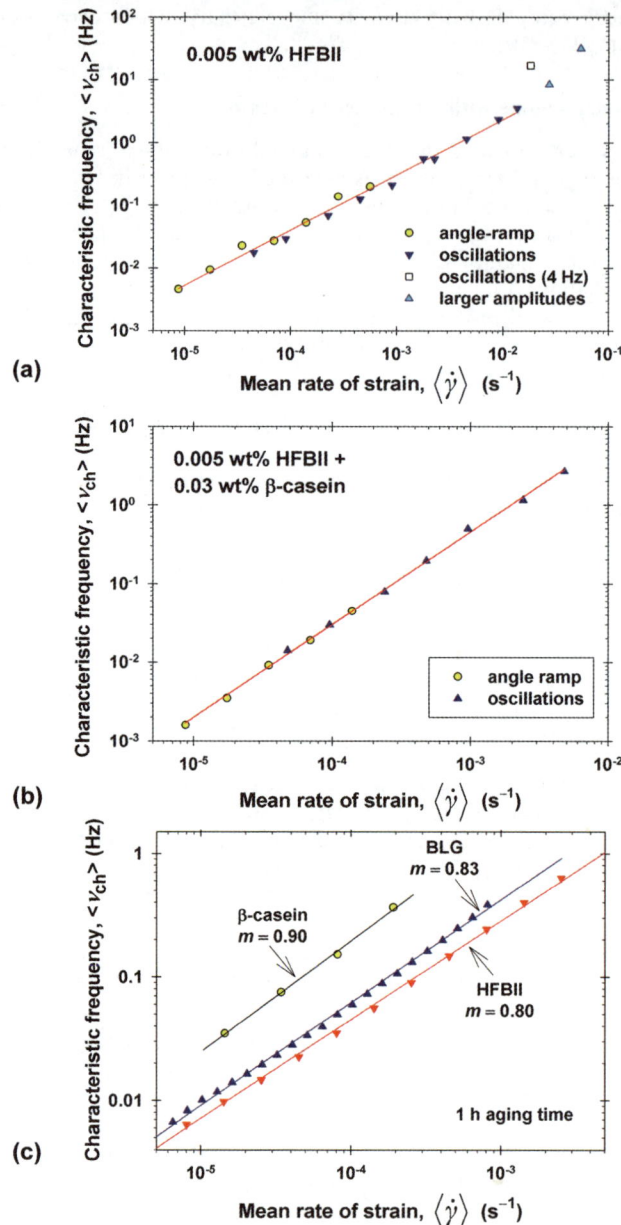

Fig. 10 Plots of the characteristic frequency of rheological response *vs.* the mean rate of strain. For the angle-ramp regime, the plot is $v_{ch} \equiv E_{sh}/\eta_{sh}$ *vs.* $\dot{\gamma}$. For the oscillatory regime, the respective mean quantities are plotted: $\langle v_{ch} \rangle \equiv \omega G''/G'$ *vs.* $\langle \dot{\gamma} \rangle$. The slope and intercept of the linear regression give m and Q; see eqn (18) and (23). (a) Data for 0.005 wt% HFBII. (b) Data for 0.005 wt% HFBII + 0.03 wt% β-casein. (c) Comparison with literature data[11] for β-casein, β-lactoglobulin and HFBII from measured in the oscillatory regime.

The data points in the *oscillatory* regime (Fig. 10) were obtained in the following way. $\langle v_{ch} \rangle = \omega G''/G'$ was calculated from the experimental data in Fig. 5 and 7 ($G' = G'_1$, $G'' = G''_1$). Then, $\langle v_{ch} \rangle$ was plotted *vs.* $\gamma_a \omega$ in a double logarithmic scale; m and Q were determined from the slope and intercept of the obtained linear dependence. The values of m and Q obtained from the fits of the data in angle-ramp and

This journal is © The Royal Society of Chemistry 2012

oscillatory regimes are very close, which confirms the applicability of the combined Maxwell–Herschel–Bulkley model to protein adsorption layers.

The values of m and Q in Table 3 were obtained by a simultaneous fit of the data from the angle-ramp and oscillatory regime with the help of numerical minimization of the respective compound merit function. From the obtained value of m, we calculated also the coefficient μ in eqn (24) which is also given in Table 3. The values of n and p refer to the asymptotic region of sufficiently small $\dot\gamma$, where the power laws in eqn (17) are obeyed; see Fig. 9.

The three points in the upper-right corner of Fig. 10a correspond to oscillatory regime with nonlinear rheological response (greater amplitude or frequency of oscillations), at which the terms with $k - 3$ in eqn (6) are not negligible. In this case, the mean characteristic frequency is calculated from the first Fourier modes: $\langle \nu_{ch} \rangle = \omega G''_1/G_1$. The respective points in Fig. 10a are close to the linear dependence that holds in the case of quasi-linear response, but still deviate from it.

As seen in Fig. 10b, the data for the adsorption layers at the surface of mixed solutions of HFBII and β-casein also comply with eqn (23) when plotted as ν_{ch} vs. $\langle \dot\gamma \rangle$. The results obtained in angle-ramp and oscillatory regimes are in excellent agreement.

Not only HFBII, but also other proteins form viscoelastic adsorption layers at the air/water interface. To check whether their behavior complies with eqn (23), in Fig. 10c we have plotted data for adsorption layers from β-casein, β-lactoglobulin (BLG) and HFBII obtained by rotational rheometer in oscillatory regime in Ref. 11. The original data[11] are in terms of G' and G'', from which we calculated $\langle \nu_{ch} \rangle = \omega G''/G'$. Fig. 10c shows that the data for β-casein and BLG excellently agree with straight lines in accordance with eqn (23). The values of the exponent m are close, in the range 0.8–0.9, for all these proteins. In other words, the combined Maxwell–Herschel–Bulkley model is applicable not only to layers from hydrophobin, but also to viscoelastic adsorption layers from other proteins. In particular, the data for HFBII obtained here and in ref. 11 are in good agreement, with a small difference in the exponent: $m = 0.88$ (Fig. 10a) vs. $m = 0.80$ (Fig. 10c). (Smaller m corresponds to more rigid layer.) This difference can be attributed to the different surface ages of the adsorption layers (before the beginning of the rheological measurements): 5 min in our experiments vs. 1 h in ref. 11.

In general, the surface rheology of proteins varies with age. For proteins such as β-lactoglobulin and ovalbumin this effect is mostly due to conformational changes, which occur with the protein molecules after their adsorption. For the rigid hydrophobin molecules conformational changes are not expected, but the number of hydrophobin aggregates that adsorb below the protein adsorption layer is increasing with time, which can also lead to age effects.

4.4. Dependences of E_{sh} and η_{sh} on the rate of strain

The comparison of theory and experiment (Fig. 10) shows that the power dependence in eqn (18), viz. $\nu_{ch} = Q| \dot\gamma|^m$, is applicable in the whole region of quasi-linear response of the protein adsorption layer. However, the comparison with the experiment indicates that the power dependencies $\eta_{sh}(\dot\gamma)$ and $E_{sh}(\dot\gamma)$ in eqn (17) can be used only at sufficiently low shear rates. To find $\eta_{sh}(\dot\gamma)$ and $E_{sh}(\dot\gamma)$ in the whole region of quasi-linear response, we will employ eqn (22), which has not been used so far.

Table 3 Parameters in eqn (17)–(18) and (23)–(24) determined from fits of experimental data.[a]

Solution	m	n	p	Q (s^{m-1})	μ
0.005 wt% HFBII	0.88	0.14	0.02	134	0.413
0.005 wt% HFBII + 0.03 wt% β-casein	1.18	0.17	0.35	1538	0.440

[a] Note: The values of n and p refer to the asymptotic region of small shear rates; $m = p - n + 1$.

Let us introduce an average shear elasticity, which is defined as follows:

$$\langle E_{sh} \rangle \equiv \frac{2}{\pi} \int_0^\pi E_{sh} \cos^2 \xi \, d\xi \tag{25}$$

For $E_{sh} = $ const. Eqn (25) yields $\langle E_{sh} \rangle = E_{sh}$. Substituting ν_{ch} from eqn (18) into eqn (22), in view of eqn (25) we obtain:

$$\langle E_{sh} \rangle = G' + (m+1) \frac{G''}{\omega} Q \langle \dot{\gamma} \rangle^m = \frac{G'^2 + (m+1)G''^2}{G'} \tag{26}$$

where eqn (21) and (23) have been used at the last step. The combination of eqn (25) and (26) gives an integral equation for determining the function $E_{sh}(\xi)$. The problem can be solved by using an appropriate empirical expression for $E_{sh}(\dot{\gamma})$. To find such an expression, one can utilize the fact that the $E_{sh}(\dot{\gamma})$ dependence is similar to the dependence of $\langle E_{sh} \rangle$ on the amplitude $\gamma_a \omega$ (we recall that $\dot{\gamma} = \gamma_a \omega \cos(\omega t)$). The latter dependence can be calculated from eqn (26) using experimental data for G' and G'', and the value of m that is known from the fit in Fig. 10 (see Table 3).

In this way, from the data for 0.005 wt% HFBII (without added β-casein), it was found that we can seek $E_{sh}(\dot{\gamma})$ in the form:

$$E_{sh} = a_0 \exp(a_1 |\dot{\gamma}| - a_2 |\dot{\gamma}|^2) \tag{27}$$

where $\dot{\gamma} = \gamma_a \omega \cos(\xi)$. The empirical parameters a_0, a_1 and a_2 are to be determined from the fit of experimental data, as follows. The points in Fig. 11a are calculated from eqn (26), where the experimental values of G' and G'' from Fig. 5b are substituted. These points, expressing the experimental $\langle E_{sh} \rangle$, are fitted with the theoretical dependence of $\langle E_{sh} \rangle$ on $\gamma_a \omega$, which is obtained by substituting eqn (27) in the integrand of eqn (25). The parameter values determined from the fit (the dashed line in Fig. 11a) are:

$$a_0 = 160 \text{ mN m}^{-1}, \ a_1 = 15.3 \text{ s}, \ a_2 = 771 \text{ s}^2 \tag{28}$$

With the above parameter values, the dependence $E_{sh}(\dot{\gamma})$ is calculated from eqn (27) with $\dot{\gamma} \equiv \gamma_a \omega$; see the solid line in Fig. 11a. One sees that the curves representing E_{sh} and $\langle E_{sh} \rangle$ are really very close, the greatest differences appearing in the zone of larger variations of E_{sh}, as it could be expected in view of eqn (25). Finally, the dependence of viscosity on the shear rate is calculated using the relationship $\eta_{sh} = E_{sh}/\nu_{ch}$, where ν_{ch} is given by eqn (18) with $\dot{\gamma} \equiv \gamma_a \omega$.

In the case of 0.005 wt% HFBII + 0.03 wt% β-casein, the data suggest that we can seek $E_{sh}(\dot{\gamma})$ in the form:

$$E_{sh} = b_0 + b_1 \exp(-b_2 |\dot{\gamma}|)[1 - \exp(-b_3 |\dot{\gamma}|)] \tag{29}$$

where $\dot{\gamma} = \gamma_a \omega \cos(\xi)$. The empirical parameters b_0, b_1, b_2 and b_3 have been determined from the fit of the data in Fig. 11b in the following way. Two sets of experimental data have been simultaneously fitted. The first one includes the data for G' and G'' from Fig. 7 obtained in oscillatory regime, which are substituted in eqn (26) to find $\langle E_{sh} \rangle$ as a function of the rate-of-strain amplitude $\gamma_a \omega$; see the points in Fig. 11b. The second set consists of the data for E_{sh} vs. $\dot{\gamma}$ from Table 2 that have been obtained in an angle-ramp regime; see the inset in Fig. 11b. The theoretical curve in Fig. 11b was computed from eqn (25) along with eqn (29). The least squares method was applied to fit the data with a merit function that represents a sum of two merit functions corresponding to the sets of data obtained in

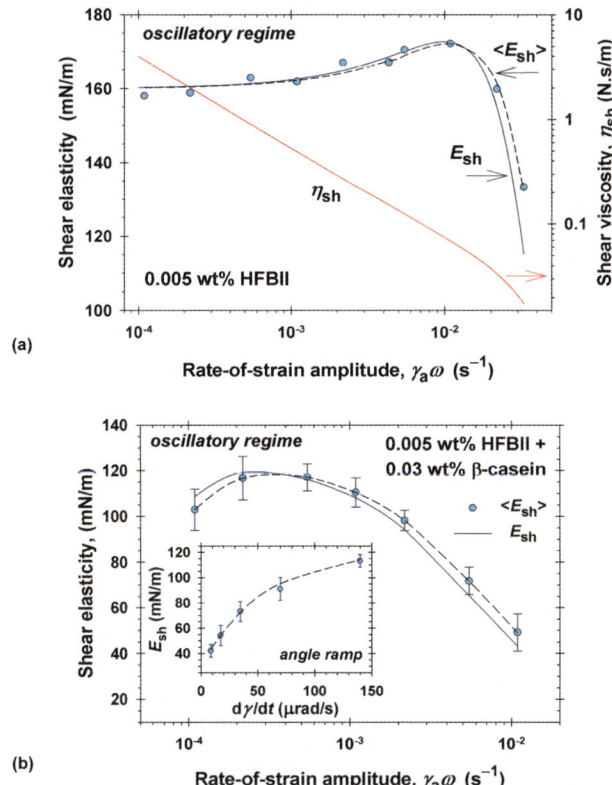

Fig. 11 Plots of shear elasticity *vs.* rate-of-strain amplitude. The points represent the mean elasticity $\langle E_{sh} \rangle$ calculated from eqn (26) using the experimental $G'(\gamma_a\omega)$ and $G''(\gamma_a\omega)$ dependences. The dashed lines represent the best fits with eqn (25), along with eqn (27) or (29). (a) Results for 0.005 wt% HFBII; E_{sh} is calculated from eqn (27) and (28) with $\dot\gamma = \gamma_a\omega$; then, $\eta_{sh} = E_{sh}/\nu_{sh}$. (b) Results for 0.005 wt% HFBII + 0.03 wt% β-casein; E_{sh} is calculated from eqn (29) and (30) with $\dot\gamma = \gamma_a\omega$; the inset shows data for E_{sh} obtained in angle-ramp regime, which are fitted simultaneously with the data for $\langle E_{sh} \rangle$ obtained in oscillatory regime.

angle-ramp and oscillatory regimes. From a statistical viewpoint, this is the most appropriate method, because the experimental error of the variables is correctly taken into account. Of course, it is possible to merge the data from the inset with the main curve in Fig. 11b by using an effective amplitude $\gamma_a\omega \equiv \dot\gamma/\mu$ along the horizontal axis for the data in angle-ramp regime; see eqn (24). The combined dependence, corresponding to eqn (29), is shown in Fig. 12a. The parameter values determined from the best fit are:

$$b_0 = 30 \text{ mN m}^{-1}, \ b_1 = 95 \text{ mN m}^{-1}, \ b_2 = 181 \text{ s}^{-1}, \ b_3 = 16978 \text{ s}^{-1} \qquad (29a)$$

Fig. 12b shows the respective dependence of viscosity on the shear rate, which is calculated using the relationship $\eta_{sh} = E_{sh}/\nu_{ch}$, where ν_{ch} is given by eqn (18) with $\dot\gamma \equiv \gamma_a\omega$, and with m and Q from Table 3.

4.5. Summary and discussion

4.5.1. The combined Maxwell–Herschel–Bulkley model. As mentioned above, it is remarkable that the modified Herschel–Bulkley law in terms of ν_{ch} *vs.* $\dot\gamma$, eqn

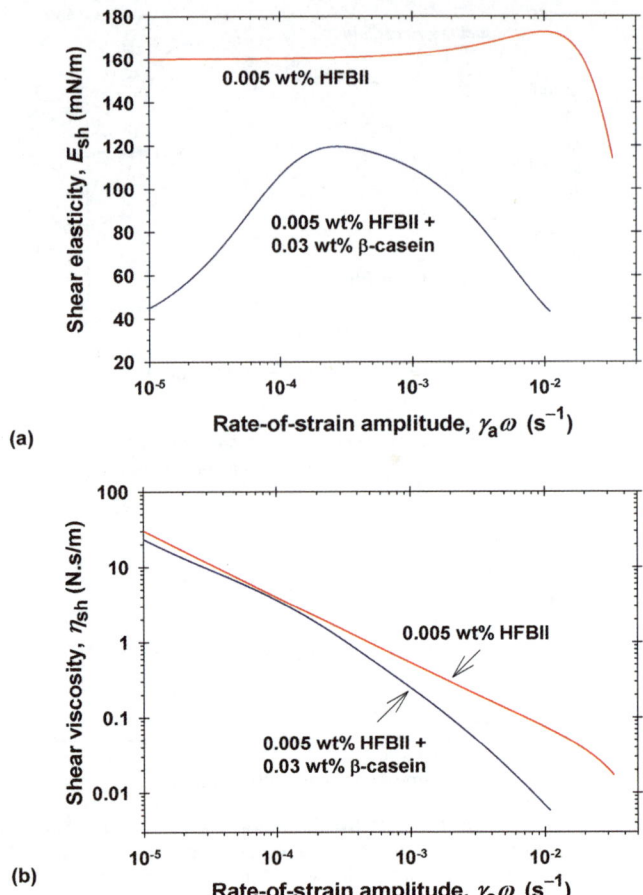

Fig. 12 Comparison of the results for HFBII adsorption layers without and with added β-casein. (a) Surface shear elasticity, E_{sh}, vs. $\gamma_a\omega$ calculated from eqn (27)–(29a) with $\dot{\gamma} \equiv \gamma_a\omega$. (b) Surface shear viscosity, $\eta_{sh} = E_{sh}/\nu_{ch}$, vs. $\gamma_a\omega$, where ν_{ch} is given by eqn (18) with $\dot{\gamma} \equiv \gamma_a\omega$, and with m and Q from Table 3.

(18), is satisfied in the whole region of quasi-linear response of the protein adsorption layers. Moreover, as seen in Fig. 10, the data obtained in the angle-ramp and oscillatory regimes collapse on a single master line, whose slope and intercept determine the parameters m and Q; see Table 3. The comparison between the two regimes is possible if the data from the oscillatory regime are plotted in terms of average values, $\langle\nu_{ch}\rangle$ vs. $\langle\dot{\gamma}\rangle$ defined by eqn (21) and (24). The full agreement between the data obtained in two very different kinetic regimes confirms the adequacy of the combined Maxwell–Herschel–Bulkley model, which is based on eqn (11), (18) and (19).

In contrast with the dependence $\nu_{ch}(\dot{\gamma})$, the dependence $\eta_{sh}(\dot{\gamma})$, and especially, $E_{sh}(\dot{\gamma})$, can considerably deviate from a power law of Herschel-Bulkley type; see Fig. 11 and 12. The power laws in eqn (17) can be used asymptotically only in the limit of low shear rates; see Fig. 9. Here, the dependence $E_{sh}(\dot{\gamma})$ has been determined by fitting the experimental data with an appropriate empirical curve; see eqn (27) and (29). The dependence $E_{sh}(\gamma_a\omega)$ turns out to be very close to the respective dependence of the average shear elasticity $\langle E_{sh}\rangle$ on the amplitude $\gamma_a\omega$; see Fig. 11. Hence, it is much easier to characterize the dependence of elasticity on the shear rate by plotting the experimental data as $\langle E_{sh} \rangle$ vs. $\gamma_a \omega$, where $\langle E_{sh} \rangle$ is calculated from the experimental G' and G'' using eqn (26), without using any fits, like those in Fig. 11. Data obtained in angle-ramp regime can be added on the same graph by plotting the determined E_{sh} vs. an effective amplitude $\gamma_a \omega \equiv \dot{\gamma}/\mu$. Having once determined E_{sh}, we can further determine the viscosity $\eta_{sh} = E_{sh}/\nu_{ch}$, where ν_{ch} has to be calculated for the same shear rate from eqn (18).

In the framework of the Maxwell model (Fig. 8b), an elastic body corresponds to $E_{sh} \to$ const., $\eta_{sh} \to \infty$, and consequently $\nu_{ch} = E_{sh}/\eta_{sh} \to 0$. In the other limit of a purely viscous body, we have $E_{sh} \to \infty$, $\eta_{sh} \to$ const., and consequently $\nu_{ch} \to \infty$. Hence, the increase of ν_{ch} indicates fluidization (softening) of the body. In other words, the value of the characteristic frequency, $0 < \nu_{ch} < \infty$, can serve as an indicator for the degree of fluidization of the viscoelastic protein layer. In this respect, the increase of ν_{ch} with the rate of strain in Fig. 10 indicates fluidization of the adsorption layer upon increasing the shear rate. Likewise, the greater slope of the line in Fig. 10b in comparison with that in Fig. 10a (see the values of m in Table 3) means that the addition of β-casein enhances the softening of the protein layer upon shearing. More detailed information for the viscoelastic behavior of the system can be obtained by calculating separately $\langle E_{sh} \rangle$ and $\langle \eta_{sh} \rangle$, as demonstrated above (see Fig. 11 and 12).

Eqn (26) indicates that $\langle E_{sh} \rangle$ represents a nonlinear combination of the storage and loss moduli G' and G''. A similar expression holds for the average viscosity

$$\langle \eta_{sh} \rangle \equiv \frac{\langle E_{sh} \rangle}{\langle \nu_{ch} \rangle} = \frac{G'^2 + (m+1)G''^2}{G''\omega} \qquad (30)$$

where eqn (21) and (26) have been used. Hence, the popular paradigm that G' and G'' characterize, respectively, the elastic and viscous response of the system (which is correct for the Kelvin model) is not applicable to the investigated viscoelastic protein layers, whose behavior obeys the combined Maxwell–Herschel–Bulkley model. Note that at constant E_{sh}, η_{sh} and $\nu_{ch} = E_{sh}/\eta_{sh}$ [i.e. at $m = 0$; see eqn (23)], eqn (26) and (30) transform into eqn (12a). One of the important conclusions from the present study is that if the rheology of the adsorption layer complies with the Maxwell model, then the surface shear viscosity has to be calculated from eqn (30), rather than by the frequently used expression $\eta_{sh} = G''/\omega$, which corresponds to the Kelvin model. Likewise, if the viscoelastic layer obeys the Maxwell model, then E_{sh} must be estimated from eqn (26), the simple relation $E_{sh} = G'$ of the Kelvin model being inapplicable.

The data obtained in angle-ramp and oscillatory regimes are mutually complementary. The main advantage of the angle-ramp regime is that it indicates the type of the model, which can provide an adequate description of the system's rheological behavior. In our case, the data obtained in the angle-ramp regime comply with the Maxwell (rather than the Kelvin) model; see Fig. 6 and the related text, as well as preceding studies.[17,18,36] For protein layers, the main disadvantage of the angle-ramp regime is that the reproducibility of the results is not so high. This demands the measurements to be carried out many times and to take average values. The most probable reason for the lower reproducibility is that the rheology of the protein adsorption layers is sensitive to the surface age, i.e. to the prehistory. Conversely, in the oscillatory regime a steady-state periodic strain is imposed, which makes inessential the prehistory of the protein layer. The main advantage of the oscillatory regime is in the enhanced reproducibility of the experimental results. However, in this regime it is difficult to identify the adequate rheological law. A combination of data obtained using the two kinetic regimes allows one to avoid the aforementioned problems.

The Maxwell model, applied to protein adsorption layers, effectively describes the simultaneous stretching, breakage and restoration of intermolecular bonds upon shearing. As seen in Fig. 12a, with the increase of the rate of strain, the elasticity,

E_{sh}, initially increases, reaches a maximum, and then decreases. The initial increase can be interpreted with a predominant effect of stretching, whereas the decrease at higher shear rates—with predominant breakage of intermolecular bonds, which leads to softening of the adsorption layer. It is remarkable that the HFBII layer (without added β-casein) has elasticity which is insensitive to the shear rate in a relatively wide range. The addition of β-casein results in a markedly lower elasticity (Fig. 12a), whereas the viscosity of the β-casein containing layers becomes smaller only at the higher rates of strain (Fig. 12b). This evidences for the intercalation of β-casein molecules in the voids of the HFBII adsorption layer that have been detected by microscopic observations.[31,45,46] Despite the voids, at the investigated concentrations the HFBII certainly forms an interconnected network at the surface. If this network is broken, then E_{sh} should be considerably lower, as for β-casein alone (see Section 2).

The generalization of the Maxwell model to the case of variable η_{sh} and E_{sh} is nontrivial. The experiment (Fig. 6) indicates that this can be achieved by postulating variable η_{sh} and E_{sh} in eqn (11), which leads to the combined Maxwell–Hershel–Bulkley model, as described above.

A different approach to the generalization of the Maxwell model could be attempted by postulating variable η_{sh} and E_{sh} in eqn (8). In such a case, instead of eqn (11), one obtains:

$$\frac{d}{dt}\left(\frac{\tau_{sh}}{E_{sh}}\right) + \frac{\tau_{sh}}{\eta_{sh}} = \frac{d\gamma}{dt} \tag{31}$$

Because $E_{sh} = E_{sh}(|\dot{\gamma}|)$, eqn (31) leads to the appearance of the derivative

$$\frac{dE_{sh}}{dt} = \frac{dE_{sh}}{d|\dot{\gamma}|}\frac{d|\dot{\gamma}|}{dt}. \tag{32}$$

The derivative $d|\dot{\gamma}|/dt$ is a discontinuous function at $\dot{\gamma} = 0$ [we recall that $\dot{\gamma}(t)$ is a sinusoid]. However, the experimental $\tau_{sh}(t)$ dependence is a continuous and smooth function, which means that the postulate used to obtain eqn (31) is in conflict with the experiment.

4.5.2. Comparison of the viscoelastic behavior of different protein adsorption layers.
Table 4 shows data for the effect of concentration of added β-casein on the rheological parameters of mixed adsorption layers from solutions containing 0.005 wt% HFBII. The values of m determined from fits like those in Fig. 10 are increasing with the rise of β-casein concentration, which indicates faster fluidization with the rise of the shear rate. The values of $\langle \nu_{ch}\rangle$, $\langle E_{sh}\rangle$ and $\langle \eta_{sh}\rangle$ in Table 4 are calculated from m, G' and G'' by means of eqn (13), (26) and (30) and correspond to amplitude $\gamma_a = 1.74$ mrad and oscillatory frequency $\nu = 1$ Hz. The characteristic

Table 4 Effect of the concentration of added β-casein on the rheological parameters of adsorption layer from a 0.005 wt% HFBII; the values of $\langle \nu_{ch}\rangle$, $\langle E_{sh}\rangle$ and $\langle \eta_{sh}\rangle$ correspond to $\gamma_a = 1.74$ mrad and $\nu = 1$ Hz

β-casein (wt%)	m	$\langle \nu_{ch}\rangle$ (s^{-1})	$\langle E_{sh}\rangle$ (mN m^{-1})	$\langle \eta_{sh}\rangle$ (N.s/m)
0.015	0.81	1.10	133	0.110
0.030	0.82	1.16	111	0.082
0.045	0.91	1.23	124	0.101
0.070	0.98	2.06	79.5	0.039

frequency $\langle v_{ch} \rangle$ increases with the rise of β-casein concentration. In addition, $\langle E_{sh} \rangle$ and $\langle \eta_{sh} \rangle$ exhibit a tendency to decrease (excluding the point at 0.045 wt% HFBII). In general, the data in Table 4 indicate an increasing fluidization (softening) of the mixed protein adsorption layer with the rise of the β-casein concentration.

The difference between the values of m at 0.030 wt% β-casein concentration in Tables 3 and 4 is due to the fact that the data in Table 4 have been obtained with another sample of β-casein, the one that was used in ref. 36.

For HFBII, G' is ca 10 and 100 times greater than for BLG and β-casein, respectively.[11] However, the difference between the values of $\langle v_{ch} \rangle$ in Fig. 10c is not so great, especially for BLG and HFBII. This is due to the fact that $\langle v_{ch} \rangle = \langle E_{sh} \rangle / \langle \eta_{sh} \rangle$, and the softening of the adsorption layer may lead to a decrease of both $\langle E_{sh} \rangle$ and $\langle \eta_{sh} \rangle$, so that the changes in their ratio $\langle v_{ch} \rangle$ can be not so significant. For this reason, the viscoelastic properties of protein adsorption layers should be characterized by the dependencies of two parameters, e.g. $\langle v_{ch} \rangle$ and $\langle E_{sh} \rangle$, on the shear rate.

Our experiments have been carried out at a fixed temperature, 25 °C. The investigation of the temperature dependence of $\langle v_{ch} \rangle$, $\langle E_{sh} \rangle$ and $\langle \eta_{sh} \rangle$ can be a subject of a subsequent study. If the increase of temperature also leads to softening of the adsorption layers, then the strain rate could be interpreted as an effective temperature, as in the studies on concentrated particle suspensions (colloidal glasses).[47,48]

4.5.3. Applications to long-term foam stability. The experiment shows that the main reason for the decay of foams formed from protein solutions is the phenomenon foam disproportionation (Ostwald ripening).[49,50] This phenomenon is related to the transfer of gas across the foam films from the smaller to the bigger bubbles driven by the higher pressure in the smaller bubbles. As a result, the smaller bubbles shrink and disappear, whereas the bigger bubbles grow. This process can be characterized by the rate of decrease of the volume V of a small bubble:[51]

$$-\frac{dV}{dt} = k_g \, A \frac{p_c}{p_a} \approx k_g A \frac{1}{p_a} \frac{2\sigma}{R} \tag{33}$$

where t is time, A is the film area, $p_a = $ const. is the atmospheric pressure; $p_c \approx 2\sigma/R$ is the capillary pressure with σ being the surface tension; R is the bubble radius, and k_g is the permeability of the film to gas. Dense viscoelastic protein adsorption layers on the bubble surfaces can suppress the foam disproportionation in two ways: (i) decrease of the surface tension, σ, and (ii) decrease of the permeability k_g; see eqn (33). Indeed, upon the bubble shrinking the protein layer on its surface is spontaneously compressed and solidifies, and can have a very low surface tension, which is evidenced by the appearance of wrinkles on the bubble surface.[9,52] Second, the solidification of the protein adsorption layer can lead to a significant decrease of the permeability of the foam films to gases, k_g. The main reason for that is the low solubility and diffusivity of the gas molecules in the respective condensed adsorption layer, which has solid rather than fluid molecular packing.[50] In this respect, better foam-stabilizing effect is expected from adsorption monolayers of higher E_{sh} and lower v_{ch} that indicate a higher rigidity of the film.

Experimentally, solidification phase transition of a HFBII adsorption layer was detected upon increase of the surface pressure π_s (and surface coverage) by means of the pendant-drop method.[53] At 25 °C, the transition occurs at $\pi_s \approx 22$ mN m^{-1}. It is registered by the sharp increase of the error of the fit of the pendant-drop profile by the Laplace equation of capillarity. This effect is explained with the fact that the solidified protein adsorption layer has an anisotropic tensorial surface tension, whereas the Laplace equation presumes isotropic surface tension.[52,54] The same method[53] could be applied to investigate the temperature dependence of the solidification phase transition for adsorption layers of various proteins.

5. Oscillatory regime with nonlinear response

5.1. Effects of the frequency and strain amplitude

Here, our goal is to investigate the behavior of the protein adsorption layers at higher values of the rate of strain $\dot{\gamma} = \gamma_a \omega \cos(\omega t)$. The amplitude $\gamma_a \omega$ can be increased by raising either the strain γ_a, or the angular frequency $\omega = 2\pi\nu$. Fig. 13a and b show plots of the experimental data for the stress $\tau_{sh}(t)$ vs. strain $\gamma(t)$ (Lissajous plot) corresponding to the same frequency, $\nu = 1$ Hz, but to two different strain amplitudes: $\gamma_a = 1.05$ and 2.10%. The respective rate-of-strain amplitudes are $\gamma_a \omega = 0.066$ and 0.132 s^{-1}. The non-elliptical (parallelogram-shaped) Lissajous curves in Fig. 13a and b indicate that the rheological response of the layer is nonlinear (see Section 3.1 above).

Fig. 13c shows a similar Lissajous plot of the experimental data from Fig. 2b obtained at a higher frequency, $\nu = 4$ Hz, but at a lower strain amplitude, $\gamma_a = 0.174\%$, so that the rate-of-strain amplitude is $\gamma_a \omega = 0.044$ s^{-1}, i.e. it is smaller than those corresponding to Fig. 13a and b. The scattered points in Fig. 13c (the lack of Lissajous curve) imply that there is no definite relation between the stress and strain in this specific case. This could be interpreted as breakage of the elastic network formed by the adsorbed hydrophobin molecules at the air/water interface. The solid line (the ellipse) in Fig. 13c represents the sinusoid that is drawn as fit of the data in Fig. 2b. The comparison of the three plots in Fig. 13 indicates that at comparable rates of strain the increase of the frequency ω damages the viscoelastic adsorption layer easier than the increase of the strain amplitude, γ_a.

5.2. Fits of the Lissajous curves with the theoretical model

Here, our aim is to demonstrate that the combined Maxwell–Herschel–Bulkley model, based on eqn (11) and (18) can provide a quantitative description of the experimental Lissajous curves in the case of nonlinear response, as in Fig. 13a and b.

In the case of nonlinear response, the use of Fourier expansions, like eqn (6), is not convenient, because of the appearance of two infinite series of coefficients, G_k and G_k', $k = 1, 3, 5, \ldots$. Instead, we could integrate the basic equation of the Maxwell model, eqn (11). Substituting $\dot{\gamma} = \gamma_a \omega \cos(\omega t)$, we bring eqn (11) in the form:

$$\frac{d\tau_{sh}}{d\xi} + \frac{\nu_{ch}}{\omega}\tau_{sh} = E_{sh}\gamma_a\cos\xi \tag{34}$$

where $\xi \equiv \omega t$. Applying the known formula for the solution of a linear first-order differential equation, from eqn (34) we obtain:

$$\tau_{sh}(\xi) = f(\xi)\left[\tau_{sh}(0) + \gamma_a \int_0^\xi \frac{E_{sh}(\hat{\xi})}{f(\hat{\xi})}\cos\hat{\xi}\,d\hat{\xi}\right] \tag{35}$$

$$\text{where } f(\xi) \equiv \exp\left[-\int_0^\xi \frac{\nu_{ch}(\hat{\xi})}{\omega}\,d\hat{\xi}\right] \tag{36}$$

$\hat{\xi}$ and $\tau_{sh}(0)$ are, respectively, integration variable and constant. For a periodic variation of the stress, we have $\tau_{sh}(0) = \tau_{sh}(2\pi)$. Then, eqn (35) gives an expression for determining the integration constant:

$$\tau_{sh}(0) = \frac{\gamma_a f(2\pi)}{1 - f(2\pi)}\int_0^{2\pi} \frac{E_{sh}(\xi)}{f(\xi)}\cos\xi\,d\xi \tag{37}$$

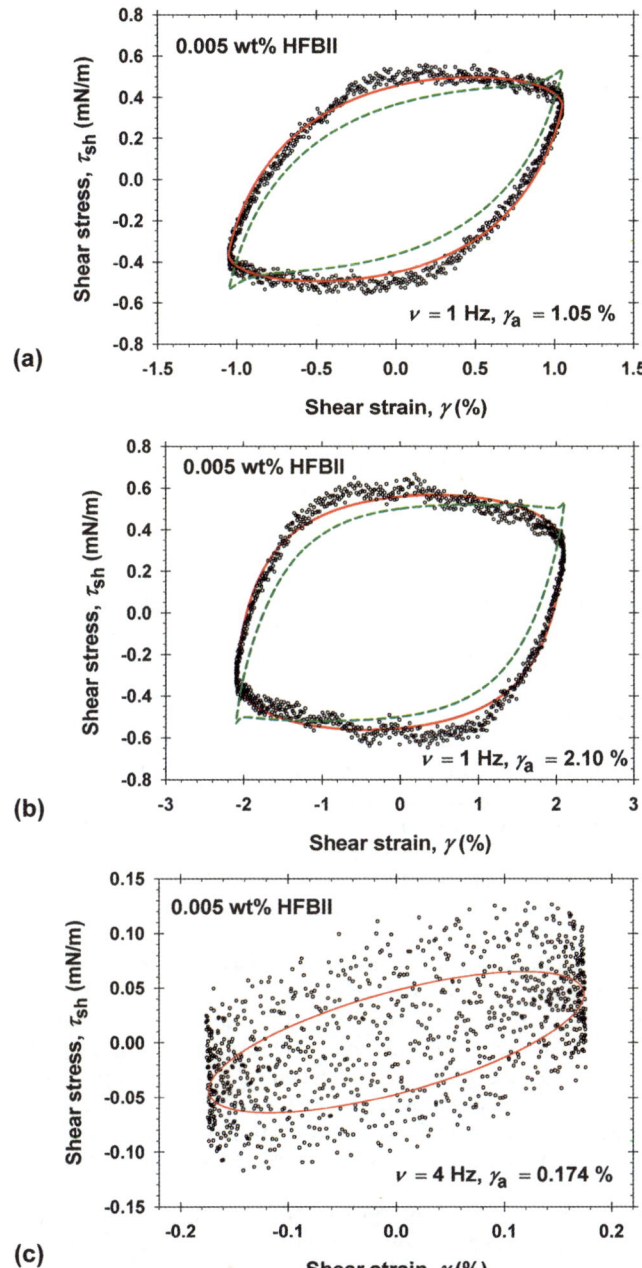

Fig. 13 Lissajous plots of experimental data for the stress *vs.* strain. The dashed lines are drawn using m, Q and $E_{sh}(\dot{\gamma})$ determined in quasi-linear regime (section 4). The solid lines are fits by means of eqn (35)–(37) using m, Q and E_{sh} as adjustable parameters. (a) Data from Fig. 3a for $\nu = 1$ Hz and $\gamma_a = 1.05\%$; (b) Data from Fig. 3b for $\nu = 1$ Hz and $\gamma_a = 2.10\%$. (c) Data from Fig. 2b for $\nu = 4$ Hz and $\gamma_a = 0.174\%$; in this special case the solid line represents the sinusoidal fit in Fig. 2b.

Table 5 Comparison of parameters determined from the fits of data in Fig. 13a and b (0.005 wt % HFBII, nonlinear response) with the respective values for quasi-linear response (the first line)

$\gamma_a\omega$ (s^{-1})	m	Q (s^{m-1})	E_{sh} (mN m^{-1})	$\langle\nu_{ch}\rangle$ (Hz)	$\langle\eta_{sh}\rangle$ (mN.s/m)
<0.033	0.88	134	$\geq$ 133	$\leq$ 3.06	$\geq$ 17.3
0.066	0.58	83	137	9.81	14.0
0.132	0.45	80	137	20.5	6.7

The dashed lines in Fig. 13a and b are calculated from eqn (35)–(37) using m, Q and $E_{sh}(\dot\gamma)$ determined in quasi-linear regime (Section 4). In particular, ν_{ch} is calculated from eqn (18) with m and Q from Table 3. In addition, E_{sh} is calculated from eqn (27) and (28) with $\dot\gamma = \gamma_a\omega\cos\xi$. As seen on Fig. 13a and b, the theoretical curves calculated in this way do not agree with the experimental data.

To fit the data in Fig. 13a and 13b, we calculated τ_{sh} from eqn (35)–(37) using m, Q and E_{sh} as adjustable parameters. In particular, $\nu_{ch}(\xi)$ was calculated from the modified Herschel–Bulkley law, eqn (18) with $\dot\gamma = \gamma_a\omega\cos\xi$. For $E_{sh}(\xi)$ we cannot use the empirical expression, eqn (27), because it has been obtained as a fit at lower shear rates. Unlike the viscosity η_{sh}, the elasticity E_{sh} is not varying too much, so that a reasonable approximation is to substitute E_{sh} = const. in eqn (35) and (37) and to determine its value as an adjustable parameter (an averaged elasticity) from the fit. The best fits are shown by solid lines in the respective figures, and the obtained parameters are given in Table 5. It is curious that the same values of E_{sh} were obtained from the fits of the data corresponding to the two different γ_a. Furthermore, substituting the obtained m and Q in eqn (23)–(24), we obtain $\langle\nu_{ch}\rangle$. Finally, the averaged viscosity is calculated from the expression $\langle\eta_{sh}\rangle = E_{sh}/\langle\nu_{ch}\rangle$. The obtained values are given in Table 5.

The first line of Table 5 shows parameter values corresponding to quasi-linear response; the maximal value of $\langle\nu_{ch}\rangle$ (from Fig. 10a) and the minimal values of of E_{sh} and η_{sh} (from Fig. 11a) are given for comparison. The last two lines of Table 5 show the parameters corresponding to the fits of the data in Fig. 13a and b. The results show that $\langle\nu_{ch}\rangle$ increases with the rise of the rate-of-strain amplitude $\gamma_a\omega$, which indicates an increasing fluidization of the adsorption layer (see Section 4.5). In addition, the last column of Table 5 indicates shear thinning, as it should be expected.

Fig. 14 compares the values of m and Q in the cases of quasi-linear and nonlinear rheological response of the HFBII adsorption layer. In the case of quasi-linear response, the values of m and Q are constant, independent of the rate-of-strain amplitude, $\gamma_a\omega$. In the case of nonlinear response, both m and Q are decreasing with $\gamma_a\omega$. This could be explained with structural changes in the adsorption layer that appear at rate-of-strain amplitudes $\gamma_a\omega$ > 0.033 s^{-1}. In Fig. 13c, the lack of a definite stress-*vs.*-strain dependence indicates that at higher frequencies the viscoelastic network of interconnected protein molecules at the interface can be destroyed.

In conclusion, the application of the combined Maxwell–Herschel–Bulkley model to the case of nonlinear response allows one to determine the elasticity and viscosity of the layer (Table 5) by fitting the Lissajous curves (Fig. 13a and b) with eqn (35)–(37). In this case the parameters m and Q of the modified Herschel–Bulkley law depend on the rate-of-strain amplitude, which indicates structural changes in the adsorption layer that precede its breakage. So, our recommendation is the comparison of different protein adsorption layers to be carried out in the regime of quasi-linear response, where the layer behaves as a viscoelastic body characterized by constant rheological parameters Q and m in a relatively wide range of shear-rate amplitudes; see Fig. 10 and 14.

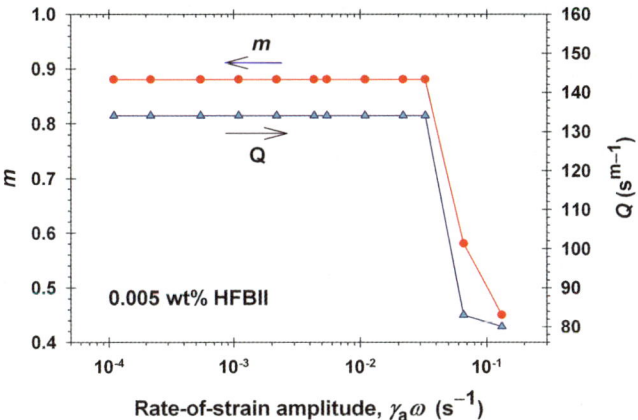

Fig. 14 Plots of m and Q from Table 5 *vs.* the experimental rate-of-strain amplitude. These two parameters are constant in the region of quasi-linear response (the line in Fig. 10a), but both of them decrease in the region of nonlinear response (Fig. 13a and b).

6.   Conclusions

A rheological model of viscoelastic protein adsorption layers is developed, which allows one to determine the interfacial shear elasticity and viscosity, E_{sh} and η_{sh}, as universal functions of the shear rate $\dot{\gamma}$, which are the same irrespective of the used kinetic regime: angle ramp or oscillations. In general, experiments in different regimes give different values of E_{sh} and η_{sh}, but the present analysis shows that the obtained experimental points must lay on universal curves $E_{sh}(\dot{\gamma})$ and $\eta_{sh}(\dot{\gamma})$. Experimental stress-*vs.*-strain dependences were obtained for adsorption layers of the protein hydrophobin HFBII and its mixture with β-casein. These protein layers exhibit a well pronounced viscoelastic behavior, which is usually characterized by the phenomenological storage and loss moduli, G' and G''.[9–11] These moduli depend on two kinetic parameters, frequency and amplitude of oscillations, and it has been unclear how they are related to the true surface shear elasticity and viscosity. Our analysis of data in both angle-ramp and oscillatory regimes revealed that the rheological behavior of the system obeys a combined Maxwell–Herschel–Bulkley model. In the angle-ramp regime, the fits of the experimental rheological curves give directly E_{sh} and η_{sh}. Their ratio, $\nu_{ch} = E_{sh}/\eta_{sh}$, represents a characteristic frequency of the system's rheological response. In oscillatory regime, these three quantities are universal functions of the rate of strain: $E_{sh}(\dot{\gamma})$, $\eta_{sh}(\dot{\gamma})$ and $\nu_{ch}(\dot{\gamma})$. Because the latter functions exhibit periodic variations, the comparison with the angle-ramp regime has to be made in terms of the average values of the respective quantities, $\langle E_{sh}\rangle$, $\langle\eta_{sh}\rangle$ and $\langle\nu_{ch}\rangle$, defined as described in the text. A complete agreement between the results obtained in the two different kinetic regimes has been achieved. It turns out, that only $\langle\nu_{ch}\rangle$ obeys a simple law of Herschel–Bulkley type, $\langle\nu_{ch}\rangle = Q\langle\dot{\gamma}\rangle^m$, in a wide range—more than three orders of magnitude; see Fig. 10. The dependences $E_{sh}(\dot{\gamma})$ and $\eta_{sh}(\dot{\gamma})$ are more complex than the simple Herschel–Bulkley power law for $\langle\nu_{ch}\rangle$. The determination of the basic rheological parameters Q and m allows comparison of different viscoelastic protein layers. Q and m can be determined from the experimental data for the moduli G' and G'', and subsequently, $E_{sh}(\dot{\gamma})$ and $\eta_{sh}(\dot{\gamma})$, and their average values, $\langle E_{sh}\rangle$ and $\langle\eta_{sh}\rangle$, can be calculated. The characteristic frequency ν_{ch} plays a central role in the rheological model. Its value, $0 < \nu_{ch} < \infty$, characterizes the softness/rigidity of the medium (0 = elastic layer; ∞ = viscous fluid layer). Despite the nonlinear character of the Herschel–Bulkley law, the protein adsorption layers exhibit a quasi-linear rheological response in a wide range of shear-rate amplitudes. However, at greater amplitudes nonlinear effects appear, which are detected by the presence of a non-negligible third harmonic in the Fourier expansion of $\tau_{sh}(t)$, and by a non-elliptic (parallelogram shaped) Lissajous plot. The combined Maxwell–Herschel–Bulkley model can be applied also to the regime of nonlinear response, but in this case the parameters Q and m become dependent on the rate-of-strain amplitude. In general, the nonlinear rheological response indicates structural changes in the protein adsorption layer caused by the more intensive shearing, which leads to breakage of the interfacial viscoelastic network at sufficiently high frequencies. For this reason, it is recommended to investigate the rheology of protein adsorption layer in the regime of quasi-linear response, for which the layer is characterized with constant Q and m, as well as with representative $E_{sh}(\dot{\gamma})$ and $\eta_{sh}(\dot{\gamma})$ dependencies. The results can be utilized for the optimization and control of the properties of fluid dispersions stabilized by protein adsorption layers. From a formal viewpoint, the developed approach can be applied to any viscoelastic continuum, not necessarily a protein layer.

Acknowledgements

The authors gratefully acknowledge the support from Unilever Research; from the National Science Fund of Bulgaria, grant No. DO-02-121/2009, and from ESF COST Action CM1101. The authors are grateful to Ms. Mariana Paraskova for her assistance in figure preparation.

References

1 A. J. Wilson, in *Foams*, ed. R. K. Prud'homme and S. A. Khan, Marcel Dekker, New York, 1996, ch. 5, pp. 243–285.
2 K. Mysels, K. Shinoda and S. Frankel, *Soap Films*, Pergamon, London, 1959.
3 D. A. Edwards, H. Brenner and D. T. Wasan, *Interfacial Transport Processes and Rheology*, Butterworth-Heinemann, Boston, 1991.
4 I. B. Ivanov, K. D. Danov and P. A. Kralchevsky, *Colloids Surf., A*, 1999, **152**, 161–182.
5 S. A. Koehler, S. Hilgenfeldt and H. A. Stone, *Langmuir*, 2000, **16**, 6327–6341.
6 K. Golemanov, S. Tcholakova, N. D. Denkov, K. P. Ananthapadmanabhan and A. Lips, *Phys. Rev. E: Stat., Nonlinear, Soft Matter Phys.*, 2008, **78**, 051405.
7 K. Tsujii, *Surface Activity: Principles, Phenomena and Applications*, Academic Press, London, 1998.
8 W. Xu, A. Nikolov, D. T. Wasan, A. Gonsalves and R. P. Borwankar, *Colloids Surf., A*, 2003, **214**, 13–21.
9 A. R. Cox, F. Cagnol, A. B. Russell and M. J. Izzard, *Langmuir*, 2007, **23**, 7995–8002.
10 A. R. Cox, D. L. Aldred and A. B. Russell, *Food Hydrocolloids*, 2009, **23**, 366–376.
11 T. B. J. Blijdenstein, P. W. N. de Groot and S. D. Stoyanov, *Soft Matter*, 2010, **6**, 1799–1808.
12 D. E. Graham and M. C. Phillips, *J. Colloid Interface Sci.*, 1980, **76**, 240–250.
13 E. Dickinson, B. S. Murray and G. Stainsby, *J. Colloid Interface Sci.*, 1985, **106**, 259–262.
14 D. Langevin, *Adv. Colloid Interface Sci.*, 2000, **88**, 209–222.
15 R. Borbas, B. S. Murray and E. Kiss, *Colloids Surf., A*, 2003, **213**, 93–103.
16 E. M. Freer, K. S. Yim, G. G. Fuller and C. J. Radke, *Langmuir*, 2004, **20**, 10159–10167.
17 I. Panaiotov, D. S. Dimitrov and L. Ter-Minassian-Saraga, *J. Colloid Interface Sci.*, 1979, **72**, 49–53.
18 P. M. Vassilev, S. Taneva, I. Panaiotov and G. Georgiev, *J. Colloid Interface Sci.*, 1981, **84**, 169–174.
19 J. Krägel, R. Wüstneck, D. Clark, P. Wilde and R. Miller, *Colloids Surf., A*, 1995, **98**, 127–135.
20 S. Roth, B. S. Murray and E. Dickinson, *J. Agric. Food Chem.*, 2000, **48**, 1491–1497.
21 J. Krägel, S. R. Derkatch and R. Miller, *Adv. Colloid Interface Sci.*, 2008, **144**, 38–53.
22 B. S. Murray and E. Dickinson, *Food Sci. Technool.*, 1996, **2**, 131–145.
23 R. Miller, R. Wüstneck, J. Krägel and G. Kretzschmar, *Colloids Surf., A*, 1996, **111**, 75–118.
24 E. Dickinson, *Colloids Surf., B*, 2001, **20**, 197–210.
25 M. A. Bos and T. van Vliet, *Adv. Colloid Interface Sci.*, 2001, **91**, 437–471.
26 J. Krägel and S. R. Derkatch, *Curr. Opin. Colloid Interface Sci.*, 2010, **15**, 246–255.
27 R. Miller, J. K. Ferri, A. Javadi, J. Krägel, N. Mucic and R. Wüstneck, *Colloid Polym. Sci.*, 2010, **288**, 937–950.

28 M. Reger, T. Sekine, T. Okamoto and H. Hoffmann, *Soft Matter*, 2011, **7**, 8248–8257.
29 M. Reger, T. Sekine, T. Okamoto, K. Watanabe and H. Hoffmann, *Soft Matter*, 2011, **7**, 11021–11030.
30 B. Niu, D. Wang, Y. Yang, H. Xu and M. Qiao, *Amino Acids*, 2012, **42**, DOI: 10.1007/s00726-011-1126-5.
31 M. B. Linder, *Curr. Opin. Colloid Interface Sci.*, 2009, **14**, 356–363.
32 J. Hakanpää, A. Paananen, S. Askolin, T. Nakari-Setälä, T. Parkkinen, M. Penttilä, M. B. Linder and J. Rouvinen, *J. Biol. Chem.*, 2004, **279**, 534–539.
33 E. S. Basheva, P. A. Kralchevsky, N. C. Christov, K. D. Danov, S. D. Stoyanov, T. B. J. Blijdenstein, H.-J. Kim, E. G. Pelan and A. Lips, *Langmuir*, 2011, **27**, 2382–2392.
34 E. S. Basheva, P. A. Kralchevsky, K. D. Danov, S. D. Stoyanov, T. B. J. Blijdenstein, E. G. Pelan and A. Lips, *Langmuir*, 2011, **27**, 4481–4488.
35 X. L. Zhang, J. Penfold, R. K. Thomas, I. M. Tucker, J. T. Petkov, J. Bent, A. Cox and R. A. Campbell, *Langmuir*, 2011, **27**, 11316–11323.
36 G. M. Radulova, K. Golemanov, K. D. Danov, P. A. Kralchevsky, S. D. Stoyanov, L. N. Arnaudov, E. G. Pelan and A. Lips, *Langmuir*, 2012, **28**, 4168–4177.
37 G. B. Bantchev and D. K. Schwartz, *Langmuir*, 2003, **19**, 2673–2682.
38 M. H. Lee, D. H. Reich, K. J. Stebe and R. L. Leheny, *Langmuir*, 2010, **26**, 2650–2658.
39 P. Erni, P. Fischer, E. J. Windhab, V. Kusnezov, H. Stettin and J. Läuger, *Rev. Sci. Instrum.*, 2003, **74**, 4916–4924.
40 K. Hyun, M. Wilhelm, C. O. Klein, K. S. Cho, J. G. Nam, K. H. Ahn, S. J. Lee, R. H. Ewoldt and G. H. McKinley, *Prog. Polym. Sci.*, 2011, **36**, 1697–1753.
41 W. Thomson, *Proc. R. Soc. London*, 1865, **14**, 289–297, DOI: 10.1098/rspl.1865.0052.
42 W. Voigt, *Ann. Phys. Chem.*, 1892, **283**, 671–693, DOI: 10.1002/andp.18922831210.
43 J. C. Maxwell, *Philos. Trans. R. Soc. London*, 1867, **157**, 49–88; URL: http://www.jstor.org/stable/108968.
44 W. H. Herschel and R. Bulkley, *Kolloid-Z.*, 1926, **39**, 291–300, DOI: 10.1007/BF01432034.
45 A. Paananen, E. Vuorimaa, M. Torkkeli, M. Penttilä, M. Kauranen, O. Ikkala, H. Lemmetyinen, R. Serimaa and M. B. Linder, *Biochemistry*, 2003, **42**, 5253–5258.
46 G. R. Szilvay, A. Paananen, K. Laurikainen, E. Vuorimaa, H. Lemmetyinen, J. Peltonen and M. B. Linder, *Biochemistry*, 2007, **46**, 2345–2354.
47 C. Eisenmann, C. Kim, J. Mattsson and D. A. Weitz, *Phys. Rev. Lett.*, 2010, **104**, 035502.
48 H. M. Vollebregt, R. G. M. van der Sman and R. M. Boom, *Soft Matter*, 2010, **6**, 6052–6064.
49 D. J. Carp, J. Wagner, G. B. Bartholomai and A. M. R. Pilosof, *J. Food Sci.*, 1997, **62**, 1105–1109.
50 S. Tcholakova, Z. Mitrinova, K. Golemanov, N. D. Denkov, M. Vethamuthu and K. P. Ananthapadmanabhan, *Langmuir*, 2011, **27**, 14807–14819.
51 H. M. Princen and S. G. Mason, *J. Colloid Sci.*, 1965, **20**, 353–375.
52 K. D. Danov, P. A. Kralchevsky and S. D. Stoyanov, *Langmuir*, 2010, **26**, 143–155.
53 N. A. Alexandrov, K. G. Marinova, T. D. Gurkov, K. D. Danov, P. A. Kralchevsky, S. D. Stoyanov, T. B. J. Blijdenstein, L. N. Arnaudov, E. G. Pelan and A. Lips, *J. Colloid Interface Sci.*, 2012, **376**, 296–306.
54 J. T. Petkov, T. D. Gurkov, B. E. Campbell and R. P. Borwankar, *Langmuir*, 2000, **16**, 3703–3711.

Elucidation of density profile of self-assembled sitosterol + oryzanol tubules with small-angle neutron scattering

Arjen Bot,[*a] Elliot P. Gilbert,[b] Wim G. Bouwman,[c] Hassan Sawalha,[de] Ruud den Adel,[a] Vasil M. Garamus,[f] Paul Venema,[d] Erik van der Linden[d] and Eckhard Flöter[ag]

Received 12th February 2012, Accepted 28th February 2012
DOI: 10.1039/c2fd20020a

Small-angle neutron scattering (SANS) experiments have been performed on self-assembled tubules of sitosterol and oryzanol in triglyceride oils to investigate details of their structure. Alternative organic phases (deuterated and non-deuterated decane, limonene, castor oil and eugenol) were used to both vary the contrast with respect to the tubules and investigate the influence of solvent chemistry. The tubules were found to be composed of an inner and an outer shell containing the androsterol group of sitosterol or oryzanol and the ferulic acid moieties in the oryzanol molecule, respectively. While the inner shell has previously been detected in SAXS experiments, the outer shell was not discernible due to similar scattering length density with respect to the surrounding solvent for X-rays. By performing contrast variation SANS experiments, both for the solvent and structurant, a far more detailed description of the self-assembled system is obtainable. A model is introduced to fit the SANS data; we find that the dimensions of the inner shell agree quantitatively with the analysis performed in earlier SAXS data (radius of 39.4 ± 5.6 Å for core and inner shell together, wall thickness of 15.1 ± 5.5 Å). However, the newly revealed outer shell was found to be thinner than the inner shell (wall thickness 8.0 ± 6.5 Å). The changes in the scattering patterns may be explained in terms of the contrast between the structurant and the organic phase and does not require any subtle indirect effects caused by the presence of water, other than water promoting the formation of sitosterol monohydrate in emulsions with aqueous phases with high water activity.

[a]Unilever Research and Development Vlaardingen, Olivier van Noortlaan 120, NL-3133 AT Vlaardingen, The Netherlands. E-mail: arjen.bot@unilever.com
[b]Bragg Institute, Australian Nuclear Science and Technology Organisation, Locked Bag 2001, Kirrawee DC, NSW 2232, Australia
[c]Department of Radiation, Radionuclides & Reactors, Faculty of Applied Sciences, Delft University of Technology, Mekelweg 15, NL-2629 JB Delft, The Netherlands
[d]Laboratory of Physics and Physical Chemistry of Foods, Department of Agrotechnology and Food Sciences, Wageningen University, Bomenweg 2, NL-6703 HD Wageningen, The Netherlands
[e]Chemical Engineering and Material Science, An-Najah National University, P.O. Box 7, Nablus, Palestine
[f]Helmholtz-Zentrum Geesthacht: Zentrum für Material- und Küstenforschung, Max Planck Strasse 1, D-21502 Geesthacht, Germany
[g]Food Process Engineering, Department of Food Technology and Food Chemistry, Technical University Berlin, Königin Luise Strasse 22, D-14195 Berlin, Germany

Introduction

In recent years, γ-oryzanol has been found to self-assemble with a range of plant sterols to form helical ribbons in triglyceride oil.[1] The helical ribbon tubules can aggregate subsequently into a firm 'organogel' network.[2] Most detailed work has been performed on the mixture of oryzanol with sitosterol, but similar assemblies are formed by oryzanol with the plant sterols ergosterol, stigmasterol, cholesterol and cholestanol. Small-angle X-ray scattering (SAXS) has been used to determine the tubule diameter and wall thickness, which were found to vary between 67 and 80 Å (diameter) and between 8 and 12 Å (wall thickness), respectively, depending on the particular plant sterol used.[3] The tubule is far in excess of 1000 Å in length.

The oryzanol + plant sterol mixtures are part of a wider class of systems that have recently been investigated in order to reduce the reliance of the structuring of edible oils on crystalline fat (*i.e.* triglycerides rich in saturated fatty acids).[4–8] Such non-triglyceride structuring systems for food-grade vegetable oils include fatty acids, fatty alcohols, monoglycerides (monoacylglycerols or MAGs), diglycerides (diacylglycerols or DAGs), phospholipids, ceramides, waxes, wax esters, or mixtures thereof[9–13] and modified biopolymers.[14] Some of these systems eliminate the relation between providing firmness to an oil-based system on the one hand and raising blood cholesterol when consumed in a food product on the other hand. This is a very desirable property since raised blood cholesterol levels are considered to be a risk factor for cardiovascular disease.[4]

The sensitivity of the sitosterol + oryzanol structurant to water was investigated recently, since most potential applications are foreseen in emulsion systems similar to margarine or butter.[15] Generally, it was found that the firmness of water-in-oil (w/o) emulsion gels based on sitosterol + oryzanol mixtures is much lower than would be expected based on the organogel properties.[16–18] This was explained by results from SAXS which demonstrated that water in emulsion gels binds to the sitosterol and promotes the formation of sitosterol monohydrate crystals. This prevents the self-assembly of sitosterol and oryzanol *via* hydrogen bonding.[19] However, the detrimental effect of water on structuring by sitosterol and oryzanol can be prevented or slowed down in two ways:[18] (i) reduction of the water activity below $\sim$0.9, *e.g.* through the addition of NaCl or sugar to the water phase to make the formation of the monohydrate crystals energetically unfavourable; or (ii) changing the organic phase to a lower polarity oil (*e.g.* decane, limonene) with a water solubility below $\sim$0.1% to reduce the rate of monohydrate crystal formation.

One of the puzzling aspects of the SAXS scattering pattern in emulsions is the splitting of the first broad peak into a, possible, overlapping double peak, as is illustrated in Fig. 1 for emulsions with high water activity (*i.e.* no salt). For the remainder, both the position and width of this double peak are very similar to the corresponding parameters for the first peak for emulsions with low water activity (Fig. 1) or, indeed, for organogels, which makes it likely that the structure of the tubules in emulsions is very similar to that of the tubules in the organogels. The emulsion data with the double peak in SAXS cannot be fitted to a hollow tube model as found for the pure organogel data.[17] Previously, however, it was found that the double peak can be constructed (qualitatively) using a combination of two hollow tubes, one with $\sim$70 and one with $\sim$100 Å diameter, with an essentially constant relative amplitude of both contributions to the SAXS data.[17] The constant relative intensity of both parts of the double peak, suggests that these are aspects of a single feature: a more complex wall structure than a tube with one wall of finite thickness. One of the more simple explanations is that the electron density of the tubule is not completely homogeneous. One possible cause is the presence of water, since the latter may interfere with the formation of a hydrogen bonds between sitosterol and oryzanol. Alternatively, the different densities of stacked androsterol groups (or sterane cores) and protruding ferulic acid moieties of the oryzanol molecules may cause a contrast difference in the tubule wall. Unfortunately, neither hypothesis

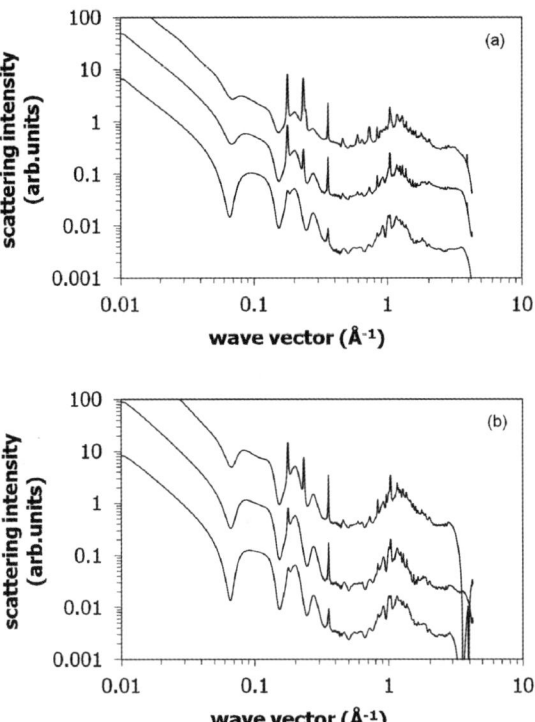

Fig. 1 SAXS for w/o emulsion gels in sunflower oil–(non-deuterated) decane mixtures, with 32% (40 : 60 sitosterol : oryzanol) as a structurant and 10% aqueous phase at a temperature of 20 °C: (a) 90 : 10; (b) 80 : 20. The aqueous phase varies in each graph from top to bottom 0, 10 and 20% NaCl. The scattering patterns have been shifted vertically for clarity.

can be tested easily in SAXS experiments since there is no way to independently vary the electron density of solvent and structurant.

As an alternative, small-angle neutron scattering (SANS) can be considered as an approach to obtain more detailed information at the nanometre length scale.[20,21] SANS provides more freedom to choose the contrast, mainly by varying the degree of deuteration of solvent and structurant. The present study, therefore, will investigate the structure of the sitosterol + oryzanol structurant mixture in more detail using solvent contrast variation enabling selected phases to be highlighted. Amongst the parameters under study are the effects of water activity (NaCl) and the effect of changing the oil in the organic phase.

Experimental

Preparation of the emulsions

In the present study, γ-oryzanol (Tsuno Rice Fine Chemicals, Wakayama, Japan) and tall oil sterol (78.5% β-sitosterol, 10.3% β-sitostanol, 8.7% campesterol and 2.5% of other minor sterols, Unilever, the Netherlands) were used as structurants. Sunflower oil (SF) (Reddy, NV Vandemoortele, Breda, the Netherlands), non-deuterated decane (>99%, Sigma-Aldrich, The Netherlands), eugenol (99%, Aldrich, The Netherlands), castor oil (Sigma, The Netherlands) and limonene (97%, Sigma-Aldrich, The Netherlands) were used as solvents. Sodium chloride, NaCl (purity >99%, Merck, the Netherlands) or sucrose (>95% (GC), Sigma, The Netherlands) dissolved in Milli-Q water solutions were used as the aqueous phase in the (w/o) emulsion. Deuterated decane was supplied by Acros Organics with D-enrichment >98.5%. Deuterated water (D_2O) was supplied by Merck. All materials were used as received.

The organic phase was prepared by dissolving the structurants in the solvent at elevated temperatures (~100 °C). The structurant concentration in the organic phase was kept constant at 32% (w/w) with a fixed γ-oryzanol to β-sitosterol ratio of 60 : 40% w/w (note that a relatively high structurant concentration was chosen to provide a clear scattering signal from the tubules. As such, organogels can be formed at much lower concentrations[2]). Aqueous solutions of different NaCl concentrations (0, 10, 20% w/w) were prepared and heated to 90 °C. All percentages in formulations are expressed on a weight basis. To prepare the emulsion, the two phases were mixed at a fixed weight fraction of the aqueous phase (10% w/w) in a closed container at 90 °C and stirred at 1300 rpm for ~2 min using a magnetic stirrer. The resulting w/o emulsion was cooled down to room temperature and stirred until gelled. The solidified emulsion was subsequently stored at 5 °C for 1 week before characterization. It should be noted that no additional emulsifiers or surfactants were added during emulsification and that the final w/o emulsion was stabilized by solidification.

Small-angle neutron scattering (SANS)

Preliminary SANS experiments were performed at the SANS-1 beamline at the Helmholtz-Zentrum Geesthacht in Geesthacht, Germany.[22] A wavelength of λ = 8.1 Å was used and a full width at half maximum of 0.81 Å. Several sample-to-detector configurations were used to cover the wave vector range $0.005 < q/Å^{-1} < 0.25$ (sample-detector distance (SDD) from 0.7 to 9.7 m) where $q = 4\pi \cdot \sin\theta/\lambda$ is the wave vector (and 2θ the scattering angle). The samples were contained in Hellma quartz cells of 2 mm path length. Samples were measured at 20 °C. The raw data were corrected for sample transmission and sample cell scattering by conventional procedures, azimuthally averaged, converted to an absolute scale and corrected for detector efficiency by dividing by the known incoherent scattering from 1 mm thick water. The scattering from solvents used for the sample preparation were subtracted as backgrounds. These experiments demonstrated that it was possible to obtain a clear tubule signature from sitosterol + oryzanol structured emulsion gels having deuterated decane as an organic phase by means of SANS (data not shown).

Final SANS experiments were performed on the 40 m Quokka instrument at the OPAL reactor (ANSTO, Sydney, Australia).[23] A wavelength λ = 5.078 Å and 14% wavelength resolution was used with source aperture diameter of 50 mm. Three instrument configurations were used to cover the wave vector range $0.004 < q/Å^{-1} < 0.7$ (SDD = 20.10, 3.20, and 1.32 m). The sample diameter used was selected by means of an automatic sample aperture changer within the instrument control system to maximise flux at the sample position but ensuring the scattering rate on the detector is within linearity in the detector response. Thus sample diameters of 15 mm, 12.5 mm and 7.5 mm were used at low, medium and high q configurations respectively. Samples were contained in demountable cells of 1 or 2 mm thickness depending on the deuteration level of the samples and contained within a thermostatically controlled 20-position automatic sample changer at 20 °C.

SANS datasets were reduced, normalized and radially averaged using a package of macros in Igor (Wavemetrics, Lake Oswego, Oregon, USA) software originally written by Kline[24] and modified to accept HDF5 data files from Quokka. Scattering curves are plotted as a function of absolute (SANS) intensity, I, versus q using the attenuated direct beam of known attenuation coefficient. A model was used that will be introduced later in this paper (eqn (1)). Preliminary fitting established ranges for the free parameters: the core diameter was limited in the range between 15 and 30 Å, the wall thicknesses of inner and outer shell between 5 and 20 Å, the scattering length density of inner and outer shell between 5×10^{-7} and $5 \times 10^{-6} Å^{-2}$, the intensity scale factor was limited to positive values. The somewhat slanting background identified for emulsions containing deuterated decane introduced a further complicating factor to the fitting procedure. Since the outer shell fitting parameters were compensating for the background for some curves, the emulsions and organogels based on a pure deuterated decane organic phase were left out in the simultaneous fit. This exclusion has little effect on the fitting parameters.

Due to their firmness, small amounts of air were introduced in the samples, affecting the absolute intensity of the SANS scattering pattern. To correct for this, the scattering intensity was normalised to a linear combination of the solvents (water, organic phases) for each sample averaged over the wave vector range $0.6 < q/\text{Å}^{-1} < 0.7$. This correction could amount to 25% of the (small) background contribution in some cases, but was usually smaller. Due to finite beam time allocation, only selected combinations of aqueous phase and organic phase systems could be measured.

Small-angle X-ray scattering (SAXS)

A number of supporting small-angle and wide-angle X-ray scattering (SAXS and WAXS, respectively) experiments were performed at the high-brilliance ID2 beamline of the European Synchrotron Radiation Facility (ESRF) in Grenoble, France.[25] An incident X-ray wavelength of $\lambda = 0.996$ Å was used. Two instrument configurations were used to cover the wave vector ranges $0.006 < q/\text{Å}^{-1} < 0.45$ (SAXS) and $0.38 < q/\text{Å}^{-1} < 3.78$ (WAXS) (SDD = 1.50 and 0.110 m, respectively). The samples were 2.2 mm thick and held in an aluminium cell with thin mica windows. The sample temperature of 20 °C was controlled by Peltier elements.

The incident and transmitted X-ray beam intensities were recorded with each SAXS pattern and used to normalise the measured SAXS intensities. The normalized two-dimensional SAXS patterns were azimuthally averaged to obtain the scattered intensity as a function of q. The corresponding background intensity was subtracted and the resulting corrected scattered intensity presented here.

Differential Scanning Calorimetry (DSC)

DSC was performed using a Perkin Elmer Diamond DSC (Perkin-Elmer Co., Norwalk, CT). 7–15 mg sample was loaded in stainless steel pans, sealed and mounted in the DSC. Scans involved a temperature cycle from 0 °C to 120 °C to 0 °C and back to 120 °C at a constant heating/cooling rate of 10 °C min⁻¹.

Results

Solvent scattering

While it would be desirable to use a deuterated analogue of sunflower oil as a solvent for conducting SANS studies, it is prohibitively expensive to acquire sufficiently large quantities. However, it has been shown that sitosterol + oryzanol tubules can be formed in a range of solvents.[18] Amongst these, decane can be readily acquired commercially in a fully deuterated form. In these experiments, therefore, a range of solvents was investigated including deuterated decane as well as mixtures of sunflower oil with both non-deuterated and deuterated decane to enable selective contrast variation studies.

Data from non-deuterated decane, deuterated decane, sunflower oil, castor oil, eugenol and a number of sunflower oil : decane mixtures (100 : 0, 90 : 10, 80 : 20 and 50 : 50) are shown in Fig. 2. The SAXS and WAXS data show pronounced peaks (see Fig. 2a) with, for example, broad features at 0.3 Å⁻¹ and 1.5 Å⁻¹ for sunflower oil (cf ref. 1). The 0.3 Å⁻¹ 'long spacing' is attributed to the bilayer thickness in the melt,[26] while the 1.5 Å⁻¹ 'short spacing' feature is generally attributed to intermolecular adjacently-packed fatty acid chains and intramolecular atom-atom

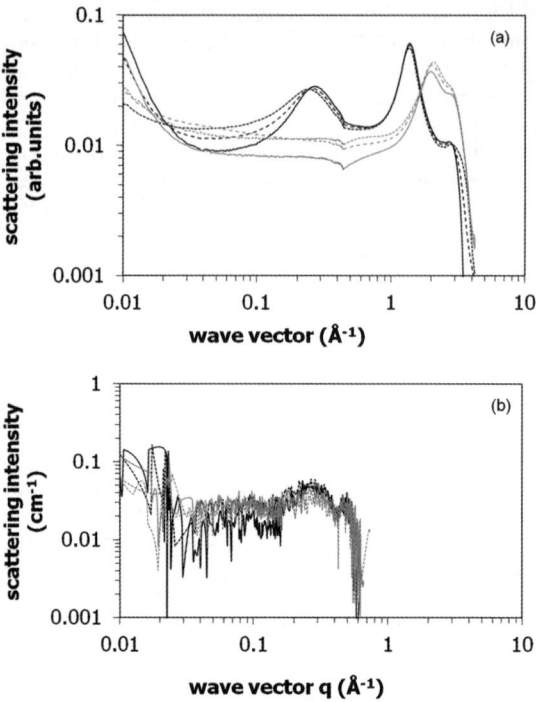

Fig. 2 Examples of (a) SAXS (and WAXS) and (b) SANS data for a selection of (non-deuterated) bulk phases. Black lines for sunflower oil : decane mixtures: (—) 100 : 0; (---) 90 : 10; (···) 80 : 20; grey lines for water : salt mixtures: (—) 100 : 0; (---) 90 : 10; (···) 80 : 20. The small cusp in the X-ray scattering data is caused by the transition between SAXS and WAXS detectors.

distances. The SANS data show only a weak feature at 0.3 Å^{-1} (Fig. 2b). In part this is explained by the dominance of incoherent scattering.

Organogels

Fig. 3 shows SANS data for the organogels with various solvents. Fig. 3a shows the effect of the polarity of the solvents, (non-deuterated and deuterated) decane, sunflower oil, castor oil and eugenol (cf reference 18). The data clearly confirm tubule formation in non-deuterated and deuterated decane. Neutron scattering length densities (SLD) for the individual components are summarised in Table 1. The structure with deuterated decane is most pronounced which may be explained by the greater contrast between the solvent and the structurant.

Amongst the higher-polarity oils, the systems with sunflower and castor oil show very weak tubular patterns. The system with eugenol no longer exhibits evidence for tubules. Since these systems have also been studied by SAXS, which *did* reveal tubular patterns for all these samples,[18] we attribute the absence of tubule scattering patterns to small contrast differences between structurant and solvent. Indeed, Table 1 indicates that the neutron scattering length densities for the solvents are all in the range 1.7×10^{-7} to 1.3×10^{-6} Å^{-2}, namely the same range as for sitosterol and oryzanol itself. In contrast, the neutron scattering length densities of non-deuterated and deuterated decane are outside this range, being -4.9×10^{-7} and 6.6×10^{-6} Å^{-2}, respectively. This highlights the value of using deuterated labelling in these studies.

Fig. 3b considers the more subtle effects of changes in solvent properties, in particular changes in the ratio between sunflower and decane, either deuterated or

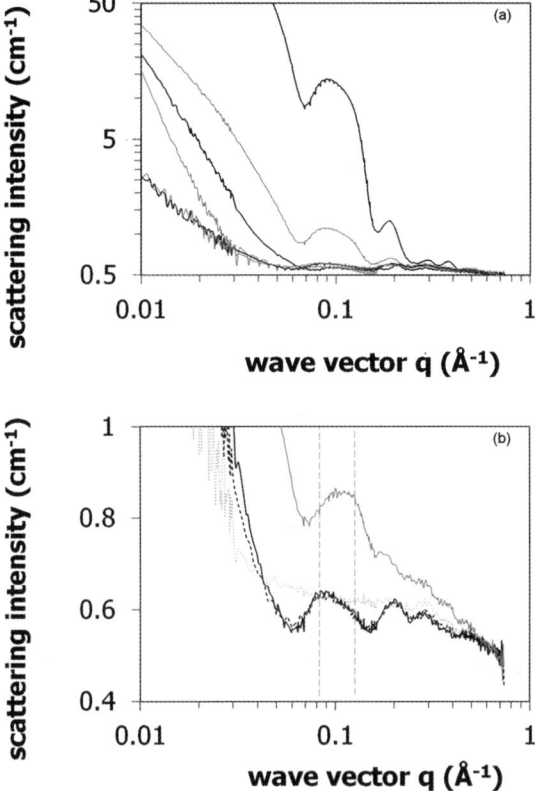

Fig. 3 SANS scattering patterns for organogels based on different organic phases: (a) deuterated decane, non-deuterated decane, eugenol, limonene, sunflower oil, castor oil (ordered from top to bottom according to intensity at small q); (b) various ratios of sunflower oil and decane (black curves are obtained from organogels containing non-deuterated decane, grey curves deuterated decane. The solid curves are obtained from organogels containing an 80 : 20 sunflower oil : decane organic phase, the dashed curves 90 : 10). The vertical dashed lines indicate the position of the maxima of the first interference maximum for the organogels with deuterated or non-deuterated decane.

non-deuterated decane. The main variation in the scattering is influenced by the amount of deuterated decane in the organic phase. This is as expected, since changes in the ratio of non-deuterated decane and sunflower oil has little effect on the overall neutron scattering length density. Obviously, the substitution of sunflower oil for deuterated decane has a much greater effect on hydrogen density. An organic phase with 10% deuterated decane is rather well contrast-matched. However, tubules can be observed for the sample with 20% deuterated decane (low hydrogen density) or the samples without deuterated decane (high-hydrogen density). For the high-hydrogen-density organic phase, the ratio of sunflower oil and decane did not affect the scattering pattern significantly.

It is interesting to note that the peak position of the first interference maximum (around 0.1 Å^{-1}) occurs at smaller wave vectors for high-hydrogen density organic phases (low neutron scattering length density, SLD) than for low-hydrogen density organic phases (high neutron SLD). Their positions at $2\pi/q_i = 73$ and 53 Å, respectively, are indicated by the dashed vertical grey line. The most likely origin for the shift in peak position is in terms of an inhomogeneous neutron scattering length density distribution in the tubules, in particular a double walled tubule with a inner

Table 1 Neutron scattering length densities (SLD) for components studied (http://www.ncnr.nist.gov/resources/sldcalc.html)

Compound name	Compound formula	Neutron SLD (Å^{-2})
Decane (H)	$C_{10}H_{22}$	-4.88×10^{-7}
Decane (D)	$C_{10}D_{22}$	6.58×10^{-6}
Limonene	$C_{10}H_{16}$	2.47×10^{-7}
Sunflower oil	$(C_{57}H_{101}O_6) = C_3H_5(C_{18}H_{32}O_2)_3$	2.25×10^{-7}
Castor oil	$(C_{57}H_{107}O_9) = C_3H_5(C_{18}H_{34}O_3)_3$	1.91×10^{-7}
Eugenol	$C_{10}H_{12}O_2$	1.29×10^{-6}
Water (H)	H_2O	-5.60×10^{-7}
Water (D)	D_2O	6.48×10^{-6}
Oryzanol	$C_{40}H_{58}O_4$	7.21×10^{-7}
Sitosterol	$C_{29}H_{50}O$	1.68×10^{-7}
Salt	NaCl	2.95×10^{-6}

high-hydrogen density (low SLD) cylinder and an outer low hydrogen density (high SLD) cylinder. This qualitative representation of the tubule structure will be developed in more detail in the Discussion section.

Emulsions

Emulsions with different organic phases. The next step is to introduce water in these systems by creating emulsions. Data obtained from corresponding emulsions are shown in Fig. 4. Again, it is clear that tubules are formed both in non-deuterated and deuterated decane emulsions. The tubule interference pattern is most pronounced in the emulsion based on deuterated decane. This is in line with the observation for the organogels (Fig. 3), and in agreement with the greater contrast difference between the solvent and the structurant (Table 1). Furthermore, tubules are still formed in limonene, although the limited contrast makes the scattering pattern less pronounced. Emulsions containing castor oil or SF (with pure water in the aqueous phase) show sharp crystallographic peaks instead of the characteristic broad tubule features.[17,18] Peaks can be found at $d = 2\pi/q_i = 35$ and 27 Å which

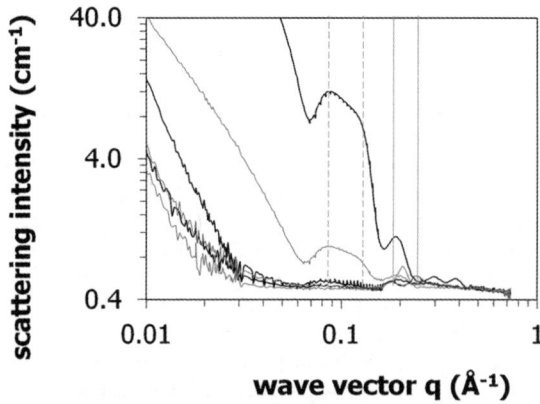

Fig. 4 SANS data for emulsions containing 10% dispersed water and an organic phase based on various solvents: non-deuterated decane, deuterated decane, limonene, castor oil, sunflower oil, eugenol oil (ordered from top to bottom according to intensity at small q). The vertical grey lines indicate the position of the sitosterol monohydrate peaks (solid lines) and of the peaks identified in Fig. 3b (dashed lines).

reflect the bilayer structure of sitosterol monohydrate.[19,27,28] The positions of these peaks are indicated by the two grey solid vertical lines, for later comparison. Sitosterol in an eugenol-based emulsion shows a different crystalline reflection at $d = 2\pi/q_i = \sim 31$ Å, indicating a different type of organisation for the sitosterol monohydrate crystals in the more polar organic solvent. The two grey dashed vertical lines reflect the position of the peak maxima in the organogels (cf Fig. 3) and it can be concluded that these positions coincide with the maxima or shoulders in the SANS pattern for these emulsion gels.

Emulsions with different organic phases and different water activity. In previous SAXS experiments, emulsions based on decane were found to form tubules, whereas emulsions based on sunflower oil, castor oil or eugenol did not form stable tubules in the presence of a pure aqueous phase.[18] This difference was attributed to the ability to form sitosterol hydrates if either the transport of water through the organic phase is fast (kinetic argument) or if the water activity of the emulsion is too high (equilibrium argument). This leaves two different routes to avoid hydrate formation and thus the intriguing possibility to study systems in which the contrast between structurant and organic phase is sufficient high to obtain the SANS signature of the tubules but, in addition, in which the water activity of the aqueous phase can be tuned to control the hydration behaviour of the sterols.

The DSC results shown in Fig. 5 demonstrate that sitosterol monohydrate formation occurs for up to ~20% decane in sunflower oil. Although this unfortunately implies a very modest contrast between structurant and organic phase, the experiments in the previous section have indicated that the contrast is sufficient to still extract valuable information from such experiments. By creating an emulsion with an organic phase that has sufficient water solubility to still allow hydrate formation but that can be stabilised by tuning the water activity of the aqueous phase, it should be possible to study the effect of water on the structure of the sitosterol + oryzanol tubules, in particular the 'double first peak' pattern that is observed in many emulsions.[17]

SANS data from emulsions prepared from sunflower oil/deuterated decane mixtures with 10 and 20% deuterated decane, respectively, are shown in Fig. 6. For each organic phase, three salt levels were assessed (0, 10 and 20% NaCl). Sharp crystallographic sitosterol monohydrate peaks are observed that were encountered previously for the salt-free emulsions with sunflower oil.[18,19] Their position is marked by solid grey vertical lines. These reflections are much stronger in emulsions with D_2O than in emulsions with H_2O which reflects the enhanced contrast of the deuterated monohydrate with the hydrogenated organic phase. Emulsions with lower water activity due to the presence of salt (10 or 20%) essentially do not exhibit

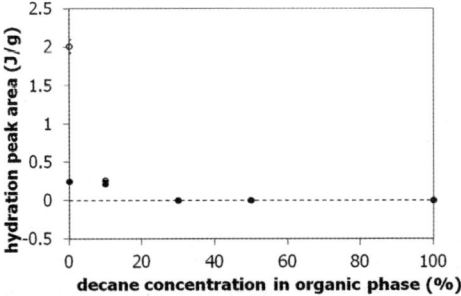

Fig. 5 Effect of (non-deuterated) decane : sunflower oil ratio on the area under the DSC hydration peak for a w/o emulsion with 10% pure water and 32% 40 : 60 sitosterol : oryzanol mixture on the organic phase: (●) first heating; (○) second heating. The data indicate that the transition vanishes at a decane concentration in sunflower oil of 20 ± 10%.

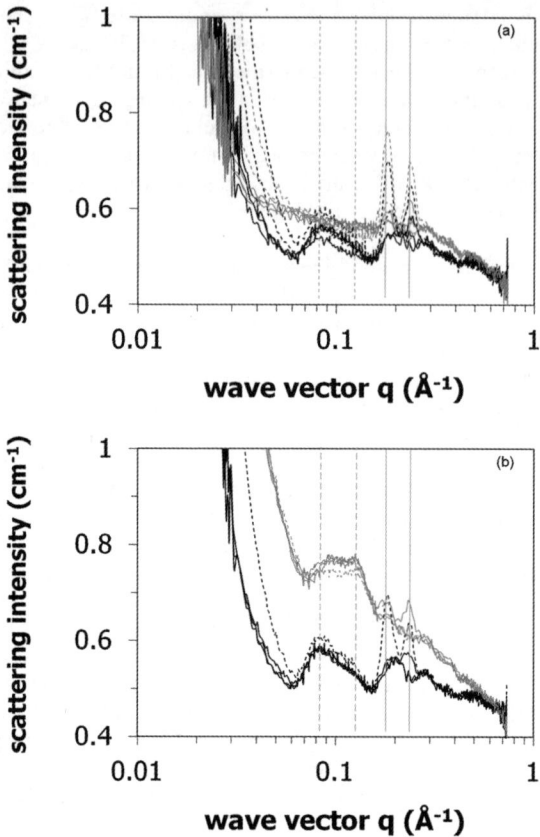

Fig. 6 Emulsions with sunflower oil : decane mixtures as the organic phase: (a) 90 : 10; (b) 80 : 20. The black curves are obtained from emulsions containing non-deuterated decane, the grey curves deuterated decane. The solid curves are obtained from emulsions containing H_2O, the dashed curves D_2O. In addition, the aqueous phase may contain 0, 10, 20% NaCl but this is only reflected in sitosterol monohydrate formation (*cf.* Fig. 1). The vertical grey lines indicate the position of the sitosterol monohydrate peaks (solid lines) and of the peaks identified in Fig. 3b (dashed lines).

sitosterol monohydrate formation with the scattering being dominated by the presence of tubules.

Furthermore, it can be seen that the high q scattering for emulsions with deuterated decane does not seem to be completely flat, but decreases with increasing wave vector. In contrast, the pure deuterated decane sample does not show this effect indicating that selective deuteration enables a smaller-scale structure to be discerned. The emulsion with 10% deuterated decane in the organic phase appears to be almost contrast-matched with any potential tubules in the emulsion (Fig. 6a)—neutron SLDs are 0.500×10^{-6} Å^{-1} and 0.219×10^{-6} Å^{-1} for the sunflower oil : decane mixture and 40 : 60 sitosterol : oryzanol, respectively.

Fig. 6 confirms the observation in Fig. 3b that the peak position of the first interference maximum at ~0.1 Å^{-1} occurs at smaller q values for the emulsion with 20% non-deuterated decane compared to the emulsion with 20% deuterated decane in the organic phase (at $2\pi/q_i = 73$ and 53 Å, respectively). Their positions are marked by dashed grey vertical lines. The effect implies, once more, the existence of two regions in the tubule wall: an inner wall with a high hydrogen density and an outer wall with a lower hydrogen density. The corresponding SAXS curves, using hydrogenated decane and water (Fig. 1), unequivocally demonstrate the presence of tubules in these emulsions, especially at lower water activities.

Except for the monohydrate peaks (position indicated by the solid vertical solid line), which become more pronounced at higher water activity, there is surprisingly little effect of the water on the SANS curves. The larger contrast in D_2O-containing emulsions increases the signal strength at small q values, but does not change the peak shape of the first interference maximum at ~ 0.1 Å^{-1}. This implies that if water affects the tubule shape directly, it can only involve very small amounts of water.

Discussion

One initial hypothesis in this work was that the double first interference maximum observed in SAXS was caused by a more complex wall structure of the tubules in emulsions, induced by the presence of water. It was anticipated that a comparison between samples that are identical except for the presence of either deuterated or non-deuterated water would mainly manifest itself by changes in the double-peaked scattering pattern around 0.1 Å^{-1}. This is not what is observed in the data, however. Changes in the 'water-contribution' to the scattering pattern always occur either in the range well *beyond* 0.1 Å^{-1} and not close to the double-peaked interference maximum, and are associated with changes in the contrast of the sitosterol monohydrate peaks, or in the range well *below* 0.1 Å^{-1}.

It can be concluded, therefore, that the double maximum observed in SAXS for the first interference peak is not directly associated with water. In particular, both the organogel and the emulsion data indicate that the two peaks represent parts of the tubule wall with different hydrogen density, the inner part being denser than the outer part. A very natural interpretation of this observation is sketched in Fig. 7, in which the tubules are considered to be double-walled structures. The low-hydrogen density of the outer tubule is caused by the protruding ferulic acid moieties of the oryzanol in the partly deuterated organic phase; note that the latter are only present in about half of the molecules that form the tubule as sitosterol does not contain such a group. This model would confirm the inference that the ferulic acid moieties are located on the outside of the tubules, as has previously been proposed based on circumstantial evidence of aggregation behaviour at various sitosterol/oryzanol ratios[4] and consideration of molecular stacking[3].

Such a qualitative model can be turned into a somewhat more quantitative comparison by translating Fig. 7 in a scattering length density profile for the tubules to form cylinders with a core–shell–shell structure and whose length is much greater than the dimensions of the cross-section. The result is shown in Fig. 8. The associated fitting equation is:

$$I(q) = \varphi \int_0^{\frac{\pi}{2}} f^2(q, \alpha) \cdot \sin \alpha \, d\alpha$$

where φ is the volume fraction of tubules and α is defined as the angle between the cylinder axis and the scattering vector, q. The integral over α averages the form factor over all possible orientations of the cylinder.

$$
\begin{aligned}
f(q,\alpha) = & \; 2 \cdot (\rho_{\text{solv}} - \rho_1) \cdot V_{\text{solv}} \cdot j_0(q(L/2) \cdot \cos\alpha) \cdot J_1(qr_{\text{solv}} \cdot \sin\alpha)/(qr_{\text{solv}} \cdot \sin\alpha) + \\
& \; 2 \cdot (\rho_1 - \rho_2) \cdot V_1 \cdot j_0(q((L/2) + (r_1 - r_{\text{solv}})) \cdot \cos\alpha) \cdot J_1(qr_1 \cdot \sin\alpha)/(qr_1 \cdot \sin\alpha) + \\
& \; 2 \cdot (\rho_2 - \rho_{\text{solv}}) \cdot V_2 \cdot j_0(q((L/2) + (r_2 - r_{\text{solv}})) \cdot \cos\alpha) \cdot J_1(qr_2 \cdot \sin\alpha)/(qr_2 \cdot \sin\alpha)
\end{aligned}
\tag{1}
$$

where the radii r_i are depicted in Fig. 8, $j_0(x) = \sin(x)/x$, and $J_1(x)$ is the first order Bessel function and ρ_i the SLD of phase i. L is the length of the cylindrical tubule.

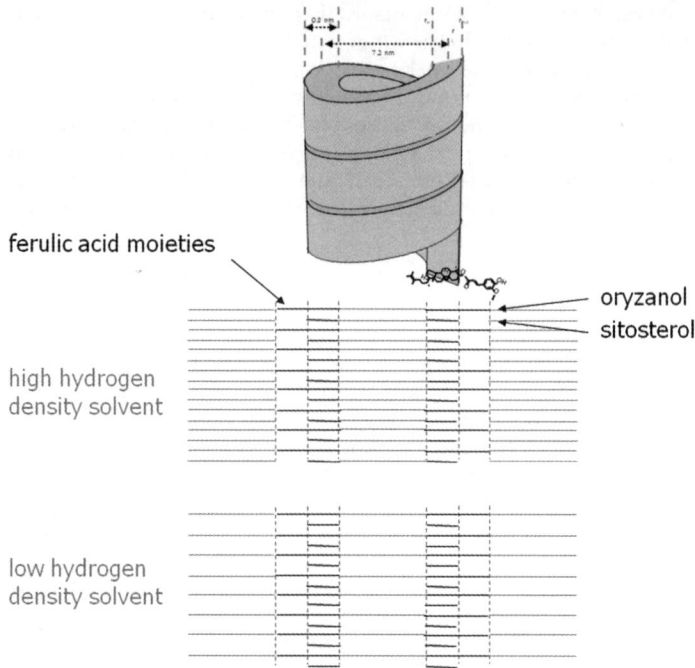

Fig. 7 Schematic representation of the high hydrogen density inside cylinder of the tubule (androsterol group in sterol and oryzanol) and the low hydrogen density outside cylinder (ferulic acid moieties of the oryzanol). The inner cylinder is revealed in a low hydrogen density solvent, and the outer cylinder in a high hydrogen density solvent.

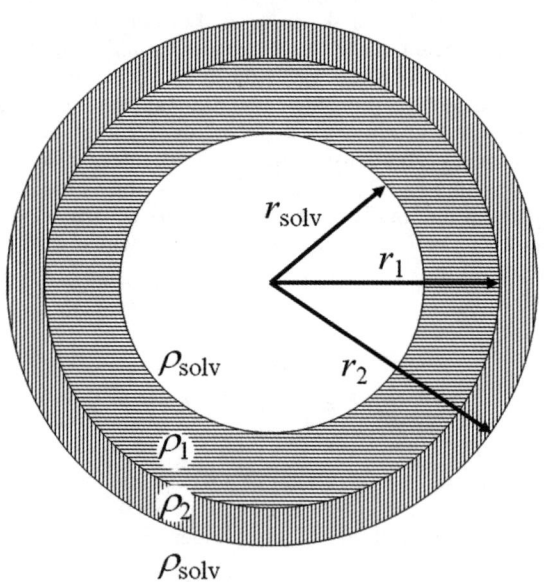

Fig. 8 Schematic representation of the double-walled tubule.

There are thus three regions defined: a core of radius, r_{solv}; an inner shell of thickness, $r_1 - r_{solv}$ and an outer shell of thickness, $r_2 - r_1$. The shell thickness is uniform over the entire tubule thus:

$$V_{\text{solv}} = \pi r_{\text{solv}}^2 L$$

$$V_1 = \pi r_1^2 (L + 2r_1)$$

$$V_2 = \pi r_2^2 (L + 2r_1 + 2r_2)$$

The equation reduces to the equation used for the analysis of previous SAXS experiments by taking $L \to \infty$ and $\rho_1 = \rho_2$ for a hollow cylinder.[29]

Eqn (1) should still be considered as an oversimplification of the real system as it cannot perfectly generate the double first peak but will constitute enough of an advance over the existing model[1,3] to warrant its use.

Eqn (1) was used to *simultaneously* fit all data for either the set of organogels or the set of emulsions over the q range from 0.04 and 0.74 Å^{-1} while fixing the neutron scattering length densities of the solvents (Table 1); this q range contains the most relevant information concerning the tubule wall structure. For fitting the set of emulsions, the q range featuring the sitosterol monohydrate peaks was excluded (range between 0.15 and 0.30 Å^{-1}). The data were used to extract the values for the diameter of the solvent-containing core and the wall thickness of the inner and outer shells. Note, therefore, that the model uses only six adjustable parameters, in addition to the scale factor and incoherent background of each curve. Some results are shown in Fig. 9.

Table 2 summarises the fitting parameters. The sum of core radius and inner shell agrees with the value that was obtained in earlier SAXS experiments:[1,3] 39.4 ± 5.6 *versus* 39.1 ± 2.2 Å (note that the present result should be compared to r_{out} in reference [3], not to r_{c}). The thickness of the inner shell is consistent with the SAXS observation (15.1 ± 5.5 *vs.* 8 ± 2 Å); despite the enhanced signal to noise ratio in SAXS, it remains difficult to extract inner shell dimensions.

The data confirm that the masking of the outer shell in a SAXS experiment is a result of insufficient contrast. While the thickness of this outer shell cannot be determined precisely in the present SANS experiment, it seems to be thinner than

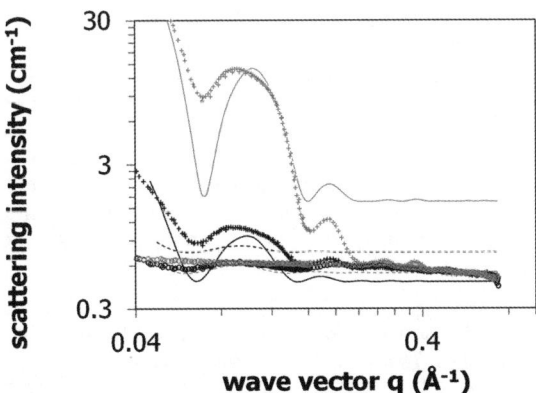

Fig. 9 Model fits (lines) to experimental data (symbols) on organogels using eqn (1). The black curves are obtained from organogels containing non-deuterated decane, the grey curves deuterated decane. The solid curves and (+)-symbols are obtained from emulsions containing pure decane as an organic phase, the dashed curves and (o)-symbols are obtained from emulsions containing 90 : 10 mixtures of sunflower oil : decane.

Table 2 Fitting parameters for eqn (1)

Parameter	Value	St. dev
Core radius, r_{solv} (Å)	24.3	0.9
Core length, L (Å)	6.7×10^3	0.2×10^3
Shell 1 thickness, $r_1 - r_{solv}$ (Å)	15.1	5.5
Shell 2 thickness, $r_2 - r_1$ (Å)	8.0	6.5
SLD shell 1, ρ_1 (Å^{-2})	7.8×10^{-7}	0.4×10^{-7}
SLD shell 2, ρ_2 (Å^{-2})	1.3×10^{-6}	0.5×10^{-6}

the inner shell: 8.0 ± 6.5 Å. Finally, the length of the tubules is of the order of a few thousand Å. For the present data, it suffices to note that the length of the tubules is *much* longer than their diameter.

Finally, the scattering length density of the inner and outer shell may provide clues whether their composition is in line with the hypothesis that the inner shell consists of androsterol groups (or sterane core) and the outer shell of ferulic acid moieties. Assuming a realistic density for the stacked molecules of approximately 1 g ml^{-1}, Table 3 indicates that the neutron scattering length density of the outer shell is very similar to that of ferulate and that of the inner shell is very close to androsterol.

Despite the success of the present analysis, it should be noted that the present model understates the true complexity of the structures formed. Indeed, the behaviour of the base line in the emulsions containing organic phases with higher deuterated decane concentration suggests more complex structures. It is anticipated that the present fit would be improved by explicitly considering alkyl groups protruding at the inside of the tubules into the organic phase (as Table 3 shows that the contrast difference between deuterated decane and the sitosterol alkyl chains is large) or by introducing a somewhat patchy distribution of ferulic acid moieties on the exterior surface of the tubules. The number of fitting parameters required would obviously increase as a result of the inclusion of pitch, roughness, polydispersity and other physically reasonable additions to the model, but might also explain the somewhat irregularly shaped first interference peak.

It would be of interest to conduct further contrast variation studies on these systems to elucidate more details of these self-assembled systems. One approach would be to focus on contrast differences but ignore any subtle effects due to the presence of water. Such an experiment would vary the contrast of the organic phase to the structurant without changing the physical properties (*e.g.* polarity) of the organic phase, for example by using mixtures of deuterated and non-deuterated decane. Hydration effects will not occur in such a system and they could be executed in organogel instead of emulsions. The second, more complex, approach would take hydration effects into account and would repeat the above experiment with mixtures of non-deuterated and deuterated triglycerides. These experiments would generate most additional information when performed on emulsions, but the first experiment could also be restricted to organogels. A deeper understanding of the structure of the tubules might also help to explain the kinetics of their formation.[30,31]

Table 3 Neutron scattering length densities (SLD) for a number of moieties in the sitosterol and oryzanol molecules (http://www.ncnr.nist.gov/resources/sldcalc.html)

Compound name	Compound formula	Neutron SLD (Å^{-2})
Androsterol	$C_{19}H_{30}O$	4.37×10^{-7}
Ferulate (minus one O)	$C_{10}H_9O_3$	1.71×10^{-6}
Alkyl chain of sitosterol	$C_{10}H_{18}$	-3.67×10^{-8}

Conclusions

The present neutron scattering experiments addressed the structure of self-assembled tubules composed of sitosterol and oryzanol that form in edible oils. By variation of the contrast between organic phase and structurant, it was possible to reveal that the cylindrical tubules are composed of an inner and an outer shell; this greatly extends our knowledge of the system which was based until now mostly on SAXS data. A core–shell–shell scattering model is necessary to describe the SANS with the scattering length densities consistent with an inner core consisting of the androsterol group in sitosterol and oryzanol for the inner shell, and with an outer shell composed of ferulic acid moieties (*i.e.* the ester part of oryzanol). The model indicates a core radius of 24.3 ± 0.9 Å, and a wall thickness of inner and outer shell of 15.1 ± 5.5 Å and of 8.0 ± 6.5 Å, respectively.

It was somewhat surprising to note that none of these changes were directly associated with the presence of water in organogels or emulsions, as had been anticipated. Instead, it seems that the more complex scattering patterns observed in SANS and in some cases in SAXS (especially for emulsions) should all be explained by changes in the contrast between organic phase and structurant.

Acknowledgements

The authors would like to thank T. Narayanan, M. Sztucki, S. Callow, J. Gummel and M. Fernandez Martinez (ID2 beam line, ESRF Grenoble, France) for their support.

References

1 A. Bot, R. den Adel and E. C. Roijers, *J. Am. Oil Chem. Soc.*, 2008, **85**, 1127–1134.
2 A. Bot and W. G. M. Agterof, *J. Am. Oil Chem. Soc.*, 2006, **83**, 513–521.
3 A. Bot, R. den Adel, E. C. Roijers and C. Regkos, *Food Biophys.*, 2009, **4**, 266–272.
4 M. Pernetti, K. F. van Malssen, E. Flöter and A. Bot, *Curr. Opin. Colloid Interface Sci.*, 2007, **12**, 221–231.
5 M. A. Rogers, *Food Res. Int.*, 2009, **42**, 747–753.
6 P. Wassell, G. Bonwick, C. J. Smith, E. Almiron-Roig and N. W. G. Young, *Int. J. Food Sci. Technol.*, 2010, **45**, 642–655.
7 A. G. Marangoni, N. Garti, Eds, *Edible oleogels: Structure and health implications*, AOCS press, Urbana, Ill, USA, (2011).
8 L. S. K. Dassanayake, D. R. Kodali and S. Ueno, *Curr. Opin. Colloid Interface Sci.*, 2011, **16**, 432–439.
9 J. Daniel and R. Rajasekharan, *J. Am. Oil Chem. Soc.*, 2003, **80**, 417–421.
10 F. G. Gandolfo, A. Bot and E. Flöter, *J. Am. Oil Chem. Soc.*, 2004, **81**, 1–6.
11 L. S. K. Dassanayake, D. R. Kodali, S. Ueno and K. Sato, *J. Am. Oil Chem. Soc.*, 2009, **86**, 1163–1173.
12 M. A. Rogers, A. J. Wright and A. G. Marangoni, *Soft Matter*, 2009, **5**, 1594–1596.
13 J. F. Toro-Vazquez, J. A. Morales-Rueda, E. Dibildox-Alvarado, M. Charó-Alonso, M. González-Chávez and M. M. Alonzo-Macias, *J. Am. Oil Chem. Soc.*, 2007, **84**, 989–1000.
14 T. Laredo, S. Barbut and A. G. Marangoni, *Soft Matter*, 2011, **7**, 2734–2743.
15 D. W. de Bruijne, A. Bot, in: *Food Texture: Measurement and Perception* (Rosenthal A. J., Editor), Aspen, Gaithersburg, MD, USA, 1999, p. 185–227.
16 A. Bot, Y. S. J. Veldhuizen, R. den Adel and E. C. Roijers, *Food Hydrocolloids*, 2009, **23**, 1184–1189.
17 A. Bot, R. den Adel, C. Regkos, H. Sawalha, P. Venema and E. Flöter, *Food Hydrocolloids*, 2011, **25**, 639–646.
18 H. Sawalha, R. den Adel, P. Venema, A. Bot, E. Flöter and E. van der Linden, *J. Agric. Food Chem.*, 2012, **60**, 3462–3470.
19 R. den Adel, P. C. M. Heussen and A. Bot, *J. Phys.: Conf. Ser.*, 2010, **247**, 012025.
20 A. Lopez-Rubio and E. P. Gilbert, *Trends Food Sci. Technol.*, 2009, **20**, 576–586.
21 A. Bot, E. Flöter, in: *Edible oleogels: Structure and health implications* (Marangoni A G, Garti N, Editors), AOCS press, Urbana, Ill, USA, chapter 3, pp. 49–79 (2011).

22 H. B. Stuhrmann, N. Burkhardt, G. Dietrich, R. Junemann, W. Meerwinck, M. Schmitt, J. Wadzack, R. Willumeit, J. Zhao and K. H. Nierhaus, *Nucl. Instrum. Methods Phys. Res., Sect. A*, 1995, **A356**, 124–132.

23 E. P. Gilbert, J. C. Schulz and T. J. Noakes, *Phys. B*, 2006, **385–386**, 1180–1182.

24 S. R. J. Kline, *J. Appl. Crystallogr.*, 2006, **39**, 895–900.

25 T. Narayanan, O. Diat and P. Bösecke, *Nucl. Instrum. Methods Phys. Res., Sect. A*, 2001, **467**, 1005–1009.

26 L. Hernqvist, *Fette, Seifen, Anstrichm.*, 1984, **86**, 297–300.

27 B. M. Craven, in: *Handbook of lipid research, volume 4, The physical chemistry of lipids – from alkanes to phospholipds* (Hanahan D. J. and Small D. M., Editors), p. 149–182 (1986).

28 G. Argay, A. Kálmán, S. Vladimirov, D. Zivanov-Stakic and B. Ribár, *Z. Kristallogr.*, 1996, **211**, 725–727.

29 J. M. Deutch, *Macromolecules*, 1981, **14**, 1826–1827.

30 M. A. Rogers, A. Bot, R. S. H. Lam, T. Pedersen and T. May, *J. Phys. Chem. A*, 2010, **114**, 8278–8285.

31 H. Sawalha, P. Venema, A. Bot, E. Flöter and E. van der Linden, *Food Biophys.*, 2011, **6**, 20–25.

General discussion

Professor Nicolai opened the discussion of the paper by Professor van der Linden: It is important to distinguish between the formation of fibrils from peptides formed by hydrolysis at pH 2 without added salt and those formed by intact proteins at higher pH or in the presence of salt. It would be interesting to explore in more detail the differences between these two types of fibrils.

Professor van der Linden answered: I am glad that you bring up this point and I fully agree about these two ways of firbil formation. Indeed it would be interesting to better compare those two ways.

Dr de Vries asked: What is the salt stability of the fibril systems at different values of pH?

Professor van der Linden replied: Yes, the stability is good, fibrils do remain stable under different salt concentrations. The salt concentration however does matter for the specific characteristics of the fibril upon formation, one should not go to too high salt concentrations.

Dr de Vries commented: To what extent is the rheology of the fiber systems determined by the fiber–fiber interactions (*i.e.* their stickiness), and to what extent by the mechanical properties of the individual fibers ?

Professor van der Linden replied: This is a very good question and has been subject of discussions between Professor Nicolai and ourselves for 10 years now. It seems still hard to think of a conclusive experiment. However, taking the results of the Mezzenga group on SDS interacting with the fibrils by means of electrostatics preferentially, and not by means of the hydrophobic tails of the SDS, I would still be inclined to believe that electrostatic repulsion and chain excluded volume do play a role in the interaction between the fibrils and thereby contribute to a critical gel concentration (as argued before in our papers by Veerman *et al.*). This then would imply a repulsive glass type of system. However, with our newest insights, given the chemical analysis results on the composition of the peptides in the fibrils, one cannot exclude hydrophobic interactions neither. In terms of numbers though, the sites giving rise to electrostatic contributions seem to outweigh the sites that give rise to hydrophobic interactions. I'd like to add that the recent findings of the Mezzenga group that report multiple twisted strands (multiplicity up to something like 7 at least) also may point towards hydrophobic sites being present, although this also still might be only attributed to electrostatic interactions. So, the short answer is: I still do not know.

Mr Hettiarachchi asked: My question is whether the fibrils are stable or do they get disintegrate at pH = 5. (According to the results of Jones *et al.*, *Biomacromolecules*, 2011, **12**, 3056–3065, shortening of the fibril structure was reported at high pHs)

Professor van der Linden answered: Our experiments showed that the fibrils stayed stable for a couple of weeks at higher pH (this was the time necessary for our experiments to be completed), but we have not studied this stabilisation as a function of pH in detail. We were at that time interested in whether we could use the fibrils at their opposite charge to make very low weight fraction gels using calcium bridging, and this worked by the way (at pH 7–8).

Dr Ettelaie commented: You mentioned that the assembly of tubules is an enthalpically driven process, whereas that of fibrils is an entropy-driven one, as of course substantiated by behaviour of these two systems under dilution or temperature changes. However, self assembly normally involves a reduction in the entropy, arising from the formation of a more structured entity. Can you please elaborate on the molecular nature of the increase in entropy, which drives the assembly of fibrils?

Professor van der Linden replied: The self assembly into a certain state in general follows from a reduction in free energy towards that state. In our case of aqueous fibril formation, indeed the process is caused by an entropic contribution as deduced from our observed temperature dependency for the critical aggregation concentration. Those experiments probe the initial aggregation of the peptides. One should note that this initial aggregation is not necessarily already into a well defined fibrillar morphology, it may still be a preceding more spherical or less defined form. Anyway, we allude that the according hydrophobic interaction to the hydrophobic moeties of the peptides, as also identified, to be present from our MALDI-TOF results. The net entropy increases due to the fact that there is an entropy increase due the release of water molecules interacting with the peptide hydrophobic sites, which overrules the entropy decrease due to the aggregation of the peptides.

Dr Velikov asked: What is the behavior of the peptide fibrils at high ionic strength? Do they display a "salting in" effect?

Professor van der Linden responded: This is a very interesting point. We have not studied this in detail though.

Dr van der Sman opened the discussion of the paper by Dr Rayner with a series of questions: What happens to the stabilized emulsion droplets in the digestive tract? How can the starch particles be desorbed? If the emulsion droplet contains functional ingredients, how do you target their release? Additionally, how can you make the content of the droplets bioavailable: the starch granules are semi-crystalline, and therefore cannot be digested (in the short intestine). How does this impair the bioavailability ?

Dr Rayner replied: This is of course very interesting and was one of the main focuses of this project in the wider perspective, as its original goal was to create a targeted delivery system for bioactive components to the colon. Starch particles can be desorbed by different mechanisms depending on if we have heat treated the emulsion or not and to what degree. Our preliminary results indicate that non-heat treated emulsions incubated in simulated intestinal fluid are easily broken—likely as the surface active bile salts displace the starch particles destabilizing the emulsion—however if we heat enough to create a cohesive layer by partial gelatinization of the starch granules at the oil–water interface we find that the emulsions remain intact after 2 h incubation in simulated intestinal fluid and centrifugation. We have seen similar increased stability to digestion in studies using lipolysis,[1] and using amylase pretreatments[2,3]. Keeping in mind this is an *in vitro* system, and we do not have published data yet, if the emulsion is first broken down in the colon with the help of gut micro-flora it would be a bonus, rather than a problem, as few systems survive so far down in the GI tract. By increasing the heat treatment we speculate one can reduce the crystallinity and thereby also reduce resistance to digestion and in turn affect the eventual bioavailability.

1. A. Timgren, M. Rayner, M. Sjöö, P. Dejmek, Starch particles for food based Pickering emulsions, *Procedia Food Science*, 2011, **1**, 95–103.
2. M. Sjöö, A. Timgren, M. Rayner, P. Dejmek, Heat treated starch granule stabilized Pickering emulsions, Poster presentation, *Food colloids: creation and breakdown of structure*, Copenhagen Denmark 16–18 April 2012.

This journal is © The Royal Society of Chemistry 2012

3. M. Sjöö, A. Timgren, M. Rayner, S. C. Emek, P. Dejmek, Starch granule stabilized emulsions with enhanced barrier properties, Oral presentation OC34: *11th International Hydrocolloid Conference*, Whistler IN, USA. 11–14 May 2012.

Professor Norton asked: Why do the unmodified starch particles go to the interface?

Dr Rayner answered: To be honest we are not completely sure why we observe this phenomenon and have previously attributed it to the small amount of surface proteins or other impurities likely to be remaining after isolation or some degree of partial dual wettability. However, there may be other explanations. This is still an open question.

Professor Norton continued: When you use a triglyceride how can you be sure that you haven't got surface active di- or monoglycerides present as a consequence of oxidative damage? This can be facilitated by light?

Dr Rayner replied: We prevent the oxidation of the oils we use by storage in a dark cold-room. If you mean that we get a certain amount of emulsifying capacity due to the impurities in the oil; we had also suspected that this was contributing to the apparent emulsifying capacity and for this reason have also carried similar studies using purified oils by pre-treating them to remove any surface active components before use as well as tests using tetradecane, paraffin, and mineral oils.

Professor Norton asked: When you report the effect of heat treated starch gel in the dry state, what is the chemical change driving the effect?

Dr Rayner replied: I am not sure I understand the question. We do not have a gel when we heat treat dry, we apply dry heat to the granules before using them as Pickering particles. We heat treat starch granules in the dry state to increase the hydrophobicity of starch, causing starch granule surface proteins to change character from hydrophilic to hydrophobic.[1] An advantage of thermal modification is that no specific labelling is required when used in food applications. Furthermore, the hydrophobic alteration is explicitly occurring at the granule surface.

1. M. Seguchi, Oil-binding ability of heat-treated wheat starch, *Cereal Chem.*, 1984, **61**, 248–250.

Professor Norton commented: It would be interesting to know the change in crystallinity of your starch samples.

Dr Rayner answered: I agree, we have done some differential scanning calorimetry (DSC) work on both the starch granules and concentrated emulsions. We have published a DSC thermograph of quinoa starch granule gelatinization.[1] We have also monitored crystallinity qualitatively using polarized light *i.e.*, do we see a birefringent structure indicating remaining crystal structure or not after heat treatments at different temperatures.[2]

1. A. Timgren, M. Rayner, M. Sjöö, P. Dejmek, Starch particles for food based Pickering emulsions, *Procedia Food Science*, 2011, **1**, 95–103.
2. M. Sjöö, A. Timgren, M. Rayner, P. Dejmek, Heat treated starch granule stabilized Pickering emulsions, Poster presentation, Food colloids: creation and breakdown of structure, Copenhagen Denmark, 16–18 April 2012.

Dr van der Sman commented: I would like to discuss other ways of making starch granules hydrophobic, as mentioned earlier in the discussion, for octenyl succinic anhydride (OSA) would not stimulate my appetite. Suppose that one can make the native starch granules porous (say *via* enzymes). These are known to absorb some oil. If you add these oil filled starch granules to the emulsion mix, would they absorb onto the interface? I guess that their surface energy is an average of the oily patches and the more hydrophilic non-etched surface.

Dr Rayner answered: This is an interesting idea. I am aware that an amylose helix can trap a fatty acid but I have not seen work using triglycerides. We are not aware of any results showing the pores in starch granules to be hydrophobic, but if they are,making oily patches could be an interesting approach to making an 'additive' free granule that was sufficiently hydrophobic. This warrants further considerations.

Dr van der Sman asked: The helix of amylose is known to have an affinity for oil.[1] I would expect that the starch granules would have some affinity for oil. Have you observed some effects of that in your experiments? Does the drop volume decrease upon adsorption of the particles?

1. J. Nuessli, B. Sigg, B. Conde-Petit. and F. Escher, Characterization of amylose—flavour complexes by DSC and X-ray diffraction, *Food Hydrocolloids*, 1997, **11**, 27–34.

Dr Rayner replied: Yes the helix has a known affinity to some types of lipids. However in the study you cited and in other work on lipid amylose complexes they are generally formed when the starch has been gelatinized and dissolved in a heated solution. I am not sure if we could make such complexes using solid granules for steric reasons. Furthermore we use triglycerides in this work. However at the same time our starch granules in the native state (0% OSA) display some affinity to oil. This can be observed in the top left plot in Fig. 4 (0% OSA) in the article as we do get some emulsion drop formation, but we have a large amount of free starch in the system as well. We are not completely sure why we observe this phenomenon and have previously attributed it to the small amount surface proteins likely to be remaining after isolation. It would be interesting to test your hypothesis.

In regard to you second question, we do not observe the drop volume to measurable decrease upon adsorption of particles as we cannot measure emulsion drop size before we make the emulsions. However we do have some longer term storage data that indicated no significant change in droplet size over an 8 week storage period.[1,2]

1. D. Marku, M. Wahlgren, M. Rayner, M. Sjöö, A. Timgren, Characterization of starch Pickering emulsions for potential applications in topical formulations, *Int. J. Pharm.*, 2012, **428**, 1–7.
2. A. Timgren, M. Rayner, M. Sjöö, P. Dejmek, Starch particles for food based Pickering emulsions, *Procedia Food Science*, 2011, **1**, 95–103.

Professor Kulozik asked: What could explain the stability of Pickering emulsions even sparsely covered by particles? Partially uncovered surfaces could also be expected to coalesce rapidly. Does this observation for Pickering emulsions also hold for long-term stability?

Dr Rayner replied: Intuitively one would conclude that drops formed at a lower concentration of starch granules are less covered and more subjected to coalescence, which is likely the case and the reason why we generally get larger drops at lower particle to oil ratios. However, Pickering emulsions have been shown to be stabilized effectively even when there is a low surface coverage, for example Vignati *et al.* in ref. 1 used silica particles 0.5–0.8 µm and while the particles coverage was only 15% of a packed monolayer at the surface they still had imparted substantial stabilization power. Another example is a work using spores particles (~25 ?µm)[2] that were highly uneven distributed at the surface of the drops, but still stabilized the Pickering emulsions. Hunter *et al.* in ref. 3 discusses some of the possible explanations and give further examples in their review article.

With respect to the long term stability, in the case of our studies using quinoa starch stabilized emulsions the droplet size does not significantly increase with storage for a wide range of starch to oil ratios. In previous studies we have tested starch to oil ratios ranging from 35 mg ml^{-1} to 3500 mg ml^{-1}. Only in the lowest concentration did we see a significant increase in mean droplet size d_{43} from 60 micron to 70 micron, after 7 days. At all other concentrations there was no statistically significant change in mean droplet size.[4] In a longer storage study at a starch to oil ratio of 214 mg ml^{-1} (note this is lowest concentration used in the present work) we observed no significant difference in mean drop diameter (d_{43}) over time (1 day to 8 weeks) or among concentrations (12.5 to 33.3% oil content).[5]

1. E. Vignati, R. Piazza, T. P. Lockhart, Pickering emulsions: interfacial tension, colloidal layer morphology, and trapped-particle motion, *Langmuir*, 2003, **19**, 6650–6656.
2. B. P. Binks, J. H. Clint, G. Mackenzie, C. Simcock, C. P. Whitby, Naturally occurring spore particles at planar fluid interfaces and in emulsions, *Langmuir*, 2005, **21**, 8161–8167.
3. T. N. Hunter, R. J. Pugh, G. V. Franks, G. J. Jameson, The role of particles in stabilising foams and emulsions, *Adv. Coll. Interface Sci.*, 2008, **137**, 57–81.
4. M. Rayner, A. Timgren, M. Sjöö, P. Dejmek, Quinoa starch granules: a candidate for stabilising food-grade Pickering emulsions, *J. Sci. Food Agric.*, 2012, **92**, 1841–1847.
5. A. Timgren, M. Rayner, M. Sjöö, P. Dejmek, Starch particles for food based Pickering emulsions, *Procedia Food Science*, 2011, **1**, 95–103.

Dr Ettelaie asked: In some experiments involving stability of bubbles stabilised by hydrophobically modified silica, we found that particles on the surface of bubbles became incorporated in large aggregates of such particles in the bulk (T. Kostakis, B. S. Murray, and R. Ettelaie, *Langmuir*, 2006, **22**, 1273–1280). In other words bubbles, through the silica particles on their surface, were becoming incorporated into large particle aggregates. Some of the micrographs in your paper seem to suggest a similar situation for your emulsion droplets. Can you comment on this possibility, such interpretation of the micrographs for your system and possible implications for the enhanced stability of emulsion droplets?

Dr Rayner answered: This bubble question is very intriguing; we really should do a study using oil fluorescence stain so that we can clearly distinguish between oil and air to quantify the existence and or amount of bubbles in our system. To follow up on your micrograph question, we did do a small test to see if the size of the measured particles were aggregates of oil drops (or not) by carefully dislodging the quinoa starch granules from the o/w interface by the addition of excess Tween 20 and observing the starch stripped emulsions in the microscope. In Fig. 1, we see that the 4.66% OSA that had the larger LS measured particle size has what appears to be smaller (or at least not significantly) larger individual oil droplets. This does not definitely answer the question if air bubbles were contributing to the overall microstructure but we can definitely say that the individual oil droplet size for the high OSA modified starch granule stabilized emulsions were not as large as measured in the LS confirming our hypotheses that they are aggregated.

Dr Bot commented: You showed that these Pickering-stabilised emulsions with starch can be heated to some extent, leading to partial aggregation of the starch particles. You also indicated that the starch layer remains stable because not all parts of the starch granules gelatinise, because a part resides in the oil phase. I have two questions: (i) are these layers stable under pasteurisation conditions? (ii) Do you think that the heating conditions in the non-gelatinised part of the starch granules will be sufficient to pasteurise these low-moisture parts of the emulsion as well?

Dr Rayner responded: In regard to your first question: if these layers are stable under pasteurization conditions, in terms of the times and temperatures required for pasteurization then the answer is yes. The time temperature required for HTST pasteurization of milk, for example is 72 °C for 15 to 20 s. We have a

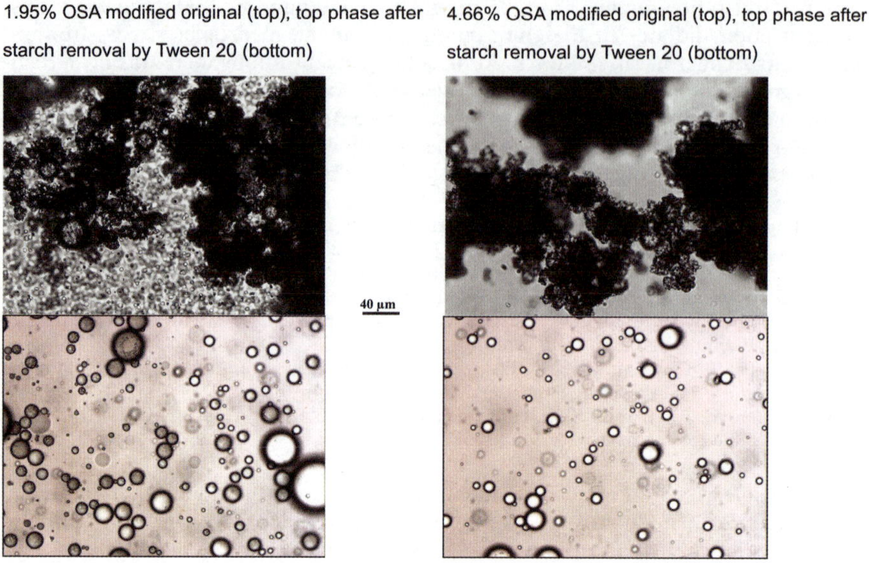

1.95% OSA modified original (top), top phase after starch removal by Tween 20 (bottom)   4.66% OSA modified original (top), top phase after starch removal by Tween 20 (bottom)

Fig. 1 Looking at actual drop sizes by removing starch.

considerable come-up time in our lab set up so 70 °C for 1 min is a greater thermal treatment than required for HTST, in addition we have used higher temperatures and still maintain droplet integrity.[1] In terms of other conditions found in industrial pasteurizers, such as the high shear and turbulent condition created during pumping, we cannot be sure at present. However, we have heated under stirred conditions using a laboratory propeller type mixer obtaining similar drop sizes obtained after heating.[2] Tests in a pilot scale HTST pasteurization process would be very interesting to conduct. In regard to your second question, heat transport at micrometer distances is really fast—if we compare starch to carbohydrate or wood with thermal diffusivity of about 0.08 mm^2 s^{-1} the characteristic time for approximately 10 μm is 1 s, if we have calculated correctly, there should not be any problems at the surface layer as heat will be easily conducted through the starch layer and oil drops during the heating process.

1. M. Sjöö, A. Timgren, M. Rayner, P. Dejmek, Heat treated starch granule stabilized Pickering emulsions, *Food Colloids: Creation and Breakdown of Structure*, Copenhagen Denmark, 16–18 April 2012.)
2. A. Marefati, M. Sjöö, M. Rayner, A. Timgren, P. Dejmek, Oil filled powders produced from starch stabilized emulsions, *11th International Hydrocolloid Conference*, Whistler IN, USA, 11–14 May 2012.)

Professor Stoyanov opened the discussion of the paper by Mr Waschatko: How you can expect the fact that oleosomes survive the 15 k g centrifugation process and do not coalesce with the air–water or solid–water interface in the centrifugation vessel, but when injected below planar air–water interface they 'spontaneously' and 'immediately' burst ? One can expect that above pI of the oleosins, oleosomes to have significant negative charge coming from the oleosins on their surface and also due to the apparent negative charge due to hydroxyl ion adsorption known to occur at almost any bare hydrophobic surface in contact with water including mineral and silicon oil, hydrocarbons and triglycerides and even air (see for example, K. G. Marinova, R. G. Alargova, N. D. Denkov, O. D. Velev, D. N. Petsev, I. B.

Ivanov, and R. P. Borwankar, *Langmuir*, 1996, **12**, 2045). The bare air–water interface will have an apparent negative charge as well due to the same reasons at arround −50 to −80mV at these conditions as well, so there should be significant repulsion between the air–water interface and 350 nm olesomes. This barrier is at least a few kT and is unlikely to be overcome by the buoyancy alone. So my questions is what screens this barrier or how oleosomes are 'tunneling' through it ?

Mr Waschatko responded: Thank you for your questions and remarks.

Firstly, we don't argue that oleosomes will not burst at the air–water interface during centrifugation, but one should keep in mind that the size and concentration ratios are completely different in the two situations. The air–water interface of the Langmuir trough is 715 cm², probably slightly larger due to the convex meniscus, and we're inserting only 22 µl (6.6 mg l⁻¹), so the interactions are driven by the behavior at the interface. In the centrifuge tube, the interface is 6 cm², with a concave meniscus and with a factor of 10^6 more oleosomes. So even if the oleosomes that are in contact with air burst, this will only be a negligible fraction of the whole amount and due to the centrifugation/tight packing we do not assume exchange between the layers.

Additionally, in the centrifuge tubes we have solutions with about 20% sugar concentration. Sucrose is known to have a strong hydrate shell and a strong interaction with the polyampholitic part of the oleosins, which will stabilize the samples. Centrifuge samples without sugar do not produce stable concentrated oleosomes (creamlayer). Furthermore, inside the creamlayer oleosomes are still surrounded by a water phase (also on a more speculative note, we assume that the hydration shell of the oleosomes plays a huge role in their stability. On the Langmuir trough this shell is probably destroyed when exposed to air as its 50% or more reaching out of the water. In the centrifuge tube the packing might be different and the hydration shell less exposed).

Dr Ettelaie remarked: Would the presence of an energy barrier against oleosomes reaching the air–water interface require a reasonable amount of charge on the air–water interface? This will only come about if there are some pre-existing charged species adsorbed on that interface.

Professor Stoyanov answered: Even the bare air–water interface has an apparent negative charge—see for example: K. G. Marinova *et al.*, *Langmuir*, 1996, **12**, 1996, 2045 or J. Cao, S. Stoyanov *et al.*, *Soft Matter*, 2012, **8**, 2194–2205. At neutral pH this is around −60 mV, which is few kT. This charge seams to occur at any hydrophobic surface in contact with water and is due to adsorption of hydroxyl ions.

Professor Wilde queried: For the extended time surface pressure isotherm (Fig. 5 of your paper), the surface pressure begins to decrease for the trypsin treated oleosomes, which is unusual. However, it looks as though the triglycerides are then forming a layer on the surface, which would explain the lower surface pressure, as the surface pressure becomes dominated by triglyceride-air interface. Is this correct? Also, as the surface is then compressed, the surface pressure increases dramatically with several phase boundaries. Does this suggest the triglyceride is being displaced from the surface by other surface active material, probably phospholipids?

Mr Waschatko responded: Thank you very much for your remark. The behavior you are describing is correct and is explained in the Results and Discussion section of our paper.

Professor Wilde asked: In response to the comment that the oleosomes are too stable to coalesce spontaneously with the air–water interface: looking at the results, the untreated oleosomes appear to be pretty stable, and probably don't fully coalesce with the interface. However, the trypsin treated oleosomes are very unstable due to the removal of the hydrophilic proteinaceous coat. This is supported by a decrease in surface pressure. Also, oleosomes are highly buoyant, despite their size and composition, and the experiment was run for 20 h, which gives sufficient time for these processes to occur.

Mr Waschatko replied: Intact oleosomes are described (in the introduction) to be very stable for long time in bulk water. At the air–water interface their stability, which corresponds to their rupture, is indeed dependent on their concentration, the time, the pH and ionic strength of the subphase. We recorded many isotherms for untreated oleosomes that show phase transitions of their former interior lipids, which demonstrates that a coalescence of intact oleosomes with the air–water interface is possible (reference 28 in the paper). However, they stay intact at the air–water interface significantly longer than the trypsin digested oleosomes.

Professor Marangoni continued the discussion of the paper by Professor van der Linden: The self assembly of proteins and peptides into fibrils could be strongly influenced by the interaction of such proteins and peptides with water. pH obviously affects this dynamic, but have you tried to induce fibril formation by judicious use of chaotropic agents? Urea could be an example of such.

Professor van der Linden responded: Yes, excellent point. Other groups have looked at the formation in alcohol (Ross-Murphy) for example. we have not gone down this avenue.

Professor Marangoni asked: Can you list the properties in an polypeptide that make it a good candidate for fibril formation?

Professor van der Linden replied: Yes, ability to form inter beta sheets, slighly hydorphobic sites, absence of proline, and slightly charged. See our paper by Akkermans *et al.* some years ago.[1]

1. Akkermans *et al.*, *Biomacromolecules*, 2008, **9**, 1474–1479.

Professor Marangoni remarked: Is there a global influence of the dipole moment of a protein or peptide on their ability to form fibrils?

Professor van der Linden responded: In principle there is a certain range of net charge necessary, but we have not determined the dipole moment of the peptides. Most likely, these will also differ a bit between one another. But in principle, there might be an influence. Analogous to the work of Pincus *et al.* on magneto–polymeric substances.

Dr Bot asked: You have demonstrated that the protein hydrolysis process results in fibrils and a fraction of material that does not seem functional in the present context. That seems wasteful from a functionality perspective. Did you establish the peptide structure that is required for protein fibril self assembly, and did you ever consider preparing these peptides *via* biotechnological routes?

Professor van der Linden answered: Yes, we know from our other work (Akkermans *et al.*) which class of peptides is built in the fibril. it needs to be partially hydrophobic, slightly charged and low in proline to have the ability to form inter-beta sheets. We indeed have considered this possibility of using only the specific peptides and acquiring them by a more efficient route, but not pursued. This would drift us away from foods too much and would be more in line with what others have been doing in this area of amyloid formation.

Professor Kulozik said: How can (and how was) the degree of hydrolysis of β-lactoglobulin be controlled to optimise the fibril formation or to reduce variability as the intended outcomes?

Professor van der Linden replied: The degree of hydrolysis was not controlled. In fact it is hard to control when at pH=2, since the hydrolysis cannot be stopped. The hydrolysis products often get further hydrolysed. Even this hydrolysis continues once the peptides are being built into the fibils (so called shaving phenomenon as found by Mishra *et al.* for lysozymes).[1] So, control in the end would only be partially relevant.

1. Mishra *et al.*, *J. Mol. Bio.*, 2007, **366**, 1029–1044.

Professor Wilde asked: SDS was added to stabilise the peptides against aggregation. Considering the interaction of SDS with proteins and peptides through both hydrophobic and/or electrostatic interactions, one would expect the SDS to affect the integrity of the fibrils, especially at higher SDS concentrations, was any such effect observed? and if so, at what SDS concentration?

Professor van der Linden answered: The surprising thing is indeed that SDS does not seem to lead to fibril breakdown. We have not observed such a thing. Chaotropic agents like urea do break the fibrils down. Perhaps the SDS does not affect the beta sheet structures like urea or guadinine chloride.

Professor Cates asked: Does the addition of SDS restore complete transparency in the slightly turbid fibril only state seen at pH 5? If so this shows that it has some direct effect on fibrils as well as free peptides.

Professor van der Linden replied: Yes indeed, this is a very good point, and indeed, the answer is yes. I agree with your conclusion that SDS is indeed interacting with the fibrils as well.

Dr Ettelaie said: How do the protein and SDS interact and associate with each other? The protein is positively charged at these low pH values, while SDS is negative, so in principle the association can be electrostatic. On the other hand the hydrophobic part of these two molecules can also associate through hydrophobic interactions. Is there any evidence to suggest which of these two mechanisms dominates in your system?

Professor van der Linden responded: This was reported by the Mezzenga group to be at low concentrations controlled by electrostatics indeed, whereas at higher concentrations the resulting higher number of hydrophobic tails stick outwards from the fibril leading to the formation of a cylindrical surfactant double layer, with the sulphate groups sticking into the aqueous phase, making it in fact an electrolyte, insensitive to pH changes. The amount of hydrophobically associated SDS at low concentrations does not therefore seem to be the dominant effect.

Dr Ettelaie continued: Following from my previous question, can a non-ionic surfactant be also used to stabilise the fibrils against aggregation. This would be of great advantage in foods, as many food grade surfactants (*e.g.* span, tween) are non-ionic.

Professor van der Linden answered: From the answer to your previous question I would think not that readily. However, we have tried to coat the fibrils with AOT and use that to make fibrils in oil systems, *i.e.* to use AOT coated fibrils to gel oil continuous systems, until now unsuccessfully however.

Dr van der Sman re-opened the discussion of the paper by Mr Waschatko: Why do you investigate the disintegration of the system at the air–water interface? Is it a model system for situations in the stomach/intestine ?

Mr Waschatko responded: Thank you for your questions. We have chosen Trypsin as digestive enzyme, because of its relevance in the human gastrointestinal tract. We could have chosen other enzymes for the cleavage, *e.g.*, proteinase K. The tryptic cleavage is just one part of the process that takes place during digestion. Nevertheless, we could show that after the cleavage of the outer part of the oleosins the concentration of free phospholipids and triacylglycerides at the air–water interface increases, which can be a hint towards the elevated bioavailability of those nutrients. Other detailed experiments with, *e.g.*, pepsin, bile acids and intestinal epithelial cells should be done to complete the picture of digestion of oil bodies. More importantly to us, was the impact of the cleavage itself on the system. Since the purification of natively folded oleosins is rather difficult, because of their required environment (PL-monolayer and TAG-matrix), the obvious alternative was their partial disintegration. Thereby we revealed that oleosins lose their surface activity (and therefore their emulsification properties) after digestion. Furthermore there were attractive forces or binding sites between the charged headgroups of the phospholipids and the cut hydrophilic parts of the oleosins, which are still present after the rupture of oleosomes and their new arrangement at the air–water interface. Additionally the digestion of the (pH-dependently) charged hydrophilic parts of the oleosins leads to a change in the pH-dependence, becoming similar to that of pure phospholipid stabilized systems.

Dr Ubbink addressed Mr Waschatko and Professor Vilgis : 1. You suggest a model to account for the shape of the phase-separated lipid domains. This model comprises an electrostatic component to the free energy and a free energy contribution arising from the line tension of the phase separated domains. a) Did you perform actual calculations on such models? b) Did you establish that the line tension contribution is indeed larger than the surface tension contribution? c) You assume quasi two-dimensional droplets in the phase-separated domains. Is this realistic? To which extent could the lipid-rich phase superpose and form multi-layer assemblies at the interface?

2) Did you observe further changes in structure and shape of the phase-separated domains? In particular at pH = 8, the surface pressure continues to increase after several hours. What would you expect the long-time structures to look like, in particular for the situations in which "star-like" domains are formed.

Mr Waschatko answered: Thank you for your questions and remarks.

1a) Calculating such models would be interesting for future work.

1b) The line tension has an influence on the free energy, *F*, which determines domain shape. In the case of measurements at pH 5, around the pI of phosphatidylethanolamine, no charged headgroups are present and therefore, *F* is only defined by the line tension, leading to circular structures. To our knowledge, surface tension may induce domain formation, but has no influence on their shape.

1c) Due to the amphiphilic nature of the lipids and TAGs, a 2D assembly seems to be the most reasonable one for structures visible at the air–water interface, as any 3D assembly (liposomes, bilayer, ...) would most likely drift into the subphase. Also, the BA micrographs show only a small contrast, compared to the previously observed oleosin aggregates, which strengthen a 2D assumption. We can safely assume that we don't have multilayer assemblies at the air–water interface, since our isotherms show typical lipid monolayer phase transitions and lipid monolayer collapse.

2) Due to the experimental set-up (shallow Langmuir trough where evaporation will lead to a change in surface pressure as well as loss of contact of the subphase with the barrier), no experiments over 20 h were systematically performed, as no reproducible results could be gained.

Dr Ettelaie asked: You have associated the changes in the surface pressure entirely to the release of phospholipids. Can protein fragments of oleosin also be present on the surface and would these have any influence on the surface pressure?

Mr Waschatko answered: Thank you for your question. For subphases with pH 2 and pH 5 we have seen large irregular and separated aggregates (> 50 μm) in the BAM micrographs, that can be attributed to protein aggregates of the hydrophobic fragment (8 kDa). The hydrophilic fragments are small and water soluble. The existence of protein aggregates at the air–water interface increases and their desorption decreases the surface pressure. However we could not observe any phase transitions of oleosins or oleosin fragments. A major difference between the oleosins and their tryptic fragments is the loss of their capability of homogeneous distribution and formation of a network-like film at the air–water interface after cleavage.

Dr van der Sman addressed Mr Waschatko, Professor Vilgis and Dr Smith: How are the oleosomes digested ? Does the oil become liberated during digestion ? This enzymatic breakdown of structure appears to be a sustainable way of liberating oil from the raw material. Would this be an alternative process/oil source for food ingredient companies like Cargill ?

Mr Waschatko responded: During the tryptic digestion oil is not liberated from the oleosomes, because those are still stabilized by the remaining phospholipids. However the droplets (trypsin digested oleosomes) are less stable and significant creaming and free oil films can be observed after a long time or by centrifugation, which might help in oil extraction/exploitation.

For further information we draw attention to the following references, in agreement with our own observations, where freeze drying, freeze-thaw or heating of enzymatically disrupted oleosomes liberate the interior triacylglycerides (oil) from oleosomes:

1. L. Towa, V. Kapchie, C. Hauck and P. Murphy, *J. Am. Oil Chem. Soc.*, 2010, **87**, 347–354.
2. L. Towa, V. Kapchie, G. Wang, C. Hauck, T. Wang and P. Murphy, *J. Am. Oil Chem. Soc.*, 2011, **88**, 1581–1591.

The authors describe an aqueous oil extraction process and recovered 90% free oil.

Dr van der Sman addressed Mr Waschatko and Dr Smith: Are the oleosomes digestible for humans? Is the oil contained liberated in the digestive tract? If this is the case, oleosomes would be useful to incorporate in food products. Do food companies like Cargill see some business in these oleosomes ?

Dr Smith answered: I am aware of research on the production of oleosomes, especially at the University of Iowa. Presumably there could be future possibilities to manufacture them on an industrial scale. I am unaware of any research on nutrition or digestion of them.

Mr Waschatko answered: Before purified oil was available for human consumption, oleosomes were the only source of plant triacylglycerides, since they are naturally contained in seeds, nuts, beans, corn *etc.* In the human digestive tract oleosomes lose their protective protein shell (due to protease digestion) and their phospholipid monolayer is replaced by bile salts. Furthermore, compared to artificial emulsions, oleosomes are much smaller and therefore have a larger surface exposed to enzymes like lipases. Many emulsions are prepared with isolated oil and isolated emulsifiers (lecithin), whereas oleosomes could be used in many food formulations (salad dressings, sauces *etc.*) directly without the isolation and emulsification steps. Food companies could easily profit by the longtime stability of oleosomes compared to phospholipid stabilized emulsions. But sadly, at least to our knowledge, companies do not use whole oleosomes for food formulations.

Dr Smith asked: There is a clear industry-wide drive for cleaner processing. The challenge is to produce cheap, healthy food ingredients. Is it possible to make such complex ingredients in a safe way?

Dr van der Sman responded: If the calculations are performed correctly, there should always be an economic gain if one does the processing more sustainably. A drawback is that in the current economic situation not all environmental costs are included in the prices of raw materials or foods. We propagate the use of exergy analysis or "equivalently", the minimization of entropy production—with given (economic) constraints for the finity of time, volume *etc.* (also named thermo-economics). Here sustainable processing is minimizing loss of exergy or entropy production, which should also give maximal economic gains.[1]

1. Y. Demirel, S. I. Sandler, Nonequilibrium thermodynamics in engineering and science, *J. Phys. Chem. B*, 2004, **108**, 31–43.

Dr van der Sman asked: With respect to sustainability, structured foods, and natural resources, do you perceive in your company, Cargill, a movement towards better use of natural microstructural building blocks, like oleosomes ? I have the impression that currently Cargill is producing food ingredients, based on their molecular identity: fats/gluten/starch. Is there something to gain in terms of business or sustainability to consider (partial) retention of the natural microstructure, and use that in the making of manufactured (structured) foods ?

Dr Smith responded: Personally I am not aware of any industry drive to produce such natural structures. There is a clear drive towards natural products so there are possibilities. However I think that storage and transport of such natural structures may be complex and expensive and have microbiological or food safety concerns.

Dr van der Sman remarked to Dr Bot: What is your view on how Unilever looks at the issue of using more natural food structures for more sustainability (*i.e.* the issues I have also raised in the question addressed to Paul Smith of Cargill)?

Dr Bot answered: I would like to refer to the Unilever strategy on reducing its environmental impact, which can be found in the Unilever Sustainable Living Plan.[1] This plan contains targets on reducing waste, water-use and greenhouse gas production. Generally, it is found that the direct contribution of manufacturing on these aspects is limited.[2] Therefore, a holistic approach is taken, addressing all aspects of the lifecycle of a product, including *e.g.* raw material production and consumer use, and focusing on the biggest potential gains. Thus, the avoidance of fractionation of raw materials as such is not a main focus, because there are much bigger potential gains elsewhere in the course of the lifecycle of a product and its raw materials. However, if certain natural structures exist that show clear benefits if used directly, be it in terms of environmental impact or in functionality, Unilever would certainly consider their use on a case-by-case basis.

1. [http://www.unilever.com/sustainable-living/uslp/ , accessed July 2nd, 2012]
2. [http://www.unilever.com/images/UnileverSustainableLivingPlan_tcm13-284876.pdf, accessed July 2nd, 2012]

Dr van der Sman addressed all the delegates: The title of this Faraday Discussion might lead to the impression that we focus on food structuring processes. However, microstructure is also important in the making of food ingredients and half materials. These materials are derived from natural, edible resources, which are microstructured by nature. Food ingredients and half materials are liberated by degrading the structure of the raw materials. This process I have named as food destructuring.[1]

Currently in our group and elsewhere, there is a clear trend in both the education and the research towards sustainable food processing.[2–5] Sustainability is equally important in food structuring as well as in food destructuring processes. In part of our research we have been investigating the fractionation of food suspensions like cow's milk into fractions, which are rich in one type of suspended colloid (fat globule, casein micelles, and globular proteins).

Quantitative analysis of sustainability For design of sustainable food processing it is required that it is assessed in a quantitative way. In our education we propagate the use of exergy analysis.[5-9] Exergy is the amount of work that becomes available if a system is brought into equilibrium with its environment. For the exergy it is important to carefully define the environment, in terms of temperature and its composition of matter. A related method is the method of minimizing entropy production.[10–11] The method incorporates exergy analysis, but combines it with specified economic constraints for the optimization. Furthermore, it does not involve a precise definition of the environment. Entropy production can also be interpreted as a loss of free energy.[7]

Biorefinery? A clear trend in the sustainable production of chemicals is the biorefinery. Here, the main objective is to degrade natural resources down to their molecular constituents in a environmentally-friendly way. Even large molecular entities can be broken down to smaller units as in the conversion of starch or cellulose to ethanol for biofuels. Several researchers in the field of food science are investigating the principles of biorefinery to the processing of edible, natural resources.[12,13]

From a pure sustainability point of view degradation of edible, natural resources down to small molecular building blocks appears most logical. These molecular building blocks, *i.e.* catabolic monomers like amino-acids, fatty-acids, carbohydrates and nucleic acids, can directly be absorbed into our body for feeding our anabolism and catabolism. But, humans like to enjoy life and thus also the pleasure of the texture and taste of food. Also, frankly speaking, the taste of catabolic monomers is awful. Hence, this explains the existence of a large industry producing structured foods, appealing to our desire our flavoursome foods. Many of the natural structures we find in the resources used for our foods, we reuse in the manufacture of processed foods, like the starch granules. They offer a whole toolbox of mesoscale building blocks,[1] and even not yet optimally used like the oleosomes one finds in seeds and beans—used and discussed in paper by Waschatko *et al.* in this Faraday Discussion.

Given the fact that humans desire to eat structured foods, and the edible natural resources contain microstructure by nature, it is not that logical/sustainable anymore to degrade the natural resources down to the molecular level. However, some degree of processing of the natural resources (cooking) is required to make the food ingredients sufficiently available for our digestion. For example, native starch granules and proteins need to denature to make these molecules bio-available to humans.[14] Of course, the breakdown of natural structures and building of man-made food structures require other resources as energy and fresh water, which are becoming more limited. Next to that, the availability of edible natural resources are not homogeneously distributed over time and space. Hence, more and more, we have to make conscious decisions on the balance between bio-availability of nutrients, logistics, food indulgence and the use of limited resources.

Free energy and sustainability Given the above arguments, I find that the microstructure of edible natural resources and manufactured foods need to be taken into account in the quantitative assessment of the sustainability of the food processing. In literature this important issue is hardly mentioned, except in the study of Charpentier,[15] who proposes to study green (chemical) product design and engineering via multiscale simulations. But, this is too much involvement for a global assessment of the sustainability of food (de)structuring. I think (loss of) free energy is a useful concept in this assessment, viewing the applicability of the method of entropy generation minimization. Free energy is also the key concept in simulations of soft matter.[4,16–19] The general expression for the Gibbs free energy of a microstructured material is:

$$G = \sum_{i,\alpha} V^\alpha n_i^\alpha \mu_i^\alpha + \sum_{\alpha > \beta} \gamma_{\alpha\beta} A^{\alpha\beta}$$

The different phases are indicated by α and β, each having a volume V^α. The area of the interface between phases α and β is indicated with $A^{\alpha\beta}$, having an interfacial tension of $\gamma_{\alpha\beta}$ (which can be a function of absorbed amphiphiles). n_i^α is the molar concentration of compound i in phase α having a chemical potential μ_i^α, which can be expressed as:

$$\mu_i^\alpha = \hat{\mu}_i^\alpha\left(T\right) + RT \ln a_i^\alpha$$

a_i^α is the activity of compound i in phase α and arises due to the mixing of the constituent compounds in phase α. For pure crystalline phases $a_i^\alpha = 1$. $\hat{\mu}_i^\alpha(T)$ is the standard chemical potentials, whose temperature dependence is calculated as:

$$\hat{\mu}_i^\alpha(T) = T\left[\frac{\Delta G_i^\alpha}{T_i^\alpha} + \Delta H_i^\alpha\left(\frac{1}{T} - \frac{1}{T_i^\alpha}\right) + \ldots\right]$$

ΔG_i^α is the free energy of formation (internal energy contained in the bonds), ΔH_i^α is the enthalpy of formation of compound i in phase α, and T_i^α is the temperature of the phase transition of compound i. It is common to take the liquid state as the reference state, with $\Delta H = 0$. where we have neglected higher order terms which are linear in the specific heat c_p, and the derivatives of the ΔH and c_p.

Hence, with the free energy one can quantify: 1) the energy embodied in the compartimentalization of food (ingredients) and their interfaces; 2) the free energy loss if they are mixed and degraded in the further processing; 3) the amount of work required to create new interfaces or separate specific food ingredients; and 4) the amount of work required for drying, freezing or heating to make food shelf stable, microbial safe or edible. I invite the delegates to critically review my statements and express their own views on this issue.

1. R. G. M van der Sman and A. J van der Goot, The science of food structuring, *Soft Matter*, 2009, **5**, 501–510.
2. G. Brans, C. G. P. H Schroen, R. G. M van der Sman, and R. M Boom, Membrane fractionation of milk: state of the art and challenges, *J. Membr. Sci.*, 2004, **243**, 263–272.
3. T. Kulrattanarak, R. G. M. van der Sman, C. G. P. H. Schroen, and R. M. Boom, Classification and evaluation of microfluidic devices for continuous suspension fractionation, *Adv. Colloid Interface Sci.*, 2008, **142**, 53–66.
4. H. M Vollebregt, R. G. M van der Sman, and R. M Boom, Suspension flow modelling in particle migration and microfiltration, *Soft Matter*, 2010, **6**, 6052– 6064.
5. R. M. Boom, Nanotechnology in food production, *Nanotechnology in the Agri-Food Sector*, pp. 37–57, 2011.
6. Y. Demirel and S. I. Sandler, Nonequilibrium thermodynamics in engineering and science, *J. Phys. Chem. B*, 2004, **108**, 31–43.
7. Y. Demirel, *Nonequilibrium thermodynamics: transport and rate processes in physical, chemical and biological systems.* Elsevier Science, 2007.

8. R. K. Apaiah, E. M. T. Hendrix, G. Meerdink, and A. R. Linnemann. Qualitative methodology for efficient food chain design, *Trends in food science & technology*, 2005, **16**, 204–214.
9. M. Aghbashlo, H. Mobli, S. Rafiee, and A. Madadlou, Energy and exergy analyses of the spray drying process of fish oil microencapsulation, *Biosyst. Eng.*, 2011, **111**, 229–241.
10. A. Bejan, Entropy generation minimization: The new thermodynamics of finite-size devices and finite-time processes, *J. App. Phys.*, 1996, **79**, 1191–1218.
11. A. Bejan, Fundamentals of exergy analysis, entropy generation minimization, and the generation of flow architecture, *Int. J. Energy Res.*, 2002, **26**, 0–43.
12. H. Ohara, Biorefinery, *Appl. Microbiol. Biotechnol.*, 2003, **62**, 474–477.
13. B. Subhadra *et al.*, Algal biorefinery-based industry: an approach to address fuel and food insecurity for a carbon-smart world, *J. Sci. Food Agric.*, 2011, **91**, 2–13.
14. R. N. Carmody and R. W. Wrangham, The energetic significance of cooking, *Journal of Human Evolution*, 2009, **57**, 379–391.
15. J. C. Charpentier, Perspective on multiscale methodology for product design and engineering. *Comp. Chem. Eng.*, 2009, **33**, 936–946.
16. M. Doi and A. Onuki, Dynamic coupling between stress and composition in polymer solutions and blends, *J. Phys. II*, 1992, **2**, 1631–1656.
17. R. G. M. van der Sman and S. van der Graaf, Diffuse interface model of surfactant adsorption onto flat and droplet interfaces. *Rheol. Acta*, 2006, **46**, 3–11, 2nd Annual European Rheology Conference (AERC 2005), Univ Grenoble, Grenoble, FRANCE, 21–23 April, 2005.
18. R. G. M. van der Sman and S. van der Graaf, Emulsion droplet deformation and breakup with Lattice Boltzmann model. *Comput. Phys. Commun.*, 2008, **178**, 492–504.
19 R. G. M. van der Sman, Simulations of confined suspension flow at multiple length scales, *Soft Matter*, 2009, **5**, 4376–4387.

Mr Waschatko answered: Thank you for mentioning our work (soybean oleosomes) in your statement about the biorefinery.

We agree with your statement of using the "mesoscale building blocks" provided from nature as food ingredients. Indeed such an approach makes sense, especially in our case, since the oleosomes could, as natural building blocks, be used for novel food formulations. They contain structural, nutritious components such as digestible proteins, triacylglycerides and phospholipids. In addition, the fat cores of the oleosomes carry high concentrations of tocopherols and other essential micronutrients, which do not need to be added as in conventionally structured (and nutrition enriched) foods.

As mentioned in the comment, starch is currently the most common "mesoscale building block". No one would think of rebuilding granules from purified amylose and amylopectin for most of the (edible) products where you use starch. As you said, it is not "logical or sustainable". However, regarding oleosomes, this is what is done for many applications: oil, lecithin and proteins are first extracted (with solvents, energy consumption) and then reunified and emulsified (energy consumption again) for processed food.

Professor Cates opened the discussion of the paper by Professor Marangoni: On what length scale in Fig.1 of your paper should I be looking for a fractal structure—*e.g.* in the case without shear is it internal to the visible 10 micron aggregates or is it on a length scale bigger than that?

Professor Marangoni responded: Fractality in these fat crystal networks is consistent with the aggregation of 100–300 nm long nanoplatelets into 20–100 μm clusters. This has been confirmed recently by USAXS experiments at Argonne National Laboratories.

Professor Cates continued: In Fig. 1 (*e.g.*, frame F), the maltese cross pattern in PLM is indicative of strong tangential ordering of the platelets, rather than any kind of fractal structure, do you agree?

Professor Marangoni replied: You bring up an excellent point. Fractality within our networks lies in the aggregation of small crystallites (~300 nm) into larger clusters (~20-50 μm). It seems that when samples are crystallized statically, the dendritic pattern observed (1C) is reminiscent of a fractal structure. Moreover, we noticed that our rheological fractal analysis yields a fractal dimension of ~2.5, in line with a 3D DLA cluster. However, under any of the two shear levels, spherical and non-fractal looking clusters are observed. What we had not noticed is that our rheological fractal analysis then yielded a fractal dimension of ~2.9–3.0, suggesting a Euclidean dimensionality. Thus, we have to conclude that you are correct, we are switching from a fractal to a near non-fractal regime. What is interesting, though, is that our rheological analysis seems to be appropriate for characterizing both regimes.

Dr Royall asked: In your paper, you describe obtaining the fractal dimension from a 2D image of a 3D system by incrementing the 2D result by 1. Is this commonly used and how is it rationalised?

Professor Marangoni responded: I share your concerns about merely adding "1" to the dimensionality of an object embedded in 2d space. It is commonly used, but that does not make it correct. In fact, in Ref. 1 below, we show that this is not true…. However, experimental difficulties make the determination of the 3d fractal dimension very difficult on a routine basis: it can only be done at low concentrations by microscopy. However, we have started using ultra-small angle X-ray scattering to determine the fractal dimension of our aggregates and we are thus not restricted to low volume fraction of solids only. As you can imagine, carrying out USAXS on a routine basis is not easy either. Hence, we resort to the "solution" of merely adding "1" to the 2d fractal dimensionality obtained. Not very satisfying indeed.

1. D. Tang, A. G. Marangoni, 3D fractal dimension of fat crystal networks, *Chem. Phys. Lett.*, 2006, **433**, 248–252.

Dr Royall asked: How many molecular species are involved in the formation of the fat crystals you study, and if it is more than one, is there a well-defined melting temperature? Can one talk of a degree of undercooling?

Professor Marangoni responded: We have now included a peak melting temperature for the systems we study. The systems we used are very homogeneous for a fat. Milk fat, for example, can have more than 300 TAG species, giving a melting range from −40 °C to +34 °C. So, the answer is no in terms of a unique melting point. In our system, hydrogenated soy oil is mostly composed of SSS and SSP (S = octadecanoic acid, P = hexadecanoic acid). Thus, the melting range is fairly narrow.

Dr van Gruijthuijsen commented: You conclude in your paper that you "have been able to correlate effectively structural and mechanical properties of the system with oil migration". I agree that you demonstrate, for instance in Fig. 8, that there are clear trends between typical length scales in the system and resulting OMR values, *for a given shearing condition*. As far as I can see, only the mechanical properties correlate with OMR values for all shearing conditions in Fig. 11A and C, but not in Fig. 11B and D. Indeed, the permeability coefficient, including both structural and mechanical properties, does not correlate with OMR values at all.

Could you comment on this apparent discrepancy? Do you have some ideas on how the results under different shearing conditions might be related to each other? Or do you think there is a discontinuous effect of shearing on the internal structure and its oil retaining capacity?

Professor Marangoni responded: The difference between Fig. 11A and C *vs.* B and D is the solids' volume fraction. Fig. 11B and D correspond to 45% solids' content, while A and B correspond to 40% solids' content. I would argue that there is a correlation, but that the exact relationship is system-specific. A higher material G' or yield stress is always inversely related to the oil migration rate. However, I do not have a clear explanation for the reason why the trend is not universal for all shear rates in the higher solids' sample. More troubling, I would think, is the correlation between the permeability coefficient and the oil migration rate reported in Fig. 12. A correlation was only observed at 30s-1 and not under static conditions or 240s-1 shearing conditions. This all suggests that are factors other than crystal size, solids' content, fractal dimension of the network and network tortuosity at play. Strengths of intermolecular interactions are not considered in the models and these could have a significant effect on oil migration. The approach suggested by Ruud van der Sman, using disjoining pressures, could be a useful approach to follow.

Professor van der Linden commented: Oil holding capacity is mentioned, however here it is measured by means of determining the amount of oil passing through the system and coming out at one end, as a function of pressure exerted on the oil coming in from the other end of the system. Oil holding capacity is a parameter referring to the amount of oil a system can hold, *i.e.* referring to a steady state situation. In contrast, your measurement is a transport parameter.

Could you please clarify to what extent this measure is the right measure for oil holding capacity?

Professor Marangoni replied: In our accelerated test, oil is pulled from the fat crystal network by the capillary forces in the underlying piece of paper, thus the network is under an external pressure. We do find the method quite sensitive to differences between systems, though. Oil leakage follows a hyperbolic pattern with a definite maximum reached. We are reporting this 'equilibrium' or steady state value. We have corrected this point in the paper: it is not an oil migration rate, but a steady state oil loss, which is related to thermodynamic factors influencing oil binding. The rate of oil loss would be given by other methods, such as MRI, dye migration, FRAP or the slope in the early stages of the oil leakage behavior. I do believe that oil migration kinetics and oil binding need to be separate phenomena. During this conference it became very clear that we need to determine both to have a better idea of what the oil binding capacity of a fat is. Kinetic and thermodynamic factors come into play, and a rapid oil loss does not necessarily mean that more oil will be lost in the end. Both need to be determined and reported, and they could be related to different structural factors.

Dr van der Sman asked: In your paper you try to relate the oil binding capacity to a transport coefficient (a permeability or diffusivity). Drawing the analogy between water holding capacity of (hygroscopic food materials)[1–3] and oil holding capacity I regard the oil binding capacity as a thermodynamic quantity and thus it is the driving force for oil migration. Whereas permeability or diffusivity is related to the migration rate. What is your view on that?

1. R. G. M. van der Sman, Moisture transport during cooking of meat: An analysis based on Flory–Rehner theory, *Meat Sci.*, 2007, **76**, 730–738.
2. R. G. M. van der Sman, Soft condensed matter perspective on moisture transport in cooking meat. *AIChE J.*, 2007, **53**, 2986–2995.
3. R. G. M van der Sman, Thermodynamics of meat proteins, *Food Hydrocolloids*, 2012, **27**, 529535.

Professor Marangoni answered: I believe now that oil binding capacity is a complex term that is influenced by kinetic and thermodynamic factors. Thermodynamic factors would influence that total amount of oil that is retained by the network, while kinetic factors would influence the rate at which the oil is lost. These two factors are not necessarily correlated. We need to report both from now on.

Dr van der Sman commented: In relation to oil binding capacity I reckon you think that capillary effects are playing a significant role. Would van der Waals forces, as might be expressed in terms of a disjoining pressure, contribute also to the oil bonding capacity? Again I base it on the analogy between water holding and oil binding. Nice papers on the contribution of capillary effects and adsorption effects of liquid retention in porous media are by Dani Or.[1,2]

1. M. Tuller, D. Or and L. M. Dudley, Adsorption and capillary condensation in porous media: Liquid retention and interfacial configurations in angular pores, *Water Resour. Res.*, 1999, **35**, 1949–1964
2. E. Shahraeeni and D. Or, Pore-Scale Analysis of Evaporation and Condensation Dynamics in Porous Media, *Langmuir*, 2010, **26**, 13924–13936.

Professor Marangoni responded: This had not been done before and I believe that relating van der Waals forces to capillary effects in terms of a disjoining pressure would help link capillary effects to diffusion effects, which seem to be the matter of controversy in this area.[1]

1. J. M. Aguilera, M. Michel, and G. Mayor, *J. Food. Sci.*, 2004, **69**, R167–R174.

Professor Norton remarked: In terms of controlling nucleation, do you have secondary nucleation in your experiment? How do you control secondary nucleation?

Professor Marangoni answered: The Peclet number suggests that we are definitely in a regime where the nanoplatelets are being affected; thus, inertial effects will come into play. I am certain that at this point we'll have crystal breakage and secondary nucleation taking place. We did not control secondary nucleation at all in these experiments. However, shear affects nucleation events, thus I believe that differences in nucleation rates are part of the effect that we are studying. Why would we want to control nucleation rates then? We wanted to see if we obtained different nucleation behavior due to shear.

Professor Norton said: As you have heterogeneous nucleation, how do you control the number of nuclei in your experiment?

Professor Marangoni answered: We did not control the nucleation rate in the experiments. However, we were interested in determining if shear affected the nucleation rate, so I believe we do not want to control the nucleation rate. It is part of the effect studied.

Professor Nicolai opened the discussion of the paper by Dr Bot: What is the connectivity between the fibrils? What is the origin of the resistance to deformation (elastic modulus); the crosslinks or the bending of the individual fibrils?

Dr Bot responded: Until now we have focused our attention mainly on the microscopic scale (structure of the tubules) and the macroscopic scale (rheology), but we have done only limited studies into the structure of the network on intermediate length scales. A small number of scanning electron microscopy images were obtained for concentrated systems (Fig. 2).[1] These images reveal a dense network of tubules, somewhat flattened because the oil inside and outside the tubules has been removed to make imaging possible. The tubules sometimes engage in lateral aggregation, and sometimes show some evidence of intertwining. Overall, the persistence length of these tubules is very long, straight tubules of several micrometers not being exceptions. The structure of the network might be revealed more clearly if images would be obtained at lower tubule concentrations. The concentrations used in the recent studies have been chosen to produce a very clear SANS or SAXS signal, but the

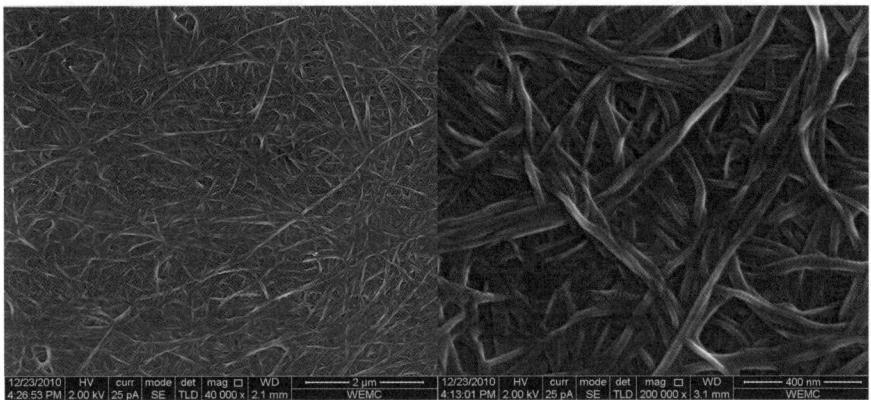

Fig. 2 Scanning electron microscopy images of a 32% plant sterol organogel (60 : 40 oryzanol:sitosterol) in sunflower oil. Image by H. Sawalha.

organogels form at much lower concentrations too. Nevertheless, the images suggest that the tubules form an enthalpic network, in which bending and ultimately breaking of the tubules accounts for most of the rheological properties[2]. The system is different from a typical enthalpic network in which the rheological properties are determined mainly by the cross-linking of the network. Kroes-Nijboer *et al.* in a paper presented in this Faraday Discussion, compared the bending rigidity of tubules and fibrils, and calculated that the tubules are stiffer than fibrils for the same amount of mass per unit of length.[3] Thus the hollow structure of the tubules directly affects the resistance to deformation as well.

1. H. Sawalha, R. den Adel, P. Venema, A. Bot, E. Flöter, E. van der Linden, *J. Agric. Food Chem.*, 2012, **60**, 3462–3470.
2. A. Bot, W. G. M. Agterof, *J. Am. Oil Chem. Soc.*, 2006, **83**, 513–521.
3. A. Kroes-Nijboer, H. Sawalha, P. Venema, A. Bot, E. Flöter, R. den Adel, W. G. Bouwman, E. van der Linden, *Faraday Discuss.*, 2012, **158**, DOI: 10.1039/c2fd20031g.

Professor De Kruif remarked: You describe the formation of self-assembled tubules of sitosterol and oryzanol in triglyceride oils. There is a similarity with the formation of so-called microtubules of the cytoskeleton, and α-lactalbunin nanotubes.[1,2] These nanotubes form a transparent gel as well which becomes liquid on shaking but re-gel on standing. Would you observe similar phenomena in your system? Also, does the growth (rate) depend on the sitosterol : oryzanol ratio?

1. J. F. Graveland-Bikker, R. Koning, H. Koerten, R. Geels, R. Heeren. and C. G. de Kruif, *Soft Matter*, 2009, **5**, 2020–2026.
2. J. F. Graveland-Bikker, C. G. de Kruif, *Trends Food Sci. Technol.*, 2006, **17**, 196–203.

Dr Bot answered: The sitosterol + oryzanol organogels are very firm (typical conditions: > 5% sterols + sterolesters, ageing for 1 day at 5 °C) and will not flow upon manual shaking. Disrupting the network requires mashing of the gel or similar large deformation mechanical treatments. We haven't really looked for conditions under which curing of these firm gels might occur, so we cannot exclude the possibility of this happening, but we typically find that damage to the network structure is irreversible at time scales of, for example, one day.

Most of our work has addressed either the macroscopic rheology of the gels or the structure of the tubules that form the network, but relatively little work has been done on the kinetics of gel and tubule formation.[1] However, there are two studies that may partially address your question.

The first study involved rheology and has addressed the macroscopic observations.[2] The study shows that a mechanical disturbance may help the solidification of the gel. In particular, it demonstrates that the moment of gelling in a quiescently cooled sample tends to happen much more erratically than gelling under a 'small' oscillatory shear deformation. If such a mechanical disturbance is applied, it is found that oryzanol-rich systems gel quicker than sitosterol-rich systems (and form firmer gels). On the other hand, oryzanol-rich systems have been shown to be much more prone to supercooling than sitosterol-rich samples under quiescent cooling conditions.[3]

The second study, involving time-resolved infrared spectroscopy, may come closer to answering your question.[4] When interpreted within the Avrami framework, the results show for a slightly different system (oryzanol + cholesterol in mineral oil) that the nucleation mechanism does not depend on the oryzanol : cholesterol ratio (sporadic nucleation), but that the activation energy does show a minimum as a function of the oryzanol : cholesterol ratio at an approximately equimolar ratio. However, the nucleation/growth mechanism does turn out to be very sensitive to the chemical details of the structurant and organic solvent, and the oryzanol and sitosterol system in mineral oil shows instantaneous nucleation instead of sporadic nucleation. The most likely explanation is the lower activation energy for growth in the latter system (~15 kJ mol^{-1} vs. ~40 kJ mol^{-1} or higher). These experiments have not been repeated yet in sunflower oil or another edible oil, so a direct comparison is not possible.

1. A. Bot and E. Flöter, in: Edible oleogels: Structure and health implications, ed. A. G. Marangoni, and N. Garti, AOCS Press, Urbana, IL, USA, 2011, ch. 3, pp. 49–79.
2. A. Bot and W. G. M. Agterof, J. Am. Oil Chem. Soc., 2006, 83, 513–521.
3. A. Bot, R. den Adel and E. C. Roijers, J. Am. Oil Chem. Soc., 2008, 85, 1127–1134.
4. M. A. Rogers, A. Bot, R. S. H. Lam, T. Pedersen and T. May, J. Phys. Chem. A, 2010, 114, 8278–8285.

Dr Ubbink commented: To what extent does the incorporation of the sitosterol in the self-assembled structures slow down the formation of sitosterol monohydrate at higher water activities?

Dr Bot replied: The hydration process was not studied in detail, but it seems that water does not immediately interfere with tubule formation. Instead, sitosterol monohydrate formation is a process that takes weeks.[1–3] Therefore, the hypothesis is that the kinetics of self-assembly of the sitosterol + oryzanol tubules is not affected much by the presence of water initially. The monohydrate formation appears to be a recrystallisation process, that requires (slow) transport of water through the oil phase to the tubules. This process proceeds more quickly in high polarity oils (higher water solubility) and high water-activity aqueous phases (greater driving force for hydration).

1. A. Bot, Y. S. J. Veldhuizen, R. den Adel and E. C. Roijers, Food Hydrocolloids, 2009, 23, 1184–1189.
2. A. Bot, R. den Adel, C. Regkos, H. Sawalha, P. Venema, E. Flöter, Food Hydrocolloids, 2011, 25, 639–646.
3. H. Sawalha, R. den Adel, P. Venema, A. Bot, E. Flöter, E. van der Linden, J. Agric. Food Chem., 2012, 60, 3462–3470.

Dr van der Sman continued the discussion of the paper presented by Professor Marangoni: Can the different behaviour of the system above and below the critical shear rate be related to inertial effects? You said that the reduction of particle size at high shear rates (240 s^{-1}) is due to collisions, which impart breakage of the (poly?) crystalline particles. That makes me draw the analogy with pastes under flow, which have been analysed in an intelligent manner by Coussot and Ancey.[1] They present a nice classification of rheological behaviour of pastes as a function of shear rate and volume fraction, based on various dimensionless numbers. On first thought I regarded the mixture of crystalline particles and oil as a "hard suspension", for which three regimes can be distinguished: 1) the friction regime, 2) the lubrication regime, and 3) the collision regime. These regimes are distinguished by ratios of contact forces, *i.e.* the Leighton (Le) number and the Bagnold (Ba) number. I would regard the transition in behaviour of the fat–oil system as a transition from the lubrication regime to the collisional regime. The transition is characterized by Ba > 1. Coussot and Ancey define the Bagnold number as follows:

$$Ba = \frac{\rho_s \dot{\gamma} r \varepsilon}{\eta_f}$$

where ρ_s is the solid density, $\dot{\gamma}$ is the shear rate, r the size of the particle, ε the surface roughness and η_f the fluid viscosity. Hence, it is linear with the Reynolds number. A quick calculation of Ba for $\rho_s \approx 900$ kg m^{-3}, $\dot{\gamma} = 240$ s^{-1}, $r = 20$ μm, $\varepsilon = 0.01r$, and $\eta_f = 0.06$ Pa s, suggests that Ba $\ll$ 1, and actual collisions seem not to occur. Another study also gives reference to a critical shear rate for fat–oil mixtures[2], which is based on the Peclet number, Pe. This number indicates the transition between the Brownian regime and the viscous regime. Can you relate your critical shear rate to any transition between regimes as discussed for pastes by Coussot and Ancey? Is this different from the critical shear rate as mentioned by Dewettinck and coworkers ?

1. P. Coussot and C. Ancey, *Phys. Rev. E*, 1999, **59**, 4445.
2. V. de Graef, B. Goderis, P. van Puyvelde, I. Foubert, and K. Dewettinck, *Eur. J. Lipid Sci. Technol.*, 2008, **110**, 521–529.

Professor Marangoni replied: This Bagnold number could be useful in characterizing transitions from mass transfer limited effects to inertial effects, however, your calculation using 20 μm spheres shows a very small Bagnold number. If we consider that the effect was also detectable at the nanoscale, with 300 nm *platelets* (not spheres), we quickly realize that this number cannot explain the observed effects. We have very small platelets experiencing these relatively low shear levels. It is unknown, to me, how such small shear fields can have an effect on 300 nm objects.

Dr van der Sman asked: The spherulite particles exhibit surface roughness. I expect that shear contributes to make its surface smoother. The roughness represents dendrites which might melt due to shear-enhanced melting. In prior research, we have introduced an effective temperature, which is linear in the viscous stress.[1,2] In the case of colloidal glasses Dave Weitz has shown that shear can induce 'melting'.[3] Dendrites have a lower melting point, and are thus more vulnerable to shear-induced melting. What are your views on that ? Would it explain the smoothness of mesoscale particles at moderate shear rates?

1. H. M. Vollebregt, R. G. M. van der Sman and R. M. Boom, Suspension flow modelling in particle migration and microfiltration, *Soft Matter*, 2010, **6**, 6052–6064.
2. M. Vollebregt, R.van der Sman and R. Boom, Model for particle migration in bidisperse suspensions by use of effective temperature, *Faraday Discuss.*, 2012, **158**, DOI: 10.1039/c2fd20035j.
3. C. Eisenmann, C. Kim, J. Mattsson and D. Weitz, Shear melting of a colloidal glass, *Phys. Rev. Lett.* 2010, **104**, 35502.

Professor Marangoni responded: This is an excellent analysis. Yes, in the absence of shear you can observe 'fractal-like' dendritic structures, which would be very sensitive to collisions with other dendrites. Melting would be a mechanism by which surface roughness is decreased. An effective temperature is a very elegant way to describe shear effects.

Dr van der Sman remarked: Continuing the line of reasoning of the previous question: if the dendrite experience shear-induced melting, can there be resolidification of molten fats inside the pores? The pores have negative curvature compared to dendrites, and the Gibbs–Thomson effect states the pores have elevated melting point. Due to shear diffusion limitation are probably also reduced enhancing the melting of dendrites and resoldification inside pores.

Professor Marangoni responded: Yes, I believe that shear induced melting would extend to the pores. Close examination of the SLM's in Fig. 1 show that even the internal morphology of the aggregates has changed. This melting and resolidification in the pores would further reduce the ability of the crystalline structures to entrap oil, effectively excluding it from the crystalline matrix. This could be the reason for the observed decrease in oil binding capacity.

Dr Royall concluded this small discussion on Professor Marangoni's paper by remarking: This follows a comment that the fractal dimension of a sphere, cut into 2d slices, was indeed equal to its value in 3D when incremented by one, which followed on from my previous question. My comment was that, in any kind of microscopy, there is an issue of resolution. This is particularly pertinent in the axial direction, where the resolution is limited in the case of 3D confocal microscopy and poorly defined in the case of "wide-field" techniques. What this means is that rather than an idealised slicing up into 2D of a 3D object, instead one should consider the nature of the image (the convolution of the specimen and the point spread function) when tackling this issues. I believe simulation is a reasonable way forward and it may be fruitful to consider as a way to aid extracting fractal dimension data from 2D images.

Dr Ettelaie opened the discussion of the paper by Professor Stoyanov: Have you monitored the graphs of characteristic frequency *vs.* mean rate of strain over a period of time? How does the aging of the interfacial film affect the graphs?

Professor Stoyanov responded: All data reported in this paper are done after 5 min of initial ageing time. This choice is a compromise between having adsorbed layers with sufficient structure build-up, but without having large yield stress developed, which in our case can lead to irreproducibility.

Dr De Vries said: You have shown that the surface rheology of both the hydrophobins and of β-lactoglobulin can be fitted with your model. How general do you believe the model that you have used is? Do you expect it to apply *e.g.* to the surface rheology of all globular proteins? Do you known of any counter examples, *i.e.* a molecule with a surface rheology that cannot be fitted with this model ?

Professor Stoyanov responded: We found that the model is applicable for other (globular) proteins as well, but is difficult to generalize if it will be generic model for all proteins. The model is not applicable for Quillaja saponins for example where we needed to use the more complex Kelvin Voigt model consisting of Maxwell plus two Kelvin elements (paper has been submitted to Langmuir).

Professor van der Linden commented: You mention stability of foams and use shear deformation, being an important parameter in this. I think that it in your case shear gives a reasonable idea for stability, since the interfaces are rather sturdy. Of course, if one has less solid like interfaces, the different deformations are not always equally important.

You ascribe the instability in Ostwald ripening as being partially related to wrinkling of the surface. This is mainly caused by a volume decrease with constant surface area. I would argue that that type of instability is more a bending contribution instead of shear or dilatation deformation. Could you comment on the importance of shear and dilatation in regards to this wrinkling? Could you also comment on the degree of elasticity of the interface at which one expects that shear and dilatation are equally important for the foam stability?

Professor Stoyanov answered: For 'typical' proteins and surfactants shear elasticity, viscosity and dilatational modulae are monotonic functions of the adsorbed amount and as we try to explain bellow, one can find apparent correlations of foam satiability and against any of them. To see which one is relevant indeed one needs to look at underlying processes and to try to understand the physics there. For example shear deformation is more relevant for the process of foam formation and film drainage and break-up occurring during foam drainage and bubble coalescence, while dilatation is more relevant for the process of disproportion. Also modulae are frequency (time) dependent and it is important to measure their values for the time scales which are relevant for the process. For example if we compress the air–water interface having adsorbed surface active species, very fast so that they could not equilibrate with the bulk or rearrange at the interface, we will measure high modulus, while if we do the same compression slowly so that the system manages to equilibrate with its bulk, then the interfacial tension will be constant, independent on the trough area and equal to the equilibrium one, thus relevant modulus $\Delta\Pi/\Delta\log A$ will be zero. This is the case with low to medium weight emulsifiers and some proteins, which can adsorb/desorb within time scales of minutes to hours. In this case if we measure the dilatational modulus in time scales of a few minutes we can get high values, but if we take into account that the time scale of disproportionation is of an order of hours, days and even months (think of shelf-life-time of the foamed food product), then the relevant modulus of such systems will be zero or very low. In order to have system, which have a high dilatational modulus at very low frequencies it should be basically irreversibly adsorbing at the interface (so that relaxation between bulk and interface be much slower than ripening process) and in addition to having weakly interacting species at the interface which are able to build strong but reversible surface networks, which can quickly form and re-arrange to accommodate other deformations and self-repair to accommodate defects (see for example R. D. Groot and S. D. Stoyanov, *Soft Matter*, 2011, **7**, 4750 and R. D. Groot and S. D Stoyanov, *Soft Matter*, 2010, **6**, 1682). If interactions are too strong, then often these molecules will have very low bulk solubility and will form strong but brittle and fragile interfaces, which will break upon mild shear deformation. An additional complication comes from the fact that deformations that occur on a single bubble surface during disporortination are very high and non-linear, while often we measure the modulus at very small liner deformations. Many protein systems can have relatively high low frequency modulae (due to irreversibility of protein adsorption) at low deformation, which however quickly decreases with increase of the deformation due to structure collapse and as result they also will fail to arrest the Oswald ripening. Taking all this into account we can say that in order to stop or slow down foam disproportionation *via* the surface elasticity route, the relevant modulus is a very low frequency, high deformation dilatational modulus, which can be difficult to measure experimentally due to the technical limitations of the equipment, processes like evaporation and deformation pre-history dependence.

For the systems, which have high dilatational modulus at mid/low frequencies (indicating a dense and close packed layer), which however can relax relatively quickly so that their very low frequency dilatational modulus is low, can still slow down the Oswald ripening but not because of 'elastic blocking', but due to their dense packing at the interface, which blocks or slows down the gas transport across the surface (and thin films) (see for example, Tcholakova *et al.*, *Langmuir*, 2011, **27**, 14807–14819). For these systems one can find apparent direct correlations between their high dilatational modulus (at mid frequency range) and the disproportionation, but in reality this will be an indirect correlation coming *via* the dependence of gas permeability on the surface coverage. For such systems one can also find that foams drain slower and are more stable against coalescence, where shear elasticity and viscosity at relatively higher frequencies (when compared with disproportionation time scales) are important. In situations where wrinkling appears, usually the Laplace equation does not hold anymore since: (i) one needs to account for bending contributions to the normal stress (see for example Kralchevsky *et al.*, *Langmuir*, 2010, **26**, 143–155); (ii) for non smooth surfaces the macroscopic curvature (think bubble radius) can be very different from the local microscopic one (due to wrinkles), while the driving force for the disproportionation comes from the capillary pressure differences, which depend on the local surface tension and the local curvature rather than on the global one; (iii) the local concentration can vary from place to place, since the surfaces can have sufficiently high 2D yield stress which can balance the surface tension gradients, thus the assumption of homogeneous interface with constant surface tension and elasticity is not valid any more as well. These three factors in combination can greatly complicate the analysis and makes it depend not only on the local deformations but on whole deformation pre-history (see for example J. K. Ferri and P. A. L. Fernandes, Axisymmetric drop and bubble shape analysis of interfaces with anisotropic stress distributions, manuscript in preparation)

Professor Cates asked: You mentioned that the model did not work so well in relaxation as in the angle ramp. Because it assumes instantaneous adaptation to strain rate, I would also expect it to have problems in a creep test (fixed stress, hence varying strain rate). Am I right?

Professor Stoyanov responded: As shown in Ref. 36, in the angle ramp regime the data obey the Maxwell–Herschel–Bulkley model, whereas the relaxation of stress at a fixed strain obeys a modified Andrade's cubic-root law. In our opinion, this difference is due to the fact that in the first case the process is *forced* (breakage and restoration of intermolecular bonds), whereas in the second case it is *spontaneous* (only bond restoration). The difference between the physical nature of the processes leads to different laws of rheological behaviour. For the time being, we have no systematic data in creep/creep-relaxation regime, **but it is quite possible to observe a difference between the rheological responses to forced shearing and free relaxation of strain (at zero applied stress).**

Professor Cates asked: The Maxwell–Herschel–Bulkley model seems to involve a Maxwell response with instantaneous dependence of parameters such as the modulus on the rate of strain. In this case I would expect the model to predict large amplitudes for the higher Fourier harmonics even for extremely small strain rates. Is that correct?

Professor Stoyanov replied: In the Maxwell–Herschel–Bulkley model, the shear stress, $\tau_{sh}(t)$, is to be determined from the Maxwell equation, eqn (11), where $v_{ch} = E_{sh}/\eta_{sh}$ is defined by the Herschel–Bulkley relation, eqn (18), and E_{sh} is a known function of the shear rate; see *e.g.* eqns (27) and (29). Because all of them are continuous functions of time, then the theory of Fourier expansions implies that the amplitude of the higher harmonics in eqn (6) decreases proportional to $1/n^2$. There cannot be resonance, because eqn (11) is a first-order differential equation (with a decaying exponent solution), rather than a second-order equation, which would have a solution corresponding to free oscillations with a given resonance frequency.

Professor Cates remarked: Can you clarify how you define the term quasi-linear. Does this refer specifically to the smallness of the higher harmonics?

Professor Stoyanov answered: The definition is as follows: the rheological response of the system is quasi-linear if at small sinusoidal oscillations of strain, the registered stress oscillates harmonically, with the same frequency (that of the applied strain), the higher-order Fourier modes in the stress being negligible; see Fig. 2 of our paper. However, at greater strain amplitudes, the higher-order Fourier modes become greater and can be experimentally detected (like G_3 in Fig. 3); see also eqn (6).

Dr van der Sman remarked: In your paper you use the melting metaphore, and you also mention that shear might act as an effective temperature. If you extend this metaphor towards the mixed system of hydrophobins and caseins, would the minor component, casein, act as a plasticizer? In glassy (arrested) systems plasticizers lower the glass transition temperature. Hence, the solidification (arrest) of the network would happen at lower shear rates—if one can identify that with an effective temperature ?

Professor Stoyanov replied: Yes, the addition of β-casein decreases the shear elasticity of the hydrophobin adsorption layers, and in this respect the β-casein can be considered as a plasticizer. However, with our experiments we could not register a transition from elastic to viscoelastic behaviour, which would be the analogue of the glass transition temperature. It seems that the yield stress is very low for these soft layers of organic macromolecules.

Dr van der Sman commented: Continuing the line of reasoning of the previous question, you said that the solidified (arrested) hydrophobin network will show some voids. In the thermal metaphor, this could be interpreted as free volume. Do you expect this free volume/voids to be filled with the plasticizer, *i.e.* casein ?

Professor Stoyanov responded: Yes, we expect that the β-casein fills the voids in the hydrophobin adsorption layer, bearing in mind the adhesive interactions between the molecules of these two proteins. A combination of data from rheological and thin-liquid-film experiments indicates that the hydrophobin is more surface active and occupies the air–water interface, whereas the β-casein fills the voids in the hydrophobin layer and forms a second adsorption layer.

Dr van der Sman said: If I continue the metaphor further, you mention that the hydrophobin networks show aging. Can the aging be interpreted in terms of change with respect to the voids in the network? Is the void space (free volume) slowly decreasing in time?

Professor Stoyanov replied: The aging can be interpreted in terms of change with respect to the voids in the network if a hydrophobin layer is spread on pure water and subjected to compression in a Langmuir trough. After the compression, the surface pressure decreases (relaxes). This can be interpreted as filling of voids. However, if the adsorption layer is formed on the surface of a hydrophobin solution (as in our case), there is another source of aging. With time, more and more rodlike hydrophobin aggregates (which are present in the solution) stick to the lower surface of the adsorption layer and form a reinforcement network that enhances the layer's rigidity.

Professor Kulozik asked: Could the measures used in this study to describe foam stability be correlated with more empirical measures such as drainage properties, or loss of overrun volume?

Professor Stoyanov replied: Yes they can: see my reply to a previous question on the foam stability and also some of these papers, which clarify some of these aspects in more details: *J. Colloid Interface Sci.*, 2012, **376**, 296–306; *Langmuir*, 2012, **28**, 4168–4177; *J. Colloid Interface Sci.*, 2012, **368**, 342–355; *Langmuir*, 2011, **27**,

4481–4488; *Langmuir*, 2011, **27**, 2382–2392; *Food Hydrocolloids*, 2010, **25**, 627–638; *Soft Matter*, 2010, **6**, 1799; *Langmuir*, 2010, **26**, 143–155.

Professor Kulozik said: Hydrophobin foams were described as super-stable. How would mixed emulsifying proteins compare against pure hydrophobins in real food systems? The data in Fig. 12a show that mixed systems seem to lose this property of superstability.

Professor Stoyanov answered: It depends very much on type and relative concentration of the other protein/emulsifier. If it is in excess when compared to hydrophobin, soon or later it will start to dominate the interface and thus determine foam properties. However since foams are non-equilibrium systems, by playing with the order of addition or processing we can shift this balance to a certain extent.

Professor Wilde asked: You presented data for hydrophobin mixtures with β-casein, but did you look at hydrophobin–β-lactoglobulin mixtures? As we found a synergistic increase in the surface elastic modulus with these mixtures, did you see a similar effect.

Comment: Care must be taken when measuring the surface dilatational modulus of hydrophobin with the pendant drop technique, as the surface becomes so rigid, it distorts the shape of the drop thus rendering the calculation of the surface tension impossible.

Professor Stoyanov answered: Yes we have also looked at other mixtures and for some we also find synergistic effects. This will be published in follow up work. We fully agree with the comment and we have recently published a paper on this: *J. Colloid Interface Sci.*, 2012, **376**, 296–306.

Dr van der Sman remarked: You have mentioned that the surface rheological behaviour of the hydrophobin network can be quite similar to that of other proteins, while some behaviour of foams stabilized by these compounds can be quite different. Thermodynamics would state that both surface tension and elasticity contribute to the surface pressure. Would differences in either surface tension and/or elasticity explain much of the differences amongst different proteins (including hydrophobins) ?

Professor Stoyanov replied: Yes, not only hydrophobin, but also other proteins (such as β-lactoglobulin and β-casein) form viscoelastic adsorption layers that comply with the Maxwell–Herschel–Bulkley model; see Fig. 10c. However, only the hydrophobin forms very stable foams. The reason for this difference is that the hydrophobin combines three favourable properties: (i) the shear elasticity of its adsorption layers is much higher than those of the other proteins; see the values of G' in Ref. 11; (ii) the adsorption layers of hydrophobin solidify much faster, within several seconds or minutes (depending on concentration), whereas for the other proteins the solidification may take from hours to days; this is important for the dynamic stage of foam generation; (iii) the 'hydrophilic' parts of the hydrophobin molecules can strongly adhere to each other (not observed with other proteins); see Ref. 34. This leads to attachment of hydrophobin aggregates to the film surfaces and Plateau borders in the foams, and to blocking of the foam drainage

Dr Gebhardt returned to the discussion of the paper by Professor Marangoni: You estimated the crystalline domain size from the width of the 001 reflection. Did you also observe a shift in the peak position at different shear rates?

Professor Marangoni replied: We analyzed the short spacing region ($2\theta \approx 19°$) and long spacing region ($2\theta \approx 4°$) and we could not detect any changes.

Dr Bot asked: You discuss the break-up of the nuclei in shear fields of ~200 s⁻¹. A quick back-of-the-envelope calculation suggests to me that such shear fields are able to break up particles with a modulus of the order of kPa's, but not crystalline particles with moduli of the order of MPa's or GPa's. Should the nanocrystals be considered much softer than normal crystals, *e.g.* due to being in a relatively unstable polymorphic state (*e.g.* the alpha polymorph), or do you have another explanation for the effect?

Professor Marangoni responded: We have carried out direct AFM measurements of the elastic modulus of spherulites, which yield a Young's modulus of about 2 MPa. So, your comments are correct in terms of crystal breakage. However, we also determined that within an aggregate (sperulite), individual crystallites are very loosely packed and the external force fields will affect/disrupt this packing. So, maybe breakage is not taking place, but aggregate breakage into individual crystallites. One word of caution with these calculations is that they mostly apply to spherical particles. Recall we have highly asymmetric platelets which are very thin (30 nm). These could be quite fragile if stressed along their width.

Dr Wyss returned the discussion of the paper by Professor Stoyanov: The suggested model implies that the amplitude of strain rate determines the characteristic frequency/characteristic time scale of the system. This is supported by the presented experimental data supplied in the paper. However, the strain rate amplitude has only been varied *via* a change in strain amplitude at fixed frequency. Did you also test experimentally if this applies for changes in frequency at fixed strain amplitude? An interesting option could be to perform 'constant-strain-rate frequency sweeps', where the strain rate amplitude is kept fixed as the frequency is varied.[1] Did you try to perform such measurements to further test if indeed the strain rate amplitude is governing the magnitude of the characteristic frequency?

1. H. M. Wyss, K. Miyazaki, J. Mattsson, Z. B. Hu, *et al.*, Strain-rate frequency superposition: A rheological probe of structural relaxation in soft materials, *Phys. Rev. Lett.*, 2007, **98**, 238303.

Professor Stoyanov answered: For the case of shear oscillations we have changed: a) the strain amplitude by fixing frequency (Fig. 5a); b) the frequency at a given fixed strain amplitude (Fig. 5b). The experimental data from both types of oscillations and from the angle ramp regime are summarized in Fig. 10a. We did not perform experiments in a 'constant-strain-rate frequency sweeps' regime. However we have also combined data from the angle ramp regime (where rate of strain amplitude is proportional to the mean shear rate with numerical factor depending on the Herschel–Bulkley exponent *n*—see eqn 24) as shown in Fig. 12a. In each case both approaches give self consistent and give same results for the shear elasticity. We have not performed constant strain-rate frequency sweep experiments, but since the angle ramp deformation can be expanded in Fourier space as a series of oscillations with different frequencies and amplitudes one can think that we have varied both strain rate and the angular frequencies simultaneously. The fact that in this case we have the same result gives us confidence that if we do constant strain rate-frequency sweep experiments we will get the same results as well.

Dr van der Sman remarked: You mention that your theory is self-consistent. Do you mean it is universal, or does it have a similar meaning in self-consistent theory?

Professor Stoyanov responded: Let us first give a counterexample. We would say that a theoretical interpretation is not self-consistent, if the rheological data obtained in the angle-ramp regime are interpreted by the Maxwell–Herschel–Bulkley model, whereas the data obtained in the oscillatory regime are interpreted by the

Kelvin model. Indeed, in both regimes we are dealing with a forced shearing of the adsorption layer, and it is to be expected that the data can be interpreted in the framework of the same theoretical model. In our paper, we have demonstrated that (i) the rheological data obtained in angle-ramp and oscillatory regime comply with the Maxwell–Herschel–Bulkley model, and (ii) the data points from the two different regimes collapse on the same master curve; see Fig. 10a and b.

Dr Zanchetta asked: The Herschel–Bulkley model includes the presence of a yield stress, *i.e.* a constant, shear rate-independent value for measured stress at low enough imposed shear rate. In your angle-ramp experiments, have you measured lower shear rates than those reported in Fig. 6? Have you observed anything like, *e.g.*, Fig. 3 in Ref 1, that is the existence of measurable dynamic- and static-yield stresses?

1. H. A. Barnes and Q. D. Nguyen, Rotating vane rheometry - a review, *J. Non-Newtonian Fluid Mech.*, 2001, **98**, 1–14.

Professor Stoyanov replied: We have measured static/dynamic yield stress for our systems in the case of aging time larger than 30 min. For large aging times the measured rheological curves were not well reproducible at one and the same experimental conditions. In the case of fresh interfaces (5 min aging time) the experiments were reproducible but the fresh interfaces do not have static/dynamic yield stress. What is known is that yield stress appears for more concentrated systems with aged monolayers (see for example *Langmuir*, 2001, **17** 4556–4563). One can argue of course that for the case of pure hydrophobin where the layer has a very high surface elasticity, which is most likely due to the presence of some perculating network structure one can expect to measure yield stress as well and we have indications for it in the conditions of our experiments for a large aging time. The results however in this case are much less reproducible and that is why we decided to work with 5 min. ageing.

New routes to food gels and glasses

Thomas Gibaud,[ab] Najet Mahmoudi,[c] Julian Oberdisse,[d]
Peter Lindner,[e] Jan Skov Pedersen,[f] Cristiano L. P. Oliveira,†[f]
Anna Stradner[g] and Peter Schurtenberger*[g]

Received 7th March 2012, Accepted 3rd April 2012
DOI: 10.1039/c2fd20048a

We describe the possibility to create solid-like protein samples whose structural
and mechanical properties can be varied and tailored over an extremely large
range in a very controlled way through an arrested spinodal decomposition
process. We use aqueous lysozyme solutions as a model globular protein system.
A combination of video microscopy, small-angle neutron and X-ray scattering
and reverse Monte Carlo modeling is used to characterize the structure of the
bicontinuous network with two coexisting phases of a dilute protein solution and
a glassy or arrested dense protein backbone at all relevant length scales.
Rheological measurements are then used to determine the complex mechanical
response of these protein gels as a function of protein concentration and quench
temperature. While in particular the origin of the dependence of the mechanical
properties on quench depth and concentration is not well understood currently, it
seems ultimately connected to the particular bicontinuous structure of the
arrested spinodal network created by the interplay between the early stage of
a spinodal decomposition and the position of the glass line. We then generalize
this behavior and discuss how this could open up new routes to prepare gel-like
food systems with adjustable structural and mechanical properties. We present
results from a first feasibility study where we use a depletion interaction caused
by the addition of small non-adsorbing polymers to suspensions of casein
micelles in order to create food gels with tunable structural and mechanical
properties through an arrested spinodal decomposition process.

1 Introduction

Gels and glasses play a very important role in today's food science and technology.[1]
Typical examples of food gels are yoghurt or cheese, where soft solid-like food materials are created through the irreversible aggregation of casein micelles, a prime
representative of food colloids. Currently, the majority of food gels are based on
sometimes quite elaborate and energy-intensive processing steps such as heat or

[a]*Brandeis University, 415 South Street, Waltham, MA 02453, USA*
[b]*Université de Lyon, Laboratoire de Physique, École Normale Supérieure de Lyon, CNRS UMR 5672 - 46 Alle d'Italie, 69364 Lyon cedex 07, France*
[c]*Sector of Biological and Soft Systems, Cavendish Laboratory, University of Cambridge, J.J. Thomson Avenue, CB3 0HE Cambridge, UK*
[d]*Laboratoire Charles Coulomb Université Montpellier 2/CNRS UMR 5221, F-34095 Montpellier, France*
[e]*Institut Laue-Langevin, B.P. 156, Grenoble, France*
[f]*Department of Chemistry, University of Aarhus, Langelandsgade 140, Aarhus C, Denmark*
[g]*Lund University, Division of Physical Chemistry, Lund, Sweden*

† Current address: Instituto de F'sica, Universidade de S'o Paulo, Caixa Postal 66318,
05314-970 S'o Paulo, Brasil

pressure treatments. It is thus quite obvious that food science and engineering could enormously profit from parallel developments in colloid science, where fluid–solid transitions such as dynamical arrest, jamming or gelation frequently encountered in colloidal suspensions have been at the center of experimental and theoretical activities during the last few years.

These investigations of dynamical arrest in colloidal suspensions have been mainly triggered by attempts to make and exploit analogies between the resulting phase or state diagrams of colloidal suspensions and atomic and molecular systems. Numerous studies have demonstrated the presence of solid-like structures such as particle gels or colloidal glasses under conditions where either repulsive or attractive interactions dominate.[2–5] Particularly interesting are also systems where mixed potentials, *i.e.* a combination of a short range attraction and a large range soft repulsion, exist.[2,6] Our current state of understanding of the phenomenon of dynamical arrest in systems with mixed potentials as a function of the strength of the attractive part and of the range of the soft repulsive part of the interparticle interaction potential is schematically summarized in Fig. 1.

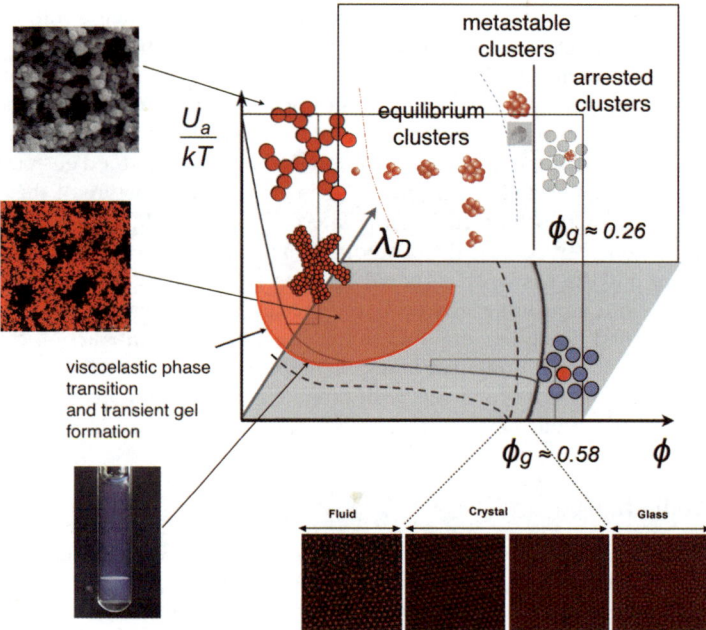

Fig. 1 Schematic state diagram of colloidal particles with mixed interaction potentials consisting of a short range attraction of strength U_a and a screened Coulomb repulsion with screening length λ_D. Shown are the freezing line (dashed line) as well as the arrest line (solid line) as a function of volume fraction ϕ, U_a and λ_D. Three cases are particularly highlighted: (1) short range attractive particles, where a metastable liquid–liquid phase separation occurs in addition to the commonly found liquid, crystalline and glassy or gel states; (2) purely repulsive systems; and (3) particles with mixed potentials where the range of the soft repulsion is significantly larger than the attraction, and where one also finds an equilibrium cluster phase and a cluster glass. Also shown are several examples from the different regions of this state diagram. The scanning electron microscopy (SEM) picture shown in the upper left part illustrates the formation of particle gels for strong attraction (casein micelles in yoghurt formation). Also shown is a confocal laser scanning microscopy (CLSM) picture of a particle gel formed by arrested spinodal decomposition (casein micelles with added polymer, left middle), a photograph of a protein solution that has undergone liquid–liquid phase separation (lower left corner), and a sequence of CLSM pictures of colloidal suspensions with increasing ϕ that shows fluid, crystalline and glassy states (adapted after ref. 1).

For ideal hard sphere particles, we observe a transition from a liquid to a disordered solid phase, a glass, at volume fractions of approximately $\phi \approx 0.58$.[7] If a weak and short-ranged attraction is now turned on, this leads to the astonishing observation of a melting of the glass, followed by a so-called re-entrant glass or solid formation at even stronger attractions.[3,8,9] In the other extreme case of very strong interparticle attraction, we reach the regime of so-called irreversible aggregation where one observes the formation of soft fractal gels already at very low volume fractions.[9,10] At intermediate strength of the attraction, the situation is even more complicated due to the fact that phase separation into a dilute (gas-like) and a concentrated (liquid-like) suspension can occur. The position of the corresponding coexistence curve can then intersect with the arrest or gel line, and the phase separation can subsequently lead to the formation of a long-lived interaction network of particles, if the attractive interactions between them are strong enough.[11]

When now adding a screened Coulomb repulsion, we see a significant move of the freezing line to lower values of ϕ, accompanied by a similar evolution of the glass line. In the extreme case of fully de-ionized colloidal suspensions, crystallization can occur at volume fractions as low as 10^{-3} or lower.[12,13] When combined with a short range attraction, new states appear, the metastable liquid–liquid phase separation is suppressed and we observe the formation of (transient) equilibrium clusters.[14] The ability of colloidal particles to form equilibrium cluster phases under conditions where the particles interact *via* such mixed potentials is well documented.[6] In this class of systems, the short range attraction drives cluster formation, whereas the increasing long-range repulsion due to the accumulating charge of the clusters limits their size. Cluster phases have been demonstrated to exist in a large range of colloidal systems[15–19] and protein solutions.[14,20–26] Particularly interesting in this context is the fact that such protein clusters were found to undergo an arrest transition from a cluster fluid to a cluster glass that occurs at a significantly lower volume fraction $\phi_g \approx 0.26$ when compared to the hard sphere glass at $\phi_g \approx 0.58$.[6]

The schematic state diagram shown in Fig.1 is however not only relevant for colloidal model systems, but has already been used to rationalize the behavior of food systems and provide guidelines for creating materials.[1] Food colloids such as casein micelles have been shown to closely follow theoretical descriptions originally developed for colloidal model systems, and the analogy between the structure of dense colloid gels used in sol–gel ceramics production and casein particle gels produced in yoghurt formation has for example allowed to understand and model this important food process quantitatively.

An important technological challenge in food technology is to optimize the mechanical properties of food gels and simplify the processing by controlling interparticle interactions between food colloids such as proteins. In an ideal case, this would then allow for creating food gels with tailored structural and mechanical properties through controlled self-assembly without invoking costly processing steps such as heat or pressure treatment. It is in this context that the widely differing solid-like states encountered with colloids offer interesting possibilities. In particular the existence of an arrest line that extends into the spinodal region for colloids with short range attractions appears very attractive. It has already been reported previously that quenching such a system deeply enough into the spinodal such that the equilibrium concentration of the dense phase would be located in the arrested region of the state diagram will lead to solid-like materials where the bicontinuous structure resembles those found during classical spinodal decomposition, with a correlation length mainly determined by the quench depth.[27] It seems quite conceivable that this could offer a mechanism that would allow us to create food gels with controllable mechanical and structural properties.

However, while the phenomenon of an arrested spinodal decomposition as a possible mechanism for gelation in attractive colloidal suspensions appears to be generally accepted, there exist contradicting findings as to the location of the arrest line.[11,28] The presence of an arrested spinodal decomposition has for example been demonstrated for concentrated solutions of the model protein lysozyme, which is known to exhibit a phase diagram that closely matches the predictions for colloids with short-ranged attractions.[11] Here the mechanism underlying the liquid–solid transition encountered when quenching a protein sample deeply below the coexistence curve for liquid–liquid phase separation has been interpreted based on a state diagram as shown in Fig.1. It has been proposed that arrest occurs because the spinodal decomposition process leads to a bicontinuous structure which gets "pinned" into a rigid self-supporting network when the concentration of the high density regions crosses the arrest line and subsequently undergo dynamical arrest. This scenario has been supported by several studies using lysozyme,[11,27,29] it is however in sharp contrast to one of the central conclusions drawn in ref. 28, where a universal phase diagram for colloids with short-range attraction had been constructed that suggested that the gelation line coincides with the phase separation boundary in the Baxter model. These authors then concluded that the origin of dynamical arrest came from the dense phase undergoing an attractive glass transition at $\phi_g \approx 0.55$, and claimed that the attractive glass line would thus not extend into the phase separation region but instead follow its high density boundary.

It is clear that the possibility to create solid-like food materials with tunable structural and mechanical properties through an arrested spinodal decomposition process will ultimately depend on which one of these scenarios is correct, as only the first scenario will lead to a large accessible range of mechanical properties. We thus further explore the nature of the arrested spinodal decomposition process and the resulting structures and their mechanical properties of lysozyme solutions as a model for food proteins. We then investigate whether the same structures can be generated for much larger particles such as casein micelles, now using depletion interaction from added non-adsorbing (bio)polymers to create the necessary short range attraction.

2   Materials and methods

We use hen egg white lysozyme (Fluka, L7651) dispersed in an aqueous buffer (20 mM Hepes, pH = 7.8) containing 0.5 M sodium chloride. Lysozyme is a monodisperse globular protein of a molecular mass of 14.4 kDa carrying a net charge of +8 electronic charges at pH = 7.8. Details of the sample preparation are given in ref. 11. Initially a suspension at $\phi \approx 0.22$ is prepared in pure buffer without added salt, and its pH is adjusted to 7.8 ± 0.1 with sodium hydroxide. Under these conditions the solution is stable and is used as stock. We then dilute it with a NaCl-containing buffer at pH = 7.8 to a final NaCl concentration of 0.5 M. Particular care is taken to avoid partial phase separation upon mixing by pre-heating both buffer and stock solution well above the coexistence curve for liquid–liquid phase separation. This procedure results in completely transparent samples at room temperature with volume fractions ranging from $\phi = 0.01$–0.18, where ϕ was obtained from the protein concentrations c measured by UV absorption spectroscopy using $\phi = c/\rho$, where $\rho = 1.351\ g/cm^3$ is the protein density. To prepare samples at even higher volume fractions up to $\phi \approx 0.34$, we take advantage of the ability of the system to phase separate into a protein-rich and protein-poor phase.

Temperature quenches to investigate samples undergoing an arrested spinodal decomposition were done as described in ref. 27: The cuvette (1 mm Hellma quartz cell or 1 mm capillary) used in the experiment was filled with a fresh lysozyme dispersion around 25 °C. The sample is then pre-quenched for a minute in an ethanol bath at the desired final temperature, T_f. It is then quickly transferred to the thermostated cuvette holder of the experiment, also at T_f. This procedure allows to obtain quench rates where we estimate the time for the sample to go from 25 °C to T_f to be around 30 s.

The feasibility tests using casein micelles were made with casein micelles free from whey proteins and from lactose (Micellar Caseins, MPI-85MC) with an average hydrodynamic radius of $R_h = 100$ nm provided by the Hungarian Dairy Research Institute. As a flexible polymer for inducing depletion interactions we used poly(ethylene oxide) (PEO) from Polymer Source, Inc. It has a weight average molar mass, M_w, of 102 000 g mol^{-1}, a polydispersity index of 1.08 and a radius of gyration, R_g, of 15.86 nm, as stated by the manufacturer. Mixtures of casein micelles at a fixed volume fraction of 0.25 and PEO at various concentrations at neutral pH and an ionic strength of 100 mM NaCl were prepared from casein and PEO stock solutions using a minishaker (MS1 Minishaker, IKA) in plastic cuvettes. The ionic strength was adjusted using a concentrated NaCl solution. The casein micelles were labelled with Rhodamine B for confocal laser-scanning microscopy (CLSM) imaging.

The microstructure of casein micelles dispersed in PEO was captured using a Leica TCS SP5 confocal laser-scanning microscope (CLSM) operated with a 63× glycerol-immersion objective. A square pixel slice was taken (512 × 512) with image dimensions varying from 246 × 246 to 25 × 25 μm. Z-stacks were acquired until a depth of approximately 30–40 μm.

Small-angle neutron (SANS) and X-ray (SAXS) scattering experiments were performed on lysozyme samples that underwent an arrested spinodal decomposition process. SANS experiments were performed at the SANS I facility at the Swiss neutron source SINQ at the Paul Scherrer Institut, Switzerland, and at the instrument D11 at the Institute Laue Langevin in Grenoble, France. We used 1 mm Hellma quartz cells and a thermostatically regulated sample holder. Combinations of different wavelengths, sample-to-detector distances and collimation lengths were used to cover a q range of 0.015 to 2 nm^{-1}. In the SANS experiments, the contrast, K, depends on the scattering length density of water, $\rho_W = -5.60 \times 10^{-7}$ Å^{-2}, the scattering length density of lysozyme, $\rho_L = 1.88 \times 10^{-6}$ Å^{-2}, the partial specific volume of lysozyme, $v = 0.74$ mL g^{-1} and Avogadro's number, N_A. It is equal to $K = v^2(\rho_L - \rho_W)/N_A = 5.41\ 10^{-8}$ m^2 mol g^{-2}. SAXS experiments were carried out with a pinhole camera (NanoSTAR, Bruker AXS) employing a rotating anode (Cu K$_\alpha$) X-ray source, a thermostatically regulated sample chamber and a two-dimensional gas detector. For the present experiments the sample to detector distance was 24 cm, providing a q range of 0.2 to 10 nm^{-1}.[30]

We also measured the rheological properties of the arrested spinodal network with a stress-controlled rheometer (MCR300 from PaarPhysica) in a cone and plate geometry, using a solvent trap to minimize sample evaporation. In a typical experiment we loaded a sample in the rheometer at $T = 25\ ^\circ$C, i.e., well above the binodal line. T was then lowered to the final temperature of the quench, T_f. Once the temperature in the rheometer reached T_f, the sample was left to rest for 300 s during which the structure stabilizes. This procedure enabled us to reach a reproducible steady state with negligible aging. Two types of tests were then carried out to characterize the viscoelastic properties of the arrested spinodal decomposition in the linear response regime: (i) frequency sweeps giving G' and G'' as a function of the frequency f, and (ii) creep experiments which yield the compliance $J(t)$ as a response to a step stress. These two measurement protocols gave consistent results.

3   Results and discussion

3.1   Arrested spinodal decomposition in lysozyme solutions

The phase diagram of lysozyme at 0.5 M NaCl has been described in detail previously.[11,27,31] It shows all the characteristic features of colloids interacting via a short range attractive potential such as liquid–liquid phase separation that is metastable with respect to the liquid–solid (crystal) phase boundary. The location of the binodal is shown again in Fig. 2a. Samples quenched below the binodal into region I show macroscopic phase separation as demonstrated by the photograph in Fig. 2a. However, for quenches below the tie line at 15 °C the phase separation becomes

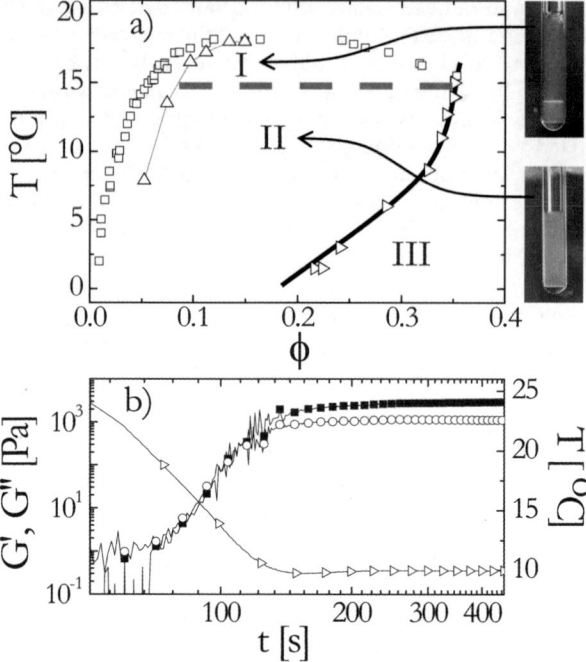

Fig. 2 a) State diagram of lysozyme. Liquid–liquid phase boundary (binodal, □), spinodal (△), glass line (▷), as determined in ref. 11, and tie line that separates complete phase separation from arrested spinodal decomposition (dashed line). Region I: phase separation. Region II: arrested spinodal decomposition. Region III: glass. b) Measurements of G' (■) and G'' (○) at 1 Hz during a temperature quench from 25 °C to $T_f = 10$ °C (▷) at $\phi = 0.11$.

arrested.[11] This can be seen in Fig. 2b, where the evolution of the storage (G') and loss (G'') moduli as a function of temperature T are shown. As T decreases from 25 °C to T_f, the sample first becomes turbid at the binodal line as a result of the spinodal decomposition (lower photograph in Fig. 2a). At $T \simeq 15$ °C, the sample then undergoes a fluid to solid transition. The bicontinuous appearance clearly visible in video microscopy no longer evolves, and the correlation length ξ linked to the characteristic mesh size of the bicontinuous network gets pinned as described in ref. 27. The corresponding evolution of the storage (G') and loss (G'') moduli with time measured at 1 Hz is shown in Fig. 2b, and we see that G' increases above G'' at $T \simeq 15$ °C.

3.2 Structure of the arrested network from SANS and SAXS

It has previously been proposed that systems that have undergone an arrested spinodal decomposition have a bicontinuous structure with a protein-poor fluid and a dense glassy protein network as sketched in Fig. 1.[27] Estimates of the local volume fraction in the dense strands have been determined by centrifugation experiments, and the resulting concentrations correspond to the volume fraction of the "homogeneous" attractive glass at the arrest line within the spinodal region separating region II and III in Fig. 2a.[11] Previous characterizations of the network structure have mainly concentrated on revealing the bicontinuous network and its correlation length ξ.[27] However, no direct and quantitative characterization of the network structure at all relevant length scales has been reported so far. A major problem here are the extremely different length scales that characterize these samples, with a correlation length of the network $\xi \sim \mu$m, while the local nearest neighbor distance in the dense glassy branches must be of order protein diameter $\sigma = 2a \approx 3.4$nm, where a is the protein radius.‡

In a next step we thus take a closer look at the local structure of the arrested spinodal network probed with SAXS and SANS. These are ideal techniques to reveal the internal structure on complementary length scales all the way to the single protein monomer. We aim at characterizing the interface between the gas-like and the glass-like phases and the local structure of the glass-like phase. The results from a corresponding combined SANS and SAXS study of the dependence of the network structure as a function of initial sample concentration and quench depth, i.e. final temperature T_{f}, are shown in Fig. 3 together with typical video micrographs of samples that underwent an arrested spinodal decomposition.

The micrographs for samples obtained at different quench depths shown in Fig. 3a indeed demonstrate that the arrested spinodal decomposition results in a bicontinuous network with decreasing mesh size, i.e. decreasing correlation length ξ. The SANS and SAXS data in Fig. 3b and c summarize the evolution of the intensity as a function of the final temperature T_{f} after a quench at $\phi_0 = 0.148$ (Fig. 3b) and the evolution of the intensity as a function of the initial volume fraction, ϕ_0, at 10 °C (Fig. 3c), respectively. At low q we observe a Porod regime, $I \sim q^{-4}$, which confirms that the gel is indeed a bicontinuous network with a sharp interface between the dilute (gas-like) and the dense glass-like phases. As temperature decreases, the intensity in the Porod regime increases, reflecting an increase of the surface to volume ratio of the glassy backbone. One can deduce the surface to volume ratio from the Porod regime:

$$I_{\mathrm{Porod}}(q) = \frac{2\pi\Delta\rho^2}{q^4}\frac{S}{V} \tag{1}$$

The excess scattering length density $\Delta\rho$ defines the contrast between the gas-like and the dense phases. The gas-like phase is characterized by its lysozyme volume fraction ϕ_1 and the relative volume of the gas-like network branch $1 - h$, while the dense phase is characterized by its lysozyme volume fraction ϕ_2 and the relative volume of the dense network branch, h, where h is the relative height of the dense phase for a macroscopically separated sample as determined using centrifugation experiments (see ref. 11 for details). For SANS $\Delta\rho$ can be approximated by

$$\Delta\rho = [\rho_{\mathrm{L}}\phi_2 + (1 - \phi_2)\rho_{\mathrm{W}}] - [\rho_{\mathrm{L}}\phi_1 + (1 - \phi_1)\rho_{\mathrm{W}}] \tag{2}$$

The results of the calculation of the surface to volume ratio S/V using eqn (1) and 2 are tabulated in Tab. 1 together with an analysis of the correlation lengths ξ from microscopy. We find that S/V increases with decreasing temperature, which seems reasonable as the characteristic length ξ decreases with decreasing temperature. For a given quench temperature, it also increases with decreasing concentration, which again agrees with the fact that both the local volume fractions ϕ_1 and ϕ_2 as well as the correlation length ξ are independent of the initial concentration and depend on the final temperature only.

3.3 Local structure

At larger q the intensity reflects the local organization of the dense phase, which depends on the local volume fraction but also on the interactions between the proteins. Given the fact that we expect the local structure to exhibit only weak variations as the system dynamically arrests when crossing the arrest line, we use liquid state theory to calculate the effective static structure factor $S(q)$ at length scales

‡ Note that these are average values, as lysozyme has an approximately ellipsoidal shape with two minor axes of about 3.3 nm and a major axis of about 5.5 nm diameter.

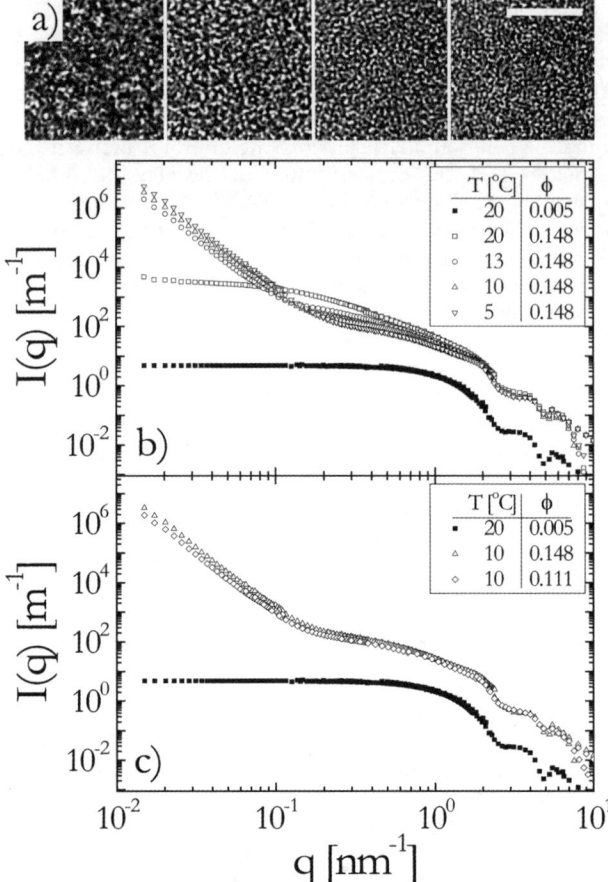

Fig. 3 a) Micrographs showing arrested spinodal decomposition for a fast quench at (from left to right): $T_f = 15\,°C$, $T_f = 10\,°C$, $T_f = 5\,°C$, $T_f = 1\,°C$. The scale bar is 20 μm. b) SANS and SAXS results for a temperature series at constant concentration ($\phi = 0.148$): 20 (□), 13 (○), 10 (△), 5 °C (▽), and for a sample at low concentration (form factor, $\phi = 0.005$) at 20 °C (■); and c) concentration series at constant $T_f = 10\,°C$: $\phi = 0.111$ (◇) and $\phi = 0.148$ (△), as well as $\phi = 0.005$, at 20 °C (■) as an effective form factor. Note that due to the different scattering contrast for SANS and SAXS, the SAXS data have been rescaled to the SANS data. Moreover, there is a slight mismatch in the overlap region due to the fact that for SAXS the electron density of the hydration shell is larger than that of bulk water, which results in a slightly larger effective radius of gyration for lysozyme in SAXS experiments than in SANS. $q = 1.5 \times 10^{-2}$–2 nm^{-1}: SANS data. $q = 0.3$–10 nm^{-1}: SAXS data.

comparable to the monomer diameter. The effective structure factor $S(q)$ is determined as the ratio between two measurements normalized with the particle concentrations:

$$S(c, \phi) = \frac{I(q, \phi)\phi_{\text{dilute}}}{\phi I(q, \phi_{\text{dilute}})} \tag{3}$$

where $I(q,\phi)$ is the intensity scattered by a solution of volume fraction ϕ and $I(q,\phi_{\text{dilute}})$ is the intensity of the dilute sample, $\phi_{\text{dilute}} = 0.005$, used as the effective form factor. Typical results for samples after an arrested spinodal decomposition are shown in Fig. 4.

We then compare the scattering results with model calculations where we use a square-well potential to approximate the protein–protein interactions. Here $S(q)$

Table 1 Variation of ϕ_1, ϕ_2, h (obtained from the centrifugation experiments[11]), ξ (obtained from the microscopy experiments shown in Fig. 7) and S/V, the surface to volume ratio (extracted from the Porod regime) as a function of T_f and ϕ_0

T_f [°C]	13	10	5	10
ϕ_0	0.148	0.148	0.148	0.111
ϕ_1	0.048	0.041	0.026	0.041
ϕ_2	0.344	0.329	0.270	0.329
h	0.342	0.381	0.520	0.242
ξ [μm]	~3.5	~2.5	~1.9	~2.5
S/V [nm^{-1}]	76	186	331	327

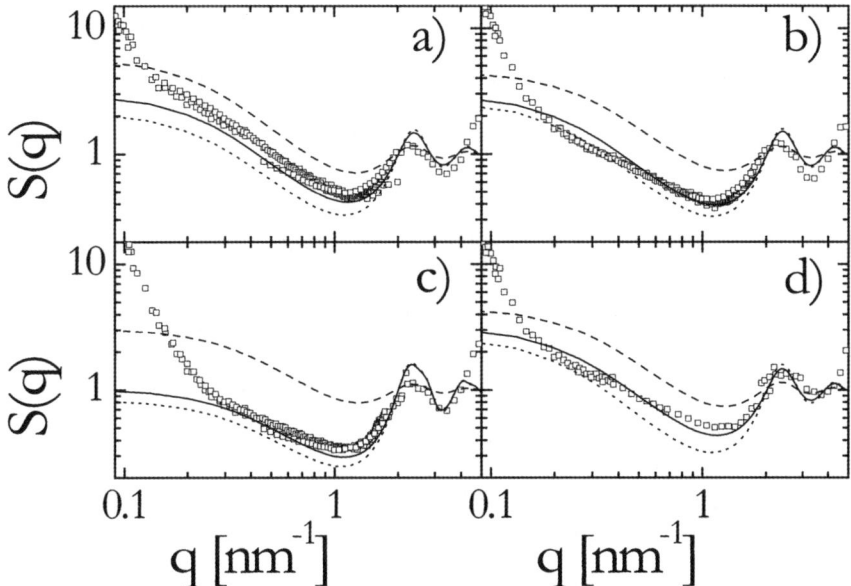

Fig. 4 Effective structure factor of the arrested spinodal decomposition. Symbols are data from SAXS experiments, lines calculated structure factors for the dilute phase $S_1(q)$ (dashed line), the concentrated phase $S_2(q)$ (dotted line) and the total effective structure factor $S_{cal}(q)$ (solid line), respectively (see text for details). a) Quench at 13 °C with $\phi_0 = 0.148$. b) Quench at 10 °C with $\phi_0 = 0.148$. c) Quench at 5 °C with $\phi_0 = 0.148$. d) Quench at 10 °C with $\phi_0 = 0.111$.

can be calculated by using integral equation theory for simple liquids with a Percus–Yevick closure relation for a model of polydisperse spheres based on the algorithm in ref. 32. The square well potential is given by

$$U_{SW}(r) = \begin{cases} \infty & 0 < r < 2a_{eff} \\ -\varepsilon & 2a_{eff} < r < 2a_{eff}(1 + \lambda) \\ 0 & r > 2a_{eff}(1 + \lambda) \end{cases}$$

where a_{eff} is an effective hard sphere radius, and ε and λ are the depth and the range of the square well, respectively. While square well potentials have unphysical shapes, they have been widely used to successfully model colloids with short range attractions since the relevant physical properties were found to depend only weakly on the shape of the potential.[33]

The parameters were taken from a fit of SANS data obtained with a temperature and a concentration series in the fluid state above the coexistence curve. Typical results are shown in Fig. 5a. We obtain a satisfying fit of the temperature dependence of the structure factor with a square well interparticle interaction. For the concentration series at a given temperature the range and the depth of the well could be maintained constant, with $\varepsilon/kT = -3.52$ for 20 °C and -3.24 for 30 °C and $\lambda = 2.2\%$ (data not shown). The study of the structure factor as a function of temperature then yields a direct relation between the depth of the short range attraction, ε and T. For a given concentration, $\phi = 0.052$, $S(q)$ could be fitted maintaining the range of the attraction constant ($\lambda = 2.2\%$) as shown in Fig. 5a. This is consistent with the presence of an isosbestic point at $q \approx 0.8$ nm^{-1}. Isosbestic points indeed appear in scattering experiments when adjusting the strength of attraction at constant range.[34–36] The analysis of the data shown in Fig. 5a provides an estimate of the temperature dependence of ε as shown in Fig. 5b. ε/kT scales linearly with temperature, which would suggest that the main temperature dependence of the interparticle interaction is simply through the temperature factor kT, while other specific temperature effects, such as changes in the hydration of lysozyme, if present, play a minor role. Note, however, that while a simple square well potential as used here is capable of reproducing the structure factors at all temperatures, it fails when simultaneously trying to quantitatively predict the corresponding virial coefficients B_2 as a function of T. In this case it is necessary to depart from a purely centrosymmetric potential and include a certain degree of patchiness, at the same strength of the attraction.[31]

Having fixed the potential values for all relevant temperatures from an extrapolation of the data shown in Fig. 5b, we can next calculate the full effective structure factor that describes the local structure in samples with an arrested spinodal decomposition. Here we have to include contributions from the gas-like phase, given by an effective structure factor $S_1(q)$, and from the dense phase, given by $S_2(q)$, respectively. Assuming that the contribution of the gas phase (1) and the glass phase (2) are uncorrelated, the resulting structure factor $S(q,\phi_0,T_f)$ of a sample with initial concentration ϕ_0 and final temperature T_f is given by

$$S_{\mathrm{cal}}(q, \phi_0, T_f) = \left[\frac{(1-h)\phi_1}{\phi_0} S_1(q, \phi_1, \varepsilon) + \frac{h\phi_2}{\phi_0} S_2(q, \phi_2, \varepsilon) \right] \qquad (4)$$

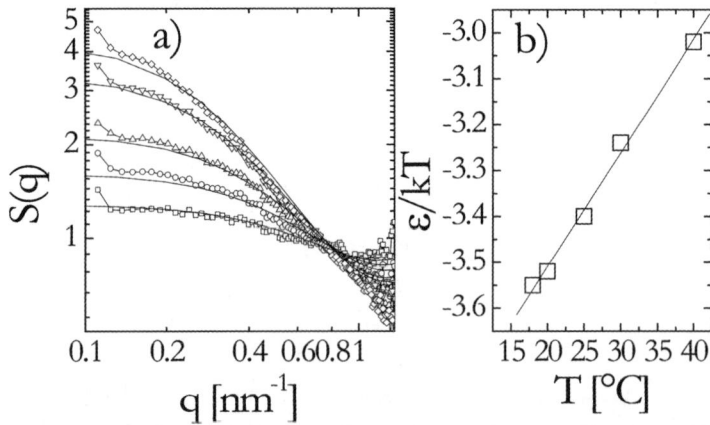

Fig. 5 a) $S(q)$ from SANS for different temperatures (from below: 18, 20, 25, 30, 40 °C) at $\phi = 0.052$. $S(q \to 0)$ increases monotonously as temperature decreases. The lines show the calculated $S(q)$. b) Evolution of ε obtained from the fit of $S(q)$ in a) as a function of temperature.

where all parameters have been determined independently from centrifugation experiments (Tab. 1) and SANS measurements in the liquid phase above the binodal (Fig. 5). This approach is motivated by the fact that in general repulsive and attractive glasses show typical fluid structure as the system goes through the dynamical arrest transition. The results of these calculations are shown in Fig. 4. Although this empirical model provides surprisingly good agreement at intermediate and high q, deviations at lower values of q show that the approach does not capture the structure on larger length scales. Therefore we extend our analysis to computer simulations using a reverse Monte Carlo method.

3.4   Russian doll reverse Monte Carlo analysis of the intensity

So far we have been able to show that we obtain quantitative and consistent information about the structure of the arrested spinodal from a combination of scattering and optical microscopy over a very large range of length scales. A remaining difficulty is the structure at intermediate length scales or wave vectors, in particular if we like to get information beyond the fact that the extended Porod region indicates the formation of well separated regions having different concentrations and a well defined interface. Therefore, we have developed a new reverse Monte Carlo (RMC) method for the analysis of the structure of the concentrated phase based on the scattering data (Fig.6).

The use of standard direct Monte Carlo would be extremely time consuming, as typically some 10^6 particles make up the biggest structures. Indeed, the experimental q-range is very large (down to 0.01 nm^{-1}), and the size of the primary lysozyme monomer quite small ($a = 1.6$ nm). To limit the computational effort, we need a course graining procedure, which we term 'Russian Doll RMC'. The idea is to describe the structure on a small length scale with few particles, regroup them into a 'meta-particle', and use it to build a higher order structure, and so on. Each level is described by an ensemble of particles, using a conventional Reverse Monte Carlo (RMC) algorithm.[37,38] The idea is to build a first-guess-structure, and improve the agreement of its $I(q)$ in the corresponding q-range with the experimentally

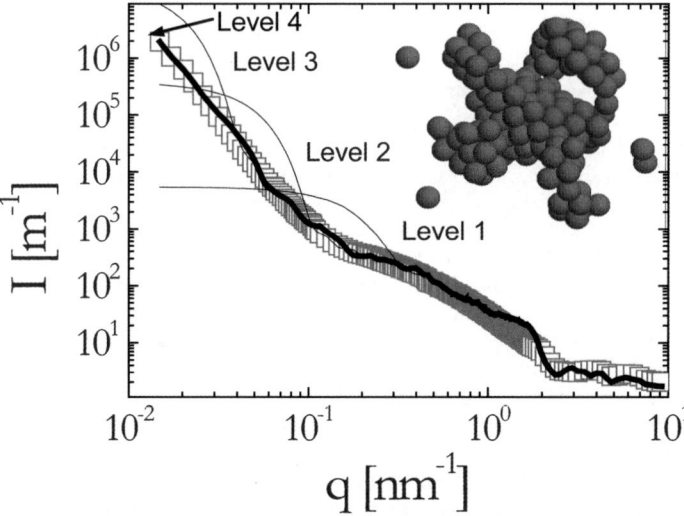

Fig. 6 Intensity scattered by a sample after an arrested spinodal decomposition (□). The thin lines show the intensity scattered by the *meta*-particles at different levels. The thick line shows the result of the simulation. The Inset shows an example of the simulation results in real space where aggregation of third level metaparticles (radius 131 nm) are shown.

measured one by randomly displacing individual subunits. These displacements are confined inside the spherical next-order metaparticles, which is a way to control internal volume fraction and interactions. Due to possible interpenetration of the less dense metaparticles, the excluded volume condition is maintained only on the smallest length scale. Structure factors can then be calculated using the Debye formula.[39] Intensities are obtained by straightforward multiplication of lysozyme form factor and structure factors for (on average) spherically symmetric metaparticles, followed by the addition of the dilute-phase intensity.

We focus on the temperature quench to 13 °C ($\phi_0 = 0.148$), but a very similar approach could be applied to the other samples. In Fig. 6, the experimental intensity is compared to the successive fits on the various length scales. At the smallest scale, 50 lysozyme particles make up a first metaparticle of radius $a_{m1} = 11.5$ nm. Some deviations can be observed, presumably due to non sphericity of the proteins. At intermediate q, 70 of these metaparticles form the next higher structure of radius $a_{m2} = 44.0$ nm, 50 of which are then regrouped on the third level ($a_{m3} = 131.0$ nm), and finally 200 of these biggest particles make up the largest structure (overall size 1000.0 nm). This structure represents 3.5×10^6 lysozyme particles which make up the dense phase (volume fraction $\phi_2 = 0.35$) inside the dilute one. For illustration, this biggest structure is plotted as an inset in Fig. 6. It seems to reproduce a microscopic phase separation, the typical length scale of which—two to three metaparticle diameters—approaches one micron (consistent with light scattering and microscopy). The scattered intensity is found to be well fitted by the RMC simulation. In particular, all major features (low-q Porod regime, break in slope, high-q lysozyme structure) are reproduced in an at least semi-quantitative manner, and the resulting local volume fraction of the dense phase $\phi_2 = 0.35$ is in very good agreement with the value $\phi_2 = 0.34$ obtained from the centrifugation experiments (Tab. 1). The Russian-Doll RMC method presented here has been tested on simulated aggregate structures with two levels, and the results are given in the appendix. It is however clear that for future work, the limits of this approach will need to be tested, and in particular the set of working parameters (size and number of metaparticles) will have to be determined.

3.5 Mechanical properties

Having established that the arrested spinodal decomposition in lysozyme solutions results in the formation of a bicontinuous network with a coexisting dilute fluid phase and an arrested glassy high density backbone with a density and correlation length that decreases with increasing quench depth, we next investigated the resulting mechanical properties of these samples. Typical results from such a series of different quench depths T_f at constant initial volume fraction $\phi_0 = 0.111$ are shown in Fig. 7a. Shown are data from creep tests as the time dependence of the inverse creep compliance $J^{-1}(t)$. The creep measurements reveal an unusual mechanical response with two plateaus at short and long times, respectively, separated by a dissipative regime at intermediate times. These plateaus can be associated with the short (G_∞) and long (G_0) time elastic moduli that have already previously been found in oscillatory shear and creep measurements that underwent an arrested spinodal.[11] We do in fact find in all of the arrested samples two elastic plateaus at high (G_∞) and low (G_0) frequencies or short and long times, with $G_\infty \gg G_0$, and the results from the creep measurements and the oscillatory shear measurements are all fully consistent.

Fig. 7b summarizes the resulting values of the moduli G_∞ and G_0 for the series of different quench depths T_f at a constant initial concentration $\phi_0 = 0.111$. We see that while the high frequency modulus G_∞ appears to be independent of T, the low frequency modulus G_0 dramatically increases with decreasing temperature over almost 3 orders of magnitude for the range of T investigated. The two moduli appear to merge close to the extrapolated position of the arrest line for $\phi_0 = 0.111$, *i.e.* at

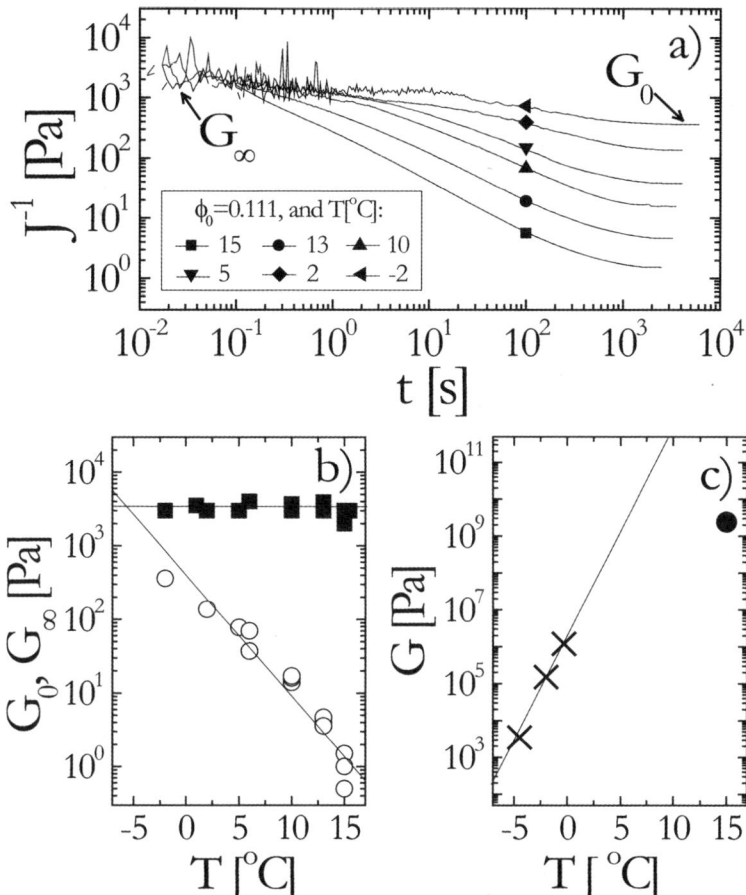

Fig. 7 a) Results from creep experiments showing the time evolution of the compliance J^{-1} in the linear regime at $\phi = 0.111$ for a series of different temperature quenches to $T_f = 15\,°C$ (■), $13\,°C$ (●), $10\,°C$ (▲), $5\,°C$ (▼), $2\,°C$ (◆), $-2\,°C$ (◄). The line shows the conversion of the corresponding compliance J^{-1} into storage and loss moduli. b) Dependence of G_∞ (■) and G_0 (○) on T_f for $\phi_0 = 0.111$. c) Intersection points of G_∞ and G_0 from the T-dependence for three different initial concentrations $\phi_0 = 0.111$, 0.148, 0.186 (x). Also shown is the theoretical estimate of the modulus of an attractive glass at the intersection point of the arrest tie line and the binodal (●, see text for details).

$T_g(\phi_0) = -5\,°C$. Given the fact that the mechanical response of the samples with two elastic moduli most likely reflects the bicontinuous structure with a coexistence of a dilute liquid and a dense solid-like phase,[11] such a coincidence of the intersection point of $G_\infty(\phi)$ and $G_0(\phi)$ with the position of the arrest line is not unexpected. We do in fact expect that for a quench to the arrest line or below, *i.e.* within region III of the state diagram in Fig. 2a, the sample is in a homogeneous attractive glass state. Here a bicontinuous texture should no longer exist as the spinodal decomposition process becomes immediately arrested upon a quench to $T_f \leq T_g$. Therefore, we should then also find a single modulus $G = G_\infty = G_0$ only. Indeed we find the same behavior for all temperature series at different initial volume fractions investigated, with the intersection points located on the extrapolated arrest line in Fig. 2a.

The resulting values of the moduli at the different intersection points are shown in Fig. 7c. These values reflect the mechanical properties of the attractive glasses at different positions on the arrest line in Fig. 2a. They indicate that an arrested spinodal decomposition indeed allows us to create soft solids with mechanical properties that may vary over many orders of magnitude. While we have so far only covered the low temperature/volume fraction part of the arrest line dividing regions II and III in Fig. 2a, we can at least estimate the maximum attainable value for G that would correspond to the intersection of the arrest line and the binodal, $i.e.$ at a value of $\phi_0 \approx 0.35$ and $T \approx 15\,°C$. Under these conditions, a prediction for attractive glasses formed by colloids with short range attractions should provide at least some guidelines. We can use a simple model for the modulus of an attractive glass G_{ag} given by $G_{ag} \approx U_a/(r_{loc}^2\sigma)$, where U_a is the value of the potential at contact, $i.e.$ $U_a = U(r = \sigma)$, and r_{loc} is the localization length of the particles trapped in the attractive well.[40,41] If we use $U_a \approx 3kT$ and $r_{loc} = \lambda a$, with $\lambda = 2.22\%$ the range of the attraction, we obtain a value of $G_{ag} \approx 2.4 \times 10^9$ Pa. While we have not explored a sufficiently large range of concentrations and temperatures yet, Fig. 7c nevertheless already indicates that an arrested spinodal decomposition for attractive proteins or food colloids could be a very interesting mechanism for forming gels with tunable properties.

4 Conclusions—lessons learned for making food gels

Our combined SANS, SAXS and rheology experiments have confirmed that quenching aqueous lysozyme solutions at sufficiently low temperatures below the binodal leads to an arrested spinodal decomposition and the formation of a bicontinuous structure with a dilute fluid gas-like phase and a dense glass-like solid backbone. The formation of a well defined interface and values of the local concentration of the dense phase in agreement with those previously obtained through ex-situ centrifugation experiments further support these previous studies. When combining the findings in Fig. 7b and 3a, we realize that we can create solid-like protein samples whose structural and mechanical properties can be varied and tailored over an extremely large range in a very controlled way through an arrested spinodal decomposition process. While in particular the origin of the dependence of the mechanical properties on quench depth and concentration is not well understood currently, it seems ultimately connected to the particular bicontinuous structure of the arrested spinodal network created by the interplay between the early stage of a spinodal decomposition and the position of the glass line. This obviously opens up new routes to prepare gel-like systems with adjustable structural and mechanical properties in materials and food science. If this is a generic behavior of colloids with short range attractions as postulated also in Fig.1, we can then try to use for example a depletion interaction caused by the addition of small non-adsorbing polymers to suspensions of food colloids such as casein micelles or emulsion droplets in order to create food gels with tunable structural and mechanical properties through a very simple mixing process.

We tested the feasibility of such an approach using casein micelle suspensions with added PEO at a concentration c_p, where PEO induces an attractive depletion interaction. These first tests are indeed encouraging, we have qualitatively observed the same phase behavior as in lysozyme, with the existence of liquid states at low c_p and a liquid–liquid phase separation at higher values of c_p that then becomes arrested at sufficiently high polymer concentrations. Moreover, the arrested samples also exhibit a bicontinuous network structure, again qualitatively similar to those already observed for lysozyme solutions, as demonstrated in Fig. 8. The figure shows confocal laser scanning microscopy pictures from casein micelle suspensions at a constant volume fraction $\phi = 0.25$ with added PEO at two different polymer concentrations c_p, $i.e.$ for different strengths of a short range depletion interaction, under conditions where phase separation is arrested. While PEO is of course not a food polymer, we use it as a simple and well defined model of a linear flexible and non-adsorbing polymer to create a well defined depletion interaction between the casein micelles. Given the casein radius of $a_{casein} \approx 100$ nm and the radius of

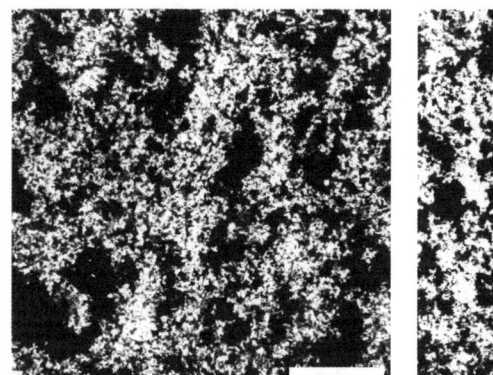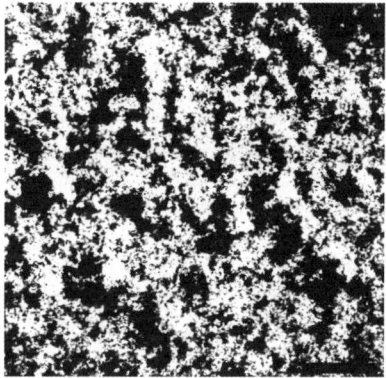

Fig. 8 Confocal images in the XY plane of arrested casein–PEO mixtures. The casein phase is stained with Rhodamine B and is shown in white. The images show suspensions with a casein volume fraction $\phi = 0.25$ and two polymer concentrations $c_p = 10$ mg mL^{-1} (left) and 12 mg mL^{-1} (right). Scale bars denote 50 μm.

gyration of PEO of $R_g \approx 16$ nm in the dilute limit, the range of the resulting depletion interaction is less than 15% of the particle radius, *i.e.* sufficiently short ranged to result in a metastable liquid–liquid phase separation scenario. For a sufficiently strong depletion attraction we indeed find arrested spinodal decomposition resulting in solid-like samples with a bicontinuous network structure such as the ones shown in Fig. 8, with a mesh size that depends on the quench depth (polymer concentration) similar to what has been observed previously for lysozyme (Fig. 3a).

We are currently performing a systematic investigation of the phase diagram and the resulting structural and mechanical properties with a particular focus on samples that underwent an arrested spinodal decomposition. It is clear that further experimental and theoretical input and more refined computer simulations will be needed in the future in order to achieve a quantitative understanding of the interplay between the local and global network structure and the resulting intricate mechanical response. However, given the enormously large range of length scales involved in these systems this will not be an easy task. Finally, one should also take into account the added possibilities that the presence of an additional Coulomb repulsion provides as shown schematically in Fig. 1. The results discussed in this paper do however clearly demonstrate the existence of a variety of mechanisms that can produce dynamically arrested states with solid-like appearance for colloids with short range attractions, mechanisms that should be possible to exploit in suspensions of food colloids through controlled application of a non-adsorbing food polymer.

Appendix

A.1 One-level reverse Monte Carlo

The aim of our implementation of the standard RMC algorithm with simulated annealing is to find a family of real-space configurations of spherical beads the scattering function of which is consistent with the experimentally observed one. In this appendix, the method is tested with model aggregates. First, an aggregate is generated, and its bead-bead structure factor $S(q)$ is calculated. Then, this structure factor is used as an input to the standard aggregate RMC algorithm for aggregates as presented in ref. 37. Note that the topology of the aggregate is conserved, in the sense that beads must always touch, and the aggregate always remains a single connected assembly of beads. The outcome is a series of real-space snapshots of aggregates, which can be compared in shape and size to the original aggregate (known by construction in this test). Naturally, the question of how to quantify good agreement arises. A close or even identical pair correlation function is fulfilled if the RMC-fit of $S(q)$ is good. Therefore, the radius of gyration of the resulting aggregate, as well as the local coordination numbers and intermediate scaling regimes (if present), are trustworthy. Here, we limit ourselves to a presentation of snapshots which show the overall robustness of the method.

In Fig. 9, the structure factor of the aggregate shown in the upper inset is compared to the final RMC-fit, and perfect agreement is found. The original aggregate was constructed by an individual diffusion process of sticky beads (bead radius 5 nm, 100 beads). Its structure factor reflects the 3D-arrangement, starting with the total aggregation number at $q \rightarrow 0$, the radius of gyration in the Guinier regime, and the internal structure with strong correlation peaks due to the excluded volume of monodisperse beads. A possible compatible 3D-arrangement found by the RMC-algorithm is depicted in the lower inset. Visual comparison with the original one shows that the algorithm produces a trustworthy configuration. Needless to say, the result is not unique: small changes obtained by displacing one bead will not affect the $S(q)$, as long as they do not induce notable changes in the local coordination or overall shape of the aggregate.

A.2 Two-level RD-RMC

A two-level RD-RMC was applied to a bigger aggregate (bead radius 5 nm, 1000 beads), following the same concept as in the previous section. Note that the generalization to higher level aggregates as shown in this article is straightforward. The aggregate structure factor has been constructed using the same algorithm as previously. It is shown in Fig. 10, with the 3D-representation of the original aggregate in the upper inset on the right.

This calculated structure factor is again the input of the RMC algorithm. We have chosen a first level of 30 beads. The standard one-level RMC-algorithm as in Fig. 10 was applied to the structure factor of the 1000-bead aggregate, in the high-q range

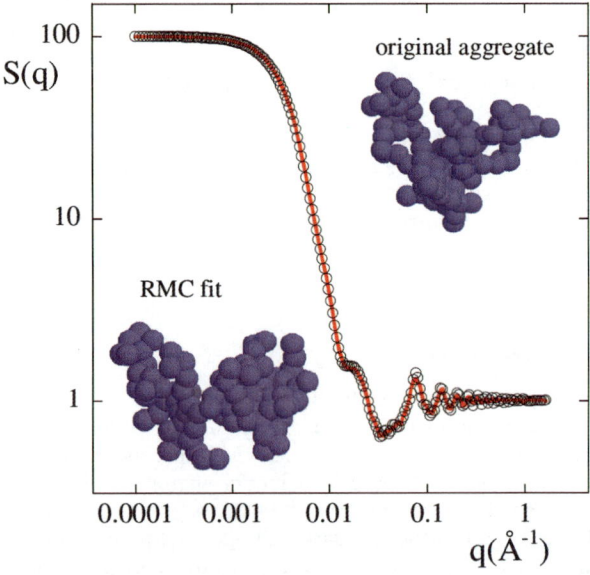

Fig. 9 Standard RMC analysis of the structure factor of a known aggregate (symbols) of 100 beads. The RMC-fit is the solid line. Upper inset: original aggregate configuration. Lower inset: RMC-snapshot compatible with $S(q)$.

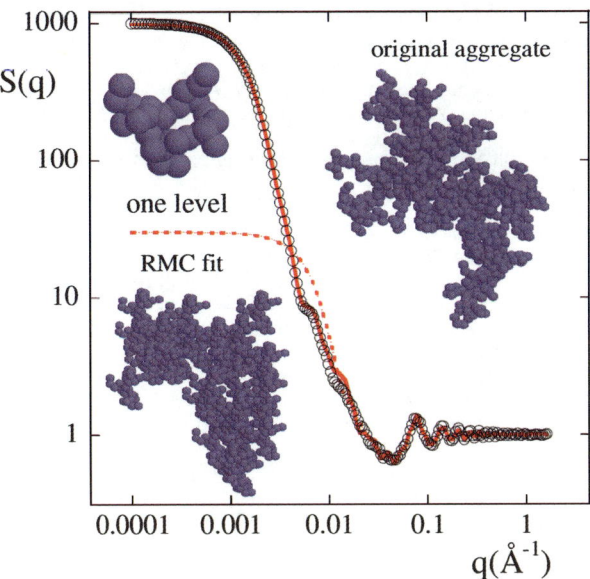

Fig. 10 Two-level RD-RMC analysis of a known aggregate of 1000 beads. The one-level (30 beads, dotted line) and two-level (33 first level aggregates, solid line) RMC-fits are compared to the total structure factor (black circles). In the upper inset on the right, the initial aggregate, on the left the first level one, and below the total one made of combinations of level one and two are shown.

($q > 0.01 \text{Å}^{-1}$). The agreement between the dotted curve and the structure factor is seen to be good in this q-range. The first level aggregate of 30 beads (shown in the upper left inset) was then used as an input for the two-level fit, of 33 aggregates. The result, now over the complete q-range, is equally satisfactory. In the lower inset, the reconstruction of the complete aggregate from level one and two is shown. It compares reasonably well with the original aggregate, shown in the top right inset.

Acknowledgements

We gratefully acknowledge financial support by the Swiss National Science Foundation, the Nestlé Research Center, Lausanne, Switzerland, the University of Fribourg, Switzerland, and the Adolphe Merkle Foundation. Julian Oberdisse thanks the Royal Society of Chemistry for an international author award financing his stay in Fribourg. Thomas Gibaud acknowledges a grant from the Agence National de la Recherche Française (ANR-11-PDOC-027). The small-angle neutron scattering experiments were performed at the instrument SANS I of the Swiss Neutron Source SINQ at the Paul Scherrer Institute (Villigen, Switzerland), and the instrument D11 at the Institute Laue-Langevin (Grenoble, France). The authors thank J. Kohlbrecher for his help during the measurements at SINQ.

References

1 R. Mezzenga, P. Schurtenberger, A. Burbidge and M. Michel, *Nat. Mat.*, 2005, **4**, 729.
2 E. Zaccarelli, *J. Phys.: Condens. Matter*, 2007, **19**, 323101.
3 K. A. Dawson, *Curr. Opin. Colloid Interface Sci.*, 2002, **7**, 218.
4 F. Sciortino, *et al.*, *Comput. Phys. Commun.*, 2005, **169**, 166.
5 W. C. K. Poon, *et al.*, *Faraday Discuss.*, 1995, **101**, 65.
6 F. Cardinaux, E. Zaccarelli, A. Stradner, S. Bucciarelli, B. Farago, S. Egelhaaf, F. Sciortino and P. Schurtenberger, *J. Phys. Chem. B*, 2011, **115**, 7227.

7 P. Pusey and W. van Megen, *Nature*, 1986, **320**, 340–342.
8 K. N. Pham, *et al.*, *Science*, 2002, **296**, 104.
9 V. Trappe and P. Sandkuhler, *Curr. Opin. Colloid Interface Sci.*, 2005, **18**, 494500.
10 S. Romer, F. Scheffold and P. Schurtenberger, *Phys. Rev. Lett.*, 2000, **85**, 4980.
11 F. Cardinaux, T. Gibaud, A. Stradner and P. Schurtenberger, *Phys. Rev. Lett.*, 2007, **99**, 118301.
12 E. B. Sirota, H. D. Ou-Yang, S. K. Sinha, P. M. Chaikin, J. D. Axe and Y. Fujii, *Phys. Rev. Lett.*, 1989, **62**, 1524.
13 P. Wette, *et al.*, *J. Chem. Phys.*, 2010, **132**, 131102.
14 A. Stradner, H. Sedgwick, F. Cardinaux, W. C. K. Poon, S. U. Egelhaaf and P. Schurtenberger, *Nature*, 2004, **432**, 492.
15 A. I. Campbell, V. J. Anderson, J. van Duijneveldt and P. Bartlett, *Phys. Rev. Lett.*, 2005, **94**, 208301.
16 H. Sedgwick, A. Kroy, K., Salonen, M. B. Robertson, S. U. Egelhaaf and W. C. K. Poon, *Eur. Phys. J. E*, 2005, **16**, 77–80.
17 C. J. Dibble, M. Kogan and M. J. Solomon, *Phys. Rev. E*, 2006, **74**, 041403.
18 V. Gopalakrishnan and C. F. Zukoski, *Phys. Rev. E*, 2007, **75**, 021406.
19 A. Shalkevich, A. Stradner, S. K. Bhat, F. Muller and P. Schurtenberger, *Langmuir*, 2007, **23**, 3570–3580.
20 F. Zhang, M. W. A. Skoda, R. M. J. Jacobs, R. A. Martin, C. M. Martin and F. Schreiber, *J. Phys. Chem. B*, 2007, **111**, 251–259.
21 W. Pan, O. Galkin, L. Filobelo, R. L. Nagel and P. G. Vekilov, *Biophys. J.*, 2007, **92**, 267–277.
22 O. Gliko, W. Pan, P. Katsonis, N. Neumaier, O. Galkin, S. Weinkauf and P. G. Vekilov, *J. Phys. Chem. B*, 2007, **111**, 3106–3114.
23 S. Boutet and I. K. Robinson, *Phys. Rev. E*, 2007, **75**, 021913.
24 N. Destainville, *Phys. Rev. E*, 2008, **77**, 011905.
25 N. Javid, K. Vogtt, C. Krywka, M. Tolan and R. Winter, *Phys. Rev. Lett.*, 2007, **99**, 028101.
26 L. Ianeselli, F. Zhang, M. W. A. Skoda, R. M. J. Jacobs, R. A. Martin, S. Callow, S. Prevost and F. Schreiber, *J. Phys. Chem. B*, 2010, **114**, 3776–3783.
27 T. Gibaud and P. Schurtenberger, *J. Phys.: Condens. Matter*, 2009, **21**, 322201.
28 P. J. Lu, E. Zaccarelli, F. Ciulla, A. B. Schofield, F. Sciortino and D. A. Weitz, *Nature*, 2008, **453**, 499.
29 T. Gibaud, F. Cardinaux, J. Bergenholtz, A. Stradner and P. Schurtenberger, *Soft Matter*, 2011, **7**, 857.
30 J. S. Pedersen, *J. Appl. Cryst.*, 2004, **37**, 369.
31 C. Gogelein, G. Nagele, R. Tuinier, T. Gibaud, A. Stradner and P. Schurtenberger, *J. Chem. Phys.*, 2008, **129**, 085102.
32 R. Klein and B. D'Aguanno, *Light Scattering, Principles and Development*, Oxford, 1996.
33 M. G. Noro and D. Frenkel, *J. Chem. Phys.*, 2000, **113**, 2941–2944.
34 M. H. G. Duits, R. P. May, A. Vrij and C. G. D. Kruif, *Langmuir*, 1991, **7**, 62.
35 X. Ye, T. Narayanan, P. Tong and J. S. Huang, *Phys. Rev. Lett.*, 1996, **76**, 4640.
36 E. Dubois, V. Cabuil, F. Boue and R. Perzynski, *J. Chem. Phys.*, 1999, **111**, 7147.
37 J. Oberdisse, P. Hine and W. Pyckhout-Hintzen, *Soft Matter*, 2007, **2**, 476.
38 R. McGreevy, *J. Phys.: Condens. Matter*, 2001, **13**, 877.
39 P. Debye, *Phys. Coll. Chem.*, 1947, **51**, 18.
40 S. A. Shah, Y. L. Chen, K. S. Schweizer and C. F. Zukoski, *J. Chem. Phys.*, 2003, **119**, 8747.
41 M. E. Cates, M. Fuchs, K. Kroy, W. C. K. Poon and A. M. Puertas, *J. Phys.: Condens. Matter*, 2004, **16**, S4861.

Protein structure and interactions in the solid state studied by small-angle neutron scattering

Joseph E. Curtis,[a] Arnold McAuley,[b] Hirsh Nanda[a] and Susan Krueger[*a]

Received 17th February 2012, Accepted 5th April 2012
DOI: 10.1039/c2fd20027a

Small-angle neutron scattering (SANS) is uniquely qualified to study the structure of proteins in liquid and solid phases that are relevant to food science and biotechnological applications. We have used SANS to study a model protein, lysozyme, in both the liquid and water ice phases to determine its gross-structure, interparticle interactions and other properties. These properties have been examined under a variety of solution conditions before, during, and after freezing. Results for lysozyme at concentrations of 50 mg mL^{-1} and 100 mg mL^{-1}, with NaCl concentrations of 0.4 M and 0 M, respectively, both in the liquid and frozen states, are presented and implications for food science are discussed.

Introduction

The role of freezing in food preservation is to prevent the growth of microorganisms and to slow chemical reactions to preserve the quality, nutrient content, texture, flavor and color of foods. The food industry must change continuously in response to consumer demands for a greater variety of high quality, convenient products at economic prices. As such, traditional methods of food preservation such as freezing need continual improvements to keep abreast of advances in technology. Thus, the freezing process and its affects on water, proteins, lipids, carbohydrates, vitamins and minerals in food must be well understood. The state diagram of foods[1] is complex because foods are complex. However, there are clear lines representing the freezing curve and glass transition that are similar to the state diagrams of simpler systems such as proteins in aqueous solutions. Thus, an understanding of the freezing process in more complex systems such as foods can be obtained by studying these simpler model systems.

During the freezing of a typical protein solution, only a fraction of the water molecules form the crystalline ice phase, whereas the remaining water molecules and other solutes present remain in the amorphous state, forming a freeze-concentrated solution, with a water concentration of around 30 percent by weight.[2] This water, confined in small areas, cannot overcome the large activation barrier that would allow it to diffuse into the ice crystal lattice. The higher the initial concentration of solute, the greater the mass fraction of water that remains unfrozen.[3] This freeze-concentration process is not new to the food science industry and, in fact, is a method used to concentrate liquid foods by allowing ice to form in the solution and then extracting the water, as ice, from the freeze-concentrated liquid phase.[4]

[a]NIST Center for Neutron Research, NIST, Gaithersburg, MD 20899, USA. E-mail: susan.krueger@nist.gov
[b]Department of Analytical and Formulation Sciences, Amgen Inc., One Amgen Center Drive, Thousand Oaks, California 91320-1799, USA

Many proteins are known to suffer inactivation upon freezing, which can occur due to dissociation, aggregation or other chemical mechanisms.[3] Proteins can also become denatured, or unfolded, as a result of freezing. The extent of denaturation of proteins upon freezing depends on several factors including the initial pH, the protein concentration, the temperature of the frozen part of the solution and the presence of other substances, such as salt or sugar, in the solution.[3] NaCl can inhibit protein denaturation down to $-30\,°C$, at which point the salt likely precipitates and, eventually, the freeze-concentrated liquid reaches the same composition as it would have in the absence of salt.[3] While structural changes in cold-denatured proteins have been studied using phosphorescence emission[5] and NMR,[6] structural studies of proteins in water ice are limited.

In this work, small-angle neutron scattering (SANS) was used to directly probe the structure and interactions of the model protein, lysozyme, in aqueous solution as a function of temperature and salt (NaCl) content. SANS probes structure and interactions on length scales of 10 Å to more than 1000 Å, making it a well-suited technique for the study of proteins in aqueous solution. Unlike X-rays, neutrons are particularly sensitive to hydrogen and the ability to substitute deuterium for hydrogen in the protein or aqueous solution makes it possible to determine the origin of features in the scattering curves. The SANS results, combined with those obtained from other techniques, were used to construct a picture of the frozen state of the lysozyme solution, including the location and aggregation state of the protein.

Materials and methods

Proteins and solutions

Hen egg lysozyme was purchased from Sigma† in powder form and used without further purification. Lysozyme solutions for SANS experiments were prepared in 99.9% D_2O (Cambridge Isotope Labs, Inc.) at protein concentrations of 50 mg mL^{-1} and 100 mg mL^{-1}. The 100 mg mL^{-1} solution was prepared with 0 mol L^{-1} (M) NaCl, whereas the 50 mg mL^{-1} solution was prepared with 0.4 M NaCl. The measured pD was near 7 for all samples.

Small-angle neutron scattering

SANS measurements were performed on the 30-meter SANS instruments[7] at the NIST Center for Neutron Research (NCNR) in Gaithersburg, MD. The neutron wavelength, λ, was 6 Å, with a wavelength spread, $\Delta\lambda/\lambda$, of 0.15. Scattered neutrons were detected with a 64 cm × 64 cm two-dimensional position-sensitive detector with 128 × 128 pixels at a resolution of 0.5 cm/pixel. The data were reduced using the IGOR program with SANS macro routines developed at the NCNR.[8] Raw counts were normalized to a common monitor count and corrected for empty cell counts, ambient room background counts and non-uniform detector response.

Data from the samples in the liquid and frozen states were placed on an absolute scale by normalizing the scattered intensity to the incident beam flux. Finally, the data were radially averaged to produce scattered intensity, $I(q)$, *versus* q curves, where $q = 4\pi\sin(\theta)/\lambda$ and 2θ is the scattering angle. A sample-to-detector distance of 1.3 m was used for measurements of lysozyme in D_2O to cover the range 0.03 $\text{Å}^{-1} \leq q \leq 0.4\ \text{Å}^{-1}$. The scattered intensities from the samples in the liquid state were then further corrected for buffer scattering and incoherent scattering from hydrogen in the samples. The buffer scattering could not be directly subtracted from the sample scattering in the frozen state due to the presence of additional

---

† Certain commercial equipment, instruments, materials, suppliers, or software are identified in this paper to foster understanding. Such identification does not imply recommendation or endorsement by the National Institute of Standards and Technology, nor does it imply that the materials or equipment identified are necessarily the best available for the purpose.

scattering at low q. In this case, the scattering from the samples in the frozen state were approximately corrected for background scattering by subtracting the scattering from the corresponding buffers at higher q values, where the buffer scattering was flat, and then subtracting a constant of the same magnitude from the scattering at lower q values.

Lysozyme solutions were loaded into demountable 2 mm path length titanium cells with titanium windows. Samples were measured at temperatures between 20 °C and −80 °C. The 0 M NaCl sample was cooled by placing it in a closed cycle refrigerator that was at 20 °C and slowly lowering the temperature, in increments, to 10 °C, 5 °C, 0 °C, −5 °C, −10 °C, −20 °C, −40 °C and −80 °C. The sample was allowed to remain at each temperature for 30 min before making a measurement and then moving to the next temperature. The actual measurement time at temperature was 5 min. The 0.4 M NaCl sample was cooled in the same way, except it was measured at −25 °C and −30 °C in addition to the temperatures above, since its freezing temperature is near −25 °C.

Data analysis

Data from samples in both the liquid and frozen states were fit using an empirical functional form for SANS data characterized by a broad scattering peak. This model function is part of the NCNR IGOR SANS data analysis package and curve fitting was accomplished using the non-linear curve fitting routine within IGOR. The scattered intensity, $I(q)$ is calculated using the relation:

$$I(q) = \frac{A}{q^n} + \frac{C}{1 + (|q - q_o|\xi)^m} + B \tag{1}$$

where A, B, and C are constants, n is the Porod exponent, m is the Lorentzian exponent and ξ is the Lorentzian screening length. The Porod exponent results mainly from the scattering at lower q values and is applicable when there is scattering in this region froFm particle aggregates. The aggregate can be a dimer, trimer or higher oligomer of the particle. Porod scattering can also result due to scattering from interfaces between co-existing phases. The Lorentzian parameters are derived mainly from the higher q region and are applicable when a broad scattering peak due to interactions between particles is present in the data. The average center-to-center distance between neighboring particles, d, also known as the d-spacing, is related to the peak position, q_o, as $d = 2\pi/q_o$. While many factors such as solution conditions can play a role in determining the actual distance between neighboring particles in solution, this approximation allows the change in particle concentration upon crowding to be followed in a simple manner. The constant, B, is a background term.

Results and discussion

SANS from 0 M NaCl lysozyme upon freezing

The SANS data from lysozyme at 100 mg mL^{-1} in D$_2$O with 0 M NaCl are plotted on a log(I) vs. log(q) scale in Fig. 1. At 20 °C, clear signs of interparticle interference, *i.e.*, interaction between the lysozyme particles, are seen in that the data show a downturn at the lowest q values. This downturn in the data at low q allows for the observation of a broad peak in the data that is centered at approximately $q = 0.1$ Å^{-1}. The presence of such a peak, and the absence of significant scattering at lower q values, suggests that the solution is fairly monodisperse and consists mainly of interacting lysozyme monomers. Such interactions can be due to the fact that proteins at higher concentration are more closely packed such that the solution has a higher degree of order. Thus, a d-spacing exists between neighboring proteins as defined above. Long-ranged, repulsive electrostatic interactions between proteins, especially in this case where there is no salt in the solution to screen them, can cause

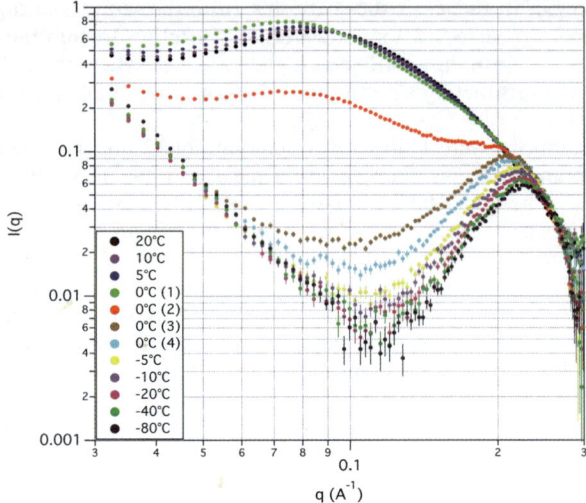

Fig. 1 SANS from lysozyme in 0 M NaCl D_2O solution as a function of temperature. Error bars represent plus and minus the combined standard uncertainty of the data collection.

a similar effect. As the temperature is slowly reduced from 20 °C to 0 °C, the data show a clear shift of the broad peak in the data to lower q values. Thus, the d-spacing becomes slightly greater as the solution cools to 0 °C. This is perhaps due to the expansion of water as it freezes, thus allowing for a greater distance between protein molecules at a given concentration. Alternatively, it could be due to temporal clustering of the protein molecules as the temperature is reduced.[9]

As the solution is equilibrated at 0 °C, the data show a reduction of the scattered intensity as a whole, as well as a shift of the broad scattering peak back to a higher q value and the appearance of an additional peak near $q = 0.2$ Å^{-1}, indicating that a subset of proteins are now much closer together in the solution. At this point, the solution is in a mixed state, with the scattering showing evidence of lysozyme with the same d-spacing as in the liquid state, as well as with a significantly smaller d-spacing. While it is assumed that this peak is correlated with the onset of freezing, this cannot be confirmed by the SANS data alone.

As the sample continues to freeze at 0 °C, the scattering shows only a well-defined peak near $q = 0.2$ Å^{-1}, along with sharply increasing scattering at lower q values. The peak near $q = 0.2$ Å^{-1} is a result of interacting lysozyme monomers that are closer together than in the liquid state, due to the freeze-concentration effect as ice forms in the solution. As the sample continues to freeze down to −80 °C, the peak continues to sharpen and shift to higher q values.

The observed peaks in the scattering occur due to the crowding of protein as the solution freezes. As ice forms, the remaining unfrozen water becomes freeze-concentrated in a separate phase. The structure of a solution of phosphate buffer saline (PBS) at a temperature of −26 °C has been described, using confocal Raman microscopy, as consisting of ice crystals surrounded by narrow channels and more rounded domains that contain unfrozen water.[10] This phase separation induces segregation of the protein, forming a region consisting of freeze-concentrated protein. While protein was observed in the ice phase of PBS containing both lysozyme and trehalose, the concentration of both protein and trehalose in the unfrozen water phase was about two orders of magnitude greater. Similarly, NMR experiments[11] have shown that a solute-rich liquid phase persists in a solution of bovine serum albumin, potassium fluoride and water down to temperatures as low as −100 °C. Fig. 1 shows that the formation of the freeze-concentrated protein regions can clearly be detected in the SANS data as the solution is frozen.

The increased scattering at lower q values arises from larger structures in the system. Such scattering was observed even in the absence of protein,[12] indicating that it is arising, at least partially, from the ice structure itself. A contrast variation series of experiments performed on $-40\ °C$ frozen water solutions of several mixtures of $D_2O : H_2O$ (by volume) in the absence of protein and salt showed an absence of scattering at a solution composition of 8% D_2O : 92% H_2O.[12] Under these conditions, the neutron scattering length density matches that of air. Thus, the scattering at low q values is due to cracks in the ice that create a contrast in neutron scattering length density between air and ice. The cracks are large features, explaining why the scattering occurs mainly at lower q values. However, a second contrast variation series performed on the same frozen water solutions with 100 mg mL^{-1} lysozyme and 0 M NaCl showed that the low q scattering was still present at a solution composition of 8% D_2O : 92% H_2O.[12] This means that there also must be scattering from large-scale protein structures in the frozen samples, in addition to the scattering from the cracks in the ice itself. It is not possible to determine from the SANS data alone whether these protein aggregates are in the ice or unfrozen water phase. Perhaps the protein is aggregating at the boundaries of the water and ice phases in addition to being forced into tight clusters due to the small space available in the water phase. If there is some protein in the ice phase, it could propagate to the edges of the cracks formed during the freezing process, where it could form large aggregates.

SANS from 0.4 M NaCl lysozyme upon freezing

The SANS data from lysozyme at 50 mg mL^{-1} in D_2O with 0.4 M NaCl are plotted on a log(I) vs. log(q) scale in Fig. 2. Unlike the 0 M NaCl data at 20 °C, there is no indication of a peak due to interparticle interference. While the concentration of this sample is one-half of the 0 M NaCl sample, a peak in the scattering should be easily visible if the solution consists of interacting lysozyme monomers.[12] Rather, the 0.4 M NaCl data show an upward slope at lower q values, which is usually an indication of the presence of higher order aggregates in the sample. Aggregation could be due to the fact that the salt is screening the long-ranged repulsive electrostatic interaction between proteins, allowing them to come closer together and form larger aggregates. The presence of aggregates in the solution means that the solution is no longer

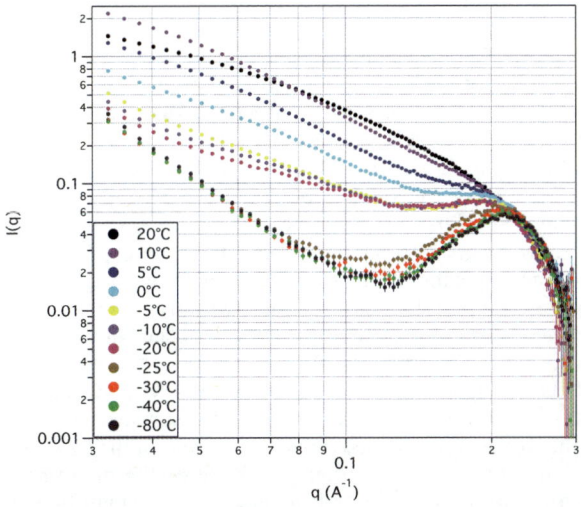

Fig. 2 SANS from lysozyme in 0.4 M NaCl D_2O solution as a function of temperature. Error bars represent plus and minus the combined standard uncertainty of the data collection.

monodisperse. This would also explain the absence of an interaction peak in the data, as there is no longer a well-defined average d-spacing between particles. As the temperature is slowly reduced from 20 °C to 5 °C, the data show a decrease in the scattering at lower q values, indicating that the population of large aggregates is decreasing, and the appearance of a peak near $q = 0.2$ Å^{-1}, indicating the existence of close-packed protein at temperatures above the freezing point of the water in the solution. The temperature at which this liquid–liquid phase separation occurs is consistent with that found in previous studies of lysozyme in 0.4 M NaCl solution.[13] However, it is uncertain what effect the existence of protein aggregates in the liquid state has on the packing of the proteins as the solution cools. The fact that a peak exists implies that there is a population of proteins in the solution that is more monodisperse than before. This correlates with the decrease in scattering at lower q values, suggesting that some of the larger aggregates originally in the solution are breaking up into smaller particles that are becoming more concentrated as the solution cools.

The low q scattering continues to decrease as the temperature is reduced to -20 °C. The shape of the scattering curve then changes dramatically below -20 °C, as the salt begins to crystallize out of solution (at -21 °C).[3] As the sample continues to freeze, by lowering the temperature down to -80 °C, the peak near $q = 0.2$ Å^{-1} continues to sharpen and shift to higher q values, indicating that the distance between proteins in the freeze-concentrated phase continues to decrease with decreasing temperature.

Modeling of protein–protein interactions

The SANS data for the 0 M NaCl and 0.4 M NaCl lysozyme solutions cooled from 20 °C to -80 °C were fit with the broad peak functional form defined in eqn (1). For the 0 M NaCl samples in the liquid state, the Porod exponent was fixed at 0.0001, since there is no scattering from large scale structures in the system. The average d-spacing between the lysozyme particles cannot be determined from the fit unless the scattering curve exhibits an interaction peak. Thus, q_o was fixed at 0 when no interaction peak was present, as was the case for the 0.4 M NaCl solution at temperatures above 5 °C. Fig. 3 shows an example of the best-fit broad peak curve along with the data for both solutions in the liquid and frozen states.

The resulting best-fit parameters for both the 0 M NaCl and 0.4 M NaCl lysozyme solutions at all temperatures are listed in Table 1. It can be seen from this information that the d-spacing decreases with decreasing temperature, as expected since the proteins are being forced closer together as the temperature is decreased. This is further illustrated in Fig. 4, which shows a plot of the d-spacing as a function of temperature for both solutions. There is an increase in d-spacing of the 0 M NaCl sample as it approaches the freezing point and a sharp decrease in d-spacing as the sample freezes at 0 °C. The d-spacing continues to decrease with decreasing temperature all the way down to -80 °C. A population of close-packed particles is not apparent in the 0.4 M NaCl solution until the temperature is lowered to 5 °C. The d-spacing has a large error above -25 °C because the interaction peak in the data is not sharp. It becomes sharper below the freezing point of the salt solution (-21 °C). Also, the d-spacing at the lowest temperature is slightly larger for the 0.4 M NaCl sample as compared to that for the 0 M NaCl sample.

The final concentration of protein in the freeze-concentrated phase is independent of the original protein concentration in the solution.[3,12] Yet, the spacing between lysozyme particles is slightly larger in the frozen 0.4 M NaCl solution all the way down to -80 °C, well below the freezing point of the salt solution. Thus, the packing of proteins in the freeze-concentrated phase is different in the two cases, perhaps due to a difference in protein shape. Either the shape of the monomer is different or perhaps there are both monomers and dimers or other small oligomers present in this phase, which may certainly be the case since there were aggregates in the system before freezing.

This journal is © The Royal Society of Chemistry 2012

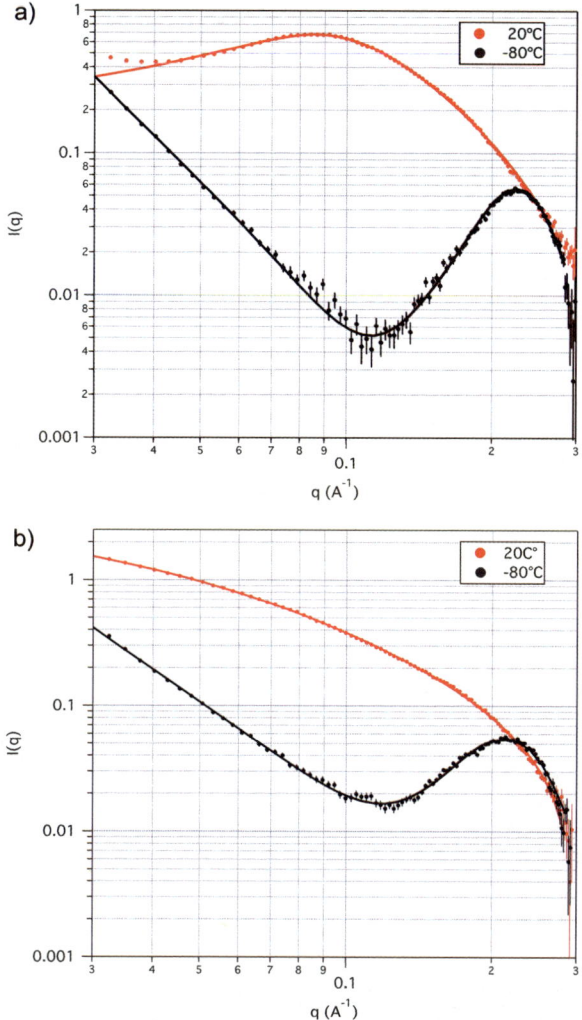

Fig. 3 SANS data from (a) 100 mg mL^{-1} lysozyme in 0 M NaCl D$_2$O solution and (b) 50 mg mL^{-1} lysozyme in 0.4 M NaCl D$_2$O solution in the liquid and frozen states. Solid lines in (a) and (b) represent best-fit curves to eqn (1). Error bars represent plus and minus the combined standard uncertainty of the data collection.

It can also be seen from Table 1 that the Porod exponent behaves differently upon freezing of the 0 M NaCl and 0.4 M NaCl solutions. This is further illustrated in Fig. 5, which shows a plot of the Porod exponent as a function of temperature for the two solutions. In this case, the Porod exponent arises due to scattering from a population of large particles in the solution and its value gives an idea of the morphology of these aggregates.[14,15] Lysozyme in 0 M NaCl solution does not significantly aggregate in the liquid state, as evidenced by the lack of significant scattering at low q values. However, lysozyme in 0.4 M NaCl solution in the liquid state has a Porod exponent around 1, which can be indicative of rod-like particles with lengths much longer than their widths. Given the lack of an interaction peak in these data, it is likely that the rod-like particles are polydisperse in length and/or width. However,

Table 1 Broad peak fitting parameters[a]

Concentration (mg mL^{-1})	NaCl (M)	T/°C	Porod Exponent	q_0 (Å^{-1})	d-spacing (Å)	χ^2
100	0	20.00 ± 0.05	0.0001	0.0859 ± 0.0001	73.14 ± 0.08	2.3
100	0	10.00 ± 0.05	0.0001	0.0830 ± 0.0002	75.7 ± 0.2	1.8
100	0	5.00 ± 0.05	0.0001	0.0793 ± 0.0001	79.2 ± 0.1	1.7
100	0	0.00 ± 0.05	0.0001	0.0738 ± 0.0001	85.1 ± 0.1	2.0
100	0	0.00 ± 0.05	0.0001	0.0736 ± 0.0002	85.4 ± 0.3	1.0
100	0	0.00 ± 0.05[b]	3	0.20 ± 0.03	31 ± 4	0.8
100	0	0.00 ± 0.05	3.13 ± 0.06	0.2062 ± 0.0002	30.47 ± 0.03	2.2
100	0	0.00 ± 0.05	3.27 ± 0.06	0.2113 ± 0.0002	29.74 ± 0.03	1.4
100	0	−5.00 ± 0.05	3.22 ± 0.06	0.2161 ± 0.0002	29.08 ± 0.03	1.3
100	0	−10.00 ± 0.05	3.11 ± 0.06	0.2181 ± 0.0002	28.81 ± 0.03	1.4
100	0	−20.00 ± 0.05	3.13 ± 0.05	0.2209 ± 0.0002	28.44 ± 0.02	1.1
100	0	−40.00 ± 0.05	3.05 ± 0.05	0.2237 ± 0.0003	28.09 ± 0.04	1.4
100	0	−80.00 ± 0.05	3.23 ± 0.05	0.2245 ± 0.0003	27.99 ± 0.04	1.5
50	0.4	20.00 ± 0.05	1.1 ± 0.2	0	—	1.3
50	0.4	10.00 ± 0.05	1.38 ± 0.02	0	—	2.5
50	0.4	5.00 ± 0.05	1.54 ± 0.01	0.2 ± 0.05	31 ± 6	3.8
50	0.4	0.00 ± 0.05	1.35 ± 0.01	0.2 ± 0.05	31 ± 6	2.0
50	0.4	−5.00 ± 0.05	1.27 ± 0.01	0.2 ± 0.05	31 ± 6	1.5
50	0.4	−10.00 ± 0.05	1.06 ± 0.01	0.2 ± 0.05	31 ± 6	1.9
50	0.4	−20.00 ± 0.05	0.83 ± 0.02	0.2 ± 0.05	31 ± 6	1.6
50	0.4	−20.00 ± 0.05[b]	2	—	—	
50	0.4	−25.00 ± 0.05	2.88 ± 0.03	0.2042 ± 0.0003	30.77 ± 0.05	1.4
50	0.4	−30.00 ± 0.05	2.81 ± 0.03	0.2086 ± 0.0003	30.12 ± 0.04	1.3
50	0.4	−40.00 ± 0.05	2.43 ± 0.06	0.2109 ± 0.0004	29.79 ± 0.05	1.2
50	0.4	−80.00 ± 0.05	2.59 ± 0.06	0.2127 ± 0.0004	29.54 ± 0.06	1.1

[a] Errors represent one standard deviation. [b] Indicates second population.

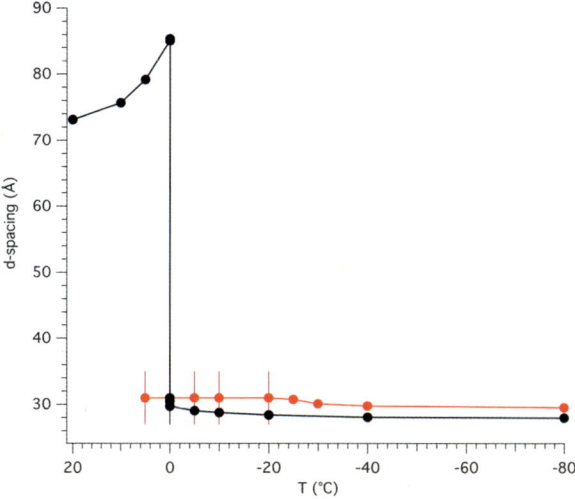

Fig. 4 Average *d*-spacing between lysozyme molecules as a function of temperature determined from fitting the data to eqn (1) for the 0 M NaCl (black) and the 0.4 M NaCl (red) samples. Error bars represent one standard deviation. The lines between points are included for clarity.

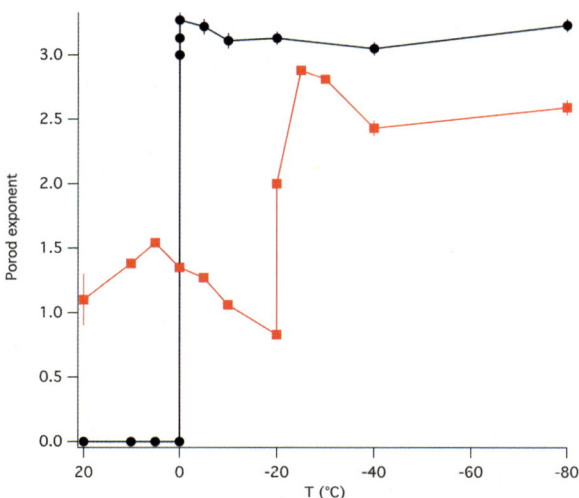

Fig. 5 Porod exponent for lysozyme as a function of temperature determined from fitting the data to eqn (1) for the 0 M NaCl (black) and the 0.4 M NaCl (red) samples. Error bars represent one standard deviation. The lines between points are included for clarity.

the fitted Porod region is based on the model assumed in eqn (1). The Porod exponent determined assuming a different fitting function could be different. Thus, a physical interpretation beyond the fact that there is scattering from large structures may not be warranted.

Upon freezing, the data from both the 0 M NaCl and 0.4 M NaCl show signs of larger aggregates in the system, in addition to the close-packed population of small particles. The Porod exponent for the 0 M solution is near 3 upon freezing. For the 0.4 M solution, the Porod exponent increases to a value between 2 and 3 upon freezing. However, as stated previously, contrast variation experiments have shown that the low q scattering in the frozen state also arises in part due to the formation of cracks in the ice structure. The differences observed in the Porod exponent are real and indicate the different morphology of the scatterers upon freezing. But, experiments performed in 8% D_2O solution, where the scattering from the cracks in the ice is negligible, would be necessary to isolate the scattering from the protein aggregates in the system.

Slow *versus* fast cooling

For the purposes of this work, slow cooling is defined as lowering the temperature in increments, from 20 °C down to −80 °C as described above, allowing the sample to equilibrate at each temperature increment for 30 min. On the other hand, fast cooling is defined as changing the temperature, in one increment, from 20 °C down to the desired frozen-state temperature (−80 °C in this case), and allowing the sample to cool directly to that temperature without stopping at any temperatures in between. Plunging the sample into liquid nitrogen or placing it directly into a freezer from room temperature would also be defined as fast cooling procedures for the purposes of this work. When the slow cooling procedure is followed, the two-dimensional SANS scattering pattern (prior to converting to $I(q)$ *vs.* q) is symmetric. However, when a fast cooling procedure is followed, the scattering pattern contains asymmetric "flares" that are indicative of an ordered structure at very long length scales relative to SANS, *i.e.*, microns or larger.[12] Such structures may arise due to strain on the sample when subjected to a large temperature change.[16]

Regardless of the cause of the asymmetric features, their main effect on the SANS data is to increase the scattering at low q values significantly. This also results in an apparent broadening of the interaction peak near $q = 0.2$ Å^{-1}. However, the location of the interaction peak, indicating the distance between the freeze-concentrated lysozyme particles, is not affected. This is illustrated in Fig. 6, which shows a $\log(I)$ *vs.* $\log(q)$ plot of the data from two 100 mg mL^{-1} lysozyme in 0 M NaCl D_2O solutions that were cooled to −80 °C using either the slow or fast cooling methods. Following the slow cooling method results in cleaner scattering curves and sharper interaction peaks in the frozen state, making it easier to fit the data to theoretical functions.

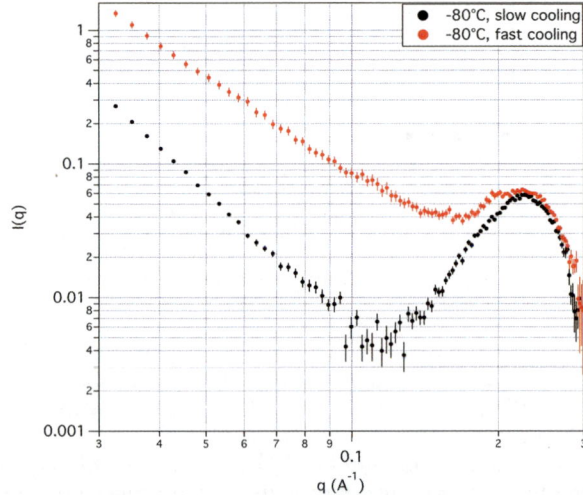

Fig. 6 SANS data from 100 mg mL^{-1} lysozyme in 0 M NaCl D_2O solution cooled to −80 °C from 20 °C by both the slow and fast cooling methods as described in the text. Error bars represent plus and minus the combined standard uncertainty of the data collection.

Multiple freezing and thawing cycles

In order to assess the effect of multiple freezing and thawing cycles on the lysozyme solutions, both solutions were cooled to −80 °C from 20 °C multiple times, using both the fast and slow cooling methods as described above. Fig. 7a shows log(I) vs. log(q) plots of the data from the 100 mg mL^{-1} lysozyme in 0 M NaCl D$_2$O solution that was cooled to −80 °C, warmed to 20 °C and then cooled again to −80 °C using the slow cooling method. It can be seen that the scattering curve is reproducible at all q values. This was found to be the case after several additional cooling cycles. Fig. 7b shows log(I) vs. log(q) plots of the data from the same system at 20 °C after several freeze–thaw cycles to final temperatures ranging from −20 °C to −80 °C using both the slow and fast cooling methods. Again, the data are essentially reproducible at all q values.

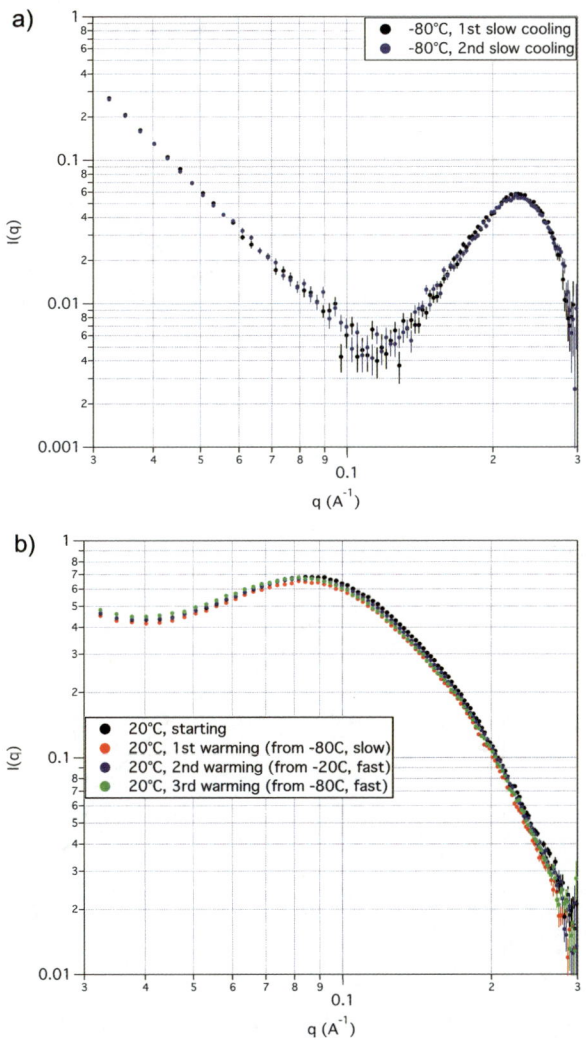

Fig. 7 (a) SANS data from 100 mg mL^{-1} lysozyme in 0 M NaCl D$_2$O solution cooled to −80 °C from 20 °C by the slow cooling method as described in the text. (b) SANS data from 100 mg mL^{-1} lysozyme in 0 M NaCl D$_2$O solution warmed to 20 °C from −80 °C by both the slow and fast methods as described in the text. Error bars in (a) and (b) represent plus and minus the combined standard uncertainty of the data collection.

On the other hand, the scattering data from the 50 mg mL^{-1} lysozyme in 0.4 M NaCl solution was not reproducible upon re-freezing. This is evident from Fig. 8a, which shows log(I) *vs.* log(q) plots of the data after the first and second cooling cycles using the slow cooling method. Fig. 8b shows log(I) *vs.* log(q) plots of the data from the same system at 20 °C after thawing the sample from −80 °C using the slow cooling method. It is clear that the original SANS curve was not reproduced after even one freeze–thaw cycle.

The SANS data obtained from the lysozyme solutions after several freeze–thaw cycles are reproducible at 20 °C and −80 °C for the 0 M NaCl solution, but not for the 0.4 M NaCl solution. This indicates that the aggregates formed during freezing of the 0 M NaCl solution are reversible upon thawing. However, the 0.4 M NaCl solution already contains aggregates before the first freeze–thaw cycle

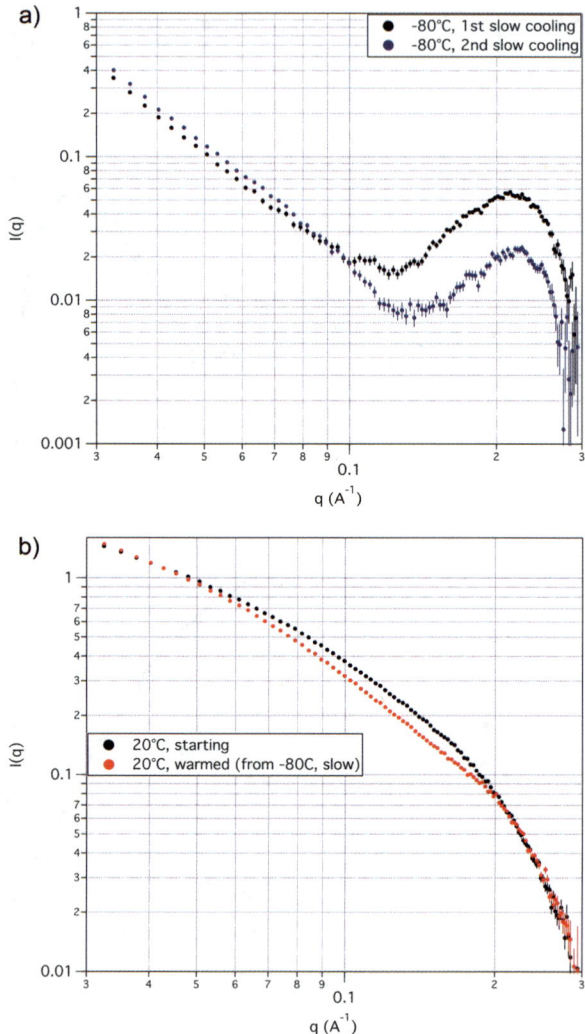

Fig. 8 (a) SANS data from 50 mg mL^{-1} lysozyme in 0.4 M NaCl D$_2$O solution cooled to −80 °C from 20 °C by the slow cooling method as described in the text. (b) SANS data from 50 mg mL^{-1} lysozyme in 0 M NaCl D$_2$O solution warmed to 20 °C from −80 °C by the slow cooling method as described in the text. Error bars in (a) and (b) represent plus and minus the combined standard uncertainty of the data collection.

and the solution is polydisperse. The nature of the polydispersity changes during the freeze–thaw cycles such that the original population of aggregates is not reproduced during subsequent freezing and thawing. However, in both cases, the aggregates formed are weakly associated with each other, as they either break up completely upon thawing, as in the 0 M NaCl solution, or they break up and reform differently upon thawing, as in the 0.4 M NaCl solution. This is consistent with size exclusion chromatography data that showed no significant aggregates in either solution before and after freeze–thawing, indicating that the aggregation is reversible.[12]

Proteins that have a larger equilibrium binding constant than lysozyme are more highly associative and may respond very differently upon freezing and thawing. Such proteins would tend to form more stable aggregates when subjected to freeze-concentration. The rates of chemical modifications would also tend to increase in such proteins, irrespective of the aggregation state. Thus, undesirable effects such as permanent aggregates, harmful chemical modifications and protein denaturation may exist upon thawing of such proteins.

Conclusions

Based on the information obtained from the SANS data of the 100 mg mL^{-1} lysozyme solution with 0 M NaCl, along with information obtained by others using Raman scattering[10] and NMR,[11] as well as a basic knowledge of the behavior of water upon freezing,[3] a cartoon of the morphology of the sample in the frozen state can be constructed. As drawn in Fig. 9, the frozen state consists mainly of ice crystals, with regions of amorphous water containing freeze-concentrated protein. The SANS data clearly show that large structures are present in the system. While some of this scattering is due to the scattering from large cracks in the ice matrix, SANS contrast variation experiments[12] showed that there is a contribution from protein aggregates. It is speculated that these aggregates form at either the ice–water or ice–air interfaces. The Raman results showed that the majority of the protein exists in the amorphous water phase.[10] Thus, while some of the protein aggregates

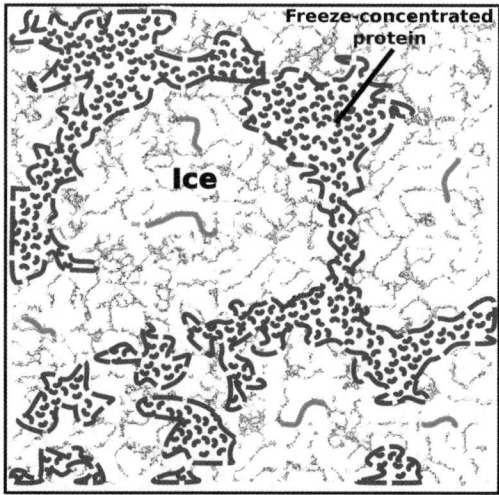

Fig. 9 Cartoon of the morphology of 100 mg mL^{-1} lysozyme in 0 M NaCl D$_2$O solution at −80 °C (based on Dong et al.[10]). The ice and freeze-concentrated protein phases are labeled. The dark grey lines represent large protein aggregates at the boundary between the ice and freeze-concentrated protein regions. The lighter grey lines in the ice phase represent possible protein aggregates trapped at the ice–air interface. Dimensions of the freeze-concentrated proteins and protein aggregates are enlarged for clarity.

may be trapped in the ice phase at the ice–air interface, the majority of the protein is drawn to be at the ice–water interface in Fig. 9. The remaining protein is shown to exist as close-packed monomers in the unfrozen amorphous water region.

The morphology of the 50 mg mL^{-1} lysozyme solution with 0.4 M NaCl is more complicated. The SANS data show that lysozyme doesn't exist as a monomer in solution. Rather, a distribution of sizes of aggregates is likely present. However, the data also show a liquid–liquid phase separation at 5 °C, well above the freezing temperature of the water component in the solution. The SANS data further indicate that the d-spacing is similar to that obtained for the 0 M NaCl solution at 0 °C and only decreases slightly below -20 °C, when the salt is frozen out of solution. This begs the question as to the mechanism of the "freeze-concentration" effect in both the 0 M NaCl and the 0.4 M NaCl samples. Is it mainly driven by a liquid–liquid phase separation as the temperature is lowered and then "frozen" into place as ice forms in the system and traps the protein in the amorphous water phase? Does this explain the slightly larger d-spacing observed for the 0.4 M NaCl sample from the onset of the phase separation all the way down to -80 °C? Further studies on the freezing of several different proteins in several different salt solutions should shed additional light on this issue.

Future directions

Neutron techniques are currently being explored for their potential contributions to food science, as evidenced by two recent meetings[17,18] and a recent paper on the subject.[19] SANS is being used to study the nanoscale structure of systems of interest to the food science community, such as food proteins, starch granules, polymers and emulsions.[19] The vast majority of these studies are performed on systems in the liquid state. However, SANS is also well-suited to the study of proteins in other bio-technologically-relevant phases. This work shows that SANS can potentially answer some important questions regarding the structure and interactions of food constituents, such as proteins, in the solid state. SANS can be used to ascertain the nature of interactions between proteins in both the liquid and solid states and under a variety of different solvent conditions. Moreover, SANS can be used to determine whether the cooling rate or multiple freeze–thaw cycles affect these interactions.

It is more difficult to ascertain the shape of the protein under conditions where it is interacting with other proteins in the solution. This hampers the ability to determine whether there is a shape change in the protein upon freezing, such as that which may occur due to denaturation, for instance. Some limited information can be obtained using alternative fitting functions that take the shape of the particle into account. However, such fitting methods are limited to those shapes that can be described in a functional form.

Work is currently underway to use real-space, all-atom computational modeling techniques to model the SANS data from close-packed protein systems. In this case, the actual structure of the protein is being used, as defined from X-ray crystallography or NMR spectroscopy. Typically, these structures are verified or refined based on SANS data from the protein in dilute solution. The solid state can then be modeled *in silico* and compared to SANS data, allowing the shape of the protein to be considered as the proteins become more concentrated and pack together more closely. To this end, it would be useful to repeat the current studies on proteins that are more asymmetric than lysozyme to determine the limits to this approach.

Contrast variation is a powerful method that can be useful for obtaining additional information about the nature of the low q scattering in the SANS data. By adjusting the $D_2O : H_2O$ ratio in the solvent, measurements can be made under conditions where either the scattering from large protein structures or from cracks in the ice are negligible. This allows the separation of the scattering from those two components to allow a better understanding of the nature of each. Contrast variation can also be used in a similar manner to eliminate or enhance the scattering contribution from many salts and sugars. Thus, further work can be done on the 0.4 M NaCl lysozyme solution to determine if the contribution to the scattering from the salt component can be separated from that of the protein and ice components.

Finally, SANS can be used to study proteins in other solid states. The scattering from freeze-dried protein samples can be measured to determine the nature of the protein interactions in the system. It is more difficult to use contrast variation in this case since there is no bulk water in the system. However, if the protein is freeze-dried in the presence of sugars, then deuterated sugars protein can be used to enhance the scattering from the protein in order to determine the effect of the sugar on the protein interactions.

Acknowledgements

This work utilized facilities supported in part by the National Science Foundation under Agreement No. DMR-0944772.

References

1 M. S. Rahman, *Trends Food Sci. Technol.*, 2006, **17**, 129–141.
2 F. Franks, *Pure Appl. Chem.*, 1997, **69**, 915–920.
3 F. Franks, *Biophysics and Biochemistry at Low Temperatures*, Cambridge University Press, London, 1985.
4 J. Welti-Chanes, D. Bermudez, A. Valdez-Fragoso, H. Mujica-Paz, and S. M. Alzamora, in *Handbook of Food Science, Technology and Engineering*, Taylor and Francis, Boca Raton, 2006, vol. 3, p. 106-1–106-8.
5 G. B. Strambini and E. Gabellieri, *Biophys. J.*, 1996, **70**, 971–976.
6 C. R. Babu, V. J. Hilser and A. J. Wand, *Nat. Struct. Mol. Biol.*, 2004, **11**, 352–357.
7 C. J. Glinka, J. G. Barker, B. Hammouda, S. Krueger, J. J. Moyer and W. J. Orts, *J. Appl. Crystallogr.*, 1998, **31**, 430–445.
8 S. R. Kline, *J. Appl. Crystallogr.*, 2006, **39**, 895–900.
9 P. Falus, L. Porcar, E. Fratini, W.-R. Chen, A. Faraone, K. Hong, P. Baglioni and Y. Liu, *J. Phys.: Condens. Matter*, 2012, **24**, 064114.
10 J. Dong, A. Hubel, J. C. Bischof and A. Aksan, *J. Phys. Chem. B*, 2009, **113**, 10081–10087.
11 J. E. Ramirez, J. R. Cavanaugh and J. M. Purcell, *J. Phys. Chem.*, 1974, **78**, 807–810.
12 J. E. Curtis, H. Nanda, S. Khodadadi, M. Cicerone, H. J. Lee, A. McAuley and S. Krueger, *J. Phys. Chem. B*, 2012, DOI: 10.1021/jp304772d.
13 V. G. Taratuta, A. Holschbach, G. M. Thurston, D. Blankschtein and G. B. Benedek, *J. Phys. Chem.*, 1990, **94**, 2140–2144.
14 J. Teixeira, *J. Appl. Crystallogr.*, 1988, **21**, 781–785.
15 O. Glatter and O. Kratky, *Small-Angle X-Ray Scattering*, Academic Press, New York, 1982.
16 Y. Rabin, P. S. Steif, K. C. Hess, J. L. Jimenez-Rios and M. C. Palastro, *Cryobiology*, 2006, **53**, 75–95.
17 Neutrons and Food Workshop, Sydney, Australia 31 October–3 November 2010, http://www.nbi.ansto.gov.au/neutronsandfood/, accessed 2012-02-07.
18 Neutrons and Food 2012, Delft, http://neutronfood.tudelft.nl/, accessed 2012-02-07.
19 A. Lopez-Rubio and E. P. Gilbert, *Trends Food Sci. Technol.*, 2009, **20**, 576–586.

The role of quench rate in colloidal gels

C. Patrick Royall[a] and Alex Malins[ab]

Received 1st March 2012, Accepted 4th May 2012
DOI: 10.1039/c2fd20041d

Interactions between colloidal particles have hitherto usually been fixed by the suspension composition. Recent experimental developments now enable the control of interactions *in situ*. Here we use Brownian dynamics simulations to investigate the effect of controlling interactions upon gelation, by "quenching" the system from an equilibrium fluid to a gel. We find that, contrary to the normal case of an instantaneous quench, where the local structure of the gel is highly disordered, controlled quenching results in a gel with a much higher degree of local order. Under sufficiently slow quenching, local crystallisation is found, which is strongly enhanced when a monodisperse system is used. The higher the degree of local order, the smaller the mean squared displacement, indicating an enhancement of gel stability.

1 Introduction

Gelation is among the most striking features of soft matter.[1–3] Although many everyday materials are readily classified as gels, from toothpastes to yoghurts, a deep understanding of the gel state remains a challenge. In particular, three questions that might be tackled are: (i) what distinguishes a gel from a glass, (ii) which materials can form gels, and (iii) for a given material, what are the requirements for gelation? This work, which explores controlled quenching in colloidal gels with Brownian dynamics simulations is relevant to question (iii).

For the first question, the identification of gelation with the crossing of a liquid–gas spinodal[1,4,5] provides a working definition to distinguish gels from other dynamically arrested states, namely glasses. For the second, it seems that an effective attraction is required. Systems with short-ranged attractive interactions (up to around 10% of the particle size) are well known to undergo gelation.[1–3,42,43] In the case of longer-ranged attractions in which the liquid state is stable, shallow quenches below the gas–liquid spinodal lead to complete phase separation.[6,7] Systems with short-ranged attractions *and* long-ranged repulsions can also exhibit similar behaviour.[1,8–10] However, the long-ranged repulsion significantly complicates the energy landscape, leading, for example, to degenerate mesophases.[11–13] Hereafter, we focus on systems without long-ranged repulsions, so the gels we consider are explicitly thermodynamically metastable.

Almost all gel literature concerns soft materials, that is to say, multicomponent systems of one or more mesoscopic component (colloids, nanoparticles, polymers) suspended in a fluid. However, gel-forming systems are often treated by integrating out the degrees of freedom of the smaller components (solvent molecules, small ions, polymers, *etc*) and considering an effective one-component system.[14] A reasonable question then is can true one-component (molecular) systems form gels? Although to our knowledge no experiment has yet been performed, C_{60} exhibits a property associated with a good gel-former, namely an attraction whose range is short

[a]School of Chemistry, University of Bristol, Bristol, BS8 1TS, UK
[b]Bristol Centre for Complexity Science, University of Bristol, Bristol, BS8 1TS, UK

compared to the molecular size. C_{60} is predicted to form a gel,[15] and even the longer-ranged Lennard-Jones model (relative to the molecular diameter) will form a gel under a sufficiently deep quench.[6]

This brings us to the third question - the requirements for gelation in a given material, which forms the subject of this article. The gels we consider are intrinsically non-equilibrium and, therefore, the means of preparation is important. Here we shall consider a system with a relatively short-ranged attraction, in which the equilibrium state is gas–crystal phase coexistence. Forming a gel, therefore, means avoiding phase separation to a gas and a crystal. Moreover, the resulting state must be dynamically arrested, which also requires that it percolates.[16] Possible routes to gelation are illustrated in Fig. 1. The most obvious route is quenching, reducing the (effective) temperature. Quenching must be carried out with sufficient speed that phase separation does not occur. For soft matter systems, this is often straightforward.

In colloids, for example, where the absolute temperature is typically held at 298 K, *effective* temperature is varied by changing the interaction strength. The effective temperature is then related to the depth of the attractive well of the interaction potential, and is often fixed for a given sample. Scanning a phase diagram then requires preparation of a considerable number of different samples, each with its own effective temperature. Following preparation of a sample, for example by shear, the system may be said to have undergone an (uncontrolled) instantaneous quench. A typical example is a colloid–polymer mixture, where the effective temperature is set by the polymer concentration (through the depletion interaction).[17] In this case, the colloid–colloid interactions are mediated by the polymer on much shorter timescales than the larger (and slower) colloids, and it is reasonable to consider an instantaneous quench as a route to gelation.

At smaller lengthscales, in nanoparticle and molecular systems, two main effects come into play. Firstly, interactions are often reasonably constant over the temperature range of interest, but temperature is used as a control parameter. A consequence then is that quench rates *cannot* be instantaneous, as is often the case for colloids. This is significant, as in the case of small nanoparticles and molecular systems the dynamics are often fast enough that quench rates that avoid phase separation can be technically challenging.[15] In this case, "crunching", where a gas of clusters is compressed, remains a route by which gels can be realised.

However, since gels are usually formed of soft matter with mescoscopic components, it is natural that most work on gelation assumes that the interactions between the particles were fixed. In other words, that quenches are instantaneous.

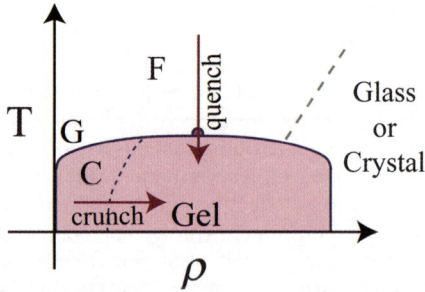

Fig. 1 Routes to gelation in short-ranged attractive systems. Gelation results upon quenching, provided the quench rate is sufficient to avoid phase separation (shaded region). Crunching of isolated clusters provides an addition route to form a gel, where accessible quench rates are too slow to prevent demixing. Here "C" denotes isolated clusters, "G" gas, "F" (supercritical) fluid.

This is typically the case in both experiments[1,3,42,43] and computer simulation.[1,10,18–20] Recent developments in controlling colloid–colloid interactions allow the interactions (and thus the effective temperature) to be changed at will, even on timescales much faster than the colloid dynamics, such that controllable quenches may be carried out.[21] It has also become possible to control attractive interactions between colloids such as temperature-dependent depletion attractions[22,23] multiaxial electric fields[24] and the critical Casimir effect.[25–27] This opens the way to consider the role of controlling effective temperature (by changing interactions) in an experiment and thus enables us to consider the role of quench rate in colloidal gelation. Varying the quench rate has been carried out in gelation in molecular systems.[15] There, using molecular dynamics simulations, a high quench rate was found to be necessary to prevent phase separation to equilibrium gas-crystal coexistence. Here we shall use Brownian dynamics simulations to model colloidal gelation, in which we shall consider the effect of quench rate upon the gels formed.

This paper is organised as follows. First we introduce the model system, which is a representation of a colloid–polymer mixture using Brownian dynamics simulations.[28] We then consider the response of the system to a conventional treatment of an instantaneous quench where the effective temperature is fixed at the outset. The system is characterised through a novel analysis of local structure we have developed, the topological cluster classification (TCC).[28,29] The effect of quench rate is then presented in both polydisperse and monodisperse systems. Finally we discuss the implications of our findings.

2 Methods

2.1 Model

We model a colloid–polymer mixture with Brownian dynamics simulations. In the experimental system upon which we base our simulations, residual electrostatic interactions are screened[28,30] and are neglected here. We found good agreement with simulations using the Morse potential.[28] The Morse potential reads:

$$\beta u(r) = \beta \varepsilon \exp(-\rho_0(\sigma - r))(\exp(-\rho_0(\sigma - r)) - 2) \tag{1}$$

where $\beta = 1/k_B T$, the thermal energy. The potential is truncated and shifted at $r = 1.4\sigma$. The effective temperature is controlled by varying the well depth of the potential ε. We set the range parameter $\rho_0 = 25.0$. Colloidal particles are typically polydisperse, and here we treat polydispersity by scaling r in eqn (1) by a Gaussian distribution in σ with 4% standard deviation (the same value as the size polydispersity in the experimental system). In addition, we consider the case of a monodisperse system. We express time in units of the Brownian time $\tau_B = (\sigma/2)^2/6D$,which is the time taken for a particle to diffuse its own radius. Here D is the diffusion constant. In the simulations, $\tau_B \approx 894$ time units. The time step is 0.03 simulation time units. Where the effective temperature is fixed throughout the simulation (Section 3.1), the runs are equilibrated for 5.0×10^6 steps and run for a further 5.0×10^6 steps. These simulation runs therefore correspond to approximately 168 Brownian times, or around 10 min, a timescale certainly comparable to experimental work. Except in the case of long quench times ($1.01 \times 10^4 \tau_B$), where 2048 particles were used, the system size was 10 000 particles. We fix the packing fraction at $\phi = \pi\rho\sigma^3/6 = 0.1$ throughout. At this packing fraction, quenching below the critical point, which is approximately located at ($\beta\varepsilon^c \approx 2.69$),[31] leads to a percolating network - i.e. a gel. The critical point gives a reasonable estimate for gelation, since the spinodal line is found to be very flat for short-ranged systems.[32] In the case of simulations where a quench is carried out, $\beta\varepsilon$ is increased linearly from unity to a chosen value, at which point the system is run for 5×10^6 time steps and analysed.

2.2 Structural analysis – topological cluster classification

To analyse the structure, we identify the bond network assuming particles closer than 1.4 σ are bonded. Having identified the bond network, we use the topological cluster classification (TCC) to determine the nature of the cluster.[29,34] This analysis identifies all the shortest path three, four and five membered rings in the bond network. We use the TCC to find clusters that are global energy minima of the Morse potential for the range we consider ($\rho_0 = 25.0$), as listed in ref. 33 and illustrated in Fig. 2. In addition we identify the thirteen particle structures that correspond to FCC and HCP, in terms of a central particle and its twelve nearest neighbours. For more details see ref. 29 and 34.

3 Results

The results section is organised as follows. We begin by presenting a brief overview of the effect of quenching. We consider the conventional case of instantaneous quenching, and provide an analysis from the perspective of local structure. Next we investigate a finite quench rate, and the effect of varying quench rate on the resulting structure and dynamics of the gel. Having established the general nature of quenching on the local structure, we examine the special case of a monodisperse system. Dynamic data is then employed to investigate relative stabilities of the two routes to gelation – instantaneous quenching and quenching at finite rates.

An overview of the system is shown in Fig. 3. These snapshots show the network structure that is developed by colloidal gels at a packing fraction $\phi = 0.1$ and attraction $\beta\varepsilon = 5.0$. While both are networks, the detail of the local structure is markedly different, as revealed by the topological cluster classification. The instantaneously quenched (polydisperse) system [Fig. 3(a)] is dominated by a 10 membered fivefold symmetric local structure, while the monodisperse system with a quench time of 671 τ_B shows a significant degree of local crystallisation but still retains a similar overall network structure. A slight degree of coarsening is apparent in the static

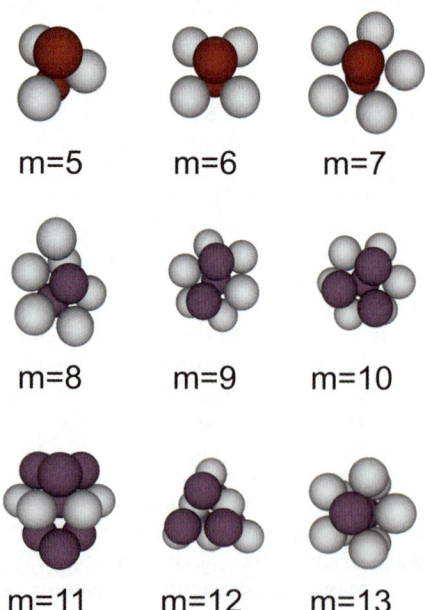

Fig. 2 Clusters detected by the topological cluster classification. These structures are minimum energy clusters of the Morse potential with $\rho_0 = 25.0$.[33]

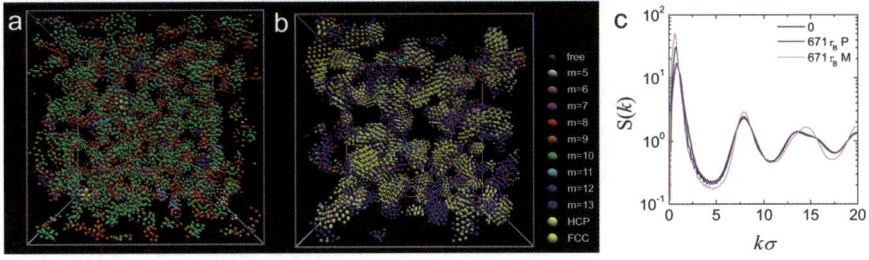

Fig. 3 Simulation snapshots labelled by the topological cluster classification. (a) Polydisperse system instantaneously quenched. (b) Monodisperse system with a quench time of 671 τ_B. Both (a) and (b) correspond to an identical state point with $\beta\varepsilon = 5.0$. The color of particles on the right of (b) denotes the cluster in which the particle is identified by the TCC as shown in Fig. 2. Note the vastly increased population of particles in crystalline environments in the case of the monodisperse system with a quench time of 671 τ_B. (c) The static structure factor $S(k)$ for state points are shown in (a) (purple line) and (b) (pink line) along with a polydisperse system with a quench time of 671 τ_B (navy line).

structure factor $S(k)$ in Fig. 3(c) relative to the instantaneous quench. We also note that, for the same quench time, a monodisperse system appears to coarsen somewhat more than a polydisperse system. However, the change in $S(k)$ appears much less dramatic than that in local structure illustrated in Fig. 3(a) and (b).

3.1 Instantaneous quenches in polydisperse systems – a local structural analysis

We now examine instantaneous quenches in more detail. A conventional structural analysis is presented in Fig. 4(a) in the form of the static structure factor $S(k)$. This illustrates the typical behaviour of systems undergoing dynamical arrest: the structure in the fluid ($\beta\varepsilon \leq 3$) responds to changes in effective temperature, in that the low k region of $S(k)$ increases strongly as phase separation is approached. Conversely, upon dynamical arrest for $\beta\varepsilon \geq 5$, no further change in the static structure is seen. Thus, one might conclude that under instantaneous quenching, all gels with $\phi = 0.1$ have the same local structure.

In fact this is not so, as shown in Fig. 4(b). Here the topological cluster classification is shown for the same data. This analysis reveals the generic nature of gelation

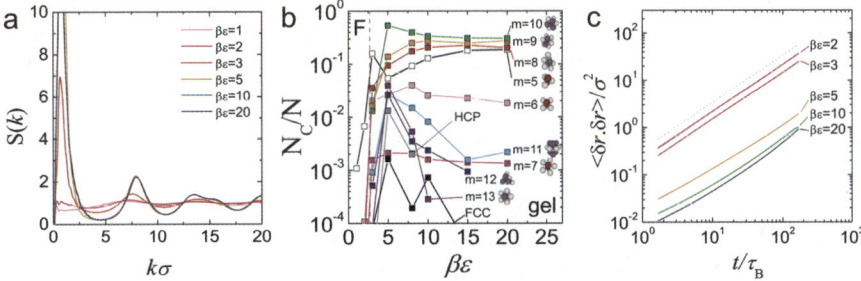

Fig. 4 The conventional situation: instantaneous quench. (a) The static structure factor ($S(k)$) for varying strengths of attraction. Where gelation occurs on the simulation timescale ($\beta\varepsilon > 3$), the $S(k)$ are indistinguishable for practical purposes. (b) The topological cluster classification reveals changes in local structure upon quenching. Dashed line denotes the strength of attraction at criticality ε^c which approximately locates gelation. N_c is the number of particles in a given cluster, N is the total number of particles. (c) The mean squared displacement for the well depth $\beta\varepsilon$ (inverse effective temperature) indicated. Dashed grey line has a slope of unity (diffusive motion).

at the single-particle level: at high effective temperatures, in the ergodic fluid phase, few clusters are found. Those few that are present are predominantly five-membered triangular bipyramids. Upon gelation, the cluster population rises very sharply. This indicates that the particles condense into clusters, which comprise the gel network [Fig. 3(a) and (b)]. However, it is the type of cluster that reveals the local structure of the gel. Although the equilibrium state is gas–crystal coexistence, even at the level of the few particles associated with a cluster, the kinetic pathway to crystallisation is arrested and very few particles are found in a local HCP or FCC environment. Upon deeper quenching, the system moves further from equilibrium, and even fewer particles in local crystalline environments are found. Rather than crystalline environments, particles are found in clusters of between 8 and 10 particles. As illustrated in Fig. 2, these are built around pentagonal bipyramids and are thus five-fold symmetric. Such five-fold symmetry has long been believed to suppress crystallisation.[35] We note that these same clusters are prevalent in dense fluids of both Morse particles and hard spheres.[36] This feature of crystallisation suppressed by five-fold symmetric clusters is found also in experiments on colloidal gels,[28] simulations of molecular gels,[15] and isolated clusters,[37] where in the case of the latter, the ground state is one of the clusters considered by the TCC. However, suppression of crystallisation is not universal, experiments in which locally crystalline clusters are found have also been reported.[38]

The change in dynamics upon gelation for instantaneous quenches is shown in the mean squared displacement (MSD) data in Fig. 4(c). Two observations are important concerning this plot. The first is that all state points are diffusive. There is little indication of a plateau that one might associate with slow dynamics. The second is that although in the fluid state, the system is rather insensitive to the effective temperature with only a slight decrease in MSD upon increasing the strength of the attraction. Upon gelation there is a sudden drop in mobility of more than an order of magnitude. Although the MSD indicates diffusive motion, the magnitude of the motion is small, only reaching the particle lengthscale after around 100 τ_B. As significant as this drop in mobility is, some restructuring cannot be ruled out, indicating that these gels age.

3.2 Finite quenches in polydisperse systems - structure

Let us consider finite quench rates. We shall use our local structural analysis, the TCC. As might be surmised from Fig. 4(a), the static structure factor is not very sensitive to quench rate. TCC analyses of data for quench times of $671\tau_B$ and $1.01 \times 10^4\,\tau_B$ are shown in Fig. 5(a) and (b) respectively. In the corresponding experimental system, the quench times are around 30 min and 8 h respectively. For comparison, instantaneous quench data is plotted in Fig. 5 (c). Here we quench from a fluid at $\beta\varepsilon = 1$ to the final value of $\beta\varepsilon$ as shown in Fig. 5(a) and (b). Each

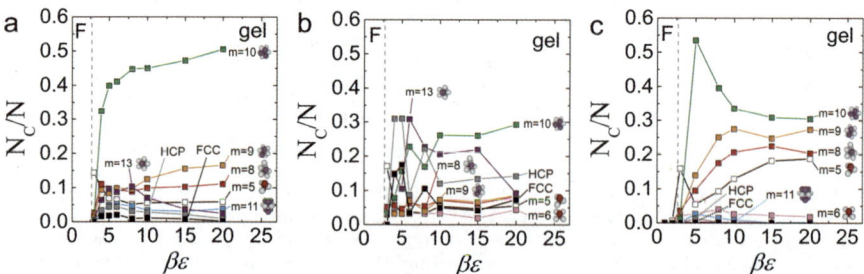

Fig. 5 The effect of quench rates on the local structure in polydisperse systems. (a) TCC analysis for a quench time of 671 τ_B. (b) TCC analysis for a quench time of $1.01 \times 10^4\,\tau_B$. (c) Data from an instantaneous quench. Dashed lines denote ε^c.

effective temperature therefore corresponds to a different simulation run. We begin by considering the moderate quench time of $671\tau_B$. The generic behaviour is similar to that of an instantaneous quench [Fig. 5(c)], with a sharp rise in cluster population upon gelation (the very small cluster population in the equilibrium fluid $\beta\varepsilon \lesssim \beta\varepsilon^c$ is of course unaffected by the quench protocol). For shallow quenches ($\beta\varepsilon \sim 5$), there is a qualitative similarity: the structure is dominated by the 10-membered cluster. However, upon quenching deeper a significant difference emerges. The effect is considerable and is summarised as follows: In the case of a moderate quench time [Fig. 5(a)], the fivefold symmetric 10-membered cluster population shows a rapid rise upon gelation up to a population of around $N_c/N \approx 0.42$ and continues to increase slowly upon deeper quenching, finally reaching a value of $N_c/N \approx 0.5$ for $\beta\varepsilon = 20$. Very different behaviour is seen in the case of the instantaneous quench [Fig. 5(c)]. Although the 10-membered cluster again exhibits strong rise upon gelation, its population *falls* upon deeper quenching, finally reaching a value around $N_c/N \approx 0.3$ for $\beta\varepsilon = 20$. Other smaller clusters, notably eight- and nine-membered clusters, form a very significant fraction of the population in the case of deep quenches.

For a long quench time ($1.01 \times 10^4 \tau_B$), Fig. 5(b) shows a further change in local structure. Higher-order clusters again are favoured. In particular, for $4 \leq \beta\varepsilon \leq 5$, the most popular cluster is the HCP crystalline environment, with a value of $N_c/N \approx 0.3$. In other words, around 30% of the system has succeeded in reaching, or at least getting close to, local equilibrium. Even at deeper quenches, although the crystal population is smaller ($N_c/N \approx 0.15$), much more of the system has crystallised compared to shorter quench times. The rise in HCP environments is accompanied by a rise in the 13-membered five-fold symmetric cluster. This bicapped pentagonal prism, which is the minimum energy structure for 13 Morse particles, is similar to – but distinct from – the icosahedron.

3.3 Finite quenches in monodisperse systems – structure

Polydispersity is known to suppress crystallisation. We therefore consider the effect of setting the polydispersity to zero (a monodisperse system), keeping all other conditions the same. The effect on the local structure is shown in Fig. 6. Except that we now consider a monodisperse system, the situation is identical to Fig. 5. Perhaps the most striking result is the instantaneous quench, shown in Fig. 6(c). The similarity to the polydisperse case is remarkable, and crystallisation is strongly suppressed. This is consistent with the idea that under instantaneous quenching, very little re-arrangement is possible; particles remain kinetically trapped in the clusters they form upon condensation, and lack the thermal energy to rearrange.

The case of the finite quenches [$671 \tau_B$ and $1.01 \times 10^4 \tau_B$ in Fig. 6(a) and (b), respectively] is very different indeed. Unlike the dominance of the five-fold

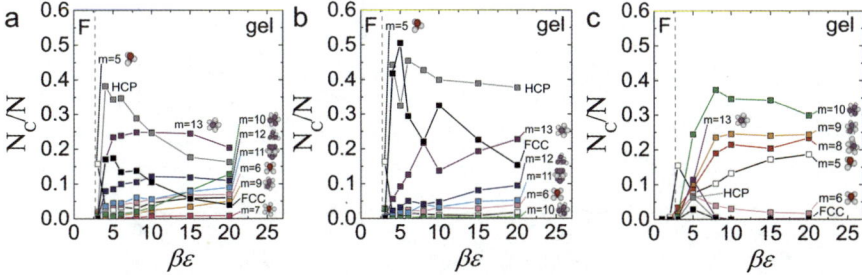

Fig. 6 The effect of quench rates on the local structure in monodisperse systems. (a) TCC analysis for a quench time of $671 \tau_B$. (b) TCC analysis for a quench time of $1.01 \times 10^4 \tau_B$. (c) Data from an instantaneous quench. Dashed lines denote ε^c.

symmetric 10-membered cluster in the polydisperse case, here the crystalline HCP and FCC clusters are the most abundant. Like the polydisperse system, the degree of crystallisation appears to be increased in the case of slow quenching $(1.01 \times 10^4\, \tau_{\mathrm{B}})$. As shown in Fig. 3(b), upon moderate quenching this monodisperse system can largely reach local equilibrium, however it retains a network structure: a crystalline gel.

3.4 Finite quenches – dynamics

The effect of quench times upon the dynamics is shown in Fig. 7. As in the instantaneous case, the mean squared displacement appears diffusive, with no indication of a plateau. However, as noted above, the magnitude of the MSD is small, with the exception of $\beta\varepsilon = 3$. For a long quench time, $(1.01 \times 10^4\, \tau_{\mathrm{B}})$, there is very considerable drop in the MSD compared to the instantaneous quench, with a reduction of more than an order of magnitude. At the very least, this suggests that the finite quench time leads to a gel with a much higher stability. Similar effects are found in the case of monodispersity, which seems consistent with the increase in crystalline order observed above.

4 Discussion

We have shown that colloidal gels quenched slowly exhibit local structural differences compared to the same system quenched instantaneously (in other words, where the particle interactions are fixed). The local structural motifs of instantaneously quenched gels are five-fold symmetric, and are also found in dense equilibrium fluids for this system.[36] It seems reasonable to suppose that, upon quenching, the system begins to phase separate, and the dense fluid phase – whose arrest drives gelation – is structurally similar to high-density fluids. In particular, the local structure is dominated by five-fold symmetric clusters whose principle motif is a five-membered ring ($7 \leq m \leq 10$ in Fig. 2). Although polydispersity is known to suppress crystallisation, the local structure upon instantaneous quenching is rather insensitive to polydispersity.

4.1 Structure and finite quench rates

Quenching at a finite rate, on the other hand, gives the system some chance to reach thermodynamic equilibrium as it is "cooled". Specifically, in equilibrium, phase separation is expected for $\varepsilon > \varepsilon^c$. As mentioned in the introduction, here we are concerned with a system with short range attractive interactions, in which the gel is explicitly metastable to crystal–gas coexistence. However, even for a quench rate

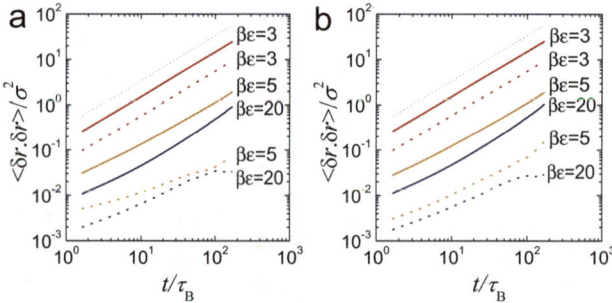

Fig. 7 The effect of quench times on the dynamics. (a) Polydisperse system, (b) monodisperse system. The mean squared displacement for quench depths is indicated, and different quench times. Solid lines – instantaneous quench, dashed lines – long $(1.01 \times 10^4\, \tau_{\mathrm{B}})$ quench time.

of $4.0 \times 10^{-4} \beta \varepsilon \tau_B^{-1}$, no such phase separation is observed, although some coarsening is observed for a moderate quench time of 671 τ_B. In the case of experiments on colloid–polymer mixtures, due to the change in polymer size with temperature, 1 °C corresponds to a change in attraction of order $k_B T$. The resulting change in effective temperature (due to the polymer-induced depletion attraction) corresponds to a quench rate of just $\sim 10^{-3}$ °C s^{-1}. Thus, the overall feature of colloidal gelation is remarkably robust to temperature quenching. Or, equivalently, colloidal model systems are very slow to equilibrate.

Although the system continues to gel, even with moderate quench rates (quench time 671 τ_B), the local structure is a strong function of cooling rate. In particular, the structure of the gel is rather constant with quench depth [Fig. 5(a)]. In other words, in the early stages of gelation, soon after the system has become metastable and begins to phase separate, the particles assemble into 10-membered clusters. Upon deeper quenching, this local structure is preserved, as the particles gradually lose their mobility. By contrast, systems subjected to instantaneous quenches are less able to self-organise into such large clusters and, particularly at deeper quenches, a substantial number of smaller clusters are present. Of particular note is the 5-membered triangular bipyramid. This is formed from two tetrahedra, and the tetrahedron is the simplest rigid assembly of spheres in 3D. In other words, instantaneous quenches lead to very simple structures, which suggest rapid formation with no further re-arrangement, such as is the case for diffusion-limited cluster aggregation.

The opposite case, then, is a slow quench ($1.01 \times 10^4 \tau_B$). Here, although full phase separation does not occur, a very considerable number of particles are able to reach local equilibrium and crystallise. We have based our discussion on a slightly polydisperse system as found in experiments.[28,30] This has two main consequences. Firstly, crystallisation should be suppressed. Although hard spheres with a polydispersity of 4% as we have employed here do crystallise, short-ranged attractive systems are known to be more sensitive to polydispersity.[2] Secondly, the relationship between polydispersity and the eventual approach to equilibrium is complex. What is known is that polydisperse hard spheres phase separate into less polydisperse daughter populations,[39] and it seems reasonable to suppose that similar behaviour might occur here. At the very least, this suggests that full equilibration would take a long time. Conversely, in the monodisperse case, a far higher degree of crystallisation is found, even though for instantaneous quenches the local structure seems insensitive to polydispersity. However, we note that previous work did indeed find some crystallisation in Brownian dynamics simulations of monodisperse attractive particles where quenching was instantaneous.[20]

4.2 Dynamics and finite quench rates

The drop in MSD induced in the same system by varying the quench rate is remarkable, and surprising in its extent. At one level, it underlines the need to use finer probes of structure than $S(k)$ in order to unravel the origins of the dynamical behaviour. We can rationalise the reduction in MSD as consistent with the system being better equilibrated, i.e. lower in its energy landscape. As intuitive as this observation is, backed up by structural observations consistent with improved equilibration, the strong dependence of the dynamics is nonetheless worthy of further investigation. Moreover, the MSD data raises additional questions.

For example, why are the MSDs diffusive? Although the sampling times are quite short, the absence of any plateau can be interpreted as indicative of further irreversible structural rearrangement. However, that all the MSDs have such similar slopes, almost regardless of quench depth and time, remains a curiosity.

A further comment on the approach to equilibrium and stability concerns aging. While this has already received attention for instantaneous quenching,[19,40,42,43] an important question concerns the link between aging and quench rate. Certainly, instantaneously quenched systems are higher in the free energy landscape and exhibit more mobility. Longer quench times produce systems lower in the energy landscape with less mobility, *i.e.* precisely the consequences of ageing.[3] Can quenching be related to aging? Is it directly equivalent? Connecting our measures of local structure with dynamical quantities such as MSD provides a means to tackle these questions.

4.3 The role of dynamics

Before closing, a few words on the role of the dynamics employed are in order. In short, this is the observation that, while here we have used Brownian dynamics and seen no phase separation, in a comparable system with molecular dynamics (MD),[15] it was hard to prevent complete phase separation. While the systems were not accurately mapped to one another (the MD system has a small region where the liquid is thermodynamically stable), the extent of the difference in behaviour seems to warrant further investigation, although some previous work did not reveal a very large difference between the two forms of dynamics.[40] Can a system with Newtonian dynamics "know" its way through the energy landscape so much better than a system with Langevin dynamics that it is hard to kinetically trap the former and apparently impossible to avoid kinetic trapping in the latter? We hope to answer this question in the near future, and also to develop the results we have presented here with larger scale simulations that can explore the role of finite size effects. Finally, we note that solvent-mediated hydrodynamic interactions have been shown to have a profound effect upon gelation and may also influence local structure.[41]

5 Conclusion

We have carried out Brownian dynamics computer simulations of a colloidal gel. By "cooling" the system from the stable fluid to a metastable gel, we have explored the role of quench rate in colloidal gels. Our results indicate that, despite its metastable nature, gelation is surprisingly robust. All quench rates we have been able to perform resulted in a gel, none exhibited phase separation. Conversely, the local structure accessed through the topological cluster classification is a strong function of quench rate. Decreasing the quench rate enables the system to access larger structures associated with lower basins in the energy landscape, and, for the slowest rates we access, a considerable part of the system can reach local equilibrium and crystallise. Monodisperse systems exhibit a much higher degree of crystallisation than only slightly polydisperse systems. This is reflected in the dynamics: the mean squared displacement drops markedly for the same state point when a slow quench is employed. We therefore conclude that local structure, as measured by the TCC, is strongly coupled to the dynamics. Our results suggest that controlled quenching, as is beginning to become possible in experiments on colloidal model systems, may be a means by which colloidal gels can be stabilised, and products based on gels may enjoy an extended shelf life.

Acknowledgements

We gratefully acknowledge stimulating discussions with Jens Eggers, Rob Jack, Hajime Tanaka and Stephen Williams. A.M. is funded by EPSRC grant code EP/E501214/1. C.P.R. thanks the Royal Society for funding. This work was carried out as part of EPSRC grant EP/H022333/1. This work was carried out using the computational facilities of the Advanced Computing Research Centre, University of Bristol.

References

1 E. Zaccarelli, *J. Phys.: Condens. Matter*, 2007, **19**, 323101.
2 W. C. K. Poon, *J. Phys.: Condens. Matter*, 2002, **14**, R859–R880.
3 L. Ramos and L. Cipelletti, *J. Phys.: Condens. Matter*, 2005, **17**, R253–R285.

4 P. J. Lu, E. Zaccarelli, F. Ciulla, A. B. Schofield, F. Sciortino and D. A. Weitz, *Nature*, 2008, **435**, 499–504.
5 N. A. M. Verhaegh, D. Asnaghi, H. N. W. Lekkerkerker, M. Giglio and L. Cipelletti, *Phys. A*, 1997, **242**, 104–118.
6 V. Testard, L. Berthier and W. Kob, *Phys. Rev. Lett.*, 2011.
7 I. Zhang, M. Faers, C. P. Royall and P. Bartlett, *Submitted*, 2012.
8 A. I. Campbell, V. J. Anderson, J. S. van Duijneveldt and P. Bartlett, *Phys. Rev. Lett.*, 2005, **94**, 208301.
9 C. L. Klix, C. P. Royall and H. Tanaka, *Phys. Rev. Lett.*, 2010, **104**, 165702.
10 A. Puertas, M. Fuchs and M. Cteas, *J. Chem. Phys.*, 2004, **121**, 2813–2822.
11 M. Tarzia and A. Coniglio, *Phys. Rev. Lett.*, 2006, **96**, 075702.
12 M. Tarzia and A. Coniglio, *Phys. Rev. E: Stat., Nonlinear, Soft Matter Phys.*, 2007, **75**, 011410.
13 A. J. Archer and N. B. Wilding, *Phys. Rev. E: Stat., Nonlinear, Soft Matter Phys.*, 2007, **76**, 031501.
14 C. Likos, *Phys. Rep.*, 2001, **348**, 267–439.
15 C. P. Royall and S. R. Williams, *J. Phys. Chem. B*, 2011, **115**, 7288–7293.
16 M. Cates, M. Fuchs, W. C. K. Kroy, K. Poon and A. M. Puertas, *J. Phys.: Condens. Matter*, 2004, **16**, S4861–S4876.
17 S. Asakura and F. Oosawa, *J. Chem. Phys.*, 1954, **22**, 1255–1256.
18 K. G. Soga, J. R. Melrose and R. C. Ball, *J. Chem. Phys.*, 1998, **108**, 6026–6032.
19 R. J. M. d'Arjuzon, W. Frith and J. R. Melrose, *Phys. Rev. E: Stat. Phys., Plasmas, Fluids, Relat. Interdiscip. Top.*, 2003, **67**, 061404.
20 A. Fortini, E. Sanz and M. Dijkstra, *Phys. Rev. E: Stat., Nonlinear, Soft Matter Phys.*, 2008, **78**, 041402.
21 L. Assoud, F. Ebert, P. Keim, R. Messina, G. Maret and H. Loewen, *Phys. Rev. Lett.*, 2009, **102**, 238301.
22 A. Alsayed, Z. Dogic and A. Yodh, *Phys. Rev. Lett.*, 2004, **93**, 057801.
23 S. Taylor, E. R. and C. P. Royall, *CondMat ArXiV*, 2012, 1205.0072.
24 N. Elsner, D. R. E. Snoswell, C. P. Royall and B. V. Vincent, *J. Chem. Phys.*, 2009, **130**, 154901.
25 C. Hertlein, L. Helden, A. Gambassi, S. Dietrich and C. Bechinger, *Nature*, 2008, **172–175**, 451.
26 H. Guo, T. Narayanan, M. Sztuchi, P. Schall and G. H. Wegdam, *Phys. Rev. Lett.*, 2007, **100**, 188203.
27 D. Bonn, J. Otwinowski, S. Sacanna, H. Guo, G. Wegdam and P. Schall, *Phys. Rev. Lett.*, 2009, **103**, 156101.
28 C. P. Royall, S. R. Williams, T. Ohtsuka and H. Tanaka, *Nat. Mater.*, 2008, **7**, 556–561.
29 S. R. Williams, *Cond. Mat. ArXiV*, 2007, ArXiv:0705.0203v1.
30 C. P. Royall, A. A. Louis and H. Tanaka, *J. Chem. Phys.*, 2007, **127**, 044507.
31 M. G. Noro and D. Frenkel, *J. Chem. Phys.*, 2000, **113**, 2941–2944.
32 J. R. Elliot and L. Hu, *J. Chem. Phys.*, 1999, **110**, 3043–3048.
33 J. P. K. Doye, D. J. Wales and R. S. Berry, *J. Chem. Phys.*, 1995, **103**, 4234–4249.
34 A. Malins, PhD Thesis, "A Structural Approach to Glassy Systems", University of Bristol.
35 F. C. Frank, *Proc. R. Soc. London, Ser. A*, 1952, **215**, 43–46.
36 J. Taffs, A. Malins, S. R. Williams and C. P. Royall, *J. Chem. Phys.*, 2010, **133**, 244901.
37 A. Malins, S. R. Williams, J. Eggers, H. Tanaka and C. P. Royall, *J. Phys.: Condens. Matter*, 2009, **21**, 425103.
38 T. H. Zhang, J. Klok, R. Hans Tromp, J. Groenewold and W. K. Kegel, *Soft Matter*, 2012, **8**, 667.
39 N. B. Wilding and P. Sollich, *Soft Matter*, 2011, **7**, 4472–4484.
40 G. Foffi, C. De Michele, F. Sciortino and P. Tartaglia, *J. Chem. Phys.*, 2005, **122**, 224903.
41 A. Furukawa and H. Tanaka, *Phys. Rev. Lett.*, 2010, **104**, 245702.
42 S. Babu, J. C. Gimel and T. Nicolai, Crystallisation and dynamical arrest of attractive hard spheres, *J. Phys. Chem.*, 2009, **130**, 064504.
43 M. Hutter, Local Structure Evolution in Particle Network Formation Studied by Brownian Dynamics Simulation, *J. Coll. Interf. Sci*, 2000, **220**, 337–350.

Delayed solidification of soft glasses: new experiments, and a theoretical challenge

Yogesh M. Joshi,[*a] A. Shahin[a] and Michael E. Cates[*b]

Received 12th January 2012, Accepted 19th March 2012
DOI: 10.1039/c2fd20005h

When subjected to large amplitude oscillatory shear stress, aqueous Laponite suspensions show an abrupt solidification transition after a long delay time t_c. We measure the dependence of t_c on stress amplitude, frequency, and on the age-dependent initial loss modulus. At first sight our observations appear quantitatively consistent with a simple soft-glassy rheology (SGR)-type model, in which barrier crossings by mesoscopic elements are purely strain-induced. For a given strain amplitude γ_0 each element can be classified as fluid or solid according to whether its local yield strain exceeds γ_0. Each cycle, the barrier heights E of yielded elements are reassigned according to a fixed prior distribution $\rho(E)$: this fixes the per-cycle probability $R(\gamma_0)$ of a fluid elements becoming solid. As the fraction of solid elements builds up, γ_0 falls (at constant stress amplitude), so $R(\gamma_0)$ increases. This positive feedback accounts for the sudden solidification after a long delay. The model thus appears to directly link macroscopic rheology with mesoscopic barrier height statistics: within its precepts, our data point towards a power law for $\rho(E)$ rather than the exponential form usually assumed in SGR. However, despite this apparent success, closer investigation shows that the assumptions of the model cannot be reconciled with the extremely large strain amplitudes arising in our experiments. The quantitative explanation of delayed solidification in Laponite therefore remains an open theoretical challenge.

1 Introduction

Soft materials such as concentrated suspensions, foams, emulsions and pastes are widely used in products such as foodstuffs, cosmetics, paints, pharmaceuticals, and ceramic precursors. Many of these systems show slow dynamics that are attributed to the trapping or jamming of mesoscopic constituents, creating barriers that the system can cross only slowly.[1,2] Such materials can fall out of thermodynamic equilibrium, evolving by slow "physical aging" towards lower energy states, with progressive slowing down of their relaxational dynamics and rheological response.[3,4] Macroscopic sample deformation can in turn promote barrier-crossing rearrangements, restoring fluidity.[5,6] The resulting interplay between aging, flow and rejuvenation in soft glasses leads to complex rheological effects including overaging,[7,8] viscosity bifurcations,[9–11] and shear banding.[12–15]

Here we present new experiments on a model soft glassy material, an aqueous suspension of Laponite,[16] under oscillatory shear of fixed stress amplitude. Building on a preliminary study,[17] we find that, despite the high fluidity present initially, the

[a]Department of Chemical Engineering, Indian Institute of Technology Kanpur, Kanpur 208016, INDIA. E-mail: joshi@iitk.ac.in; Fax: +91 512 259 0104; Tel: +91 512 259 7993
[b]SUPA, School of Physics & Astronomy, University of Edinburgh, James Clerk Maxwell Building, The King's Buildings, Mayfield Road, Edinburgh EH9 3JZ, UK. E-mail: mec@ph.ed.ac.uk; Fax: +44 131 650 5902; Tel: +44 131 650 5296

material abruptly solidifies after a certain critical time, t_c. Such "delayed solidification" could have serious consequences if it arose, for instance, during the sustained vibratory stresses that arises during transportation (*e.g.*, in road tankers) of soft materials, or in some cases during their manufacture. The sudden solidification of a supposedly fluid formulation risks catastrophic failure of expensive equipment – a situation comparable to the issue of silo rupture, caused when granular materials cease suddenly to flow.[18] The problem is all the more serious because of its apparent unpredictability: nothing obvious about the initial sample indicates that its fluidity will later be lost in this way.[17] The phenomenon of delayed solidification is thereby reminiscent of (and yet almost opposite to) that of "delayed sedimentation" in which an apparently stable colloidal gel suddenly collapses after sustained exposure to gravitational stress.[19] There are some reports in the literature where different soft glassy materials have been observed to undergo delayed solidification in creep flow.[9,11,20] The major difference between these reports and our work is the nature of the applied deformation field. We employ a stress-controlled oscillatory flow field, in which the strain induced in the material remains bounded (and therefore decreases with increasing viscosity and/or elasticity). On the other hand, in creep flow the strain induced in the material never decreases with time.

In this Discussion Paper we attempt to shed light on the physics of delayed solidification, by performing new quantitative experiments to explore the dependence of the delay time t_c on the stress amplitude, frequency, and the (age-related) loss modulus of the sample prior to the oscillatory stress being applied. Alongside this we develop a simple yet semi-quantitative theory that apparently relates these dependences directly to the distribution of energy well depths in soft glasses. Initially we will present this theory at face value, in tandem with our new experimental results (Sections 2 and 3). However, in Section 4 we will show that the model's assumptions cannot be reconciled with the extremely large strain amplitudes that arise in our experiments throughout the prolonged delay period prior to final solidification. In Section 5 we discuss delayed solidification in the context of food materials. Our reluctant conclusion (Section 6) is that, at least in its present form, our theory is not quantitatively credible in the context of the Laponite system studied here, although it may well be applicable to delayed solidification in other soft glasses. Thus our data poses an open challenge: to create a consistent quantitative theory of delayed solidification in Laponite.

2   Experimental methods and results

In this work we have used aqueous suspensions of 3.2 and 2.8 wt% Laponite RD® (Southern Clay Products, Inc.), prepared as described elsewhere.[16] Each Laponite suspension was stored in a sealed polypropylene bottle at 30 °C for a predetermined "idle time" t_i (7–28 days), then loaded in a Couette cell (bob diameter 28 mm, gap 1 mm) of an AR1000 stress controlled rheometer, and shear-melted for 15 min by applying oscillatory shear stress of amplitude 80 Pa at frequency 0.1 Hz. Immediately after this shear melting, an oscillatory stress of amplitude (in the range 5–40 Pa) and frequency f was applied, and the strain evolution was recorded. Experiments were performed at 25 °C; the free surface of the sample was covered with a thin layer of inviscid silicone oil to avoid water evaporation.

Our Laponite suspensions generally have a paste-like consistency. Aging in these systems increases both the elasticity and the relaxation time. Fig. 1 shows the elastic modulus G' as a function of time elapsed (after the shear-melting pre-treatment ends) for experiments at various idle times. Although the strain induced in the material is large, the characterization of the aging data in terms of the linear storage modulus G' in Fig. 1 is justified, as the third strain harmonic is less than 20% of the first.[17] The modulus G', initially too small to detect, shows a sudden and dramatic increase (delayed solidification) beyond a critical time t_c. This arises despite the very large

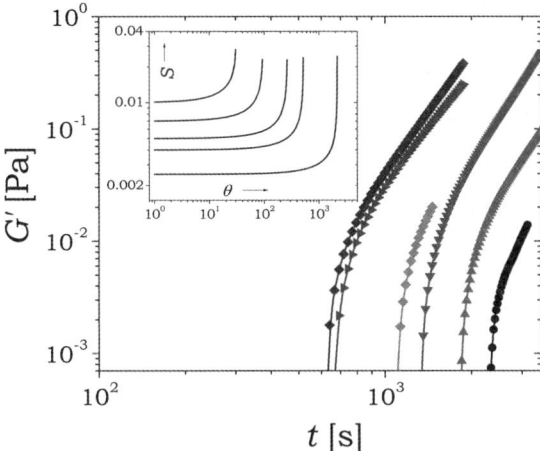

Fig. 1 Evolution of G' for various idle times. Evolution of G' subsequent to shear melting as a function of time for various idle times (from right to left: $t_i = 7, 10, 13, 15, 24, 28$ days) under application of oscillatory shear stress of 30 Pa at 0.1 Hz frequency for 3.2 wt% suspension. Inset shows prediction of eqn (1)–(3) for $\lambda/E_{min} = 0.01$ and $y = 3$, with various initial solid fractions $S(0)$.

fluidizing strains, in the range 10–200, experienced by the material in the first few (post-shear-melting) cycles.

Interestingly, t_c decreases markedly as the idle time is raised. Thus, in contrast to many soft glasses, the aging of Laponite is partly irreversible: subsequent shear melting does not rejuvenate all samples to the same state.[16] Although the microscopic details of this are debatable,[21,22] an aging Laponite suspension at rest clearly crosses some barriers too high to be reversed by our shear-melting protocol, allowing faster delayed solidification in older samples (Fig. 1). However, the evolution of G'

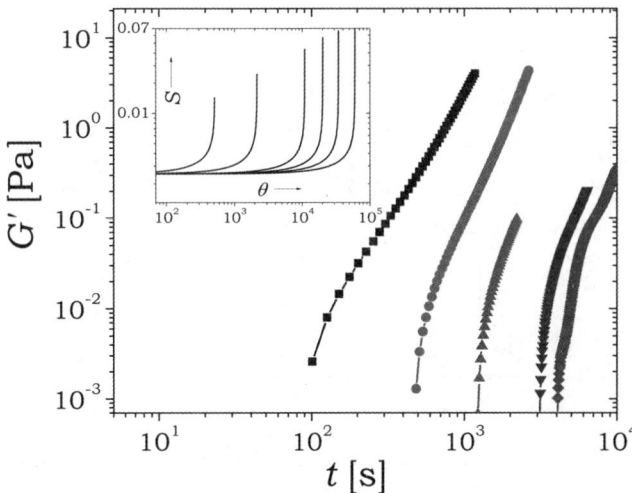

Fig. 2 Evolution of G' at various stresses. Evolution of G' with respect to time under oscillatory flow field having various magnitudes of stresses (from left to right: 20, 25, 30, 35, 40 Pa) at 0.1 Hz frequency for idle time $t_i = 13$ days for 3.2 wt% suspension. Inset shows prediction of eqn (1)–(3) for $y = 3$ (from left to right: $\lambda/E_{min} = 0.005, 0.01, 0.023, 0.031, 0.04, 0.053$).

after t_c is similar in all cases, suggesting that idle time and aging time are partly interchangeable.[16] Fig. 2 shows how the evolution of G' depends on σ at fixed idle time t_i. In this set of experiments each initial state is completely equivalent before the final oscillatory stress is applied. We see that t_c gets larger as the stress is increased. In addition, Fig. 3 shows that t_c is reduced as f is raised.

3 A simple model and its predictions

The microstructure of Laponite suspensions is variously argued to be a repulsive glass or an attractive gel;[21,22] under our conditions (without added salt), repulsions and attractions probably both are important.[16] Nonetheless, whether caging or bonding dominates locally, mesoscopic elements can be considered trapped in energy wells of various depths. These elements are forced out of their traps by macroscopic deformation, whereupon they form new cages or bonds that in turn present new barriers to rearrangement.

Our model for this process is essentially a soft-glassy-rheology (SGR) model,[5,7,23,24] considered for simplicity in the noise-free limit whereby elastic elements cross barriers only when their local mechanical yield threshold is exceeded. One important simplification within the SGR approach, which our model inherits, is the assumption that all elements strain affinely with the imposed flow between one jump and the next. Allied with the further simplifying assumptions of harmonicity within each trap and a uniform elastic constant for all traps,[23] this represents a picture in which the intra-jump elastic deformation is that of a parallel mechanical circuit. The opposite assumption would be to suppose equal stress in all elements, *i.e.*, a series mechanical circuit. The real distribution of local elastic deformations and stresses must lie somewhere between these extremes; we return to this point in Section 4.

Within the SGR framework, the arrested state (an amorphous solid) of a soft glass is described, as indicated above, in terms of mesoscopic elements trapped in energy wells created by neighbors.[5] A crucial postulate of SGR is that these well depths

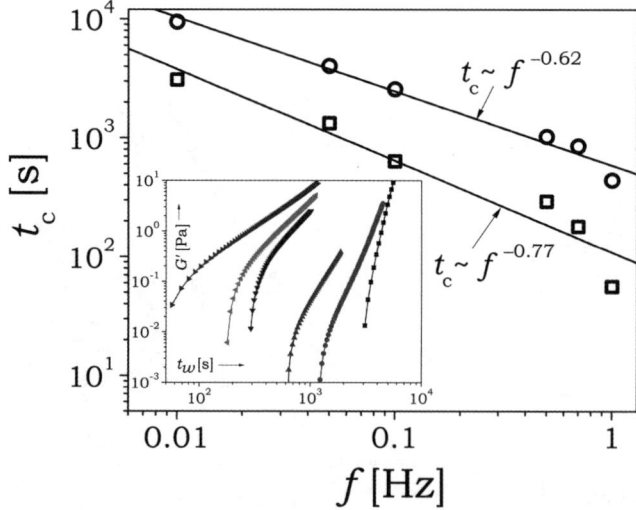

Fig. 3 Evolution of G' at various frequencies and dependence of critical time on frequency. The critical time for delayed solidification plotted against frequency of oscillations (open squares: 3.2 wt% suspension, $t_i = 28$ days, $\sigma_0 = 30$ Pa; open circles: 2.8 wt% suspension, $t_i = 21$ day, $\sigma_0 = 20$ Pa). Inset shows corresponding evolution of G' as a function of time for various frequencies (from right to left: 0.01, 0.05, 0.1, 0.5, 0.7, 1 Hz) for 3.2 wt% suspension.

(or barrier heights) are distributed broadly. As shown by Bouchaud,[4] if the *a priori* (prior) distribution of well-depths varies as $\rho(E) \sim \exp[-E/\langle E \rangle]$, ergodicity is lost at a glass transition temperature T_g, obeying $k_B T_g = \langle E \rangle$. All other forms of $\rho(E)$ lie either in the glass or the fluid according to whether their decay is slower or faster than exponential. The SGR model[23,24] further allows for deformations, and replaces the thermal energy $k_B T$ by a nonequilibrium noise amplitude. Assuming exponential $\rho(E)$, the SGR model offers a unified phenomenological model of soft-glass rheology, whose predictions include power-law fluid and Herschel–Bulkley behaviours.[5]

Although theoretical arguments suggest it asymptotically,[4,24] there is so far no direct experimental test of whether the well-depth distribution in soft glasses is indeed exponential, as SGR assumes. (Other forms might still support an arrest transition, but only if more complicated cooperative dynamics are considered.[23,24]) It would therefore be useful to gain clearer experimental insight into the true form of $\rho(E)$.

The prior distribution of well depths, $\rho(E)$, is not their occupancy probability $P(E)$ since deeper wells have lower escape rates and are more likely to be occupied. However $\rho(E)$ is the distribution from which E is drawn once a rearrangement is made and a new barrier height chosen. If oscillatory shear of amplitude σ is imposed, creating a strain amplitude γ_0, then assuming affine intra-jump deformation (as SGR does) each individual element will gain an energy $k\gamma_0^2/2 \equiv E_0$, where k is a spring constant, at the extremes of each cycle. (As in SGR, k is here taken independent of well-depth E;[5] we partially relax this assumption below.) We assume that any element of well-depth $E < E_0$ is rejuvenated during the given strain cycle, while all those occupying deeper wells are not.

We now distinguish a liquid fraction F and solid fraction $S = 1 - F$, representing in turn the fractions of elements occupying wells shallower or deeper than E_0. Taking for convenience a time coordinate $\theta = ft$, we propose F to obey:

$$\frac{dF}{d\theta} = -FR \tag{1}$$

where R is the fraction of jumps into the solid state:

$$R = \frac{\displaystyle\int_{E_0(\gamma_0)}^{\infty} \rho(E)dE}{\displaystyle\int_0^{\infty} \rho(E)dE} \tag{2}$$

Eqn (1) models the random events by which 'fluid' elements—those that cross their barriers by strain-induced dynamics during one cycle—become 'solid' elements (those that don't) in the next. In a material having solid fraction S and thus modulus GS, the strain induced by the stress of amplitude σ is then taken to be:

$$\gamma_0 = \frac{\sigma}{SG} \tag{3}$$

The threshold energy at given stress σ and solid fraction S then obeys $k\gamma_0^2/2 = E_0$, or equivalently $E_0 = \lambda/S^2$ where $2\lambda = k\sigma^2/G^2$.

Eqn (1)–(3) can now be solved to give $S(\theta)$ in terms of $\rho(E_0)$ and the initial solid content which we denote as $S(0) = \varepsilon$. We consider two cases: first the exponential distribution $\rho(E) = e^{-\alpha E}/\alpha$ for which $R(E_0) = e^{-\alpha E_0}$, and secondly a power law, $\rho(E) = AE^{-y}$ for $E > E_{min}$, with a cutoff E_{min}. The cutoff is defined so that

$\rho(E < E_{min}) = 0$. By normalization we then have $A = E_{min}^{1-y}/(1 - y)$; the resulting R is given by $R(E_0) = (E_0/E_{min})^{1-y}$.

Fig. 1 (inset) shows for the power-law case the time evolution of the solid fraction $S(\theta)$, for various initial solid contents ε, but the same values of λ/E_{min} (equivalently, the same stress σ). Our simple model captures both the sudden solidification at t_c, and the decrease in t_c with increasing initial solid content. Fig. 2 (inset) shows $S(\theta)$ for various values of λ/E_{min} (or equivalently various σ) but the same initial solid content ε. As observed experimentally, the delayed solidification time t_c is predicted to increase with stress. Since $\theta = ft$, eqn (1) directly implies $t_c \propto f^{-1}$. Fig. 3 (inset) shows a decreasing trend, although the best fit power laws (albeit with limited data) are $t_c \propto f^{-0.77}$ for the 3.2% sample and $t_c \propto f^{-0.62}$ at 2.8%. Given the simplicity of the model, these are close enough to the f^{-1} prediction to be broadly encouraging.

Moreover, integrating eqn (1) gives the time at which a given solid fraction is attained:

$$\theta(S) = \int_{\varepsilon}^{S} \frac{1}{(1 - S')R(S')} \, dS' \qquad (4)$$

The critical value θ_c identifies the time where S "ceases to be small". The precise definition of this quantity is clearly somewhat arbitrary, but once it is no longer small, $S(\theta)$ increases so steeply that the details of the definition barely matter. Thus, it suffices to ignore the saturation that occurs as $S \to 1$ in the denominator, and then set $S = 1$ as the upper limit of the integral above; these two simplifications give the following expression for the solidification time $\theta_c = ft_c$:

$$\theta_c \approx \int_{\varepsilon}^{1} \frac{1}{R(S')} \, dS' \qquad (5)$$

For our exponential and power law distributions the results are respectively:

$$\theta_c = \frac{2\varepsilon^3 G^2}{\alpha k \sigma^2} \exp\left[\frac{\alpha k \sigma^2}{2\varepsilon^2 G^2}\right] \qquad (6)$$

$$\theta_c \propto \varepsilon^{3-2y} \sigma^{2(y-1)} \qquad (7)$$

Fig. 4 shows t_c, now experimentally defined by $G'(t_c) = 5$ mPa, as a function of G'', the initial loss modulus measured directly after shear melting. Strictly speaking, our model relates the solid fraction only to G', which is too small to measure in this early time regime. Therefore we assume that the loss modulus is a similar indicator of solid content: $G'' \propto \varepsilon = S(0)$. If so, eqn (6) demonstrates a poor fit to the experimental data but the power law result (eqn (7)) fits well, with $y = 3.15 \pm 0.13$ at 3.2% and $y = 3.3 \pm 0.18$ at 2.8%.

Fig. 5 shows how t_c depends on σ for the 3.2% data of Fig. 2, and for two datasets at 2.8%. Here stress was varied at fixed idle time, so we expect ε to be fixed, giving $t_c \sim \sigma^{2(y-1)}$ by eqn (7) (note however that the measured initial G'' decreases with σ; see the Appendix for a discussion). Fig. 5 confirms this power law, and again contrasts with the prediction for exponential $\rho(E)$. A fit gives $y = 3.4 \pm 0.3$ for the 3.2% sample and $y = 3.6 \pm 0.1$ for both 2.8% samples.

To summarize, fitting separately the dependences of t_c on initial solid content and on stress leads to very similar y values for three different samples, suggesting a fairly robust power law. This is interesting from a glass physics viewpoint: any power law

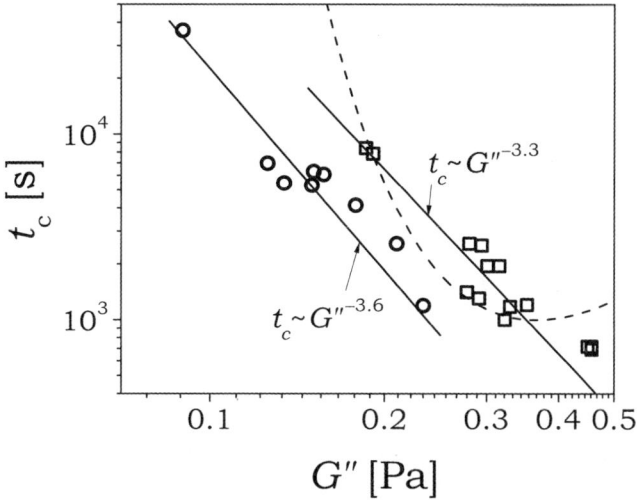

Fig. 4 Dependence of critical time on initial G''. The critical time for delayed solidification plotted against G'' measured directly after the shear-melting protocol is complete (open squares: 3.2 wt% suspension, $\sigma = 30$ Pa, $f = 0.1$ Hz; open circles: 2.8 wt% suspension, $\sigma = 20$ Pa, $f = 0.1$ Hz). Solid lines represent fit of eqn (7) while the dashed line represents fit to eqn (6) with $G'' \propto \varepsilon$ assumed in both cases. The stated exponents for the modulus dependence translate into the y values quoted in the text.

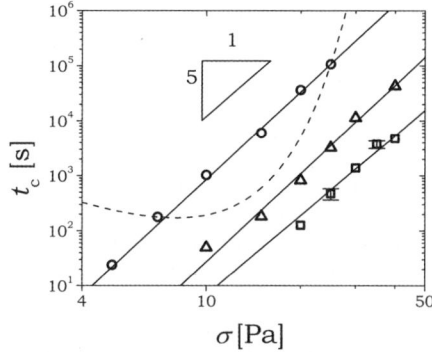

Fig. 5 Dependence of critical time on stress. The critical time for delayed solidification plotted as a function of applied stress amplitude σ (circles: 2.8 wt% suspension, $t_i = 9$ days, $f = 0.1$ Hz; triangles: 2.8 wt% suspension, $t_i = 15$ days, $f = 0.1$ Hz (data from), squares: 3.2 wt% suspension, $t_i = 13$ days, $f = 0.1$ Hz.) Solid lines represent fit of eqn (7) to the experimental data; the dashed line is a fit of eqn (6) for the uppermost dataset only.

distribution should, by the arguments of Bouchaud,[4] lie deep in the aging glass regime. In the Appendix, we generalize our simple model to allow the elastic constant k of an element to vary as $k \propto E^p$. This gives the same results as above, but with y replaced by $y_{eff} = (y + p)/(1 + p)$; the same conclusion applies. Thus we have found that, when interpreted within the precepts of the SGR-inspired model presented above, our quantitative delayed solidification measurements for Laponite indicate a power law distribution of the energy well depths. This apparently direct connection between macroscopic rheological observations and the mesoscopic energetics is tantalizing: it holds out the prospect for "spectroscopic" analysis of the energy landscape through careful study of the nonlinear rheology. In turn this might provide new insight into aging and rejuvenation in this important class of materials.

4 What's wrong with this picture?

Before answering this question, it is interesting to speculate where power laws in the local yield energy and/or elastic constant might come from. At the low volume fractions present in our Laponite samples, bonding might lead to some sort of percolation transition. Near such a transition the elastic elements comprise clusters of all sizes, whose moduli are controlled by power laws (coinciding in the scalar limit with those for resistivity.[25]) The corresponding yield energy distribution is model-dependent, but seemingly can itself exhibit power laws over one or two decades, or more in some limits,[26,27] although not enough is yet known to suggest specific values for the relevant exponents.

However the emerging picture of a buildup of the solid fraction S (from an initially minimal level) by a percolation-like process gives pause for thought. For if solidification is initiated by the formation of relatively rigid clusters within a sea of fluidized material, SGR's assumption of a parallel mechanical circuit becomes highly suspect. One could not expect such aggregates to deform affinely under any type of flow: solid objects floating in a fluid develop only small deformations before achieving stresses that match those of their continuously deformable surroundings. At first sight, this objection might not appear fatal to our model since, in its many other predictions, the SGR approach is empirically successful although in practice local deformations are never affine. Nonetheless, the percolation viewpoint argues for a model that lies much nearer the series-circuit (equal stress on all elements) end of the spectrum than the parallel-circuit assumption embodied in SGR.

A serious blow is struck by noting that, in our Laponite studies, the enormous initial strains to which the initially fluidized sample is subjected (of order 10–200, *i.e.*, 1000% to 20,000%) are maintained almost throughout the incubation period prior to the final solidification event. It is scarcely credible that *any* mesoscopic element of the type envisaged by SGR could have high enough internal or external energy barriers to sustain an affine deformation of this magnitude. By completely destroying the structures that SGR requires to survive from one cycle to the next, such strain amplitudes preclude the slow buildup of a solid component which is an essential precursor to the final dramatic solidification. (Recall that the latter occurs when the feedback between the slowly building elasticity and decreasing strain amplitude finally takes over.) The observation of delayed solidification at these initial strain amplitudes in Laponite therefore means we must look for a mechanism involving pockets of solidity that do not deform affinely, and, for that reason alone, can grow from one cycle to the next. Whatever its merits in other context, by assuming affine deformation, SGR precludes a consistent description of any such mechanism.

We can however, speculate a mechanism by assuming that the solid pockets, and the fluidized suspension surrounding these, share the same stress (series mechanical circuit). Consequently the strain induced in the solid region would be very small. Such a scenario may give rise to an apparent boundary layer around the solid pocket, wherein strain magnitude changes from practically zero at the surface to a very large value away from it. In this small strain region very near the solid surface, the liquid suspension may undergo aging following a very similar dynamics mentioned in the previous section. Owing to this aging of the liquid suspension near the solid surface, the solid region is expected to grow, which in turn will enhance the fraction of liquid suspension undergoing smaller strain. Moreover the enhanced solid fraction will also reduce the bulk strain magnitude. Through a forward feedback mechanism, the growing solids will fill the space causing jamming of the system as a whole. This picture preserves some of the physical features of our SGR-based approach, but would require a different mathematical description from the one we present above.

5 Delayed solidification in the context of food materials

Interestingly, there are many food materials that have paste like consistency and are expected to demonstrate soft glassy rheological behavior. These include: fruit jams, mustard, jellies, mayonnaise, cheese, ice-creams, tomato and chocolate puree, toothpaste (not exactly a food material, but edible), *etc.* Among these both mustard[28] and mayonnaise[29] have been reported to demonstrate physical aging (time dependent enhancement in viscosity and elastic modulus), which is a signature of soft glassy behavior. Rheological behavior is one of the most important characteristic features of food materials. The effects of time and deformation, which respectively tend to enhance and reduce viscosity and elasticity, is a very important consideration when designing food processing equipment and determining the shelf life of a food product. The present work suggests that under application of a sustained oscillatory deformation field, food materials may in some cases transform from an apparently fluid like state to solid state. Such materials are subjected various kinds of deformation fields during preparation, transportation and handling. The present work suggests that such deformation fields can delay but may not stop the process of aging which leads to solidification.

In addition to physical aging, which is a reversible process, food materials are also prone to undergo partly irreversible changes in their rheological behavior. Particularly enzymes, heating or acidification can induce gelation in certain food products causing an increase in viscosity and elastic modulus. Thus flowing liquids get converted to soft solids as a function of time.[30] The solidification that we discuss in this manuscript is essentially reversible. However, Laponite suspensions are also known to show partial irreversible aging behavior and therefore interestingly mirror what happens in certain types of food materials.

Besides time-dependent irreversible phenomena, irreversible aggregation induced by an applied deformation field is also possible and is particularly observed for proteins. It is known that mis-folded proteins tend to aggregate because of inter molecular hydrophobic associations.[31] In addition, one of the proposed mechanisms for spider silk formation also suggests deformation induced self-assembly of proteins which is irreversible in nature.[32–34] In colloidal suspensions, shear-induced aggregation has also been reported.[35] Interestingly, shear-induced enhancement in elasticity is also observed in some soft glassy materials[7] and in Laponite suspensions in particular.[8] However, there seems to be no direct connection between shear-induced solidification (for instance in steady shear) and the delayed solidification addressed in our work under sustained oscillatory shearing.

6 Conclusions

We have studied in detail the sudden and dramatic enhancement in elastic modulus at late times seen in aqueous Laponite suspensions undergoing stress-controlled oscillatory shear. The critical time t_c for this delayed solidification is reduced for older samples, despite our use of a vigorous pre-shear protocol, whereas application of higher stress amplitude, or lower frequency, increases t_c. We have proposed a simple SGR-type model wherein a liquid fraction of fluidized elements are rejuvenated every cycle. At each such event, there is a probability R of jumping into a solid fraction of deep wells, that do not rejuvenate; R is determined by the current strain amplitude, and the prior distribution of well depths. The ever-increasing solid content slowly decreases the induced strain and increases R; this positive feedback leads eventually to sudden jamming of the whole sample. Taken at face value, the model offers a semi-quantitative explanation of our Laponite experiments, with strong evidence that a power law distribution of energy well depths must be chosen in preference to the exponential form normally adopted in an SGR context. This offers a tantalizing glimpse of how macroscopic rheology might be directly relatable to the barrier distribution for rearrangements—a quantity that has previously eluded direct experimental characterization in soft glasses. But unfortunately this is no more than a glimpse, because on closer inspection the large strain amplitudes arising in the Laponite system cannot credibly be reconciled with one of the model's central approximations: that local deformations between rearrangement events follow affinely the macroscopic flow.

Where does this critique leave the model? Unless its foundations can be reinterpreted or repaired (and we have not managed to do this so far), then despite its semi-quantitative success at fitting the data, our SGR-based approach clearly does not offer a secure starting point to understand delayed solidification in Laponite suspensions. Accordingly, the evidence that it seemingly offers for a power-law rather than exponential barrier height distribution must now be set aside.

On the other hand, the qualitative physical predictions of our SGR-inspired model remain intact for systems that are credibly approximated by its assumption of affine deformation between rearrangements. We can see nothing to prevent the existence of soft glasses for which a delayed solidification scenario arises at much more modest strains, of order unity: our model, even if it must be rejected for Laponite, offers a ready-made description for such systems. Meanwhile it remains an open theoretical challenge to develop a more suitable quantitative model for delayed solidification in Laponite itself. Only when such a model exists can one know how much, if any, of the qualitative physics embodied in our model is also relevant to the Laponite system.

Appendix

A.   Stress dependence of viscous modulus at fixed idle time

For experiments carried out on 2.8 wt% Laponite suspension on day 9, the initial G'', measured in the first cycles of oscillatory shear after shear melting, was found to decrease with stress amplitude. Since all the samples have equivalent histories prior to this point, we believe that this decrease represents shear thinning of the rejuvenated suspension, not a stress-dependent initial solid fraction. To verify this we measured viscosity (η) of a sample immediately after the rejuvenation stage. The Fig. 6 shows G''/ω and η vs. stress. Both the variables show the same dependence

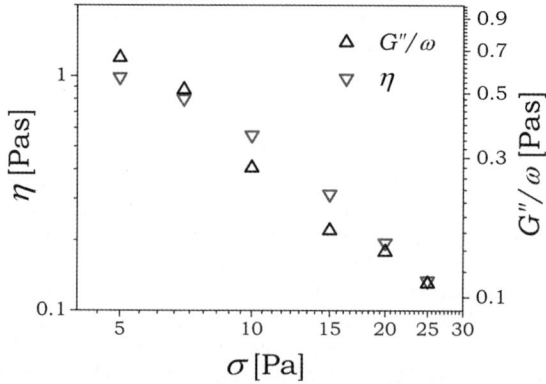

Fig. 6   Initial G''/ω for an oscillatory test (angular frequency $\omega = 0.628$ rad s^{-1}) plotted as a function of magnitude of stress for 2.8 wt% 9 days old Laponite suspension. The viscosity measured from a stress controlled shear experiment is also plotted against stress for the same system.

on shear stress, confirming that the decrease in G'' is indeed due to shear thinning of the rejuvenated suspension.

B. Modified exponents for power law traps with variable local elastic constant

Here we consider conditions where $k = k(E)$, so that the elastic modulus of a trap depends on its depth. For simplicity we assume $k(E) = E^p$. Eqn (2) for the liquid-solid conversion factor R is unaffected, where now $E_0/k \, (E_0) = \gamma_0^2/2$, so that $\gamma_0 \propto E_0^{(1-p)/2}$. Also the modulus of the solid material G in eqn (3) now obeys:

$$
G \propto \frac{\int\limits_{E_0}^{\infty} \rho(E)k(E)\mathrm{d}E}{\int\limits_{E_0}^{\infty} \rho(E)\mathrm{d}E} \propto E_0^p \tag{8}
$$

In other words, the solid fraction populates wells of depth $E > E_0$ with the prior distribution, and the modulus of the solid phase is fixed by the appropriate weighted average of the elastic constants k of the individual wells. This gives $S \propto \sigma \, E_0^{-(1-p)/2}$, while $R \propto E_0^{1-y}$ as before, and therefore $\theta_c \propto \varepsilon^{(3-2\,y+p)/(1+p)} \, \sigma^{2\,(y-1)/(1+p)}$.

Acknowledgements

We thank the Royal Society of Edinburgh—Indian National Science Academy International Exchange Programme for sponsoring YMJ's visit to Edinburgh (2010). YMJ also acknowledges the IRHPA scheme of the Department of Science and Technology, Government of India. MEC is supported by the Royal Society, by EPSRC/EP/EO30173, and by EPSRC/EP/J007404; he thanks KITP Santa Barbara for hospitality, where this research was supported in part by the National Science Foundation (USA) under Grant no. NSF PHY05-51164.

References

1 L. Cipelletti and L. Ramos, *Curr. Opin. Colloid Interface Sci.*, 2002, **7**, 228–234.
2 P. G. Debenedetti and F. H. Stillinger, *Nature*, 2001, **410**, 259–267.
3 L. C. E. Struik, *Physical Aging in Amorphous Polymers and Other Materials*, Elsevier, Houston, 1978.
4 J. P. Bouchaud, *J. Phys. I*, 1992, **2**, 1705–1713.
5 S. M. Fielding, P. Sollich and M. E. Cates, *J. Rheol.*, 2000, **44**, 323–369.
6 L. M. Falk and J. S. Langer, *Phys. Rev. E: Stat. Phys., Plasmas, Fluids, Relat. Interdiscip. Top.*, 1998, **57**, 7192–7205.
7 V. Viasnoff and F. Lequeux, *Phys. Rev. Lett.*, 2002, **89**, 065701.
8 R. Bandyopadhyay, H. Mohan and Y. M. Joshi, *Soft Matter*, 2010, **6**, 1462–1466.
9 P. Coussot, Q. D. Nguyen, H. T. Hyun and D. Bonn, *J. Rheol.*, 2002, **46**, 573–589.
10 C. Derec, A. Ajdari and F. Lequeux, *Eur. Phys. J. E: Soft Matter Biol. Phys.*, 2001, **4**, 355–361.
11 C. Christopoulou, G. Petekidis, B. Erwin, M. Cloitre and D. Vlassopoulos, *Philos. Trans. R. Soc. London, Ser. A*, 2009, **367**, 5051–5071.
12 P. Coussot, J. S. Raynaud, F. Bertrand, P. Moucheront, J. P. Guilbaud, H. T. Huynh, S. Jarny and D. Lesueur, *Phys. Rev. Lett.*, 2002, **88**, 218301.
13 F. Varnik, L. Bocquet, J. L. Barrat and L. Berthier, *Phys. Rev. Lett.*, 2003, **90**, 095702.
14 M. L. Manning, J. S. Langer and J. M. Carlson, *Phys. Rev. E: Stat., Nonlinear, Soft Matter Phys.*, 2007, **76**, 056106.
15 S. M. Fielding, M. E. Cates and P. Sollich, *Soft Matter*, 2009, **5**, 2378–2382.
16 A. Shahin and Y. M. Joshi, *Langmuir*, 2010, **26**, 4219–4225.
17 A. Shukla and Y. M. Joshi, *Chem. Eng. Sci.*, 2009, **64**, 4668–4674.
18 J. M. Rotter, *Guide for the economic design of circular metal silos*, Taylor and Francis, London 2001.

19 L. Starrs, W. C. K. Poon, D. J. Hibberd and M. M. Robins, *J. Phys.: Condens. Matter*, 2002, **14**, 2485–2505.
20 B. M. Erwin, D. Vlassopoulos and M. Cloitre, *J. Rheol.*, 2010, **54**, 915–939.
21 P. Mongondry, J. F. Tassin and T. Nicolai, *J. Colloid Interface Sci.*, 2005, **283**, 397–405.
22 B. Ruzicka and E. Zaccarelli, *Soft Matter*, 2011, **7**, 1268–1286.
23 P. Sollich, F. Lequeux, P. Hebraud and M. E. Cates, *Phys. Rev. Lett.*, 1997, **78**, 2020–2023.
24 P. Sollich, *Phys. Rev. E: Stat. Phys., Plasmas, Fluids, Relat. Interdiscip. Top.*, 1998, **58**, 738–759.
25 D. Stauffer, *Phys. Rep.*, 1979, **54**, 1–74.
26 Y. M. Strelniker, S. Havlin, R. Berkovits and A. Frydman, *Phys. Rev. E: Stat., Nonlinear, Soft Matter Phys.*, 2005, **72**, 016121.
27 A. G. Hunt, *Transp. Porous Media*, 1998, **30**, 177–198.
28 P. Coussot, H. Tabuteau, X. Chateau, L. Tocquer and G. Ovarlez, *J. Rheol.*, 2006, **50**, 975–994.
29 F. Da Cruz, F. Chevoir, D. Bonn and P. Coussot, *Phys. Rev. E: Stat. Phys., Plasmas, Fluids, Relat. Interdiscip. Top.*, 2002, **66**, 051305.
30 D. Eric, *Food Hydrocolloids*, 2012, **28**, 224–241.
31 K. M. N. Oates, W. E. Krause, R. L. Jones and R. H. Colby, *J. R. Soc. Interface*, 2006, **3**, 167–174.
32 A. K. Lele, Y. M. Joshi and R. A. Mashelkar, *Chem. Eng. Sci.*, 2001, **56**, 5793–5800.
33 H. J. Jin and D. L. Kaplan, *Nature*, 2003, **424**, 1057–1061.
34 C. Holland, J. S. Urbach and D. L. Blair, *Soft Matter*, 2012, **8**, 2590–2594.
35 A. Zaccone, D. Gentili, H. Wu, M. Morbidelli and E. Del Gado, *Phys. Rev. Lett.*, 2011, **106**, 138301.

Slow dynamics and structure in jammed milk protein suspensions

Peggy Thomar, Dominique Durand, Lazhar Benyahia and Taco Nicolai*

Received 7th February 2012, Accepted 5th April 2012
DOI: 10.1039/c2fd20014g

The dynamic mechanical properties and the structure of dense suspensions of sodium caseinate were investigated using oscillatory shear rheology and confocal laser scanning microscopy, respectively. Caseins are the most abundant milk proteins and form in the absence of calcium phosphate small star-like particles with radii of about 10 nm. The viscosity increases strongly with increasing protein concentration above $\sim$80 g L^{-1} due to jamming of the particles. The viscosity increase is stronger at lower temperatures, caused by a strong decrease in the terminal relaxation time with increasing temperature. Addition of calcium ions introduces an attractive interaction that induces phase separation above a critical calcium concentration. Increasing the $CaCl_2$ concentration leads to an increase in the terminal relaxation time, but to a decrease in the high frequency elastic modulus. The effect of adding $CaCl_2$ is stronger at higher temperatures.

Introduction

Many food systems behave like soft solids, but still contain a large fraction of water. Such systems are often called gels and are easily recognized even though it is notoriously difficult to give an unambiguous definition of the concept gel. There are two fundamentally different ways to give aqueous food systems the texture of a soft solid.

The first way is to form a system spanning network of connected particles that can be proteins, polysaccharides or even small oil droplets. It does not matter whether the connections are covalent or physical bonds, as long as they are sufficiently strong and long lived so that the system does not flow or break easily.

The second method is to fill up the space with particles that do not bind to each other. In this case, the system stops flowing, because the particles are jammed by an excluded volume interaction sometimes reinforced by electrostatic repulsion. Suspensions of non-interacting monodisperse hard spheres stop flowing close to the volume fraction of random close packing ($\phi = 63\%$).[1] In food systems soft solids of this kind are usually formed by soft particles such as oil droplets or microgels. Because such particles can be deformed and compressed up to a certain extent, higher volume fractions are generally needed to obtain solid like behavior. In the case of microgels, the particles may themselves contain a large fraction of water, which means that even if the particles are closely packed, the water content may still be high. Dense suspensions of monodisperse spheres spontaneously form a phase with crystalline order, but a small amount of polydispersity or asymmetry is enough to keep the system disordered. Therefore crystalline order is generally not found in food systems. Disordered solids of this kind are often called glasses and dense

LUNAM Université du Maine, IMMM UMR-CNRS, 72085 le Mans Cedex 9, France. E-mail: taco.nicolai@univ-lemans.fr suspensions of hard spheres have extensively been studied as model systems for glasses in general.[2]

For the first type of solid, the transition between a liquid-like and a solid-like behavior can be induced by increasing the number of strong bonds between particles leading to more and larger aggregates until at a critical amount one aggregate percolates the whole system. This process is often called the sol–gel transition and can be described by the percolation model.[3] For the second type, the liquid-solid transition can be induced by increasing the effective volume fraction of particles (ϕ_e). The theoretical description of this process, which is often called the glass transition, is not yet complete even for the simplest case of hard spheres.[2] ϕ_e can not only be increased by adding more particles, but one can also increase the effective volume of the particles, *e.g.* by swelling or increasing the electrostatic repulsion.

The situation is more complex if there are attractive forces between the particles in addition to excluded volume interactions. The influence of short range attraction on the glass transition of hard spheres has been studied intensively and the term attractive glass has been coined to describe such systems.[4] Strong attraction may lead to phase separation between a high density and a low density phase.[5] The conditions at which phase separation occurs depends not only on the strength of the interaction, but also on the number of interaction sites on the particles.[6]

Even if strong attraction will finally lead to phase separation, transient gel-like behavior may occur, because initially the particles aggregate and form a system spanning network with long lived bonds that subsequently coarsens and evolves towards macroscopic phase separation.[7,8] If the coarsening is slow compared to the time of observation, such systems will behave as stable soft solids.

All these phenomena may be observed also in food systems. Here we investigate the liquid–solid transition for dense aqueous suspensions of caseinate. Caseins are by far the most abundant proteins in milk.[9] There are four major types of caseins each with a molar mass of around 20 Kg mol^{-1}. In milk they are present in the form of spherical particles, so-called casein micelles, with a radius of about 100 nm that are held together by calcium phosphate. The calcium phosphate can be removed by reducing the pH to the iso-electric point of the caseins where they precipitate. After washing, the caseins can be resolubilized by adding sodium hydroxide resulting in solutions of sodium caseinate (NaCas).[10]

When 0.1 M or more NaCl is added, NaCas was found to spontaneously form small aggregates with a radius of about 11 nm containing approximately 15 caseins, depending somewhat on the pH and the temperature.[11–13] Without added salt, similar aggregates are formed, but only at higher concentrations.[14] The aggregation is most likely driven by hydrophobic interactions and it has been suggested that it is akin to micellisation of surfactants[15] as caseins contain relatively long chain sections with predominantly hydrophobic amino acids.[16]

The rheology of dense NaCas suspensions has been studied in some detail.[14,17–20] Above a concentration of about C = 80 g L^{-1} the viscosity increases rapidly with increasing concentration, which may be attributed to jamming of the caseinate particles. The viscosity has a remarkably strong temperature dependence that is fully reversible. Pitkowski *et al.*[14] suggested that this temperature dependence could be explained by a weak decrease of the effective volume fraction with increasing temperature. In jammed suspensions a weak variation of ϕ_e causes a strong variation of the viscosity.

Added calcium ions bind specifically to caseins and can lead to aggregation of the caseinate particles. Alvarez *et al.*[21] reported changes in the intrinsic viscosity, fluorescence and circular dichroism spectra when $CaCl_2$ was added to dilute NaCas suspensions, which they attributed to the formation of increasingly dense aggregates. At higher $CaCl_2$ concentrations an increasing fraction of the proteins precipitates with increasing $CaCl_2$ concentration.[22,23] The soluble fraction still consisted of small aggregates similar to those formed in the presence of NaCl.[22] Cuomo *et al.*[24] reported that for a given Ca^{2+} concentration the fraction of precipitated proteins increased with increasing temperature. The effect of adding calcium ions on the rheology of dense caseinate suspensions has not yet been investigated systematically.

Dense NaCas suspensions are a good example of a soft solid food system formed by jamming of predominantly repulsive soft particles. The aim of the present investigation was to study in more detail the rheology of dense NaCas suspensions and the influence of attractive interaction of varying strength caused by adding different amounts of calcium ions. We will investigate the effects on the structure and the dynamic mechanical properties as a function of the protein concentration up to 300 g L^{-1} and the temperature (10–60 °C) keeping the pH fixed at 6.7. It will be shown that addition of more than a critical amount of calcium ions drives phase separation. The results will be compared to those obtained for suspensions of very similar caseinate particles obtained by addition of polyphosphates to native casein micelles that form soft solids by aggregating into a homogeneous percolating network.[25–27]

Materials and methods

Materials

The sodium caseinate powder used for this study was provided by Armor protein (Saint-Brice, France). The water content was about 5 wt%. The powders were dissolved in deionised water (Millipore) containing 3 mM sodium azide as a bacteriostatic agent. The pH was adjusted to pH 6.7 by slow addition of 0.1 M HCl or 1 M NaOH under continuous stirring. Addition of $CaCl_2$ leads to a small decrease of the pH, which was compensated by addition of NaOH. The casein concentration (C) was determined by measuring the nitrogen content (Kjeldahl) and the UV absorption at 280 nm using a UV-Visible spectrometer Varian Cary-50 Bio (Les Ulis, France). Consistent results were obtained using an extinction coefficient of 0.85 L g^{-1} cm^{-1}. The suspensions were homogenized by stirring at 80 °C for about 30 min.

Rheology

Continuous and oscillatory shear measurements were made with a stress-imposed rheometer (TA Instruments Rheolyst AR2000) using a cone - plate geometry (40 mm, 0.58° or 20 mm, 4.0°). For $C \leq 165$ g L^{-1} the measurements were done with another stress-imposed rheometer (MCR 301, Anton-Paar) using a Couette geometry (inner and outer diameters: 26.6 and 28.9 mm). The temperature was controlled by a Peltier system and the geometry was covered with a mineral oil to prevent water evaporation.

Confocal laser scanning microscopy

CLSM was used in the fluorescence mode. Observations were made with a Leica TCS-SP2 (Leica Microsystems Heidelberg, Germany). A water immersion objective lens was used HCx PL APO 63× NA = 1.2 with a theoretical resolution of about 300 nm in the x–y plane. Caseins were labelled with the fluorochrome rhodamine B isothiocyanate (Rho), by adding a small amount of a concentrated rhodamine solution to the casein solutions. Rho was excited using a helium–neon laser with a wavelength of 543 nm and the fluorescence was detected with a photomultiplier. Care was taken not to saturate the fluorescence signal and it was verified that the amplitude of the signal was proportional to the protein concentration.

Turbidity

The turbidity was determined as a function of the wavelength using a UV-Visible spectrometer Varian Cary-50 Bio (Les Ulis, France). The path length through the cells was varied between 1 and 10 mm in order to obtain an optimum signal. The temperature was controlled with a thermostat bath.

Results and discussion

Dynamic mechanical properties of caseinate particle suspensions

NaCas suspensions are transparent at low concentrations and translucent at higher concentrations. The latter is caused by a small weight fraction of particles with a radius of about 60 nm (<5 wt%). CLSM images of NaCas suspensions are homogeneous at all concentrations. Small angle X-ray scattering of dense suspensions were compatible with that of close packed disordered particles with a diameter of about 20 nm.[14]

Here we focus on the zero shear viscosity and the frequency dependent shear moduli in the limit of linear response, but it was shown elsewhere that concentrated suspensions of NaCas are shear thinning.[14,19] Fig. 1a shows the zero shear viscosity normalized by that of water (η_r) of NaCas suspensions as a function of the

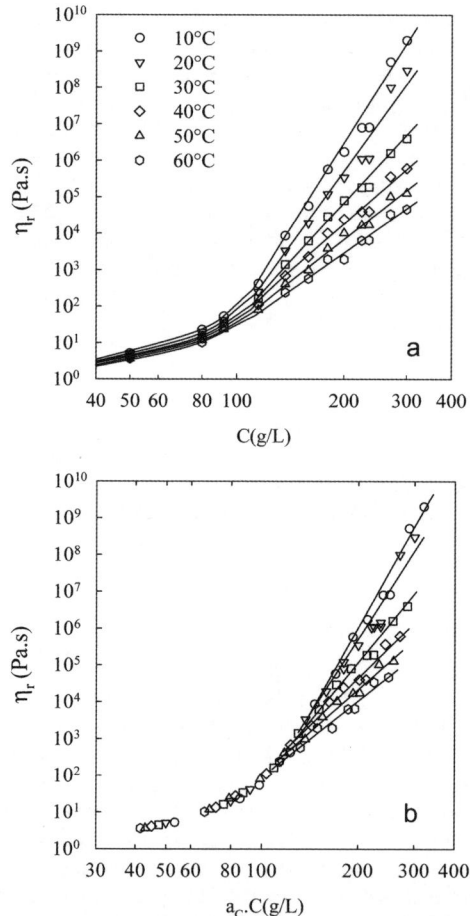

Fig. 1 (a) Dependence of the relative zero shear viscosity as a function of the NaCas concentration for different temperatures indicated in the figure. (b). Same data as in Fig. 1a after superposition of the data up to about 120 g L⁻¹ horizontal shifts. The shift factors used are: 1.07, 1.0, 0.95, 0.9, 0.86 and 0.83 for 10, 20, 30, 40, 50 and 60 °C, respectively.

concentration for a range of temperatures. η_r rises very steeply with increasing concentration above about 80 g L^{-1}, but it does not diverge because the particles can interpenetrate.

The increase for $C > 120$ g L^{-1} can be approximated by very steep power laws with exponents that decrease with increasing temperature from about 15 at 10 °C to 7 at 60 °C. It was suggested in ref. 14 that the strong temperature dependence of η_r can to a large extent be explained by a small temperature dependence of the effective volume fraction. However, the more detailed investigation reported here shows that the data cannot be superimposed within the experimental error by simple horizontal shifts over the whole concentration range, see Fig. 1b. When we superimpose the initial increase of the viscosity by shifting the concentrations, the data clearly deviate for $C > 120$ g L^{-1}. This means that in the jammed state the interaction between the particles depends on the temperature. We note that there is little effect of added NaCl (0–250 mM) and the pH (5.4–6.7) on the viscosity,[14] implying that electrostatic interactions are not very important.

The viscosity is determined by a combination of the high frequency elastic shear modulus (G_{el}) and the terminal relaxation time (τ): $\eta \propto G_{el} \cdot \tau$. At higher protein concentrations these parameters could be determined separately by oscillatory shear measurements that showed an elastic response at high frequencies and a viscous response at low frequencies as reported earlier.[14,19,20] Oscillatory shear measurements made at different temperatures could be superimposed by frequency-temperature superposition. In this way master curves at $T_{ref} = 20$ °C were obtained at different concentrations, see Fig. 2, that show that the stress relaxation is characterized by a broad relaxation time distribution. The terminal relaxation shifted rapidly to lower frequencies as the concentration was increased, while the high frequency elastic modulus increased only weakly. We note that the dynamic viscosity at low frequencies was equal to the low shear viscosity obtained from flow measurements.

The terminal relaxation time was calculated from the frequency (f_c) where G' and G'' cross: $\tau = (2\pi f_c)^{-1}$. τ increased strongly with increasing concentration and decreasing temperature following a power law dependence, see Fig. 3a. The elastic modulus was calculated as 10 times the value of G' at f_c. This choice is rather arbitrary, but the relative variation of G_{el} does not depend on this choice as the frequency dependence is almost independent of the temperature and the concentration. The concentration dependence of G_{el} is plotted in Fig. 3b. and shows that G_{el} is approximately proportional to C^2. It is clear that the increase in G_{el} with increasing concentration is much weaker than that of τ. G_{el} also increased systematically with

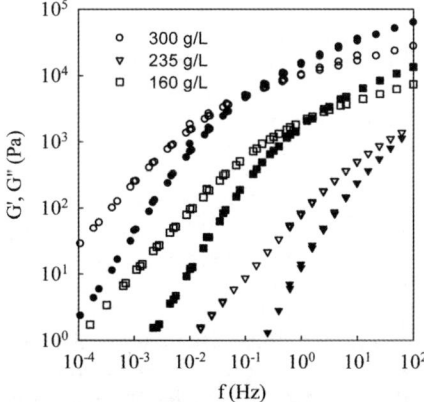

Fig. 2 Master curves of the storage (closed symbols) and loss (open symbols) shear moduli obtained by temperature–frequency shifts at $T_{ref} = 20$ °C for different concentrations of NaCas as indicated in the figure.

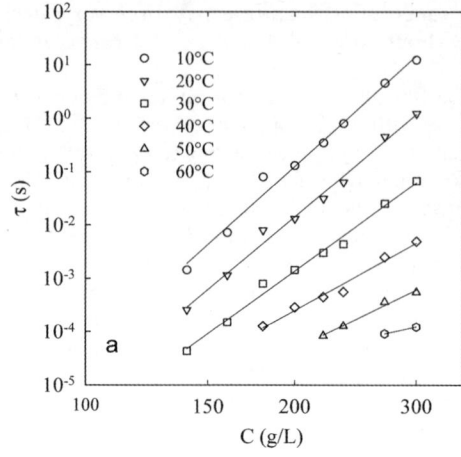

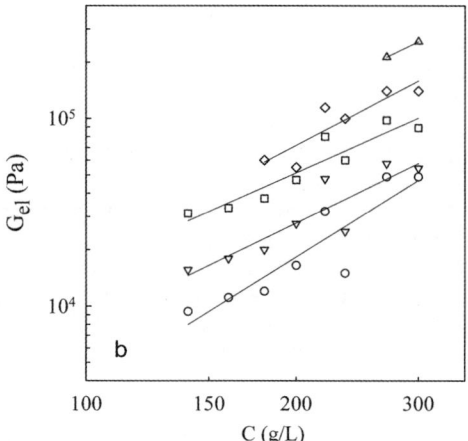

Fig. 3 Dependence of the terminal relaxation time (a) and the high frequency elastic modulus (b) as a function of the NaCas concentration for different temperatures indicated in the figure. The solid lines were obtained from linear least square fits.

increasing temperature, which taken by itself would lead to an increase in η_r with increasing temperature. However, this effect is overcompensated by the much stronger decrease in τ with increasing temperature. We may conclude that the main cause for the strong concentration and temperature dependence of η_r is the variation of τ.

On one hand, the increase of G_{el} with increasing temperature shows that the jammed particles are more difficult to deform at higher temperatures. On the other hand, the decrease of the relaxation time suggests that the particles can more easily escape from the cage formed by their neighbours. The increase in G_{el} with increasing temperature can be understood by the increase of the aggregation number of the particles and a small increase of their density that was observed in dilute solutions.[12] In the presence of 0.1 M NaCl, both the hydrodynamic radius and the molar mass of the particles were found to increase with increasing temperature. As mentioned in the Introduction we may suppose that the aggregates are formed by association of hydrophobic parts at the centre with a corona formed by the more hydrophilic chain sections. Within this scenario the increase in the aggregation number with increasing temperature is caused by an increase in the hydrophobic interactions. It is likely that the aggregation number and the density of the particles increase with increasing temperature also in the jammed state although the effect will be quantitatively different. Increased density renders the particles less deformable which explains qualitatively the temperature dependence of G_{el}.

An increase in the density of the particles implies a decrease in their volume fraction, which explains why the suspensions jam at a higher volume fraction when the temperature is increased. However, it does not explain the weaker dependence of the relaxation time on the concentration. On the contrary, one would expect that in the jammed state the relaxation time increases if the repulsion between the particles is stronger. For star polymers and polymeric micelles one observes a stronger concentration dependence of the viscosity when the aggregation number is higher.[28,29] As hydrophobic interactions become stronger with increasing temperature, it is very unlikely that the strong decrease in τ can be explained by an increase in the exchange rate of individual caseins between the caseinate particles. We speculate that in the case of caseinate particles, weaker repulsion between the particles with decreasing temperature leads to stronger interpenetration of the coronas and that this leads to stronger friction between the particles and thus a slower relaxation. Alternatively, there might be some hydrogen bond formation between the particles. Hydrogen bonding is weaker at higher temperatures, which could explain the decrease in the relaxation time.

It is clear that even in the relatively simple situation of densely packed predominantly repulsive caseinate particles the dynamics cannot be explained solely in terms of escape from the cage formed by surrounding particles. Interactions on the molecular level between overlapping particles also need to be considered.

The influence of adding calcium ions on the structure

Addition of $CaCl_2$ led to an increase of the turbidity above a critical amount (C_s^*). At room temperature the increase was slow close to C_s^*, but a stationary value was obtained rapidly at higher C_s and higher temperatures. Interestingly, we observed a strong reversible increase in the turbidity with increasing temperature, see Fig.4. C_s^* increased somewhat with increasing protein concentration and decreasing temperature. At 20 °C, C_s^* was situated between 5 and 10 mM at $C = 20$ g L^{-1} and between 10 and 15 mM for $C = 40$ g L^{-1} and $C = 80$ g L^{-1}. The system became turbid only when heated for $C = 20$ g L^{-1} at $C_s = 5$ mM and for $C = 40$ g L^{-1} at $C_s = 10$ mM.

At lower protein concentrations, a fraction of the proteins precipitated slowly for $C_s > C_s^*$. The strong increase of the viscosity inhibited sedimentation under gravity for $C > 100$ g L^{-1} at room temperature, but at elevated temperatures precipitation was also observed at higher concentrations. The fraction of soluble proteins (F) was quantified as a function of the $CaCl_2$ concentration for $C = 80$ g L^{-1} and $C = 20$ g L^{-1} at 20 °C and 40 °C. F was determined after incubating overnight by measuring the protein concentration in the supernatant after centrifugation (2 h at 5.10^4 g). F decreased sharply for $C_s > 20$ mM at $C = 80$ g L^{-1} and for $C_s > 5$ mM at $C = 20$ g L^{-1}. The drop was more important at the lower protein concentration and the higher temperature, see Fig. 5. The reversibility was tested by incubating overnight at 40 °C and subsequently raising the temperature to 20 °C. The same results were obtained as when the suspensions were incubated at 20 °C, see Fig. 5, demonstrating again that the temperature dependence was reversible.

The increase of the turbidity is caused by the formation of dense domains of proteins that at higher $CaCl_2$ concentrations become sufficiently large to be visible in the CLSM images, see Fig. 6. Between 80 g L^{-1} and 200 g L^{-1} the domains were visible when 30 mM or more $CaCl_2$ was added, while at $C_s = 20$ M or lower, the images were homogeneous. For $C = 235$ g L^{-1} the images showed dense domains for $C_s = 40$ mM and higher and were homogeneous for $C_s = 30$ mM and lower.

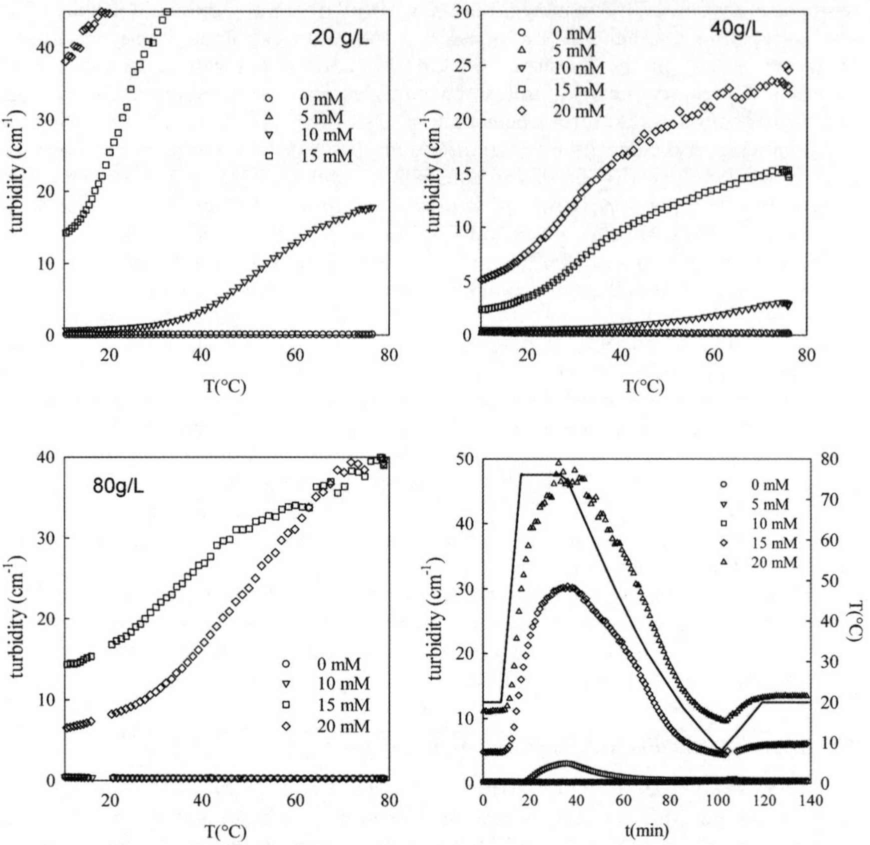

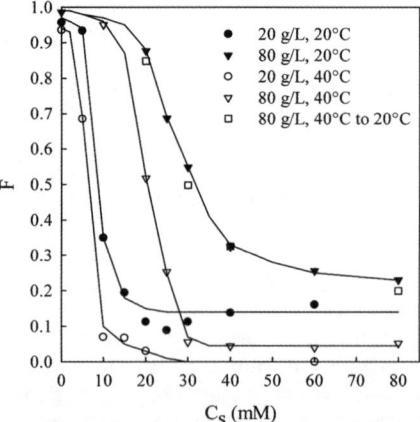

Fig. 4 Temperature dependence of the turbidity at $\lambda = 800$ nm at different protein and $CaCl_2$ concentrations indicated in the figure. The reversibility is illustrated for $C = 40$ g L^{-1} in the lower right panel that shows the turbidity as a function of the time during a heating and a cooling ramp. The sample temperature is indicated by the solid line.

Fig. 5 Fraction of soluble protein as a function of the $CaCl_2$ concentration at two protein concentrations and two temperatures. The effect of incubating at 40 °C and subsequently cooling at 20 °C is also shown.

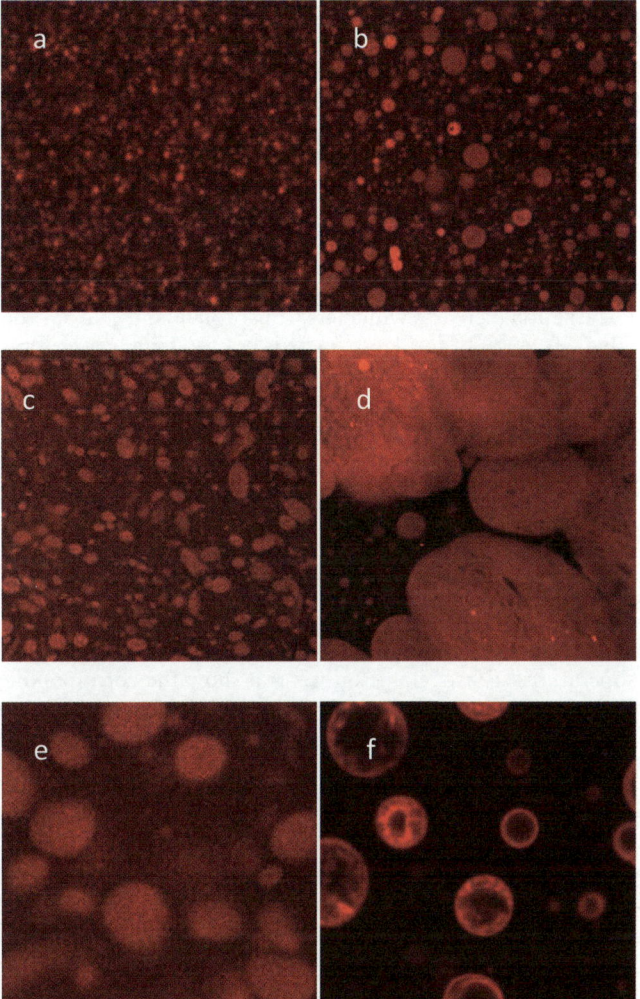

Fig. 6 CLSM images of NaCas suspensions at different protein and $CaCl_2$ concentrations: a) $C = 80$ g L^{-1}, $C_s = 30$ mM; b) $C = 140$ g L^{-1}, $C_s = 30$ mM; c) $C = 200$ g L^{-1}, $C_s = 30$ mM; d) $C = 200$ g L^{-1}, $C_s = 80$ mM; e) $C = 200$ g L^{-1}, $C_s = 30$ mM; f) $C = 80$ g L^{-1} $C_s = 12$ mM obtained by dilution with pure water from $C = 200$ g L^{-1}, $C_s = 30$ mM. Images a–d represent 150×150 μm and images e and f represent 75×75 μm. For clarity, in all images the contrast between the phases has been artificially increased.

At lower protein and $CaCl_2$ concentrations, small spherical domains were formed (see Fig. 6a) that merged and took up a larger volume fraction at higher concentrations. If the protein and $CaCl_2$ concentrations were high enough the dense phase was continuous (see Fig. 6d). By comparing the fluorescence intensity of the domains with the average intensity, we can estimate that the protein concentration within the domains is about 250 g L^{-1}. Suspensions containing 300 g L^{-1} caseinate remained homogeneous after addition of $CaCl_2$, because in this case the dense phase filled up the whole space.

We investigated the reversibility of the phase separation by diluting a phase separated system at $C = 200$ g L^{-1} and $C_s = 30$ mM (Fig. 6e) with pure water to give $C = 80$ g L^{-1} and $C_s = 12$ mM (Fig. 6f). The solutions directly prepared under the latter conditions were homogeneous. We observed that after dilution most of the proteins in the dense domains redispersed, but the outer shell of the domains remained. In some cases we also observed an internal structure that looked like connected shells of several smaller domains that had merged. This observation shows that strong bonds are formed specifically at the domain surfaces that do not break spontaneously when the concentration of calcium ions is reduced below the critical value.

The results discussed above together with those reported in the literature lead to the following overall picture of the effect of adding calcium ions on the structure of NaCas suspensions. Calcium ions bind specifically to caseins which induces an attraction between the caseinate particles that increases with increasing $CaCl_2$ concentration and temperature. Above a critical calcium concentration the system phase separates. The protein concentration in the dense phase is about 250 g L^{-1} and its volume fraction increases with increasing protein and calcium concentration. When the volume fraction of the dense phase is not too high it is present in the form of spherical particles of a few microns. At the surface of the dense domains stable bonds are formed that can withstand dilution in pure water.

The influence of adding calcium ions on the dynamic mechanical properties

The effect of adding $CaCl_2$ on the visco-elastic properties of dense NaCas suspensions was studied in detail for selected concentrations between 80 and 300 g L^{-1}. All measurements were done starting at 10 °C and progressively increasing the temperature. The stability of the suspensions was tested by repeating the measurement at 20 °C at the end of the experiment.

Fig. 7 shows the dependence of the viscosity on the $CaCl_2$ concentration at different temperatures. For comparison we also show the corresponding values of the ratio (R) between the molar concentration of $CaCl_2$ and caseinate. At $C = 80$ g L^{-1} addition of calcium ions caused a decrease in the viscosity. With increasing temperature the decrease became more important and started at lower calcium concentrations. At higher C_s, precipitation of proteins occurred rapidly especially at higher temperatures so that the range of C_s that could be investigated at this concentration was limited.

At $C = 140$ g L^{-1} and 165 g L^{-1} we found a weak increase followed by a strong drop at higher C_s. In a rare investigation of the effect of adding $CaCl_2$ on the viscosity of NaCas suspensions Carr et al.[30] also reported a weak maximum for a NaCas suspension at 14 wt%. At $C = 200$ g L^{-1} the viscosity was relatively insensitive to the addition of $CaCl_2$ except at higher temperatures where η_r decreased at larger C_s. Finally, at $C = 235$ g L^{-1} and 300 g L^{-1} the viscosity increased initially with increasing C_s and then remained approximately constant at all temperatures. The suspensions with $C \leq 165$ g L^{-1} were not stable at higher temperatures when R was larger than about 5, i.e. when the viscosity dropped sharply. This instability showed itself either by visible precipitation or by a lower viscosity at 20 °C when it was remeasured after the experiments at higher temperatures.

As was mentioned above, the dependence of the viscosity on C_s is caused by the combined effects on G_{el} and τ that could be studied separately for $C \geq 165$ g L^{-1}. Again master curves of the frequency dependent shear moduli could be formed by frequency-temperature superposition. The shape of the master curves did not change significantly with the addition of $CaCl_2$, but the values of τ and G_{el} depended on C_s. The concentration dependence of τ and G_{el} is shown in Fig. 8. We note that the spread of the data is much more important than in the absence of calcium ions, which is most likely caused by the heterogeneous structure seen in the CLSM images. The effect of phase separation on the structure may vary between different preparations and this will influence the visco-elastic properties. The absence of structural heterogeneity explains why the results at $C = 300$ g L^{-1} are less scattered.

At all concentrations except at $C = 300$ g L^{-1}, G_{el} decreased with increasing $CaCl_2$ concentration. The decrease was stronger at higher temperatures and could even lead to an inversion of the temperature dependence from increasing with increasing

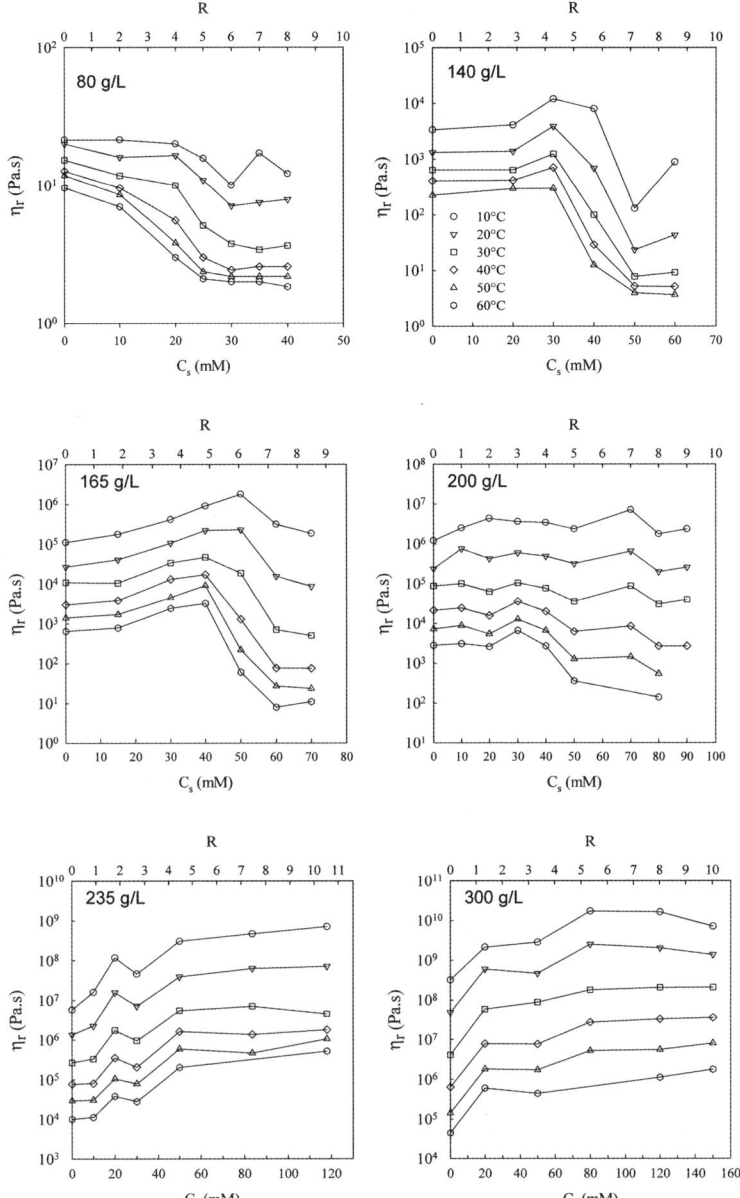

Fig. 7 Dependence of the zero shear viscosity on the $CaCl_2$ concentration at different temperatures and different caseinate concentrations indicated in the figure. The top horizontal axis shows the corresponding values of the ratio (R) between the molar concentration of $CaCl_2$ and caseinate.

temperature to decreasing with increasing temperature. At higher $CaCl_2$ concentrations and temperatures, the drop of the shear modulus became so strong that it was no longer possible to make useful oscillatory shear measurements.

At $C = 165$ g L^{-1} the terminal relaxation time shows a clear maximum for all temperatures at about 50 mM $CaCl_2$. Above this concentration the relaxation time dropped sharply. At higher protein concentrations adding $CaCl_2$ led to an increase in the relaxation time.

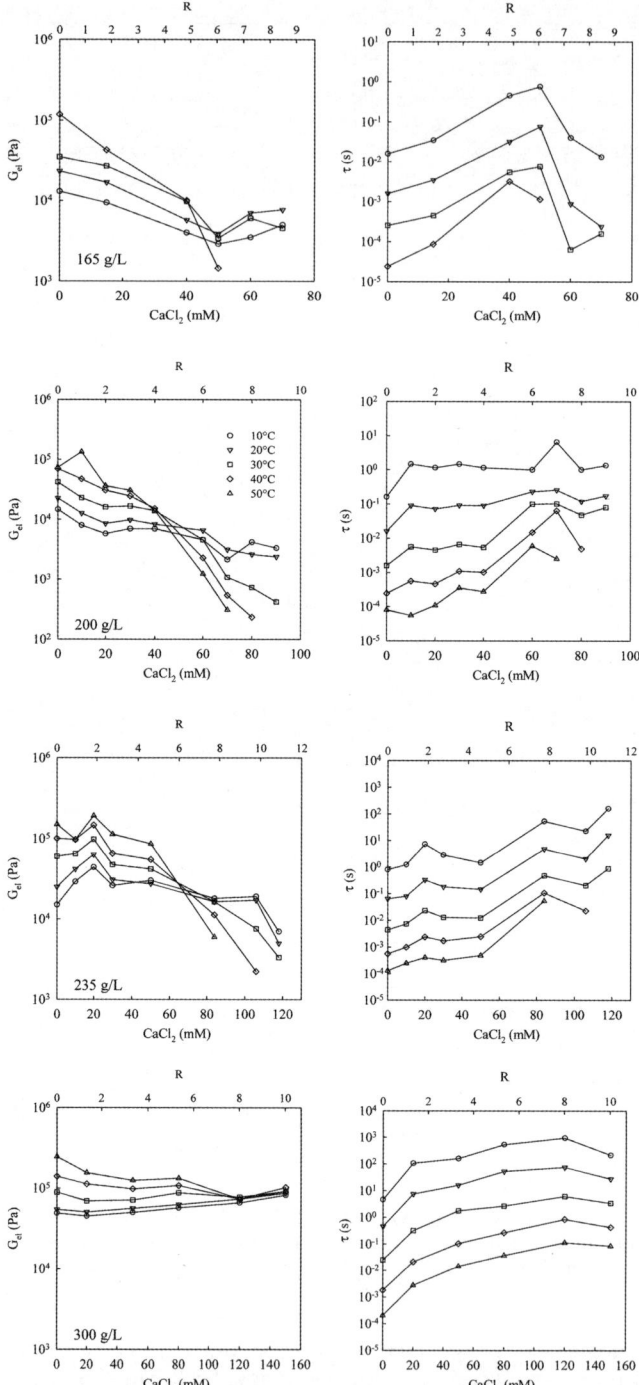

Fig. 8 Dependence of the terminal relaxation time and the high frequency elastic modulus on the $CaCl_2$ concentration at different temperatures and caseinate concentrations indicated in the figure. The top horizontal axis shows the corresponding values of the ratio (R) between the molar concentration of $CaCl_2$ and caseinate.

This journal is © The Royal Society of Chemistry 2012

As discussed above, adding calcium ions leads to an attractive interaction between the particles. Such an interaction is not expected to modify G_{el}, but increases the friction between the jammed particles and thus the terminal relaxation time. This may account for the behavior at 300 g L^{-1}. At lower protein concentrations we also need to consider the effect of phase separation which leads to a less dense continuous phase interspersed with dense domains. A decrease of the concentration of the continuous phase will lead to a decrease of G_{el}. A stronger phase separation at higher temperatures explains the stronger decrease of G_{el}.

Phase separation would also lead to a decrease in τ in the absence of any effect of attractive interaction. The net effect of calcium ions on the relaxation time is thus determined by a balance between a reduction of the caseinate particle concentration in the continuous phase and an increase of the attractive interaction in this phase. At higher protein concentrations the volume fraction of the dense phase becomes large and may percolate through the system. Clearly, in this case the structure of the dense phase will also become important for its contribution to G_{el}.

Comparison with networks of caseinate particles

When polyphosphate is added to an aqueous solution of native casein micelles, the latter disintegrate by chelation of the calcium inside the casein micelles. As a consequence, caseinate particles are formed with almost the same size as the sodium caseinate particles.[25] With increasing concentration the viscosity rises steeply similarly to NaCas. Also for this system the viscosity of dense suspensions decreases strongly with increasing temperature. An essential difference with the suspensions studied here, however, is that the particles irreversibly aggregate and form self-supporting gels at concentrations as low as 40 g L^{-1}. The aggregation is very slow at room temperature, but the rate increases with increasing temperature. Clearly, the addition of polyphosphate together with the colloidal calcium phosphate causes the formation of strong bonds between the particles by a mechanism that has not yet been elucidated. Importantly, no phase separation was observed, though at low concentration the network was too weak to support its own weight leading to precipitation of large flocs.

Dynamic mechanical measurements showed that the gelation process occurs through percolation of bound caseinate particles.[27] No relaxation was observed showing that the bonds were irreversible, contrary to the bonds formed by adding CaCl$_2$ to NaCas. This explains why in this system no phase separation occurred as densification is inhibited by the elasticity of the network. The elastic modulus increased with the square of the protein concentrations between 40–200 g L^{-1}, similarly to G_{el} for NaCas suspensions at $C > 140$ g L^{-1}, but the absolute values were somewhat smaller. Interestingly, no significant effect of the temperature on G_{el} was observed. This indicates that the elasticity for this system is determined by the connectivity of the network and not by the repulsion between jammed particles even at higher concentrations, where the connected particles are also clearly jammed.

Conclusion

Jamming, phase separation and network formation of the same caseinate particles can be induced in aqueous suspension by subtle variations of the mineral composition.

Sodium caseinate forms small particles with a radius of about 10 nm in aqueous solution by association of hydrophobic chain sections. Above a concentration of about 80 g L^{-1} the particles jam causing a sharp increase of the viscosity. In the jammed state the viscosity increases following a power law with an exponent that decreases with increasing temperature.

For $C > 140$ g L^{-1} the visco-elastic properties of the suspensions can be characterized by oscillatory shear measurements. Master curves of the frequency dependence can be obtained by temperature–frequency superposition. They show a mainly elastic response at high frequencies and a relaxation characterized by a broad relaxation time distribution. The elastic modulus increases with increasing protein concentration and temperature. The latter can be understood by an increase of the aggregation number and the density of the caseinate particles with increasing temperature. The terminal relaxation time increases very strongly with increasing concentration and decreasing temperature and is the dominant cause of the variation of the viscosity. The dynamics of the jammed caseinate particles cannot be explained solely in terms of the escape of soft particles from the cage formed by neighbouring particles. Possibly, the increase of the relaxation time upon lowering the temperature is caused by increased interpenetration of the particles and thus stronger friction. Alternatively, it could be caused by hydrogen bond formation between particles.

Addition of calcium ions causes an attractive interaction between the caseinate particles leading to phase separation above a critical concentration. Spherical dense protein domains of a few microns in diameter are formed that merge at higher protein concentrations. The volume fraction of the dense phase increases with increasing protein concentration until $C = 300$ g L^{-1} and higher, the dense phase occupies the whole volume. A small fraction of the particles in the dense domains form strong bonds that can withstand dilution in pure water.

The addition of calcium ions causes a reduction of G_{el} which is more pronounced at higher temperatures. The decrease of G_{el} is probably caused by the decrease of the protein concentration in the continuous phase when dense protein domains are formed. At higher protein and calcium concentrations the dense phase may percolate the system and contribute significantly to G_{el}, which implies that the structure of the dense phase also becomes an important parameter. At $C = 300$ g L^{-1} the effect of adding calcium ions is very small because the system remains homogeneous. The attraction between the particles caused by addition of calcium ions leads to an increase of the relaxation time. However, this effect is off-set by the reduction of the density of the continuous phase in more dilute solutions leading to a decrease in τ at high CaCl$_2$ concentrations. Clearly, the effect of adding calcium ions on the dynamics of dense caseinate particle suspensions is determined by a complex interplay of increased attractive interaction and formation of a heterogeneous structure.

References

1 W. B. Russel, D. A. Saville, W. R. Schowalter, *Colloidal dispersions*, Cambridge Univ Pr, 1992.
2 F. Sciortino and P. Tartaglia, *Adv. Phys.*, 2005, **54**, 471–524.
3 D. Stauffer and A. Aharony, *Introduction to percolation theory*, Taylor & Francis, London, 1992.
4 K. A. Dawson, *Curr. Opin. Colloid Interface Sci.*, 2002, **7**, 218–227.
5 G. Foffi, C. De Michele, F. Sciortino and P. Tartaglia, *Phys. Rev. Lett.*, 2005, **94**, 078301.
6 F. Sciortino and E. Zaccarelli, *Curr. Opin. Solid State Mater. Sci.*, 2011, **15**, 246–253.
7 S. Babu, J.-C. Gimel and T. Nicolai, *J. Chem. Phys.*, 2006, **125**, 184512.
8 W. C. K. Poon, L. Starrs, S. P. Meeker, A. Moussaïd, R. M. L. Evans, P. N. Pusey and M. M. Robins, *Faraday Discuss.*, 1999, **112**, 143–154.
9 P. F. Fox, in *Advanced Dairy Chemistry vol1: Proteins 3rd editions*, ed. K. A. P. F. Fox and P. L. H. McSweeney, 2003, pp. 427–435.
10 C. R. Southward, in *Development in Dairy Chemistry-4*, ed. F. P. Fox, Elsevier Applied Science, London, 1989, pp. 173–244.
11 B. Chu, Z. Zhou, G. Wu and H. M. J. Farrell, *J. Colloid Interface Sci.*, 1995, **170**, 102–112.
12 A. HadjSadok, A. Pitkowski, L. Benyahia, T. Nicolai and N. Moulai-Mostefa, *Food Hydrocolloids*, 2008, **22**, 1460–1466.
13 J. A. Lucey, M. Srinivasan, H. Singh and P. A. Munro, *J. Agric. Food Chem.*, 2000, **48**, 1610–1616.
14 A. Pitkowski, D. Durand and T. Nicolai, *J. Colloid Interface Sci.*, 2008, **326**, 96–102.
15 C. G. de Kruif and V. Y. Grinberg, *Colloids Surf., A*, 2002, **210**, 183–190.
16 D. S. Horne, *Curr. Opin. Colloid Interface Sci.*, 2002, **7**, 456–461.

17 D. Farrer and A. Lips, *Int. Dairy J.*, 1999, **9**, 281–286.
18 J. Fichtali, F. R. van de Voort and G. J. Doyon, *J. Food Eng.*, 1993, **19**, 203–2011.
19 S. M. Loveday, M. A. Rao, L. K. Creamer and H. Singh, *J. Food Sci.*, 2010, **75**, N30–N35.
20 A. Bouchoux, B. Debbou, G. Gasan-Guiziou, M. H. Famelart, J. L. Doublier and B. Cabane, *J. Chem. Phys.*, 2009, **131**, 165106.
21 E. M. Alvarez, P. H. Risso, M. A. M. Canales, M. S. Pires and C. A. Gatti, *Colloids Surf., A*, 2008, **327**, 51–56.
22 A. Pitkowski, T. Nicolai and D. Durand, *Food Hydrocolloids*, 2009, **4**, 1164–1168.
23 C. A. Zittle and E. S. Dellamonica, *Arch. Biochem. Biophys.*, 1958, **76**, 342–353.
24 F. Cuomo, A. Ceglie and F. Lopez, *Food Chem.*, 2011, **126**, 8–14.
25 M. Panouillé, L. Benyahia, D. Durand and T. Nicolai, *J. Colloid Interface Sci.*, 2005, **287**, 468–475.
26 M. Panouillé, D. Durand, T. Nicolai, N. Boisset and E. Larquet, *J. ColloidInterface Sci.*, 2005, **287**, 85–93.
27 A. Pitkowski, T. Nicolai and D. Durand, *J. Rheol.*, 2007, **52**, 971–986.
28 J. Roovers, *Macromolecules*, 1994, **27**, 5359–5364.
29 D. Vlassopoulos, G. Fytas, S. Pispas and N. Hadjichristidis, *Phys. B*, 2001, **296**, 184–189.
30 A. J. Carr, P. A. Munro and O. H. Campanella, *Int. Dairy J.*, 2002, **12**, 487–492.

Arrested coalescence of viscoelastic droplets with internal microstructure

Amar B. Pawar,[ab] Marco Caggioni,[a] Richard W. Hartel[b] and Patrick T. Spicer[*a]

Received 18th February 2012, Accepted 24th April 2012
DOI: 10.1039/c2fd20029e

There are many new approaches to designing complex anisotropic colloids, often using droplets as templates. However, droplets themselves can be designed to form anisotropic shapes without any external templates. One approach is to arrest binary droplet coalescence at an intermediate stage before a spherical shape is formed. Further shape relaxation of such anisotropic, arrested structures is retarded by droplet elasticity, either interfacial or internal. In this article we study coalescence of structured droplets, containing a network of anisotropic colloids, whose internal elasticity provides a resistance to full shape relaxation and interfacial energy minimization during coalescence. Precise tuning of droplet elasticity arrests coalescence at different stages and leads to various anisotropic shapes, ranging from doublets to ellipsoids. A simple model balancing interfacial and elastic energy is used to explain experimentally observed coalescence arrest in viscoelastic droplets. During coalescence of structured droplets the interfacial energy is continuously reduced while the elastic energy is increased by compression of the internal structure and, when the two processes balance one another, coalescence is arrested. Experimentally we observe that if either interfacial energy or elasticity dominates, total coalescence or total stability of droplets results. The stabilization mechanism is directly analogous to that in a Pickering emulsion, though here the resistance to coalescence is provided *via* an internal volume-based, rather than surface, structure. This study provides guidelines for designing anisotropic droplets by arrested coalescence but also explains some observations of "partial" coalescence observed in commercial foods like ice cream and whipped cream.

1 Introduction

Coalescence is a process of merging two or more droplets into a single, usually spherical, droplet. The driving force for coalescence to proceed, once initiated, is the reduction in interfacial energy by a minimization of the surface-to-volume ratio. Coalescence is a common instability mechanism in emulsions and its initiation is often sterically opposed by the addition of species like surfactants and polymers that adsorb to droplet surfaces.[1-4] Adsorbed colloids can also prevent coalescence initiation in Pickering emulsions[5-7] by forming a sufficiently continuous surface layer on droplets that provides steric stabilization. While stabilizing emulsions against the initiation of coalescence is a common goal, coalescence can also be initiated but then halted before formation of a single spherical droplet, a phenomenon known as

[a]Complex Fluid Microstructures, Corporate Engineering, Procter and Gamble Co., 8256 Union Center Blvd., West Chester, Ohio, USA. E-mail: spicer.pt@pg.com
[b]Department of Food Science, University of Wisconsin Madison, 105 Babcock Hall, 1605 Linden Drive, Madison, WI, USA

arrested coalescence.[8-13] Although arrested coalescence in food processes has been widely studied,[14-17] the exact mechanisms of resistance to coalescence and the stabilization of arrested structures are not yet fully understood because they typically occur in complex systems that do not permit the isolation of key phenomena.[8,9,18,19]

More fundamental studies [11,20] have shown that arrested structures require a dynamic balance between the forces driving shape relaxation and an opposing force. Coalescence can be stably arrested by droplet elasticity at the droplet surface[14] or within its volume.[21] While coalescence arrested by interfacial elasticity has led to fabrication of novel materials such as bijels[22,23] and armored droplets,[24,25] volume elasticity is often crucial for fabrication of structured food emulsions such as ice cream and whipped cream.[9,18,21] Through similar, but more complex, mechanisms non-spherical biological shapes can be formed by bacteria and viruses during dynamic processes like cell division and protrusion.[26] Recent work has also created simple cell analogues with emulsion drops containing elastic networks of actin.[27]

Arrest at flat interfaces [28,29] and on the surface of liquid doublets has only recently been studied for model systems.[20,30] Pawar et al.[20] found that the elasticity of the droplet surface determined the degree of arrest and, thus, the shape of the resultant liquid doublets or ellipsoids. In addition to surface structures, the elasticity of an entire droplet volume can also arrest coalescence to form novel colloidal structures but such effects have not been studied in model systems. In this article we make in situ observations of coalescing model viscoelastic droplets to isolate contributions of interfacial tension and elasticity to arrest. When elastic and interfacial forces possess similar magnitude, stable arrested droplet structures are found, while if either elastic or interfacial forces dominate then total stability or total coalescence, respectively, is observed.

2 Experimental details

Partially crystalline oil droplets are prepared by making oil-in-water (O/W) emulsions using oil containing anisotropic wax crystals.[10] The emulsions are then diluted to study droplet coalescence behavior by micromanipulation.

2.1 Emulsion preparation and characterization

Emulsions are prepared by mixing equal volumes (5 ml each) of an oil and aqueous phase. The dispersed oil phase consists of a mixture of hexadecane (99%, Sigma Aldrich) and wax (Petrolatum, Unilever) while the continuous aqueous phase consists of a 0.5 wt% microfibrous cellulose (MFC, CP-Kelco) dispersion ($\sigma_y \sim 0.17$ Pa) and 10 mM sodium dodecyl sulfate (SDS) (99%, Fluka) surfactant solution. The small yield stress of the continuous phase avoids sedimentation and uncontrolled movement of the small droplets during manipulation. The emulsion is made by combining the oil and the aqueous phases, heating to 75 °C, then shaking by hand for 10 s. The dispersed phase is initially a homogeneous solution but, as the emulsion cools down to room temperature, wax (mp ~ 40–60 °C) crystallizes inside the droplets. The hexadecane completely wets the crystals, containing them within the emulsion drop volume. The wax crystals are anisotropic solids with an aspect ratio ~10, as seen in Fig. 1, and are weakly attractive as gelation of the wax crystals is sometimes observed within droplets. The elastic modulus (G') of the wax–oil dispersion is measured in the linear viscoelastic regime by oscillatory shear experiments (AR 2000) at a strain of 0.1%.

Emulsions are fabricated to produce partially crystalline droplets with variable solid content (10%–50%). Emulsions are further diluted ~10× with an MFC dispersion without surfactant for the micromanipulation experiments.

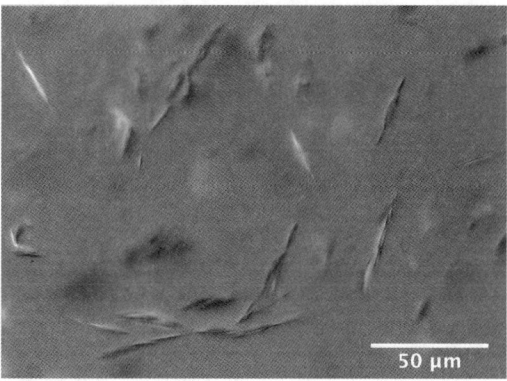

Fig. 1 Dispersion of anisotropic wax crystals in hexadecane (∼20% w/w wax).

2.2 Droplet coalescence study

Coalescence of two partially crystalline droplets is induced and observed *in situ* by micromanipulation experiments.[31-33] One of the droplets is held at the tip of a pre-pulled capillary by applying suction and is then manually contacted with the other droplet to study their coalescence behavior.

Tapered capillaries are fashioned from standard borosilicate glass capillaries (1 mm OD and 0.5 mm ID, Sutter Instruments) with a Micropipette Puller (Model P-97; Sutter Instruments). The tip of a pulled capillary is flattened using a Microforge (Model MF-830; Narishige Int'l. USA). The other end of the capillary is connected to a water reservoir (10 ml open syringe) by rubber tubing. Changing the height of the water reservoir adjusts the hydrostatic pressure applied at the flattened tip of the capillary, enabling manipulation of droplets. The capillary is mounted on a 3-axis coarse manipulator (Narishige Int'l USA) which is attached to a microscope stage (Zeiss axioplan-2). A diluted drop of emulsion (∼1 ml) is placed on a glass slide and the tip of the mounted capillary is aligned with one of the partially crystalline droplets using the micromanipulator. The droplet is drawn toward the capillary tip by applying suction (negative hydrostatic pressure). Applied suction is adjusted such that it holds the droplet stationary at the tip of the capillary without squeezing out any oil. The captured droplet is aligned and manually contacted with a second partially crystalline droplet suspended in the continuous phase and their coalescence is observed. Microscopic observation of coalesced droplets after 15 min is used to differentiate between total coalescence, arrested coalescence, and total stability behavior.

3 Results and discussion

As two droplets contact, coalescence is initiated when a liquid neck forms between them. The drops are then pulled together into a single spherical drop *via* flow of oil through the liquid neck. During the coalescence process, the instantaneous deformation of the shape is a function of the emulsion interfacial forces, as described by the Laplace pressure,[34] $P \propto \gamma/R$, where γ is the oil–water interfacial tension and R is droplet radius. The resulting change in shape of the coalescing structure is then a function of the instantaneous strain:

$$\varepsilon = \frac{\Delta L}{L_0} \tag{1}$$

where L_0 is the initial length before coalescence is initiated, $L_0 = 4 \times R$, and ΔL is the linear deformation. Fig. 2 illustrates coalescence of two hexadecane drops

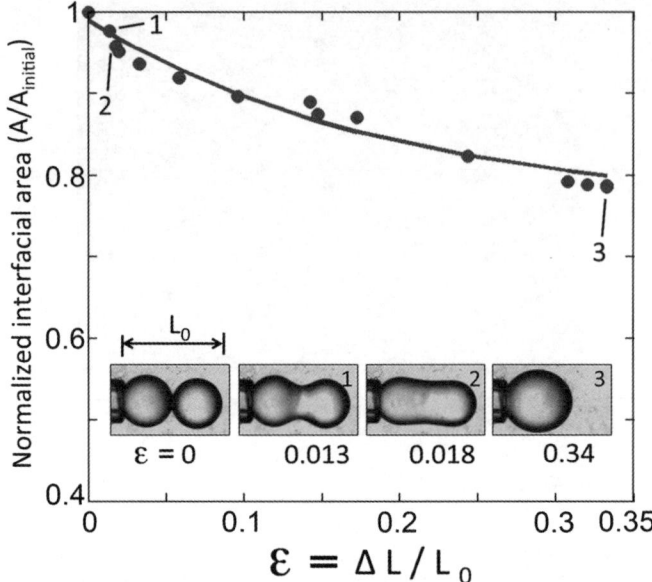

Fig. 2 Change in interfacial area as a function of strain during coalescence of hexadecane drops. Micrographs 1–3 show coalescing droplets at intermediate stages corresponding to the points labeled above.

without wax crystals and plots the change in interfacial area with strain. For coalescence of the drops in Fig. 2 only the drop viscosity resists the Laplace pressure-driven oil flow. The neck radius grows with coalescence and eventually interfacial tension drives the drop into a spherical shape.[35] As coalescence progresses , the interfacial area continuously decreases while ΔL, and thus the strain, increases accordingly. A reduction to ~79%, or $2^{-1/3}$, of the original interfacial area is achieved during total coalescence, corresponding to a strain of ~0.37, or $1-2^{-2/3}$, for identical drops.

Evolution of coalescing shapes thus depends on interfacial tension and the continuous and dispersed phase rheology. Fig. 2 indicates that the interfacial area decays exponentially with strain and can be fit using:

$$A = A_0 + A_1 \exp\left(-\frac{\varepsilon}{A_2}\right) \qquad (2)$$

where $A_0 = 0.76$, $A_1 = 0.23$, and $A_2 = 0.19$ correspond to the data in Fig. 2. Coalescing structures progress through various non-spherical shapes in Fig. 2 before reaching a final spherical one. Arresting coalescence at any of these intermediate shapes, and thus strains, is possible if the droplets offer enough elastic resistance to balance the Laplace pressure gradient.[20]

In contrast to our earlier study of interfacial elasticity-driven arrested coalescence, here we study coalescence of droplets whose interior volumes are elastic as a result of varying solid content. Fig. 3 reports the elastic modulus of wax–hexadecane dispersions for several solid levels. The partially crystalline oil phase elastic modulus grows as a power law with exponent ~4.8 with increasing solid content and, similar to our earlier findings for Pickering emulsions, we expect coalescence to be arrested at different stages depending on these parameters.

The coalescence behavior of partially crystalline droplets with increasing solid content is illustrated in Fig. 4. Fig. 4 a.1–e.1 indicate two partially crystalline droplets before coalescence is initiated with $\phi = 0.15$, 0.30, 0.40, 0.45, and 0.50 solid

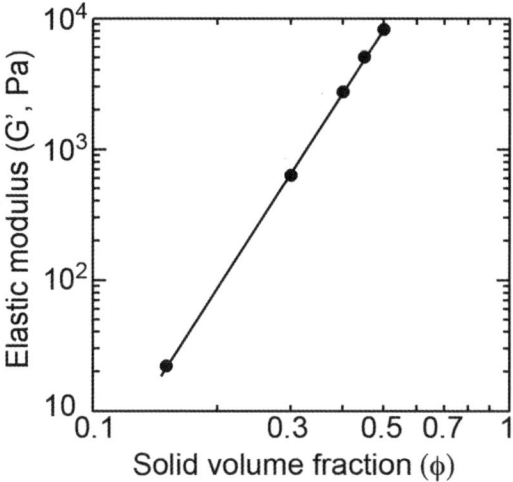

Fig. 3 Elastic modulus of wax–hexadecane dispersion as the volume fraction of wax is varied.

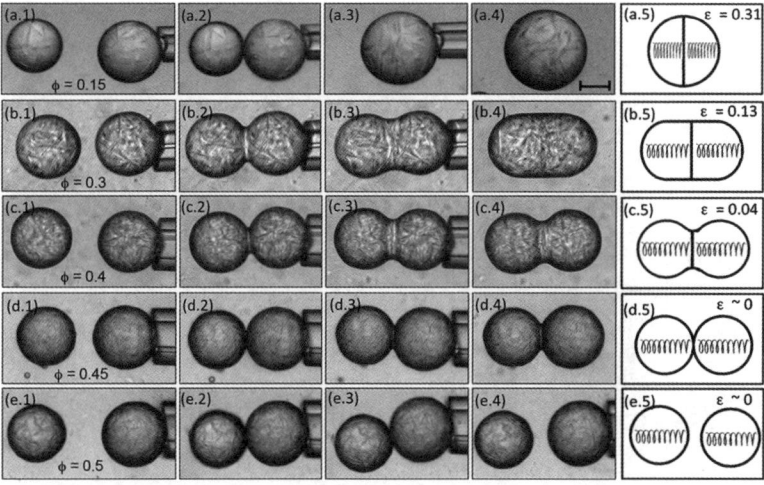

Fig. 4 Micrographs showing coalescence behavior of viscoelastic droplets, left to right, as the solid volume fraction is increased from top to bottom. Also shown are schematics of the final strain of each droplet pair. Scale bar = 50 μm.

volume fraction, respectively. As droplets are contacted coalescence begins and liquid oil flows from droplets to the coalescing neck (Fig. 4 a.2–d.2). At low solids fractions, around $\phi = 0.15$, droplets behave like weak gels that deform easily because of their low elastic moduli (Fig. 3). As a result, during coalescence the interfacial energy dominates elastic energy and the drop completely relaxes into a final spherical shape (Fig. 4 a.4). As the solids fraction is increased, the droplet elastic modulus increases and for $\phi = 0.30$–0.45 the particle network elasticity balances the Laplace driving force at an intermediate shape and arrested coalescence occurs (Fig. 4 b.4–d.4). It is clear that increasing the droplet elasticity arrests coalescence at earlier stages, changing the resulting shape. Further increasing solids fraction to $\phi = 0.50$ raises the droplet elasticity enough to completely prevent coalescence initiation (Fig. 4 e.4), at least to the limits of our observations. The progression of shapes and arrest in Fig. 4 is similar to our results for Pickering droplets[20] and the stabilization at high solids fractions is directly analogous to the interfacially-based stabilization seen in Pickering emulsions.[7]

Clearly the formation of stable arrested doublets results from a balance of the interfacial driving force and the elastic reaction force of the droplet microstructure. The partially crystalline droplets studied here contain an elastic network of solid crystals saturated with liquid oil: a poroelastic colloidal gel. During coalescence, Laplace pressure gradients exert stress on and deform the solid network. As coalescence progresses the drop area and, thus, the interfacial energy, $E_{\text{interfacial}}$, is continuously reduced:

$$E_{\text{interfacial}} = A\gamma \tag{3}$$

where γ is 10 mN m^{-1}, and A is the total interfacial area of two coalescing droplets. While coalescence decreases interfacial energy, the deformation or strain of the elastic droplet microstructure leads to an increase of elastic energy, E_{elastic}, stored within the microstructure:

$$E_{\text{elastic}} = \frac{3}{2}G'\varepsilon^2 V \tag{4}$$

where G' is the droplet shear modulus that has been multiplied by three to yield the Young's modulus [36] and V is total volume of the coalescing droplets. The decreasing interfacial energy, eqn (3), and increasing elastic energy, eqn (4), can be balanced at any strain in the range between total stability, $\varepsilon = 0$, and total coalescence, $\varepsilon = 0.37$, stabilizing doublet shapes at various minimum energy states. A simple energy balance can be used to explore the minima corresponding to stable arrested coalescence states.

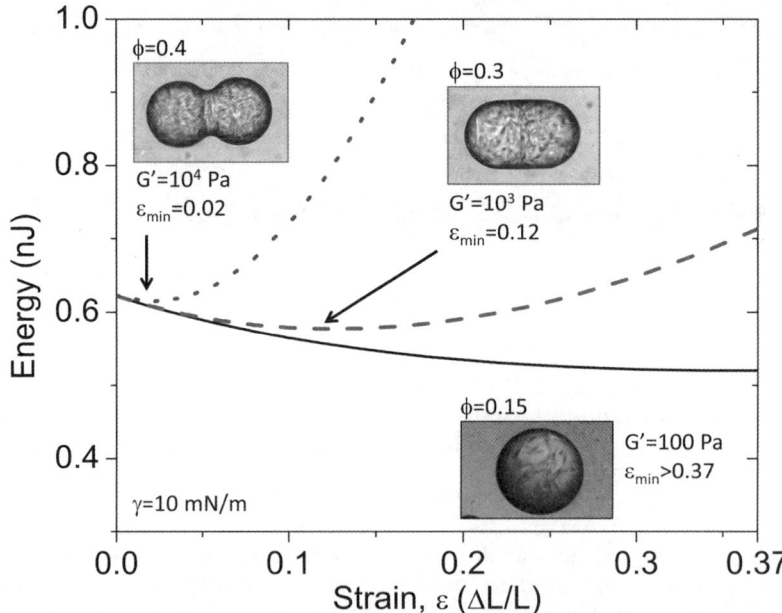

Fig. 5 Comparison of change in total energy during arrested and total coalescence of droplets. A zero strain corresponds to droplets before coalescence and a maximum strain of 0.37 represents total coalescence into a spherical droplet. The two arrows indicate the location of energy minima for each of the arrested coalescence cases.

We calculate the change in energy during coalescence of droplets with three relevant elastic moduli, $G' = 100$ Pa, 10^3 Pa, and 10^4 Pa, and plot the total energy against droplet deformation in Fig. 5. Here we assume the interfacial energy is solely a function of interfacial area during coalescence, that the droplet elastic modulus is a constant with strain, and that the shape evolution in Fig. 2 approximates the behavior of partially crystalline droplets. We also neglect any minimum deformation required to overcome disjoining pressures and initiate coalescence.[37,38] Accordingly, the interfacial area decays exponentially with strain (eqn (2)). However, the elastic energy (eqn (4)) varies as a power law with strain and is proportional to the elastic modulus. Combining both energy terms, a minimum in the total energy occurs when the stored elastic energy increases to attain the same magnitude as the interfacial energy lost by drop compression during coalescence.

In Fig. 5 for droplets with low elastic modulus, $G' = 100$ Pa, the bottom solid curve indicates the change in elastic energy is negligible *versus* the change in interfacial energy. As a result, the total energy is dominated by the exponential decay of interfacial energy and relaxes to a minimum value at a strain corresponding to a single sphere, $\varepsilon = 0.37$. For an intermediate elastic modulus of $G' = 10^3$ Pa the middle dashed curve in Fig. 5 has a minimum at an intermediate $\varepsilon = 0.12$, approximating that of the non-spherical shape shown. Further increasing the droplet elastic modulus increases the rate of elastic energy storage during coalescence and enables arrest at much smaller strains, moving the minimum down to $\varepsilon = 0.02$ in the top dotted curve of Fig. 5 and qualitatively matching the results in Fig. 4.

While this simple model correctly captures the mechanism underlying the arrested coalescence state, it is not complete. For example, our model predicts a minimum in the total energy, even at high values of G' and at vanishingly low strains, though experimentally we see no coalescence initiation for such high modulus structures. This discrepancy could arise because the model does not include coalescence initiation even though experimentally our drops required a minimum critical strain, $\varepsilon_{critical} \sim 0.01$, to trigger coalescence. A more detailed model quantifying the condition for coalescence onset appears necessary to fully describe an elastic stabilization mechanism but the model provides a starting point for understanding arrested coalescence and for mapping the expected behavior of droplets in practical food processes.

Using the above conceptual model, we predict that total coalescence and arrested coalescence can be obtained by appropriately tuning the droplet elastic modulus and interfacial tension. Adding eqn (3) and eqn (4) to obtain the total energy of two coalescing droplets, taking the derivative with respect to strain, then equating the result to zero allows us to locate minima for a range of curves like those plotted in Fig. 5:

$$\gamma = \frac{A_2}{A_1} \exp\left(\frac{\varepsilon_{min}}{A_2}\right) G' R \varepsilon_{min} \qquad (5)$$

We can use eqn (5) to calculate, for a droplet of radius R, the elastic moduli and interfacial tensions yielding either coalesced or arrested structures and we can speculate on the existence of a region of elastic stabilization as well. Plotting lines of constant minimum strain in Fig. 6 for the limiting cases of complete coalescence, $\varepsilon_{min} = 0.37$, and total stability against coalescence, $\varepsilon_{min} = \varepsilon_{critical} \sim 0.01$, produces a map of the three regions for identical size elastic droplets of a given radius. The left region corresponds to the states of complete relaxation to a spherical shape and is separated from the middle arrested coalescence region by the $\varepsilon_{min} = 0.37$ line. The right region represents the states where we experimentally observe total stability against coalescence and is bounded on its left side by the dashed $\varepsilon_{min} = 0.01$ line. Within the arrested coalescence region the structures can vary from ellipsoidal to doublet shapes. If further validated, such a map could provide a useful design basis for emulsion formulation and processing depending on the desired microstructure and bulk rheology of a product. Ultimately more experiments are

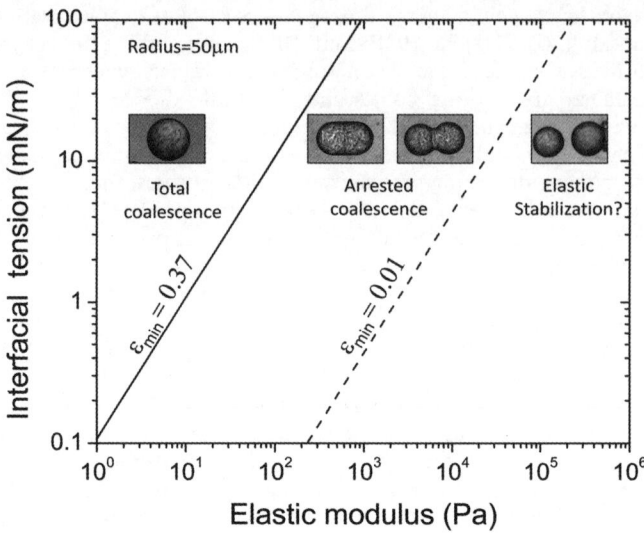

Fig. 6 Coalescence behavior map for identical size 50 μm droplets showing a conceptual basis for designing droplet structures using physical properties like interfacial tension and droplet elasticity.

required to test these ideas given the very basic nature of the proposed model, but its consistency with our experimental data is encouraging.

Here, as in the case of Pickering emulsions,[20] arrested coalescence occurs as the result of a balance between interfacial and elastic forces. By moving the arresting elastic microstructure from the drop surface to its volume, we change the mechanism of arrest from a jamming due to increased volume fraction to a reaction due to elastic deformation. While for Pickering droplets the solids surface concentration increases during coalescence, the solids volume fraction in the drops examined here remains constant. Thus the arrest is not due to irreversible jamming but to a reversible elastic reaction. The distinction is important as a completely jammed interface, like those seen in armored droplets [24] and other arrested Pickering emulsions, [20] usually will not permit subsequent addition of droplets to form a larger network. However, partially crystalline droplets can form unique supracolloidal materials with multi-unit structures *via* sequential arrested coalescence, as shown in Fig. 7. Furthermore, in certain food emulsions like ice cream and whipped

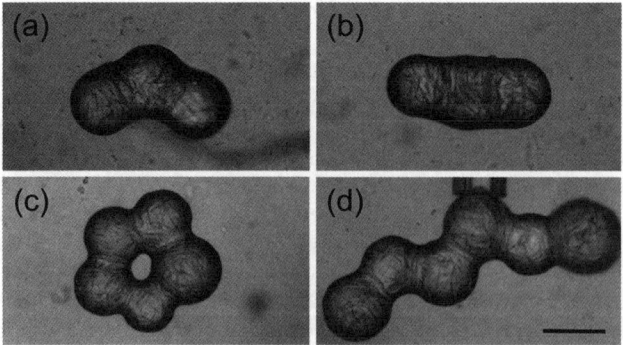

Fig. 7 Anisotropic structures of partially crystalline shapes fabricated by arrested coalescence of three or more droplets. Scale bar = 100 μm.

toppings, volume-spanning rheological networks of clustered fat droplets are formed during processing, providing structure and influencing physical and sensorial properties.[17] Our study may provide a means to design and characterize such applied systems using commonly measured parameters.

Acknowledgements

We gratefully acknowledge discussions of these results with Eric Furst (U. Delaware), Steven Hudson (NIST), and Véronique Trappe (U. Fribourg). We also sincerely thank an anonymous referee for valuable commentary on the manuscript.

References

1 F. Leal-Calderon, V. Schmitt and J. Bibette, *Emulsion Science: Basic principles*, Springer, 2007.
2 K. S. Birdi, *Surface and colloid chemistry: Principles and applications*, CRC Press Taylor and Francis group, 2010.
3 P. C. Hiemenz and R. Rajagopalan, *Principles of Colloid and Surface Chemistry*, Marcel Dekker, Inc., 1997.
4 B. P. Binks, *Modern Aspects of Emulsion Science*, The Royal Society of Chemistry, 1998.
5 S. U. Pickering, *J. Chem. Soc. Trans.*, 1907, **91**, 2001–2021.
6 N. D. Denkov, I. B. Ivanov, P. A. Kralchevsky and D. T. Wasan, *J. Colloid Interface Sci.*, 1992, **150**, 589–593.
7 B. P. Binks, *Curr. Opin. Colloid Interface Sci.*, 2002, **7**, 21–41.
8 E. Fredrick, P. Walstra and K. Dewettinck, *Adv. Colloid Interface Sci.*, 2010, **153**, 30–42.
9 P. Walstra, *Physical Chemistry of Foods*, Marcel Dekker Inc., 2003.
10 J. Giermanska, F. Thivilliers, R. Backov, V. Schmitt, N. Drelon and F. Leal-Calderon, *Langmuir*, 2007, **23**, 4792–4799.
11 A. R. Studart, H. C. Shum and D. A. Weitz, *J. Phys. Chem. B*, 2009, **113**, 3914–3919.
12 A. B. Subramaniam, M. Abkarian and H. A. Stone, *Nat. Mater.*, 2005, **4**, 553–556.
13 J. Giermanska-Kahn, V. Laine, S. Arditty, V. Schmitt and F. Leal-Calderon, *Langmuir*, 2005, **21**, 4316–4323.
14 K. Boode, P. Walstra and A. E. A. de Groot-Mostert, *Colloids Surf., A*, 1993, **81**, 139–151.
15 K. Boode and P. Walstra, *Colloids Surf., A*, 1993, **81**, 121–137.
16 D. Rousseau, *Food Res. Int.*, 2000, **33**, 3–14.
17 F. Thivilliers-Arvis, E. Laurichesse, V. Schmitt and F. Leal-Calderon, *Langmuir*, 2010, **26**, 16782–16790.
18 H. D. Goff, *J. Dairy Sci.*, 1997, **80**, 2620–2630.
19 J. Benjamins, M. H. Vingerhoeds, F. D. Zoet, E. de Hoog and G. A. van Aken, *Food Hydrocolloids*, 2009, **23**, 102–115.
20 A. B. Pawar, M. Caggioni, R. Ergun, R. W. Hartel and P. T. Spicer, *Soft Matter*, 2011, **7**, 7710–7716.
21 F. Leal-Calderon, F. Thivilliers and V. Schmitt, *Curr. Opin. Colloid Interface Sci.*, 2007, **12**, 206–212.
22 E. M. Herzig, K. A. White, A. B. Schofield, W. C. K. Poon and P. S. Clegg, *Nat. Mater.*, 2007, **6**, 966–971.
23 F. Thivilliers, N. Drelon, V. Schmitt and F. Leal-Calderon, *Europhys. Lett.*, 2006, **76**, 332–338.
24 A. B. Subramaniam, M. Abkarian, L. Mahadevan and H. A. Stone, *Nature*, 2005, **438**, 930–930.
25 A. B. Subramaniam, M. Abkarian, L. Mahadevan and H. A. Stone, *Langmuir*, 2006, **22**, 10204–10208.
26 T. J. Mitchison, G. T. Charras and L. Mahadevan, *Semin. Cell Dev. Biol.*, 2008, **19**, 215–223.
27 M. Claessens, R. Tharmann, K. Kroy and A. Bausch, *Nat. Phys.*, 2006, **2**, 186–189.
28 H. Xu, S. Melle, K. Golemanov and G. Fuller, *Langmuir*, 2005, **21**, 10016–10020.
29 B. Madivala, S. Vandebril, J. Fransaer and J. Vermant, *Soft Matter*, 2009, **5**, 1717–1727.
30 H. L. Cheng and S. S. Velankar, *Langmuir*, 2009, **25**, 4412–4420.
31 E. A. Evans and R. Skalak, *Mechanics and Thermodynamics of Biomembranes*, CRC Press, 1980.
32 R. Kwok and E. Evans, *Biophys. J.*, 1981, **35**, 637–652.
33 S. Lee, D. H. Kim and D. Needham, *Langmuir*, 2001, **14**, 5537–5543.

34 H. J. Butt, K. Graf and M. Kappl, *Physics and Chemistry of Interfaces*, Wiley VCH GmbH and Co. KGaA, 2003.
35 M. Wu, T. Cubaud and C. M. Ho, *Phys. Fluids*, 2004, **16**, L51–L54.
36 M. A. Biot, *J. Appl. Phys.*, 1941, **12**, 155–164.
37 J. Bibette, D. Morse, T. Witten and D. Weitz, *Phys. Rev. Lett.*, 1992, **69**, 2439–2442.
38 D. Filip, V. Uricanu, M. Duits, W. Agterof and J. Mellema, *Langmuir*, 2005, **21**, 115–126.

General discussion

Professor Bolhuis opened the discussion of the paper by Dr Royall: In Section 4.3 of the discussion you report that simulations employing Brownian dynamics show no sign of a phase separation, while for a molecular dynamics simulation such a phase separation occurs easily and is even hard to prevent. Now, you speculate that the Newtonian dynamics somehow "knows its way through the energy landscape". Can you elaborate on this rather mystical statement?

Dr Royall responded: Not really. As it stands, it is simply an observation. One can argue that in molecular dynamics, a system can find its way smoothly through the energy landscape to the equilibrium state, however in Brownian dynamics, each particle experiences random noise, which tends to "kick" it off its trajectory to the equilibrium state. What is needed, is a direct comparison of the same system at the same state point with both types of dynamics.

Professor Bolhuis continued: Clearly, the inertial terms of the dynamics are important to explain this behavior. Inertia might be at the origin of correlated jumps between minima in the rugged energy landscape, effectively enhancing the dynamics. The overdamped Brownian dynamics looses all memory after one time step.

Dr Royall responded: Agreed.

Dr Velikov asked: What is the possible impact of local particle order in the individual clusters on the gel rheological/mechanical properties?

Dr Royall answered: Our mean squared displacement measurements indicate that the system is less mobile upon crystallisation. Therefore, one expects that a crystalline gel should be stronger and stiffer (less flexible) than its non-crystalline counterpart. This is because the strands of the gel are thicker, and each particle has more bonds to near neighbours in a crystalline arrangement.

Dr van der Sman inquired: In your paper you mention an effective temperature. Is this a normalization against kT, or do you view it as a characterization of the out-of-equilibrium situation, similar to our paper presented (in the first session of the Faraday Discussion, by Vollebregt *et al.*)?

Dr Royall responded: We fix kT here. The simulations are somewhat inspired by recent developments in colloidal model systems, see for example Taylor *et al.*, 2012, arXiv:1205.0072. In this class of experiment, absolute temperature is fixed, or varies only to a small extent. The behaviour originates from the change in (attractive) interaction strength, so cooling is achieved by an increase in well depth, heating by a decrease in well depth. We believe this is broadly equivalent to (*e.g.* molecular) systems where interactions are typically fairly constant over the temperature regime of interest but temperature is varied.

Professor de Kruif opened the discussion of the paper by Dr Krueger: In Fig. 1–3 it seems as if the minimum of the formfactor is reached at 3 nm^{-1}. Assuming a spherical particle its radius would be 4.49/3 nm^{-1} = 1.5 nm, which is a reasonable value and would correspond to a d-spacing of about 3 nm as is found for the freeze concentrated phase In a low salt, charge stabilized, colloidal–protein dispersion, the d-spacing is maximal due to the repulsive interaction. On increasing concentration the d-spacing decreases to the minimum value which is the diameter of the particles.

In Fig. 4 the d-spacing decreases from about 7.5 nm to about 3.0 nm. This would correspond to a concentration increase by a factor $(2.5)3 = 15.6$. The original concentration is 100 mg ml^{-1} and it would increase to 1560 mg ml^{-1} in the freeze concentrated phase, a rather high value. Is there an explanation?

Dr Krueger answered: The concentration calculated is higher than what would be expected for theoretical dense packing of proteins (about 1380 mg ml^{-1}). Thus, there is clearly something that doesn't add up here. The d-spacing estimated in eqn 1 of the paper, based on the q value of the peak maximum, is an approximation that is derived from diffraction theory, i.e., $d = 2\pi/q$. Although the solution is frozen, the protein isn't undergoing crystal packing. Also, other solutes in the solution can have an effect on the actual distance between nearest neighbors. If these data are fit using model $S(q)$ functions that take these issues into account, the maximum volume fractions obtained do not suggest such a high concentration. Furthermore, our Monte Carlo analysis of the proteins under these conditions provides a visual description of this more reasonable scenario (we describe fits to alternative functions in reference 12 of the paper; J. E. Curtis, H. Nanda, S. Khodadadi, M. Cicerone, H. J. Lee, A. McAuley, and S. Krueger, *J. Phys. Chem. B*, DOI: 10.1021/jp304772d.) Thus, the fitting approach we chose to use in this paper is just a convenient way to monitor the changes in the protein concentration using a single parameter.

Dr van der Sman commented: In your final figure you show a cartoon of the protein distribution in the frozen system. You have concluded that proteins are attracted to interface between water and ice. I know of a hypothesis that states that antifreeze proteins are absorbed on these interfaces,[1] because they lower the interfacial tension similar to surfactants absorbed on emulsions. Do you think similar hypothesis holds for the adsorption of lysozyme on the water–ice interface ?

1. Y. Mastai, J. Rudloff, H. Cölfen, M. Antonietti, Control over the structure of ice and water by block copolymer additives, *ChemPhysChem*, 2002, **3**, 119–123.

Dr Krueger answered: While we cannot say for sure that the same mechanism is operating here for lysozyme, we can say that the nature of the scattering at low q suggests this type of behavior. The paper you cite may offer an explanation of our data.

Dr Bot asked: You seem to be interested in the large scale protein aggregate structures, but it seems that you have restricted yourself to a relatively narrow q range. Have you considered investigating a wider q range, or over extending the window by using e.g. Spin-echo SANS (SESANS) for this purpose?[1]

1. R. H. Tromp and W. G. Bouwman, *Food Hydrocolloids*, 2007, **21**, 154–158.

Dr Krueger answered: We did, in fact, make measurements to lower q values for some of our runs. The scattering that is seen in the lower q region of the data shown in the paper exhibits a power law behavior that extends to the lowest q values that we were able to measure with SANS, i.e., around 0.003 Å^{-1}. We also made USANS measurements to extend the q range to lower values and to determine whether we could see evidence of larger protein clusters. Unfortunately, the scattering from the titanium windows used in our sample cells dominated the scattering in the USANS range and the signal-to-noise from the sample was less than ideal. Quartz windows are generally used in USANS cells, but they cracked upon freezing of the samples, except for the samples with higher salt content. We are certainly still interested in exploring the USANS range if we can find a suitable replacement for the titanium windows. We haven't considered using SESANS at this time.

Professor Cates asked: If the contrast between air and water is matched at 8% D_2O, then I would expect that adding protein would cause the low q scattering from cracks to reappear (at least for the 8% composition). This is because at low q you would now have a finite contrast between air and the water–protein mixture. Hence, even if the latter has no interesting low q structure of its own, you would still pick up a low q signal.

Dr Krueger responded: Once protein is introduced into the system, then the scattering from the system is no longer completely matched at 8% D_2O. Thus, we agree that some scattering is expected under these conditions. If 8% D_2O matches the scattering of air, then it won't simultaneously match that of the protein, since the scattering length densities of air and protein are very different. The fact that we observed scattering at 8% D_2O when protein was present was indeed expected. In fact, the low q scattering observed at 40% D_2O, *i.e.*, the protein match point, was very similar to that seen in the absence of protein, except near the air match point (the contrast variation experiments we performed are described in reference 12 in the paper). Thus, the scattering due to the cracks looked the same whether protein was present or not. However, the scattering at the other contrasts, including 8% D_2O, in the presence of protein was very different, indicating to us that the protein was contributing significantly to the low q scattering, in addition to scattering in the high q region (where expected due to the dense packing upon freezing).

Dr van der Sman asked: Can you give further details on your definition of cracks in the ice phase? Are these grain boundaries between different ice crystals (do you assume polycrystalline ice)? Why do you assume they are/become filled with air ?

Dr Krueger replied: Yes, we are assuming polycrystalline ice. We have some preliminary X-ray diffraction data that suggests hexagonal ice in the system. The scattering that is coming from the ice in absence of protein is definitely due to a contrast between the ice and air. While there may be some air at grain boundaries between different ice crystals, what we are seeing could also be due to air in the solvent that forms bubbles that get trapped when the ice forms. We do see such scattering at low q from macroscopic air bubbles in D_2O solution. While we do degass our buffers prior to measurement, we may not be completely removing all of the air. The reason I used the term cracks, rather than bubbles, is that the low q scattering that we see in the ice is somewhat different than what we see in the D_2O solution and reminded me more of the type of scattering we often see at interfaces between grain boundaries.

Dr van Gruijthuijsen addressed Dr Krueger and Professor Schurtenberger: I was surprised to see the differences in the d-spacing (or peak position) of the lysozyme interaction peak in the frozen samples in the presence or absence of salt. Intuitively and from my experience with colloidal gels, I would expect the average binding distances to be smaller for a purely attractive system, *i.e.* in the presence of salt, and larger for a system with probably a mixed potential, *i.e.* in the absence of salt. Your findings exactly oppose this.

I'd like to raise a tentative argument, that the higher salt concentration in the non-frozen water phase in the protein-dense regions increases the local osmotic pressure, thus retaining more liquid water and creating a less dense protein-concentrated phase. The absence of salt would then allow for a closer packing of the lysozymes, albeit charged and barely screened. In line with this, did you perform or think of performing any measurements on the dynamical properties of the proteins? And to further quantify things, could you make some estimates of the osmotic and elastic pressures in the system, notably in the protein-concentrated phase?

Professor Schurtenberger responded: This observation is indeed surprising at first. However, it is also important to realize that the two systems undergo most likely a different freezing scenario caused by the different state diagrams. For 0 M NaCl the sample directly goes from a cluster fluid to freezing, which allows for a much more pronounced freeze-concentration process then in the case of 0.4 M NaCl, where the sample first forms an arrested spinodal network that is much more difficult to further compact *via* the freezing process. When interested in the osmotic pressure of the concentrated protein solutions as a function of ionic strength, you could look at Gögelein *et al.*,[1] where a detailed model for the Helmholtz free energy of concentrated lysozyme solutions has been developed that results in an almost quantitative theoretical phase diagram for different salt concentrations.

1. C. Gögelein, G. Nägele, R. Tuinier, T. Gibaud, A. Stradner, and P. Schurtenberger, *J. Chem. Phys.* , 2008, **129**, 085102.

Dr Krueger added: We postulate in the paper that the reason for the larger *d*-spacing, and hence the larger packing distance, in the presence of salt is due to a change in shape of the protein. In other words, the protein is bigger, so the distance between them is larger in the non-frozen water phase. 'Bigger' can mean that the lysozyme monomer has changed shape due to unfolding or that lysozyme perhaps exists as a mixture of monomers and other small, higher order oligomers, such as dimers or trimers. We favored the latter because we didn't find evidence for cold denaturation of lysozyme in the literature and because we already observe aggregates of lysozyme in the liquid state.

We find your postulate interesting in that the salt almost certainly is in the non-frozen water phase and its concentration is quite high. Thus, it certainly must have some effect on the water, whether it be an effect on the hydration layer around the protein or even an effect on the amount of liquid water retained in this phase, as you suggest. We agree that performing measurements of the type you suggest might help shed some light on this question. However, we have not thought about performing such measurements in detail. We do hope to be able to calculate some relevant physical quantities based on our Monte Carlo studies to further quantify the nature of the non-frozen water phase (we describe our initial MC studies in our reference 12). This work is ongoing and the goal is to reproduce our scattering results in our model system and then use the model system to calculate useful—and hopefully insightful—physical quantities.)

Dr Frith continued by discussing the paper by Professor Peter Schurtenberger: Considering the lysozyme phase diagram: when the sample is quenched into the spinodal regime, the system arrests and gels because the concentrated phase is continuous and becomes solid. If this is the case, as the concentration is reduced and the amount of phase volume of concentrated phase is reduced, there should come a point where the concentrated phase no longer percolates, and gelation does not happen. Is this correct or does the spinodal nature of the phase separation alter this behaviour in some way?

Professor Schurtenberger answered: We would indeed expect a different behavior at much lower initial concentrations. Under these conditions a nucleation and growth mechanism should take place, leading to individual (arrested) droplets that could then sediment and form a jammed arrested phase.

Dr van der Sman remarked: if you add salt to the lysosyme–water–ice systems, what happens? At water activities of $a_w < 0.75$, salt can crystallize. How would that change the signal of the neutron scattering ?

Dr Krueger answered: The salt changes the scattering at low q values significantly, both in the absence and in the presence of protein in the system. The salt is forming large structures even in the absence of protein and this could certainly be due to the fact that the salt is crystallizing.

Professor Schurtenberger asked: When comparing the data sets shown in Fig. 1–4 for lysozyme solutions in the absence of added salt and with added salt, it is important to realize that temperature quenches in these systems will result in very different freezing scenarios. There is in fact a considerable amount of published data on these two extreme cases available (and not quoted in the paper, see for example references 1–7 below and references therein) that would help to better interpret the data given in this paper. When looking at Fig. 1 in the paper of Gibaud *et al.* also presented in this Faraday Discussion,[8] it becomes apparent that a deep quench at high salt concentrations will initially lead to an arrested spinodal decomposition followed by freezing of the water-rich domains, while such a quench in the absence of salt will directly lead to freezing of the lysozyme cluster fluid as in this case the (metastable) binodal is suppressed by the weakly screened Coulomb repulsion between the proteins and clusters. The dependence of the binodal and arrest line on the salt concentration is given in ref. 6, while ref. 7 provides a detailed summary of the behavior of lysozyme solutions at low ionic strength. This will also explain the data in Fig. 5 of this paper, where the T-dependence of the so-called d-spacing appears counter-intuitive at first. However, when looking at the data for 0 M NaCl, this value corresponds to the so-called cluster peak.[1–7] The fact that the d-spacing first increases upon decreasing T simply reflects the growing cluster size with decreasing T (see ref. 7). The fact that it then sharply drops upon freezing and indicates smaller nearest neighbor distances than for 0.4 M NaCl most likely is due to the fact that for 0 M NaCl the sample directly goes from a cluster fluid to freezing, which allows for a much more pronounced freeze-concentration process then in the case of 0.4 M NaCl, where the sample first forms an arrested spinodal network that is much more difficult to further compact *via* the freezing process.

1. A. Stradner, H. Sedgwick, F. Cardinaux, W. C. K. Poon, S. U. Egelhaaf, and P. Schurtenberger, Equilibrium cluster formation in concentrated protein solutions and colloids, *Nature*, 2004, **432**, 492–495.
2. A. Stradner, F. Cardinaux and P. Schurtenberger, A small-angle scattering study on equilibrium clusters in lysozyme solutions, *J. Phys. Chem. B*, 2006, **110**, 21222–21231.
3. F. Cardinaux, A. Stradner, P. Schurtenberger, F. Sciortino and E. Zaccarelli, Modeling equilibrium clusters in lysozyme solutions, *Eur. Phys. Lett.*, 2007, **77**, 48004.
4. F. Cardinaux, T. Gibaud, A Stradner, and Schurtenberger, The interplay between spinodal decomposition and glass formation in proteins exhibiting short range attractions, *Phys. Rev. Lett.*, 2007, **99**, 118301.
5. T. Gibaud and P. Schurtenberger, A closer look at arrested spinodal decomposition in protein solutions, *J. Phys. Cond. Mat.*, 2009, **21**, 322201.
6. T. Gibaud, F. Cardinaux, J. Bergenholtz, A. Stradner, and P. Schurtenberger, Phase separation and dynamical arrest for particles interacting with mixed potentials—the case of globular proteins revisited, *Soft Matter*, 2011, **7**, 857.
7. F. Cardinaux, E. Zaccarelli, A. Stradner, S. Bucciarelli, B. Farago, S. U. Egelhaaf, F. Sciortino, and P. Schurtenberger, Cluster-driven dynamical arrest in concentrated lysozyme solutions, *J. Phys. Chem. B*, 2011, **115**, 7227
8. T. Gibaud, N. Mahmoudi, J. Oberdisse, P. Lindner, J. S. Pedersen, C. L. P. Oliveira, A. Stradner, and P. Schurtenberger, New routes to food gels and glasses, *Faraday Discuss.*, 2012, **158**, DOI: 10.1039/C2FD20048A.

Dr Krueger replied: Thank you for your comment. We also noted in our paper that the decrease in d-spacing in the 0 M NaCl lysozyme solution upon cooling was likely due to clustering (reference 9 in our paper: P. Falus, L. Porcar, E. Fratini, W.-R. Chen, A. Faraone, K. Hong, P. Baglioni and Y. Liu, *J. Phys.: Condens. Matter*, 2012, **24**, 064114), so we agree with your explanation.

Our data do seem to suggest two different freezing scenarios, as the 0.4 M lyso-zyme solution clearly showed phase separation prior to freezing. We do state in our paper that these results are consistent with previous studies of lysozyme as a function of salt concentration (our reference 13; V. G. Taratuta, A. Holschbach, G. M. Thurston, D. Blankschtein and G. B. Benedek, *J. Phys. Chem.*, 1990, **94**, 2140–2144). The question is whether the same is true for the 0 M NaCl concentra-tion. We clearly saw phase separation at 0 °C, which we stated that we assumed correlated with the onset of freezing (based on our reference 13 above, which sug-gested that the cloud point for a 0 M NaCl lysozyme solution was below the freezing point of the solution). However, this cannot be determined from the SANS data alone and the work in our reference 13 was done on lysozyme in an H_2O-based buffer whereas we used a D_2O-based buffer. So, the question remained as to whether two freezing scenarios do, in fact, exist. Thus, the information you provided is very helpful.

Professor Bolhuis returned the discussion to the paper by Dr Royall: In Fig. 6 you show the structural behavior of a monodisperse system for different quench times. While the FCC crystal is most likely the stable state for your potential, you clearly see more HCP formed. Can you explain this fact? Also, while you observe more FCC structure forming when lowering the temperature the FCC structure in Fig. 6b shows a peak around $\beta\varepsilon = 10$, and then decreases again. Can you explain the nature and the origin of this peak?

Dr Royall replied: We agree the FCC is likely to be the crystal structure. There-fore, presumably the formation of HCP is related to the fact that it may be easier to access. The fluid is dominated by fivefold symmetric structures, so forming the crystal involves bond breaking. Given that for hard spheres, the free energy of HCP and FCC is so similar, one might imagine that the pathway by which either is formed could dominate the outcome. In these gels, we expect little transition between HCP and FCC once HCP is formed. It would therefore be sensible to carry out an analysis of the pathway to crystallisation at the particle level, as we have done for nearly hard spheres in J. Taffs, S. R. Williams, H. Tanaka, C. P. Royall, 2012, arXiv:1206.5526.

For the second question, one generically finds an optimum in the effective temper-ature for crystallisation. This is the case in 'instantaneous quenches', see C. P. Royall, S. R. Williams, T. Ohtsuka and H. Tanaka, *Nat. Mat.*, 2008, **7**, 556. Here, our quench time is fixed, so, as we quench deeper, the quench rate is increased. Therefore in the case of the deeper quenches, the system spends less time in the 'good assembly' regime where the well depth is around 4–5 kT, which leads to poorer crys-tallisation for deeper quenches.

Professor Nicolai resumed the discussion of the paper by Peter Schurtenberger: One way that the phase separated system can coarsen is by evaporation of individual proteins into the low density phase and redeposition. The fact that no coarsening is observed suggests that the particles at the surface are in practice irreversibly bound as soon as they have a few neighbours. Given this situation, what is the molecular origin of the high frequency modulus, the slow relaxation process, and the low frequency modulus? Two elastic moduli separated by a slow relaxation process are not observed in heat-set protein gels at similar concentrations.

Professor Schurtenberger responded: When looking at the absence of coarsening, it is important to realize that due to the arrested spinodal decomposition process there is still a chemical potential gradient in place that has the opposite sign as the concentration gradient, *i.e.* we have a situation where 'uphill diffusion' would continue to occur if the dense phase would not actually arrest. While individual proteins at the surface are relatively weakly bound (the contact value of the attractive potential is only a few kT), this results in a quasi-stationary situation as long as the system is not subjected to an external stress above the yield stress of the arrested continuous backbone. We believe that the intriguing and unusual viscoelastic properties arise from the large difference between the protein size and the characteristic dimensions (strand length and thickness) of the network. We have recently developed a porous medium model that is indeed capable of reproducing the viscoelastic properties of these systems as a function of quench depth and volume fractions.[1] It demonstrates that the high and low frequency elastic moduli can be respectively attributed to stretching and bending modes. The unexpected decoupling of the two modes in the frequency domain is attributed to the length scale involved: while stretching mainly relates to the relative displacement of two particles, bending involves the deformation of a strand with a thickness of the order of hundreds of particle diameters.

1. T. Gibaud, A. Zaccone, E. del Gado, V. Trappe and P. Schurtenberger, manuscript in preparation.

Dr Frith remarked: When compared with a colloidal system undergoing an analogous process of gelation by arrested spinodal decomposition, it seems that the ratio of the sizes of the structures developed to their constituent molecules or particles is very different. You mention a characteristic length-scale (correlation length) that is 2–300 times that of the molecule, whereas, in a colloidal system, this ratio might be 10 or more times smaller. Does this difference in the ratios of the length scales to molecular and colloidal systems arise because the spinodal structure is able to ripen more in molecular systems than in colloidal systems prior to arrest occurring, or does it reflect the different systems having different characteristic length scales from the moment spinodal decomposition starts?

Professor Schurtenberger responded: The characteristic domain size in the initial stage of spinodal decomposition is determined by the interplay between particle diffusion and interfacial tension. It is interesting to note that for colloidal suspensions, the characteristic domain size in the initial state of spinodal decomposition after a deep quench is comparable for proteins, small nanoparticles and large colloids.[1–4] This implies that while for proteins and small nanoparticles strands are hundreds of particle diameters thick, for large colloids they encompass a few particles only. This difference between the arrested structures in colloids and proteins is thus not because of a different ripening (as arrest should always occur in the early stage of spinodal decomposition), but because of the characteristic domain size that seems quite independent of particle size.

1. T. Gibaud and P. Schurtenberger, *J. Phys.: Condens. Matter*, 2009, **21**, 322201.
2. M. H. G. M. Penders and A. Vrij, *Adv. Colloid Interface Sci.*, 1991, **36** 185–217.
3. R. Tuinier, J. K. G. Dhont, C. G. de Kruif, *Langmuir*, 2000, **16**, 1497.
4. A. E. Bailey *et al.*, *Phys. Rev. Lett.*, 2007, **99**, 205701.

Dr Wyss commented: How is the arrest line between phase II and III determined experimentally (density of arrested dense phase)? In colloidal systems this is not straightforward as, for instance, for deep quenches the network strands can be as thin as 2–3 particles across. In this case even the definition of a volume fraction for the dense phase is not straightforward.

A follow-up question: It is not obvious to me that the density of the arrested dense phase could be accessed from centrifugation experiments without running into problems with either stress-induced compactification or uncomplete packing of the broken up strands/clusters. Could one alternatively access this density from static scattering methods, thereby avoiding the additional influence of gravity/centrifugal forces?

Professor Schurtenberger responded: The position of the arrest line within the spinodal has initially been determined using centrifugation experiments that rely on the fact that the arrested network with its very large mesh size of order μm has a much lower yield stress than the dense protein glass that forms its backbone, in line with the finding of two well separated elastic plateaus that differ by at least two orders of magnitude. In ref. 1 you find a detailed account of test experiments that indeed demonstrate that centrifugation at a given quench temperature results in the formation of a sharp interface between a liquid phase on top and a homogeneous glass phase with a density that provides the position of the glass line at this particular temperature. Ref. 1 also demonstrates that the height of the meniscus for a given sample is independent of centrifugation time and acceleration value. We have subsequently also reconfirmed these values using scattering experiments, where the protein concentration in the arrested backbone can be extracted from the scattering data in the Porod regime.[2] It is however important to realize that this is only possible due to the very different yield stress values of the network and the protein glass as a result of the very large difference in length scales. This is very different for large colloids, where the characteristic domain size in the early stage of a spinodal decomposition process after a deep quench is also a few micrometers only,[3] and where the strands thus encompass a few particle diameters only. Moreover, elastic moduli and thus yield stresses will depend strongly on the particle size, and therefore be much smaller for large colloids. In this context it is interesting to look at some early data for spinodal decomposition in suspensions of sterically stabilized silica particles with much smaller radii of about 25–40 nm. Here the authors also present data that provide evidence for an arrest line that extends far into the spinodal, quite similar to the situation found for lysozyme.[4,5]

1. F. Cardinaux, T. Gibaud, A Stradner, and P. Schurtenberger, The interplay between spinodal decomposition and glass formation in proteins exhibiting short range attractions, *Phys. Rev. Lett.*, 2007, **99**, 118301.
2. T. Gibaud, N. Mahmoudi, J. Oberdisse, P. Lindner, J. S. Pedersen, C. L. P. Oliveira, A. Stradner, and P. Schurtenberger, New routes to food gels and glasses, *Faraday Discuss.*, 2012, **158**, DOI: 10.1039/C2FD20048A.
3. A. E. Bailey *et al.*, Spinodal decomposition in a model colloid–polymer mixture in microgravity, *Phys. Rev. Lett.*, 2007, **99**, 205701.
4. M. H. G. M. Penders and A. Vrij, Spinodal decomposition in a sterically stabilized colloidal silica dispersion following from quench experiments, *Adv. Colloid Interface Sci.*, 1991, **36**, 185–217.
5. H. Verduin and J. K. G. Dhont, Phase diagram of a model adhesive hard-sphere dispersion, *J. Colloid Interface Sci.*, 1995, **172**, 425.

Dr Wyss said: Coming back to the possible influence of gravity. It appears that protein solutions can in some cases exhibit a behavior different form that of colloids, for instance very thin strands of just 2–3 particles across are generally not observed. If you could density match the protein solutions, would you expect the phase diagram to change significantly?

Professor Schurtenberger replied: I believe that the difference between the findings for lysozyme and large colloids is not caused by gravity, but by the particle size. As shown in a number of papers on spinodal decomposition in colloidal suspensions, the characteristic domain size in the initial state is comparable for proteins, small nano particles and large colloids. This implies that while for proteins and small nano particles strands are hundreds of particle diameters thick, for large colloids they encompass a few particles only. Moreover, yield stresses for attractive glasses are strongly dependent on the particle size. A simple and approximate relationship between the strength of the attraction, U_a, and the elastic modulus of the attractive glass, G_{el}, has been derived as $G_{el} \approx U_a/(r^2_{loc}d)$, where r_{loc} is the localization length of the particles trapped in the attractive well of the potential and d the particle diameter.[1] This will result in much smaller values of the yield stress for large colloids, presumably the reason why arrested spinodal networks for large colloids do not withstand gravitational stress and thus exhibit transient gel formation and collapse. For the small proteins, the influence of gravity is negligible, and we thus do not expect any significant change of the phase diagram under zero or microgravity conditions.

1. S. A. Shah, Y. L. Chen, K. S. Schweizer, C. F. Zukoski, Viscoelasticity and rheology of depletion flocculated gels and fluids, J. Chem. Phys., 2003, **119**, 8747–8760.

Dr Royall remarked: This follows a comment in a discussion where Peter Schurtenberger stated that the 'arms' in gels of proteins are typically 50 particles thick.

This contrasts with colloidal gels in which the 'arms' are typically three particles thick. I presume this is related to the different diffusion times of colloids and proteins. In the case of the former (particularly those micron-sized particles used for confocal microscopy), the time to diffuse a diameter can reach 30 s. Since this time scales with the cube of the diameter, protein molecules diffuse very much faster—many orders of magnitude.

By analogy, proteins would be expected to be able to proceed further with phase separation and thus the 'arms' may become thicker.

Professor Schurtenberger replied: The characteristic domain size in the early stage of spinodal decompositions is indeed governed by the combination of the diffusion of the particles and the interfacial tension between the dilute and dense phase. Both depend on the particle size, and precise measurements of the interfacial tension of a phase separated colloid-polymer mixture has for example demonstrated that the interfacial tension γ goes as $\gamma \propto kT\phi/d^2$, where ϕ is the volume fraction of the dense phase and d the particle diameter.[1] A detailed calculation of the optimal size of concentration fluctuations that is related to the peak in the q-dependent scattering intensity at q_{m} for spinodal decomposition in colloidal suspensions can be found in ref. 2. It is in fact interesting to compare the characteristic domain size in the early stage of spinodal decomposition for small particles such as whey protein aggregates[3] or sterically stabilized silica particles[4], with particle radii of about 25–30 nm, with that found for larger PMMA particles with a radius of 160 nm.[5] In all three cases the domain size is of order a few µm, independent of particle size. Therefore it is not surprising that for comparable quench depths the characteristic domain size and strand thickness is of similar overall size for all systems, which means that it will correspond to many particle diameters for proteins or small colloids, but only a few particle diameters for large colloids.

1. E. de Hoog and H. Lekkerkerker, Measurement of the interfacial tension of a phase-separated colloid–polymer suspension, *J. Phys. Chem. B*, 1999, **103**, 5274–5279.
2. J. K. G. Dhont, Spinodal decomposition of colloids in the initial and intermediate stages, *J. Chem. Phys.*, 1996, **105**, 5112.
3. R. Tuinier; J. K. G. Dhont; C. G. de Kruif, Depletion-induced phase separation of aggregated whey protein colloids by an exocellular polysaccharide, *Langmuir*, 2000, **16**, 1497.
4. M. H. G. M. Penders and A.Vrij, Spinodal decomposition in a sterically stabilized colloidal silica dispersion following from quench experiments, *Adv. Colloid Interface Sci.*, 1991, **36**, 185–217.
5. A. E. Bailey *et al.*, Spinodal decomposition in a model colloid–polymer mixture in microgravity, *Phys. Rev. Lett.*, 2007, **99**, 205701.

Dr van der Sman remarked: In your paper you suggest that the phase diagram can be constructed by detailed simulations. How would you represent the lysozyme in these simulations? I gather it can be an elaborated sticky hard sphere picture. How do you account for the electrostatic interactions in this picture ?

Professor Schurtenberger replied: We have in fact made a detailed calculation (not simulation) of the phase behavior of lysozyme at different salt concentrations that resulted in near quantitative reproduction of the liquid–solid phase boundary and the metastable liquid–liquid binodal using a patchy sphere model.[1] Here the electrostatic interactions were treated by using a screened Coulomb interaction potential between the proteins, assuming that the net charges are smeared out homogeneously over the spherical protein surfaces. In this one-component macro-ion-fluid potential we have also included the finite size and concentration of the colloidal macroions.

1. C. Gögelein, G. Nägele, R. Tuinier, T. Gibaud, A. Stradner, and P. Schurtenberger, *J. Chem. Phys.*, 2008, **129**, 085102.

Dr Menut asked: In the conclusion of your paper, your extend your demonstration to the case of casein micelles, showing that an arrested bicontinuous structure can also be obtained by addition of PEO. Casein micelles are soft objects, due to their compressibility they can be compared at least in some aspects with microgel particles. In such systems, the glass line (difference between region II and III in your Fig. 2) shifts to higher volume fraction values. In comparison with non-deformable colloidal particles, do you think that the particle softness should also modify the shape of the binodal curve, especially when considering the micelle rich phase?

Professor Schurtenberger replied: While for casein micelles in the absence of attractions the glass transition is indeed shifted towards higher volume fractions, we would expect that this effect will become less important for attractive glasses and gels as here attractions play a major role. The same should hold for the binodal, as for the relevant volume fractions caseins can be modeled as effective hard spheres (see also general comment earlier by C. G. de Kruif). A comparison between the theoretical predictions for the binodal for casein–xanthan mixtures using a generalized free volume theory[1,2] and the experimentally determined phase boundary results in almost quantitative agreement, except for very high polymer concentrations, where presumably the viscoelastic properties of the concentrated xanthan solutions and the relatively high persistence length play a major role.

1. S. Bhat, R. Tuinier and P. Schurtenberger, *J. Phys.: Condens. Matter*, 2006, **18**, L339–L346.
2. K. van Gruijthuijsen, V. Herle, R. Tuinier, P. Schurtenberger and A. Stradner, *Soft Matter*, 2012, **8**, 1547.

Dr van der Sman commented: How would shear rate fit in the state diagram of Fig. 1? Experiments in our group have shown one obtains anisotropic structures. Hence, this would add a new dimension! Would other external fields act in similar way p.e. electric fields?

Professor Schurtenberger replied: Here I can not even speculate, as this would very much depend on the nature and strength of the shear (or any other external field) applied. It would certainly add another dimension, similar to the jamming phase diagram proposed by Liu and Nagel.[1]

1. A. J. Liu and S. R. Nagel, *Nature*, 1998, **396**, 21.

Dr Trappe resumed the discussion of the paper by Dr Royall: Could you comment on the differences between the pathways for the arrest of spinodal decomposition *via* a glassy and crystal arrest of the dense phase? Is the nucleation rate of crystals simply enhanced by the densification process during spinodal decomposition and coarsening just stops when the entire dense phase has transformed into a crystal during the quench? Or do you first fully densify in a glassy configuration before crystal-formation occurs?

Dr Royall replied: Crystals likely nucleate in a two-step process. The simulations are run at a packing fraction of 0.1, where crystallisation is extremely slow. Therefore, the system must phase separate prior to crystallisation. During phase separation, the local density for the vast majority of particles which end up in the gel network changes from 0.1 to the density of the liquid that the system would phase separate to if it were not arrested.

Whether the system crystallises or not depends how long the particles spend at a volume fraction where crystallisation is possible, namely ~ 0.5 to 0.58. Too low and crystallisation is too slow; too high and the dynamic arrest means crystallisation doesn't happen on the simulation timescale.

Dr van der Sman remarked: In your simulation of a quench you can perform *via* a decrease in temperature. How do you cope with the fluctuation dissipation theorem (FDT)? You take into account the decreasing temperature. Furthermore, do you take the local viscosity into account in the local thermal fluctuations? The local viscosity is dependent on the local volume fraction of particle, *cf.* Krieger–Dougherty. In the mean-field approximation the friction coefficient (which enters the FDT) is inversely dependent on the local viscosity (which is a function of local volume fraction)?[1]

1. H. M. Vollebregt, R. G. M. van der Sman and R. M. Boom, Suspension flow modelling in particle migration and microfiltration, *Soft Matter*, 2010, **6**, 6052–6064.

Dr Royall replied: For us, Brownian dynamics refers to the overdamped limit where the colloids carry no momentum. The timescale on which the colloids carry momentum is vanishingly small for our system. In the experiments we seek to model, temperature changes very little (283–298 K), this small change is neglected. Clearly, as the system falls out of equilibrium, there are questions about whether FDT is satisfied, however this is also true of the experimental system. At the level we are operating at, we believe the Brownian dynamics is reasonable.

Dr Frith inquired of the paper by Dr Krueger: My question relates to the apparently anomalous high concentrations that are estimated based upon the structural information you have (from the *d* spacing?), as mentioned by one of the other questioners. It seems likely that if your lysozyme was used as received from the supplier, then it would contain a significant amount of hydration water. It seems possible that freezing to −80 °C may reduce the amount of hydration water, thus making the effective concentration of lysozyme appear higher than at ambient. Do you think this is possible?

Dr Krueger responded: This is certainly a possibility. The simple approach we took to fitting the data was only intended as an approximation that provides a convenient, one parameter way to track the changes that occur in protein packing during freezing. It doesn't include the effect of protein shape, other solutes, or hydration water. Yet, this simple method does allow us to conclude that salt definitely has an effect on the average nearest neighbor distance at −80 °C.

Dr van der Sman said: During multiple freezing/thawing cycles for some proteins systems you expect some irreversible protein denaturation. This is evident in a traditional Japanese dish: kori-tofu, where freezing is used to obtain a porous structure. You state that in your systems the aggregation of proteins is reversible. When do you expect that in your system irreversible protein denaturation and subsequent aggregation will occur? Would salt or protein concentration have an influence on that? What makes proteins to undergo reversible *cq.* irreversibe cold naturation?

Dr Krueger replied: These are all very good questions. It looks from our data that lysozyme isn't undergoing irreversible denaturation—at least in the protein and salt concentration ranges that we examined. As we stated in the paper, proteins with a large equilibrium binding constant would be more highly associative, and more prone to chemical modifications, than lysozyme. We have not yet repeated our experiment on such a protein to determine whether we can detect protein denaturation and/or irreversible aggregation. However, based on our lysozyme data, I think that such changes can be detected using SANS.

Dr van der Sman commented: How can NaCl inhibit protein denaturation?

How would making a state diagram of protein–water improve the understanding of your results? Recent theories I have developed can predict those for several types of proteins.[1]

1. R. G. M. van der Sman *et al.*, Thermodynamics of meat proteins, *Food Hydrocolloids*, 2012, **27**, 529–535.

Dr Krueger answered: Thank you very much for alerting us to this work. The question you pose is indeed of interest to us and this work may be of help to us in choosing some appropriate additional proteins for additional freeze/thaw studies.

Dr Royall commenced a new discussion of the paper by Professor Cates: Have you a picture in mind (perhaps within SGR) of how the structure of the Laponite changes over time to lead to this solidification?

Professor Cates replied: There is a continuing debate in the literature regarding whether the structure of Laponite suspension is an attractive gel or a repulsive glass. This will depend strongly on salinity and other factors. In the experimental conditions addressed in our paper, we feel that both attractive and repulsive contributions are at play, but it is difficult to comment on the precise nature of microstructure. The SGR model was originally intended for caged systems such as dense emulsions where the wells represent confinement by repulsive forces, but one could also interpret these wells as representing a bonded environment. Thus our model does not point to a particular picture of the nature of the solid.

Dr Ettelaie communicated: You have studied your model in the low temperature limit where thermal hopping is negligible (*i.e.* the noise free limit, as you mentioned in the paper). Would inclusion of such a thermal noise make a qualitative difference to the predictions of the model?

Professor Cates communicated in reply: The thermal hopping would allow aging to proceed even in the quiescent state, but through most of the delay period we expect such hopping to be negligible compared to the shear-induced hopping term. In the final solidified state, the model as it stands predicts that eventually all elements are in wells deep enough to not undergo shear-induced hopping, beyond which point further evolution of the structure will cease altogether. In this late time regime there would again be a significant correction from the intrinsic hops which would allow slow aging to continue within the solid.

Dr Trappe communicated: Could you give a practical example other than a gel that would show delayed solidification? Should not a glassy state, where arrest is governed by overcrowding, just be a solid after rejuvenation? A solid that ages, but nonetheless a solid even if the elastic modulus is only a high frequency modulus. If there is a fluid–solid transition for glasses that occur in time, how would we distinguish between aging of an intrinsically solid state and a fluid–solid transition after rejuvenation?

This journal is © The Royal Society of Chemistry 2012

Professor Cates communicated in reply: Physical aging is a phenomenon where the free energy of a system decreases as a function of time in search of the equilibrium state. This aspect is generic irrespective of whether we talk about a glass or a gel, or whether aging leads to a liquid–solid transition or occurs within an already solid state. We agree that materials having very high filled volume are less likely to be in a fully liquid state after shear rejuvenation, although mode coupling theories (MCT) for hard spheres colloidal glasses do predict exactly that outcome (J. M. Brader, M. E. Cates and M. Fuchs, *Phys. Rev. Lett.*, 2008, **101**, 138301). Our paper suggests that a sudden solidification may be seen under application of a very high oscillatory shear stress causing flow that (irrespective of glass or gel) creates a state resembling one of complete rejuvenation. This might happen in glasses as well as gels if particles can get trapped in deeper wells owing to localized aging. However, we are not aware of any direct experimental observation of this in glasses as yet.

Dr Bot opened the discussion of the paper by Dr Spicer by communicating: I consider this to be a very elegant study on arrested coalescence, and it immediately raises a question with me: You have studied the coalescence of two isolated droplets. In foods, the situation usually involves emulsions with a high amount of dispersed phase. The effect of interdroplet distance due to dispersed phase volume on partial coalescence has been investigated empirically in temperature-cycled emulsions.[1] I wonder whether your model would be able to make any predictions on the chances of neck formation between two droplets as a function of the distance between the droplets, or do your experiments only show (arrested) coalescence if the droplets are brought in direct contact?

1. S. Kiokias and A. Bot, *Food Hydrocolloids*, 2006, **20**, 245–252.

Dr Spicer communicated in reply: Our model does not include the initiation of coalescence and the experiments always forced droplet contact to initiate coalescence. The probability of arrested structure formation is increased when the elastic modulus is lowered, but not entirely removed, by complete melting. So our model does at least predict arrested coalescence when raised temperatures lower the modulus and meet the conditions that promote arrested coalescence. However, our model assumes the droplets are brought into contact by some transport mechanism and does not describe the effect of separation distance between droplets on coalescence.

Dr Royall communicated: In your discussion of the coalescence of Pickering emulsions, it is very clear that there is a change in surface area and this can drive dynamical arrest, and stop coalescence. However, in the case you present here, there is no analogous change in volume. Therefore, given that the glass transition is continuous on the kind of timescales accessed here, might one not expect some slow coalescence over one or ten hours, say, rather than arrested coalescence?

Dr Spicer communicated in reply: Perhaps, but if the droplets are truly glasses then we would expect relaxation to a spherical shape of any arrested oil droplet structures with internal fat crystals (even the structures in ice cream). We see some evidence of attractive interactions between the crystals inside the oil and consider the internal structure more of a colloidal gel than a glass. So it is possible that over a characteristic time the structure will relax but on the time scale of several days we observe no relaxation. More quantitative information on the aging time scales of these structures is needed before we can be certain.

Professor Schurtenberger commented: While the assumption that the final state is determined by the balance between the elastic modulus and the interfacial tension only appears to be in good agreement with your experimental data, I was wondering whether this is sufficient to describe the boundary between total coalescence and arrested coalescence shown in your coalescence map in Fig. 6. Could one not imagine that for low solid volume fractions the yield stress of the wax–hexadecane dispersion is low enough so that total coalescence occurs in this case because the internal network yields under the stress imposed by interfacial tension?

Dr Spicer replied: We do indeed feel that at lower solid levels the elasticity and yield stress will be too low to arrest coalescence. Whether complete coalescence results from a large deformation of the network, elasticity, or breakage of the network, yielding, is an important distinction of the phenomenon that goes beyond our current simple model and data set. One distinction would be whether energy continues to be stored within the droplet structure or has been dissipated by breakage. As we have observed some structural recoil upon relaxation, though not total recovery, we suspect the answer may be some mixture of the two, as we indicate above.

Professor van der Linden opened the discussion of the paper by Professor Nicolai by asking: How does the so-called critical calcium concentration depend on the concentration of casein?

Professor Nicolai answered: As far as we can tell there is a critical ratio of about 5 calcium ions per protein.

Professor Schurtenberger communicated: In your manuscript you write that the viscosity rises steeply with increasing concentration above about 80 g L^{-1}, but that it does not diverge because the particles can interpenetrate. I was wondering whether you have additional information that demonstrates that the particles can indeed interpenetrate. There would in fact be other possibilities that would also result in a finite viscosity such as a concentration-dependent deswelling of the particles. This would in turn lead to an effective volume fraction that would increase less steeply than the weight concentration.

Professor Nicolai communicated in reply: We cannot exclude that the particles deswell rather than interpenetrate. However, we believe that the caseinate particles resemble polymeric micelles and it is well known that star polymers interpenetrate above the overlap concentration.

Dr van der Sman inquired: During heating do the caseinates not denature? Or rather, if they are more random coil like, can they change their conformation during heating? Can they become more hydrophobic?

Professor Nicolai communicated in reply: Caseins are so-called rheomorphic proteins like gelatin. They do not show a clear process of denaturation. The effect of heating is indeed most likely that hydrophobic interactions become more important. As mentioned in the article this leads to a larger aggregation number of the small particles into which the caseinate is organized.

Dr van der Sman continued the discussion of the paper by Dr Spicer by communicating: In your model you use pure thermodynamic potentials to describe arrest. Normally arrest is used for describing a out-of-equilibrium metastable state. Is it (physically) correct that a model with quantities describing equilibrium conditions (surface tension and elastic energy) can describe non-equilibrium conditions? Or is the naming of arrest incorrect? Is arrest also not affected by viscoelastic/plastic behaviour?

Dr Spicer responded: We have used the word 'arrested' to mean halted or stopped. Part of this is to distinguish from different literature meanings of the terms 'partial coalescence'.

In food science, partially fused structured droplets are described as partially coalesced. In soft matter physics, partial coalescence generally indicates a droplet, without internal solids, incompletely merged with a liquid reservoir during a collision. For this reason we use 'arrest' to mean the same sort of coalescence halting seen by Studart *et al.* (2009) for Pickering-type droplets.[1] From a purely thermodynamic point of view, the droplet arrest studied here is an equilibrium state based on our energy minimization description. So, if arrest is formally an out-of-equilibrium state we do not exactly match.

1. A. R. Studart *et al.*, *J. Phys. Chem. B*, 2009, **113** 3914-3919.

Professor Cates continued the discussion: If the spheres were simply elastic (or a viscoelastic solid) then they would form reversible adhesive contacts with each other to minimize surface area. The local deformation is maximized in the contact region, so if there is a yield stress one expects melting here. This would create local plastic deformation and a permanent bond in the contact zone, while retaining a purely elastic deformation across the central part of each droplet (so that a reduction in tension would cause partial elastic recovery). So I believe your very interesting data could be more fully explained by a model with a finite yield stress.

Dr Spicer replied: While our simple model describes the droplet structure as an ideal elastic spring, we agree that local yielding likely occurs during coalescence initiation. As a result, there is partial dissipation of the interfacial energy and the remainder is then stored in the deformed elastic matrix of the droplet. We believe the dissipation is not complete because we often observe partial recoil of the arrested doublets upon reducing interfacial tension. The yielding behavior of the structure is certainly important to determine the stability of such arrested structures and should be included in further refinements of the model.

Professor Wilde communicated: Regarding the mechanical strength of a network of coalesced, partially solidified droplets: considering the reduction in overall surface area following arrested coalescence, *versus* the strength/size of the adhesion force between droplets, what do you think would be the optimum solid content of the droplets to maximise the strength of the network? In addition, and in response to the previous questions, the droplets should not be regarded as viscoelastic solids, but a mixture of liquid oil within a porous network of solid fat/wax. The liquid oil can percolate through the solid network, allowing coalescence between droplets, and the presence of the solid network maintains the integrity of individual droplets and preventing full coalescence.

Dr Spicer answered: We have not measured the mechanical properties of a network of these droplets, but Fig. 3 is an indication of the crystal network strength within the oil. Approximating a network of structured droplets as a colloidal gel, we expect its strength to be a function of the attractive interactions between droplets and the connectivity of the network. Using Fig. 4 as an example, we see that the maximum contact between droplets occurs at intermediate solids levels: high enough to allow arrest, but low enough that contact area is maximized. More research is needed to determine the bulk mechanical properties as a function of relevant variables like solids level, connectivity, and the ratio of total oil to total water volume.

Dr van der Sman asked: You said that you let the droplets approach quite slowly. Just before coalescence did you observe fat crystals sticking out of interface?

Dr Spicer responded: We did not observe fat crystals sticking out of the droplets at any stage of the experiments. The interfacial disposition of the crystals is dependent on the relative wettability of the crystals by the oil, as pointed out by Boode and Walstra (1993).[1] For these experiments we used crystals that are entirely wetted by the oil in this particular aqueous surfactant system.

1. K. Boode and P. Walstra, *Coll. Surf. A.*, 1993, **81**, 121–137.

Dr Mayama inquired: Please show the equation for the theoretical profiles in Fig. 5 although you show eqn (3) and (4) in the text.

Dr Spicer communicated in reply: The profiles plotted in Fig. 5 are obtained by plotting the sum of eqn (3) and (4).

Professor Stoyanov communicated: Is there a possible link between the delayed solidification of Laponite dispersions that you observe and shake gels (mixtures of Laponite and large Mw PEO), please see: J. Zebrowski, V. Prasad, W. Zhang, L. M Walker, D. A Weitz, *Colloids Surf., A*, 2003, **213**, 189–197, 2003.

Professor Cates communicated in reply: In shake gels, made up of Laponite and PEO, upon shaking PEO molecules bridge the Laponite particles thereby forming a network. We do not believe that the presented work on delayed solidification is directly connected with shake gels, which undergo shear induced network formation. In delayed solidification the first effect of shear is to melt the system; despite the shear it eventually manages to solidify. In shear induced networks there is initially no structure but shear allows one to form. These appear to be fundamentally different mechanisms although they might have some physics in common.

Professor van der Linden commented to Professor Cates: In your paper you mention pocket growth as an aspect of the model you propose. Could it be that there need not be growth of the pockets (in size that is), but growth in number of pockets, and the threshold is related to percolation defined by a critical number of pockets?

Professor Cates communicated in reply: It could be so, but it does not seem likely to us that there is only an increase in the number of pockets without increasing the size of those already present. Nonetheless, an increase in the number of pockets, leading to a percolation type process, may indeed play a role.

Professor Schurtenberger returned to the discussion of the paper by Professor Nicolai by communicating: In your manuscript you write: "*As discussed above, adding calcium ions leads to an attractive interaction between the particles. Such an interaction is not expected to modify G_{el}, but increase the friction between the jammed particles and thus the terminal relaxation time.*" I don't understand the basis for this general statement, as a number of theoretical treatments of attractive colloidal glasses have demonstrated a direct relationship between the strength of the attraction as given by the value of the interaction potential at contact, U_a, and the elastic modulus G_{el}. A simple and approximate relationship between U_a and G_{el} has been derived as $G_{el} \approx U_a/(r^2_{loc}d)$, where r_{loc} is the localization length of the particles trapped in the attractive well of the potential and d the particle diameter.[1,2]

1. S. A. Shah, Y. L. Chen, K. S. Schweizer, C. F. Zukoski, Viscoelasticity and rheology of depletion flocculated gels and fluids., *J. Chem. Phys.*, 2003, **119**, 8747–8760.
2. M. E. Cates, M. Fuchs, K. Kroy, W. C. K. Poon, A. M. Puertas, Theory and simulation of gelation, arrest and yielding in attracting colloids. *J. Phys.: Condens. Matter*, 2004, **16**, S4861–S4875.

Professor Nicolai replied: This argument is correct as long as the particles do not jam. When particles jam the elastic modulus is determined by that of the particles themselves. This is the case for the caseinate particle suspensions for which the effective volume fraction was larger than unity.

Dr Ettelaie addressed Professor Cates: Would a cut off in the distribution of energy wells imply that there is a critical stress amplitude above which solidification cannot happen? If so, does the existence of shear melting in Laponite systems studies here imply such a cut-off for these dispersions?

Professor Cates answered: Within our simplified model, the answer to the first question is yes: if conditions are such that the deepest possible well undergoes a strain-induced hop every cycle at the initial strain amplitude, then the solid fraction S must remain zero forever and solidification is ruled out. Experimentally however, it is difficult to say with certainty that above critical stress amplitude solidification will not happen. We can only say that with increase in stress magnitude the time to solidification will increase. Shear melting in steady shear does not in itself imply any cutoff: under those conditions every element will eventually jump, however deep the well it occupies.

Dr Frith commented: Do you think that the same phenomena would be observed if the applied stress were steady instead of oscillatory? For example, if a suitable steady stress were applied to the sample after the shear melting, would the observed shear-rate decrease to zero after a certain delay time?

Professor Cates responded: The phenomenon of gradual solidification below a threshold of steady stress and yielding above is indeed observed in many systems. This includes systems showing a viscosity bifurcation but is not limited to these cases. Indeed, the standard SGR model predicts no viscosity bifurcation but a logarithmic increase of strain with time at small stresses. This does imply a shear-rate that falls to zero with time, but the decrease is gradual not sudden. This could be sharpened in materials with a viscosity bifurcation (see P. Coussot et al., Phys. Rev. Lett., 2002, **88**, 175501) where there is not merely a yield stress but also a minimum stable flow rate in steady state. An SGR variant showing the viscosity bifurcation is reported in S. M. Fielding, P. Sollich and M. E. Cates, Soft Matter, 2009, **5**, 2378–2382.

Experimentally, one advantage of applying the oscillatory stress field is that the strain induced in the material always remains bounded, whereas in the case of constant stress it does not.

Professor de Kruif communicated: In your experiments on the clay particles an oscillatory shear stress is applied. Would a constant shear stress lead to similar phenomena of solidification? Is there a parallel with the experiments and theoretical description by A. Zaccone, H. Wu, D. Gentili and M. Morbidelli, Theory of activated-rate processes under shear with application to shear-induced aggregation of colloids, Phys. Rev. E, 2009, **80**, 051404.

Professor Cates communicated in reply: Application of constant shear stress results in solidification below a threshold of stress (the yield stress) with continuous flow at higher stresses. The solidification is not caused by the stress in that case; aging is always faster without stress, and beyond the yield stress aging stops altogether. Therefore we do not think there is any close parallel between the phenomenon discussed in our paper and the shear induced aggregation discussed by Zaccone et al., in which the structuring of the material, and hence the viscosity increase, is caused directly by the flow itself.

Dr van der Sman enquired: In systems where food-like platelets are used, for example fat crystals, do you expect similar effects as described in the paper (delayed solidification)?

Professor Cates replied: The present work suggests that the deformation field can delay but may not stop the process of aging (leading to solidification) for those materials that demonstrate aging (or solidification) under quiescent conditions. Therefore, if systems with food platelets do undergo aging under quiescent conditions, this work indeed may be applicable to them.

Dr Bot resumed the discussion of the paper by Professor Nicolai: In Figure 6f, you show these intriguing protein 'shells'. They remind me of swollen modified starch particles. Do these features turn out to be very fragile? Do they contribute to the viscosity of the aqueous phase in which they are dispersed? You also refer to strong bonds preserving this feature–do you mean covalent bonds, or just very persistent physical bonds?

Professor Nicolai communicated in reply: They are not very fragile as they precipitate and can be redispersed by shaking without breaking up. They do not contribute significantly to the viscosity, but it should be realized that they were formed by diluting a much more concentrated solution. We do not know for sure, at the present moment, if the bonds are covalent or just very persistent physical bonds, but my guess is the latter.

Dr van der Sman addressed Professor Nicolai: In the paper you have made master curves of G' via shifting them. What is the physical hypothesis behind this shifting? In the literature I find hardly any clues why this is allowed. At times I think shear or oscillation frequency can be translated into an effective temperature, cf. references 1 and 2 below, which can be added to the regular temperature effect. Then G' and G'' will be a function of $T_{eff} - T_g$ or T_{eff}/T_g, with T_g the glass/gel transition (where $G' = G''$). Can you elucidate the physical reasoning behind the shifting of G' and G'' curves?

1. H. M. Vollebregt, R. G. M. van der Sman R. M. Boom, Suspension flow modelling in particle migration and microfiltration, Soft Matter, 2010, 6, 6052–6064.
2. H. M. Vollebregt, R. G. M van der Sman, R. M. Boom, Particle migration with effective temperature, *Faraday Discuss.*, 2012, **158**, DOI: 10.1039/C2FD20035J.

Professor Nicolai communicated in reply: Frequency–temperature shifts are standard procedure for polymer solutions and melts, but are also successfully used for many other complex liquids. If such a shift leads to correct superposition it means that the relaxation process is basically the same, but that the relaxation times are systematically shifted, e.g. because the friction decreases with increasing temperature or because the relaxation processes of the system are controlled by the same activation energy.

Professor Stoyanov continued the discussion of the paper by Professor Cates: Does the vessel (walls) play a role in the process—what will happen if you use glass instead of plastic?

Is it possible that Laponite clays are not fully exfoliated and this happens during the intensive shearing? If so your model still can be valid but in a way reversed—you have compact and non exfoliated fraction of stacked clays with poor structuring capacity and fully exfoliated fraction which structures well (via house of cards) and shear re-balances fraction distribution via the same mechanism you propose, while reducing the non-exfoliated fraction at the expense of exfoliated one, until critical concentration is reached. If so then your concern about the model disappears , while you still can have same scaling behavior, which seems to agree with the experiment.

Professor Cates communicated in reply: Firstly, we do not think vessel walls have any role to play as delayed solidification appears to be a bulk phenomenon. Secondly, yes there is a possibility that not all the clay discs are exfoliated. However shear-induced exfoliation leading to structure formation does not seem to be happening in our experiments. These show formation of a solidified structure that the applied shear field seems initially capable of preventing, but ultimately can only delay.

Dr Ettelaie addressed Professor Cates: If I am correct, your model still predicts the delayed solidification, even for low oscillatory shear stress amplitudes (small values of E_0 in eqn 2). It is interesting that some foods, notably tinned concentrated milk, stored in a fridge and therefore subject to gentle oscillation, do solidify over a long period of a few months (the so called age gelation of milk). There is no apparent change in the chemical composition or any biological degradation. I was wondering if this phenomenon of age gelation of milk can have its origins in the mechanism captured by the SGR model.

Professor Cates communicated in reply: From what you describe, then yes, the SGR model put forward in our paper could offer an explanation for this. In cases where aging would not occur without shear being applied, the neglect of intrinsic or thermal hopping may be justified even for small amplitude oscillations (though we don't believe that to hold in the laponite system).

Professor Nicolai asked: Is the oscillatory shear simply accelerating a process that would happen also by thermal motion, by allowing it to explore the energy landscape more quickly?

Professor Cates communicated in reply: In an oscillatory flow field particles that experience the strain escape the shallow wells over a unit timescale that is the inverse of frequency. This causes the delay time for solidification to scale approximately as the inverse of frequency as shown in Fig. 3. In that sense, the oscillations are allowing the system to explore energy landscape more quickly. Indeed, in the simple SGR-type model we use, the temperature is effectively zero and without the straining, no exploration at all would occur. This is a good approximation at the very large initial strains that arise in our experiments, where strain-induced jumps dominate spontaneous ones (for a related discussion see: V. Viasnoff and F. Lequeux, *Phys. Rev. Lett.*, 2002, **89**, 065701) and V. Viasnoff, S. Jurine and F. Lequeux, *Faraday Discuss.*, 2003, **123**, 253–266). Nonetheless, it is misleading in the laponite system to think of the main effect of the flow as accelerating the aging process: in practice the system would solidify faster without the flow. To model that quiescent aging one would of course have to restore the spontaneous jumps as is done in the standard SGR treatments.

Dr Royall communicated: You stated that the glass transition is driven by the hard core, even in Lennard-Jones type glasses, and that attraction-dominated arrest is gelation.

But what about silica? There I understand the glass transition is driven by attraction—should we think of silica as a gel?

Professor Cates communicated in reply: It is preferable to distinguish three types of glass: repulsive and attractive glasses in which the particle interactions are isotropic, and network glasses in which bonds can only form at specified orientation between the particles. In conventional atomic and molecular systems, so long as one excludes network glasses, the hard core repulsions are normally thought to dominate the physics of glass transition, with the longish range attractions essentially providing what would for hard spheres be a confining pressure. In colloids the attractions can be much shorter range, which gives the possibility of an attractive glass whose arrest is caused by non-directional bonding rather than caging. Recent work on 'patchy' colloids is instead more akin to the network glass case (F. Sciortino and E. Zaccarelli, *Curr. Opin. Solid State Mater. Sci.*, 2011, **15**, 246–253).

Professor Tanaka commented: For laponite suspensions under constant shear stress, it is known that viscosity bifurcation phenomena take place. That is, below a threshold stress a laponite suspension is shear melted, whereas above the threshold it continues to age. It is known that the aging of a laponite suspension at a quiescent state is characterized by the initial exponential growth of the structural relaxation time followed by the power-law growth as a function of the waiting time (see, *e.g.*, H. Tanaka, S. Jabbari-Farouji, J. Meunier, and D. Bonn, *Phys. Rev. E*, 2005, **71**, 021402). The elasticity evolves abruptly around this crossover aging time. I wonder how this solidification behaviour is related to the steep evolution of G' under oscillatory shear.

Professor Cates responded: In a freshly prepared quiescent suspension of Laponite the transition from initial exponential growth to power law growth of the structural relaxation time, which also accompanies sharp evolution of G', is a different physical phenomenon from the one addressed in our paper. In the literature the first phenomenon is represented as cage formation (Y. M. Joshi, *J. Chem. Phys.*, 2007, **127**, 081102). Interestingly an exponential increase in relaxation time is also observed for an aged Laponite suspension right after shear melting; this eventually undergoes a transition to a power law increase. However, unlike for fresh Laponite suspensions, a steep evolution of G' at the point of transition is not observed. In aged samples G' and G'' instead cross over in the neighborhood of this transition (A. Shahin and Y. M. Joshi, *Phys. Rev. Lett.*, 2011, **106**, 038302). The focus of the present work is however different. This work explores a region where a sharp increase in G' is observed under application of oscillatory shear stress. The time at which this increase occurs is a strong function of the magnitude of stress applied. The oscillatory case thus appears to be quite distinct from the viscosity bifurcation.

Viscoelastic phase separation in soft matter and foods

Hajime Tanaka*

Received 18th February 2012, Accepted 25th April 2012
DOI: 10.1039/c2fd20028g

Phase separation is a fundamental phenomenon that produces spatially heterogeneous patterns in soft matter and foods. We argue that phase separation in these materials generally belongs to "viscoelastic phase separation", where the morphology is determined by the mechanical balance of not only the thermodynamic force (interface tension) but also the viscoelastic force. The origin of the viscoelastic force is dynamic asymmetry between the components of a mixture, which can be caused by either a size disparity or a difference in the glass transition temperature between the components. Such dynamic asymmetry quite often exists in foods, which are typically mixtures of big molecules (polymers, proteins, etc.) and liquids (water, oil, etc.). We show examples of mechanically driven pattern formation in foods, in which dynamic asymmetry plays crucial roles, including the formation of network and cellular patterns in foods (e.g., breads, sponge cakes, butter, chocolates, etc.) and crack pattern formation (dried foods, cooked meat, etc.). Collapsing of these structures upon heating or moisture uptake is also discussed. We also argue that heterogeneous gels are in general formed as a consequence of dynamical arrest of the viscoelastic phase separation. Finally we mention an intimate link of viscoelastic phase separation, where deformation fields are spontaneously generated by phase separation itself, to mechanical instability and fracture induced by externally imposed strain fields. Such mechanical instability and nonlinear rheology may be relevant to food processing and also to separation and fracture of foods. We propose that all these phenomena can be understood as mechanically driven inhomogeneization with the concept of dynamic asymmetry in a unified manner.

1 Introduction

Phase-separation phenomena are commonly observed in various kinds of condensed matter including metals, semiconductors, simple liquids, soft materials such as polymers, surfactants, colloids, biological materials, and food materials. The phenomena play key roles in pattern evolution of immiscible multi-component mixtures of any materials. The resulting patterns are linked to optical, electrical, and mechanical properties of materials and also to taste, appearance, and sensory properties of foods. Thus, phase-separation dynamics has been intensively studied from both fundamental and applications viewpoints.[1,2]

On the basis of the concept of dynamic universality of critical phenomena,[3] phase separation were classified into a few groups. Phase separation in each group of condensed matter is described by a specific set of basic equations describing its dynamic process. For example, phase separation in solids is known as "solid model

Institute of Industrial Science, University of Tokyo, 4-6-1 Komaba, Meguro-ku, Tokyo 153–8505, Japan. E-mail: tanaka@iis.u-tokyo.ac.jp; Fax: +81-3-5452-6126; Tel: +81-3-5452-6125

(model B)", whereas phase separation in fluids as "fluid model (model H)".[1,3] For the former the local concentration can be changed only by material diffusion, whereas for the latter by both diffusion and flow. The universal nature of critical phenomena in each model and the scaling concept based on the self-similar nature of domain growth have been established.[1,3] In all classical theories of critical phenomena and phase separation, however, the same dynamics for the two components of a binary mixture, which we call "dynamic symmetry"[4,5] between the components, has been implicitly assumed. This assumption can always be justified very near a critical point, where the order parameter fluctuations are far slower than any other internal modes of a system (see Fig. 1). However, this is not the case far from a critical point, where most of practical phase separation takes place, for a mixture having strong dynamic asymmetry between the components. The presence of dynamic asymmetry means that there is also a large separation between the soft matter mode and the microscopic mode of a system. Furthermore, there is another gross variable of a system, the velocity field, whose relevance in dynamics comes from the momentum conservation law. Thus, dynamic asymmetry leads to complex couplings between the slow critical fluctuation mode, the slow soft matter mode, and the velocity field (see Fig. 1).

Nearly two decades ago we found unusual phase separation behaviour,[4–8] which is markedly different from phase separation of a fluid mixture (model H).[2,5–10] In the normal phase separation observed in dynamically symmetric mixtures (model H), the phase separation morphology is determined by the balance between the thermodynamic (interfacial) force and the viscous force, while satisfying the momentum conservation. In viscoelastic phase separation, on the other hand, the self-generated mechanical force also plays a crucial role in its pattern selection, in addition to the thermodynamic and viscous force. We named this type of phase separation "viscoelastic phase separation". In addition to the solid and fluid model, thus, we need the third model for phase separation in condensed matter, i.e., the "viscoelastic model".[11,12] This model is actually a general model of phase separation, which includes the solid and fluid models as its special cases.[12]

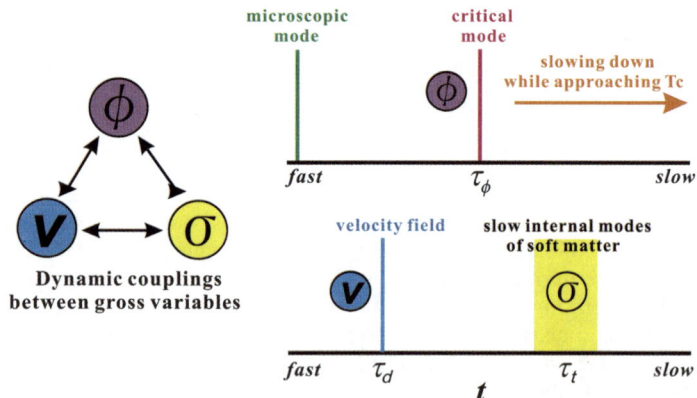

Fig. 1 Schematic figure showing dynamical couplings among the three gross variables, the composition ϕ, the velocity field $\vec{v}$, and the stress field σ, The relation among these modes and the microscopic mode are also shown. When approaching the critical point T_c, the order parameter fluctuation mode τ_ϕ should eventually become the slowest mode in principle. In this limit, the relaxation of σ does not play any role and thus the dynamic universality should hold. However, this situation may not be practically realized for a system of strong dynamic asymmetry. In phase separation, we should also consider the characteristic time of deformation τ_d. If the deformation rate τ_d is faster than the relaxation rate of the slow soft matter mode τ_t, the viscoelastic effects have a drastic influence on phase separation.

Intuitively, viscoelastic phase separation can be explained as follows. When there is a large difference in the dynamics between the components of a mixture, phase separation tends to proceed in a speed between that of the fast and slow components. Then, the slow component cannot catch up with a deformation rate spontaneously generated by phase separation itself, τ_d, and thus starts to behave as an elastic body, which switches on the elastic mode of phase separation. Thus, this phenomenon can be regarded as "viscoelastic relaxation in pattern evolution", which is the reason why we named it viscoelastic phase separation.[6] Unlike ordinary mechanical relaxation experiments, the mechanical perturbation is characterized by the rate of deformation induced by phase separation, τ_d, and the relaxation rate is that of the slowest mechanical relaxation, τ_t, in a system (see Fig. 1). Without dynamic asymmetry, the deformation rate is always slower than the relaxation rate. Thus, phase separation in such a mixture can always be described by the fluid model, no matter how slow the dynamics of the components. For example, this is the case for a mixture of two polymers having similar molecular weights and glass transition temperatures. We emphasize that dynamic asymmetry, which is prerequisite to viscoelastic phase separation, often exists in materials, particularly in soft materials and food materials.

In this article, we review the basic physics of viscoelastic phase separation,[11,13,14] including fracture phase separation,[15] and discuss its importance in food science. We show that with an increase in the ratio of the deformation rate of phase separation to the slowest mechanical relaxation rate the type of phase separation switches from fluid phase separation, viscoelastic phase separation, to fracture phase separation. We point out that there is a physical analogy of this to the transition of the mechanical behaviour of materials under shear from liquid fracture, ductile fracture, to brittle fracture. This allows us to discuss phase separation and shear-induced instability of disordered materials,[16,17] including soft matter[10,18–23] and foods, on the same physical ground. As examples of mechanically driven pattern formation in foods, we also consider the formation of network and cellular patterns in foods (*e.g.*, breads, sponge cakes, butter, chocolates, *etc.*) and crack pattern formation (dried foods, cooked meat, *etc.*) as well as collapsing of these structures upon heating or moisture uptake.

2   Importance of phase separation and rheological instability in foods

Processability, texture, flavour, taste, and stability of foods are controlled not only by chemical composition, but also by how the various ingredients are spatially organized, what is the topology, and what is the characteristic lengthscale. Food structures have rich varieties ranging from macroscopically homogeneous liquids to complex, multiphase solids containing water, salts, fats, proteins, and polysaccharides in the form of droplets, fibres, and networks and in the state of gases, liquids, liquid crystals, crystals, and glasses. Such a food structure is an important factor that affects visual impressions, sensory properties, product stability, and even digestion. These inhomogeneous structures are usually generated by nonequilibrium phenomena such as phase separation, emulsification, and crystallization. Such pattern formation has many features in common with that in soft matter. The common physics between pattern evolution in these two systems, soft matter and foods, have recently been addressed in a convincing manner.[24–33] Along this line, we focus on the fact that in most cases there is strong dynamic asymmetry between the components of foods. We argue that the basic concept of viscoelastic phase separation that we established in soft matter is relevant to pattern evolution in food materials.

There are also many situations where foods are in a strongly nonequilibrium state and exhibit nonlinear rheology such as shear thinning and thickening. For example, phase separation and emulsion stability are major issues for food structures and the effects of externally applied strain fields are also crucial for their processing. The strongest nonequilibrium situation for foods is seen in the processes of being violently mixed in cooking and being chewed in the mouth. In these processes, shear-induced instability and mechanical fracture of foods are key physical phenomena.

Here, we consider pattern formation in food materials and their mechanical instability under shear flow, on the basis of the knowledge of soft matter, putting a special emphasis on the concept of dynamic asymmetry.[11]

3   Theory

3.1   Two-fluid model of polymer solution and stress-diffusion coupling

Shear effects on complex fluids have attracted much attention because of their unusual nature known as "Reynolds effect". For example, shear flow that intuitively helps the mixing of the components actually induces phase separation in polymer solutions.[2,18] This is caused by couplings between the shear velocity fields and the elastic internal degrees of freedom of polymers. To explain this unique feature of polymer solution, there have been considerable theoretical efforts.[18-22] Doi and Onuki[10] established a basic set of coarse-grained equations describing critical polymeric mixtures, based on a two-fluid model whose original form was developed by de Gennes and Brochard[34-36] for polymer solutions and by Tanaka and Filmore[37] for chemical gels.

Later, we proposed that an additional inclusion of the strong concentration dependence of the bulk stress, which is not important in shear-induced instability, is necessary for describing viscoelastic phase separation of dynamically asymmetric mixtures, more specifically, the volume shrinking behaviour of the slow-component-rich phase.[12,38,39] We also argued its generality beyond polymer solutions to particle-like systems such as colloidal suspensions, emulsions, and protein solutions.[23] That is, we showed that the internal degrees of polymer chains and entanglement effects peculiar to polymer systems are not necessary for viscoelastic phase separation to take place and strong dynamic asymmetry between the components of a mixture is the only necessary condition. A main difference between shear-induced phase separation and viscoelastic phase separation is that the velocity fields are induced by external shear fields in the former whereas they are self-induced by phase separation itself in the latter.

The dynamic equations for polymer solutions are given as follows:[10]

$$\frac{\partial \phi}{\partial t} = -\vec{\nabla} \cdot (\phi \vec{v}) + \vec{\nabla} \cdot \frac{\phi(1-\phi)^2}{\zeta} \vec{\nabla} \cdot [\mathbf{\Pi} - \mathbf{\sigma}] \tag{1}$$

$$\vec{v}_p - \vec{v} = -\frac{(1-\phi)}{\zeta} \vec{\nabla} \cdot [\mathbf{\Pi} - \mathbf{\sigma}] \tag{2}$$

$$\rho_0 \frac{\partial \vec{v}}{\partial t} = -\vec{\nabla} \cdot [\mathbf{\Pi} - \mathbf{\sigma}] - \nabla p + \eta_s \nabla^2 \vec{v} \tag{3}$$

$$\vec{\nabla} \cdot \vec{v} = 0 \tag{4}$$

Here $\vec{v}_p(\vec{r},t)$ and $\vec{v}_s(\vec{r},t)$ are, respectively, the average velocities of polymer and solvent at point $\vec{r}$ and time t, and then the average velocity of a mixture $\vec{v}$ is given by $\vec{v} = \phi \vec{v}_p + (1-\phi)\vec{v}_s$. $\phi(\vec{r},t)$ is the composition of the polymer. $\mathbf{\Pi}$ is the osmotic stress tensor, which is related to the thermodynamic force $\vec{F}_\phi$ as shown in eqn (5):

$$\vec{F}_\phi = -\vec{\nabla} \cdot \mathbf{\Pi} = -\phi \nabla (\delta \mathscr{F}/\delta \phi) \tag{5}$$

where $\boldsymbol{\sigma}$ is the mechanical stress tensor, ρ_0 is the average density, η_s is the solvent viscosity, and ζ is the friction constant per unit volume. Here, p is a part of the pressure, which is determined to satisfy the incompressible condition $\vec{\nabla} \cdot \vec{v} = 0$. The free energy $\mathscr{F}(\phi)$ is given by the following Flory-Huggins-de Gennes form:

$$\mathscr{F}(\phi) = k_B T \int d\vec{r} \left[f(\phi) + \frac{C(\phi)}{2} (\nabla \phi)^2 \right]$$

$$f(\phi) = \frac{1}{N} \phi \ln \phi + (1 - \phi)\ln(1 - \phi) + \chi \phi (1 - \phi)$$

where N is the degrees of polymerization of polymer and χ is the interaction parameter between polymer and solvent. The terms containing the mechanical stress tensor cause couplings between the composition and the stress fields *via* the velocity fields. The above equations clearly tell us that the relative velocity of polymers to the average velocity is determined not only by the thermodynamic osmotic force but also by the mechanical force. To close these equations, we need a constitutive equation, which describes the time evolution of $\boldsymbol{\sigma}$.

Here it is worth noting that in eqn (3) the inertia term is not relevant for the description of viscoelastic phase separation in ordinary situations, since viscoelasicity suppresses the development of velocity fields. However, this is not necessarily the case for a shear problem, and even a nonlinear velocity term plays an important role for high Reynolds number flow. This, however, is out of the scope of this article.

In the above, we consider a case of polymer solution, where only polymers can support viscoelastic stress, for simplicity. However, for a more general case, where viscoelastic stress is not supported only by one of the components, we need a more general set of equations.[12] In such a case, the constitutive relation may also become more complex.

For a later discussion, here we just note that effects of gravity can be included by replacing $\vec{\nabla} \cdot \boldsymbol{\sigma}$ in eqn (1)–(3) by $\vec{\nabla} \cdot \boldsymbol{\sigma} + \Delta \rho \phi g \vec{i}_z$, where $\Delta \rho$ is the density difference between the components of a mixture, g is the gravitational acceleration, and $\vec{i}_z$ is the unit vector along the gravitational direction.

Finally, we mention a fundamental remaining problem of the two-fluid description. In the above derivation, the dissipation in a mixture is separated into the two contributions: one is viscous dissipation of the liquid component, and the other comes from the friction between the two components. This intuitively looks OK, however, the hydrodynamic couplings between the slow components are not considered in a systematic manner in the coarse-gaining procedure. This makes the validity of the above separation a bit obscure. Thus, we need theoretical justification for the treatment of dissipation, which remains a subject for future investigation.

3.2 What is the relevant form of the constitutive relation for a mixture of components having large size disparity?

We note that for mixtures composed of a large particle (or molecule) component, component 1, and a simple fluid (liquid or gas), component 2, the stress division becomes almost perfect ($\alpha_1 \cong 1$ and $\alpha_2 \cong 0$), reflecting the large size disparity and the resulting large difference in the friction constant. Stress is selectively supported almost by the large component 1 alone. This is the case for polymer solutions,[10] and suspensions of colloids, proteins, and emulsions.[23] So, the velocity relevant to the description of viscoelastic stress is the average velocity of component 1 ($\vec{v}_p$ for a polymer solution).

As an example of this type of mixture, here we consider how the mechanical stress, $\boldsymbol{\sigma}$, should be expressed in the case of a polymer solution. In principle, we can incorporate any constitutive equation into the above two-fluid model, depending upon materials. Doi and Onuki[10] employed the upper-convective Maxwell equation as a constitutive relation describing its time evolution for polymer solution:[40]

$$\frac{D}{Dt}\boldsymbol{\sigma}_S = \boldsymbol{\sigma}_S \cdot \vec{\nabla}\vec{v}_p + (\vec{\nabla}\vec{v}_p)^T \cdot \boldsymbol{\sigma}_S - \frac{1}{\tau_S(\phi)}\boldsymbol{\sigma}_S + G_S(\phi)\left\{\vec{\nabla}\vec{v}_p + (\vec{\nabla}\vec{v}_p)^T\right\} \quad (6)$$

where $\frac{D}{Dt} = \frac{\partial}{\partial t} + \vec{v}_p \cdot \vec{\nabla}$, and τ_S and G_S are the relaxation time and the modulus of the shear stress, respectively. Note that $(\vec{\nabla}\vec{v}_p)_{ij} = \partial_i v_{pj}$. To make the shear stress a traceless tensor, $\boldsymbol{\sigma}_s$ was defined as $\boldsymbol{\sigma}_S = \boldsymbol{\sigma}_S - \frac{1}{d}(Tr\boldsymbol{\sigma}_S)\mathbf{I}$, where $\mathbf{I}$ is the unit tensor and d is the space dimensionality.

We proposed to introduce the bulk stress, to describe the volume shrinking behaviour of the viscoelastic phase separation.[12,38,39] Since the bulk stress is isotropic, it can be expressed by a scalar variable, namely, $\tilde{\sigma} = \frac{1}{d}Tr\boldsymbol{\sigma}_B$. Then, the bulk stress obeys the following equations:

$$\frac{D}{Dt}\tilde{\sigma} = -\frac{1}{\tau_B(\phi)}\tilde{\sigma} + G_B(\phi)\vec{\nabla}\cdot\vec{v}_p \quad (7)$$

Here, τ_B and G_B are the relaxation time and the modulus of the bulk stress, respectively.

Here, we discuss the rheological functions in the above constitutive equations. In the case of polymer solutions, $G_S(t)$ was estimated[10,19,20,22] on the basis of rheological theories of polymer solution including the reptation theory[40,41] for good and θ solvents. The bulk stress related to $G_B(t)$ was not regarded to be important, since the longitudinal relaxation along a tube is much faster than the shear relaxation by reptation.[40] This is true locally, however, even in a good solvent, there may be elasticity associated with entanglements for large scale volume deformation at a high concentration. The elastic modulus in such a case is scaled as[34,35] $E \sim k_B T/\xi_e^3 \sim \phi^{2.25}$, where ξ_e is the characteristic length of entanglement. Since such a modulus does not exist below the overlap concentration, ϕ^*, there may be a steep concentration dependence of E on ϕ. Thus, the bulk modulus G_B may be mimicked by a step-like function, $\Theta(x)$ ($\Theta(x) = 1$ for $x \geq 1$ and $\Theta(x) = 0$ for $x < 0$),[12,38,39] as $G_B(\phi) = G_B^0\Theta(\phi - \phi^*)$. Furthermore, elastic effects associated with the volume deformation may be even more pronounced for polymer solutions under a poor solvent condition.[12,38] It should be stressed that phase separation of polymer solutions always occurs in a poor-solvent condition. Thus, we cannot apply theories for polymers in good and θ solvents to our problem. In a poor solvent, there exist attractive interactions between polymer chains. Thus, we expect that there are temporal crosslinkings of energetic origin between the polymer chains, as schematically drawn in Fig. 2.

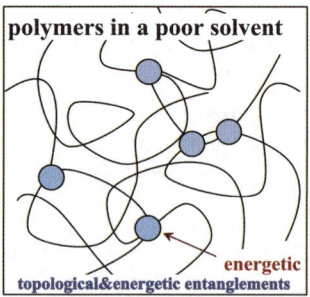

polymers in a poor solvent

energetic
topological&energetic entanglements

Fig. 2 Schematic figure showing the topological and energetic entanglements between polymer chains in a poor solvent. Attractive interactions between polymer chains probably form temporal entanglement points between close segments, whose lifetime, τ_x, increases with a decrease in the temperature. Thus, the system behaves as a gel in a short time scale. We expect that, in addition to the shear relaxation modulus, $G_S(t)$, the system has the bulk mechanical relaxation modulus $G_B(t)$, which steeply depends upon the local concentration.

The most natural model for polymer solutions under such a poor-solvent condition may be a transient gel model, in which the interpolymer attractive interactions produce temporal contact (crosslinking) points between polymer chains. If we assume that the lifetime of temporal contacts between chains is τ_x, we expect that the bulk relaxational modulus $G_B(t)$ has the relaxation time of the order of τ_x. Then, the deformation described by $\vec{\nabla} \cdot \vec{v}_p$, which accompanies a change in the volume occupied by polymer chains, causes bulk stress if the characteristic time of the deformation, τ_d, is shorter than τ_x. However, since polymer dynamics in a poor solvent is far from being completely understood, we need further theoretical studies on this problem. We point out that this type of attractive interaction between molecules of the same component may commonly exist in the unstable region of a mixture, which may generally result in the formation of a transient gel in dynamically asymmetric mixtures.

We cannot estimate $G_B(t)$ and $G_S(t)$ on a quantitative level since we do not have any reliable theory for polymer dynamics in a poor solvent yet. However, we may use knowledge of gels to estimate their magnitudes. According to a standard theory of gels, the mechanical bulk and shear modulus, G_B and G_S, are given by the following relations:[2]

$$G_B = \frac{k_B T}{\xi_c^3} \left[B\left(\frac{\phi}{\phi_0}\right) - \frac{1}{3}\left(\frac{\phi}{\phi_0}\right)^{1/3} \right] \tag{8}$$

$$G_S = \frac{k_B T}{\xi_c^3} \left(\frac{\phi}{\phi_0}\right)^{1/3} \tag{9}$$

where ξ_c is the characteristic length of crosslinking in the relaxed state, ϕ_0 is the volume fraction in the relaxed state, and B is a dimensionless parameter.

We argued[23] that the same physics may be applied to particle-like systems, such as colloidal suspensions, emulsions, and protein solutions, on noting that under the action of attractive interactions particles tend to form a transient network with the help of hydrodynamic interactions.[13,42,43] To include the effects of transient gel formation, and the resulting transient elasticity due to the gel-like connectivity, on an intuitive level, we introduced a steep ϕ-dependence of G_B,[12,29,38] as described above.

Besides the above origin, there is a possibility that for particle suspensions the slow bulk stress relaxation may originate from hydrodynamic interactions under the incompressible condition: hydrodynamic squeezing effects.[23] The relative importance of the energetic and hydrodynamic origins in the bulk stress relaxation remains a problem for future investigation. This is related to the treatment of dissipation in the two-fluid description (see section 3.1).

3.3 Constitutive relation for a more general case

The above perfect stress division only applies to a mixture of large size disparity. In polymer blends[10,44] or in a system where the glass transition has very different T_g's,[12] stress is supported by both of the two components. In this case, the dynamical eqn (1)–(7) must be generalized.[12]

Here, we briefly discuss a general rule of the stress division in such a case. First, we introduce the rheologically relevant velocity $\vec{v}_r$, which appears in the constitutive relation. It is defined as $\vec{v}_r = \alpha_1 \vec{v}_1 + \alpha_2 \vec{v}_2$, with $\alpha_1 + \alpha_2 = 1$.[12,44] Here $\vec{v}_k$ is the relative motion of component k, which has the average velocity of $\vec{v}_k$, to the mean-field rheological environment, which has the velocity of $\vec{v}_r$, and α_k is the stress division parameter. For simplicity, we neglect the transport and rotation of the stress tensor, which does not affect the pattern evolution so much since the transport and rotation are very slow in viscoelastic phase separation. In a linear-response regime, then, the most general expression of σ_{ij} is formally written by introducing the time dependence of bulk and shear moduli in the theory of elasticity,[45] as shown in eqn (10):

$$\sigma_{ij} = \int_{-\infty}^{t} dt' [G_S(t - t')k_r^{ij}(t') + G_B(t - t')\vec{\nabla}\cdot\vec{v}_r(t')\delta_{ij}], \tag{10}$$

where

$$\kappa_r^{ij} = \frac{\partial v_r^j}{\partial x_i} + \frac{\partial v_r^i}{\partial x_j} - \frac{2}{d}(\vec{\nabla}\cdot\vec{v}_r)\delta_{ij} \tag{11}$$

Here, $\vec{v}_r$ is the velocity relevant to rheological deformation and for polymer solutions $\vec{v}_r = \vec{v}_p$. $G_S(t)$ and $G_B(t)$ are material functions, which we call the shear and the bulk relaxation modulus, respectively. It should be noted that the rheological relaxation functions, $G_S(t)$ and $G_B(t)$, are functions of the local composition $\phi(\vec{r})$. We note that $G_B(t)$ is a purely mechanical modulus and is different from the bulk osmotic modulus, $G_{os} = \phi^2(\partial^2 f/\partial\phi^2)$. We have the relation $\eta = \int_0^{\infty} G(t)dt$, where η is the viscosity of a material.

The second term of eqn (10) was introduced to incorporate the effect of volume change into the stress tensor.[12,39] In a two-component mixture, the mode associated with $\vec{\nabla}\cdot\vec{v}_r$ can exist as far as $\vec{v}_r \neq \vec{v}$, even if the system is incompressible, $\vec{\nabla}\cdot\vec{v} = 0$. We proposed that this term plays a crucial role in viscoelastic phase separation[12,39] (see below), although it is not so important when we consider shear-induced demixing.[10,19,20,22]

Now we consider the stress division for the above general case. The friction force is given by $\zeta_k(\vec{v}_r - \vec{v}_k)$, where ζ_k is the average friction of component k and the mean-field rheological environment at point $\vec{r}$, where the volume fraction of component k is $\phi_k(\vec{r})$. Here, $\zeta_k = \phi_k\zeta_k^m$, and ζ_k^m is proportional to the friction between an individual molecule of component k and the mean-field rheological environment, which we call the generalized friction parameter. Because of the physical definition of the mean-field rheological environment, the two friction forces should be balanced. This fact guarantees that the rheological properties can be described only by $\vec{v}_r$. Thus, we have the following relation, in general:

$$\zeta_1(\vec{v}_r - \vec{v}_1) + \zeta_2(\vec{v}_r - \vec{v}_2) = 0. \tag{12}$$

Then, the general expression of the stress division parameter, α_k, is obtained as eqn (13).

$$\alpha_k = \frac{\phi_k\zeta_k^m}{\phi_1\zeta_1^m + \phi_2\zeta_2^m} \tag{13}$$

The above relation is consistent with a simple physical picture, where the friction only is the origin of the coupling between the motion of the component molecules and the rheological medium. We expect that this relation holds, irrespective of the microscopic details of rheological models, and, thus, we can apply it to a mixture of any material where the motion of both components is described by a common mechanism. However, for the theoretical estimation of friction coefficients, we need microscopic rheological theories, which are not generally available, unfortunately. More importantly, as mentioned in section 3.1, there is obscurity associated with the treatment of hydrodynamic couplings in the coarse-graining procedure of the two-fluid model.

3.4 Roles of dynamic asymmetry linked to glass transition in diffusion

Dynamic asymmetry affects not only the constitutive relation of a system, but also the kinetics of diffusion.[8,12] This is particularly important in a mixture whose components have very different T_g, since there is drastic slowing down of the dynamics towards T_g. This gives rise to an extremely strong dependence of the diffusion coefficient, D, on the composition, ϕ, near T_g. Furthermore, the formation of a transient gel is not expected for a mixture that has little size disparity between its components. The formation of a transient gel may be specific to a mixture with large size disparity. Thus, for a mixture whose components have very different T_g, we do not expect a significant role of the bulk stress in suppressing the diffusion, unlike the case of a mixture of large size disparity (see above), since there may be no strong ϕ-dependence of G_B. Even in this case, the strong ϕ dependence of D can cause an effect similar to the bulk stress, as shown below.

The ϕ dependence of D near (colloid) glass transition can be expressed by the following empirical Vogel-Fulcher-Tammann (VFT) relation: $D(\phi) = D_0 \exp(A\phi/(\phi_0 - \phi))$, where ϕ_0 is the VFT volume fraction and A is the fragility index. Thus, we have to take into account this ϕ-dependence of D, or the friction coefficient ζ. Effects of a steep ϕ-dependence of $D(\phi)$ were studied by numerical simulations.[46] It should be noted that large bulk stress in the slow-component-rich phase (see above) and slow diffusion in the phase that is rich in the high T_g component play similar roles in phase separation: they both suppress the rapid growth of the composition fluctuations and slow down the composition change in the more viscoelastic phase. Accordingly, the rate of the material transport between the two phases is limited or controlled by that in the slower phase. In this manner, a disparity in the diffusion coefficient, D, between the two components of a mixture, i.e. a steep ϕ-dependence of $D(\phi)$, has similar effects on phase separation as that in the bulk relaxation modulus $G_B(\phi)$.

Relevant examples of this type of dynamic asymmetry in foods can be found in many water soluble polymers and proteins,[47–49] water/sugar mixtures,[50] and meat proteins.[51] It is widely known that glass transitions and water plasticization strongly affect food quality, safety, and stability.[47] Water acts as a ubiquitous plasticizer of natural and fabricated amorphous food ingredients and products. Water-compatible food polymers include polysaccharides, starch, amylose, amylopectin, gluten, glutenin, gliadin, and gelatin. The strong composition (ϕ) dependence of the glass transition of water-soluble ingredients (polymers, proteins, sugars) leads to a steep ϕ-dependence of $D(\phi)$,[50] $D(\phi) = D_0 \exp(AT_0(\phi)/(T - T_0(\phi)))$ and slow dynamics of the ingredients-rich phase, which are prerequisites for asymmetric stress division and the resulting viscoelastic phase separation.

Finally, we note that there is a decoupling between viscosity and translational diffusion in a supercooled liquid,[52] which results in the violation of the Stokes–Einstein relation. Since the crystal growth rate is controlled by translational diffusion rather than viscosity, it is faster than that expected from the viscosity, which may solve the so-called Kauzmann paradox.[53] This decoupling may even allow crystallization below T_g. This fact may be important when we want to keep amorphous foods while avoiding crystallization,[54] i.e. in food storage.

3.5 Roles of the steep ϕ dependence of bulk stress and/or diffusion in viscoelastic phase separation

Here we briefly discuss the roles of the steep ϕ dependence of bulk stress and diffusion. According to the continuity equation:

$$\frac{\partial \phi}{\partial t} = -\vec{\nabla} \cdot \left(\phi \vec{v}_p \right) \tag{14}$$

we can see it is $\vec{\nabla} \cdot \vec{v}_p$ that causes the composition change. The bulk stress caused by the deformation type of $\vec{\nabla} \cdot \vec{v}_p$, thus, suppresses the growth of composition fluctuations if τ_d is shorter than τ_x. In this way, the bulk stress is directly coupled with the composition change and the volume shrinking.[12,38] Note that the volume change of the polymer-rich (slow-component-rich) phase is directly associated with the deformation described by $\vec{\nabla} \cdot \vec{v}_p$. So, the volume shrinking behaviour peculiar to viscoelastic phase separation is a consequence of (i) the slow bulk stress relaxation due to the connectivity of a transient gel formed by the large component and/or hydrodynamic squeezing effects, or (ii) a steep composition dependence of the diffusion constant D.

3.6 Beyond the simple constitutive relation

As will be discussed later, viscoelastic phase separation accompanies mechanical fracture of the slow-component-rich phase and thus nonlinearity of rheology may also play an important role in the pattern evolution. In the above, we assume a simple Maxwell-type constitutive relation. However, in some cases, we need more complicated constitutive relations to describe the rheology of materials. For example, McLeish and Larson[55] showed that strain hardening of branched polymers upon large deformation is related to the fact that the backbone can readily be stretched in an extensional flow since the branches are entangled with the surrounding molecules. Two key deformation types in viscoelastic phase separation are elongation (uniaxial stretching) and extension (biaxial extension). The former is important in a network-forming viscoelastic phase separation whereas the latter is important in a cellular one. The importance of strain hardening in the formation of cellular patterns has been recognized for both synthetic polymers[56] and food polymers (e.g., breads).[57,58] Qualitatively, strain hardening makes the more viscoelastic phase mechanically more resistive to fracture, or makes the morphological selection due to mechanical force balance more robust (see eqn (22)). It is highly desirable to incorporate these features into the constitutive relation for characterizing these nonlinear effects on a quantitative level.

3.7 The early stage of viscoelastic phase separation

First we consider viscoelastic effects for a case of a shallow quench, where a transient gel is not formed and the characteristic deformation rate is slower than the viscoelastic relaxation rate ($\sim 1/\tau$). This also applies to the early stage of viscoelastic phase separation.[59,60] For simplicity, here we do not consider a difference in the relaxation time between shear and bulk stress and assume $\tau_B = \tau_S = \tau$. Using the relation $\vec{\nabla} \cdot \vec{v}_p = -\dfrac{1}{\phi}\dfrac{\partial \phi}{\partial t}$ we obtain the linearized equation for $z_q = [\vec{\nabla} \cdot \vec{\nabla} \cdot \sigma_p]_q$:

$$\frac{\partial Z_q}{\partial t} \cong -\frac{Z_q}{\tau} + \frac{2G}{\phi}q^2 \frac{\partial \phi_q}{\partial t}$$

where $G = G_B + \dfrac{4}{3}G_S$. Here, ϕ_q is the Fourier component of the deviation from the initial composition ϕ_0, and it obeys, to linear order,[10,59] eqn (15).

$$\frac{\partial \phi_q(t)}{\partial t} \cong -\Gamma_q \phi_q(t) - \frac{2LGq^2}{\phi^2}\int_0^t dt' e^{-\frac{t-t'}{\tau}}\frac{\partial \phi_q(t')}{\partial t'} \tag{15}$$

Here we use the Ginzburg-Landau-type free energy $f = k_B T\left[\dfrac{r_0}{2}(\phi - \phi_c)^2 + \dfrac{u}{4}(\phi - \phi_c)^4\right]$. This form of the free energy is reasonable as far as we concern only a shallow quench near a critical point. Then, $\Gamma_q = Lq^2(r_\phi + cq^2)$, where $L = \phi^2(1 - \phi)^2/\zeta(\phi)$, is the decay rate in the absence of viscoelastic couplings. $r_\phi = r_0 + 3u(\phi_0 - \phi_c)^2$, where $r_0 = a(T - T_c)$ (a: a positive constant) and T_c and ϕ_c are the critical temperature and composition, respectively. The correlation length is given by $\xi = \left[\dfrac{C}{|r_\phi|}\right]^{1/2}$. For a case when the time scale of ϕ_q change is slower than τ, we can set $\dfrac{\partial \phi_q(t')}{\partial t'} = \dfrac{\partial \phi_q(t)}{\partial t}$ in eqn (15) and, thus, the growth rate of ϕ_q is given by:

$$A(q) = L|r_\phi|q^2(1 - \xi^2 q^2)/(1 + \xi_{ve}^2 q^2) \qquad (16)$$

where $\xi_{ve} = (2\eta L/\phi^2)^{1/2}$ is the so-called viscoelastic length.[34,35,61] This ξ_{ve} gives us the length scale above which the dynamics are dominated by diffusion and below which they are dominated by viscoelastic effects. We can also say that this length scale is the length up to which the shear stress can transmit. Furukawa showed that the viscosity of a polymer solution has the following wavenumber (k) dependence associated with the viscoelastic length:[62,63]

$$\eta(k) = \eta_s + \frac{\eta_m}{1 + \xi_{ve}^2 k^2} \qquad (17)$$

where η_m is the macroscopic viscosity and η_s is the solvent viscosity. This nonlocal nature of the viscous transport is a manifestation of the temporal hierarchical structure of dynamically asymmetric systems. Without viscoelastic couplings, the relation $A(q) = L|r_\phi|q^2(1 - \xi^2 q^2)$ should hold as the Cahn's linear theory[1] predicts. It was shown[60] that the early stage of phase separation of a polymer solution is well explained by the above Onuki-Taniguchi theory.[59]

We emphasize that the early stage of phase separation in dynamically asymmetric mixtures, including soft matter and foods, should be analysed by this theory. Applications of the Cahn's theory without considering viscoelastic effects may not be appropriate in many cases since ξ_{ve} can easily become mesoscopic in dynamically asymmetric mixtures. In relation to this, we note that the above relation [eqn (16)] well explains the unusual q-dependence of $A(q)$ experimentally observed in colloid phase separation.[23] This suggests the relevance of the viscoelastic model to phase separation not only in polymer solutions, but also in colloidal suspensions, emulsions, and protein solutions, which further indicates the importance of viscoelastic effects in any dynamically asymmetric mixtures, including food materials.[11]

3.8 The late stage of viscoelastic phase separation

In ordinary phase separation, the late stage phase separation is discussed on the basis of the scaling concept, which relies on the fact that there is only one characteristic length scale, *i.e.* the domain size, in a system. For viscoelastic phase separation, however, such a scaling concept is not valid because of the volume shrinking of the slow-component-rich phase during phase separation. Because of this difficulty, there has been no analytical theory on domain coarsening so far. In the following, thus, we describe pattern evolution in phase separation on a qualitative level.

4 Pattern evolution in viscoelastic phase separation

4.1 Initiation of phase separation: quench

In ordinary physical experiments, phase separation is usually initiated by an almost instantaneous temperature change from the initial to the final target temperature. Depending upon the type of phase diagram, phase separation can be induced either by cooling or heating. However, phase separation in foods can also be induced by changes in pressure, pH, and salt concentrations, and also by polymerization, drying (removal of solvents), and mixing of an insoluble component. For example, gas-liquid phase transition can be initiated by a pressure drop: spontaneous cavitation of bubbles can be regarded as phase separation and if the growth of bubbles exceeds the relaxation rate of the slow-component-rich phase, the phase separation can be regarded as viscoelastic phase separation. Changes in pH, salt concentration, and mixing of insoluble components all modify the interaction potential of a system, which may result in phase separation. The polymerization reaction leads to the reduction of mixing entropy (note that mixing entropy is inversely proportional to the degree of polymerization, N, according to the Flory-Huggins theory). Removal of a solvent by drying may also lead to phase separation if a solvent is rather poor. All these changes in physical parameters may be very slow. Even for heating or cooling, the rate of the temperature change can be very slow as in the case of baking of breads. However, the key rate here is not the rate of the change of a physical variable, but the rate of deformation induced by phase separation itself. If this deformation rate is faster than the relaxation rate of the slow-component-rich phase, viscoelastic effects should play a crucial role in the resulting phase separation. We emphasize that the key is the Weissenberg number defined for the self-generated deformation rate (see below).

Furthermore, in the pattern formation of foods, inhomogeneization is not necessarily induced by phase separation, but may involve much more complex nonequilibrium processes, such as crystallization of fats (fat crystal networks in butters, margarines, and chocolates), ice crystallization (freeze dry foods and ice creams). Some of these cases will also be discussed later.

4.2 General features

First we emphasize that pattern evolution in viscoelastic phase separation is essentially the same between the two types of dynamically asymmetric mixtures:[11,13,14] one is a system like polymer solutions,[5–7,64] colloidal suspensions,[65] and protein solutions,[66] where the strong dynamic asymmetry comes from a large difference in the molecular size and topology between the components, and the other is a system whose components have a large difference in the glass transition temperature.[8]

In both cases, a mixture first becomes cloudy just after the temperature quench, then, after some incubation time, small solvent holes start to appear (see Fig. 3, left). We call this incubation period the "frozen period", which is the initial stage of viscoelastic phase separation. The number and the size of solvent holes increase with time. The slow-component-rich matrix phase expels the fast liquid component and shrinks its volume and becomes networklike or sponge-like with the growth of holes made of the fast-component-rich phase (see Fig. 3, middle). In this volume-shrinking process, the bulk mechanical stress plays a crucial role.[38,39] Thin parts of a networklike structure are elongated and eventually broken. In this network-forming process, the pattern is dominated by the mechanical shear force balance condition and thus the shear stress plays a major role.[39] In the final stage, a networklike structure tends to relax to a structure of rounded shape and the domain shape starts to be dominated by the interface tension as in usual fluid-fluid phase separation (see Fig. 3, right). Domains finally become spherical. If the concentration of the slow-component-rich phase reaches the glass transition composition, a structure is eventually dynamically arrested. This may be regarded as the general scenario for

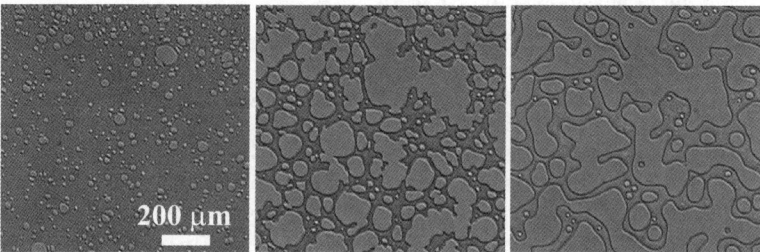

Fig. 3 The phase-separation process observed with phase contrast microscopy in a polymer solution of PS (molecular weight: 1.90×10^5) and diethyl malonate (8.53 wt% PS) at 0.0 °C, which is 11.6 K below the phase-separation temperature of 11.6 °C. Patterns are observed after 16 s, 103 s, and 2100 s after the quench from left to right. We can see a gradual transition in morphology from network to droplet, reflecting the crossover from the mechanical-stress-dominated to interface-tension-dominated regime.

formation of colloidal gels (see below).[23,67,68] When the slow-component-rich phase is the minority phase, then there is a phase inversion during phase separation. This phase inversion is a unique feature specific to viscoelastic phase separation.

According to the common sense of ordinary phase separation, after the formation of a sharp interface between the coexisting phase (namely, in the so-called late stage) the concentration of each phase almost reaches the final equilibrium one and, thus, there should be no change in the volume and concentration of each phase.[1,2] We pointed out[11] that the volume decrease of the more viscoelastic phase with time, after the formation of a sharp interface, is essentially the same as the volume shrinking of gels during volume phase transition.[69–71] The physical reason of this similarity to gels will be discussed later.

The scaling law, established in ordinary phase separation, is a direct consequence of the conservation of the volumes of the two phases after the formation of a sharp interface and the resulting self-similar growth of domains. The volume shrinking of the slow-component-rich phase inevitably leads to the absence of self-similarity during the viscoelastic phase separation and, thus, the absence of an extended scaling regime. Nevertheless, we observe a transient scaling law (the characteristic domain size $R \sim t^{1/2}$) in the intermediate coarsening stage for a few systems,[65,66,72] although its physical mechanism remains elusive.

In sum, the whole pattern evolution process can be clearly divided into three regimes: the initial, intermediate, and late stages. The crossovers between these regimes can be explained by viscoelastic relaxation in pattern evolution and the resulting switching of the primary order parameter, as will be described below.

Besides the early stage, we do not have any reliable analytical predictions and thus numerical simulations based on the viscoelastic model play a crucial role in its understanding.[13,39,72–77] We showed that a steep composition dependence of the bulk modulus or the diffusion constant is the key to volume shrinking and the resulting phase inversion and a rather smooth ϕ^2-dependence of the shear modulus is responsible for the formation of a network-like structure.[13,39,75]

We also showed that the mechanical stress accumulated in a network structure leads to its coarsening by repeating the following sequence: stress concentration on a weak part of the network, its break up and the resulting stress relaxation,

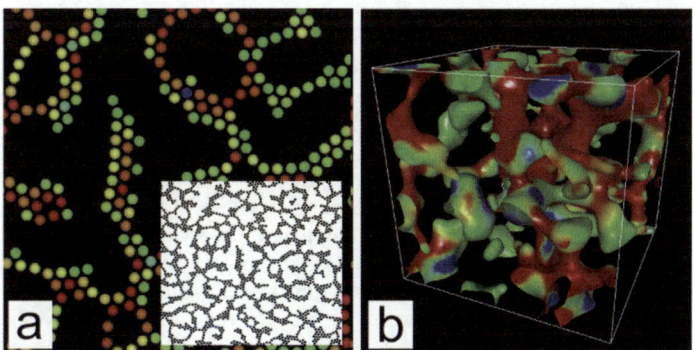

Fig. 4 Phase separation processes of colloidal suspensions interacting with the Asakura-Oosawa potential, whose range is characterized by $R = d_p/D_p$, where D_p is the particle diameter and d_p is the range of the potential. (a) 2D pattern (the volume fraction is 0.248 and $R = 0.7$). The inset is the overall structure of the colloidal network. (b) 3D (coarse-grained) pattern evolution (the volume fraction is 0.100 and $R = 0.6$), where we coarse-grained structures by replacing a particle by a Gaussian field and extracting the interface by applying a black&white operation to the field. The details of the simulations are described in Ref. 72. In both (a) and (b), red particles are stretched and in a high energy state, whereas blue particles are in a low energy state. The most significantly stretched part, due to stress concentration, eventually breaks up, which allows the decrease in elastic energy and results in stress relaxation. This process, which is an elementary process of coarsening, is repeated.

and structural rearrangements towards a lower interface energy structure.[13,39,75] Such examples can be seen in 2D and 3D colloid simulations, as shown in Fig. 4. We stress that this process can proceed without any thermal activation. Actually, the simulations in Fig. 4 were performed without any thermal noises, namely, at $T = 0$. This indicates that the coarsening of network-type viscoelastic phase separation can proceed purely mechanically: mechanically driven coarsening. This is markedly different from a conventional picture based on the activation-type coarsening process. We emphasize that mechanically driven coarsening cannot be characterized by the strength of attractive interactions measured by the thermal energy $k_B T$ alone.

4.3 Examples of viscoelastic phase separation in foods

4.3.1 Ordinary viscoelastic phase separation and gelation. Foods can generally be regarded as multi-component mixtures whose components have a large difference in their elementary dynamics. Thus, phase separation in foods should basically belong to viscoelastic phase separation. Here we mention a few examples. Globular proteins and polysaccharides are two major components of many food products and are often used to control the structure, texture, and stability of the products. There is often a competition between phase separation and gelation process in such systems, *e.g.*, a mixture composed of a globular protein, bovine serum albumin and an anionic polysaccharide, low-methoxyl pectin.[78] Milk proteins are also known to exhibit phase separation due to deletion interactions induced by polymers[79] or by addition of salt or acid. Very often network-like phase separation patterns are observed in phase separated milk,[80–82] acid skim milk gels,[83] pressure-induced gelation of whey proteins,[84] and phase separation in ice creams.[85–87] Similar phase separation behaviour is also observed in confectionery gels.[88]

Phase separation in emulsions is also an important issue in food science. For example, sodium caseinate is widely used as an emulsifying agent in many dairy products, and it imparts stability to emulsions by a combination of steric and electrostatic mechanisms. However, despite excellent coalescence stability above a certain critical protein concentration, caseinate-based emulsions can exhibit pronounced creaming or serum separation due to depletion flocculation induced by excess unadsorbed proteins in the aqueous continuous phase. Upon destabilization, network-forming phase separation is observed and can be interpreted as viscoelastic phase separation.[89] Similar behaviour is also observed in soy protein systems.[90] This type of pattern formation is basically the same as the viscoelastic phase separation observed in polymer solutions[6] and colloidal suspensions[65] and protein solutions,[66] which is supported by a striking similarity between the phase separation patterns observed.

Under a competition between phase separation and gelation, the final spatial pattern is determined by the stage at which the pattern is frozen by gelation. This is crucially dependent on the quenching condition: if phase separation can proceed sufficiently before gelation starts, a phase separation structure with large characteristic length scale can be formed. In the opposite case, a rather homogeneous pattern is formed. We also note that for the stability of a gel, the mechanical stress generated by viscoelastic phase separation has to be supported by the yield stress of the gel formed. On noting these points, we can say that slow quenching (slow temperature change, slow change in other external variables such as pH and salt concentration, or slow chemical reaction) generally leads to a phase separation structure with a large domain size, since phase separation can proceed before being arrested by glass transition or gelation.

4.3.2 Dough formation. The importance of thermodynamic aspects in dough formation has been pointed out.[31,91,92] It was shown that milk proteins and flour proteins have similar structures and functionalities and the phase behaviour of skimmed milk protein-polysaccharide systems can be used to model that of wheat dough. The underlying commonality may be phase separation induced by depletion interaction.[93] During the mixing of flour with water, albumins, globulins, water-soluble starch (from damaged starch granules) and pentosans form a liquid aqueous phase. This is immiscible with glutelins and gliadins, which form a separated gluten phase. Thus, one phase is the concentrated protein viscoelastic phase containing gliadins and glutelins, called gluten. The other co-existing phase is a viscous mixed solution of albumins, globulins, neutral and charged polysaccharides, which is treated as the liquid phase. Thus, the two phases have strong dynamic asymmetry. This means that phase separation between these two phases may be classified into viscoelastic phase separation.

Applying shear in the process of kneading[94,95] may lead to the formation of complex phase-separation morphologies under dynamical couplings between stress and diffusion.[96] The liquid phase acts as a lubricant and the composition heterogeneity further enhances viscoelastic heterogeneity. In this regime it was found that in a steady state the characteristic domain size is inversely proportional to the average shear stress for various shear rates. In the Newtonian liquid, it is known that the domain size R is determined by a balance between the surface energy density γ/R and the viscous shear stress $\eta\dot{\gamma}$. In dynamically asymmetric mixtures, the stress is of elastic origin rather than viscous origin. During a mixing process, gluten particles deform and make crosslinkings *via* covalent di-sulphide bonds, which makes the process even more complicated. Later we will also discuss phenomena like shear-induced separation of starch and glutens.[97]

4.3.3 Formation of fat crystal networks.
Plastic fat products such as shortening, margarine, butter, lard, and chocolate are characterized by their solid-like nature. The yield stress of these products is the consequence of the presence of a three-dimensional network of fat crystals.[98–101] The relation between the fractal nature of the networks and the shear elastic modulus has been suggested.[102] The formation of fat crystal networks involves at least two processes: the formation of microcrystallites and their aggregation towards the formation of networks. It is believed that the second aggregation process is driven by van der Waals attraction between crystallites.[98] Thus, the process after the formation of microcrystallites is essentially the same as the viscoelastic phase separation of colloidal suspensions.[23] However, if there is no time separation between crystal growth and aggregation, their coupling makes the process far more complicated than ordinary viscoelastic phase separation.

Here we note that the aggregation process must be affected by hydrodynamic interactions,[42,43] as far as the viscosity of the surrounding liquid is not so high. Hydrodynamic interactions have significant effects on the formation of the network structure, such as its fractal dimension.[23] The interplay between the shape of particles and the hydrodynamic interactions and their effects on the network morphology are also an interesting topic for future study. The stabilization of the network formed by microcrystallites might be induced by secondary crystallization, in addition to van der Waals attractions. This is related to the above mentioned degree of separation between the crystal growth and aggregation processes.

4.4 Collapsing of network or foam structures

As described above, in the final stage of viscoelastic phase separation, due to a dynamical crossover between the deformation rate induced by phase separation and the structural relaxation of the slow-component-rich phase, a network structure transforms into a droplet structure (see Fig. 3). This process is physically very similar to the collapsing of network or foam structures upon a loss of elasticity under gravity. We note that the collapsing behaviour in the absence of gravitational fields has successfully been simulated by the viscoelastic model.[13,39,75] The essential behaviour should be captured by the viscoelastic model including the gravitational effects (see section 3.1). Here we mention a few examples in food products.

4.4.1 Collapsing of fat crystal networks. Here we mention the melting process of fat crystal networks. Upon melting, the solid network transforms into a fluid network, which is then destabilized by interfacial tension and breaks into droplets. We note that a one-dimensional fluid tube is unstable for pinching off, which is known as Rayleigh instability, or tube hydrodynamic instability. This process is the same as the late stage of viscoelastic phase separation, where a network structure transforms into droplet structures and accordingly the yield stress also disappears (see Fig. 3). Under gravitational fields, collapsing dynamics should obey the viscoelastic model with gravity effects (see section 3.1).

4.4.2 Formation of spongy structures in cryogels and freeze-drying products and their collapsing. The processes of cryotropic gelation of polymeric systems occur in the non-deep freezing, storage in the frozen state, and thawing of the solutions or colloidal dispersions containing monomeric or polymeric precursors potentially capable of producing gels.[103–106]

Polymeric materials formed under these conditions were termed as cryogels, which we often see also in freeze-drying foods[107–110] such as kori-tofu and dried vegetables and fruits. Similar phenomena are also observed for freezing colloidal suspensions,[111,112] although there is a difference in the mechanism of exclusion between polymers and colloids. When the initial solution or colloidal sol is frozen non-deeply, *i.e.* not lower than several tens of degrees from the crystallization point of the pure solvent, the resulting system is composed of the crystallized solvent (ice in the case of aqueous systems) and the unfrozen liquid, where the gel-forming components are concentrated: cryoconcentration. Cryoconcentration is the consequence that the noncrystallizable component is expelled from crystals into the surrounding liquid. Although the physics behind the formation of spongy structures is very different, spongy patterns formed in this way have many similarities to those formed by foam-like structures that are formed in viscoelastic phase separation. The commonality comes from nucleation of solvent-rich holes or crystals and the continuous increase of the concentration of shrinking or expelled polymeric components during pattern evolution (phase separation or crystallization, respectively). Unlike viscoelastic phase separation, however, the mechanical force balance does not play any role in the formation of cryogels due to the solid nature of crystals, but after thawing the stress is divided quite asymmetrically between the polymer and solvent phase: the stress is supported exclusively by the spongy polymer structure.

Fruits and vegetables are cellular tissues containing gas-filled pores that tend to collapse upon dehydration.[113] The collapsing in the drying process involves tissue shrinkage, cellular shrinkage, and then cell collapse. If the system size is small enough, the system may homogeneously shrink without the formation of pores. For a large enough system, however, the overall shrinkage of a system is very slow and thus internal mechanical instability spontaneously takes place inside the system, which leads to the formation of porous spongy structures. The physics of this phenomenon is the same as that of mechanical instability (solvent hole formation) upon volume shrinking of the slow-component-rich phase in viscoelastic phase separation (see Fig. 3) or in gels undergoing volume shrinking phase transition.[64]

As in the case of fat crystal networks, the reverse process, *i.e.*, melting of spongy structures due to heating or absorption of solvents, should be essentially the same as the late stage break-up process of spongy or network structures into droplets in viscoelastic phase separation: the process is dominated by interface-tension driven flow (see Fig. 3). For example, when freeze-dried cake is heated to a certain temperature, a change in the structure called collapse generally takes place. The cause of this shrinkage has been attributed to a reduction in the elasticity and viscosity of the matrix, to a point where the viscosity is too low to support the matrix weight. This is essentially the same as the viscoelastic relaxation process of patterns in the final stage of viscoelastic phase separation (see Fig. 3). This may further be related to delayed sedimentation of gels under gravity (see, *e.g.*, Ref. 23,65,114–116). Here transient gels formed by viscoelastic phase separation collapse under the gravitational field, when the gravitational force exceeds the yield stress of the gel network.

4.4.3 Collapsing of foam structures of ice creams. Ice cream has a foam structure composed of a fat globule network, ice crystals, a serum phase, and air cells. This foam structure is formed in the freezing process, which both freezes a portion of the water and adds air to increase the volume of the product. Although this process of the formation of foam-like structure is quite different from viscoelastic phase separation, the collapsing of foam structures is similar to the above cases. Ice cream foams can be destabilized when ice crystals melt into liquid water. So, the basic behaviour of collapsing should be the same as the above cases.

5 Pattern evolution in fracture phase separation

Here we show a special case of viscoelastic phase separation, where phase separation proceeds accompanying mechanical fracture of a mixture.

5.1 Physical mechanism

The above-described mechanical nature of viscoelastic phase separation implies a close analogy to the mechanical response of materials. Indeed, we recently found novel phase-separation behaviour accompanying mechanical fracture ("fracture phase separation") in polymer solutions[15] (see Fig. 5(a)). Surprisingly, mechanical fracture becomes the dominant coarsening process in this phase separation. This type of phase separation is observed when the deformation rate of phase separation becomes much faster than the slowest mechanical relaxation time of a system. In this sense, the transition from viscoelastic to fracture phase separation corresponds to the "liquid-ductile-brittle transition" in the fracture of materials under shear deformation[17] (see Fig. 5(b)). The only difference between fracture phase separation and material fracture is whether the deformation is induced internally by phase separation itself or externally by loading.

We argue that fracture phase separation is the process of mechanical fracture of a transient gel against *self-generated shear deformation*, which is caused by volume shrinking of the slow-component-rich phase. For slow shear deformation, a transient gel behaves as viscoelastic matter and exhibits liquid fracture behaviour for shear deformation: *viscoelastic phase separation*. A network is stretched continuously under stress, elongated along the stretching direction, and eventually breaks up. This process resembles the process of liquid fracture of a material under a stretching force (see Fig. 5(c)).[117–120] For fast shear deformation, a transient gel should behave in a solid-like manner, and exhibit brittle (or ductile) fracture behaviour: *fracture phase separation* (see Fig. 5(d)). At this moment, it is not so clear whether crack formation in fracture phase separation belongs to ductile or brittle fracture, since we are not able to visualize the deformation field in the coarse of phase separation. We speculate that cracks are formed perpendicular to the stretching direction (see Fig. 5(b)). This fracture behaviour is a manifestation of solid-like (or, elastic) behaviour[119,120] of a transient gel.

The physical mechanism of this mechanical instability is basically the same as shear-induced fracture of a viscoelastic matter: self-amplification of density fluctuations under shear.[16,17] In our view, a steep composition dependence of the bulk stress leads to instability of the interaction network for the volume deformation of type $\vec{\nabla} \cdot \vec{v}_p < 0$, whereas that of the shear stress leads to its instability for shear-type deformation, which should be the origin of fracture-like behaviour. In fracture phase separation, elastic couplings between cracks also play a crucial role in pattern formation. We studied this problem by using a simple spring model,[76] but further detailed studies are necessary to elucidate the roles of spatio-temporal elastic coupling.

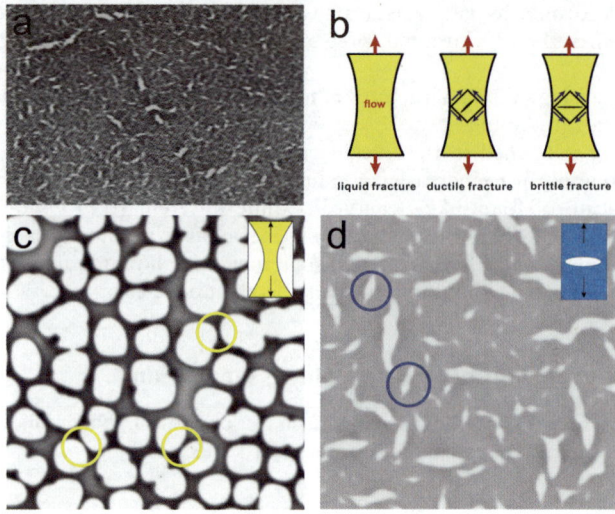

Fig. 5 (a) Crack formation in the initial stage of fracture phase separation. Fracture phase separation observed for a polystyrene(PS)/diethylmalonate mixture (4 wt% PS) after a quench to 22 °C. Crack formation is clearly observed. Cracks are solvent-rich domains. The sample thickness is 5 μm. The width of the image corresponds to 0.5 mm. (b) Schematic figure showing liquid, ductile and brittle fracture of a material under elongational deformation. For ductile fracture a crack is formed along 45° from a stretching direction, whereas for brittle fracture it is formed perpendicular to a stretching direction. Brittle fracture is also characterized by crack formation just after the linear Hookian regime. On the other hand, liquid and ductile fracture occur after large nonlinear deformation. Viscoelastic phase separation accompanies liquid or ductile fracture for self-generated shear deformation, whereas fracture phase separation accompanies brittle fracture. (c) Viscoelastic phase separation and (d) fracture phase separation simulated on the basis of the viscoelastic model.[15] We can see typical patterns of liquid and solid fracture in (c) and (d), respectively.

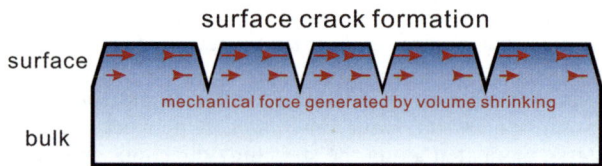

Fig. 6 Schematic figure showing the localization of mechanical stress near the surface of drying soft matter and foods. The mechanical stress is a consequence of volume shrinking induced by solvent evaporation.

For a situation of fracture phase separation, the break-up of bonds is required not only for volume deformation, but also for shear deformation of the network. To represent such a strongly nonlinear behaviour, we introduce a steep (actually, step-like) composition dependence also for the shear modulus:[15] $G_S(\phi) = G_S^0 \Theta(\phi - \phi_0^S)$, where ϕ_0^S is the threshold polymer composition for the shear modulus. ϕ_0^S may be material specific, reflecting its constitutive relation. We speculate that $\phi_0^S < \phi_0^B$ since the instability occurs for volume deformation before it occurs for shear deformation. This is because only volume deformation can induce a composition change and shear deformation cannot. We confirmed that the introduction of a step-like ϕ dependence for the relaxation time τ_S has a similar effect.

Fracture phase separation also provides a mechanism for the formation of shrinkage crack patterns in both nature (tectonic plates, dried mud layers, and cracks on rocks) and materials (cracks in concretes and coatings and grazes on a ceramic mug). This mechanism may also be relevant to crack formation in foods upon shrinking, which we discuss below.

5.2 Examples of shrinkage crack formation in foods

Here we briefly discuss a few examples of fracture phase separation in foods. It is not easy to find such examples in the narrow meaning of fracture phase separation. However, pattern formation induced by mechanical fracture self-induced by volume shrinking can be found in many situations.

The most obvious examples are surface crack formation upon drying of foods, *i.e.*, evaporation of a liquid component (*e.g.* water) from them. Crack formation should occur when the deformation induced by evaporation exceeds the mechanical relaxation rate of a material. Thus, we propose that the control of the evaporation rate allows us to control surface crack formation or avoid it. We also note that in the process the mechanical boundary condition plays a crucial role. This issue will be discussed later in more detail.

An interesting example can also be seen in the structural change occurring *post mortem* in meats.[121] This phenomenon is induced by slow fibre shrinkage, which leads to the formation of gaps between the fibre bundles. Lateral shrinkage of myofibrils, which is induced by pH drop after death, occurs while accompanying expelling water, which resembles viscoelastic phase separation or the volume shrinkage of gels. After the formation of gaps between fibre bundles, further shrinkage of fibres eventually leads to gap formation between fibres. Thus, gap formation occurs in a two step process, which leads to two-level gap patterns. This can be explained by two pre-existing boundaries, which have different strengths: weaker boundaries between fibre bundles and stronger boundaries between fibres. We note that even without such boundaries, mechanical instability generally takes place in a longer length scale and develops towards shorter length scale upon further shrinking. in this particular case, fracture takes place from weaker boundaries. In our viscoelastic model, we do not have any spatial heterogeneity in elastic and viscous properties in the initial state besides thermal fluctuations. However, if there is heterogeneity, fracture should first take place in a weak part due to stress concentration. This feature can be incorporated into the viscoelastic model by introducing the spatial dependence of the elastic modulus as an initial condition. After the detachment of fibre bundles, further shrinkage of the fibres creates mechanical stress. However, since at this stage each fibre bundle is already mechanically isolated, it has a free surface boundary condition. Since the transport of water from the surface of fibre bundles is limited, however, the mechanical stress is generated inside a fibre bundle, which leads to secondary fracture at boundaries between fibres. We note that the same mechanical instability also happens in cooking meats.[122] When we heat meats in cooking, lateral shrinkage of fibres also takes place and leads to gap formation, which leads to the formation of a peculiar texture like surface crack patterns. The basic mechanism is the same as the above and may be classified into shrinkage-induced crack patterns.

In principle, similar cracking or porosity formation occurs when there is a significant volume shrinking in the drying process of foods such as vegetables and fruits.[113,123,124] The basic physics should be the same. Whether volume shrinking leads to the formation of cracks or pores should depend upon the shrinking rate and the rheological relaxation rate of the matrix. If the relevant Weissenberg number is much larger than 1, fracture-type cracking should take place and otherwise a porous structure should be formed. Even in the fracture mode, significant shrinking may eventually lead to a porous structure (but without smooth interfaces).[15] As in the above cases, the initial mechanical heterogeneity pre-existing in foods should significantly affect the spatial characteristics of initial mechanical instability. This feature is absent in ideal viscoelastic or fracture phase separation, besides very weak inhomogeneity due to thermal fluctuations, but can easily be incorporated into the model.

6 Switching of the order parameter during viscoelastic phase separation

6.1 Concept of order-parameter switching

Here we show that the dynamic behaviour of viscoelastic phase separation can be explained by the concept of "order-parameter switching". Phase separation is usually driven by the thermodynamic force and the resulting ordering process can be described by the temporal evolution of the relevant order parameter associated with the thermodynamic driving force. The primary order parameter describing phase separation of a binary mixture of isotropic components is only the composition difference between the two phases in the solid or fluid model of phase separation. In the viscoelastic model, on the other hand, the phase-separation mode can be switched between "fluid mode" and "elastic gel mode". This switching is caused by a change in the coupling between the stress and the velocity fields, which is described by eqn (10): eqn (10) tells us that these two ultimate cases, namely, (i) fluid model ($\kappa_{ij}^{p} \sim$ constant) and (ii) elastic gel model ($G_S(t)$ and $G_B(t) \sim$ constant), correspond to $\tau_{ts} \gg \tau_d$ and $\tau_{ts} \ll \tau_d$, respectively. For $\tau_d \gg \tau_{ts}$ the primary order parameter is the composition as in usual classical fluids, whereas for $\tau_d \leq \tau_{ts}$ it is the deformation tensor as in elastic gels. The deformation tensor $\mathbf{u}_p$ is defined as

$$\mathbf{u}_{pij} = \frac{1}{2}\left(\frac{\partial u_{pi}}{\partial x_j} + \frac{\partial u_{pj}}{\partial x_i}\right) \tag{18}$$

It is well known[2,71] that the free energy of gel, f, can be expressed only by the local deformation tensor as $f(\mathbf{u}_p)$. Thus, we can say that the order-parameter switching is a result of the competition between two time scales characterizing the domain deformation τ_d and the rheological properties of domains τ_{ts}. As mentioned above, thus, this can be regarded as viscoelastic relaxation in pattern evolution. Here it should be noted that the above two order parameters are related with each other in a gel state as[2]

$$\frac{\phi_0}{\phi} = \text{Det}\left[\frac{\partial u_{pi}}{\partial x_j}\right] \tag{19}$$

where ϕ_0 is the volume fraction in the relaxed state.

6.2 Crossovers between the characteristic timescales

We next consider how τ_{ts} and τ_d change with time during phase separation. To simplify the problem, we estimate the temporal change of τ_d and τ_{ts}, provided that they are independent of each other. Under this crude assumption, we can estimate the velocity field determining the deformation rate, neglecting the contribution of $\vec{\nabla}\cdot\boldsymbol{\sigma}$, from the relation

$$\vec{v} = -\int d\vec{r}''\mathbf{T}(\vec{r} - \vec{r}')\cdot\vec{\nabla}\cdot C\nabla^2\phi\nabla\phi \tag{20}$$

where $\mathbf{T}(\vec{r})$ is the so-called Oseen tensor given by

$$\mathbf{T}(\vec{r}) = \frac{1}{8\pi\eta_s r}\left(\mathbf{I} + \frac{\vec{r}\vec{r}}{r^2}\right) \tag{21}$$

According to the above equation, in the initial stage the velocity fields should grow[125] as $|\vec{v}| \sim k_B TC/3\eta\xi\Delta\phi^2$, where $\Delta\phi$ is the composition difference between the two phases, and ξ is the correlation length, or the interface thickness. Since $\Delta\phi$ approaches to $2\Delta\phi_e$ (ϕ_e: the equilibrium composition) with time, this expression of $|\vec{v}|$ reduces to the well-known relation $|\vec{v}| \sim \gamma/\eta$ (γ: interface tension) in the late stage [note that $\gamma \sim k_B T C (2\phi_e)^2/3\xi$]. Thus, the characteristic deformation time τ_d changes with time as $\tau_d \sim R(t)/V(t) \sim R(t)/\Delta\phi(t)^2$. In the initial stage, the domain size does not grow so much with time whereas $\Delta\phi$ rapidly increases with time; and, accordingly, τ_d decreases rapidly. On the other hand, τ_{ts} increases steeply with an increase in $\Delta\phi$, reflecting the increase in the polymer concentration in the polymer-rich domain. Thus, τ_{ts} becomes comparable to τ_d in this intermediate stage of phase separation. Once τ_d exceeds τ_{ts}, the slower phase cannot follow a deformation speed and behaves as an elastic body: the mechanical force balance dominates a coarsening process in the intermediate stage. Next, we consider what happens in the late stage. Since $\Delta\phi$ approaches the value of $2\phi_e$ and becomes almost constant in the late stage, τ_d ($\sim R\eta/\gamma$) increases with an increase in R whereas τ_{ts} becomes almost constant. Thus, τ_d becomes longer than τ_{ts} again. This results in the fluid-like behaviour in the final stage of phase separation. We may regard $Wi = \tau_{ts}/\tau_d$ as the Weissenberg number for self-generated deformation rate. The viscoelastic effects become significant when this Wi significantly exceeds 1.

In short, $\tau_d \gg \tau_{ts}$ in the initial stage, $\tau_d \lesssim \tau_{ts}$ in the intermediate stage, and $\tau_d \gg \tau_{ts}$ in the late stage again. Accordingly, the order parameter switches from the composition to the deformation tensor, and then switches back to the composition again. When phase separation accompanies an ergodic-to-nonergodic transition such as glass transition, phase separation ends up with a dynamically arrested state, which can freeze network-like and sponge-like structures: gelation (see below).

Here we consider possible effects of a difference in the two types of origins of dynamic asymmetry on pattern evolution: size disparity and the difference in T_g between the two components. In the above, the domain deformation rate is related to the interfacial tension γ, or the coefficient c. It is known that γ is inversely proportional to ξ^2, $\gamma \sim 0.1 k_B T/\xi^2$, according to the two scale-factor universality.[2] Since the interfacial thickness, or the correlation length, ξ, is the size of a component, the interface tension, γ, is known to be extremely small for systems of macromolecules, emulsions, and colloids simply because of their large sizes.[126–128] However, the large size of the slow component also results in the slow relaxation in proportion to ξ^3. Thus, the above Weissenberg number, Wi, can become very large even for a system with size disparity.

7 Viscoelastic selection of phase-separation morphology

7.1 What physical factors determine the shape of pattern?

Since the deformation tensor $\mathbf{u}_p$ has an intrinsic coupling to the mechanical stress, a pattern in the elastic regime is essentially different from that of usual phase separation in fluid mixtures, which is dominated by the balance between the thermodynamic and the viscous force. The domain shape during viscoelastic phase separation is determined by which of the mechanical and interface force is more dominant in the momentum conservation equation. Roughly, the elastic energy is scaled as $G_S e^2 R^d$ (e: strain and d: spatial dimensionality) for a domain of size R, since it is the bulk energy. On the other hand, the interface energy is estimated as γR^{d-1}. For macroscopic domains, thus, the elastic energy is much more important than the interface energy in the intermediate stage where $\tau_d \lesssim \tau_{ts}$.

The momentum conservation tells us that the domain shape is generally determined by the mechanical shear force balance condition:[11]

$$\partial_i \left[C(\phi) \left\{ \partial_i\phi\partial_j\phi - \frac{1}{d}(\partial_i\phi)(\partial_j\phi)\delta_{ij} \right\} - \sigma_{ij} \right] = 0 \qquad (22)$$

This leads to networklike or spongelike morphology. In two dimensions, this force balance condition favours a three-armed treelike structure where the angles between the arms are about $120°$, whereas in three dimensions a four arm (tetrapod-like) structure around its junction point are favoured. This is consistent with what is observed in Fig. 3 and 4. In the late stage of phase separation where $\tau_d \gg \tau_{ts}$, on the other hand, the interface energy dominates a domain shape since the stress becomes very weak.

Here we note a possible difference between a system with size disparity and a system with disparity in T_g. As mentioned above, a system with large size disparity is characterized by ultra-low interface tension, γ. For such a system, the above force balance condition can approximately be given by $\partial_i \sigma_{ij} = 0$. For a system of disparity in T_g, on the other hand, the interface tension plays a more important role when the mechanical stress is about the same. In the final stage of viscoelastic phase separation, where the Weissenberg number, Wi, decreases and becomes smaller than 1, the interface tension leads to the breakage of a network structure, which transforms the morphology from network-like to droplet-like. This process may take place more slowly for a system with size disparity than for a system with disparity in T_g.

7.2 Crucial roles of the boundary condition for a system in viscoelastic and fracture phase separation

As described above, the mechanical force balance plays a crucial role in pattern selection in viscoelastic phase separation. A transient gel always tends to shrink to reduce the elastic energy, as a gel undergoing volume-shrinking transition does.[69,70] This means that the entire network tends to shrink its volume. This stress leading to volume shrinking of a whole sample must be supported by the boundary to have only internal mechanical instability, leading to solvent-hole formation and crack formation. In simulations, the employment of a periodic boundary condition automatically allows us to avoid long-wavelength instabilities. The rate of volume shrinking is controlled by the rate of the transport of the fluid component under the stress fields. In many experimental situations, this elastic stress is supported by the boundary, which prevents the shrinking of the overall volume of a transient gel. This can be realized by wetting or adsorption of the slow-component-rich phase to walls confining a sample. In our experiments using a quasi two-dimensional sample for microscopic observation, or for an anisotropic confinement of a sample, the volume shrinking in the lateral direction is strongly suppressed by a large friction of the sample to the walls. This boundary effect is the very origin of the mechanical stress acting against concentration diffusion ($\vec{\nabla} \cdot \vec{v}_p$). Even if there is no fixed boundary condition, for a very large sample there is a clear separation between the time scale of the volume shrinking of the entire sample and that of the local development of the mechanical instability. For a small sample, however, the volume shrinking takes place rather rapidly and thus affects or interferes the internal mechanical instability. We reported such volume shrinking behaviour of the whole transient gel accompanying mechanical instability in a macroscopic length scale in Ref. 64.

This special role of the boundary condition in phase separation is a manifestation of the mechanical nature of phase separation, which is common to both viscoelastic and fracture phase separation. We note that surface crack formation is also affected by such a boundary condition. During evaporation of the liquid component, the volume shrinking of the surface part takes place much faster than the bulk part far from the surface. Thus, the bulk part plays the same role as a fixed boundary condition and supports the mechanical stress, which leads to the formation of surface crack patterns.

Surface crack patterns can also be induced by bulk expansion: the slow (or solid-like) surface layer cannot catch up with the expansion of the bulk. This is, for example, the case of surface crack formation of chocolate loafs. Surface crack formation during freezing of foods may also share the same origin: volume expansion due to ice crystallization in the bulk may lead to surface crack patterns, if crystallization near the surface is more suppressed than bulk due to partial drying or any other reasons. We also note that surface crack formation can also be caused by cooling of a glassy material from its surface. This is because surface cooling leads to a larger volume shrinking near the surface. This causes the extensional mechanical stress on the surface, which may induce brittle fracture of the surface region that becomes solid-like near and below the glass transition upon cooling. This may be the case for formation of grazes on ceramic or glass mugs.

8   Arrest of viscoelastic phase separation

8.1   Gelation as dynamically arrested viscoelastic phase separation

Gels and glasses are important nonergodic states of condensed matter, both of which are dynamically arrested nonequilibrium states.[129] Unlike crystals, their static elasticity does not come from translational order. These states are particularly important in soft matter and foods. In particular, a gel can sustain its shape under gravity, but is still soft enough to eat; because of these features, it is a major form of foods. Thus, we briefly consider the nature of gel and the mechanism of its formation.

We show a schematic state diagram for colloidal suspensions in Fig. 7, which shows that a transient gel is a consequence of viscoelastic phase separation and a permanent gel is a consequence of viscoelastic phase separation dynamically arrested by glass transition.[23] Recently, by combining careful experiments and simulations, Lu *et al.*[68] showed evidence that colloidal gelation is spinodal decomposition dynamically arrested by glass transition. Here it is worth pointing out that spinodal decomposition is not the necessary condition, but phase separation including nucleation-growth type is enough to cause gelation if the slow-component-rich phase is the majority phase.[23] In this scenario, there is an intimate relation between gels and glasses, since the source of dynamic arrest for these two nonergodic states is the same. However, there are many distinct differences in both structures and dynamics between them (see, *e.g.*, Ref. 130). Locally the dynamic arrest is a consequence of glass transition. However, since gelation is a consequence of phase separation, it

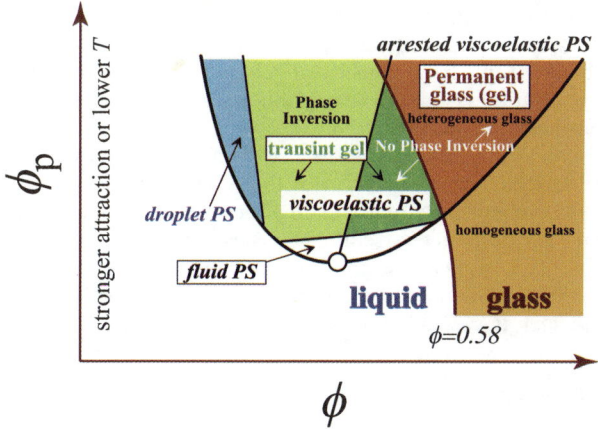

Fig. 7   Schematic state diagram for colloidal suspensions, emulsions, and protein solutions. Whether phase inversion takes place or not is determined by the static symmetry line on which the two separated phases occupy the same volume. In the left-hand side of this line, a network pattern is formed, whereas in the right-hand side a sponge-like structure is formed. Whether viscoelastic phase separation is arrested or not is determined by the glass-transition line. The timing when viscoelastic phase separation is arrested by glass transition, or the degree of coarsening of a phase-separated structure, is dependent on the quenching condition (the composition, the effective temperature, and the quench speed).

intrinsically has macroscopic spatial heterogeneity. This is always the case if a gel is formed by ordinary attractive interactions between particles.

In some cases, however, gelation involves specific interactions such as strong hydrogen or covalent bonding and microcrystallite formation (*e.g.*, gelatin gels and agarose gels). We note that the mechanism of gelation in these cases is different from the above scenario, reflecting the difference in the mechanism of local dynamic arrest. For example, in gelatin and agarose crosslinking points are formed by micro-crystallites of polymers. In some biopolymers hydrogen bonding between polymers is responsible for gelation, whereas in gluten di-sulphide bonds are responsible. The difference in the physical interactions stabilizing a gel network leads to the difference in the stability and yield stress. Upon viscoelastic phase separation, mechanical stress is always generated in the polymer-rich phase but the formation of crosslink-ings leads to an increase in the yield stress, which results in the stabilization of the gel under the mechanical stress. We emphasize that this mechanical stress is induced by many body effects (the sum of attractions between many molecules) and thus can well exceed the interaction strength per bond, which is often measured in the unit of k_BT. Thus, even for strong attractions ($\gg k_BT$), coarsening can proceed upon phase separation accompanying gelation (see the discussion in section 4.2) if there is a strong driving force for volume shrinking, although stronger bonds of course tend to increase the yield stress and make a gel more stable. The level of coarsening can also crucially depend upon the stage at which gelation takes place upon phase separation (see the state diagram and the caption of Fig. 7).

8.2 Ageing of gels and glasses

The scenario that gelation is viscoelastic phase separation dynamically arrested by glass transition immediately tells us a crucial difference in the ageing mechanism between gelation and vitrification. In viscoelastic phase separation, the coarsening is driven by elastic stress associated with volume shrinking and interfacial tension. These features are absent in the ageing of glass transition, at least in colloidal suspensions, due to the conservation of the composition. In ordinary glass transi-tion, which takes place under a condition of constant pressure, the volume of a sample decreases (or, the density increases) while ageing, since the ageing accom-panies the densification due to attractive interactions. We can say that the ageing of gels proceeds under the momentum conservation while satisfying eqn (22).[72] The intrinsic macroscopic heterogeneity of gels, which comes from its link to visco-elastic phase separation, leads to strong inhomogeneity of particle mobility, *i.e.*, faster dynamics near the interface.[131]

Whether viscoelastic phase separation is dynamically arrested or not may be determined by whether the connectivity of the slow-component-rich phase remains when the system reaches a nonergodic state or not. Once the volume shrinking stops, the driving force for domain coarsening becomes only the interfacial tension. If the yield stress of a gel is higher than the force exerted by this interfacial tension, the system is basically frozen and only exhibits slow ageing towards a lower free-energy configuration, which is basically the same as that of glasses.

8.3 Moving droplet phase

Here we mention another type of viscoelastic phase separation, which takes place in a mixture with a low volume fraction of the slow component. At such a low volume fraction, the slow-component-rich phase immediately forms droplets, which shrink their size by expelling a solvent. This shrinking process is finished rather quickly because of the small size of the droplets. Thus, the volume fraction inside droplets may rapidly reach a glassy state if the final volume fraction is higher than the glass transition volume fraction. If the collision timescale is shorter than the structural relaxation time, droplets behave as elastic glassy balls. This may also be expressed as follows. If the contact time during droplet collision due to Brownian motion is shorter than the material transport between droplets, droplets do not coalesce and stay without growing for a fairly long time. We refer to this interesting metastable state due to the elastic or glassy nature of droplets as "moving droplet phase".[4,5,7,11]

This phenomenon may be used to make rather monodisperse particles whose size is in the order of sub microns to microns. Recently, it has been shown that even random nonionic amphiphilic copolymers can form stable aggregates, a mesoglobular phase between individual collapsed single-chain globules and macroscopic precipitation.[132] The monodisperse nature is a direct consequence of the formation of droplets due to the growth of concentration fluctuations with a characteristic wavenumber and little coarsening after that. So, this phenomenon may provide us with a new very simple and low cost method to make particles with a desired size, which may be useful in both soft materials and foods industries. For example, we speculate that the formation of elastic particle gels of proteins[133] may share a common mechanism with the moving droplet phase.

We also note that if the droplet concentration becomes too high, droplets are no longer stable and tend to aggregate to form networks.[65] After the formation of networks, the behaviour is similar to viscoelastic phase separation. This may be regarded as a two-step viscoelastic phase separation: the formation of elastic gel particle followed by network formation.

8.4 Arrest of viscoelastic phase separation by smectic order

So far we have considered only phase separation of a mixture of isotropic disordered materials. However, there is a possibility that viscoelastic phase separation accompanies other ordering phenomena. Here we show an interesting example in which viscoelastic phase separation is arrested by smectic ordering.[134] We studied phase separation of an ordered phase (lamella) of a lyotropic liquid crystal (tri-ethyleneglycol mono n-decyl ether ($C_{10}E_3$)/water mixtures) into the coexistence of an ordered (lamella) and a disordered (sponge) phase upon heating. When phase separation cannot follow the heating rate, viscoelastic phase separation is observed. The slow lamella phase, which has internal smectic order and anisotropic elasticity, cannot catch up with the fast domain deformation, and thus it transiently behaves like an elastic body and supports most of the mechanical stress. On the other hand, the less viscous sponge phase, which is an isotropic Newtonian liquid, cannot support any stress. This dynamic asymmetry leads to the formation of a well-developed network structure of the lamella phase.

If phase separation is slow enough to satisfy the quasi-equilibrium condition, then membranes can homeotropically align along the interface between the two phases while keeping their connectivity, to lower the elastic energy. This leads to the formation of a cellular structure (see Fig. 8). The lamellar films forming closed polyhedra cannot exchange material with layers in neighbouring polyhedra, except by permeation (*i.e.*, diffusion of material normal to the layers). For slow permeation, which should be the case in our system, the lamellar films can exhibit dilational elasticity, which leads to a (quasi-)stable cellular structure. This is markedly different from a soap froth, where the stretching fluid film merely pulls in material from the others without any elastic cost. Thus, a high degree of smectic order in the cell walls and borders stabilises the cellular structure: any deformation of the smectic order increases elastic energy and thus the structure is selected by the elastic force balance condition.

We also note that this is an interesting example showing that the heating rate can be used to control the type of phase separation, covering from droplet phase separation, to network phase separation, to foam-like phase separation. This may be relevant in pattern formation in foods, where the change in a physical variable, such as temperature, is not instantaneous (see also section 4.1).

This phenomenon may be applied to phase separation of systems with smectic order, such as lyotropic and thermotropic smectic liquid crystals and block

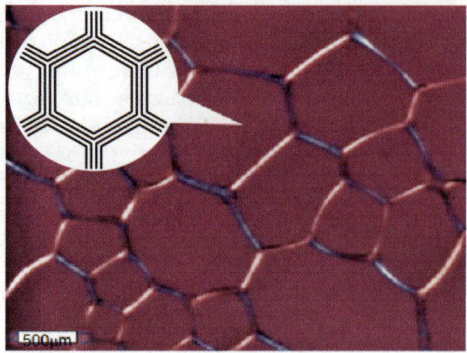

Fig. 8 Cellular pattern formation in a lyotropic liquid crystal (a $C_{10}E_3$/water mixture of 19.9 wt% $C_{10}E_3$) observed at 42.15°with polarizing microscopy. The inset schematically shows how membranes are organized in a cellular structure. In a border region, there is a disclination line of strength $-1/2$.

copolymers.[135] The basic physical strategy may also be used for various types of soft matter and foods, which have other internal order that can support elastic stress. Here it may be worth noting that similar stabilization of foam-like structures by introducing lamellar order in gels has been known as α-gels in food science.[136–138] This phase is often thermodynamically unstable and further transforms into the coa-gel phase, where monoglycerides are crystallized into plate-like crystals. These states are applied to dressings, mayonnaises, sauces, processed cheese, meat products and fat spreads.[136] The establishing lamellar order without crystallization would lead to extremely stable foam structures. We also note that the kinetics of phase separation is a key factor for attaining lamellar order in the cell wall.

9 General mechanism of the formation of cellular, foam-like, or sponge-like structures in materials

Next we discuss the universal nature of a spongelike morphology and its physical origin. It is known that gel undergoing volume-shrinking phase transition forms a bubble-like structure.[69–71]

We argue that the physical origin of the appearance of a honeycomb structure in plastic foams (*e.g.* polystyrene and urethane foams) and breads is also similar to that of a network structure in viscoelastic phase separation. When we consider the mechanical force balance equation in the formation of network patterns, the pressure, p, plays only a minor role: p is determined to satisfy the incompressibility condition. However, the formation of foam structures is usually induced by the liquid-to-gas transformation of one of the components of a mixture (see below), which accompanies its large volume expansion. This expansion creates high internal pressure in gas bubbles, and thus the gas pressure, p, plays a key role in the morpho-logical selection in the foam formation. To describe this phenomenon we need to use the dynamic equations for compressible liquids. The force balance can be satis-fied only when a gas bubble is surrounded by the matrix phase: the internal gas pressure is balanced with the mechanical stress created by the stretched matrix phase surrounding the gas bubble. It is this feature that leads to the formation of cellular foam structures. As in the case of network formation in viscoelastic phase separation, we can say that the foam structure formation is a mechanically selected pattern formation, and thus can be regarded as a special case of viscoelastic phase separation.

Besides the above-mentioned difference in the morphological selection, all these processes have a common feature that holes of a less viscoelastic fluid phase (gas in plastic foams, water in gels, solvent in polymer solutions, and so on) are nucleated in a phase-separation process to balance the force associated with the formation of a heterogeneous structure in an elastic medium. Then, the more viscoelastic phase decreases its volume with time (only relatively in the case of foams). This volume shrinking process is dominated by the transfer (diffusion or flow) of the more mobile component under stress fields, from a more viscoelastic phase to the less viscoelastic phase. The limiting process of material transport between the two phases is that in the slower phase. The above picture suggests that a spongelike structure is the *universal morphology* for phase separation in systems in which one of the components asymmetrically has (visco)elasticity stemming from either topological connectivity or long-range interactions.

9.1 Selection principles of patterns for viscoelastic and elastic phase separation

It is worth noting that this pattern selection differs from phase separation of elastic solid mixtures (*e.g.* metal alloys). We note that elasticity, which is static, does not involve any time scales (or velocity fields). For elastic phase separation, thus, the momentum conservation, or force balance condition, is irrelevant for the selection of morphology and the pattern evolves solely to lower the elastic energy, while obeying the diffusion equation alone. We emphasize that the momentum conservation is relevant only when a mixture contains a fluid as its component.

Elastic effects often originate from a lattice mismatch between the two atomic components in solid alloys. First of all, solid phase separation accompanies little volume change of each phase. Furthermore, the softer phase always forms a network-like continuous phase to minimize the total elastic energy,[2] in contrast to our case. This is because it is energetically more favourable to deform the soft phase than the hard phase. In solid mixtures, the elastic energy minimization determines pattern formation, whereas in liquid mixtures the momentum conservation (or the force balance) determines the phase-separation morphology.

Concerning the momentum conservation, we note that hydrodynamic degrees of freedom play a significant role in the initial and final stage of phase separation. For example, network formation in colloid phase separation is significantly influenced by the hydrodynamic interactions between colloids.[13,42,43] In the final stage, hydrodynamic effects are important to describe Rayleigh instability of tubes (or networks). In the intermediate stage, on the other hand, hydrodynamic effects are not so significant and only the force balance plays an important role in pattern evolution. To describe this regime, thus, we may use Langevin (Brownian) dynamics[139–141] or Newton dynamics.

9.2 Pattern formation in plastic foam

A typical formation process of plastic foams is as follows (see, *e.g.*, Ref. 142). First, a polymer matrix absorbing a low-boiling-point solvent is prepared. Then, its temperature is raised above the boiling point of the solvent, which induces bubble formation in the polymer matrix. These bubbles nucleate and grow as the result of evaporation of the solvent from the polymer matrix. The total volume of the sample expands as a result of the liquid-gas transformation of the solvent. In this process, a pattern is dominated by the mechanical force balance condition with the contribution of the gas pressure, p. This is caused by the strongly asymmetric stress division: gas bubbles cannot support any mechanical stress besides the hydrostatic pressure and only the polymeric phase can support it. In this way, a cellular pattern is formed. As mentioned above, it was pointed out that strain hardening plays an important role in the formation of cellular patterns.[56] This may be because strain hardening prevents the liquid-type rupture of cell walls. Thus, it can be viewed as the enhancement of the importance of mechanical stress over interfacial tension in the force balance condition (see, *e.g.*, eqn (22)). Finally, foam structures are stabilized by glass transition or crystallization.

9.3 Foam structure formation in foods such as breads, cakes, snacks

Here we consider the formation mechanism of a foam structure of foods on the basis of the concept of viscoelastic phase separation. We note that a diverse range of foods are aerated, using a similarly varied assortment of processing methods. The physical mechanism is essentially the same as that for plastic foams.

First we consider the foam structure formation in breads. A bread is basically comprised, at a macroscopic level, of the gas and the solid (cell wall material) phase. When viewing the final structure of a bread crumb, we can see that the solid wheat phase is connected and the gas cells are often isolated, but sometimes partially connected.[143] The volume fraction of the phases and the nature of their connectivity and topology determines the structure, and consequently the mechanical properties of breads. Thus, it is crucial to understand how the two phases are formed in a bread-making process.[31] First we prepare a dough, which is made from a mixture of wheat flour, yeast, and water. In the mixing processes, proteins are hydrated, small gas (CO_2) bubbles are formed by yeast fermentation and then their sizes are reduced by kneading. In this process, gluten particles are crosslinked by di-sulfide bonding and a gluten network is formed. In the heating process, small CO_2 bubbles may act as nucleation centres for water vapour droplets formation. This latter process creates a large mechanical stress upon their volume expansions and is responsible for the formation of a cellular structure. To maintain the vapour pressure, it is important that surface heating makes the dough surface dense enough to prevent transport of water vapour. Then the structure is frozen by glass transition[48] or crystallization, namely, the transformation from a viscoelastic matter to a solid. An example of a cellular structure of a bread is shown in Fig. 9 (see also Ref. 144). Thus, this entire process is markedly similar to the above-explained formation process of plastic foams. As in plastic foams, the importance of strain hardening in the formation of cellular patterns are also pointed out for food polymers such as breads.[57,58,145,146] We note that the formation of foam structures in other spongy foods such as cakes[147,148] and baked starch foams[149] is also basically the same as breads.

Here, we also mention the application of supercritical fluids for foaming of polymers and foods.[150,151] A supercritical fluid, which is in a region above its critical temperature and pressure, exhibits interesting behaviour by combining the properties of conventional liquids and gases. Its liquid-like density allows for solvent power

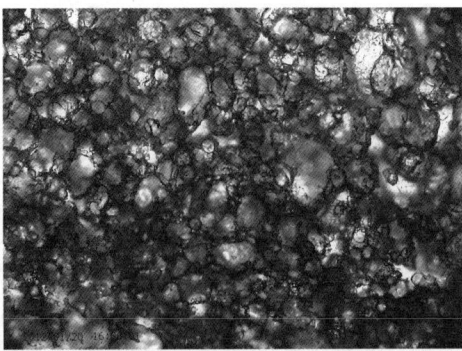

Fig. 9 A cellular pattern of a thin bread formed on a temperature-controlled hot stage. The pattern is observed with optical microscopy. Coexisting cellular and network-like structures are clearly observed.

of orders of magnitude higher than gases, while gas-like viscosity leads to a high rate of diffusion. These facts combine to ensure rapid swelling of polymers by supercritical fluids to equilibrium values comparable to liquid solvents. In addition, supercooled fluids can readily plasticize glassy polymers. A pressure quench from supercritical conditions at constant temperature ensures that no vapour-liquid boundary is encountered during the process of solvent removal. This helps avoiding damaging the delicate cellular structure. Foam formation using supercritical liquid can be triggered simply by changing pressure. The mechanism of foam structure formation is essentially the same as the above plastic foams and breads.

10   Shear-induced composition fluctuations and demixing

10.1   Basic mechanism of shear-induced instability

Finally, we consider shear-induced composition fluctuations and demixing (or flocculation) in polymer solutions[10,18–22,152–156] as well as in colloidal suspensions and emulsions.[23] Shear-induced composition fluctuations are induced by a steep increase of η with ϕ. An intuitive explanation was given as follows.[152,153] Shear-induced demixing is caused by a certain mechanism to store elastic energy under shear. This elastic energy effectively leads to a change in the free energy functional, which results in an effective shift of the phase diagram and destablizes a thermodynamically stable system. However, this picture has turned out to be too simplistic. It was shown by intensive theoretical studies[10,18–22] that we need to treat dynamical couplings between the composition and stress fields properly to explain shear-induced demixing of polymer solutions. This phenomenon is now widely known as "shear-induced demixing" in polymer solutions under shear.[18]

Some time ago we considered whether similar phenomena can be observed in colloidal suspensions, emulsions, and protein solutions or not.[23] In polymers, the conformational degrees of freedom of chains and entanglement effects play a crucial role in shear-induced instability. Since such internal degrees of freedom are absent in suspensions of particle-like objects, the mechanism to store elastic energy under shear in colloidal suspensions should be essentially different from that in polymer solutions.[23] At first sight, shear effects seem less pronounced for colloidal suspensions than for polymer solutions. Thus, this problem is far from being obvious.

In the following, we briefly discuss shear effects on colloidal suspensions on an intuitive level.[23] Under thermal fluctuations, local shear stress is stored inhomogeneously due to a strong nonlinear and asymmetric dependence of $G_S(\phi)$ and $\tau_S(\phi)$ on ϕ. Note that the stress relevant to a shear problem is the "shear" stress, σ_c^S. The linear stability analysis tells us that this enhances composition fluctuations along the extension axis of the flow, since this stress moves colloidal particles towards a more concentrated region. This positive feedback process results in shear induced instability in a self-catalytic manner.

In the linear Newtonian regime under the condition $\dot{\gamma}\tau_s \ll 1$, where $\dot{\gamma}$ is the shear rate, σ_c is given as

$$\sigma_c \sim \eta(\phi)(\vec{\nabla}\vec{v} + (\vec{\nabla}\vec{v})') \sim \eta(\phi)\dot{\gamma}. \tag{23}$$

Then, one can straightforwardly obtain the following expression for the relaxation rate of the composition fluctuations convected by shear flow:[18]

$$\Gamma_{\rm eff} = L\left[q^2(r_0 + Cq^2) - 2\left(\frac{\partial\eta}{\partial\phi}\right)_T \phi^{-1}\dot{\gamma}q_xq_y\right]\Big/\left[1 + \xi_{\rm ve}^2q^2\right] \tag{24}$$

It is important to note that if $(\partial\eta/\partial\phi)_T > 0$, $\Gamma_{\rm eff}$ can be negative even for positive r_0 for $\dot{\gamma} > \dot{\gamma}_c$, indicating the growth of composition fluctuations even in a thermodynamically stable region. Compare this equation for shear-induced instability with that for thermodynamic instability, eqn (16). The critical shear rate $\dot{\gamma}_c$ is thus obtained, using r_ϕ defined in section 3.7, as

$$\dot{\gamma}_c \sim r_\phi \phi/(\partial \eta/\partial \phi)_T. \tag{25}$$

Recently it was demonstrated by Furukawa and Tanaka[16] that this condition can be rewritten by using the osmotic pressure, Π, as follows:

$$\dot{\gamma}_c \sim (\partial \eta/\partial \Pi)_T^{-1}. \tag{26}$$

For a general implication of this relation and its relevence to single-component glassy systems, please refer to Ref. 16 and 17.

10.2   Shear-induced instability and fracture in foods

Shear flow is often applied to foods in the processing. Because of intrinsic dynamic asymmetry between the components of foods, shear flow often enhances or induces phase separation.[18] At the same time, shear gradient deforms and breaks up domains formed by phase separation.[2,96] As mentioned above, such examples can be seen in dough formation by kneading.[94] Unlike the formation of a rather narrow domain size distribution in simple shear, shear induced migration may lead to macroscopic phase separation under shear gradient or curve linear flow.[157] Such migration phenomena may also been described in terms of the two fluid model similar to the viscoelastic model[158] by considering effects of normal stress differences.[159]

Furthermore, flow can generate anisotropic structures such as layered structures and fibrous structures, which provide anisotropic mechanical properties sometimes useful for food products known as anisotropic protein-rich foods.[160] In polymer mixtures[18,161] and colloidal suspensions,[162] at a high shear rate string-like phase separated structures are formed. We note that string-like phase separation is observed for a system with rather weak dynamic asymmetry between the two phases. For strongly dynamically asymmetric cases, more chaotic and disordered structures are formed.[96] This indicates that string-like domain formation is of hydrodynamic origin and the interplay between shear deformation and interface tension may play a primary role in the selection of the string structure. We also note that stringlike morphology, more precisely, leek-like structures, can also be formed along the flow direction by shear flow in a lyotropic lamellar phase.[163]

Lamella-like layered structures are often ascribed to so-called shear banding, which is a consequence of nonlinear rheology accompanying non-monotonic stress-strain rate relation.[164,165] Such nonlinearity may come from a coupling between shear flow and internal degrees of freedom of slow components, e.g., orientation of polymer chains.[166] A constitutive relation such as the nonlocal Johnson-Segalman (JS) model can describe rheological instability,[164,165] which is very similar to the upper-convective Maxwell relation, besides additional inclusion of the slippage effects and the so-called stress diffusion term in the nonlocal JS model. Thus, the viscoelastic model may describe rich pattern evolution in a nonlinear flow regime at least on a phenomenological level. Unlike a single-component description, the two-fluid model provides a coupling between shear, stress, and concentration fields, which is crucial in multi-component systems such as foods. In relation to this, it is worth noting that in the two-fluid model, the nonlocal constitutive relation may not be required to have stable shear banding since similar nonlocal effects are expected to be produced by the concentration gradient and their couplings to stress and strain fields.[18,96,167,168] For theoretical analysis, we need to treat nonlinear effects properly, including the spatial variation not only in the concentration field, but also the stress and strain fields, and their couplings. This is a difficult theoretical task. Whether we fix the total stress or strain rate applied is also crucial for the selection of nonequilibrium steady states, *e.g.*, gradient and vorticity banding, if they exist.[164,165,169] In previous studies of shear instability, the steep dependence of the transport coefficient, the structural relaxation time, and the elastic modulus on the order parameter such as the composition, ϕ, has not been considered carefully. However, as emphasized above, it may induce instability of a different mechanism and thus play a crucial role in shear-induced phenomena.[10,16–23,152–156,170] This problem needs further study in the future.

11 General nature of viscoelastic phase separation and classifications of rheological behaviour of materials, phase separation, fracture

First we consider the general nature of the basic equations describing viscoelastic phase separation:[12] (i) if we set $G_S(t - t') = G_S(\phi(\vec{r},t))$, where G_S is the shear modulus, $G_B(t - t') = G_B(\phi(\vec{r},t))$, where G_B is the bulk modulus, and the absence of the velocity fields ($\vec{v} = 0$), a viscoelastic model reduces to the elastic solid model.[171] (ii) If we further assume that G_S and G_B do not depend on the composition, ϕ, it reduces to the solid model (model B[3]). (iii) If we assume dynamic symmetry between the two components of a mixture in the viscoelastic model, it reduces to a new model of symmetric viscoelastic model.[12] If we further assume slow enough deformation, then, it reduces to the fluid model (model H[3]). (iv) If we assume only $G_S(t) = G_S$ and $G_B(t) = G_B$, the viscoelastic model reduces to the "elastic gel model"[2,71] that describes phase separation in elastic gels. Note that the time integration of the velocity becomes the deformation $\vec{u}_p$, and

$$\sigma_{ij} = G_S \left[\frac{\partial u_{pj}}{\partial x_i} + \frac{\partial u_{pi}}{\partial x_j} - \frac{2}{3}(\vec{\nabla} \cdot \vec{u}_p)\delta_{ij} \right] + G_B(\vec{\nabla} \cdot \vec{u}_p)\delta_{ij}.$$

Thus, the viscoelastic model is the general model of phase separation that can describe any type of phase separation in mixtures of isotropic condensed matter, as its special cases.[12]

The viscoelastic model in the classification of isotropic phase separation corresponds to viscoelastic matter in the classification of isotropic condensed matter. Viscoelastic matter includes any condensed matter ranged from solid to fluid. The

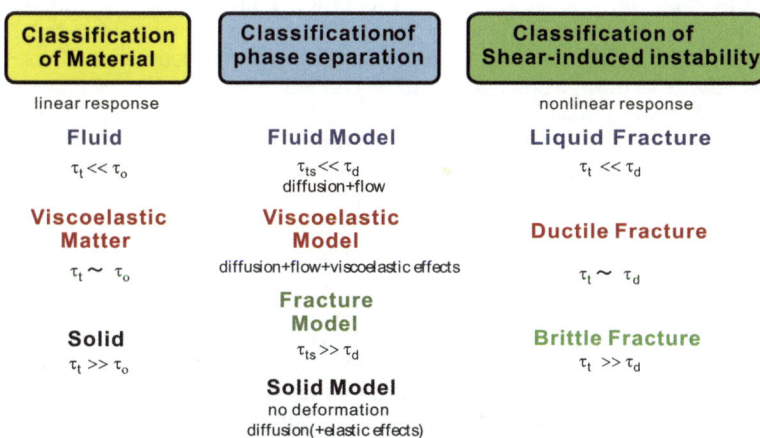

Fig. 10 Schematic figure explaining the classification of phase separation of isotropic matter and its relation to the classification of materials and mechanical fracture. In the classification of materials (left), the ratio of the structural relaxation time, τ_t, to the observation time, τ_o, which is known as the Deborah number, is a key number. In the classification of phase separation (middle), the ratio of τ_{ts} to τ_d, plays a crucial role, as discussed in section 6.2, and is regarded as the Weissenberg number for deformation self-induced by phase separation. On the classification of fracture (right), please refer to Ref. 17.

key factor for the classification of materials is the relation between the characteristic internal rheological time, τ_t, and the characteristic observation time, τ_o. Corresponding to this, the key physical factor for the classification of isotropic phase separation is the relation between the characteristic time of phase separation (domain deformation), τ_d, and the characteristic rheological time of the slower phase, τ_{ts}. The above analogy is schematically summarized in Fig. 10.

Furthermore, this classification may also be common to that of mechanical fracture,[17] which is determined by the relation between the time when instability set in, the mechanical relaxation rate, and the deformation rate (see Fig. 10). The only difference between the two is whether the deformation is induced by phase separation or externally imposed. See Ref. 17 on the details of mechanical fracture.

12 Summary

In summary, we show that viscoelastic phase separation and the concept of dynamic asymmetry are very useful for understanding not only phase separation and gelation in soft matter and food materials but also their mechanical instability under shear deformation. We demonstrated that the viscoelastic model including both bulk and shear stress contributions is a very general model that can universally describe phase separation of condensed matter.

We also demonstrate that the formation of heterogeneous network or cellular structures in foods and their collapsing may be regarded as mechanically driven pattern evolution and can be understood in the framework of viscoelastic phase separation. Dynamic asymmetry may be a key to the physical understanding of not only phase separation but also mechanical instability of materials under deformation. These phenomena of mechanically driven inhomogeneization can be understood in a unified manner on the basis of the concept of dynamic asymmetry. Besides these direct applications, finally we mention another interesting possibility of applications of viscoelastic phase separation: recently it was shown that a spatially heterogeneous pattern formed by protein phase separation causes a Bragg reflection of light, which is an origin of a colour of bird feathers.[172] We believe that this phase separation should also belong to viscoelastic phase separation. This phenomenon may be used to put beautiful colours to foods without using (toxic) dye molecules.

At this moment, viscoelastic phase separation and shear-induced mechanical instability can be studied analytically only in their linear regimes. Thus, numerical simulations play a major role in our understanding of these phenomena. Thus, simulations based on the phenomenological viscoelastic two-fluid model may be very useful in studying nonequilibrium and nonlinear dynamical behaviour of foods, including phase separation and flow-induced phenomena.

In relation to this, we finally mention some fundamental remaining problems of the current viscoelastic model.[1] The dissipation in a dynamically asymmetric mixture may not be given by a simple sum of friction due to the relative motion of the components and hydrodynamic dissipation. Here the nonlocal nature of the transport, which is characterized by the viscoelastic length, ξ_{ve}, should also be considered properly in the process of the coarse-graining.[2] The phenomenological constitutive relation crucially depends on the composition dependence of the elastic moduli and the mechanical relaxation times. However, there is no firm basis for the physical description of these quantities. For more quantitative understanding of viscoelastic phase separation it is crucial to overcome these difficult problems.

We hope that this article will contribute to better understanding of pattern formation and mechanical instability of foods.

Acknowledgements

The author is indebted to the reviewer for his or her invaluable suggestions on possible connections between viscoelastic phase separation and many interesting phenomena in foods. He is also grateful to T. Araki, Y. Iwashita, T. Koyama, and Y. Nishikawa for their collaboration on viscoelastic phase separation and A. Furukawa for his collaboration on mechanical instability and fracture of materials. He also thanks M. Leocmach and K. Murata for their kind help in making experiments on cellular pattern formation in breads. This work was partly supported by a Grant-in-Aid from the Ministry of Education, Culture, Sports, Science and Technology, Japan and Aihara Project, the FIRST program from JSPS, initiated by CSTP.

References

1 J. D. Gunton, M. San Miguel and P. S. Sahni, *Phase transitions and Critical phenomena, Vol. 8*, Academic, London, 1983.
2 A. Onuki, *Phase Transition Dynamics*, Cambridge University Press, Cambridge, 2002.
3 P. C. Hohenberg and B. I. Halperin, *Rev. Mod. Phys.*, 1976, **49**, 435.
4 H. Tanaka and T. Nishi, *Jpn. J. Appl. Phys.*, 1988, **27**, L1787.
5 H. Tanaka, *Macromolecules*, 1992, **25**, 6377–6380.
6 H. Tanaka, *Phys. Rev. Lett.*, 1993, **71**, 3158–3161.
7 H. Tanaka, *J. Chem. Phys.*, 1994, **100**, 5323.
8 H. Tanaka, *Phys. Rev. Lett.*, 1996, **76**, 787–790.
9 H. Tanaka and T. Miura, *Phys. Rev. Lett.*, 1993, **71**, 2244–2247.
10 M. Doi and A. Onuki, *J. Phys. (Paris) II*, 1992, **2**, 1631–1656.
11 H. Tanaka, *J. Phys.: Condens. Matter*, 2000, **12**, R207.
12 H. Tanaka, *Phys. Rev. E*, 1997, **56**, 4451.
13 H. Tanaka and T. Araki, *Chem. Eng. Sci.*, 2006, **61**, 2108–2141.
14 H. Tanaka, *Adv. Mater.*, 2009, **21**, 1872–1880.
15 T. Koyama, T. Araki and H. Tanaka, *Phys. Rev. Lett.*, 2009, **102**, 65701.
16 A. Furukawa and H. Tanaka, *Nature*, 2006, **443**, 434–438.
17 A. Furukawa and H. Tanaka, *Nat. Mater.*, 2009, **8**, 601–609.
18 A. Onuki, *J. Phys.: Condens. Matter*, 1997, **9**, 6119.
19 A. Onuki, *Phys. Rev. Lett.*, 1989, **62**, 2472–2475.
20 E. Helfand and G. H. Fredrickson, *Phys. Rev. Lett.*, 1989, **62**, 2468–2471.
21 H. Ji and E. Helfand, *Macromolecules*, 1995, **28**, 3869–3880.
22 S. T. Milner, *Phys. Rev. E*, 1993, **48**, 3674–3691.
23 H. Tanaka, *Phys. Rev. E*, 1999, **59**, 6842.
24 A. M. Donald, *Rep. Prog. Phys.*, 1994, **57**, 1081.
25 E. Dickinson, *Colloids Surf.*, 1989, **42**, 191–204.
26 E. Dickinson and M. Golding, *Food Hydrocolloids*, 1997, **11**, 13–18.
27 E. Dickinson, *Curr. Opin. Colloid & Interface Sci.*, 2010, **15**, pp. 40–49.
28 R. Mezzenga, P. Schurtenberger, A. Burbidge and M. Michel, *Nat. Mater.*, 2005, **4**, 729–740.
29 R. Mezzenga, *Food Hydrocolloids*, 2007, **21**, 674–682.
30 J. Ubbink, A. Burbidge and R. Mezzenga, *Soft Matter*, 2008, **4**, 1569–1581.
31 R. G. M. Van der Sman and A. J. Van der Goot, *Soft Matter*, 2009, **5**, 501–510.
32 S. Bhat, R. Tuinier and P. Schurtenberger, *J. Phys.: Condens. Matter*, 2006, **18**, L339.
33 T. Gibaud and P. Schurtenberger, *J. Phys.: Condens. Matter*, 2009, **21**, 322201.
34 P. G. de Gennes, *Macromolecules*, 1976, **9**, 587–593.
35 P. G. de Gennes, *Macromolecules*, 1976, **9**, 594–598.
36 F. Brochard, *J. Phys.*, 1983, **44**, 39–43.
37 T. Tanaka and D. J. Fillmore, *J. Chem. Phys.*, 1979, **70**, 1214–1218.
38 H. Tanaka, *Prog. Theor. Phys., Suppl.*, 1997, **126**, 333–338.
39 H. Tanaka and T. Araki, *Phys. Rev. Lett.*, 1997, **78**, 4966–4969.
40 M. Doi and S. Edwards, *The Theory of Polymer Dynamics*, Oxford University Press, USA, 1988, vol. 73.
41 P. G. de Gennes, *Scaling Concepts in Polymer Physics*, Cornell University Press, 1979.
42 H. Tanaka and T. Araki, *Phys. Rev. Lett.*, 2000, **85**, 1338–1341.
43 A. Furukawa and H. Tanaka, *Phys. Rev. Lett.*, 2010, **104**, 245702.
44 A. Onuki, *J. Non-Cryst. Solids*, 1994, **172–174**, 1151–1157.
45 L. D. Landau and E. M. Lifshitz, *Theory of Elasticity*, Butterworth-Heinenann, Oxford, 1975.
46 D. Sappelt and J. Jäckle, *Europhys. Lett.*, 1997, **37**, 13.
47 L. Slade and H. Levine, *Adv. Food Nutr. Res.*, 1991, **38**, 103–269.
48 B. Cuq, J. Abecassis and S. Guilbert, *Int. J. Food Sci. Technol.*, 2003, **38**, 759–766.
49 R. G. M. van der Sman and M. B. J. Meinders, *Soft Matter*, 2011, **7**, 429–442.

50 X. He, A. Fowler and M. Toner, *J. Appl. Phys.*, 2006, **100**, 074702.
51 R. G. M. van der Sman, *Food Hydrocolloids*, 2011, **27**, 529–535.
52 M. D. Ediger, *Annu. Rev. Phys. Chem.*, 2000, **51**, 99–128.
53 H. Tanaka, *Phys. Rev. E*, 2003, **68**, 011505.
54 R. W. Hartel, R. Ergun and S. Vogel, *Compr. Rev. Food Sci. Food Saf.*, 2011, **10**, 17–32.
55 T. C. B. McLeish and R. G. Larson, *J. Rheol.*, 1998, **42**, 81.
56 P. Spitael and C. W. Macosko, *Polym. Eng. Sci.*, 2004, **44**, 2090–2100.
57 B. Dobraszczyk, *J. Non-Newtonian Fluid Mech.*, 2004, **124**, 61–69.
58 B. J. Dobraszczyk and C. A. Roberts, *J. Cereal Sci.*, 1994, **20**, 265–274.
59 A. Onuki and T. Taniguchi, *J. Chem. Phys.*, 1997, **106**, 5761–5770.
60 M. Takenaka, H. Takeno, T. Hashimoto and M. Nagao, *J. Chem. Phys.*, 2006, **124**, 104904.
61 F. Brochard and P. de Gennes, *Macromolecules*, 1977, **10**, 1157–1161.
62 A. Furukawa, *J. Phys. Soc. Jpn.*, 2003, **72**, 209–212.
63 A. Furukawa, *J. Phys. Soc. Jpn.*, 2003, **72**, 1436–1445.
64 T. Koyama and H. Tanaka, *Europhys. Lett.*, 2007, **80**, 68002.
65 H. Tanaka, Y. Nishikawa and T. Koyama, *J. Phys.: Condens. Matter*, 2005, **17**, L143.
66 H. Tanaka and Y. Nishikawa, *Phys. Rev. Lett.*, 2005, **95**, 78103.
67 M. E. Cates, M. Fuchs, K. Kroy, W. C. K. Poon and A. M. Puertas, *J. Phys.: Condens. Matter*, 2004, **16**, S4861.
68 P. Lu, E. Zaccarelli, F. Ciulla, A. B. Schofield, F. Sciortino and D. A. Weitz, *Nature*, 2008, **453**, 499–503.
69 E. S. Matsuo and T. Tanaka, *J. Chem. Phys.*, 1988, **89**, 1695.
70 E. S. Matsuo and T. Tanaka, *Nature*, 1992, **358**, 482–485.
71 K. Sekimoto, N. Suematsu and K. Kawasaki, *Phys. Rev. A*, 1989, **39**, 4912–4914.
72 H. Tanaka and T. Araki, *Europhys. Lett.*, 2007, **79**, 58003.
73 T. Taniguchi and A. Onuki, *Phys. Rev. Lett.*, 1996, **77**, 4910–4913.
74 A. Bhattacharya, S. D. Mahanti and A. Chakrabarti, *Phys. Rev. Lett.*, 1998, **80**, 333–336.
75 T. Araki and H. Tanaka, *Macromolecules*, 2001, **34**, 1953–1963.
76 T. Araki and H. Tanaka, *Phys. Rev. E*, 2005, **72**, 041509.
77 J. Zhang, Z. Zhang, H. Zhang and Y. Yang, *Phys. Rev. E*, 2001, **64**, 051510.
78 L. Donato, C. Garnier, B. Novales, S. Durand and J. L. Doublier, *Biomacromolecules*, 2005, **6**, 374–385.
79 R. Tuinier, J. K. G. Dhont and C. G. De Kruif, *Langmuir*, 2000, **16**, 1497–1507.
80 P. W. de Bont, G. M. P. van Kempen and R. Vreeker, *Food Hydrocolloids*, 2002, **16**, 127–138.
81 L. F. van Heijkamp, I. M. de Schepper, M. Strobl, R. H. Tromp, J. R. Heringa and W. G. Bouwman, *J. Phys. Chem. A*, 2010, **114**, 2412–2426.
82 S. Bourriot, C. Garnier and J. Doublier, *Food Hydrocolloids*, 1999, **13**, 43–49.
83 C. Sanchez, R. Zuniga-Lopez, C. Schmitt, S. Despond and J. Hardy, *Int. Dairy J.*, 2000, **10**, 199–212.
84 J. S. He, H. Yang, W. Zhu and T. H. Mu, *J. Phys.: Conf. Ser.*, 2010, **215**, 012169.
85 C. Vega and H. D. Goff, *Int. Dairy J.*, 2005, **15**, 249–254.
86 H. D. Goff, *Curr. Opin. Colloid Interface Sci.*, 2002, **7**, 432–437.
87 H. D. Goff, D. Ferdinando and C. Schorsch, *Food Hydrocolloids*, 1999, **13**, 353–362.
88 P. Burey, B. R. Bhandari, R. P. G. Rutgers, P. J. Halley and P. J. Torley, *Int. J. Food Prop.*, 2008, **12**, 176–210.
89 T. Moschakis, B. S. Murray and E. Dickinson, *J. Colloid Interface Sci.*, 2005, **284**, 714–728.
90 J. Renkema, *Food Hydrocolloids*, 2004, **18**, 39–47.
91 V. Tolstoguzov, *Food Hydrocolloids*, 1997, **11**, 181–193.
92 V. Tolstoguzov, *Food Hydrocolloids*, 2003, **17**, 1–23.
93 C. G. De Kruif and R. Tuinier, *Food Hydrocolloids*, 2001, **15**, 555–563.
94 D. Peressini, S. H. Peighambardoust, R. J. Hamer, A. Sensidoni and A. J. Van Der Goot, *J. Cereal Sci.*, 2008, **48**, 426–438.
95 S. H. Peighambardoust, A. J. Van der Goot, T. Van Vliet, R. J. Hamer and R. M. Boom, *J. Cereal Sci.*, 2006, **43**, 183–197.
96 T. Imaeda, A. Furukawa and A. Onuki, *Phys. Rev. E*, 2004, **70**, 051503.
97 E. E. J. van der Zalm, A. van der Goot and R. M. Boom, *J. Food Eng.*, 2009, **95**, 572–578.
98 J. M. Deman and A. M. Beers, *J. Texture Stud.*, 1987, **18**, 303–318.
99 S. S. Narine and A. G. Marangoni, *Food Res. Int.*, 1999, **32**, 227–248.
100 S. S. Narine and A. G. Marangoni, *Adv. Food Nutr. Res.*, 2002, **44**, 33–145.
101 G. Alejandro and S. E. McGauley, *Cryst. Growth Des.*, 2003, **3**, 95–108.
102 S. S. Narine and A. G. Marangoni, *Phys. Rev. E*, 1999, **59**, 1908.

103 V. I. Lozinsky, I. Y. Galaev, F. M. Plieva, I. N. Savina, H. Jungvid and B. Mattiasson, *Trends Biotechnol.*, 2003, **21**, 445–451.

104 V. I. Lozinsky, *Russ. Chem. Bull.*, 2008, **57**, 1015–1032.

105 F. M. Plieva, I. Y. Galaev and B. Mattiasson, *J. Sep. Sci.*, 2007, **30**, 1657–1671.

106 F. M. Plieva, H. Kirsebom and B. Mattiasson, *J. Sep. Sci.*, 2011, **34**, 2164–2172.

107 M. C. Gutiérrez, M. L. Ferrer and F. Del Monte, *Chem. Mater.*, 2008, **20**, 634–648.

108 G. Petzold and J. M. Aguilera, *Food Biophys.*, 2009, **4**, 378–396.

109 L. Qian and H. Zhang, *J. Chem. Technol. Biotechnol.*, 2011, **86**, 172–184.

110 H. Kiani and D. W. Sun, *Trends Food Sci. Technol.*, 2011, **22**, 407–426.

111 H. Zhang, I. Hussain, M. Brust, M. Butler, S. P. Rannard and A. I. Cooper, *Nat. Mater.*, 2005, **4**, 787–793.

112 S. S. L. Peppin, M. G. Worster and J. S. Wettlaufer, *Proc. R. Soc. London, Ser. A*, 2007, **463**, 723–733.

113 F. Prothon, L. Ahrne and I. Sjöholm, *Crit. Rev. Food Sci. Nutr.*, 2003, **43**, 447–479.

114 W. C. K. Poon, L. Starrs, S. P. Meeker, A. Moussaid, R. M. L. Evans, P. N. Pusey and M. M. Robins, *Faraday Discuss.*, 1999, **112**, 143–154.

115 W. C. K. Poon, *J. Phys.: Condens. Matter*, 2002, **14**, R859–R880.

116 P. Bartlett, L. J. Teece and M. A. Faers, *Phys. Rev. E*, 2012, **85**, 021404.

117 F. Spaepen, *Acta Metall.*, 1977, **25**, 407–415.

118 A. S. Argon, *Acta Metall.*, 1979, **27**, 47–58.

119 A. E. Carsson and E. R. Fuller Jr., *Fracture: Instability Dynamics, Scaling, and Ductile/Brittle Behavior*, Materials Research Society, Boston, 1996.

120 M. L. Falk and J. S. Langer, *Phys. Rev. E*, 1998, **57**, 7192–7205.

121 G. Offer and T. Cousins, *J. Sci. Food Agric.*, 1992, **58**, 107–116.

122 T. Astruc, P. Gatellier, R. Labas, V. S. Lhoutellier and P. Marinova, *Meat Sci.*, 2010, **85**, 743–751.

123 M. S. Rahman, *Drying Technol.*, 2001, **19**, 1–13.

124 L. Mayor and A. M. Sereno, *J. Food Eng.*, 2004, **61**, 373–386.

125 H. Tanaka and T. Araki, *Phys. Rev. Lett.*, 1998, **81**, 389–392.

126 I. T. Norton and W. J. Frith, *Food Hydrocolloids*, 2001, **15**, 543–553.

127 D. G. A. L. Aarts, M. Schmidt and H. N. W. Lekkerkerker, *Science*, 2004, **304**, 847–850.

128 C. P. Royall, D. G. A. L. Aarts and H. Tanaka, *Nat. Phys.*, 2007, **3**, 636–640.

129 E. Zaccarelli, *J. Phys.: Condens. Matter*, 2007, **19**, 323101.

130 H. Tanaka, J. Meunier and D. Bonn, *Phys. Rev. E*, 2004, **69**, 031404.

131 T. Ohtsuka, C. P. Royall and H. Tanaka, *Europhys. Lett.*, 2008, **84**, 46002.

132 C. Wu, W. Li and X. X. Zhu, *Macromolecules*, 2004, **37**, 4989–4992.

133 L. E. van Riemsdijk, J. H. B. Sprakel, A. J. van der Goot and R. J. Hamer, *Food Biophys.*, 2010, **5**, 41–48.

134 Y. Iwashita and H. Tanaka, *Nat. Mater.*, 2006, **5**, 147–152.

135 Y. L. Yan, X. G. Jia, M. Meng and C. T. Qu, *Chem. Lett.*, 2011, **40**, 261–263.

136 I. Heertje, E. C. Roijers and H. Hendrickx, *LWT–Food Sci. Technol.*, 1998, **31**, 387–396.

137 D. Van de Walle, P. Goossens and K. Dewettinck, *Food Res. Int.*, 2008, **41**, 1020–1025.

138 H. D. Batte, A. Wright, J. W. Rush, S. H. J. Idziak and A. G. Marangoni, *Food Res. Int.*, 2007, **40**, 982–988.

139 B. M. Whittle and E. Dickinson, *Mol. Phys.*, 1997, **90**, 739–758.

140 M. T. A. Bos and J. H. J. van Opheusden, *Phys. Rev. E*, 1996, **53**, 5044.

141 S. Babu, J. C. Gimel and T. Nicolai, *J. Chem. Phys.*, 2006, **125**, 184512.

142 J. J. Crevecoeur, J. F. Coolegem, L. Nelissen and P. J. Lemstra, *Polymer*, 1999, **40**, 3697–3702.

143 M. Scanlon and M. Zghal, *Food Res. Int.*, 2001, **34**, 841–864.

144 P. M. Falcone, A. Baiano, F. Zanini, L. Mancini, G. Tromba, F. Montanari and M. A. D. Nobile, *J. Food Sci.*, 2004, **69**, FEP38–FEP43.

145 L. E. van Riemsdijk, P. J. M. Pelgrom, A. J. van der Goot, R. M. Boom and R. J. Hamer, *J. Cereal Sci.*, 2011, **53**, 133–138.

146 L. E. van Riemsdijk, A. J. van der Goot, R. J. Hamer and R. M. Boom, *J. Cereal Sci.*, 2011, **53**, 355–361.

147 T. K. Berry, X. Yang and E. A. Foegeding, *J. Food Sci.*, 2009, **74**, E269–E277.

148 D. Kocer, Z. Hicsasmaz, A. Bayindirli and S. Katnas, *J. Food Eng.*, 2007, **78**, 953–964.

149 R. L. Shogren, J. W. Lawton, W. M. Doane and K. F. Tiefenbacher, *Polymer*, 1998, **39**, 6649–6655.

150 S. K. Goel and E. J. Beckman, *Polym. Eng. Sci.*, 1994, **34**, 1137–1147.

151 M. Sauceau, J. Fages, A. Common, C. Nikitine and E. Rodier, *Prog. Polym. Sci.*, 2010, **36**, 749–766.

152 B. A. Wolf and H. Krämer, *J. Polym. Sci., Polym. Lett. Ed.*, 1980, **18**, 789–794.

153 B. A. Wolf, *Macromolecules*, 1984, **17**, 615–618.

154 X. L. Wu, D. J. Pine and P. K. Dixon, *Phys. Rev. Lett.*, 1991, **66**, 2408–2411.
155 T. Hashimoto and K. Fujioka, *J. Phys. Soc. Jpn.*, 1991, **60**, 356–359.
156 E. Moses, T. Kume and T. Hashimoto, *Phys. Rev. Lett.*, 1994, **72**, 2037–2040.
157 A. J. Van der Goot, S. H. Peighambardoust, C. Akkermans and J. M. van Oosten-Manski, *Food Biophys.*, 2008, **3**, 120–125.
158 H. M. Vollebregt, R. G. M. van der Sman and R. M. Boom, *Soft Matter*, 2010, **6**, 6052–6064.
159 J. F. Morris, *Rheol. Acta*, 2009, **48**, 909–923.
160 J. M. Manski, A. J. van der Goot and R. M. Boom, *Trends Food Sci. Technol.*, 2007, **18**, 546–557.
161 T. Hashimoto, K. Matsuzaka, E. Moses and A. Onuki, *Phys. Rev. Lett.*, 1995, **74**, 126–129.
162 D. Derks, D. G. A. L. Aarts, D. Bonn and A. Imhof, *J. Phys.: Condens. Matter*, 2008, **20**, 404208.
163 H. Miyazawa and H. Tanaka, *Phys. Rev. E*, 2007, **76**, 011513.
164 M. E. Cates and S. M. Fielding, *Adv. Phys.*, 2006, **55**, 799–879.
165 P. D. Olmsted, *Rheol. Acta*, 2008, **47**, 283–300.
166 H. Tanaka, *J. Phys. Soc. Jpn.*, 2000, **69**, 299–302.
167 L. Jupp, T. Kawakatsu and X. F. Yuan, *J. Chem. Phys.*, 2003, **119**, 6361.
168 L. Jupp and X. F. Yuan, *J. Non-Newtonian Fluid Mech.*, 2004, **124**, 93–101.
169 J. K. G. Dhont and W. J. Briels, *Rheol. Acta*, 2008, **47**, 257–281.
170 R. Besseling, L. Isa, P. Ballesta, G. Petekidis, M. E. Cates and W. C. K. Poon, *Phys. Rev. Lett.*, 2010, **105**, 268301.
171 H. Nishimori and A. Onuki, *Phys. Rev. B*, 1990, **42**, 980–983.
172 E. R. Dufresne, H. Noh, V. Saranathan, S. G. J. Mochrie, H. Cao and R. O. Prum, *Soft Matter*, 2009, **5**, 1792–1795.

Kinetic model for the mechanical response of suspensions of sponge-like particles

Markus Hütter,[*a] Timo J. Faber[ab] and Hans M. Wyss[a]

Received 16th February 2012, Accepted 17th April 2012
DOI: 10.1039/c2fd20025b

A dynamic two-scale model is developed that describes the stationary and transient mechanical behavior of concentrated suspensions made of highly porous particles. Particularly, we are interested in particles that not only deform elastically, but also can swell or shrink by taking up or expelling the viscous solvent from their interior, leading to rate-dependent deformability of the particles. The fine level of the model describes the evolution of particle centers and their current sizes, while the shapes are at present not taken into account. The versatility of the model permits inclusion of density- and temperature-dependent particle interactions, and hydrodynamic interactions, as well as to implement insight into the mechanism of swelling and shrinking. The coarse level of the model is given in terms of macroscopic hydrodynamics. The two levels are mutually coupled, since the flow changes the particle configuration, while in turn the configuration gives rise to stress contributions, that eventually determine the macroscopic mechanical properties of the suspension. Using a thermodynamic procedure for the model development, it is demonstrated that the driving forces for position change and for size change are derived from the same potential energy. The model is translated into a form that is suitable for particle-based Brownian dynamics simulations for performing rheological tests. Various possibilities for connection with experiments, e.g. rheological and structural, are discussed.

1   Introduction

Suspensions and pastes of soft, deformable particles are remarkably common in foods, cosmetics, paints, biological materials and a wide range of other industrial products. These particles have a sponge-like character; they can change both their shapes as well as their volume in response to deformation and flow, or in response to a change in particle concentration.[1-4] While the open-porous structure is elastic, swelling or shrinking of particles also requires them to take up viscous solvent or expel it from their interior, which leads to rate-dependent behavior. Surprisingly, despite their technological importance, the mechanical behavior of these materials still remains poorly understood.

Most experimental and theoretical studies of suspensions have focused on hard, spherical particles of equal size, suspended in a Newtonian liquid. The behavior of these hard sphere systems is well understood and numerical models based on Brownian dynamics or Stokesian dynamics methods accurately predict the behavior of such suspensions, even at high concentrations[5] or at large shear rates.[6] The

[a]Eindhoven University of Technology, Mechanical Engineering, Materials Technology (MaTe), P.O. Box 513, 5600 MB Eindhoven, The Netherlands. E-mail: M.Huetter@tue.nl; Fax: +31 40 244 7355; Tel: +31 40 247 2486
[b]FrieslandCampina, Stationsplein 4, P.O.Box 1551, 3800 BN Amersfoort, The Netherlands

behavior seen in hard spheres is indeed surprisingly general. Even soft spherical particles[7] or particles that deviate significantly from a spherical shape, as well as strongly repulsive particles still obey the general behavior seen in hard spheres. Indeed, their behavior can often be directly matched onto that of hard spheres with a larger, "effective", particle size.[8]

However, this generality does not extend to suspensions of sponge-like particles, as they can exhibit behavior qualitatively different from that observed for hard particles. For example, these materials can exhibit significant shear thinning even at very high Péclet numbers, where this is not observed for hard spheres.[1] Moreover, with increasing particle softness, the viscosity and the structural relaxation processes in these materials become significantly less sensitive to concentration changes, a fact that has recently been used to extend the concept of glass fragility to colloidal suspensions.[9] This becomes intuitively clear if we consider the effect of concentration for soft and hard particles, respectively. For the latter there is clearly a limit to the suspension's concentration, the volume fraction of random close packing, where the system arrests into a solid, jammed state. In contrast, for sponge-like particles the concentration can be increased far beyond the point of random close packing, as particles can be compressed in volume as well as deformed in their shape to accommodate more particles within the same volume.[1,2,9,10] In this densely packed regime, where particles are compressed and deformed, these systems form elastic pastes that behave essentially like yield stress fluids, often described adequately by a Herschel–Bulkley type flow behavior. The flow behavior in this regime has also been shown to depend sensitively on the single particle elastic modulus.[1]

These examples illustrate that for suspensions of sponge-like particles information on the properties at the single particle level is key towards understanding and predicting the macroscopic response. Thus, the macroscopic mechanical properties of these systems can only be interpreted in a meaningful way if the behavior at the single particle level can also be experimentally accessed. The application and validation of numerical models that could provide a link between the behavior at the particle and the macroscopic scale thus requires experimental access to adequate properties at both length scales.

While the experimental characterization of these materials at the macroscopic scale is relatively straightforward and can be achieved using standard rheological techniques, accessing the single particle properties is more difficult.[11] One method that can be employed is atomic force microscopy, which is able to characterize the typical forces required to test the properties of microscopic soft particles, on the order of pN.[12,13] However, this requires an assumption to be made on the Poisson ratio of the particles, as shear and compressive moduli cannot be measured independently. The most direct way to measure the compressive elastic modulus of sponge-like particles is to measure the change in particle volume in response to an increase in osmotic pressure, mediated by the addition of non-adsorbing macromolecules or nano-particles to the background fluid.[9,14] Important here is that the size of the species used to apply the osmotic pressure is larger than the typical pore size or mesh size of the sponge-like particles. If this is the case, the osmotic species cannot enter the particle, which results in an isotropic, purely compressive external stress. For large enough particles the resulting volume change can be quantified directly by imaging in a microscope, while for smaller particles it can be measured using dynamic light scattering measurements. This osmotic compression method thus provides an accurate and well-defined measure for the compressive elastic modulus. However, it does not enable the measurement of the elastic shear modulus of the particles, associated with a change in particle shape. For micrometer-sized particles, this can be achieved using the recently developed capillary micromechanics technique,[15,16] which is based on quantifying the pressure-induced deformation of single particles in tapered glass capillaries. The single particle properties that can be measured directly in experiments should ideally be the starting point for modeling the macroscopic response of these sponge-like particle systems.

Various modeling and simulation efforts have been undertaken to describe the effect of the particle softness on the mechanical properties of suspensions, gels, and glasses. For example, the effective viscosity of concentrated suspensions has been calculated in the limit of low and high frequencies,[17,18] where the particles either have an elastically soft outer layer and hard core, or for soft particles, without a core.[19] For particles with neo-Hookean elastic behavior, immersed in a viscous fluid, the particle deformation has been studied under Stokes flow conditions.[20] With respect to the mechanical properties of pastes[2] and gels[21-24] of soft particles, particle-based simulations have been performed. Mode-coupling theory has been employed to examine glassy dynamics of soft particles.[25,26] Also the nonlinear Langevin equation theory of activated glassy dynamics has been used to study the activated relaxation and the mechanical properties of dense fluids of soft particles.[27,28]

However, it must be pointed out that all these efforts consider only the elastic origin of particle softness. Specifically, one resorts e.g. to neo-Hookean elasticity,[20] or soft interaction potentials[2,21-26,29] for describing the softness of the particle. Instead for modeling sponge-like particles, and suspensions and gels made thereof, a further ingredient is required. Since the viscous background fluid flows through the pores of the particle, which results in swelling or shrinking of the particle, viscous rather than only elastic effects should be accounted for to describe the change in particle size and shape. In analogy to potential-based and friction forces leading to particle motion, there are elastic and viscous contributions to the size-change of the sponge-like particles. In turn, this suggests treating the particle positions and their shapes on an equal footing, i.e., to include both types of degrees of freedom when constructing a model for rate-dependent material properties.

In this article, we present a new model that describes the mechanics and dynamics of materials consisting of sponge-like soft particles in a liquid. The proposed model description pays attention to changes in the particle volume or radius, while changes in the particle shape are neglected at this stage. The manuscript is organized as follows. In Sec. 2, the thermodynamics framework is presented, that is used in Sec. 3 to develop the consistent two-scale model. This is followed by detailed analysis of the developed model in Sec. 4, where also a formulation is presented that is suitable for direct particle-based numerical simulations, akin to Brownian dynamics simulations. The article is concluded with a discussion, Sec. 5.

2 Methodology

Nonequilibrium thermodynamics can be used as a guard-rail, that helps to set up a model that is in agreement with certain principles of thermodynamics. Particularly, when microstructural variables are involved simultaneously with macroscopic ones, doing so makes sure that the mutual couplings between the macroscopic and microscopic levels is taken into account consistently. Many different approaches to nonequilibrium thermodynamics modeling exist; the relations between many of them have been established.[30,31] In this article, we use a derivative of the general equation for the nonequilibrium reversible–irreversible coupling (GENERIC) framework by Grmela and Öttinger,[32-34] as introduced earlier.[35] The GENERIC method for closed systems emphasizes to study the energy and entropy of the system as separate model ingredients, rather than only their composite Helmholtz free energy. The framework then makes very specific statements about how these two building blocks evolve in time, as a result of reversible and irreversible dynamics, respectively. Making for two generators, two statements each leads to four conditions on the dynamics of the system under consideration, as will be elaborated in the following.

Let us assume that one has decided for a specific set of variables, called x, that describe the system in a non-redundant way to the desired level of detail. Once the thermodynamic properties are specified by an appropriate choice of the energy $E[x]$ and the entropy $S[x]$, conditions on the evolution of these functionals amount to conditions on the evolution of the variables x, by way of the chain rule. For closed systems, we have introduced the following conditions (1) as the "weak formulation" of GENERIC:[35]

- The reversible dynamics does not affect the total energy and entropy of the system,

$$\frac{d}{dt} E\bigg|_{rev} = 0,$$ (1a)

$$\frac{d}{dt} S\bigg|_{rev} = 0.$$ (1b)

- The irreversible dynamics leaves the total energy unaffected, while the entropy change is non-negative,

$$\frac{d}{dt} E\bigg|_{irr} = 0,$$ (1c)

$$\frac{d}{dt} S\bigg|_{irr} \geq 0.$$ (1d)

Note that only in the case of isothermal conditions, the two criteria for the reversible and irreversible dynamics, respectively, can be merged into one for the Helmholtz free energy $E - TS$ with T being the constant temperature. In the following, we shall use the conditions (1) for analyzing the dynamics of sponge-like particles in a suspension.

It is mentioned in passing that the condition (1d) is closely related to the so-called Clausius–Duhem inequality.[36-38] The latter is used frequently in continuum mechanics for expressing the second law of thermodynamics.

The conditions (1) can be translated as follows into conditions on the evolution equations of x. Consider a general functional A of the variables x. Using the chain-rule of variational calculus, one obtains

$$\frac{d}{dt} A[x] = \sum_i \int \frac{\delta A}{\delta x_i(z)} \partial_t x_i(z) dz,$$ (2)

where the summation runs over all variables in the set x, and z stands for any set of integration variables (e.g. three-dimensional space or microscopic configuration space). The reversible and irreversible contributions to dA/dt are obtained by inserting in (2) the reversible and irreversible contributions of $\partial_t x_i$, respectively. In (2), we have tacitly assumed that the integration domain is not changing in time, which is sufficient for our present purpose. Furthermore, since we consider closed systems, there are also no fluxes through the physical boundary into or out of the system.

To exploit the conditions (1), one needs two ingredients. On the one hand, the functionals of energy $E[x]$ and entropy $S[x]$ must be specified. On the other hand, one can make use of some characteristics of the evolution equations. This two-step procedure is illustrated in the following, for the example of sponge-like particle suspensions.

3 Development of the model

The main goal of this article is the formulation of a dynamic model that describes the (transient) mechanical behavior of suspensions of sponge-like particles. In order to lessen the amount of phenomenological assumptions but still arrive at a tractable model, a two-level approach is taken. Particularly, the macroscopic level of hydrodynamics is coupled to dynamics on the microscopic level of sponge-like particles. The mutual interplay of these levels will be of special interest, namely, (i) the effect of deformation on the microstructure, and (ii) the effect of the microstructure on the macroscopic mechanical behavior, *i.e.*, the constitutive relation for the stress tensor.

The formulation of constitutive relations for the macroscopic mechanical response in terms of the micro- or mesostructure is superior to purely macroscopic phenomenological ansatzes, because physical insight can be captured in the model in a more direct way. Prominent examples are the elastic dumbbell model,[39] the temporary network model,[40] and the reptation model for polymer melts.[41,42]

3.1 Dynamic variables of interest

Macroscopically, we consider the solvent-particle suspension effectively as a fluid. Since compressibility and thermal effects may play a role, we choose nonisothermal hydrodynamics of a compressible fluid as the suitable macroscopic description. In terms of variables, this means that we are interested in the densities of mass and momentum per unit volume, respectively, $\rho(r)$ and $u(r)$, and in addition the temperature field $T(r)$, to be able to discuss nonisothermal effects. Note that, the choice of the momentum density instead of the velocity field is one of convenience. Similarly, instead of the temperature, one could have chosen the internal energy density (or only a specific contribution thereto). While the latter may be useful in some circumstances,[43] we here assume that the concept of temperature is well defined. It will also be evident further below, that the temperature is more suitable for the physical interpretation of the effects.

The microscopic state of the system is captured as follows. First, it is important to emphasize that we are interested only in mechanical deformations that are slow and moderate with respect to internal relaxations, while extremely rapid deformations are not considered. The main simplification arising from this restriction is that one can consider overdamped dynamics on the microscopic level, while inertia on that scale is not relevant. More specifically, the following argument applies. The typical time scale for the velocity to be equilibrated is on the order of $\tau_v = m/\zeta$ with m the mass of the particle and ζ the friction coefficient. For example, for sub-micron sized particles immersed in water and assuming Stokes drag, $\tau_v \sim 10^{-7}$s. If the applied deformation rates are significantly smaller than τ_v^{-1}, the particle velocities will not be perturbed significantly by the applied deformation field. Instead, the particle velocities will relax to the equilibrium distribution on time scales much shorter than the time the flow requires to significantly affect the microstructure. As a result, it is not only an admissible, but rather a physically meaningful consequence to omit the particle velocities from the description. Therefore, we will describe the instantaneous state of the system of N sponge-like particles by their positions $\{Q_i\}_{i=1,...,N}$ and sizes $\{R_i\}_{i=1,...,N}$. For later convenience, we introduce ξ to denote all microscopic degrees of freedom,

$$\xi = (Q_1, R_1, ..., Q_N, R_N),\tag{3}$$

that is a $4N$-dimensional vector. This level of description is sometimes also called "mesoscopic" in the following, because the constituent particles are not atomistic, and due to their size experience Brownian motion.[44,45]

To conveniently calculate averages over different realizations of such states, we use the distribution function as a dynamic variable for the microstructure. However, to describe macroscopically inhomogeneous situations, this distribution function must vary in the macroscopic 3-dimensional space. In order to realize that, we use the function $p(r;\xi)$, with the interpretation that the microscopic positions Q_i are measured relative to r. The average of a quantity $h(r;\xi)$ over microscopic states, $\langle...\rangle$, is then calculated as

$$\langle h \rangle(r) \equiv n(r)^{-1} \int h(r; \xi)\, p(r; \xi)\, d^{4N}\xi. \tag{4a}$$

The symbol n denotes the number of N-particle systems per unit volume,

$$n(r) = \int p(r; \xi)\, d^{4N}\xi, \tag{4b}$$

which hence specifies the normalization of p. In turn, n^{-1} stands for the representative volume of the N-particle system at the macroscopic position r.

In summary, the full set of dynamic variables is given by the hydrodynamic variables, supplemented by the microstructure,

$$x = (\rho(r), u(r), T(r), p(r; \xi)). \tag{5a}$$

To simplify notation in the sequel, we denote with

$$x' = (\rho(r), T(r), p(r; \xi)), \tag{5b}$$

the set of variables excluding the momentum density.

3.2 Functionals

To keep the treatment as concise as possible, we refrain from specifying the functionals of energy and entropy. Doing so, the general thermodynamic scheme remains to be executed only one, while the system specification is postponed to the later steps only. The only assumption we would like to make for technical simplification is that both functionals can be written in the form

$$E[x] = \int \frac{u^2}{2\rho}\, d^3r + E_{\mathrm{int}}[x'] = \int \frac{u^2}{2\rho}\, d^3r + \int e(x')\, d^3r, \tag{6a}$$

$$S[x] = \int s(x')\, d^3r, \tag{6b}$$

with E_{int} the internal energy, and where the internal energy density per unit volume e and the entropy density per unit volume s depend on the variables x'. The momentum density hence only plays a role in the kinetic energy, $i.e.$, in the first term on the right hand side (r.h.s.) of (6a).

3.3 General form of the evolution equations

Some parts of the evolution equations for the variables (5a) are already known because they are valid for a large class of systems, such as the general form of the mass balance equation, or the structure of the momentum balance. In contrast, the constitutive relations deserve our special attention. For example, the relation for the stress tensor in terms of the microstructure is one of the key questions, that we are addressing below.

First, we start by taking into account general aspects of the evolution equations. In particular, the partial time-derivatives of the variables x can be written in the form

$$\partial_t \rho = -\nabla_r \cdot (v\rho), \tag{7a}$$

$$\partial_t u = -\nabla_r \cdot (vu) + \nabla_r \cdot \sigma, \tag{7b}$$

$$\partial_t T = -v \cdot \nabla_r T + \Theta, \tag{7c}$$

$$\partial_t p = -\nabla_r \cdot (vp) - \nabla_\xi \cdot (\dot{\xi} p),$$ (7d)

with

$$v = u/\rho$$ (8)

for the velocity field, and σ the total macroscopic stress tensor. The latter, as well as the temperature change Θ in the temperature equation are specified further below. In (7d), the symbol $\dot{\xi}$ represents in terms of $\dot{Q}_i$ the rate of change of the internal coordinate Q_i, and in terms of $\dot{R}_i$ the change in size of particle i.

To interpret (7d), it may be convenient to split it into two evolution equations, one for the number density n, defined in (4b), and one for the probability density

$$\tilde{p} = p/n$$ (9)

that is normalized to unity,

$$\partial_t n = -\nabla_r \cdot (vn),$$ (10a)

$$\partial_t \tilde{p} = -v \cdot \nabla_r \tilde{p} - \nabla_\xi \cdot (\dot{\xi} \tilde{p}).$$ (10b)

In this equivalent formulation, the number density n of N-particle systems obeys the same evolution equation as the mass density. However, this does not mean that the chosen set of variables contains redundant information. Rather, in general, the density of microscopic particles may change on macroscopic scales e.g. due to gravity, electromagnetic fields, diffusion, and thermophoretic effects.[46,47] Including such effects would exemplify the different nature of the mass density ρ and the number density n of N-particle systems. However, in this article we choose to neglect such effects to simplify the model.

The evolution (10b) of the reduced distribution function $\tilde{p}$ contains two contributions. The first term on the r.h.s. represents the advection of the particles (and hence of the distribution function) with the macroscopic flow field. The second term accounts for the changes in particle positions and sizes. As will be shown below in Sec. 4, the eqn (10b) for $\tilde{p}(\mathbf{r};\xi)$ eventually assumes the form of a diffusion equation, that can be interpreted in terms of the Fokker–Planck equation.[48–50] While the latter is an evolution equation for transition probabilities, it can be translated into evolution equations for the trajectories ξ in the form of stochastic differential equations.[48–50] This equivalence justifies the form of the second term on the r.h.s. of (7d) and (10b).

In the above calculations and throughout the entire article, we assume that boundary terms with respect to the ξ-integrals vanish. This will apply particularly also when using the thermodynamic conditions (1).

3.4 Reversible dynamics

Considering the evolution of the variables x, the dynamics of each variable can be split into a reversible and an irreversible contribution in the following sense. As reversible contributions to the dynamics we denote those terms that are invariant upon time-reversal, i.e. upon $t \rightarrow -t$ and reversal of all time-odd quantities such as v. All other contributions are called irreversible. Note that this definition of reversible and irreversible is one on the level of the time-evolution law $\mathcal{T}x = G(x)$ with $\mathcal{T}$ representative of time-derivatives, where a current state x relates to the changes, $\mathcal{T}x$. It is well possible that the current state x is a result of irreversible dynamics (e.g., viscous flow leads to dissipation in the evolution of the velocity v). However, our definitions of reversible and irreversible do not ask how a certain state x came about (*i.e.*, the history), but they are rather based on how the current state x determines the further evolution of the system, $\mathscr{T}x$.

In the following, we are exploiting the thermodynamic conditions (1a,1b) for the functionals (6) of the variables (5a), with evolution equations of the form (7).

3.4.1 Temperature equation.
To elaborate on the consequences of the condition (1b), we first discuss our understanding of the dynamics of the particle positions and particles sizes. It is reasonable to assume that the reversible change in particle positions is exclusively due to the flow field v. If the velocity field varies slowly on the scale of the N-particle system, and keeping in mind that the positions Q_i are measured relative to a frame of reference moving with the entire volume element, we find for purely affine deformation

$$\dot{Q}_i|_{\text{rev}} = (\nabla_r v)^{\text{T}} \cdot Q_i, \tag{11a}$$

where the superscript T denotes the matrix transpose. The change in particle size is more subtle. If the particle i was perfectly rigid, its size R_i would be constant in time. Contrary, if the particle surface deforms under flow, the change of particle size relative to its present size must be proportional to the divergence of the velocity field, $\nabla_r \cdot v$. All other aspects of $\nabla_r v$ than its trace are irrelevant for the evolution of R_i, because the particle shape is only quantified by the scalar R_i that can not couple to the tensorial structure of $\nabla_r v$. We hence find

$$\dot{R}_i|_{\text{rev}} = \alpha(\nabla_r \cdot v)R_i, \tag{11b}$$

with a prefactor α. While $\alpha = 0$ for rigid particles, one finds $\alpha = 1/3$ for the special case of vanishing modulus, *i.e.*, when the particle surface deforms affinely with the flow field. In general, α is a complicated function of the properties of the particles, with $0 \leq \alpha \leq 1/3$.

With (11), the condition (1b) for the conservation of entropy in reversible dynamics can be worked out. Particularly, one finds that Θ in the temperature evolution (7c) can be written in the form

$$\Theta = (\nabla_r v)^{\text{T}} : Y, \tag{12}$$

with the second-rank tensor Y given by

$$\left(\frac{\delta S}{\delta T}\right) Y = \left(-s + \rho\frac{\delta S}{\delta \rho} + n\left\langle\frac{\delta S}{\delta p}\right\rangle\right)\mathbf{1} - n\left\langle\sum_i\left(\nabla_{Q_i}\frac{\delta S}{\delta p}\right)Q_i\right\rangle$$

$$-\alpha n\left\langle\sum_i\left(\partial_{R_i}\frac{\delta S}{\delta p}\right)R_i\right\rangle\mathbf{1}, \tag{13}$$

with the average $\langle...\rangle$ and the number density of systems n introduced in (4). For deriving this relation we have made use of

$$\frac{\delta A}{\delta \rho}(\nabla_r \rho) + \frac{\delta A}{\delta T}(\nabla_r T) + \int\int\frac{\delta A}{\delta p}(\nabla_r p)d^{3N}Q d^N R = \nabla_r a, \tag{14}$$

with $A = S$ and $a = s$.

3.4.2 Non-dissipative contribution to the stress tensor.
Using the conservation of energy during reversible dynamics, (1a), one obtains the following result for the stress tensor in the momentum balance (7b),

$$\sigma^F = ((e - Ts) - \rho F_{;\rho} - n\langle F_{;p}\rangle)\mathbf{1} + n\left\langle \sum_i (\nabla_{Q_i} F_{;p})Q_i \right\rangle + \alpha n\left\langle \sum_i (\partial_{R_i} F_{;p})R_i \right\rangle \mathbf{1} \cdot$$

$$\tag{15}$$

The symbols $F_{;\rho}$ and $F_{;p}$ are defined as to abbreviate the following combination of derivatives,

$$F_{;\rho} \equiv \frac{\delta E_{\mathrm{int}}}{\delta \rho} - T\frac{\delta S}{\delta \rho}, \tag{16a}$$

$$F_{;p} \equiv \frac{\delta E_{\mathrm{int}}}{\delta p} - T\frac{\delta S}{\delta p}. \tag{16b}$$

The letter 'F' has been chosen since these specific combinations of internal energy E_{int} and entropy S are akin to a Helmholtz free energy. However, we refrain from introducing a Helmholtz free energy functional for two reasons. First, the entire procedure above essentially rests on using E_{int} and S as separate ingredients. And second, it is only on the level of the ρ- and p-derivatives that a Helmholtz free energy-like combination occurs. A Helmholtz free energy functional as such is not required.

In deriving (15) we have made use of (14) with $A = E_{\mathrm{int}}$ and $a = e$. Furthermore, it has been assumed that the consistency relation

$$\frac{\delta E_{\mathrm{int}}}{\delta T} = T\frac{\delta S}{\delta T} \tag{17}$$

between the internal energy E_{int} and the entropy S holds. If that relation did not hold, all prefactors T in (15,16) would have to be replaced by $(\delta S/\delta T)^{-1}(\delta E_{\mathrm{int}}/\delta T)$.

3.5 Irreversible dynamics

In the following, we are exploiting the thermodynamic conditions (1c,1d) for the functionals (6) of the variables (5a), with evolution equations of the form (7). Three effects are considered, namely the conduction of heat, viscous flow, and the over-damped particle dynamics. In the most general scenario, these effects can be coupled. To keep the current treatment concise, we neglect cross-couplings between these three effects.

3.5.1 Heat flow. With regard to the flow of heat, the only variable of x that is affected is the temperature. Using the conservation of energy during irreversible dynamics, (1c), one observes that the temperature change must be of the form

$$\partial_t T\Big|_{\mathrm{irr}} = -\left(\frac{\delta E_{\mathrm{int}}}{\delta T}\right)^{-1} \nabla_r \cdot \boldsymbol{q} , \tag{18}$$

where, for physical reasons, $\boldsymbol{q}$ assumes the meaning of a heat flux. Using the consistency relation (17), non-negative change in total entropy (1d) results in

$$\int \left[\left(\nabla_r \frac{1}{T}\right)\cdot\boldsymbol{q} - \nabla_r\cdot\frac{\boldsymbol{q}}{T}\right]d^3r \geq 0. \tag{19}$$

While the second contribution on the left hand side (l.h.s.) is the exchange of entropy between different volume elements, the first term represents the actual production of entropy. It is hence the first term in (19) only that is to be used as a constraint on the constitutive relation for the heat flux.

3.5.2 Viscous flow. Viscous flow affects neither the mass balance, nor the evolution of the microstructure, but only relates to the change in momentum and, because of dissipation, also the temperature. The conservation of energy, (1c), thus implies

$$\partial_t T\big|_{irr} = \left(\frac{\delta E_{int}}{\delta T}\right)^{-1} (\nabla_r \boldsymbol{v})^{\mathsf{T}} : \boldsymbol{\sigma}^{\mathsf{v}}, \tag{20}$$

with $\boldsymbol{\sigma}^{\mathsf{v}}$ the viscous contribution to the stress tensor. With the aid of the consistency relation (17), the non-negativity of the entropy production (1d) results in

$$\int \frac{1}{T} (\nabla_r \boldsymbol{v})^{\mathsf{T}} : \boldsymbol{\sigma}^{\mathsf{v}} d^3 r \geq 0. \tag{21}$$

3.5.3 Overdamped particle dynamics. In the absence of inertial effects, *i.e.* in the overdamped limit, the forces acting on the particles result directly in the rate-of-change of particle positions and sizes, rather than in accelerations. The ratio between force and rate-of-change has units of an inverse friction coefficient. Overdamped particle dynamics must hence be part of the irreversible dynamics.

The overdamped dynamics of the particle positions and particle sizes affects neither the mass balance, nor the momentum balance, but only relates to the evolution of the distribution p and, because of dissipation, to the change in temperature. In view of the conditions (1c,1d) and the chain rule (2), we may thus use $\partial_t \rho = 0$, $\partial_t \boldsymbol{u} = 0$, $\partial_t T \neq 0$, and for the irreversible contribution to the particle dynamics we write in view of (7d) the form

$$\partial_t p\big|_{irr} = -\nabla_\xi \cdot (\dot{\xi}\big|_{irr} p). \tag{22}$$

The conservation of energy, (1c), implies

$$\partial_t T\big|_{irr} = -\left(\frac{\delta E_{int}}{\delta T}\right)^{-1} n \left\langle \left(\nabla_\xi \frac{\delta E_{int}}{\delta p}\right) \cdot \dot{\xi}\big|_{irr} \right\rangle, \tag{23}$$

while the condition of non-negative entropy production, (1d), leads finally to

$$\int \frac{n}{T} \left\langle (-\nabla_\xi F_{,p}) \cdot \dot{\xi}\big|_{irr} \right\rangle d^3 r \geq 0, \tag{24}$$

for general nonisothermal situations, with $F_{,p}$ defined in (16b).

3.6 Summary of the model

Summarizing all of the above results, one obtains the following model for nonisothermal hydrodynamics with heat conduction, viscous flow, and overdamped dynamics of the particle evolution:

$$\partial_t \rho = -\nabla_r \cdot (\boldsymbol{v}\rho), \tag{25a}$$

$$\partial_t \boldsymbol{u} = -\nabla_r \cdot (\boldsymbol{v}\boldsymbol{u}) + \nabla_r \cdot (\boldsymbol{\sigma}^F + \boldsymbol{\sigma}^{\mathsf{v}}), \tag{25b}$$

$$\partial_t T = -\boldsymbol{v} \cdot \nabla_r T + \left(\frac{\delta E_{int}}{\delta T}\right)^{-1} \left[(\nabla_r \boldsymbol{v})^{\mathsf{T}} : (\boldsymbol{\sigma}^S + \boldsymbol{\sigma}^{\mathsf{v}}) - \nabla_r \cdot \boldsymbol{q} - n \left\langle \left(\nabla_\xi \frac{\delta E_{int}}{\delta p}\right) \cdot \dot{\xi}\big|_{irr} \right\rangle \right], \tag{25c}$$

$$\partial_t p = -\nabla_r \cdot (vp) - \sum_i \nabla_{\boldsymbol{Q}_i} \cdot \left((\nabla_r v)^T \cdot \boldsymbol{Q}_i p \right) - \sum_i \partial_{R_i} (\alpha (\nabla_r \cdot v) R_i p) - \nabla_\xi \cdot (\dot{\xi}|_{irr} p), \quad (25d)$$

with the velocity field

$$v = u/\rho. \quad (26)$$

The non-viscous stress contributions are given by

$$\boldsymbol{\sigma}^F = \left((e - Ts) - \rho F_{;\rho} - n\langle F_{;p} \rangle \right) \mathbf{1} + n \left\langle \sum_i (\nabla_{\boldsymbol{Q}_i} F_{;p}) \boldsymbol{Q}_i \right\rangle + \alpha n \left\langle \sum_i (\partial_{R_i} F_{;p}) R_i \right\rangle \mathbf{1} \quad (27a)$$

$$\boldsymbol{\sigma}^S = T \left(-s + \rho \frac{\delta S}{\delta \rho} + n \left\langle \frac{\delta S}{\delta p} \right\rangle \right) \mathbf{1} - nT \left\langle \sum_i \left(\nabla_{\boldsymbol{Q}_i} \frac{\delta S}{\delta p} \right) \boldsymbol{Q}_i \right\rangle$$

$$- \alpha nT \left\langle \sum_i \left(\partial_{R_i} \frac{\delta S}{\delta p} \right) R_i \right\rangle \mathbf{1}, \quad (27b)$$

with the abbreviations $F_{;\rho}$ and $F_{;p}$ defined in (16).

The constitutive relations for the heat flux q, the viscous stress $\boldsymbol{\sigma}^v$, and the irreversible particle dynamics $\dot{\xi}|_{irr}$ are constrained by the inequalities

$$\int \left(\nabla_r \frac{1}{T} \right) \cdot q d^3 r \geq 0, \quad (28a)$$

$$\int \frac{1}{T} (\nabla_r v)^T : \boldsymbol{\sigma}^v d^3 r \geq 0, \quad (28b)$$

$$\int \frac{n}{T} \langle (-\nabla_\xi F_{;p}) \cdot \dot{\xi}|_{irr} \rangle d^3 r \geq 0. \quad (28c)$$

Violation of these inequalities results in a negative entropy production, and hence in unphysical behavior of the model.

Since our main interest is in the dynamics of the particle positions and their sizes, as well as their coupling to the macroscopic deformation, we neglect in the remainder of this article the heat flux and the viscous flow effects.

In the relation (28c), one observes that $\dot{\xi}|_{irr}$ plays the role of thermodynamic fluxes, while $-\nabla_\xi F_{;p}$ are thermodynamic forces.[34,51] A sufficient condition to satisfy (28c) is given by the local criterion

$$(-\nabla_\xi F_{;p}) \cdot \dot{\xi}|_{irr} \geq 0, \quad (29)$$

if we assume, on physical grounds, that $n > 0$ and $T > 0$. A general solution is then given by the quasi-linear relation

$$\dot{\xi}|_{irr} = \boldsymbol{\mu} \cdot (-\nabla_\xi F_{;p}), \quad (30)$$

with $\boldsymbol{\mu}$ a positive semi-definite and symmetric $4N \times 4N$ mobility tensor. In view of modeling hydrodynamic interactions and the like, it is essential to point out that the mobility tensor in general can be an arbitrary function of the particle configuration

$\dot{\xi}$ and of the fields x, as long as it is positive semi-definite and symmetric. The constitutive rule (30) demonstrates that, in principle, all particle positions and particle sizes are affected by all thermodynamic forces. Hence, the change in position of particle i, $\dot{Q}_i|_{\mathrm{irr}}$, is in principle affected by all forces $-\nabla_{Q_k}F_{;p}$ and $-\nabla_{R_f}F_{;p}$. A completely analogous statement holds for the change in the size of particle i, $\dot{R}_i|_{\mathrm{irr}}$.

4 Application of the model

4.1 Ansatz for the energy and entropy

The first step in making the above model more specific consists of making an ansatz for the non-kinetic energy E_{int} and the entropy S of the system. In the absence of immersed particles, the internal energy density ε and the entropy density η per unit volume of the solvent are functions of the mass density ρ and of the temperature T. When the sponge-like particles are present, the following two additional contributions arise. The first represents the sum $\hat{\phi}$ of all interaction energies in the N-particle system. Since the immersed particles are not of molecular size but rather mesoscopic objects, the interactions between them are coarse-grained (effective) interaction potentials,[52] i.e., a form of Helmholtz free energies themselves that depend in general on both the temperature and the mass density of the system. The energetic and entropic contributions to the effective interaction energy of the N-particle system are denoted by $\hat{\phi}^E$ and $\hat{\phi}^S$ with[34]

$$\hat{\phi} = \hat{\phi}^E + \hat{\phi}^S. \tag{31}$$

Secondly, there is a particle-related contribution to the entropy that accounts for the (Boltzmann) configurational entropy. Altogether, we have the following expressions for the thermodynamic functions,

$$E_{\mathrm{int}}\left[x'\right] = \int \varepsilon(\rho(r), T(r))d^3r + \int \hat{\phi}^E(\rho(r), T(r); \xi)p(r, \xi)d^{4N}\xi d^3r, \tag{32a}$$

$$S\left[x'\right] = \int \eta(\rho(r), T(r))d^3r - \int \frac{1}{T}\hat{\phi}^S(\rho(r), T(r); \xi)p(r, \xi)d^{4N}\xi d^3r$$
$$- k_B \int p(r, \xi)\ln\frac{p(r, \xi)}{p_0}d^{4N}\xi d^3r, \tag{32b}$$

where the constant p_0 has been introduced for dimensional reasons. For example, the interaction potential $\hat{\phi}$ can, but does not have to, be a sum over pairwise interactions V,

$$\hat{\phi}(\xi) = \sum_i \sum_{j>i} V(Q_i, R_i, Q_j, R_j), \tag{33}$$

where V may stand for Derjaguin–Landau–Verwey–Overbeek (DLVO)-type contributions for non-contacting particles,[53,54] and contact contributions for deformable particles e.g. the Hertzian potential.[3,55] An example for a temperature-dependent interaction potential is the electrostatic double-layer repulsion in the DLVO-theory.[56,57]

4.2 Non-dissipative contribution to the stress tensor

The ramifications of the choice (32) for the entire model are manifold, as can be seen in the summary in Sec. 3.6. While the general couplings between the microscopic scale and the macroscopic scales have been described in detail in the above, we now concentrate on the evolution of the microstructure and on the stress tensor, for the specific functionals (32). After calculating the functional derivatives $\delta E_{\text{int}}/\delta x'$ and $\delta S/\delta x'$, one obtains according to the definitions (16)

$$F_{;\rho} = \frac{\partial}{\partial \rho} \left[(\varepsilon - T\eta) + n\langle \hat{\phi} \rangle \right], \tag{34a}$$

$$F_{;p} = \hat{\phi} + k_{\text{B}} T \left(\ln \frac{p}{p_0} + 1 \right). \tag{34b}$$

As a result, the stress tensor expression (27a) becomes

$$\boldsymbol{\sigma}^F = \left[-p_{\text{sol}} - n\rho \frac{\partial}{\partial \rho} \langle \hat{\phi} \rangle - n(1 + N + \alpha N)k_{\text{B}} T \right] \mathbf{1} + n \left\langle \sum_i (\nabla_{\boldsymbol{Q}_i} \hat{\phi}) \boldsymbol{Q}_i \right\rangle$$

$$+ \alpha n \left\langle \sum_i (\partial_{R_i} \hat{\phi}) R_i \right\rangle \mathbf{1}, \tag{35}$$

with the pressure of the solvent

$$p_{\text{sol}} = -(\varepsilon - T\eta) + \rho \frac{\partial}{\partial \rho} (\varepsilon - T\eta). \tag{36}$$

The second contribution on the r.h.s. of (35), $-n\rho\partial\langle\hat{\phi}\rangle/\partial\rho\mathbf{1}$, is due to the fact that the effective interaction between the mesoscopic particles may depend on the density of the system. Furthermore, there are contributions akin to an ideal gas due to the degrees of freedom, while the last two contributions to the stress are due to the interaction forces.[34,39] Specifically, $-\nabla_{\boldsymbol{Q}_i}\hat{\phi}$ is the interaction force on particle i, while $-\partial_{R_i}\hat{\phi}$ is a thermodynamic force, that can be interpreted analogously to an effective pressure on the particle volume.

4.3 Particle dynamics, extended Brownian dynamics scheme

The dynamics of the particles for the ansatz (32) is given as follows. Using the force-flux relations (30), one can calculate the evolution of p, eqn (25d), and in turn the evolution of the normalized distribution function $\tilde{p}$ introduced in (9),

$$\partial_t \tilde{p} = -\boldsymbol{v} \cdot \nabla_r \tilde{p} - \sum_i \nabla_{\boldsymbol{Q}_i} \cdot \left((\nabla_r \boldsymbol{v})^{\text{T}} \cdot \boldsymbol{Q}_i \tilde{p} \right) - \sum_i \partial_{R_i} (\alpha(\nabla_r \cdot \boldsymbol{v}) R_i \tilde{p}) - \nabla_\xi \cdot (\boldsymbol{\mu} \cdot [(-\nabla_\xi \hat{\phi}) \tilde{p}$$

$$- k_{\text{B}} T \nabla_\xi \tilde{p}]). \tag{37}$$

To study the behavior of the microstructure under deformation, it is often cumbersome to do that in terms of the distribution function $\tilde{p}$. Instead, it may be beneficial to resort to simulation of several trajectories $\xi(t)$, the ensemble average of which also properly describes the system response. To that end, one interprets the diffusion eqn (37) as a Fokker–Planck equation for the transition probability, that in turn can be re-expressed equivalently in terms of stochastic differential equations.[48–50] Particularly, for the Fokker–Planck eqn (37) one obtains the stochastic differential equations in the Itô-interpretation[48–50]

$$d\boldsymbol{Q}_i = (\nabla_r \boldsymbol{v})^{\text{T}} \cdot \boldsymbol{Q}_i dt + \sum_k \boldsymbol{\mu}_{\boldsymbol{Q}_i \boldsymbol{Q}_k} \cdot \left(-\nabla_{\boldsymbol{Q}_k} \hat{\phi} \right) dt + \sum_k \boldsymbol{\mu}_{\boldsymbol{Q}_i R_k} \cdot \left(-\partial_{R_k} \hat{\phi} \right) dt$$

$$+ k_{\text{B}} T [\nabla_\xi \cdot \boldsymbol{\mu}]_{\boldsymbol{Q}_i} dt + [\boldsymbol{B} \cdot d\boldsymbol{W}_t]_{\boldsymbol{Q}_i}, \tag{38a}$$

$$dR_i = \alpha(\nabla_r \cdot \mathbf{v})R_i dt + \sum_k \boldsymbol{\mu}_{R_i Q_k} \cdot \left(-\nabla_{Q_k}\hat{\phi}\right)dt + \sum_k \mu_{R_i R_k}\left(-\partial_{R_k}\hat{\phi}\right)dt + k_B T[\nabla_\xi \cdot \boldsymbol{\mu}]_{R_i} dt$$

$$+ [\mathbf{B} \cdot d\mathbf{W}_t]_{R_i},$$

(38b)

with a matrix $\mathbf{B}$ that satisfies

$$\mathbf{B} \cdot \mathbf{B}^T = 2k_B T\boldsymbol{\mu}.$$

(39)

The terms involving $d\mathbf{W}_t$ represent the fluctuating Brownian contributions, that add the stochastic nature to the dynamics. The symbol $d\mathbf{W}_t$ stands for the increments of Wiener processes,[48–50] with average $\langle\!\langle d\mathbf{W}_t \rangle\!\rangle = 0$ and variance $\langle\!\langle d\mathbf{W}_t d\mathbf{W}_{t'}^T \rangle\!\rangle = \delta_{t,t'}dt\mathbf{1}$. In other words, the noise is uncorrelated in time (called "white" noise), and the components of $d\mathbf{W}_t$ are mutually uncorrelated. The condition (39) for $\mathbf{B}$ and the occurrence of the contributions proportional to $\nabla_\xi \cdot \boldsymbol{\mu}$ in (38) follow rigorously from (37), and ensure that the trajectories $\xi(t)$ obey the proper statistics, in particular (42) at equilibrium. The relation (39) is a manifestation of the so-called fluctuation–dissipation theorems.[34,51,58,59]

The system of eqn (38) serves as a basis for numerical simulations. In the absence of the particle sizes, one recovers the common Brownian dynamics for the evolution of particle positions Q_i.[34]

The interpretation of the individual contributions in (38) is as follows. The terms involving $\mathbf{v}$ describe the effect of the macroscopic velocity field, discussed earlier. The symbols $\boldsymbol{\mu}_{Q_i Q_k}$, $\boldsymbol{\mu}_{Q_i R_k}$, $\boldsymbol{\mu}_{R_i Q_k}$, and $\boldsymbol{\mu}_{R_i R_k}$ stand for the respective parts of the symmetric and non-negative $4N \times 4N$ mobility matrix. Since $-\nabla_{Q_k}\hat{\phi}$ is the force on particle k, many-particle hydrodynamic interactions[60] can be included by the choice $\boldsymbol{\mu}_{Q_i Q_k} \neq 0$ for $i \neq k$. In the absence of many-particle hydrodynamic interactions, single-particle drag can be described by

$$\boldsymbol{\mu}_{Q_i Q_k} = \mu_{Q_i Q_i}\delta_{ik}\mathbf{1} = \frac{1}{\zeta_{\text{eff},i}}\delta_{ik}\mathbf{1},$$

(40)

with $\zeta_{\text{eff},i} > 0$ the effective friction coefficient of the single particle. It is in principle possible that dynamic cross-couplings with respect to the force $-\nabla_{R_k}\hat{\phi}$ are possible due to non-vanishing $\boldsymbol{\mu}_{Q_i R_k}$, and $\mu_{R_i R_k}$ for $i \neq k$; however, this asks for a detailed discussion of the hydrodynamic effects, which is beyond the scope of this article. Instead, we briefly comment on the physics contained in the single-particle coefficients $\mu_{R_i R_i}$. It is the nature of the sponge-like particles that they have an open-porous structure that is deformable and that can be penetrated by the solvent with viscosity η_{sol}. For a specific thermodynamic force $-\nabla_{R_k}\hat{\phi}$, the rate of change in particle size should be inversely proportional to η_{sol}. For dimensional reasons, we propose to write, akin to Darcy's law,[61,62]

$$\mu_{R_i R_i} = \frac{\chi}{R_i \eta_{\text{sol}}},$$

(41)

where χ is a dimensionless measure of the pore structure of the particle. The relation (41) together with the evolution eqn (38b) also makes clear that the change in particle size is irreversible, since the viscosity of the solvent enters. However, the irreversible nature of particle-size change is also clear on purely dimensional grounds, since the mobility $\mu_{R_i R_i}$ has the same units as the inverse of a friction coefficient.

It is noteworthy that the two driving forces for position-change and size-change, $-\nabla_{Q_i}\hat{\phi}$ and $-\partial_{R_i}\hat{\phi}$, are derived from the same potential, $\hat{\phi}$. The reason for that is that we have started out with a thermodynamic formulation. The latter ensures that the equilibrium distribution of microscopic states ξ is given by the Boltzmann distribution, as desired. Particularly, solving the diffusion eqn (37) for the equilibrium distribution, *i.e.* in the absence of flow, leads to

$$\tilde{p} = \frac{1}{Z}\exp\left(\frac{-\hat{\phi}}{k_B T}\right), \tag{42}$$

with Z being fixed by the normalization condition $\int \tilde{p}\,d^{4N}\xi = 1$. In other words, in order to recover the correct limiting distribution in equilibrium, one is forced to use the force expressions as described above. Therefore, modeling the system of (sponge-like) particles indeed starts with the specification of the energy and entropy, which includes effective interaction potentials between the particles. If the latter depend both on relative particle positions and on particles sizes (e.g., Hertzian potential[3,55]), the driving forces for position-change and size-change are a direct consequence. There is no room for additional modeling of the forces. However, there is clearly plenty of room in modeling how the forces, $-\nabla_\xi\hat{\phi}$, give rise to changes $\dot{\xi}|_{irr}$, namely through specification of the mobility matrix $\boldsymbol{\mu}$.

5 Discussion

This article has presented a method to develop a two-scale model for suspensions of sponge-like particles. The development rests on the thermodynamic conditions (1) for the changes in energy and entropy. As a consequence, the non-kinetic energy E_{int} and the entropy S are ubiquitous in the entire model. The thermodynamic properties of the entire suspension, consisting of solvent and sponge-like particles, are contained in these two functionals. The other kind of system specification concerns the formulation of constitutive relations for the thermodynamic fluxes, namely for the heat flux, the viscous stress, and for the irreversible particle dynamics. In this article, we have focused our attention on the irreversible particle dynamics, which finally amounts to specification of the mobility matrix $\boldsymbol{\mu}$. Once the system of interest is specified in these terms, the mutual coupling between microscopic and macroscopic scales can be studied in terms of the particle dynamics in flow, eqn (25d), and the resulting stress tensor (27a). While (25d,27a) is the general form of our main result, for practical applications the realization in terms of the stress tensor (35) and the stochastic differential eqn (38) is more relevant.

The developed general model offers many opportunities for application and comparison to experimental systems. In practice, it is desirable to represent in the model as much of the experimentally gained knowledge about the system as possible. Specifically, experiments on the particle level can be used to render the model (e.g. in the form of the numerical simulations based on (38)) system specific. For example, the elastic properties of the particles and their interactions can be captured in terms of the potential $\hat{\phi}$. The force for changing particle positions due to interactions, $-\nabla_{Q_i}\hat{\phi}$, and the driving force for changing the particle size, $-\nabla_{R_i}\hat{\phi}$, are then two sides of the same coin. Hence, separate experimental investigations about both kinds of forces can be used to infer the structure of the potential $\hat{\phi}$. The structure of the general model gives room to also account for the, experimentally observed, rate-dependent size change of the sponge-like particles, namely through appropriate choice of $\mu_{R_iR_i}$, see e.g. (41). Through these contributions to the mobility tensor, one can incorporate the effective permeability of the particles. There are also several ways to confront the model predictions with experimental data. While the comparison of the mechanical response in terms of stresses, flow curves, and the like is most obvious, also the structure can be compared. For example, based on the numerical solutions of the particle-based dynamics (38), the pair-correlation function and the distribution of particle sizes can be determined and compared with experiments.

A formulation of the model has been presented that is suitable for particle-based simulations, in terms of stochastic differential eqn (38). Their numerical implementation leads to an extension of common Brownian dynamics schemes, in the sense that the particle-sizes enter as additional degrees of freedom. For given flow field

$\mathbf{v}$, trajectories $\xi(t)$ can be determined by numerical integration of (38). Based on these trajectories, the average mechanical response can be determined with the constitutive relation (35) for the macroscopic stress $\boldsymbol{\sigma}^{\mathrm{F}}$.

The microstructure can depart significantly from equilibrium if the applied flow is sufficiently strong. In common Brownian dynamics schemes (for particle positions only), one makes use of a characteristic time scale τ_{Br}, which in turn is compared with the inverse of the applied rate of deformation $\dot{\gamma}$, *i.e.*, a Péclet number is introduced as $Pe = \tau_{\mathrm{Br}}\dot{\gamma}$. Significant distortion of the microstructure and non-Newtonian behavior set in for $Pe \geq 1$. In the case of single-particle Stokes drag (40), the Brownian time scale can be defined as $\tau_{\mathrm{Br}} = \delta_Q^2/(6D_Q)$, with the diffusion coefficient $D_Q = k_{\mathrm{B}}T\mu_{Q_i Q_i}$ and δ_Q^2 a characteristic mean square displacement, leading to

$$Pe = \dot{\gamma}\frac{\delta_Q^2}{6k_{\mathrm{B}}T\mu_{Q_i Q_i}}, \tag{43}$$

and with the mobility $\mu_{Q_i Q_i} = (6\pi\eta_{\mathrm{sol}}R_i)^{-1}$. A similar procedure could be followed for the analysis of the evolution of particle sizes. One would then arrive at a definition of a dimensionless number that characterizes the onset of significant change in particle size due to strong flow. However, an alternative analysis seems more appropriate. As can be seen in (38b), the particle size is only affected by the flow field if the latter is compressible. Since the dispersing fluid of the suspension, that also penetrates the sponge-like particles, is nearly incompressible (*e.g.* water), the direct effect of the flow field on the particle size is negligible in most situations. Neglecting cross-coupling by setting $\boldsymbol{\mu}_{R_i Q_k} = 0$ and $\mu_{R_i R_{k \neq i}} = 0$ in (38b), the change in particle size originates primarily from the term $\mu_{R_i R_i}(-\partial_{R_i}\hat{\phi})$. In other words, the evolution of the particle size R_i is informed about the flow indirectly through the changed positions Q_k ($k = 1,...,N$), that enter the R_i-evolution through the interaction potential $\hat{\phi}$. The crucial question is thus whether the particle positions evolve so slowly, that the sizes have got enough time to adjust to this new spatial arrangement. This can be assessed by considering the ratio Rm of the mobilities for particle position and size,

$$Rm = \frac{\mu_{Q_i Q_i}}{\mu_{R_i R_i}}. \tag{44}$$

For $Rm \ll 1$, the diffusion-related change in particle positions is slow enough so that the particle sizes can continuously equilibrate with respect to these changes. In contrast, for $Rm \gg 1$, the particles sizes lag behind the ongoing diffusion-related change in particle positions. Note that for single-particle Stokes drag, one obtains $Rm = (6\pi\chi)^{-1}$, with χ being the measure of the porosity of the particle (see (41)). The magnitudes of the two dimensionless parameters, Pe and Rm, dictate the characteristics of the mechanical response. If the evolution of particle positions is dominated by the flow contribution, the product $Pe\,Rm$ can be used to discuss whether the particle sizes manage to continuously adjust to the ongoing particle rearrangements. Detailed computer simulations based on (38) are required to get a more detailed view of the rheological response of the suspension.

In this article, we have quantified the geometry of the particles only through their radius. While this is a first natural step for accounting for the sponge-like character of the particles, there are substantial short-comings. Particularly, strong flow or the interaction with other particles in dense suspensions leads to a change also in the particle shape. Therefore, future steps should aim at replacing the scalar R_i by a tensorial quantity $\boldsymbol{T}_i$ for each particle. Tensorial characterizations of morphology have been introduced in studies about polymer blends[63-65] and particularly for ellipsoidal droplets,[66] and more generally in integral geometry.[67] When choosing tensorial measures of the particle geometry, the change in size and particularly shape can be accounted for more naturally.

References

1 S. Adams, W. J. Frith and J. R. Stokes, *J. Rheol.*, 2004, **48**, 1195–1213.
2 J. R. Seth, M. Cloitre and R. T. Bonnecaze, *J. Rheol.*, 2006, **50**, 353–376.
3 R. T. Bonnecaze and M. Cloitre, *Adv. Polym. Sci.*, 2010, **236**, 117–161.
4 A. Bouchoux, G. Gésan-Guiziou, J. Pérez and B. Cabane, *Biophys. J.*, 2010, **99**, 3754–3762.
5 J. F. Brady, *J. Chem. Phys.*, 1993, **99**, 567–581.
6 J. F. Brady and J. Morris, *J. Fluid Mech.*, 1997, **348**, 103–139.
7 F. d. J. Guevara-Rodríguez and M. Medina-Noyola, *Phys. Rev. E: Stat. Phys., Plasmas, Fluids, Relat. Interdiscip. Top.*, 2003, **68**, 011405.
8 R. Buscall, *Colloids Surf., A*, 1994, **83**, 33–42.
9 J. Mattsson, H. M. Wyss, A. Fernandez-Nieves, K. Miyazaki, Z. B. Hu, D. R. Reichman and D. A. Weitz, *Nature*, 2009, **462**, 83–86.
10 M. Cloitre, R. Borrega and L. Leibler, *Phys. Rev. Lett.*, 2000, **85**, 4819–4822.
11 H. M. Wyss, J. Mattsson, T. Franke, A. Fernández-Nieves and D. A. Weitz, *Microgel Suspensions: Fundamentals and Applications*, Wiley VCH, 2011, pp. 311–324.
12 O. Tagit, N. Tomczak and G. J. Vancso, *Small*, 2008, **4**, 119–126.
13 S. M. Hashmi and E. R. Dufresne, *Soft Matter*, 2009, **5**, 3682–3688.
14 A. Fernández-Nieves, A. Fernández-Barbero, B. Vincent and F. J. de las Nieves, *J. Chem. Phys.*, 2003, **119**, 10383.
15 M. Y. Guo and H. M. Wyss, *Macromol. Mater. Eng.*, 2011, **296**, 223–229.
16 H. M. Wyss, T. Franke, E. Mele and D. A. Weitz, *Soft Matter*, 2010, **6**, 4550–4555.
17 C. I. Mendoza, *J. Chem. Phys.*, 2011, **135**, 054904.
18 H. Ohshima, *Langmuir*, 2009, **26**, 6287–6294.
19 H. Ohshima, *Colloids Surf., A*, 2009, **347**, 33–37.
20 T. Gao and H. H. H. P. P. Castañeda, *J. Fluid Mech.*, 2011, **687**, 209–237.
21 M. Whittle and E. Dickinson, *J. Chem. Soc., Faraday Trans.*, 1998, **94**, 2453–2462.
22 M. Whittle and E. Dickinson, *J. Chem. Phys.*, 1997, **107**, 10191–10200.
23 A. A. Rzepiela, J. H. J. van Opheusden and T. van Vliet, *Comput. Phys. Commun.*, 2002, **147**, 303–306.
24 A. A. Rzepiela, J. H. J. van Opheusden and T. van Vliet, *J. Rheol.*, 2004, **48**, 863–880.
25 L. Berthier, E. Flenner, H. Jacquin and G. Szamel, *Phys. Rev. E: Stat., Nonlinear, Soft Matter Phys.*, 2010, **81**, 31505.
26 L. Berthier, A. Moreno and G. Szamel, *Phys. Rev. E: Stat., Nonlinear, Soft Matter Phys.*, 2010, **82**, 060501.
27 J. Yang and K. S. Schweizer, *J. Chem. Phys.*, 2011, **134**, 204908.
28 J. Yang and K. S. Schweizer, *J. Chem. Phys.*, 2011, **134**, 204909.
29 J. R. Seth, L. Mohan, C. Locatelli-Champagne, M. Cloitre and R. T. Bonnecaze, *Nat. Mater.*, 2011, **10**, 838–843.
30 M. Grmela, *J. Non-Newtonian Fluid Mech.*, 1997, **69**, 105–107.
31 R. J. J. Jongschaap and H. C. Öttinger, *J. Non-Newtonian Fluid Mech.*, 2001, **96**, 1–3.
32 M. Grmela and H. C. Öttinger, *Phys. Rev. E: Stat. Phys., Plasmas, Fluids, Relat. Interdiscip. Top.*, 1997, **56**, 6620–6632.
33 H. C. Öttinger and M. Grmela, *Phys. Rev. E: Stat. Phys., Plasmas, Fluids, Relat. Interdiscip. Top.*, 1997, **56**, 6633–6655.
34 H. C. Öttinger, *Beyond Equilibrium Thermodynamics*, Wiley, Hobroken, 2005.
35 M. Hütter and L. C. A. van Breemen, *J. Appl. Polym. Sci.*, 2012, **125**, 4376–4389.
36 C. Truesdell and W. Noll, *The Non-Linear Field Theories of Mechanics*, Springer, Berlin, 1992.
37 J. F. Besseling and E. Van der Giessen, *Mathematical Modelling of Inelastic Deformation*, Chapman and Hall, London, 1994, vol. 5.
38 M. Šilhavý, *The Mechanics and Thermodynamics of Continuous Media*, Springer, 1997.
39 R. B. Bird, C. F. Curtiss, R. C. Armstrong and O. Hassager, *Dynamics of Polymeric Liquids*, John Wiley & Sons, New York, 2nd edn, 1987, vol. 2: Kinetic Theory.
40 M. S. Green and A. V. Tobolsky, *J. Chem. Phys.*, 1946, **14**, 80–92.
41 M. Doi and S. F. Edwards, *J. Chem. Soc., Faraday Trans. 2*, 1978, **74**, 1789.
42 M. Doi and S. F. Edwards, *J. Chem. Soc., Faraday Trans. 2*, 1978, **74**, 1802.
43 M. Hütter, C. Luap and H. C. Öttinger, *Rheol. Acta*, 2009, **48**, 301–316.
44 S. Chandrasekhar, *Rev. Mod. Phys.*, 1943, **15**, 1–89.
45 E. Nelson, *Dynamical Theories of Brownian Motion*, Princeton University Press, Princeton, 1967.
46 L. Waldmann, *Z. Naturforsch. A*, 1959, **14**, 589–599.
47 M. Hütter and M. Kröger, *J. Chem. Phys.*, 2006, **124**, 044511.
48 W. Gardiner, *Handbook of Stochastic Methods for Physics, Chemistry and the Natural Sciences*, Springer, Berlin, 2nd edn, 1985.

49 P. E. Kloeden and E. Platen, *Numerical Solution of Stochastic Differential Equations*, Springer, Berlin, Heidelberg, 1992.
50 H. C. Öttinger, *Stochastic Processes in Polymeric Fluids*, Springer, Berlin, 1996.
51 S. R. de Groot and P. Mazur, *Non-equilibrium Thermodynamics*, North-Holland, Amsterdam, 1962.
52 C. N. Likos, *Phys. Rep.*, 2001, **348**, 267–439.
53 W. B. Russel, D. A. Saville and W. R. Schowalter, *Colloidal Dispersions*, Cambridge University Press, 1989.
54 J. Israelachvili, *Intermolecular and Surface Forces*, Academic Press, London, 2nd edn, 1991.
55 K. L. Johnson, *Contact Mechanics*, Cambridge University Press, Cambridge, 1985.
56 B. V. Derjaguin and L. D. Landau, *Acta Physicochim. USSR*, 1941, **14**, 633.
57 E. J. W. Verwey and J. T. Overbeek, *Theory of the Stability of Lyophobic Colloids*, Elsevier, Amsterdam, 1948.
58 R. Kubo, M. Toda and N. Hashitsume, *Nonequilibrium Statistical Mechanics, vol. II of Statistical Physics*, Springer, Berlin, 2nd edn, 1991.
59 D. J. Evans and G. P. Morriss, *Statistical Mechanics of Nonequilibrium Liquids*, Academic Press, London, 1990.
60 S. Kim and S. J. Karrila, *Microhydrodynamics: Principles and Selected Applications*, Butterworth-Heinemann, Stoneham MA, 1991.
61 H. P. G. Darcy, *Les fontaines publiques de la ville de Dijon*, Victor Dalmont, Paris, 1856.
62 S. Irmay, *Trans. Am. Geophys. Union*, 1958, **39**, 702–707.
63 M. Doi and T. Ohta, *J. Chem. Phys.*, 1991, **95**, 1242–1248.
64 E. D. Wetzel and C. L. Tucker, *Int. J. Multiphase Flow*, 1999, **25**, 35–61.
65 C. L. Tucker and P. Moldenaers, *Annu. Rev. Fluid Mech.*, 2002, **34**, 177–210.
66 N. E. Jackson and C. L. Tucker, *J. Rheol.*, 2003, **47**, 659–682.
67 C. Beisbart, R. Dahlke, K. Mecke and H. Wagner, in *Lecture Notes in Physics*, Springer, Berlin, 2002, vol. 600, ch. Vector- and tensor-valued descriptors for spatial patterns, pp. 238–260.

Nanoscale characteristics of triacylglycerol oils: phase separation and binding energies of two-component oils to crystalline nanoplatelets

Colin J. MacDougall,[a] M. Shajahan Razul,[a] Erzsebet Papp-Szabo,[b] Fernanda Peyronel,[c] Charles B. Hanna,[d] Alejandro G. Marangoni[c] and David A. Pink[*ac]

Received 28th February 2012, Accepted 26th March 2012
DOI: 10.1039/c2fd20039b

Fats are elastoplastic materials with a defined yield stress and flow behavior and the plasticity of a fat is central to its functionality. This plasticity is given by a complex tribological interplay between a crystalline phase structured as crystalline nanoplatelets (CNPs) and nanoplatelet aggregates and the liquid oil phase. Oil can be trapped within microscopic pores within the fat crystal network by capillary action, but it is believed that a significant amount of oil can be trapped by adsorption onto crystalline surfaces. This, however, remains to be proven. Further, the structural basis for the solid–liquid interaction remains a mystery. In this work, we demonstrate that the triglyceride liquid structure plays a key role in oil binding and that this binding could potentially be modulated by judicious engineering of liquid triglyceride structure. The enhancement of oil binding is central to many current developments in this area since an improvement in the health characteristics of fat and fat-structured food products entails a reduction in the amount of crystalline triacylglycerols (TAGs) and a relative increase in the amount of liquid TAGs. Excessive amounts of unbound, free oil, will lead to losses in functionality of this important food component. Engineering fats for enhanced oil binding capacity is thus central to the design of more healthy food products. To begin to address this, we modelled the interaction of triacylglycerol oils, triolein (OOO), 1,2-olein elaidin (OOE) and 1,2-elaidin olein (EEO) with a model crystalline nanoplatelet composed of tristearin in an undefined polymorphic form. The surface of the CNP in contact with the oil was assumed to be planar. We considered pure OOO and mixtures of OOO + OOE and OOO + EEO with 80% OOO. The last two cases were taken as approximations to high oleic sunflower oil (HOSO). The intent was to investigate whether phase separation on a nanoscale took place. We defined an "oil binding capacity" parameter, $B(Q,Q')$, relating a state Q to a reference state Q'. We used atomic scale molecular dynamics in the NVT ensemble and computed averages over 1–5 ns. We found that the probability of the OOE phase separating into a layer on the surface of the CNP compared to being retained randomly in an OOO + OOE mix were approximately equal. However, we found that it was probable that the EEO component of an OOO + EEO mix would phase separate and coat the surface of the CNP. These results suggest a mechanism whereby

[a]Physics Department, St.Francis Xavier University, Antigonish, NS, Canada
[b]Physics Department, University of Guelph, Guelph, ON, Canada
[c]Guelph-Waterloo Center for Graduate Work in Physics, Department of Food Science, University of Guelph, Guelph, ON, Canada
[d]Department of Physics, Boise State University, Boise, ID 83725-1570, USA

many-component oils undergo phase separation on a nanoscale so as to create a transition oil region between the surface of the CNP and the bulk major oil component (OOO in the case considered here) so as to create the appropriate oil binding capacity for the use to which it is put.

Introduction

Triacylglycerol (TAG) polycrystalline particles in TAG oils find extensive use in the food industry. Although modelling studies of TAG systems have been carried out,[1-4] none have attempted to address a problem of central concern to food science and technology: the "oil binding capacity" of a system. In a many-component oil the existence of a solid fat surface could induce phase separation on a nanoscale in the oil. Here, we investigate whether the surface of a crystalline fat particle can induce such nanoscale phase separation (nanophase separation) and we relate this to oil binding capacity.

The oils with which we are concerned are those composed of a variety of triacylglycerols. Accordingly, the possibility that separation of their components into many coexisting phases—possibly microphases, depending upon the environment—separated by interfaces of different characteristics and complexities, cannot be passed over. Indeed it is possible that it is exactly this phenomenon which makes different complex oils suitable for a range of different food preparation techniques.[5]

Recently, Acevedo and Marangoni[6-8] identified the seemingly-fundamental components making up solid fat structures at ∼300 K. These appear to be highly-anisoptropic crystalline nanoplatelets (CNPs) with dimensions approximately 500 nm × 200 nm × 50 nm. If these CNPs are indeed the fundamental components from which all larger scale fat structures are constructed, then it is important to understand how they interact with the surrounding oils and with each other.

These interactions are all Coulombic of which the major components are van der Waals interactions though in cases where hydrogen bonding might play a role, electrostatic interactions become important.

The question of exactly what determines oil binding capacity is related to that concerning what are the essential aspects which drive the self-assembly of microscopic fat structures from nanoscale CNPs. Much work has been done on the self-assembly of membranes, vesicles and micelles from phospholipids[9-15] and work has been done on the related area of polar food lipids.[16] However, the question of what structures arise from the interaction of CNPs with each other and with the oils in which they are embedded, and the pathways through which self-assembly comes about, is essentially completely unresolved.

Understanding the effects of the Coulombic interactions will enable advances in obtaining a picture of exactly what it is that determines self-assembly of fat particles and "oil binding capacity", and how one can use this picture to design oils in which the *trans* components are reduced. In this paper we are concerned with oil binding capacity as defined by the "binding energy" (below) of a given mix of oils. Here we present results obtained using atomic scale molecular dynamics to compute the chemical potential of a mixed triacylglycerol oil phase interacting with a solid triacylglycerol CNP. We chose an oil comprising ∼80% of triolein, which we refer to as the "majority component" or "majority phase", together with ∼20% of a dielaidin-olein or a diolein-elaidin oil which we refer to as the "minority component" or "minority phase". We refer to this as an "80 : 20 mix". This approximates high oleic sunflower oil (HOSO) which has a minimum of 80% oleic acid.[17] Our intent was 2-fold: (1) to establish whether phase separation of an 80 : 20 mix was likely at ∼300 K and we studied this by comparing the chemical potentials of a random mix of the two molecules and a phase-separated system in which the minority component was located at the surface of the CNP. (2) If phase separation did occur then we wanted to know what is the binding energy of the dielaidin-olein and diolein-elaidin oils at the solid fat surface.

Here we shall denote triolein by OOO, diolein-elaidin by OOE and dielaidin-olein by EEO. The particular realizations are shown in Fig. 1. The approach that we shall use is to calculate the chemical potentials of 80 : 20 mixes of OOO–OOE and OOO–EEO in two cases: (a) a random mix of the two molecules and (b) a completely phase-separated system in which the minority phase is located at the surface of the CNP. We shall then compute the chemical potential of a pure OOO oil. We shall then interpret the differences in the chemical potentials to be the binding energies of the systems. We shall take the CNPs to be composed of fully hydrogenated canola oil (FHCO) which is ~80% tristearin.

Theory

The model

Fig. 1 shows schematic diagrams of a majority phase triolein molecule (Fig. 1A), and the two minority-phase molecules, diolein-elaidin (Fig. 1B) and dielaidin-olein (Fig. 1C). There we see the *trans* or *cis* double bond at C_9 on the elaidic and oleic chains. We chose the configurations, B and C, in order to examine whether there is any difference in having the two elaidic or two oleic chains in the 1,2 positions. Our hypothesis is that molecules with elaidic chains will be more tightly bound to the CNP surface than those with only oleic chains and that the binding energy of EEO molecules will be stronger than that of OOE molecules.

Fig. 2 shows schematic diagrams of the three cases that we simulated. There we see a single fat CNP composed of tristearin molecules in contact with an oil composed of pure triolein (OOO) (2A), (B) a random mix of OOE or EEO with OOO (2B), and the case in which complete phase separation has taken place with OOE or EEO molecules in contact with the surface of the CNP and OOO filling the remainder of the space. The CNP possess a flat surface lying in the $x - y$ plane. The z-axis is perpendicular to this plane and the CNP occupies the half-space $-\infty < x,y < \infty$ and $-\infty < z \leq 0$. Fig. 2 also shows the simulation space comprising two regions: one of length S and a shorter one of length L. The simulation was carried out in the space of length S. The surface on the right hand end is defined by only a short-range repulsive interaction in order to keep the lipid molecules inside the simulation volume. However, that surface will introduce correlations not characteristic of a volume of infinite extent. In order that these correlations play no role in the averaging procedures, we restricted the volume inside which averages would be computed to that enclosed by length L. The lateral dimensions of the simulation volume in the $x - y$ plane are $d \times d$ with a cross section area, $A = d^2$.

Interaction energies

The Lennard-Jones (L-J) 6-12 potential, $u_{AB}(r)$, describing the van der Waals interaction between two spherically-symmetric atomic moieties, A and B, a centre-to-centre distance r apart is,[18,19]

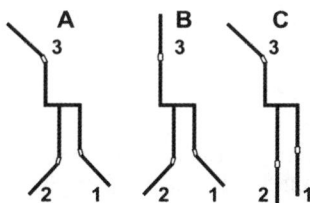

Fig. 1 Three triacylglycerols modelled. A: OOO. B: OOE with an elaidic chain in the 3 position. C: EEO with an oleic chain in the 3 position.

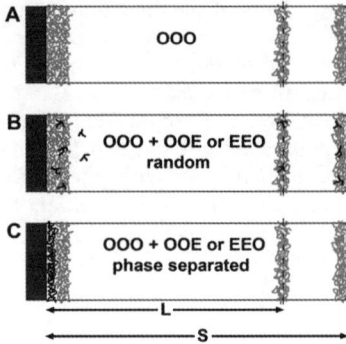

Fig. 2 Model system of a flat CNP surface in contact with an oil. The solid black region indicates the tristearin CNP. A: pure triolein (pale grey). B: triolein containing randomly distributed OOE or EEO molecules (dark grey). C: triolein with OOE or EEO molecules (dark grey) located at the CNP surface. S and L indicate the regions in which the simulations were carried out, and in which averaging took place, respectively.

$$u_{AB}(r) = -\frac{C_6^{AB}}{r^6} + \frac{C_{12}^{AB}}{r^{12}} \qquad (1)$$

We represented the CNP as a solid continuum possessing the density of crystalline fully hydrogenated coconut oil (FHCO) as measured by us (below). The CNP interacted with the molecules of the oil with the coefficients of eqn (1), C_6^{AB} and C_{12}^{AB}, averaged over the atomic moieties of the CNP. The averaged coefficients describing the interactions between an infinitesimal volume of a solid CNP and a B atomic moiety in the liquid oil was

$$\langle C_m^B \rangle = \sum_A n_A C_m^{AB} \quad m = 6,12 \qquad (2)$$

where n_A is the fraction of A atomic moieties of the molecules in the CNP. This enabled us to represent the interaction energy, $U_B(S,z)$, between the CNP and an atomic moiety, B, of the oil, possessing a z-coordinate z ($0 < z < S$) as

$$U_B(S,z) = \left[-\frac{\pi \langle C_6^B \rangle}{6z^3} + \frac{\pi \langle C_{12}^B \rangle}{45} \left(\frac{1}{z^9} + \frac{1}{(S-z)^9} \right) \right] \Phi_1 A \qquad (3)$$

Here A is the area (in nm²) of the face of the CNP, Φ_1 is the number density of atomic moieties nm⁻³ in the crystalline solids, and the total energy of the solid–oil interaction can be obtained by summing over all oil atomic moieties, B. The short-range repulsive term in $1/(S-z)^9$ of eqn (3) represents the right hand edge of the simulation box. Here we have simply represented it as a soft wall which has no attractive interaction with the oil molecules and we shall choose S sufficiently large so that the effect of the wall has vanished over 1–2 nm. Accordingly, we choose L so that it is sufficiently smaller than S. Below we refer to the S- and L-regions as those volumes AS and AL respectively.

Oil binding capacity

This concept is associated with the ability of solids to retain liquid oil so that it does not leak away. One can quantify it by relating it to the volume of oil that is retained by the solids in a system, normalized by the volume of oil originally in the system. Here, however, we need to define a parameter that we can relate to the properties of the system. The amount of oil retained by the solids in the system is determined by the chemical potential, μ_Q, of the system labelled as Q. We label the five systems that we have identified above as follows: $Q = [OOO]$ represents the system with pure OOO, $Q = [OOO,E,c,1]$ and $Q = [OOO,E,c,2]$ represents the mix of OOO and concentration c of OOE in a random mix (1) or phase separated (2), and $Q = [OOO,EE,c,1]$ and $Q = [OOO,EE,c,2]$ represents the mix of OOO and concentration c of EEO in a random mix (1) or phase separated (2).

We define the oil binding capacity parameter, $B(Q,Q')$, of system Q compared to that of system Q' as,

$$B(Q, Q') = \exp[-\Delta\mu(Q, Q')/k_B T]|$$
$$\Delta\mu(Q, Q') = \mu_Q - \mu_{Q'} \tag{4}$$

where k_B is Boltzmann's constant and T is the absolute temperature. It is clear the following inequality holds,

$$0 \leq B(Q,Q') < \infty \tag{5}$$

In our cases we shall compare the chemical potentials of the four mixed oils with that of OOO, which we take as the reference system, Q'.

Computer simulation

Atomic scale molecular dynamics (AMD)

Atomic scale molecular dynamics (AMD) is concerned with simulating the motion of atomic nuclei, or atoms as a whole, in contrast to other methods, such as the Car–Parrinello technique[20] which includes electronic degrees of freedom. In AMD, electronic "motion" is assumed to take place on a time-scale much faster than the motion of the atom as a whole so that an average electronic distribution can be assumed. AMD is concerned with weak physical interactions and is not concerned with chemical reactions involving energies that can break or form covalent bonds. The technique defines sets of forces ("forcefields") that act between pairs of atoms or small atomic moieties. The atoms or atomic moieties are characterized by fixed (average) point partial electric charges located inside a spherical volume which defines the size of the atomic moiety. Distinction is made between "bonded" and "non-bonded" atoms: a pair of bonded atoms possess a permanent covalent bond between them. All other pairs of atoms are "non-bonded". The forces to be used do not act between pairs of "bonded" atoms. The net force, $\vec{F}(t)$, acting at time t on an atom with mass m defines its acceleration, $\vec{a}(t)$, via Newton's equation, $\vec{a}(t) = \vec{F}(t)/m$, which is then integrated to obtain the velocity, $\vec{v}(t + \Delta t)$, and position, $\vec{r}(t + \Delta t)$, after a pre-selected elapsed "time-step", Δt. From the new position, $\vec{r}(t + \Delta t)$, one calculates the net force, $\vec{F}(t + \Delta t)$, acting on the atom at time $t + \Delta t$ and repeats the procedure. The magnitude of Δt used in AMD is typically ~ 1–10×10^{-15} s (1–10 fs). Simulations are generally carried out under one of the following conditions ("ensembles"): microcanonical (constant number of particles, N, constant volume, V, constant energy, E), canonical (constant N, constant V, constant temperature, T) or the isothermal-isobaric (constant N, constant T, constant pressure, p).

Here we report AMD simulations performed using NVT ensembles with periodic boundary conditions employing the force field given by Berger et al.[21] and using the V-rescaled thermostat.[22] We used eqn (1) and represented the solids as continua, but modelled the liquid as a molecular fluid using AMD. We modelled the radial interactions between 'non-bonded' triacylglycerol (TAG) atomic moieties using GRO-MACS.[23,24] TAGs possess only CH, CH_2, CH_3, O, and C=O moieties, and Tables are available which list the coefficients C_6^{AB} and C_{12}^{AB} that define the L-J potential between given pairs of these atomic moieties.

We used the *NVT* ensemble rather than the *NpT* ensemble because we wished to use the experimentally known density for HOSO oil of 0.9016×10^{-21} gm nm^{-3} as measured by us using a Mettler Toledo DE 40 density meter. This gave an oil number density of $\Phi_2 \approx 38.5$ moieties nm^{-3}. We used a pyctometer, to measure the density of FHCO and obtained a value of 1.03×10^{-21} gm nm^{-3} which yielded a CNP number density of $\Phi_1 \approx 44.0$ moieties nm^{-3}.

Choice of parameters

We chose the lateral dimension of the simulation volume to be $d = 5.662$ nm so that the area is $A = 32.058$ nm^2. We chose $S = 10.00$ nm and $L = 6.00$ nm. The volume of the *S*-region accommodated a total of 200 molecules of which 40 were OOE or EEO. For case B where the minority phase molecules are randomly distributed, we have ~24 OOE or EEO molecules in the *L*-region, on the average. When we studied case C, we placed all ~24 such molecules from the *L*-region onto the CNP surface and allowed them to move. We ran the simulation for 10 ns and carried out averaging only in the *L*-region. Whether a molecule was to be included in the averaging was determined by where its "central" CH moiety was located: if it was located inside the *L*-region, then all of the molecule was included in the averaging procedure.

Results and discussion

After initialization, the simulation was run for 10 ns and Fig. 3 shows the free-potential-energy-per-molecule, which we define to be the chemical potential, in kJ mole^{-1}, computed over the final n nanoseconds of the simulation, where $n = 1, 2, 3, 4, 5$. We denote this elapsed time by Δt. The chemical potentials are as defined above, $\Delta\mu(Q,Q') = \mu_Q - \mu_{Q'}$, for cases $Q = [OOO,E,c,1]$ (squares, a), $Q = [OOO,E,c,2]$ (circles, b), $Q = [OOO,EE,c,1]$ (triangles, c) and $Q = [OOO,EE,c,2]$ (triangles, d) with $c = 0.20$. The reference state Q' is that with pure OOO. Error bars are shown for 1, 3 and 5 ns (e). There are two points to note: (1) the chemical potentials for the two cases involving the mix of OOO and OOE are within the RMS fluctuations and so we conclude that OOE molecules are essentially equally likely to be found randomly mixed with OOO, or expelled from the OOO "sea" and attached to the surface of the solid CNP. (2) The difference in the chemical potentials of the random mix of OOO and EEO is ~1–2 kJ mole^{-1} higher than that of the case in which the EEO molecules are expelled from the OOO and coat the surface of the solid CNP.

The latter result reflects the fact that two extended E-chains can align themselves with the CNP surface, whereas a molecule with a single E-chain is equally at home with the surface or the OOO.

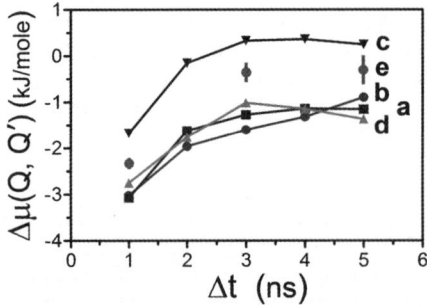

Fig. 3 Chemical potentials, $\Delta\mu(Q,Q') = \mu_Q - \mu_{Q'}$, for cases $Q = [OOO,E,c,1]$ (squares, a) and $Q = [OOO,E,c,2]$ (circles, b), and $Q = [OOO,EE,c,1]$ (triangles, c) and $Q = [OOO,EE,c,2]$ (triangles, d). The state Q' is that with pure OOO. Error bars for 1, 3 and 5 ns are inserted (e).

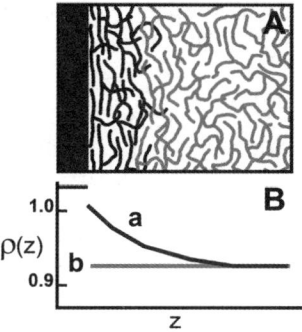

Fig. 4 Schematic diagram of a mechanism to create high oil binding capacity. A. The many-component oil (shown in grey scale) undergoes nanophase separation at the surface of the solid (dark grey). B. The density of the layers (a) exhibits an approximately continuous decrease towards the density of the bulk oil (b).

The results shown in Fig. 3 enable us to answer the questions that we raised in the introduction: (1) to establish whether phase separation of an 80 : 20 mix was likely at ~300 K and (2) if phase separation did occur then what is the binding energy of the dielaidin-olein and diolein-elaidin oils at the solid fat surface. We see that phase separation is likely only for the OOO + EEO mix so that the CNP surface becomes coated with EEO molecules. The oil-binding capacity for such an oil compared to that of pure OOO can be calculated using eqn (4). For the system, $Q = $ OOO, we obtain, of course, $B(Q',Q') = 1.00$. When $Q = $ [OOO,EE,c,2] with a binding free energy of $\Delta\mu(Q,Q') = -1.363$ kJ mole^{-1} and using that $k_B T = 2.494$ kJ mole^{-1}, we obtain $B(Q,Q') = 1.73$. That is, the oil binding capacity with a layer of EEO on the surface of the solid CNP is 1.73 times greater than if the oil is pure OOO and there is no EEO present.

The results of our simulation suggests a mechanism by which fats could exhibit the appropriate "oil binding capacity": the oil components undergo phase separation at the nanoscale in the neighbourhood of the surfaces of solid fat CNPs. This nano-phase separation arises in a layer-like arrangement with those oil components which possess hydrocarbon chains that best "fit" the local CNP surface being closest to it and the other components "coating" the deposited layers, until those components which exhibit the worst "fit" to the surface is reached. This is shown schematically in Fig. 4A. Fig. 4B shows the approximate density (a) as a function of distance, z, from the CNP surface. The horizontal line (b) shows the density of the "bulk" oil and it can be seen that there is a mismatch in density at the interface.

We have shown elsewhere[25] that a system composed of tristearin (SSS) CNPs and pure OOO oil possesses a low oil-binding capacity and that CNPs can aggregate into needle-like structures with essentially no oil between them. This is brought about by the mismatch in the density of OOO and solid SSS CNPs. Accordingly, with the results obtained here, it is reasonable to propose the system of nanophase separation shown in Fig. 4.

Conclusions

We have used atomic scale molecular dynamics (AMD) to model triacylglycerol (TAG) oils in contact with a solid planar crystalline nanoplatelet (CNP) surface. We employed GROMACS and carried out simulations for 5 oil systems: a single CNP composed of tristearin (SSS) molecules in contact with an oil composed of pure triolein (OOO), a random mix of OOE or EEO with OOO, and the case in which complete phase separation has taken place with OOE or EEO molecules in contact with the surface of the CNP and OOO filling the remainder of the space.

We employed the NVT ensemble together with periodic boundary conditions in the $x - z$ and $y - z$ planes where the z-axis is perpendicular to the planar surface of the CNP.

We found that:

(1) At 300 K, a 20 : 80 mix of EEO and OOO will phase separate with the former coating the surface of the SSS CNP whereas a similar mix of OOE and OOO has approximately equal probabilities of phase separating or not.

(2) The binding energy of the nanophase separated 20 : 80 mix of EEO and OOO at 300 K, defined as the difference in the chemical potentials of the phase separated phase and the phase comprising a random mix of EEO and OOO, is $\sim$1.36 kJ mole^{-1}. This should be compared to thermal energies at 300 K of $\sim$2.5 kJ mole^{-1}.

Our results have led us to propose a mechanism by which oils could exhibit the appropriate "oil binding capacity" for a given food system: the oil components undergo phase separation on a nanoscale in the neighbourhood of the surfaces of solid fat CNPs. This nanophase separation could arise as a layer-like arrangement with those oil components which possess hydrocarbon chains that best "fit" the local CNP surface being closest to it and the other components "coating" the deposited layers, until those components which exhibit the worst "fit" to the surface is reached. This scenario is under investigation.

Acknowledgements

This work was supported by NSERC of Canada and ACEnet.

References

1 Z.-Y. Yan, S. D. Huhn, L. P. Klemann and M. S. Otterburn, *J. Agric. Food Chem.*, 1994, **42**, 447.
2 A. K. Sum, M. J. Biddy and J. J. dePablo, *J. Phys. Chem. B*, 2003, **107**, 14443.
3 A. Hall, J. Repakova and I. Vattulainen, *J. Phys. Chem. B*, 2008, **112**, 13772.
4 W.-D. Hsu and A. Violi, *J. Phys. Chem. B*, 2009, **113**, 887.
5 L. H. Wesdorp, J. A. van Meeteren, S. de Jong, R. V. D. Giessen, P. Overbosch, P. A. Grootscholten, M. Struik, E. Royens and A. Don. 2005. In *Fat Crystal Networks*, ed. Marangoni, A. G., New York, Marcel Dekker Inc., pp 481–709.
6 N. C. Acevedo and A. G. Marangoni, *Cryst. Growth Des.*, 2010, **10**, 3327.
7 N. C. Acevedo and A. G. Marangoni, *Cryst. Growth Des.*, 2010, **10**, 3334.
8 N. C. Acevedo, F. Peyronel and A. G. Marangoni, *Curr. Opin. Colloid Interface Sci.*, 2011, **16**, 374.
9 A. R. Patel and K. P. Velikov, *LWT–Food Sci. Technol.*, 2011, **44**, 1958.
10 D. J. McClements and Y. Li, *Adv. Colloid Interface Sci.*, 2010, **159**, 213.
11 M. Fathi, M. R. Mozafari and M. Mohebbi, *Trends Food Sci. Technol.*, 2012, **23**, 13–27.
12 J. Weiss, S. Gaysinsky, M. Davidson and J. McClements In *Global Issues in Food Science and Technology*, 2009, Chapter 24, 425.
13 F. E. Antunes, E. F. Marques, M. G. Miguel and B. Lindman, *Adv. Colloid Interface Sci.*, 2009, **147–148**, 18.
14 K. Larsson, *Curr. Opin. Colloid Interface Sci.*, 2009, **14**, 16.
15 L. Sagalowicz and M. E. Leser, *Curr. Opin. Colloid Interface Sci.*, 2010, **15**, 61.
16 M. E. Leser, J. Sagalowicz, M. Michel and H. J. Watzke, *Adv. Colloid Interface Sci.*, 2006, **123–126**, 125.
17 British Pharmacopoeia Commission, "Ph Eur monograph 1371", *British Pharmacopoeia*, 2005, Norwich, England, The Stationery Office, ISBN 0-11-322682-9.
18 V. A. Parsegian *Van der Waals Forces*, Cambridge University Press, 2005.
19 J. N. Israelachvili, *Intermolecular and Surface Forces*, Academic Press, 2006.
20 R. Car and M. Parrinello, *Phys. Rev. Lett.*, 1985, **55**, 2471. See also http://www.cpmd.org/.
21 J. Berger, O. Edholm and F. Jähnig, *Biophys. J.*, 1997, **72**, 2002.
22 B. Hess, C. Kutzner, D. van der Spoel and E. Lindahl, *J. Chem. Theory Comput.*, 2008, **4**, 435. See also www.gromacs.org, GROMACS users manual version 4.0, gromacs4_manual.pdf.
23 H. J. C. Berendsen, D. Van der Spoel and R. Vandrunen, *Comput. Phys. Commun.*, 1995, **91**, 43.

24 C. Kutzner, D. Van der Spoel, M. Fechner, E. Lindahl, U. W. Schmitt, B. L. De Groot and H. Grubmuller, *J. Comput. Chem.*, 2007, **28**, 2075.
25 D. A. Pink, M. S. G. Razul, C. J. MacDougall, F. Peyronel, C. B. Hanna and A. G. Marangoni, Nanoscale Characteristics of Molecular Fluids in Confined Spaces: Triacylglycerol Oils, unpublished.

Soft matter approaches as enablers for food macroscale simulation

Ashim K. Datta,[*a] Ruud van der Sman,[b] Tushar Gulati[a] and Alexander Warning[a]

Received 2nd March 2012, Accepted 14th May 2012
DOI: 10.1039/c2fd20042b

Macroscopic deformable multiphase porous media models have been successful in describing many complex food processes. However, the properties needed for such detailed physics-based models are scarce and consist of primarily empirical models obtained from experiment. Likewise, driving forces such as swelling pressure have also been approached empirically, without physics-based explanations or prediction capabilities. Soft matter based prediction of properties will provide an additional avenue to obtaining properties and also provide a deeper and critical understanding of how these properties change with composition, temperature and other process variables.

1 Introduction

Physics-based (as opposed to empirical) models of food processes have tremendous potential in terms of providing insight into complex processes and the developing capabilities to pre-optimize in the design of product, processes and equipment. For solid and semi-solid foods, a modeling framework[8,9] that considers homogenized macroscale multiphase transport in the food as a deformable/swellable porous medium has been successful in modeling a number of important processes including drying, rehydration,[83] baking,[49,86] frying,[24,85] meat cooking,[9] microwave heating,[48] and microwave puffing.[59] This approach recognizes that foods are commonly structured materials. The multiple phases are assumed to be homogenously mixed, and are represented *via* their volume fractions. Transport modes of each phase are related to clearly identifiable driving forces and transport parameters, such as molecular diffusion in the gas phase, capillarity in the liquid phase.

One of the difficulties in large scale use of the above macroscale homogenized porous-media based approach is the lack of material properties and closure relations such as driving forces and equilibrium relationships that are also based on thermodynamic rigor. Here, soft matter approach[45,68,74] can come to the rescue. Soft matter is a branch of physics dealing with materials having dispersed phases of micron size, and which yield easily if external force is applied (which makes them soft). Almost any food can be classified as soft matter. The soft matter approach can be characterized by (1) the conception that physics of soft matter at and beyond the length scale of the dispersed phase is largely independent of the chemical details of the molecules, and (2) the rigorous application of thermodynamics. Soft matter scientists predominantly focus on the mesoscale of the dispersed phase, where physics is governed by a balance between thermal fluctuations and external applied fields, which explain the rigor in thermodynamics. Application of the soft matter approach

[a] Riley Robb Hall, Cornell Univ., Ithaca, NY, USA. E-mail: akd1@cornell.edu; Fax: +1 607 255 4449; Tel: +1 607 255 2482
[b] Agrotechnology Food Sciences Group, Wageninen University, the Netherlands

has been successful in modeling food structuring[34,70] and predicting the properties and closure relations needed in the macroscale porous media models. For example, the Free Volume Flory–Huggins (FVFH) theory has been successfully applied to explain moisture sorption and phase transitions in mixtures of water and polysaccharides and proteins[75,76] and its extension to compute the water holding capacity in meat proteins.[71,72,76]

However, the porous media approach cannot be readily coupled to the results obtained by food scientists following the soft matter approach. The tools developed and the results produced in soft matter approach have been intended for broader understanding, not necessarily geared toward (or restricted to) complementing macroscale porous media transport models. The purpose of this article is to show how to bridge between the macroscale porous media and the soft matter approaches, which will end in a complementary and synergistic relation between the two, in the context of foods. This will be accomplished by providing specific examples of transport parameters and closure relations that can be developed from a soft matter approach, showing relationships between the macroscale and the soft matter approach, and by applying the combined macroscale and soft matter framework to a real food process.

The article is organized as follows: a short introduction is provided of the macroscale porous media modeling framework. Generalized fluid pressure as a driving force (a closure relation) in macroscale porous media is developed from synthesis of work in many disciplines and broad classes of materials, relating it to the thermodynamic quantities recognized by the soft matter approach, and making it broadly applicable to food processes. How the fluid pressure as a driving force can be included in the food process modeling framework is conjectured. The soft-matter approach to the prediction of three of the transport parameters needed in macroscale models, intrinsic permeability, molecular diffusivity and capillary diffusivity, is developed, showing preliminary prediction results. A proposed framework for the prediction of mechanical properties during food processing is provided next. Finally, examples of the application of combined soft matter and macroscale porous media modeling approaches to two food processes (rehydration of freeze-dried vegetables and cooking of meat) are provided.

2 Formulations and alternatives

2.1 Macroscale multiphase porous media formulations

Food process models that are based on multiphase transport in a porous medium have typically used the common volume averaged equations,[84] although the linkage to averaging process may not always be made explicit. The food matrix is mostly considered rigid, although deformable porous media has been considered; the relevant equations are provided in detail in Datta.[6,7,9] The phases considered for a solid food are solid, liquid (e.g., water, oil), and gas (e.g., water vapor, carbon dioxide, nitrogen, ethylene). Evaporation is considered as either distributed throughout the domain or at an evaporating interface and is dictated by the local equilibrium between the liquid and vapor phases. The transport mechanisms considered are capillarity and gas pressure (due to evaporation) for liquid transport, and molecular diffusion and gas pressure for vapor and air transport. The pressure driven flow is modeled using Darcy's law when the permeability is small (pores are small, including possible Knudsen effects) or its more general Navier–Stokes analog when the matrix is very permeable.[30] A local thermal equilibrium, where all phases share the same temperature at a location, is assumed, leading to one energy equation. The final governing equations for a rigid matrix consist of one energy equation, one mass balance equation and either the Darcy's law or the Navier–Stokes for the momentum equation for each of the fluid phases. In addition, there will be transport equations for each solute component, such as flavor.

2.2 Multiphase, rigid porous media continuum equations

An enormous range of food processes can be viewed as involving the transport of heat and mass through deformable porous media (Fig. 1). Examples include drying, frying, microwave heating, meat roasting, puffing, and rehydration of solids. Most solid food materials can be treated as hygroscopic and capillary-porous. A porous media formulation homogenizes the real porous material and treats it as a continuum, where the pore scale information is no longer available. Since this framework can cover a large range of situations and can be simplified to various levels, the general set of equations are presented first, followed by various adaptations that are the simplified versions.

A general version of the conservation equations for rigid porous media are provided below for the individual phases of water, vapor, air, and a combined energy equation for all phases:

$$\frac{\partial c_w}{\partial t} + \nabla \cdot (\vec{n}_w) = -\dot{I} \tag{1}$$

$$\frac{\partial c_v}{\partial t} + \nabla \cdot (\vec{n}_v) = \dot{I} \tag{2}$$

$$\frac{\partial c_a}{\partial t} + \nabla \cdot (\vec{n}_a) = 0 \tag{3}$$

$$\left(\rho c_p\right)_{\text{eff}} \frac{\partial T}{\partial t} + \left(c_{pv}\vec{n}_v + c_{pa}\vec{n}_a + c_{pw}\vec{n}_w\right) \cdot \nabla T = \nabla\left(k_{\text{eff}}\nabla T\right) - \lambda\dot{I} + \dot{Q} \tag{4}$$

Food as a porous medium and its processing

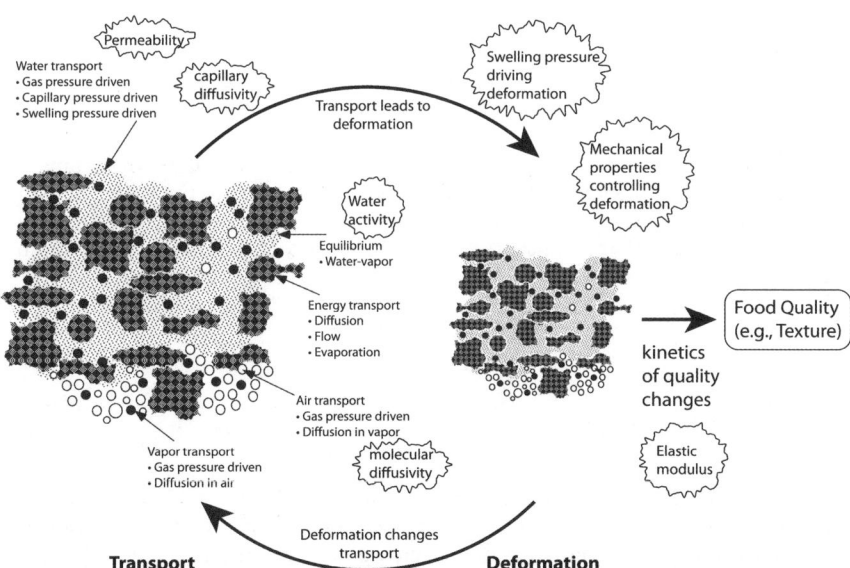

Fig. 1 Schematic showing food treated as a deformable porous medium and the processing of food studied in terms of the transport of various components and phases due to their respective driving forces. The circled quantities are some of the input parameters needed for such a porous media based model that can be potentially predicted using soft matter approach. These quantities are a function of the material and process parameters.

In the above equations, c_i is the concentration of phase i, where $i = w$, v or a, $\vec{n}_i$ are the fluxes, $\dot{I}$ is the volumetric rate of evaporation, ρ is the density, c_p is the specific heat, T is the temperature, k is the thermal conductivity, $\lambda(T)$ is the latent heat of evaporation, and $\dot{Q}$ is any volumetric rate of heating, such as from microwaves. The food (porous) structure is often created *via* phase transitions, like boiling/evaporation, melting/solidification of crystals, and protein denaturation. For example, during baking, one heavily relies on starch gelatinization for the structuring. Like other phase transitions, this can precisely be described *via* thermodynamics.[75] In general, phase transitions produce or consume significant amounts of heat, and this has to be accounted for in the energy equation. The fluxes in the above equations are given by:

$$\vec{n}_w = -\rho_w(\kappa_w/\mu_w)\nabla p_w = -\rho_w(\kappa_w/\mu_w)\nabla p_g - D_{w,c_w}\nabla c_w - D_{w,T}\nabla T \qquad (5)$$

$$\vec{n}_v = -\rho_v(\kappa_g/\mu_g)\nabla p_g - (c^2/\rho_g)M_vM_aD_{bin}p_v/p_g^2\nabla p_g - D_{v,c_w}\nabla c_w - D_{v,T}\nabla T \qquad (6)$$

$$\vec{n}_a = -\rho_a(\kappa_g/\mu_g)\nabla p_g - (c^2/\rho_g)M_vM_aD_{bin}p_a/p_g^2\nabla p_g - D_{v,c_w}\nabla c_w - D_{v,T}\nabla T \qquad (7)$$

Here κ_i and μ_i are the permeabilities and dynamic viscosities, respectively, of the ith phase. Flux of water, $\vec{n}_w$, is due to liquid pressure and is discussed at length in Section 4. In many food process models, with the implicit assumption that the material is unsaturated, the liquid pressure is assumed to be equal to $p_g - p_c$, where p_c is the capillary pressure. Moisture dependence of capillarity and temperature dependence of capillarity, respectively, leads to the last two terms in eqn (5). The appropriate water pressure to use in various situations is discussed in more detail in Section 4.1. The less common formulation of liquid pressure in terms of swelling pressure is discussed in Section 4.2. Flux of vapor, $\vec{n}_v$, is due to gradients in gas pressure and vapor pressure (decomposed into three separate effects representing the last three terms: binary diffusion, driven by liquid concentration and driven by temperature).[3] There is no distinct momentum equation as the Darcy's law used in water, vapor and air transport equations is a replacement for the fluid flow or momentum equation.

As the solid matrix of food materials is hygroscopic, since they are commonly constituted of biopolymers, for some moisture transport situations, the moisture bound to the solid matrix has been considered as a separate phase that transports by diffusion only.[56] If the concentration of bound water is c_b, the corresponding mass balance becomes:

$$\frac{\partial c_b}{\partial t} + \nabla \cdot (\vec{n}_b) = -\dot{J} \qquad (8)$$

with

$$\vec{n}_b = -D_b\nabla c_b \qquad (9)$$

where D_b is the diffusion coefficient for bound water. When considering this bound water in a formulation, eqn (1) is rewritten for transportable water with a separate source term that corresponds to the bound water becoming transportable water. Since the magnitude of this term is also $\dot{J}$, in such formulations, equations for transportable water, bound water and water vapor are all added together to develop an equation for total moisture, removing the need to estimate $\dot{J}$.

Eqn (1)–(4) have one extra variable than the number of equations (since the rate of evaporation, $\dot{I}$, is unknown). To complete the system, an additional equation is needed that provides the rate of evaporation; this can be formulated in one of two ways:[25]

$$p_v = p_{v,eq}(M,T) \tag{10}$$

$$\dot{I} = K(\rho_{v,eq} - \rho_v)S_g\phi \tag{11}$$

Here M is the moisture content at a spatial location and ϕ is the porosity. For the purposes of describing the system, these two equations are equivalent. Eqn (10) is the equilibrium relation for the material relating the vapor pressure to moisture and temperature given by the absorption isotherms. For a variety of biopolymer, polysaccharide and animal-based proteins, water activity can be predicted using the soft matter (thermodynamic) approach.[69] Eqn (11) is a non-equilibrium formulation that approaches eqn (10) for large values of K. A number of food processes (e.g., drying, baking, frying and microwave heating) have been modeled by several researchers, mostly using eqn (10) but some with eqn (11).

Variations of the continuum porous media formulation are available, the most common one being frontal approach to evaporation or sharp interface phase change problem.[20] The liquid water and water vapor transport equations can also be combined, leading to the simple diffusion equation with an effective diffusivity. There are also phenomenological approaches[41] to multiphase transport in porous media whose origin in terms of averaging have not been demonstrated and many of the transport coefficients in this model cannot be traced to standard properties. Among the detailed mechanistic approaches to the modeling of food processes, the multiphase porous media based approach at the macroscale appears to be the most popular one as it has been used to model a number of food processes mentioned in the introduction.

2.3 Inclusion of deformation (shrinking/swelling) in these formulations

A deforming (shrinking/swelling) porous medium is essentially handled by treating all the fluxes, discussed in Section 2.2 for a rigid porous media, to be those relative to the solid skeleton, and combining this with a velocity of the solid skeleton that comes from deformation obtained from solid mechanical stress-strain analysis (also assuming macroscale continuum). Since the solid has a finite velocity, $\vec{v}_{s,G}$, due to deformation, the mass flux of a species, i, with respect to a stationary observer, $\vec{n}_{i,G}$, can be written as the sum of flux with respect to solid and flux due to the movement of solid with respect to stationary observer:

$$\vec{n}_{i,G} = \vec{n}_{i,s} + c_i\vec{v}_{s,G} \tag{12}$$

Thus, $\vec{n}_w$ in eqn (1), which is really $\vec{n}_{w,s}$, will be replaced by:

$$\vec{n}_{w,G} = \vec{n}_{w,s} + c_w\vec{v}_{s,G} \tag{13}$$

Movement of solid, $\vec{v}_{s,G}$, in turn, is obtained from stress-strain analysis. If $\bar{\sigma}'$ is the effective stress on the solid skeleton, it can be written following the well-known Terzhagi's principle[2] as:

$$\bar{\sigma} = \bar{\sigma}' - p_f\mathbf{I} \tag{14}$$

where p_f is the volume averaged fluid pressure of the gas and liquid phases, given by:

$$p_f = S_g p_g + S_w p_w \tag{15}$$

where S_g and S_w are the saturations of the gas phase and water phase, respectively. Eqn (15) is the well known extension of Terzhagi's principle, useful for unsaturated porous media (for a saturated porous media, $p_f = p_w$).

The solid matrix of the food material can be treated as elastic, viscoelastic (or following other material models) and the corresponding strain energy function can be used with the linear momentum balance equation for the deforming solid:

$$\nabla \cdot \bar{\sigma} = 0 \tag{16}$$

leading to the final equation as:

$$\nabla \cdot \bar{\sigma}' = \nabla p_f \tag{17}$$

Here p_f, the fluid pressure, is obtained differently for saturated and unsaturated medium, respectively, depending on the nature of the physicochemical forces present, as discussed in Section 4. These sets of equations are readily implementable[9,58] in a commercial software capable of combining various physics, such as COMSOL Multiphysics (COMSOL, Inc., Burlington, MA, USA).

3 Other multiscale formulations of the porous media

Multiscale formulations of transport in food as porous media exist and can be distinguished based on whether a real structure is included or it is a hypothetical continuum (smeared) model. In contrast with the standard porous media formulation that assumes a hypothetical continuum, the actual geometry can be acquired at various scales (see also discussion under prediction of mechanical properties) and the problem can be formulated to include the detailed structure at these scales. This is continuum multiscale formulation using real geometry. In the hypothetical continuum model that is multiscale, one general transport theory for biopolymers that captures the fluid transport behavior in glassy, rubbery and transition states of food has been developed starting from thermodynamics—it leads to multiscale fluid transport equations.[63,64] It is referred to as the hybrid mixture theory and has been applied to food processes such as in prediction of stresses in corn drying.[66] During processing, a food material may transition from rubbery to a glassy state and, during the transition, non-Fickian diffusion can occur that has been captured by this theory. It considers phenomena at three spatial scales: micro-, meso-, and macro-scales. At the microscale, the vicinal fluid and the solid matrix exist as separate phases. At the mesoscale, the solid matrix and the vicinal fluid are considered together as a homogenous mixture representing a particle; and the liquid(s) and/or gaseous bulk fluid(s) are separated from this particle. At the macroscale, the particle and the bulk fluid(s) form overlaying continua. Complexities arise in the implementation of such a three-scale model and a simplified two-scale model along the same line has also been presented.[88] These models, however, still need material properties at some scale (see also discussion under mechanical properties).

4 Obtaining the liquid pressure using soft matter approaches

The liquid pressure, p_w, in eqn (14) is the driving force—it is part of the closure relations needed to solve the governing transport equations. Thermodynamic theories from the field of soft matter physics can provide such a valuable closure relation. These theories show a remarkably high degree of universalism, thus, they are applicable to a wide variety of food components, and require only a few fitting parameters. Using the knowledge of the food structure and composition, these theories can even be predictive,[78] as is shown in this section as well as in the section for predicting properties.

The liquid pressure in a capillary hygroscopic porous media is written in general as:[46]

$$p_w = p_{gas} - p_{capillary} + p_{swelling} + z \qquad (18)$$

Here p_g is gas pressure that can be due to evaporation of water or gas release, as for carbon dioxide in baking. The term $p_{capillary}$ represents capillary pressure in an unsaturated solid (e.g., many food materials); it would be a function of the temperature and moisture content of the food material. The term $p_{swelling}$ stands for swelling pressure, present in a shrinkable/swellable system. Swelling pressure itself is broken down into three components:[76]

$$p_{swelling} = p_{mix} - p_{elastic} + p_{ion} \qquad (19)$$

If we think of food as biopolymers, $p_{elastic}$ (also known as p_{matrix}) is the elastic contribution and represents the resistance to swelling of the polymeric matrix. It can be related to the effective stress (see more discussion in Section 2.3). The term p_{mix} (also known as $p_{osmotic}$) is the mixing or osmotic pressure, which represents the interaction of the solvent (i.e., water) with the polymer. This is due to the tendency for the flow of solvent from their region of higher concentration to their region of lower concentration. The term p_{ion} (also referred to as $p_{electrostatic}$) is the ionic contribution of the dissolved salts, which can be computed using the Raoults law or the Pitzer equation.[69] Under equilibrium conditions, the matrix pressure balances the other two and the swelling pressure, $p_{swelling}$, is zero. Thus, loosely speaking, one can think of swelling pressure as the excess osmotic pressure. The contribution due to electrostatic forces seems to have been studied more extensively in the soil literature but has been either ignored or presumed lumped into other terms in the food and biomaterial literature. The soft matter approach, $p_{osmotic}$, that has also been called the mixing contribution, can be derived from FVFH theory.[76] However, the FVFH theory only predicts sorption, and cannot explain the hysteresis between desorption and adsorption. It is expected that hysteresis can be explained via viscoelastic relaxation, which definitely plays a role in moisture sorption.[51]

Eqn (18), in its form as shown, has been sporadically mentioned in soil literature (e.g. ref. 46) but as far as can be seen, this is the first mention of the generalized liquid pressure formulation in the context of transport in deformable biomaterials that includes both the terms due to capillary pressure and swelling. If the literature for transport in soil, clay, concrete, wood, and food are scanned, two formulations for liquid transport appear in an ad-hoc manner without necessarily discussing the rationale or the assumptions. Either the capillary pressure term, $p_{capillary}$, is present, or the swelling pressure term, $p_{swelling}$, is present, but not both, i.e.,

$$p_w = \begin{cases} p_{gas} - p_{capillary} & \text{for unsaturated medium, with gas generation} \\ -p_{capillary} & \text{for unsaturated medium} \\ p_{swelling} & \text{for saturated medium} \end{cases} \qquad (20)$$

where the gravity term, z, has been ignored. The rationale for the mutually exclusive contributions of capillarity and swelling is likely to be that when the porous medium is unsaturated, the capillary pressures are quite large and thus swelling pressure can probably be ignored. On the other hand, in a saturated porous medium (where the gas phase is absent), the capillary pressure is considered to be zero but swelling pressure can be significant, depending on the relative values of p_{matrix}, $p_{osmotic}$, and $p_{electrostatic}$.

As a general rule, however, it may not always be possible to separate the two pressure terms as is shown in eqn (20). In a drying or wetting (rehydration) process, there can be a significant zone in which the saturation changes from very high to very low.

In such a region, it is not clear from the literature whether it can be assumed that one of the two pressures, $p_{capillary}$ or $p_{swelling}$, dominate. The only estimate[72] of swelling pressure in the food literature is for meat and it is in the range of 0.03–0.3 MPa. A crude estimate of capillary pressure in meat, as discussed later and presented in Fig. 4, shows that the capillary pressures are generally much higher and can be as high as 300 MPa, depending on saturation. Thus, it is entirely possible that for some food processing situations, the capillary and swelling pressure contributions can be comparable in some spatial regions or at some time during a process.

A resolution of the question of the correct formulation of the liquid pressure in unsaturated, swelling porous media is to assume that water bound to biopolymers forms a different phase than the capillary water, which are at local equilibrium. This approach is taken in the example of the rehydration of freeze-dried vegetables in Section 9.1.

4.1 Obtaining the liquid pressure: capillary pressure dominating

The most common approach in modeling transport in various biomaterials uses the capillary pressure formulation, generally in the context of drying or drying-like processes such as baking and frying. It is assumed that the material is unsaturated and capillary pressure dominates over swelling pressure. Capillary pressure is taken to be function of moisture content and temperature. The flux of water can be written as:

$$\vec{n}_w = -\rho_w \frac{\kappa_w}{\mu_w} (\nabla p_w)$$

$$= -\rho_w \frac{\kappa_w}{\mu_w} \nabla (p_g - p_c(M, T)) \tag{21}$$

$$\vec{n}_w = -\rho_w \frac{\kappa_w}{\mu_w} \left(\nabla p_g - \frac{\partial p_c}{\partial M} \nabla M - \frac{\partial p_c}{\partial T} \nabla T \right)$$

$$= -\rho_w \frac{\kappa_w}{\mu_w} \left(\nabla p_g - \frac{\partial p_c}{\partial M} \frac{\partial M}{\partial c_w} \nabla c_w - \frac{\partial p_c}{\partial T} \nabla T \right) \tag{22}$$

$$\vec{n}_w = -\rho_w \frac{\kappa_w}{\mu_w} \nabla p_g - D_{w,c_w} \nabla c_w - D_{w,T} \nabla T \tag{23}$$

where diffusivity due to moisture gradient, D_{w,c_w}, and diffusivity due to temperature gradient, $D_{w,T}$, are defined as:

$$D_{w,c_w} = -\rho_w \frac{\kappa_w}{\mu_w} \frac{\partial p_c}{\partial M} \frac{\partial M}{\partial c_w}$$

$$D_{w,T} = -\rho_w \frac{\kappa_w}{\mu_w} \frac{\partial p_c}{\partial T} \tag{24}$$

Using the thermodynamic relation given by eqn (36), the capillary pressure from water activity data can be estimated, thus relating the macroscopic formulation to soft matter approach. There are, however, difficulties with this approach that are described in Section 7.3.

4.2 Obtaining the liquid pressure: Swelling pressure dominating

The other approach was used to model meat cooking, first empirically[52] and later[72] relating to the fundamental concept of swelling pressure and thermodynamics (soft matter approach). The soft matter approach assumes that the material (meat) stays saturated. As meat is saturated, capillary pressure is zero. When meat is heated, the term given by p_{matrix} in eqn (19) gets weaker due to the denaturation of proteins. This results in the expulsion of water from the protein matrix, leading to a non-equilibrium and a non-zero swelling pressure (or excess osmotic pressure). Since heating is never uniform over the entire volume, there will be a gradient of this swelling pressure in the meat that will drive the flow of liquid water. Starting with the well-known Flory–Rehner theory,[15] a statistical thermodynamic model used in elastic polymer and polyelectrolyte gels, it was shown[72] that the swelling pressure can be written in terms of a difference between the volume fraction of protein at any moment and the same at equilibrium at the same temperature. It was argued[72] that instead of the volume fraction of protein, it is more natural to speak in terms of moisture content. The moisture content in meat at equilibrium is postulated to be equal to the readily available information of water holding capacity of meats. Thus,

$$p_{swelling} = E(c - c_{eq}(T)) \tag{25}$$

where c is the moisture content of meat at any location in it at any time and $c_{eq}(T)$ is the water holding capacity of the meat at the same location and time, when temperature at the location is T. The flux of water is written in terms of this swelling pressure as:

$$\vec{n}_w = -\rho_w \frac{\kappa_w}{\mu_w} \nabla p_{swelling} \tag{26}$$

$$\vec{n}_w = -\rho_w \frac{\kappa_w}{\mu_w} E \left(\nabla c - \nabla c_{eq}(T) \right) \tag{27}$$

$$\vec{n}_w = -\rho_w \frac{\kappa_w}{\mu_w} E \left(\nabla c - \frac{\partial c_{eq}}{\partial T} \nabla T \right) \tag{28}$$

Note that the flux equation given by eqn (28) is equivalent to eqn (23) (there are two terms relating to concentration and temperature gradients, respectively), when gas pressure is insignificant.

5 Obtaining the material properties using soft matter approaches

As would be expected in any physics-based approach, the macroscale porous media-based approaches outlined above require many material properties, both equilibrium properties such as specific heats and enthalpies related to phase transitions, and transport properties such as permeability and diffusivity, several of which are hard to obtain. Various ways of estimating them using empirical and semi-empirical approaches have been summarized.[23] In a macroscale model, the properties are also macroscale (*i.e.*, homogenized or effective) and typically obtained from experimentation (that are macroscale). It is fair to say that with some recent exceptions, all of the property estimation equations are either completely empirical or, at best, semi-empirical.[7,23] A good example is the effective moisture diffusivity in food solids, where, with very few exceptions (*e.g.* ref. 77), the property models are primarily empirical. The available properties typically do not provide anisotropic information.

Estimating properties is where the soft matter approach can play a very significant role and shows a lot of promise. The transport coefficients in macroscale equations of the porous medium are homogenized properties, and thus depend on the food

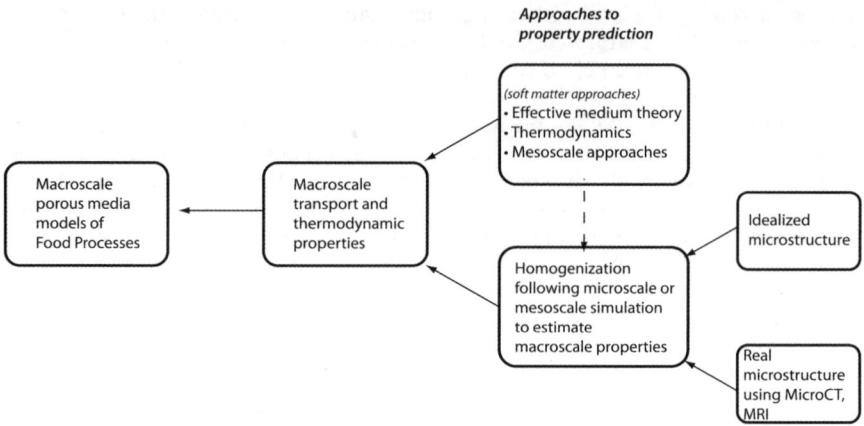

Fig. 2 Framework of property prediction and their relationship to the macroscale porous media model shown in Fig. 1.

microstructure at the lower mesoscale. The soft matter approach also can be thought of as part of multiscale modeling approaches.[50] As shown in Fig. 2, predictions of properties (or the closure relations needed in the macroscale equations) have followed several approaches: (1) Here, properties are predicted using the effective medium theory of Maxwell-Garnett and its extensions,[32] where the material is considered a two-phase media (a matrix with inclusions). This has been applied to predict thermal conductivity of frozen meat products;[73] (2) a thermodynamics-based approach such as the one applied to predict water activity;[69] (3) various mesoscale approaches, and (4) homogenization starting from microstructure. Here, the two possibilities are using idealized microstructure or obtaining real microstructure using procedures such as MicroCT and MRI. The homogenized transport coefficient for a periodic cell (in the case of idealized microstructure) or a real microstructure can be obtained, for example, performing direct numerical simulation at the mesoscale using Lattice Boltzmann,[53] using pore network models[12,13] or using continuum equations at this scale and volume averaging (homogenizing) to obtain the macroscale properties.[57] This approach (homogenization) still requires properties at a smaller scale that may be obtainable using the first three approaches mentioned. The last approach is also an example of multiscale computation,[50] where the coupling is serial—computations at the mesoscale provide closures for the homogenized properties at the macroscale.

6 Prediction of equilibrium properties using soft matter approach

Although the details of predictions of equilibrium properties are not included in this article due to space limitations, it is important to note that a number of important equilibrium properties needed for porous media models, such as water activity, partition coefficients, state diagram, *etc.*, have been predicted using soft matter approach. For example, the prediction of water activity in meat has been discussed in detail[69] and a state diagram for starches has been predicted.[75] Solute partition coefficients are needed for the transport modeling of food processes when, for example, two different materials are involved, as in the case of the food and the package in food packaging. Many regulations in this area urge the development of predictive approaches on the migration of chemicals into the food. Partition coefficients of n-alkanes, n-alcohols, volatiles, and typical antioxidants between a low density polyethylene and several alcohols (methanol and ethanol) were predicted using a soft matter (off-lattice Flory–Huggins) approach[21] that the authors plan to generalize into wider systems.

7 Prediction of transport properties using soft matter approach

The mass transport properties needed in the above macroscale models are mass diffusivity and permeability. Using soft matter and thermodynamic approaches, these properties can potentially be predicted (and used in the macroscale models).

7.1 Permeability

Permeability quantifies the material's ability to allow pressure driven (or bulk) flow of fluids through it. Permeability is defined by Darcy's law of transport through a porous medium:

$$\vec{q} = -\frac{\kappa}{\mu}\vec{\nabla}p \qquad (29)$$

The extent and nature of pores in the material and the transport properties of the fluid are the most important factors governing the permeability of foods. Permeability, κ, is further broken into intrinsic permeability, κ_i, and relative permeability, κ_r, given by the following expression:

$$\kappa = \kappa_i \kappa_r \qquad (30)$$

The intrinsic permeability, κ_i, represents the permeability of a liquid or gas in a fully saturated state. Permeability data is perhaps the most difficult to obtain among the properties included here. There is very little measured data available (a small amount of data are available for potato, meat[5] and bread[18] and only one prediction equation is available (for apple) with porosity).

Thus, the prediction of permeability is a critical need for these macroscale transport models. Using the soft matter approach, permeability can be determined using the Lattice Boltzmann (LB) method combined with a real geometry generated from X-ray microtomography. An example of this is the preliminary calculation of permeability using LB simulation, using the geometry shown in Fig. 3. The calculated permeability is 6.75×10^{-11} m^2, while a value of 4.57×10^{-11} m^2 has been reported

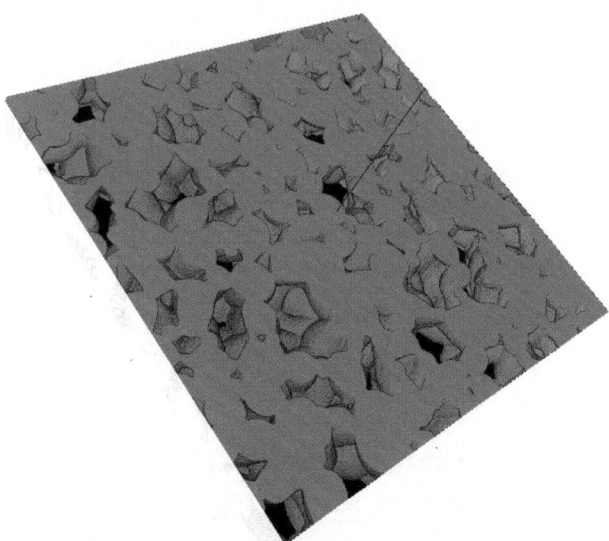

Fig. 3 Example of real geometry acquired for an apple by the Nicolai group (Some related details are provided in ref. 90). In the images, black is pore and white is solid.

in the literature for the same material.[14] The LB method is also versatile enough to be able to model the molecular forces between the fluid and the solid in the case of soaking or to determine relative permeability between two fluids, such as in frying for oil and water. The boundary conditions have been reviewed elsewhere.[35,42] Further details on this method can be seen in standard textbooks.

Also, if one approximates the food microstructure with an idealized geometry, there exist analytical correlations and theories to compute the permeability. For capillaries with a cross section having the shape of a convex irregular polygon, the (relative) permeability is readily computed *cf.* ref. 11, 47, 54, 55 and 60. These capillaries are showing universal behavior, with their shapes being characterized by the compactness, equal to $C = P_w^2/A_w$, with P_w the wetted perimeter, and A_w the wetted cross sectional area. It builds on the earlier concept of hydraulic diameter in engineering sciences. Seeking universal behavior is a natural trait of a soft matter physicist, and food science stands to benefit from it.

However, foods are still challenging because they are hygroscopic and during moisture sorption, they will swell. Consequently, the porosity and permeabilities will change in time. This will require real-time 3D X-ray microtomography, combined with LB simulations and finite element analysis of deformation, to predict the change of permeability as a function of moisture content and temperature.

7.2 Molecular diffusivity

Mass diffusivity is commonly defined by Fick's law:

$$\vec{n}_A = -D_{AB}\vec{\nabla}c_A \tag{31}$$

Here $\vec{n}_A$ is the flux of component A, D_{AB} is the diffusivity of A in B, and c_A is the concentration of A. Although describing diffusion with Fick's law is common for researchers in food engineering, the thermodynamic driving force for diffusion is the chemical potential, as expressed by the Stefan–Maxwell relation. Both Fick's law and the Stefan–Maxwell relation have diffusion coefficients that are related, but different in their nature. In the framework of thermodynamics, the diffusion coefficient in Fick's law is called the mutual diffusion coefficient, D_m, and the diffusion coefficient in the Stefan–Maxwell relation is called a self-diffusion coefficient, D_s, which is related to the friction solubilized compounds experience in the liquid phase. The Stefan–Maxwell theory defines the diffusive flux relatively to the velocity of the solid phase $\vec{u}_s$[79]

$$\vec{n}_b = \rho_w\phi_w(1 - \phi_w)(\vec{u}_w - \vec{u}_s) = -M\nabla(\mu_w - \mu_s) \tag{32}$$

where M is the so-called mobility, which also enters the Onsager relation, ρ_w is the bulk density of water, and ϕ_w is the volume fraction of water relative to that of the solid phase. Hence, $c_b = \rho_w\phi_w$, and $D_b = D_m$. Note that the diffusion is in terms of the difference in chemical potential of water and the solid phase (biopolymers): $\mu_w - \mu_s$. The mobility can be written in terms of the self diffusion coefficient, D_s, in the Stefan–Maxwell relation:

$$M = \phi_w^2(1 - \phi_w)\frac{D_s}{RT} \tag{33}$$

Hence, the relation between mutual and self diffusivity is:

$$D_m = \phi_w^2(1 - \phi_w)\frac{D_s}{RT}\frac{\partial(\mu_w - \mu_s)}{\partial\phi_w} = \phi_w(1 - \phi_w)D_s\frac{\partial\ln(a_w)}{\partial\ln\phi_w} = \phi_w(1 - \phi_w)D_sQ \tag{34}$$

where Q is the so-called thermodynamic factor. This relation is well-known in polymer physics[1,10] and is identical to that for shear-induced diffusion of particulate suspensions.[79] For the self diffusion, there exist predictive theories in polymer physics that have been shown to hold for moisture diffusion in food matrices[31] (which will be reported in detail in a forthcoming paper[77]). For dilute polymer solutions, D_s can be obtained from the solute self-diffusivity following a generalized Stokes–Einstein relation, and for relatively dry solid matrices, D_s can be obtained from the free volume theory of Vrentas and Duda,[80] which has recently been shown to hold for simple sugars and polyols.[26] This has been shown to hold for larger polysaccharides.[77] In these extreme cases of dilute solutions and concentrated, dry solid matrices, it holds that $Q \approx 1$ and $D_m \approx D_s$. In the intermediate regime of semi-dilute solutions, the variation in Q is much smaller than in D_s, and D_s can be computed as a weighted mean of the results of the generalized Stokes–Einstein and free volume theory. There, it has been shown that one of the premises of soft matter physics indeed holds for water diffusion in biopolymer matrices, namely, there is universal behavior of physics at the mesoscale independent of the chemical identity of the biopolymer. Due to the separation of length scale of the water molecule and the sugars/biopolymer monomers, these compounds appear to the water molecule much alike, namely a hydrogen bonded-network.

7.3 Moisture capillary diffusivity

Although Fick's law was originally formulated for molecular diffusion, diffusivity, D_{AB}, can be due to many different mechanisms of transport that are analogous to molecular diffusion. For water transport in an unsaturated solid, D_{AB} stands for capillary diffusivity of water in the solid. Often, many different modes of transport are lumped into D_{AB} without clear identification of the modes, leading to D_{AB} being referred to as effective diffusivity (denoted as D_{eff}).

Moisture diffusivity is a strong function of the moisture content, temperature and structure (void fraction, pore structure and distribution) of the material. Moisture diffusivity is also a function of electrolyte concentration such as calcium in the case of drying plant materials (like tomatoes). Calcium ions interact with pectin molecules increasing the stiffness of the membrane and cell wall, reducing shrinkage of cells and ultimately affecting diffusivity.[39] Moisture diffusivity in the food literature often lumps together one or more transport mechanisms (capillary diffusion of liquid water, molecular diffusion of vapor) and separating the effective diffusivity into distinct contributions from any one transport mechanism is generally not possible.

In a moist food, water transport can be in both liquid and vapor form. Often both these transport modes are combined in an effective diffusivity value. A large number of empirical models have been developed based on drying kinetics, sorption or desorption kinetics and moisture profile analysis.[43,89] However, generic models for estimating moisture diffusivity as a function of moisture and temperature values are rare.

When the food has a significant amount of moisture, effective diffusivity is essentially the capillary diffusivity of liquid water. Theoretically, it should be possible to predict capillary diffusivity from data on capillary pressure and permeability (discussed in Section 7.1) for the material using the following equation from porous media literature:[7]

$$D_w = \frac{\rho_w^2 g \kappa}{\mu_w} \frac{\partial h}{\partial c} \tag{35}$$

where D_w is capillary diffusivity, ρ_w is density of water, κ is intrinsic permeability of the matrix, μ_w is viscosity of water, and h is the matric potential for water in the food matrix. The matric potential for food materials is difficult to measure directly but its estimation from water activity data has been made using the Kelvin's equation:

$$h = \frac{RT}{\rho_w g V_m} \ln a_w \qquad (36)$$

where a_w is the water activity. This equation has been used to model water transport in soaking of tea leaves using the Richard's equation.[83] A formulation equivalent to using effective diffusivity has been used to predict the transport of water in food as a capillary solid using the Richard's equation, where eqn (36) is used to predict capillary pressure and the transport parameter hydraulic conductivity as a function of capillary pressure.[82]

Prediction of capillary diffusivity directly, using eqn (35), has not been reported. This could very well be due to the difficulties in obtaining (experimentally or using further predictive equations) the matric pressure, h, and the intrinsic permeability, κ. An attempt made to predict diffusivity using eqn (35) and (36) is now provided. Fig. 4a shows experimental water activity data for meat as a function of water content from the literature.[61] Using the standard equation relating permeability to porosity, ε, given by:

$$\kappa_{in} = 10^{-19} \frac{\varepsilon^3}{1 - \varepsilon^2} \text{ in m}^2 \text{ s}^{-1} \qquad (37)$$

the permeability is obtained as a function of moisture content. The definition of porosity is important in the context of permeability because its definition depends on the type of tissue. For fibrous materials, porosity is the volume fraction of all components between the fibers (gases and liquids). For cellular materials (like apples and potatoes), the porosity is more specifically defined as all fluids between the cells or fibers and not any intracellular fluids. Therefore, fluid inside the cells or meat fibers is considered "bound" until it permeates into the pore spaces. Using eqn (36), capillary pressure as a function of water activity is obtained using the permeabilities just calculated—this is shown in Fig. 4b. Finally, using eqn (35), diffusivity is predicted as a function of moisture content as shown in Fig. 4c. The above prediction process for diffusivity, however, uses measured water activity data. Using the soft matter approach, it has been shown[69] to be possible to predict water activity and thus moisture diffusivity should be predictable without utilizing the measured water activity data.

8   Prediction of mechanical properties

When modeling the transport processes in food as a deformable system, mechanical properties are needed. Even outside of process modeling, mechanical properties such as texture are critically linked to food quality and thus the prediction of mechanical properties can play a very significant role in food process applications. In almost all cases, such mechanical properties are measured and input to the model. Such properties, however, can be potentially predicted, as has been commonplace in the field of micromechanics.

8.1   Approach based on micromechanics

In one of the most common approaches, mechanical analysis is performed on the actual microstructure that is obtained through various imaging methods, and effective properties (that can be used in macroscale deformation analysis) can be obtained, a process commonly referred to as homogenization. This is quite analogous to the process of obtaining permeability from an actual geometry at the microscale, as described earlier. This is only beginning to be performed in a food processing context.[22,40]

The effective properties estimated using the micromechanical models generally require the properties of the dense, non-porous material. For example, in estimating mechanical properties of cereal solid foods[22] as open cellular solids (such as regular honeycombs), the modulus of elasticity, E, was expressed as a function of density, ρ, as:

$$E/E_s = C(\rho/\rho_s)^n \text{ and } \nu = 0.3 \tag{38}$$

where E_s and ρ_s are the elastic modulus and density, respectively, of the dense, *i.e.*, non-porous material, n is an exponent that varies with the type of deformation, and the cellular structure, C, is a constant that depends on the cell geometry and is generally closer to unity, and ν is the Poisson's ratio. The quantity, E_s, is either measured directly or is a parameter that matches the homogenized property estimated by the model to experimentally measured data. Since it is hard to measure such a "foam wall" property, it is suggested[22] that the mechanical property is computed from the composite morphology of the biopolymer blend that constitutes the "foam wall." Thus, soft matter approach in predicting mechanical properties should be able to come to the rescue, following the example of recent work.[4]

8.2 Approach based on rubber elasticity applied to meats: framework for meat texture prediction

Soft matter approaches to understand texture development in meats have been recently explored.[36–38] Meat texture is a complex interplay of several quality attributes such as hardness, springiness, cohesiveness and chewiness. Factors such as protein structures, connective tissue framework, fat and carbohydrate components affect meat texture critically.[65] In a series of papers, Lepetit[36–38] has shown a direct relationship between changes in connective tissue morphology and meat toughness and have provided a framework for texture prediction based on the theory of rubber elasticity.

Meats are primarily composed of muscle fibers and connective tissues. Connective tissue is a composite network of collagen (96%) and elastin (4%) fibers embedded in a matrix of proteoglycan covering individual muscle fibers and fiber bundles. In their raw state, collagen and elastin fibers have elastic moduli between 0.5–1.0 GPa and 0.1–0.41 GPa, respectively,[19,62] whereas the elastic moduli of myofibrils typically lie in the range of 0.3–0.5 MPa.[28] Since collagen is the prime component in connective tissue, properties such as its size, inter muscle variations, presence of Type I (more thermostable) and Type III (less thermostable) collagen, number of cross-links between collagen molecules and fibrils have shown to affect meat toughness–an important quality attribute.[37] Collagen has a quasi-crystalline structure and when heated to temperatures of 58 °C–65 °C, it undergoes a helix to coil transition. As a result, it gets denatured and contracts and, in this state, its behavior is governed by the theory of rubber-like elasticity.[16,17,33,37,44] The stresses generated by a collagen fiber on individual muscle fibers and fiber bundles is given by

$$\bar{\sigma} = \frac{F}{S} = g(\varphi)\frac{V_c}{V}\chi RT\nu^{\frac{1}{3}}\left(\alpha - \frac{1}{\alpha^2}\right) \tag{39}$$

where V_c is the total volume of collagen in a meat sample of volume, V, χ is the number of chains per dry volume of collagen (which in turn is related to the number of cross-links per unit volume), R is the universal gas constant, T is the temperature in °K, α is the degree of deformation of the collagen fiber and $g(\varphi)$ is a function of the angle φ between the direction of measurement and the direction of collagen fibers, expressions for which are available.[27] Stresses generated by the collagen fibers generate pressure, P, on the muscle fibers. The total pressure, P, thus generated depends upon the morphology of the connective tissue, M, and the angle that the collagen fibers make with the central axis of the muscle fibers, $\theta \approx 55°$,

$$P = \frac{V_c}{V}g(\varphi)\chi MTR\nu^{\frac{1}{3}}\left(\alpha - \frac{1}{\alpha^2}\right)\sin^2\theta \tag{40}$$

The pressure developed by connective tissues is opposed to the resistance of muscle fibers and fiber bundles. The balance between this pressure and the resistance leads to a final value of collagen contraction state, α, which determines the elastic modulus of collagen fibers and fibrils. The elastic modulus of collagen fibers is thus obtained as:

$$G(\varphi) = \frac{P}{\left(\alpha - \dfrac{1}{\alpha^2}\right)} = \frac{V_c}{V} g(\varphi) \chi M R T \nu^{\frac{1}{3}} \sin^2(\theta) \tag{41}$$

As mentioned previously, mechanical properties such as elastic moduli can be very good measures for predicting texture and hence evaluating the overall quality of food stuffs including meats. Collagen contraction affects meat toughness in a critical way owing to its large modulus of elasticity when compared with muscle fibers. Hence, its elastic modulus can be a good indicator of meat texture. Dynamic prediction of its elastic modulus can be made possible by coupling the above framework with the energy balance equation (eqn (4)) together with the kinetics of collagen denaturation in meats during cooking (e.g. ref. 3). Soft matter approaches to predict mechanical properties such as shown above thus provide a fundamental basis to better understand texture development in foods.

9 Examples of integration between macroscopic porous media based model and soft matter approach

Integration of the porous media based macroscale model with soft matter approach will now be demonstrated by applying to two food processes: (1) the rehydration of freeze-dried vegetables, and (2) meat cooking including deformation.

9.1 Rehydration of vegetables

In this example, we consider the rehydration of freeze-dried vegetables. At the product scale, freeze-dried samples will not exhibit swelling and we will disregard this in the model. Freeze-dried vegetables are highly porous and are hydrated very quickly. We are interested in the effect of structure on the rehydration, which will be analyzed *via* multiscale simulations. At the mesoscale level, the moisture transport in a collection of capillaries (as a representative elementary volume, REV) was investigated. As a first step in the development of this model, the moisture transport at the scale of a single capillary was studied. This was used to present the soft matter approach to porous media modeling. The pore structure was idealized to a long, straight capillary, with a square cross section. The model is presented in terms of 1-D macroscale equations, meaning that the vegetables are regarded as having an ideal periodic array of these capillaries. The following phases have been considered: a solid phase consisting of biopolymers (cell wall material) and bound water, a liquid phase with capillary water and a gas phase containing water vapor. The pressure in the gas phase will be assumed constant because it is connected to the environment. The freeze-dried vegetable is placed in an environment at high temperature (95 °C) and 100% relative humidity. The various mass densities are reformulated in terms of the components bulk density and the volume fraction it occupies: $c_i = \rho_i \phi_i$. All condensed matter are considered incompressible. Combined with the constant gas pressure, this means $\sum_i \phi_i = 1$. Constant temperature and local equilibrium are assumed, meaning that the water activity of all phases is equal, and that moisture transport can be described by a single equation. If the solid phase absorbs bound water, it will swell at the expense of the volume of the pore space.

Each phase allows its own mode of moisture transport: diffusion of bound water, pressure driven transport in the liquid phase, and vapor diffusion in the gas phase. Thus, the following single governing equation is obtained:

$$\rho_w \frac{\partial \phi_{tot}}{\partial t} = -\nabla \cdot \vec{n}_b - \nabla \cdot \vec{n}_w - \nabla \cdot \vec{n}_v \quad (42)$$

with $\phi_{tot} \approx \phi_b + \phi_w$, since the contribution of water vapor to the total amount of moisture is always negligible (as follows from thermodynamics and the assumption of local equilibrium). The flux relations are written more in line with previous models taking the porous media approach, and these are as follows:

$$\vec{n}_b = -(\phi_b + \phi_s)D_m\nabla\tilde{c}_b \quad (43)$$

$$\vec{n}_w = -\rho_w(\kappa_w/\eta_w)\nabla p_w = \rho_w\vec{u}_w \quad (44)$$

$$\vec{n}_v = -\phi_g D_v\nabla\tilde{c}_v \quad (45)$$

where $\tilde{c}_b = \rho_w\phi_b/(\phi_b + \phi_s) = \rho_w\tilde{\phi}_b$ is the density of water in the solid phase, and $\tilde{c}_v = \rho_v\phi_v/\phi_g$ is the mass density of water vapor in the gas phase. The values of the driving forces follows from the equality of the water activity/chemical potential:

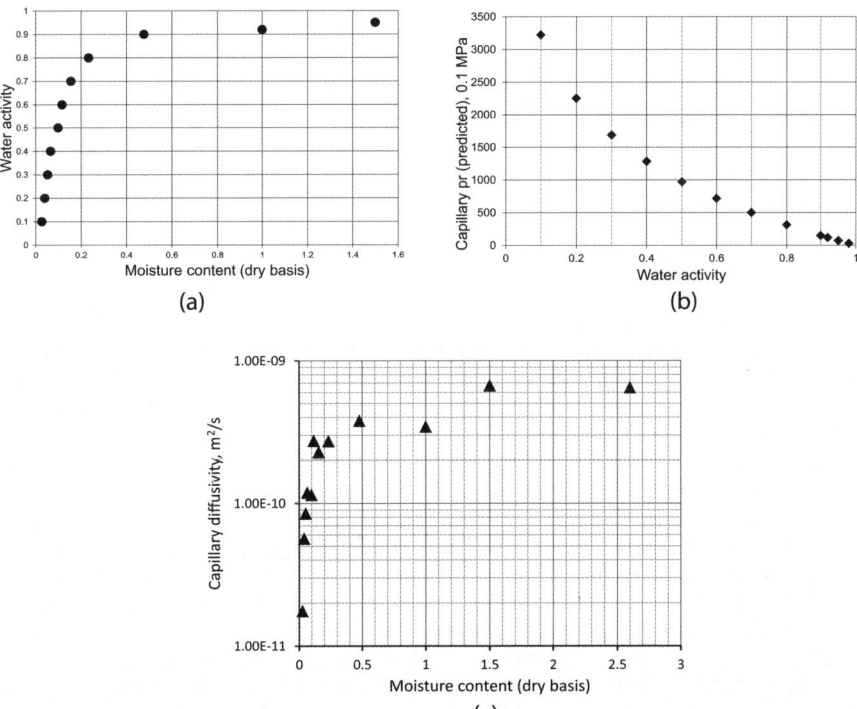

Fig. 4 (a) Measured water activity, (b) predicted capillary pressure and (c) Predicted diffusivity.

$$a_w = \frac{\tilde{c}_v}{c_{sat}(T)}$$

$$RT \ln(a_w) = (p_g - p_w) v_w \tag{46}$$

$$RT \ln(a_w) = \mu_{mix}(\tilde{\phi}_b) + \mu_{elas}(\tilde{\phi}_b) + \mu_{FV}(\tilde{\phi}_b)$$

with μ_{mix} (the mixing contribution) as follows from Flory–Huggins, μ_{elas} (the elastic contribution) as follows from Flory–Rehner and μ_{FV} (the free-volume contribution) as follows from Vrentas and Vrentas.[75,76,81] Following the Flory–Rehner hypothesis, it is assumed that individual contributions to the total chemical potential of water in the solid phase are additive.[72,76] The Flory–Huggins contribution is calculated on the basis of composition, which is classified into mono-, disaccharides and biopolymers (present in cytoplasm and cell wall material). The interaction with water of all non-crystalline biopolymers is assumed to behave similarly to amorphous polysaccharides and animal proteins, resulting in a Flory–Huggins interaction parameter to be dependent on the volume fraction, cf. ref. 75 and 86. The individual contributions of compounds to the effective interaction parameter is computed via volume averaging, as suggested in ref. 67 and 87. The free volume contribution also takes the moisture-dependent glass transition temperature, which is computed via volume averaging of the dry solutes glass transition temperatures, as suggested in ref. 75. This is allowed due to the universal value of the $\Delta C_{p,s}$ parameters, as appears in the Couchman–Karasz equation. The parameters present in the elastic contribution, μ_{elas}, are estimated via determination of water holding capacity via centrifugation of freeze-dried carrots, which are reported in a forthcoming paper. The capillary follows from Young–Laplace law. The biopolymers are assumed to be perfectly wetting (which holds in the limit of fully hydrated solid phase, $a_w > 0.99$). Hence,

$$p_c = p_g - p_w = \frac{\gamma}{r} \tag{47}$$

with γ the interfacial tension between water and air, and r the curvature of the capillary water. The capillary water is contained in the square corners of the pore, and the amount of capillary water per unit of length is then equal to $\rho_w A_w = \rho_w(1 - \pi/4)r^2$, as shown in Fig. 5a. It is assumed that the pores retain their square cross sections if the cell walls are swelling due to the sorption of water. The resistance for liquid flow follows as discussed above from the wetted perimeter and cross sectional area:

$$\kappa = \frac{A_w^2}{2P_w^2} \tag{48}$$

with $A_w = (4 - \pi)r^2$ and $p_w = 8r$.

In the 1D simulation model, the capillary is divided into several control volumes, each containing a certain amount of moisture and cell wall material. The moisture will be partitioned between the solid and liquid phases, based on the constraint of equal water activity. Knowing the water activity, and the amount of bound and capillary water, the fluxes can be calculated, and the mass balance for total moisture content can be integrated with respect to time. The model has been implemented using Finite Volume. The length of the capillary is taken as $L = 120$ μm. The cross section of the dry pore is $H = 20$ μm. The dry cell wall material, representing the wall of the square capillary, has an initial thickness of 0.37 μm. The mutual diffusion coefficient is taken equal to the self-diffusion of water, which is a good approximation if the cell wall material has a low degree of hydration ($a_w < 0.9$).

In Fig. 5a, the idealized geometry of the pore space in the freeze-dried vegetable has been sketched, with the solid phase, consisting of cell wall materials, as the wall of the capillary, and the capillary water as a film in the corners of the capillary. Furthermore, Fig. 5b shows the change of the porosity, $\varepsilon = 1 - \phi_b$, and the saturation, $S = \phi_w/\varepsilon$, as a function of the water activity. A given amount of moisture will be

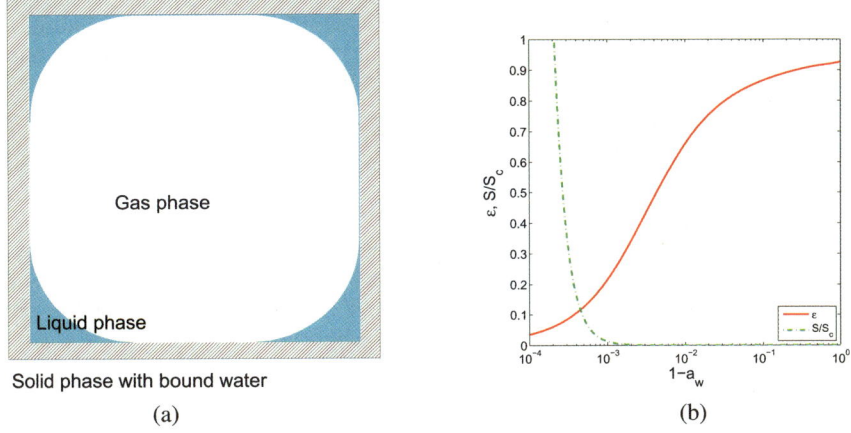

Fig. 5 (a) Idealized microstructure of a freeze-dried vegetable with a capillary wall consisting of cell wall materials, and capillary water in the corners. (b) Porosity, ε, and saturation, S, as functions of water activity, a_w, as determined by the local equilibrium condition.

partitioned between the solid phase and the liquid phase, whose relative size is thus represented by ε and S. Due to water sorption, the cell walls will swell, leading to a reduction of the porosity. Only if $a_w \approx 1$ is there a significant amount of capillary water. Notice that in this range, there is hardly any swelling of the solid phase

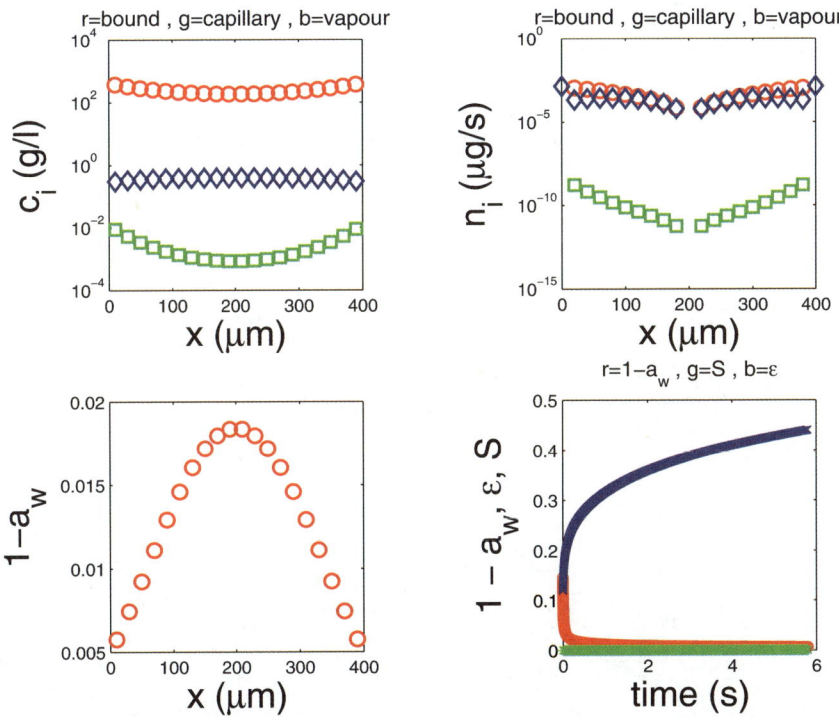

Fig. 6 Sample output of the model of rehydration of freeze-dried vegetables, showing various profiles as a function of distance, x, at the end of the simulation, and the evolution of water activity, a_w, porosity, ε, and saturation, S, in the first control volume, next to the inlet. The shown profiles are c_i, the water density in each phase, n_i, the water flux in each phase, and the water activity, a_w. Different colors indicate the phases as shown in the legends.

because of the action of elastic deformation, *i.e.* p_{elas} is resisting further swelling. In wood science, this situation is known as the fiber saturation point. If the saturation exceeds a critical value ($S > S_c$, where the radius $r = H/2$), the meniscus becomes unstable (snap-off), and the capillary will become fully saturated. Because here the modeling is limited to the unsaturated regime, the transport in the saturated region is not discussed.

In Fig. 6, evolution of the moisture in time and space is shown. It is observed that in the initial few seconds, moisture is transported mainly *via* the vapor phase, and the water activity quickly rises to $a_w \approx 1$. Thereafter, there is little driving force for vapor diffusion and diffusion *via* the solid phase, which becomes hydrated and swells. Transport *via* the capillary phase is still negligible. Practically all water is retained in the swollen solid phase. Only after a long time does this become important; finally, the capillary water in the first cell will snap off, trapping the enclosed gas.

9.2 Cooking of meats

This study is based on single-sided contact heating of hamburger patties, in which the patties are heated at a fixed-plate temperature of 140 °C. Initially, the patty is in a wet rubbery state, *i.e.*, the pores are completely saturated. As the temperature of the patty rises, denaturation of muscle proteins occurs, which leads to a decrease in the water holding capacity (WHC) of the meat, thereby releasing water from in between the muscle fibers. The patty stays in a soft and rubbery state during its entire period of cooking. According to the discussion in Section 4.2, the swelling pressure is the driving mechanism for moisture transport during meat cooking. It is obtained by linearizing the Flory–Rehner expression near equilibrium as shown in eqn (25).[72] Since the pores are completely saturated, capillary pressure is zero and there is no gas generation. The flux of water is written in terms of swelling pressure and is given by eqn (28). The energy balance equation that solves for temperature is given by eqn (4). The patty shrinks by 30% or more of its initial volume during contact heating, which necessitates the use of large deformation analysis. Since the patty stays saturated for its entire period of cooking, in eqn (17), for linear momentum balance,

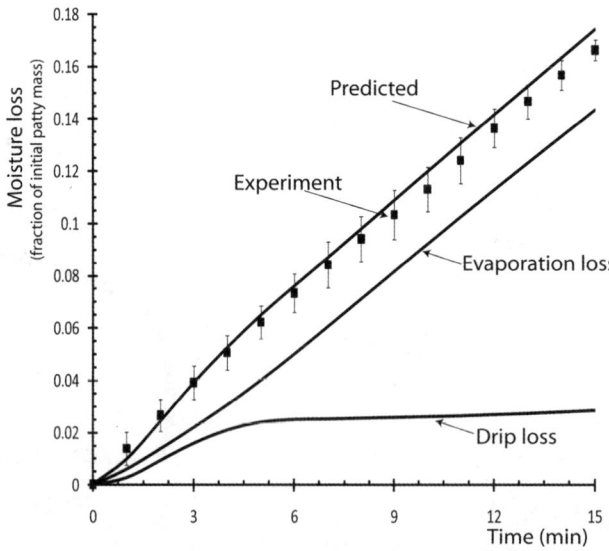

Fig. 7 Cumulative total moisture loss (predicted and experimentally observed), evaporation moisture loss (predicted) and drip loss (predicted) for single-sided contact heating of hamburger patties. The model has swelling pressure as the driving force that is related to the water holding capacity.[9]

the pore pressure is zero and the shrinkage is solely due to the amount of water lost. Stress and strain are related using a modified neo-Hookean model (which is an extension of Hooke's law) to account for the large changes in volume due to moisture loss.

Fig. 7 shows a comparison between the predicted and experimentally observed total moisture loss history of a hamburger patty heated from one side for 15 min. Evaporation loss from the surface is calculated using the standard surface convective mass transfer equations. Drip loss is considered to occur only when surface moisture concentration is more than the water holding capacity and is equal to the total moisture flux reaching the surface subtracted by the amount of surface evaporation. Rate of evaporation loss keeps increasing with time, although very slightly. Rate of drip loss, on the other hand, approaches zero at around five minutes as moisture concentration at the patty surface falls below equilibrium concentration. Evaporation loss is always higher than the drip loss. Spatial profiles (not shown here) demonstrate that the moisture gradients dominate along the thickness of the patty. Also, even at the end of heating, the minimum moisture content (near the surface being heated) is still at 0.89 dry basis. Moisture content at the unheated top surface actually rises from an initial value of 2.6 to 2.73. Since the meat stays rubbery during the process, local volume change at any point is equal to the volume of moisture loss. This being a local effect, distributing the total moisture loss over the entire patty volume, and thus predicting its diameter as cooking progresses, does not work and solid deformation equations are needed for a more accurate prediction of the change in diameter.

10   Other benefits of soft matter approach in relation to macroscale porous media models

The macroscale porous media models are smeared models that do not include the detailed structural heterogeneities at the small scale. Therefore, it should be possible to combine these with soft matter approaches to obtain multiscale models analogous to the multiscale work with continuum models at both scales using results from macroscale as boundary conditions for microscale (localization).[29]

11   Conclusions

Fundamental physics-based models are critical to the understanding of various food processes and their optimization, leading to the design of new products, processes and equipment. Soft matter approaches can nicely complement the macroscale porous media transport models that are the staple of physics-based food process modeling, by providing the material properties and other auxiliary relations that are critical to such models and are often impossible to obtain through direct experimentation. The relationships between macroscale porous media models and the predictions possible through soft matter approach were clearly identified in this study. Specifically, soft matter approach can provide the equilibrium properties (*e.g.*, water activity), transport properties (*e.g.*, permeability, mass diffusivity) and driving forces (*e.g.*, swelling pressure). These were discussed in detail and included the synthesis of developments in related areas such as soils and of theories in soft matter physics and poromechanics.

Soft matter approach provides the understanding that is an enabler in design and optimization in two ways, as follows.

(1) Directly: through physics-based and broader understanding of complex processes and transformations that a food material undergoes, by showing how a transport or other property will change with product and processing conditions. Such understanding reduces repeated experimentation and provides a more rational approach to design. As an example, food literature has often described moisture transport in food systems by lumping all modes of transport into an "effective

diffusivity" term, obtained as a fitting parameter from experimental data. As a result, little understanding is developed and the predictive power is minimal, since fresh experiments would have to be conducted for each product–process combination to obtain this "effective diffusivity." Soft matter based approach provides a more fundamental understanding of the complex moisture transport process by relating it to the state of the material and the process parameters in a more direct and explicit way, leading to greater predictability in terms of how moisture transport will change under different situations. Such understanding also makes learning easier as we can draw from similar physics studied in large classes of non-food materials such as soil, clay, concrete and wood.

(2) Indirectly: the soft matter approach is also an enabler in computer-aided or predictive food process engineering. This is the primary focus of this article. Physics-based food process models, needed for computer-aided food process engineering, require properties and related parameters that change with food and its processing condition; an exhaustive collection of such data is simply unattainable. Reliable property prediction makes computer-aided food process engineering possible, and soft matter approach to property prediction, although still in an emerging state, can be that enabler.

Acknowledgements

Author Datta greatly acknowledges the preliminary work on diffusivity prediction by Ashish Dhall of the Dept. of Biological and Environmental Engineering at Cornell University. Author van der Sman acknowledges support by the Dutch Food and Nutrition Delta programme (project FND080078U).

References

1 A. Z. Akcasu, The "fast" and "slow" mode theories of interdiffusion in polymer mixtures: Resolution of a controversy, *Macromol. Theory Simul.*, 1997, **6**(4), 679–702.
2 L. S. Bennethum, M. A. Murad and J. H. Cushman, Modified Darcy's law, Terzaghi's effective stress principle and Fick's law for swelling clay soils, *Comput. Geotech.*, 1997, **20**(3–4), 245–266.
3 N. Bertola, A. Bevilacqua and N. Zaritzky, Heat treatment effect on texture changes and thermal denaturation of Proteins in beef muscle, *J. Food Process. Preserv.*, 1994, **18**, 3146.
4 R. T. Bonnecaze and M. Cloitre, Micromechanics of soft particle glasses, *High Solid Dispersions*, 2010, **vol. 236**, 117–161.
5 A. K. Datta, Hydraulic permeability of food tissues, *Int. J. Food Prop.*, 2006, **9**(4), 767–780.
6 A. K. Datta, Porous media approaches to studying simultaneous heat and mass transfer in food processes. I: Problem formulations, *J. Food Eng.*, 2007, **80**, 80–95.
7 A. K. Datta, Porous media approaches to studying simultaneous heat and mass transfer in food processes. II: Property data and representative results, *J. Food Eng.*, 2007, **80**, 96–110.
8 A. K. Datta and A. Dhall, Modeling food process, quality and safety: Frameworks and practical aspects, *11th International Congress on Engineering and Food*, May 22–26, 2011, Athens, Greece. Peer-reviewed proceedings published in: A. K. Datta and A. Dhall, *Procedia Food Sci.*, 2011, **1**, 1202–1208.
9 A. Dhall and A. K. Datta, Transport in deformable food materials: A poromechanics approach, *Chem. Eng. Sci.*, 2011, **66**, 6482–6497.
10 M. Doi and A. Onuki, Dynamic coupling between stress and composition in polymer solutions and blends, *J. Phys. II*, 1992, **2**(8), 1631–1656.
11 M. Dong and I. Chatzis, The imbibition and flow of a wetting liquid along the corners of a square capillary tube, *J. Colloid Interface Sci.*, 1995, **172**(2), 278–288.
12 D. C. Esveld, R. G. M. van der Sman, M. M. Witek, C. W. Windt, H. van As, J. P. M. van Duynhoven and M. B. J. Meinders, Effect of morphology on water sorption in cellular solid foods. Part I: Pore scale network model, *J. Food Eng.*, 2012, **109**(2), 301–310.
13 D. C. Esveld, R. G. M. van der Sman, M. M. Witek, C. W. Windt, H. van As, J. P. M. van Duynhoven and M. B. J. Meinders, Effect of morphology on water sorption in cellular solid foods. Part II: Sorption in cereal crackers, *J. Food Eng.*, 2012, **109**(2), 311–320.
14 Tang Feng and Cavalieri Plumb, Intrinsic and relative permeability for flow of humid air in unsaturated apple tissues, *J. Food Eng.*, 2004, **62**, 185–192.

15 P. J. Flory and J. Rehner Jr, Statistical mechanics of cross-linked polymer networks II. swelling, *J. Chem. Phys.*, 1943, **11**, 521.

16 P. J. Flory and R. R. Garrett, Phase transitions in collagen and gelatin systems, *J. Am. Chem. Soc.*, 1958, **80**(18), 4836–4845.

17 P. J. Flory and O. K. Spurr, Melting equilibrium for collagen fibers under stress - elasticity in amorphous state, *J. Am. Chem. Soc.*, 1961, **83**(6), 1308–1316.

18 D. L. Goedeken and C. H. Tong, Permeability measurements of porous food materials, *J. Food Sci.*, 1993, **58**, 1329–1331.

19 J. M. Gosline, Structure and mechanical-properties of rubber-like proteins in animals, *Rubber Chem. Technol.*, 1987, **60**(3), 417–438.

20 M. Farid, The moving boundary problems from melting and freezing to drying and frying of food, *Chem. Eng. Process.*, 2002, **41**(1), 1–10.

21 G. Gillet, O. Vitrac and S. Desobry, Prediction of partition coefficients of plastic additives between packaging materials and food simulants, *Ind. Eng. Chem. Res.*, 2010, **49**(16), 7263–7280.

22 S. Guessasma, L. Chaunier, G. Della Valle and D. Lourdin, Mechanical modeling of cereal solid foods, *Trends Food Sci. Technol.*, 2011, **22**, 142–153.

23 T. Gulati and A. K. Datta, Food property prediction equations for enabling computer-aided food process engineering, *J. Food Eng.*, 2012, submitted.

24 A. Halder, A. Dhall and A. K. Datta, An improved, easily implementable, porous media based model for deep-fat frying. Part I: Problem formulation and input parameters, *Food Bioprod. Process.*, 2007, **85**(C3), 209–219.

25 A. Halder, A. Dhall and A. K. Datta, Modeling transport in porous media with phase change: applications to food processing, *J. Heat Transfer*, 2011, **133**, 031010.

26 X. He, A. Fowler and M. Toner, Water activity and mobility in solutions of glycerol and small molecular weight sugars: Implication for cryo-and lyopreservation, *J. Appl. Phys.*, 2006, **100**, 074702.

27 C. T. Herakovich, *Mechanics of fibrous composites*, 1998, Wiley, New York.

28 W. Herzog, V. Joumaa and T. R. Leonard, The force-length relationship of mechanically isolated sarcomeres, in: D. E. Rassier (Ed.), *Muscle Biophysics: From Molecules to Cells*, 2010, Springer, New York, pp.141–161.

29 Q. T. Ho, P. Verboven, B. E. Verlinden, E. Herremans, M. Wevers, J. Carmeliet and B. M. Nicolai, A three-dimensional multiscale model for gas exchange in fruit, *Plant Physiol.*, 2011, **155**, 1158–1168.

30 M. L. Hoang, P. Verboven, M. Baelmans and B. M. Nicolai, A continuum model for airflow, heat and mass transfer in bulk of chicory roots, *Transactions of the ASAE*, 2003, **46**(6), 16031611.

31 X. Jin, R. G. M. van der Sman and A. J. B. van Boxtel, Evaluation of the free volume theory to predict moisture transport and quality changes during broccoli drying, *Drying Technol.*, 2011, **29**(16), 1963–1971.

32 J. R. Kalnin, E. A. Kotomin and J. Maier, Calculations of the effective diffusion coefficient for inhomogeneous media, *J. Phys. Chem. Solids*, 2002, **63**, 449–456.

33 J. Kopp and M. Bonnet, Stress-strain and isometric tension measurements in collagen, in: A. M. Pearson, T. R. Dutson and A. J. Bailey (Ed.), *Advances in meat research: Vol. 4*, 1987, Van Nostrund Reinhold, New York, pp.163–185.

34 J. Kromkamp, A. Bastiaanse, J. Swarts, G. Brans, R. G. M. Van Der Sman and R. M. Boom, A suspension flow model for hydrodynamics and concentration polarisation in crossflow microfiltration, *J. Membr. Sci.*, 2005, **253**(1–2), 67–79.

35 J. Latt, B. Chopard, O. Malaspinas, M. Deville and A. Michler, Straight velocity boundaries in the lattice Boltzmann method, *Phys. Rev. E: Stat., Nonlinear, Soft Matter Phys.*, 2008, **77**(5), 056703.

36 J. Lepetit, A. Grajales and R. Favier, Modelling the effect of sarcomere length on collagen thermal shortening in cooked meat: consequence on meat toughness, *Meat Sci.*, 2000, **54**(3), 239–250.

37 J. Lepetit, A theoretical approach of the relationships between collagen content, collagen cross-links and meat tenderness, *Meat Sci.*, 2007, **76**(1), 147–159.

38 J. Lepetit, Collagen contribution to meat toughness: Theoretical aspects, *Meat Sci.*, 2008, **80**(4), 960–967.

39 Michaluk Lewicki, Drying of tomato pretreated with calcium, *Drying Technol.*, 2004, **22**(8), 1813–1827.

40 Z. Liu and M. G. Scanlon, Predicting mechanical properties of bread crumb, *Food Bioprod. Process.*, 2003, **81**, 224–238.

41 A. V. Luikov, Systems of differential equations of heat and mass transfer in capillary-porous bodies (review), *Int. J. Heat Mass Transfer*, 1975, **18**(1), 1–14.

42 Malaspinas, Lattice Boltzmann Method for the Simulation of Viscoelastic Fluid Flows, 2009, Ecole Polythecnique Federale de Lausanne.

43 D. Marinos-Kouris, Z. B. Maroulis, *Thermophysical properties for the drying of Solids*, in: A.S. Majumdar (Ed.), *Handbook of Industrial Drying*, 1995, Marcel Dekker, New York.

44 P. E. Mcclain, E. Kuntz and A. M. Pearson, Application of stress-strain behavior to thermally contracted collagen from epimysial connective tissues, *J. Agric. Food Chem.*, 1969, **17**(3), 629–632.

45 R. Mezzenga, P. Schurtenberger, A. Burbidge and M. Michel, Understanding foods as soft materials, *Nat. Mater.*, 2005, **4**(10), 729–740.

46 J. K. Mitchell, Components of pore water pressure and their engineering significance, *Clays Clay Miner.*, 1960, **9**, 162–184.

47 N. A. Mortensen, F. Okkels and H. Bruus, Reexamination of Hagen-Poiseuille flow: Shape dependence of the hydraulic resistance in microchannels, *Phys. Rev. E.*, 2005, **71**(5), 057301.

48 H. Ni, A. K. Datta and K. E. Torrance, Moisture transport in intensive microwave heating of wet materials: a multiphase porous media model, *Int. J. Heat Mass Transfer*, 1999, **42**, 1501–1512.

49 H. Ni and A. K. Datta, Heat and moisture transfer in baking of potato slabs, *Drying Technol.*, 1999, **17**, 2069–2092.

50 B. M. Nicolai, Multiscale modeling in food engineering, *J. Food Eng.*, to be submitted.

51 L. Oliver and M. B. J. Meinders, Dynamic water vapour sorption in gluten and starch films, *J. Cereal Sci.*, 2011, **54**(3), 409416.

52 Z. Pan, R. P. Singh and T. R. Rumsey, Predictive modeling of contact-heating process for cooking a hamburger patty, *J. Food Eng.*, 2000, **46**(1), 9–19.

53 C. Pan, M. Hilpert and C. T. Miller, Lattice-Boltzmann simulation of two-phase flow in porous media, *Water Resour. Res.*, 2004, **40**(1), W01501.

54 T. W. Patzek and D. B. Silin, Shape factor and hydraulic conductance in noncircular capillaries: I. One-phase creeping flow, *J. Colloid Interface Sci.*, 2001, **236**(2), 295–304.

55 T. W. Patzek and J. G. Kristensen, Shape factor correlations of hydraulic conductance in noncircular capillaries: II. Two-phase creeping flow, *J. Colloid Interface Sci.*, 2001, **236**(2), 305–317.

56 P. Perré and I. W. Turner, A 3-D version of TransPore: a comprehensive heat and mass transfer computational model for simulating the drying of porous media, *Int. J. Heat Mass Transfer*, 1999, **42**(24), 4501–4521.

57 M. Quintard and S. Whitaker, Two-phase flow in heterogeneous porous media: The method of large-scale averaging, *Transp. Porous Media*, 1988, **3**(4), 357–413.

58 V. Rakesh and A. K. Datta, Microwave puffing: Determination of optimal conditions using a coupled multiphase porous media - Large deformation model, *J. Food Eng.*, 2011, **107**, 152–163.

59 V. Rakesh and A. K. Datta, Transport in deformable hygroscopic porous media during microwave puffing, *AIChE J.*, 2012, DOI: 10.1002/aic.13793.

60 T. C. Ransohoff and C. J. Radke, Laminar flow of a wetting liquid along the corners of a predominantly gas-occupied noncircular pore, *J. Colloid Interface Sci.*, 1988, **121**(2), 392–401.

61 E. Roca, V. Guillard, B. Broyart, S. Guilbert and N. Gontard, Effective moisture diffusivity modelling *versus* food structure and hygroscopicity, *Food Chem.*, 2008, **106**, 1428–1437.

62 F. H. Silver, Y. P. Kato, M. Ohno and A. J. Wasserman, Analysis of mammalian connective tissue: Relationship between hierarchical structures and mechanical properties, *J. Long Term Eff. Med. Implants*, 1992, **2**, 165–198.

63 P. P. Singh, J. H. Cushman and D. E. Maier, Multiscale fluid transport theory for swelling biopolymers, *Chem. Eng. Sci.*, 2003, **58**, 2409–2419.

64 P. P. Singh, J. H. Cushman and D. E. Maier, Three scale thermomechanical theory for swelling biopolymeric systems, *Chem. Eng. Sci.*, 2003, **58**, 4017–4035.

65 M. B. Solomon, J. S. Eastridge, E. W. Paroczay, B. C. Bowker and M. Liu, Measuring Meat Texture, in: L. Nollet and F. Toldra (Ed.), *Handbook of Muscle Foods Analysis*, CRC Press, Boca Raton, FL, p. 479–502.

66 P. S. Takhar, Hybrid mixture theory based moisture transport and stress development in corn kernels during drying: Coupled fluid transport and stress equations, *J. Food Eng.*, 2011, **105**, 663–670.

67 J. Ubbink, M. I. Giardiello and H. J. Limbach, Sorption of water by bidisperse mixtures of carbohydrates in glassy and rubbery states, *Biomacromolecules*, 2007, **8**(9), 2862–2873.

68 J. Ubbink, A. Burbidge and R. Mezzenga, Food structure and functionality: a soft matter perspective, *Soft Matter*, 2008, **4**(8), 1569–1581.

69 R. G. M. van der Sman and E. Boer, Predicting the initial freezing point and water activity of meat products from composition data, *J. Food Eng.*, 2005, **66**, 469–475.

70 R. G. M. van der Sman and S. van der Graaf, Diffuse interface model of surfactant adsorption onto flat and droplet interfaces, *Rheol. Acta*, 2006, **46**(1), 3–11.

71 R. G. M. van der Sman, Moisture transport during cooking of meat: An analysis based on Flory–Rehner theory, *Meat Sci.*, 2007, **76**, 730–738.

72 R. G. M. van der Sman, Soft condensed matter perspective on moisture transport in cooking meat, *AIChE J.*, 2007, **53**, 2986–2995.

73 R. G. M. van der Sman, Prediction of enthalpy and thermal conductivity of frozen meat and fish products from composition data, *J. Food Eng.*, 2008, **84**, 400–412.

74 R. G. M. van der Sman and A. J. van der Goot, The science of food structuring, *Soft Matter*, 2009, **5**, 501–510.

75 R. G. M. van der Sman and M. B. J. Meinders, Prediction of the state diagram of starch water mixtures using the Flory–Huggins free volume theory, *Soft Matter*, 2011, **7**(2), 429–442.

76 R. G. M. van der Sman, Thermodynamics of meat proteins, *Food Hydrocolloids*, 2012, **27**, 529–535.

77 R. G. M. van der Sman and M. B. J. Meinders, Moisture diffusion in food materials, *Food Chemistry*, 2012, in press.

78 R. G. M. van der Sman, S. Khalloufi, and X. Jin, Moisture Sorption in fruits and vegetables, 2008, in preparation.

79 H. M. Vollebregt, R. G. M. van der Sman and R. M. Boom, Suspension flow modelling in particle migration and microfiltration, *Soft Matter*, 2010, **6**(24), 6052–6064.

80 J. S. Vrentas and J. L. Duda, Diffusion in polymer-solvent systems. I. Reexamination of the free-volume theory, *J. Polym. Sci.*, 1977, **15**(3), 403–416.

81 J. S. Vrentas and C. M. Vrentas, Sorption in glassy polymers, *Macromolecules*, 1991, **24**(9), 2404–2412.

82 R. Wallach, O. Troygot and I. S. Saguy, Modeling rehydration of porous food materials: II. The dual porosity approach, *J. Food Eng.*, 2011, **105**, 416–421.

83 A. H. Weerts, G. Lian and D. R. Martin, Modeling the hydration of foodstuffs: Temperature effects, *AIChE J.*, 2003, **49**(5), 1334–1339.

84 S. Whitaker, Simultaneous Heat, Mass and Momentum Transfer in Porous Media: A Theory of Drying, *Advances in Heat Transfer*, Vol. 13, 1977, Academic Press, New York.

85 R. Yamsaengsung and R. G. Moreira, Modeling the transport phenomena and structural changes during deep fat frying-Part 1: model development, *J. Food Eng.*, 2002, **53**(1), 1–10.

86 J. Zhang, A. K. Datta and S. Mukherjee, Transport processes and large deformation during baking of bread, *AIChE J.*, 2005, **51**(9), 2569–2580.

87 J. Zhang and G. Zografi, Water vapor absorption into amorphous sucrose-poly (vinyl pyrrolidone) and trehalose–poly (vinyl pyrrolidone) mixtures, *J. Pharm. Sci.*, 2001, **90**(9), 1375–1385.

88 H. Zhu, A. Dhall, S. Mukherjee and A. K. Datta, A model for flow and deformation in unsaturated swelling porous media, *Transp. Porous Media*, 2010, **84**(2), 335–369.

89 N. P. Zogzas and Z. B. Maroulis, Effective moisture diffusivity estimation from drying data. A comparison between various methods of analysis, *Drying Technol.*, 1996, **14**(7–8), 1543–1573.

90 P. Verboven, G. Kerckhofs, H. Mebatsion, Q. Ho, K. Temst, M. Wevers, P. Cloetens and B. Nicola, Three-dimensional gas exchange pathways in pome fruit characterized by synchrotron X-ray computed tomography, *Plant Physiol.*, 2008, **147**(2), 518–527.

Numerical study of the effect of thiol–disulfide exchange in the cluster phase of β-lactoglobulin aggregation

Rosanne N. W. Zeiler and Peter G. Bolhuis*

Received 20th February 2012, Accepted 17th April 2012
DOI: 10.1039/c2fd20030a

We report a numerical study of β-lactoglobulin aggregation using grand canonical Monte Carlo simulations of a simple lattice model in which the proteins are represented by a single lattice point and interact *via* a sum of a short-ranged attraction, and a long-ranged screened electrostatic repulsion. For certain values of the potential parameters we observe the so-called cluster phase, in which protein aggregates of finite size repel each other. The properties of the cluster phase are dependent on the salt concentration, the charge of the protein, and the strength of the short-ranged attraction. Disulfide bridges are modeled by covalent bonds between the lattice points, and can exchange with free thiols. Allowing the thiol–disulfide exchange leads to a severe lowering in the chemical potential of the cluster transition, or equivalently, a lower monomer density. Moreover, we find that the disulfide bridges (or the free thiol groups) are not uniformly distributed over the aggregate. The free thiol groups are significantly more abundant on the surface than in the core of the aggregate, making the surface more reactive than the inner core. This finding might explain why films made of β-lactoglobulin by cold gelation, after resolution, reconstitute finite aggregates rather than a monomer solution.

1   Introduction

Whey proteins are commonly used in the food industry as thickening agents.[1] More recently, these proteins have shown potential as components of edible protein films.[2] As such, their aggregation and gelation properties are of particular interest. While whey proteins are the subject of extensive research, a fundamental understanding of the underlying mechanisms of their aggregation is still lacking.[3,4] Not only do the whey-protein aggregate and gel properties depend on the pH, salt and protein concentration, but also on the temperature used for the initial aggregation and the time the proteins are kept at that temperature, as well as the mechanism used to obtain a gel (*e.g.* change in pH or salt concentration).[2]

Gelation can be induced in a one- or two-step procedure. In the one-step hot gelation process the sample is heated until the denatured protein forms a gel. In cold gelation small heat induced aggregates undergo a subsequent cold gelation process induced by, for example, changes in the pH, salt concentration or drying of the sample.[1] During both aggregation procedures the formation of intermolecular covalent bonds *via* thiol–disulfide exchange plays an important role. The most abundant whey protein, β-lacto-globulin (β-lac), contains both a free thiol group and two disulfide bridges. Therefore, a system containing only β-lac can undergo thiol–disulfide exchange by itself. In this exchange a free thiol group reacts with an existing disulfide bridge, breaking it, and

van't Hoff Institute of Molecular Sciences, Universiteit van Amsterdam, PO Box 94157, 1090 GD Amsterdam, Netherlands. E-mail: p.g.bolhuis@uva.nl

transferring its hydrogen atom to one of the sulfur atoms in the bridge, while reforming a bridge with the other sulfur atom. This mechanism preserves the total number of free thiol groups. The net result is an exchange of the disulfide bridge, which can lead to intermolecular covalent bonds, strengthening the aggregate. The relative importance of the thiol–disulfide exchange for the aggregation process as opposed to the other interactions, such as the hydrophobic effect or salt bridges, strongly depends on the experimental conditions, see *e.g.* ref. 5. Several studies have shown that irreversible aggregation only occurs when thiol–disulfide exchange is possible.[6–8]

In the cold gelation process, films formed by drying of heat induced aggregates of β-lac are stabilized by covalent disulfide bridges between the aggregates. Strikingly, upon resolubilisation with surfactants such as SDS the films fall apart into aggregates of the same size as the initial aggregates.[2] It is unclear why these aggregates are stable, whereas the films do fall apart. Here we will address this question by investigating the effect of thiol–disulfide exchange on the β-lac aggregates with a Monte Carlo simulation approach.

More generally, protein aggregation is driven by several short-ranged interactions. In addition to the above mentioned hydrophobic interactions, hydrogen bonds, salt-bridges and hydrophilic interactions can induce attraction. Upon denaturation at high temperature, buried residues are exposed and interact with neighboring proteins. These strongly attractive interactions are balanced by electrostatic long-ranged (screened) repulsions due to the net charge on the proteins (away from the iso-electric point). Given sufficiently strong attractive interactions, the aggregation process will result in macroscopically large aggregates. The structure and stability of such aggregates will often be determined by kinetics in addition to the underlying thermodynamic driving forces.

As in a theoretical study one needs to separate kinetics from equilibrium behavior, we focus in this paper first on equilibrium properties of β-lac aggregation. We employ a simple model system in which each protein is represented by a single particle interacting *via* a potential consisting of a short-range attraction and long-range repulsion. Computer simulations of such simple models have been previously employed and were able to reproduce aggregation and gelation properties of proteins[9] (see for a short review[10]). Particularly noteworthy is the work of Sciortino and coworkers, who showed the existence of cluster phases in colloidal and protein systems.[11,12] Protein aggregates keep on growing due to their gain in short ranged attractive energy until the repulsive Coulomb energy of the cluster becomes sufficiently large to inhibit further growth. The electrostatic repulsion between clusters stabilizes this phase. The process is thus akin to microphase separation of micelles. Indeed, the latter process is also analogous to a electrostatic charging process, limiting the cluster growth.[13]

In this work we specifically investigate the effect of the thiol–disulfide exchange reaction on the equilibrium properties of β-lac aggregation. To make the computations tractable we perform grand canonical Monte Carlo (GCMC) of the protein system on a cubic lattice. On this lattice each lattice point can hold one protein. Chemical bonds such as a disulfide bridge can be formed between neighboring lattice sites. Cubic lattice models have been used before to study polymerization reactions and colloidal gel formation.[14–16] Previous simulation studies of aggregation have also incorporated the formation of bonds, both irreversible and reversible, for example by adding a tunable infinitely narrow barrier at the edge of an attractive well.[17] Such an addition would influence the dynamics only, and not the thermodynamics. None of the aggregation modeling studies take the inherent restrictions of thiol–disulfide exchange into account. Specifically, the fact that a covalent bond can only be exchanged between a free thiol group, and a sulfur atom participating in a disulfide bond.

In the first part of the paper we focus solely on the aggregation behavior of proteins without the thiol–disulfide exchange. We investigate the influence of the salt concentration, the pH, and the aggregation strength (related to the heating temperature) on the phase behavior. In the second part we introduce the thiol–disulfide exchange move. We study the effect on the phase behavior, as well as the internal structure. In Section 4 we discuss these results and make a connection with experiments. We end with concluding remarks.

2 Methods

2.1 The model

The protein system is modeled by a cubic lattice in which each lattice point can be occupied by only one protein particle (excluded volume). An empty site represents the solvent. The protein particles interact *via* an effective potential consisting of a short-ranged attraction together with a long-ranged repulsion. The short-ranged part is set to a constant attractive energy ε, representing the average sum of the hydrophobic interactions, hydrogen bonds, salt-bridges and hydrophilic and short-ranged electrostatic interactions. Naturally, the value of ε will depend on the system parameters as well as the experimental conditions. Nevertheless, for specific experimental conditions this term can be considered a simple average of all fluctuating interactions. The long-ranged repulsion due to the charged proteins is described by a screened electrostatic potential.[18] The total potential as a function of the distance r between two protein particles is thus:

$$V(r) = \begin{cases} -\varepsilon & 1 \leq r \leq \sqrt{3} \\ A\exp(-\kappa(r-1)) & r > \sqrt{3} \end{cases} \tag{1}$$

Here, $\kappa = [(\varepsilon_0\varepsilon_r k_B T)/(2N_A e^2 I)]^{-1/2}$ is the inverse Debye length, $A = 2\pi R\varepsilon_0\varepsilon_r \Psi_0^2$, with R the radius of the particle, ε_0 and ε_r the vacuum and relative permittivity, N_A Avogadro's number, e the elementary charge, Ψ_0 the surface potential, k_B Boltzmann's constant, T the temperature and $I = \frac{1}{2}\sum_i c_i z_i^2$ the ionic strength. A salt concentration of 10 mM NaCl corresponds to a κ of 3 nm, which corresponds to 1 lattice unit, since the radius of β-lac is about $R \approx 1.5$ nm. At pH = 7, the surface potential[19] is $\Psi_0 \approx -17.6$ mV, yielding $A \approx 0.5k_B T$. In the following we use reduced variables, energies are in units of $k_B T$. All distances are measured in lattice units (1 lattice unit is around 3 nm).

Each protein particle has an attractive interaction ε with each of its 26 neighbors on the cubic lattice. The neighboring particles do not feel the electrostatic repulsion, to avoid lattice artifacts. If the repulsive potential would be used for all values of r, this would result in an effectively stronger attraction between particles separated by $\sqrt{3}$, leading to unrealistic preferential growth along the diagonal. To avoid this effect the screened repulsive electrostatic potential only acts on non-neighboring proteins, which means that short-ranged repulsions are effectively included in the short-ranged attraction ε. Although the above effect could be avoided by using an off-lattice model, for this particular study, in which we screen several important parameters in the grand canonical ensemble, the efficiency of a lattice model is preferable. We chose a value of ε around unity, leading to a net attractive energy per particle of around 26 $k_B T$, which does not seem unreasonably strong. Note that this value is the sum of the attractive and the repulsive potentials.

We note that the repulsive screened electrostatic potential is modeled as an exponential function rather than a Yukawa type potential, as this is more appropriate for macromolecules.[18] Nevertheless, the use of a Yukawa potential will not lead to qualitatively different results.

2.2 The thiol–disulfide exchange

The general thiol–disulfide exchange reaction can be written as:

$$S_i H + S_j S_k \rightleftharpoons S_i S_j + S_k H$$

where the indices i, j and k refer to a specific protein. For example, $i = j = k$ refers to a internal exchange, whereas $i \neq j \neq k \neq i$ means that the reaction occurs between three different proteins. We assume that the reaction is reversible, and has a standard Gibbs free energy $\Delta G = 0$ kJ mol^{-1} on average due to symmetry. However, this is a severe approximation as it assumes that all five cysteines are equal in reactivity, whereas in reality the position of the cysteines in the protein gives them a different reactivity.[20–23] Nevertheless, as a first estimate this assumption is useful, and will highlight the effect of bond entropy.

We further assume that the total number of disulfide bonds and free thiol groups in the system is constant. The reduction of two thiol groups resulting in a disulfide bond is therefore not taken into account. However, this reaction does occur during aggregation, and around 15% of the free thiols is lost by oxidation.[20,24]

In the simulation we do not describe the thiol–disulfide exchange reaction itself, but rather the effect on the distribution of covalent links between proteins, and the distribution of free thiols. The main effect of the thiol exchange is that the covalent bond will make a lattice particle immobile with respect to its partner. That is, particles that are covalently bonded cannot be moved apart in a Monte Carlo step, nor deleted in a grand canonical removal step. To describe the covalent bond formation due to the thiol–disulfide exchange we employ a very simple model. The only information needed for each protein is the binary state of each cysteine residue. Either a cysteine is in a free thiol state, or it is bound in a disulfide bridge to a specific cysteine residue on a specific protein. An exchange move is assumed only to occur between neighboring proteins.

2.3 Grand canonical Monte Carlo

The grand canonical Monte Carlo simulations incorporate standard displacement moves, cluster rotations, as well as translations, and insertions and deletions. To avoid kinetic trapping and to enable faster equilibration we implemented the cluster algorithm by Bhattacharyay and Troisi.[25] In this algorithm a (partial) cluster is recruited for a translation and/or rotation by adding particles to this cluster one by one. Each particle j that is a neighbor a particle i already belonging to cluster, is recruited with a probability $P_{i,j} = 1 - \exp(\beta' \min\{0, E_{ij}\})$ that depends on the pairwise energy E_{ij} and a tunable parameter β'. This step is repeated until no more particles can be recruited. The finally selected cluster is then translated or rotated, and the move is accepted with a modified Metropolis probability:

$$P_{\mathrm{acc}} = \min[1, \exp(-[\beta \Delta E - \beta' \Delta E^-])] \qquad (2)$$

where ΔE^- is the energy difference between the initial and the final state taking only the negative contributions to the energy into account. For the simulations with exchange, a regular cluster translation/rotation move was used.

The grand canonical ensemble was sampled using standard insertion and deletion moves.[26] We consider each possible protein cysteine configuration as a different species. The 12 possible configurations are depicted in Fig. 1. In the simulations with the thiol–disulfide exchange move only native monomers with two internal disulfide bridges and one free thiol group (C_120) can be inserted or deleted. While in principle also proteins with three thiol groups and one internal bridge (C_310), and proteins with five free thiols could be exchanged (C_500), it is convenient to only impose the chemical potential of a single species. The other species will find their equilibrium concentration through the thiol–disulfide exchange reaction. The reference system for the chemical potential for all simulations is thus an ideal gas of C_120 particles.

The thiol–disulfide exchange move was implemented as follows. Each β-lac protein has five cysteines. Each cysteine can either be in a disulfide bond, or it can be 'free'. For the exchange move two cysteines are randomly chosen from the total

This journal is © The Royal Society of Chemistry 2012

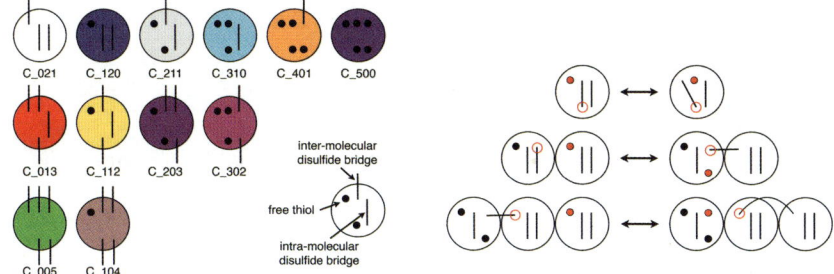

Fig. 1 Cartoon representations of the twelve possible protein configurations (left). The dots symbolize the free cysteines, while the lines represent the disulfide bridge. In the configuration code C_xxx, the first digit denotes the number of free cysteines, the second digit the number of intramolecular disulfide bonds, and the last digit the number of intermolecular disulfide bonds. C_120 is the native species with one free cysteine and two intramolecular disulfide bonds. Three different examples of an acceptable exchange move (and their reverse move) (right). In the reaction, the red dots denote the randomly chosen free cysteines, while the other randomly chosen cysteine, participating in a disulfide bond, is indicated by a red circle.

amount of cysteines in the system. If one of these is a free cysteine and the other is part of a disulfide bond, the bond is exchanged, provided that the two particles are neighbors of each other. The bound cysteine switches its disulfide bond to the free cysteine, while releasing its previously bound partner. Fig. 1 shows three examples of allowed exchange moves. Because we assume a vanishing ΔG for the exchange reaction there is no energy criterion associated with this move, *i.e.* when a move is allowed, it will be accepted with probability 1 in the Metropolis rule. The exchange move obeys detailed balance.

2.3.1 Computational details. The cubic lattice size was set to $30 \times 30 \times 30$ in all simulations. The cut-off of the potential was 15 lattice units (half the box length). The cluster algorithm parameter β' was set to 0.1. The maximum cluster translation was 10 lattice units for both normal and recruiting cluster translation. A run consisted of 90 000 Monte Carlo (MC) cycles for equilibration, followed by 100 000 production cycles. Each cycle consisted of 6000 monomer insertions/deletions, 100 recruiting cluster translations and 100 recruiting cluster rotations. For the exchange simulations we included 300 000 thiol–disulfide exchange attempts, and used normal instead of recruiting cluster moves.

The Metropolis rule acceptance ratios varied widely for different setting of the parameters. For insertions/deletions the range was 0.024 to 0.9, depending on the chemical potential μ. For cluster translations and rotations the acceptance ratios were between 0.0001 to 0.9. The lowest acceptances were either caused by very low chemical potentials (no particles to move) or very high chemical potentials (no space to move). The influence of the other interaction parameters was small. For the exchange moves acceptance ratios ranges between 0.0001 to 0.16. Here, the low values are caused by the intrinsically very high probability to choose non-neighboring cysteines.

2.4 Prediction of the cluster phase transition

The GCMC needs an input for the chemical potential range at which the clustering is expected to take place. In an equilibrium finite cluster phase the chemical potential per particle is equal throughout the system. That is, the chemical potential μ_1 of the monomers is equal to the chemical potential of an aggregate μ_n divided by the number of particles n:

$$\mu_1 = k_B T \ln \rho_1 + \mu_1^{ex} = \frac{1}{n}\mu_n = \frac{1}{n}k_B T \ln \rho_n + \frac{1}{n}\mu_n^{ex} \qquad (3)$$

Here ρ is the density, and μ_{ex} denotes the excess chemical potential. This leads to the well-known law of mass action equation:

$$\rho_n = \exp[-\beta\mu_n^{ex} + n\beta\mu_1] = \rho_1^n \exp[-\beta\Delta F(n)] = \exp[-\beta\Delta F(n) + n\beta\mu_1^{id}] \qquad (4)$$

where $\Delta F(n) = \mu_n^{ex} - n\mu_1^{ex}$ is the free energy of formation of a cluster of size n, and $\beta = 1/k_B T = 1$. This expression only applies at low density. To make an estimate for the cluster transition we thus need the free energy of the system. As a rough estimate, one can approximate the cluster free energy as the potential energy of a solid spherical cluster of size n (see also ref. 12). In Fig. 2 we plot the potential energy as a function of the size of the aggregates for different parameters A and κ. The expected equilibrium distribution of the clusters is given by applying eqn (4). From this distribution the range of chemical potentials for the monomer μ_1 can be deduced.

One can estimate the expected distribution ρ_n for a certain μ from a simulated cluster size distribution at another μ' from eqn (4), assuming the free energy $\Delta F(n)$ does not depend on μ.

$$\rho_n = \rho_n' \exp[-n(\mu - \mu')] \qquad (5)$$

This expected histogram can be used to investigate the influence of cluster–cluster interactions, as it is only valid for dilute solution. In Fig. 2 we also show a prediction of $-\ln\rho_n$, for several μ values. The minimum corresponds to the predicted cluster size.

3 Results

3.1 Grand canonical Monte Carlo without exchange move

To investigate the aggregation behavior of β-lac we first focus on the situation without exchange. We performed GCMC on a lattice size of $30 \times 30 \times 30$ for several values of ε, κ, A and μ. We chose $\varepsilon = 1, 1.2, \kappa = 0.9, 1.0, 1.2$ and $A = 0.4, 0.5, 0.6, 0.7$. As mentioned above, these values are around the estimated reasonable experimental conditions. For several combinations of parameters we ran a series of simulations at a range of μ, estimated using eqn (5). We analyze the simulations in terms of the total number of particles N, the aggregation fraction ϕ of particles involved in clusters, the number of large clusters, and the cluster size distribution. These numbers are computed using a cluster algorithm.[27]

We start with the case of $\varepsilon = 1.0$, $\kappa = 1.0$, $A = 0.5$. Here the cluster transition is expected between $\mu = -5$ and -7. As hysteresis can occur we conducted the simulations for $\varepsilon = 1.0$, $\kappa = 1.0$, $A = 0.5$, by increasing as well as decreasing the chemical

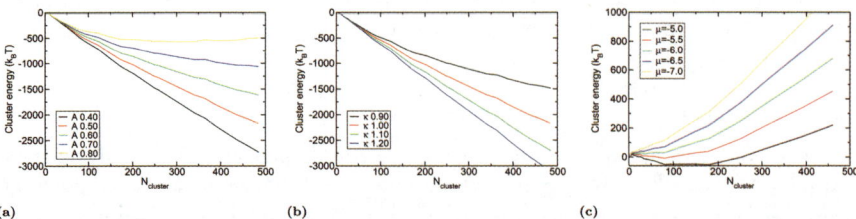

(a)   (b)   (c)

Fig. 2 Potential energy of spherical clusters as a function of the number of particles in the cluster. The influence of A, with $\varepsilon = 1.0$ and $\kappa = 1.0$ (a). The influence of κ, with $\varepsilon = 1.0$ and $A = 0.5$ (b). The energy at $\kappa = 1.0$, $\varepsilon = 1.0$ and $A = 0.5$ after applying eqn (4) (c).

potential from $\mu = -8$ to -5.4 in steps of 0.1. In Fig. 3 we plot the cluster size distribution, the aggregation fraction, the number of clusters and the cluster size as a function of μ. Fig. 3a clearly shows that at low chemical potential only monomers or small aggregates are present. Around $\mu = -6.0$ the aggregation fraction suddenly increases, indicating the formation of larger clusters. This jump is reminiscent of a phase transition and is indeed akin to the microphase separation in surfactant micelles. The hysteresis between ascending and descending along the chemical potential is only slight. The number of clusters in Fig. 3b shows a jump, although a bit more gradual. The observed hysteresis is also slightly larger, and is mainly caused by the difficulty of nucleating a new cluster from the gas phase close to the phase boundary. As the effective cluster–cluster interactions influence the cluster size, the average cluster size curves in Fig. 3c show hysteresis, but, surprisingly perhaps, not very much. Finally, the cluster size histograms in Fig. 3d show the finite size of the aggregate. Also here, hysteresis is apparent as a shift in both the position and the height of the peaks. Note that clusters of sizes between 10 and 40 hardly exists, as the (free) energy of such clusters is unfavorable. We can thus conclude that hysteresis, while present, is relatively small. We therefore only report the results obtained with descending chemical potential below.

To further investigate the effect of μ on the cluster size we extend the series of simulations for $\varepsilon = 1.0$, $\kappa = 1.0$, $A = 0.5$ by varying μ from -4.25 to -5.4. Fig. 4a shows the observed cluster size histograms. While the clusters on average increase in size with μ, this increase is not as large as expected. In Fig. 4b we show the normalized histograms for $\mu = -6$ to -5.4, together with the expected histograms from reweighting the $\mu = -6$ histogram, which is just above the phase transition, using eqn (5). The simulated cluster size grows much slower than the expected cluster size, due to repulsive inter-cluster interactions.[11] As the number of

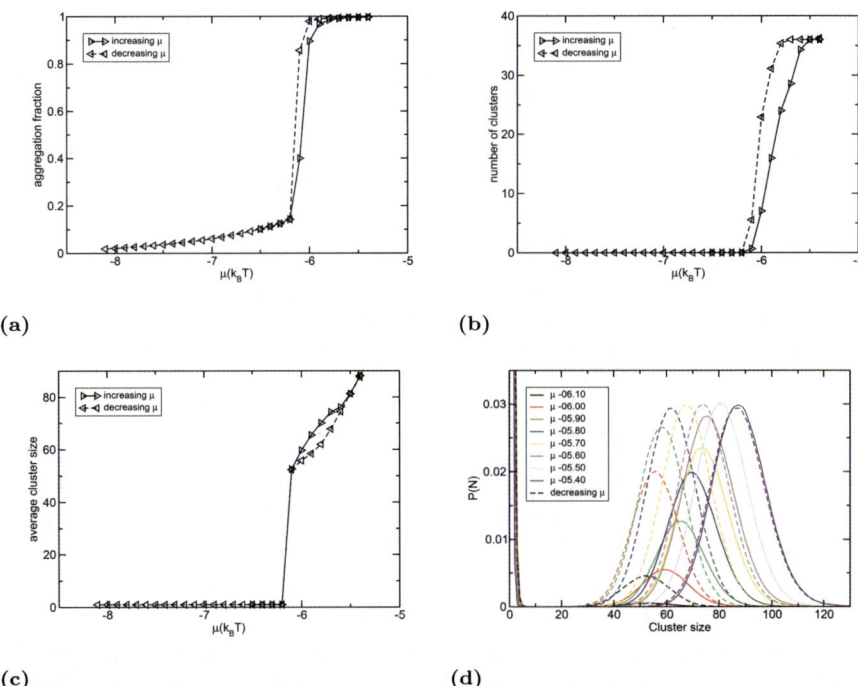

(a)   (b)

(c)   (d)

Fig. 3 Results for the GCMC simulations as a function of μ for $\varepsilon = 1.0$, $\kappa = 1.0$ and $A = 0.5$. (a) The aggregation fraction. (b) The number of clusters with a size larger than 7 monomers. (c) The average cluster size. (d) Cluster size distribution. To show the strength of hysteresis the simulations were performed by increasing as well as decreasing μ.

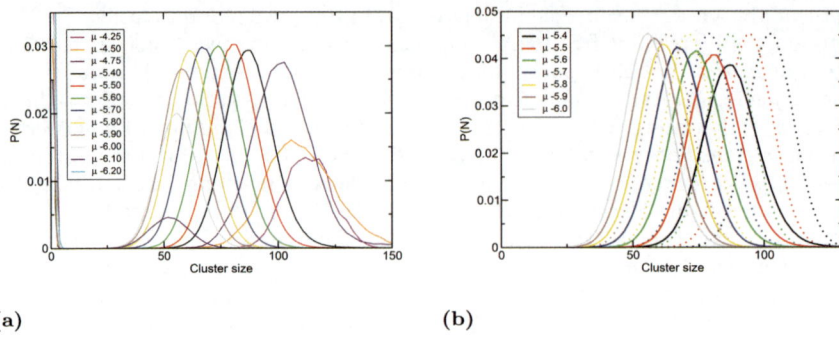

Fig. 4 (a) Cluster size histograms at various values of μ for $\varepsilon = 1.0$, $\kappa = 1.0$ and $A = 0.50$. (b) The histograms for $\mu = -6$ to -5.4 (solid lines) are compared to the expected histograms (dotted lines) obtained from reweighting the $\mu = -6$ histogram, using eqn (5).

clusters increases with μ additional repulsion arises between the aggregates, which increases the free energy, thus making a larger cluster less favorable. In addition, as μ increases the measured cluster size histograms reduce again in height. This is due to the appearance of elongated, joint clusters, or 'ribbons' (see Fig. 5c), which count as a single much larger cluster, and contribute to the histograms beyond $n = 200$ (not shown). The clusters also form cluster solids at high μ, for example at $\varepsilon = 1.0$, $A = 0.5$ cluster solids appear around $\mu = -5$. At $\varepsilon = 1.2$, $A = 0.6$ already at $\mu = -6.5$ a very clear cluster solid can be observed (see Fig. 5e). Even lamellar phases appear at $\varepsilon = 1.2$, $A = 0.4$ and $\mu = -6.0$. (see Fig. 5d). While very interesting, we do not investigate these phases further here, as these phases probably suffer from finite size effects, as their size is comparable to the box size. In contrast, the cluster phase consists of clusters that are relatively small compared to the box size, and therefore we expect no finite size effects for these phases. Moreover, the cluster–cluster interaction at long distances is still an exponential decaying function and hence vanishes completely at the cutoff. Indeed, a simulation of large box size ($50 \times 50 \times 50$) did not show different behavior besides much longer equilibration times.

Next, we investigated the influence of the ε, A and κ parameters on the cluster phase transition. Fig. 6 shows the aggregation fraction and the morphology diagrams for several combinations of ε, A and κ, as a function of μ. Introducing more attraction by increasing ε gives a sharper phase transition, more sensitive to μ, shifts the transition chemical potential to lower values and yields larger aggregates. Raising κ also leads to low μ values for the transitions, as well as larger clusters. In contrast, introducing a higher repulsion by increasing A shifts the transition to higher μ values, and smaller clusters.

Note that, effectively, changing ε to ε' is equivalent to multiplying A with a factor ε'/ε, sampled at a different temperature $\beta\varepsilon'/\varepsilon$. This scaling follows directly from the Boltzmann factor $\exp(-\beta U(r))$, with $U(r)$ the configurational energy. This scaling explains the opposite trends for the phase behavior between varying A and varying ε. The scaling of the effective temperature $\beta\varepsilon'/\varepsilon$ also largely explains the shift in μ for the transition when changing ε.

3.2 Grand canonical Monte Carlo with exchange

To study the influence of the thiol–disulfide exchange reaction, we include the exchange move as explained in Section 2.3. As in the simulations without exchange, we vary ε, κ, A and μ. We allow the exchange move between all cysteine neighbors within $r = \sqrt{3}$. This yields $2 \times 5 - 1 = 134$ possible exchange partners, including cysteines on the same protein. This number of exchange partners is probably larger than is possible in reality. Therefore, we compare these 27-neighbor simulations to

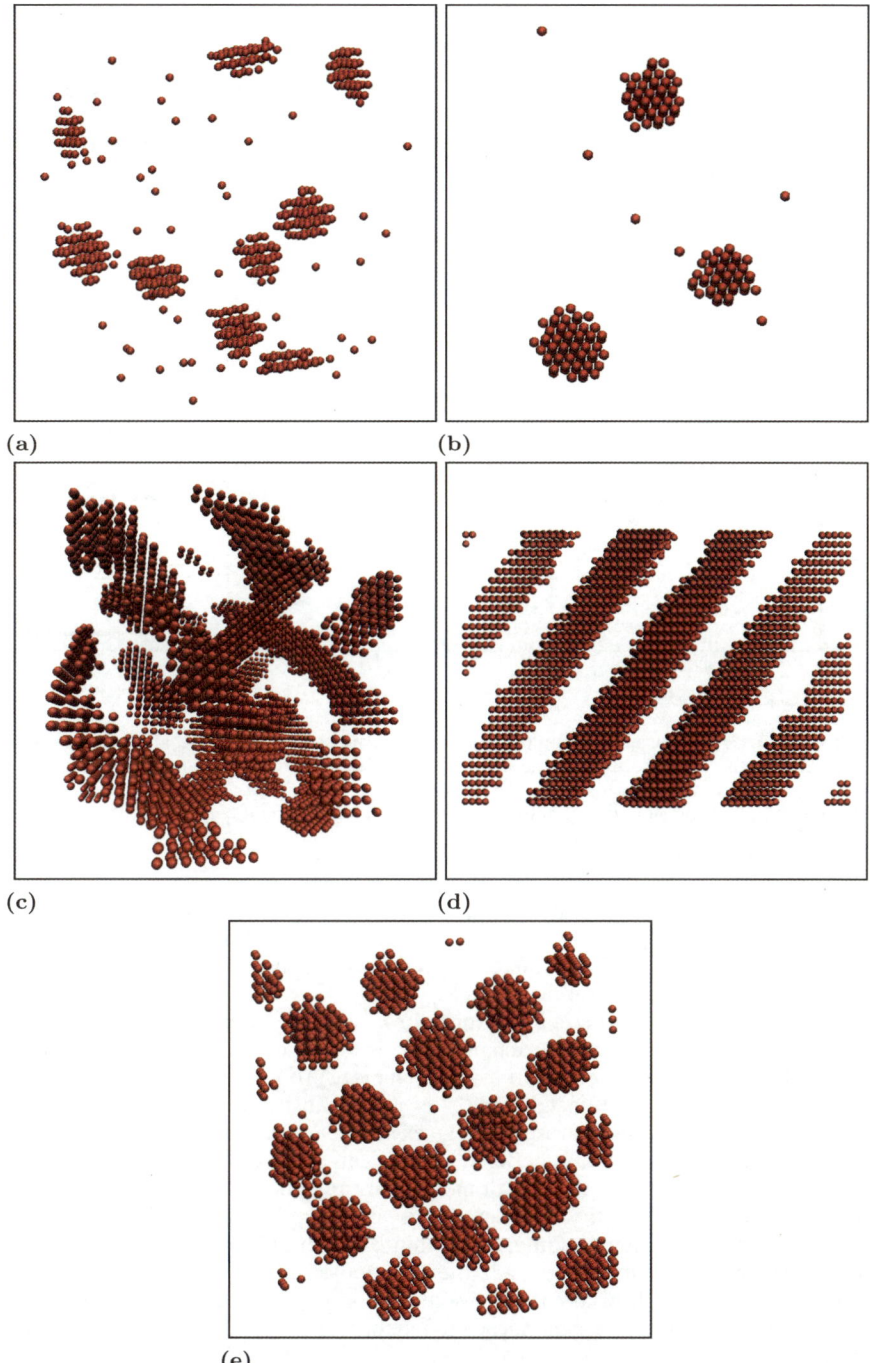

Fig. 5 Simulation snapshots of GCMC simulation without exchange. (a) $\kappa = 1.0$, $\varepsilon = 1.0$, $A = 0.50$ and $\mu = -6$. (b) $\kappa = 1.0$, $\varepsilon = 1.2$, $A = 0.50$ and $\mu = -6.8$. (c) Ribbons at $\kappa = 1.2$, $\varepsilon = 1.0$, $A = 0.50$ and $\mu = -6.6$. (d) Sheets at $\kappa = 1.0$, $\varepsilon = 1.2$, $A = 0.40$ and $\mu = -6$. (e) A cluster solid at $\kappa = 1.0$, $\varepsilon = 1.2$, $A = 0.60$ and $\mu = -6.5$, two to three clusters are superimposed in the direction into the paper.

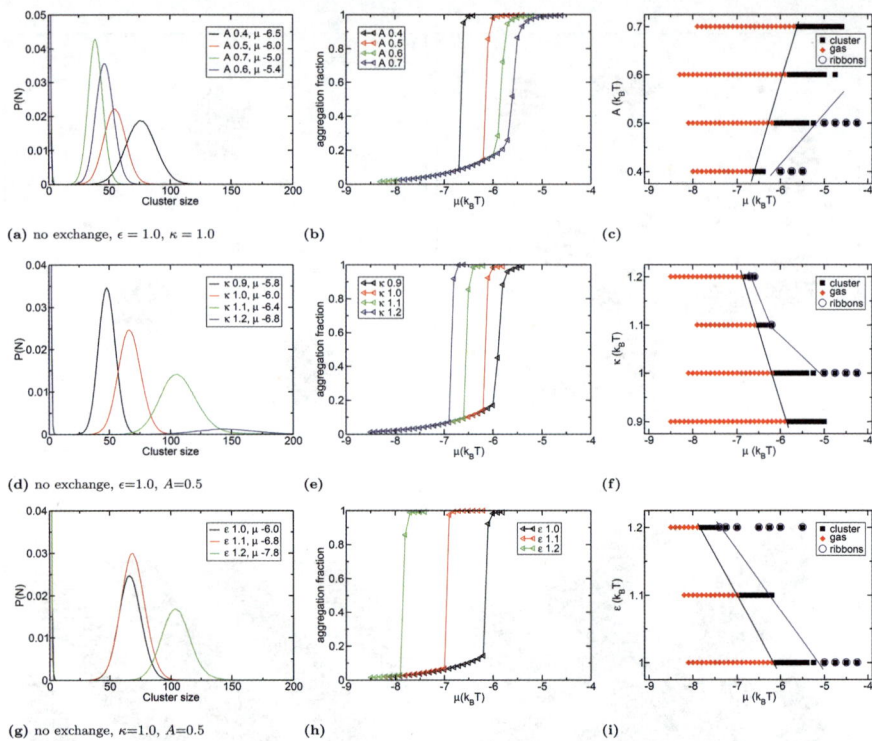

Fig. 6 Overview of the aggregation behavior without the exchange move. $\varepsilon = 1.0$, $\kappa = 1.0$, $A = 0.5$, unless otherwise specified in the graph. Varying A (top row). Varying κ (middle row). Varying ε (bottom row). Cluster size distribution close to the phase transition (left column). Aggregation fraction (middle column). Morphology diagrams (right column). Lines are added to guide the eyes.

a 7-neighbor situation in which we allow only exchange between $r = 1$ neighbors, which leads to $7 \times 5 - 1 = 34$ possible exchange partners. In reality the number of possible neighbors is somewhere between these extremes, most likely closer to 7-exchange than to the 27-exchange case.

Fig. 7 shows the cluster size histograms for $\kappa = 1.0$, $\varepsilon = 1$ and $A = 0.5$. The most striking difference with the no-exchange case is the formation of clusters at much lower chemical potential. This shift in μ is even more pronounced when the exchange is allowed between $r = \sqrt{3}$ neighbors (27-exchange) instead of only neighbors with $r = 1$ (7-exchange). This suggests that the exchange move introduces an additional effective attractive interaction between the proteins, as two proteins that are covalently bound by the exchange cannot move apart anymore, thus effectively keeping particles close to each other. However, this effect is very different from making a bond by putting a high, infinitesimally thin barrier in the potential,[17] as this will only change the dynamics, and not the thermodynamics, as we find here. That reversible bonds can lead to an effective attraction has also been found in ref. 28.

The strong effective interaction leads to kinetic trapping and slower equilibration. As a consequence of the low chemical potential at the cluster phase transition, there are almost no monomers in the system (see Fig. 8a). Indeed, the lower μ means the transition occurs at much lower protein concentration.

To get a feeling for the magnitude of the effective interaction, we investigated the behavior of the system when ε is lowered considerably. If the exchange is included, values as low as $\varepsilon = 0.2$ are enough to show aggregation, albeit with small clusters. In fact, even at $\varepsilon = 0$, aggregation is observed (data not shown). Fig. 7 shows for

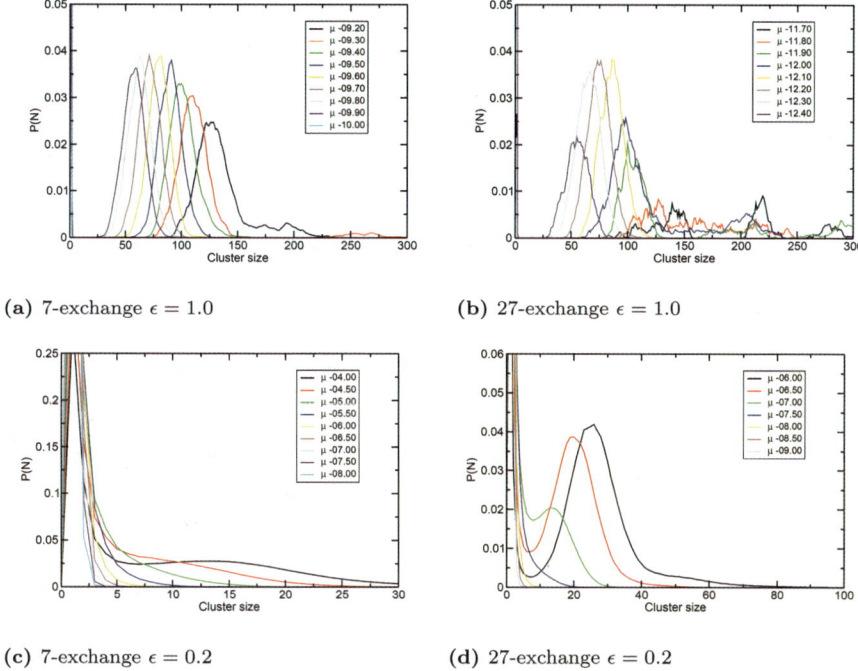

(a) 7-exchange $\epsilon = 1.0$

(b) 27-exchange $\epsilon = 1.0$

(c) 7-exchange $\epsilon = 0.2$

(d) 27-exchange $\epsilon = 0.2$

Fig. 7 Cluster size histograms for the two different exchange simulations, varying values of μ. In all simulations $\kappa = 1.0$ and $A = 0.5$.

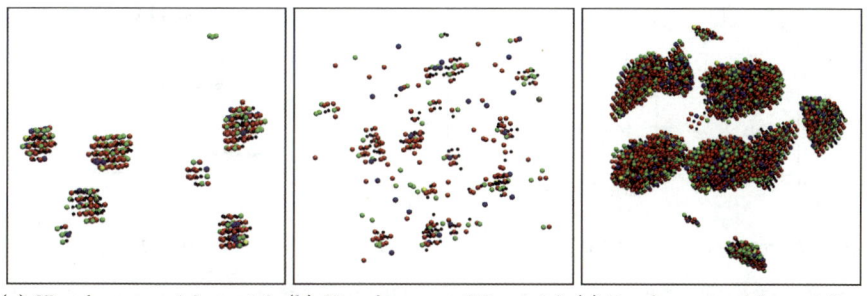

(a) 27-exchange $\epsilon = 1.0$, $\kappa = 1.0$ **(b)** 27-exchange $\epsilon = 0.2$, $\kappa = 1.0$ **(c)** 7-exchange $\epsilon = 1.0$, $\kappa = 1.2$

Fig. 8 Snapshots of simulations with the exchange move included. Proteins are colored according to the amount of free cysteines they have: 0 free cysteines - black, 1 free cysteine - red, 2 free cysteines - green, 3 free cysteines - blue, 4 free cysteines - yellow, 5 free cysteines - brown. (a) 27-exchange $\varepsilon = 1.0$, $\kappa = 1.0$, $A = 0.5$, $\mu = -12.2$. (b) 27-exchange $\varepsilon = 0.2$, $\kappa = 1.0$, $A = 0.5$, $\mu = -7.0$. (c) 7-exchange $\kappa = 1.2$, $\varepsilon = 1.0$, $A = 0.5$, $\mu = -10.5$.

$\kappa = 1.0$, $A = 0.5$ and $\varepsilon = 0.2$ the cluster size distribution for different μ values. In the 27-exchange case, we still observe a cluster phase, as can be deduced from the minimum in the histograms at low cluster size ($n \approx 10$), although the transition from gas to cluster is broader than at high ε. The aggregates are quite small, as can be seen in the snapshot in Fig. 8b. For the 7-exchange aggregation takes place (with an aggregation fraction of 0.97 for $\mu = -4$), but no clear cluster phase can be seen in the histograms.

In Fig. 9 we plot the morphology diagrams as a function of A, and of κ. For $\varepsilon = 1.0$ the morphology diagrams are qualitatively the same for the no-exchange,

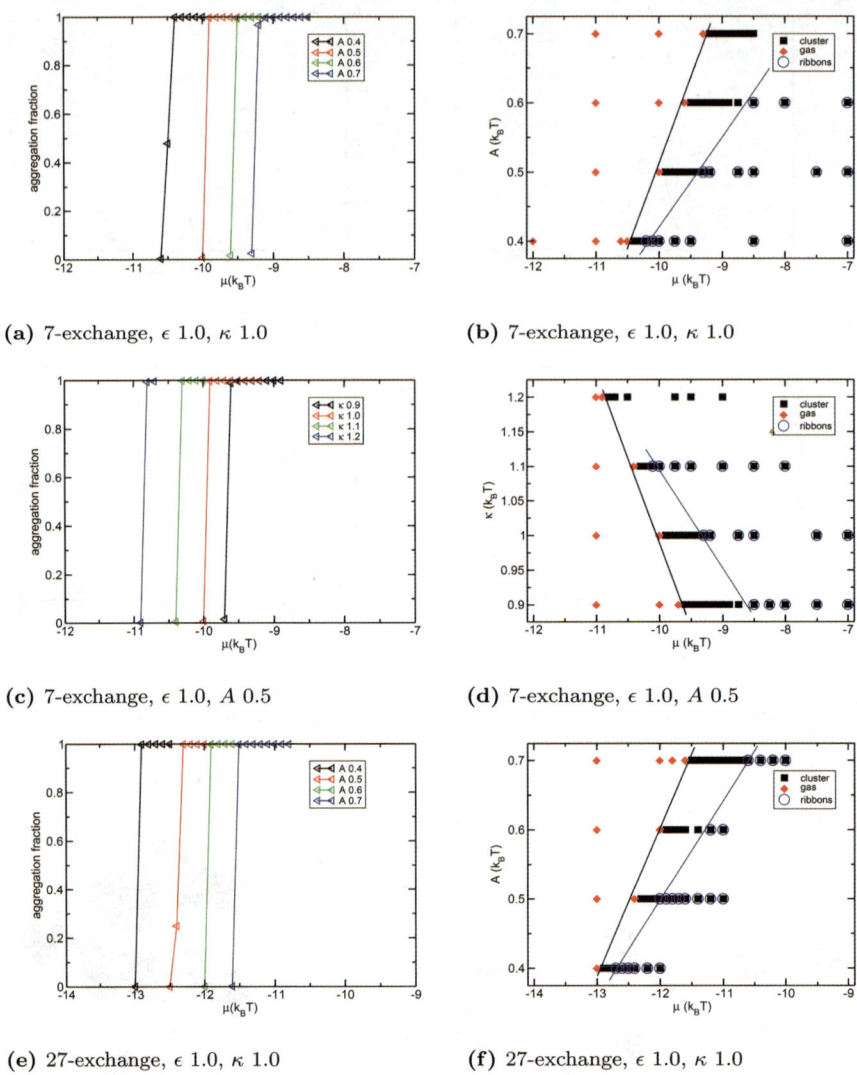

(a) 7-exchange, ϵ 1.0, κ 1.0

(b) 7-exchange, ϵ 1.0, κ 1.0

(c) 7-exchange, ϵ 1.0, A 0.5

(d) 7-exchange, ϵ 1.0, A 0.5

(e) 27-exchange, ϵ 1.0, κ 1.0

(f) 27-exchange, ϵ 1.0, κ 1.0

Fig. 9 Overview of the aggregation behavior with the exchange move. $\varepsilon = 1.0$, $\kappa = 1.0$ and $A = 0.5$ unless otherwise specified in the graph. Aggregation fraction (left), morphology diagrams (right). Lines are added to guide the eyes.

7-exchange and 27-exchange simulations. Except for a shift in the chemical potential, the phase transition line between the gas and the cluster phase shows the same dependence on A and κ for all simulation cases, with the same relative changes in μ.

Apart from the shift in μ, and hence the monomer density, the thiol exchange has an effect on the composition of the clusters. Fig. 10b shows a bar graph of the system composition, in which the length of the colored bar represents the mole fraction of each possible protein configuration. The composition depends only marginally on A and κ (data not shown), however, it depends greatly on ε. For $\varepsilon = 0.2$ the transition is more gradual and takes place at a high chemical potential compared to $\varepsilon = 1.0$. For $\varepsilon = 1.0$, we observe that the system composition of the 7-exchange differs from that of the 27-exchange simulation (see Fig. 10b). However, when we disregard the

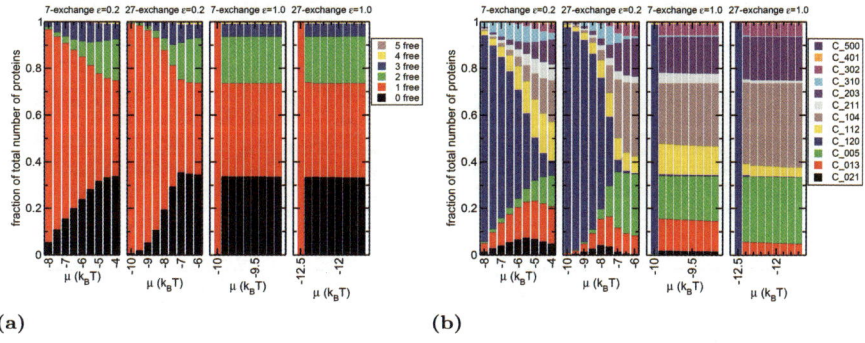

Fig. 10 Bar graphs depicting the system composition. (a) System composition according to the number of free cysteines on a protein. (b) Total fractions of all cysteine configurations. See Fig. 1 for an overview of the configurations in the same color code.

differences in intra- and intermolecular bonds the composition is the same (see Fig. 10a). Focusing on the bond types, the fraction of intermolecular disulfide bonds is higher for 27-exchange than the 7-exchange: there is relatively more of the C_005, C_104, C_203 and C_302 species present in the aggregates. This is expected as there are more available exchange partners in the 27-exchange than in the 7-exchange simulations. The same conclusion holds at $\varepsilon = 0.2$. In the transition region, intramolecular bonds are favored above intermolecular bonds, because proteins have simply fewer neighbors if there is less aggregation.

An interesting observation is that the cluster transition in the 7-exchange case at $\varepsilon = 0.2$ occurs around the same chemical potential, $\mu \approx -6$, as in the simulations without exchange at $\varepsilon = 1.0$. This means that $\varepsilon = 0.2$ with exchange has roughly the same effective interaction as $\varepsilon = 1.0$ without exchange, although there is no cluster phase in the former case.

3.3 Distribution of the free thiol groups

The thiol exchange has a pronounced influence on the value of the chemical potential, and hence the monomer concentration, at the transition, not on the topology of the morphology diagrams. However, the exchange reaction might also have an influence on the distribution of free thiol groups inside the aggregates. To investigate this we measure the composition in the clusters as a function of the distance to the center of mass of the cluster, in terms of the 12 possible cysteine configurations of each protein as identified in Fig. 1. The average number of proteins with a certain configuration, *e.g.* C_120, is calculated during the GCMC sampling, and binned according to the distance r to the center of mass of the cluster the proteins are part of. In Fig. 11a,b the number density $\rho_i(r)$ for each configuration type i is plotted as a function of r, for the 7-exchange and 27-exchange cases. The normalization is done by simply dividing by the expected number of proteins at a distance r, *i.e.* $\sim r^2$. Note that the density tails off around $r = 3$, the average radius of the clusters. The most abundant types are C_005, with all 5 cysteines involved in intermolecular bridges, and C_104, with one free thiol, but the remainder of the cysteines involved in intermolecular bridges. In Fig. 11c,d the composition fractions are shown. The C_104 type is still most frequent in the cores, as one might expect based on entropic arguments. However, towards the surface the C_005 fraction drops from over 30% to about 8% for the 7-exchange case, and in the 27-exchange case even from 40% to 10%. At the same time there is a steady increase of the configurations with multiple free thiols. This is made more clear in Fig. 11e,f where the composition is plotted in terms of configurations with a number of thiol groups. Here, the configurations with 2, 3 and 4 thiol groups are clearly more abundant closer to the surface. Strikingly, the fraction of configurations with 1 free thiol remains roughly constant. The

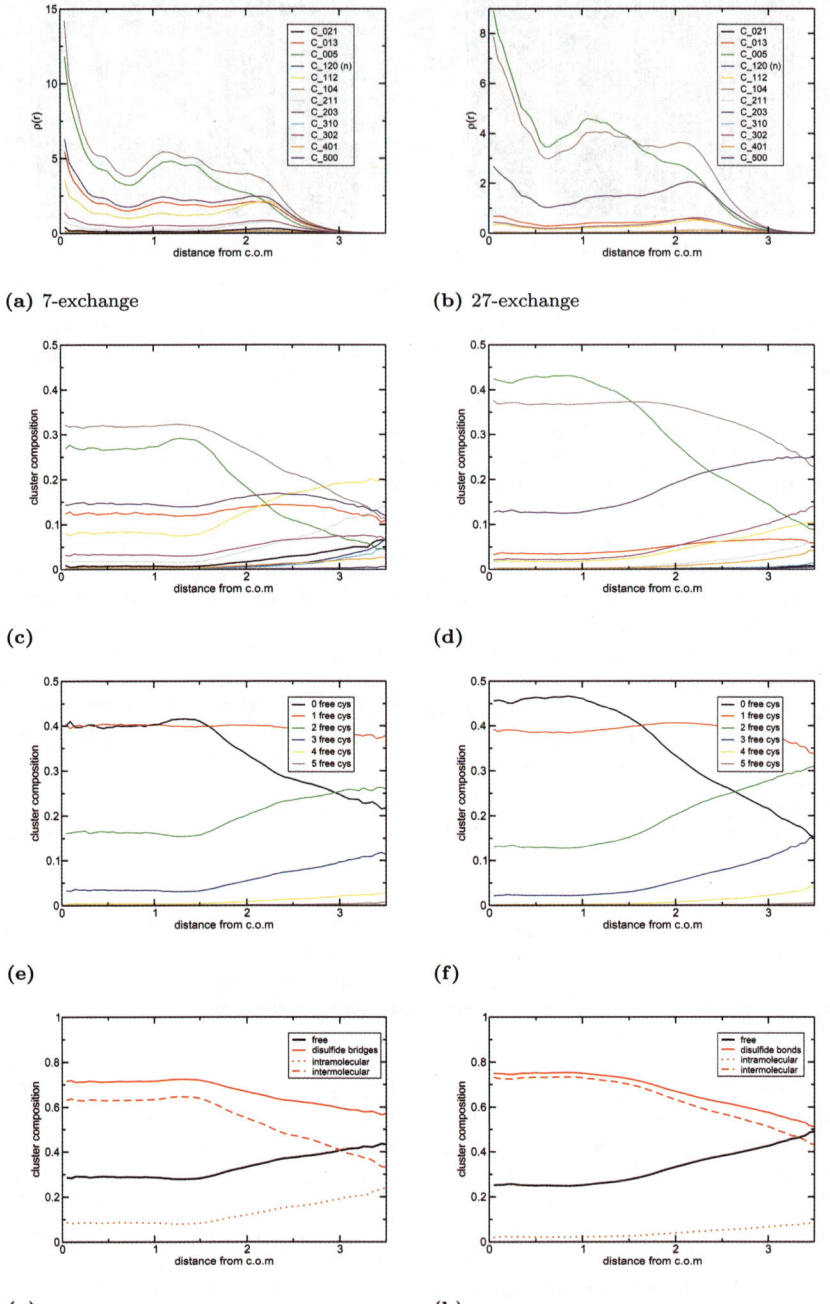

Fig. 11 Cluster composition as a function of the distance from the center of mass of the cluster. 7-exchange, $\varepsilon = 1.0$, $\kappa = 1.0$, $A = 0.5$, $\mu = -9.7$ (left). 27-exchange, $\varepsilon = 1.0$, $\kappa = 1.0$, $A = 0.5$, $\mu - 12.2$ (right). The top graphs show the number density for all possible configurations. The graphs below show the fractional composition in terms of the 12 possible configurations, the amount of free cysteines on particular protein and the total amount of thiols and disulfide bonds (subdivided into intra- and intermolecular) respectively.

reactivity of the aggregate depends on the total density of free thiol groups which is plotted in Fig. 11g,h. The concentration of free thiol groups almost doubles with respect to concentration in the center. On average the thiol to bridge fraction is 1 : 2 as the total number of free thiols is conserved. We also separate the fraction of bridge cysteines in inter- and intramolecular bridges. Here, the influence of the available partners for the exchange move is strongest. In the 7-exchange case around one quarter to a third of cysteines involved in a bridge, are part of an intramolecular bridge. In the 27-exchange case, the fraction of intramolecular bridges has significantly dropped. This difference is due to the number of available partners.

4 Discussion

In this discussion we interpret the simulation results in the light of previously published experimental data. In ref. 29, aggregates of finite size were observed after heating a 0.5 mmol L^{-1} β-lac solution for five hours at neutral pH, moderate temperature (68.5 °C) and a NaCl concentration of 0–0.02 M. The finite aggregates had a hydrodynamic diameter of around 30 nm. Such a size corresponds to roughly $r = 5$ in our model. At similar conditions another study found aggregates consisting of around 110 monomers.[24] While the precise correspondence of the experimental systems with our model is far from established, these numbers are in fact not too far from the cluster sizes we observed in the simulations. At the cluster transition the simulations predict aggregates of around 70 monomers, with a radius of 2.6. The difference in radius might well be due to a larger effective size of a denatured protein. In addition we assume a dense compact cluster, whereas in reality the cluster might be much more dilute. Since we are conducting equilibrium simulations, a compact cluster is the most stable state. Fractal-like clusters as experimentally found by, for example, Schmitt et al.,[30] might be found by employing dynamical simulation schemes.

The fact that the experiment finds a higher number of proteins in the cluster than we predict might be explained by different parameter settings of A, κ and ε. The parameter settings of A and κ in our simulations should be reasonably realistic, as they depend solely on the protein charge (pH) and the salt concentration. However, the ε parameter is not well defined, and depends on the sum of several interactions. The $\varepsilon = 1$ choice does not seem unreasonable as this amounts to a total net interaction per particle of $26k_{\mathrm{B}}T$, which is not excessively large for protein systems. Note that the effect of experimental temperature is also included in the ε parameter, as a higher experimental temperature will lead to more denaturation and hence a stronger attraction. In the simulations, a slightly stronger interaction already leads to higher cluster sizes (see Fig. 6). As mentioned in Section 3.1, increasing ε is equivalent to lowering A, simulated at a lower simulation reciprocal temperature β. However, in reality, a stronger interaction will also lead to more kinetic trapping, and hence fractal clusters. Still, the cluster sizes we find are not unreasonable in the light of the experimental results. Increasing the salt concentration leads to smaller Debye length, and a higher κ value, which reduces the chemical potential and hence the monomer concentration at which the transition takes place. Indeed, in experiment it is found that a higher (but still moderate) salt concentration gives faster aggregation, which suggests a higher thermodynamic driving force, i.e. larger chemical potential difference.[4]

Another way to compare experiments with simulation is through the fraction of non-aggregated proteins. In the above mentioned experiments the protein concentration is in the dilute regime, i.e. less than a few volume percent. This means if the experiments reach equilibrium there is a coexistence between the cluster phase and gas phase. The experimental concentration in the gas phase can thus be compared to the ideal gas density at the phase transition in the simulations. For instance in ref. 29 at low salt concentration the final non-aggregated monomer concentration fraction is found to be less than 0.005, which is equivalent to 0.0125 mM of protein. Our simulations without the exchange move predict for

$\kappa = 1$, $A = 0.5$ and $\varepsilon = 1$, a coexistence chemical potential of $\mu \approx -6$. This corresponds to a number density of $\rho = e^{\beta\mu} = 0.0024$, or equivalently, a concentration of $c = 0.153$ mM of protein. When we do include the exchange move we find a cluster coexistence chemical potential between $\mu \approx -10$ and $\mu \approx -12$, for the 7-exchange and 27-exchange case respectively. This corresponds to a number density range $\rho = e^{\beta\mu} = 4.5 \times 10^{-5}$ to 6.1×10^{-6}, which is about $c = 0.003$ mM to 0.0004 mM. The range of values for the exchange simulation are clearly more realistic, being only a factor 3 lower than the experiments. The difference in chemical potential upon introducing the exchange is so large, that neglecting the exchange move gives results that are clearly not in line with experimental findings.

One might still object that the settings of potential parameter $\kappa = 1$, and $\varepsilon = 1$ most likely do not correspond to the experiments, as the experiments were conducted with no salt added, and the ε value cannot be directly determined. However, lowering the salt concentration leads to a lower κ value, which leads in turn to higher values of the coexistence chemical potential μ (see Fig. 6) for the exchange simulations, more in line with the experiments. In principle one could also raise ε (for example at $\varepsilon = 1.5$ the cluster phase appears at $\mu \approx -10.4$) or, equivalently, lower A to explain the low aggregation density without the exchange move. However, this larger attraction/lower repulsion will quickly lead to bulk aggregation, rather than a cluster phase.

The large difference in the coexistence chemical potential between the exchange and no-exchange simulations might also explain the profound effect of inhibiting thiol–disulfide exchange by blocking the thiol groups at different values of the pH[5]. At pH $= 8$ the aggregation was significantly slower when the thiol group was blocked compared to native β-lac aggregation at pH $= 8$, while at pH $= 6$ the aggregation was sped up when the thiol groups were blocked. This suggests indeed that at pH $= 8$ aggregation is dominated by the disulfide bridge formation, whereas at pH $= 6$ non-covalent interaction is most important.

The effect of the thiol distribution is more difficult to connect to experiments. Although some measurements were done to determine the amount of free cysteines on the surface of whey protein aggregates,[31] these results cannot be interpreted correctly without knowing the density of the aggregates and how far the reagents used for determining free thiol groups can penetrate the aggregates. However, the activity of thiol–disulfide exchange is determined by the availability of both thiol groups and disulfide bonds. Hence, the finding that the surface contains as many thiols groups as bridges might explain why films made by the cold gelation process redissolve into finite aggregates. Upon the film formation the thiol–disulfide exchange covalently links the aggregates. Because the thiol group concentration is higher at the aggregate surface than in the core, resolution might also be easier achieved.

5   Conclusions

In this work we have investigated the finite cluster phase of aggregating β-lac proteins, using a simple lattice model. In line with previous work by other authors (*e.g.* ref. 11), we found that the finite cluster phase transition is dependent on the strength of the short ranged attraction ε representing the denatured protein's attraction, on the parameter A representing the protein charge, which in turn depends on pH, as well as on the Debye length κ^{-1} set by the salt concentration. Increasing ε or κ, or decreasing A lowers the transition chemical potential, and leads to large cluster sizes. Eventually the aggregates form ribbons, and even lamellar phases and cluster solids.

Incorporating the thiol–disulfide exchange yields very similar morphology diagrams but for much lower chemical potential, or equivalently lower protein concentration. Also, the fraction of free cysteines is around twice as high at the surface of the aggregates compared to the core. This finding might have consequences for the reactivity of the aggregates in β-lac film formation, and the experimentally observed resolution.[2]

Because the model is generic and computationally very efficient it can be easily extended to more complex systems that aggregate by forming intermolecular disulfide bridges *via* the thiol–disulfide exchange reaction. In future work we will employ this model to simulate for example mixtures of whey proteins.

Acknowledgements

The authors thank R. Floris and I. Bodnár for stimulating discussions on the aggregation of β-lactoglobulin. This work is part of the research programme of the Foundation for Fundamental Research on Matter (FOM), which is financially supported by the Netherlands Organisation for Scientific Research (NWO).

References

1 C. M. Bryant and D. J. McClements, *Trends Food Sci. Technol.*, 1998, **9**, 143–151.
2 R. Floris, I. Bodnár, F. Weinbreck and A. C. Alting, *Int. Dairy J.*, 2008, **18**, 566–573.
3 J. N. de Wit, *Trends Food Sci. Technol.*, 2009, **20**, 27–34.
4 T. Nicolai, M. Britten and C. Schmitt, *Food Hydrocolloids*, 2011, **25**, 1945–1962.
5 M. A. M. Hoffmann and P. J. J. M. van Mil, *J. Agric. Food Chem.*, 1997, **45**, 2942–2948.
6 Y. J. Cho, W. Gu, S. Watkins, S. P. Lee, T. R. Kim, J. W. Brady and C. A. Batt, *Protein Eng., Des. Sel.*, 1994, **7**, 263–270.
7 T. V. Burova, N. V. Grinberg, R. W. Visschers, V. Y. Grinberg and C. G. de Kruif, *Eur. J. Biochem.*, 2002, **269**, 3958–3968.
8 D. Jayat, J. C. Gaudin, J. M. Chobert, T. V. Burova, C. Holt, I. McNae, L. Sawyer and T. Haertle, *Biochemistry*, 2004, **43**, 6312–6321.
9 A. Giacometti, D. Gazzillo, G. Pastore and T. K. Das, *Phys. Rev. E: Stat., Nonlinear, Soft Matter Phys.*, 2005, **71**, 031108.
10 T. Nicolai and D. Durand, *Curr. Opin. Colloid Interface Sci.*, 2007, **12**, 23–28.
11 F. Sciortino, S. Mossa, E. Zaccarelli and P. Tartaglia, *Phys. Rev. Lett.*, 2004, **93**, 055701.
12 S. Mossa, F. Sciortino, P. Tartaglia and E. Zaccarelli, *Langmuir*, 2004, **20**, 10756–10763.
13 L. Maibaum, A. R. Dinner and D. Chandler, *J. Phys. Chem. B*, 2004, **108**, 6778–6781.
14 E. Del Gado, A. Fierro, L. de Arcangelis and A. Coniglio, *Europhys. Lett.*, 2003, **63**, 1–7.
15 E. Del Gado, A. Fierro, L. de Arcangelis and A. Coniglio, *Phys. Rev. E: Stat., Nonlinear, Soft Matter Phys.*, 2004, **69**, 051103.
16 E. Del Gado and W. Kob, *Europhys. Lett.*, 2005, **72**, 1032–1038.
17 I. Saika Voivod, E. Zaccarelli, F. Sciortino, S. V. Buldyrev and P. Tartaglia, *Phys. Rev. E: Stat., Nonlinear, Soft Matter Phys.*, 2004, **70**, 041401.
18 J. Israelachvili, *Intermolecular and Surface Forces: Revised Third Edition*, Academic press, 2011.
19 Shlomo Magdassi, Yelena Vinetsky and Perla Relkin, *Colloids Surf., B*, 1996, **6**, 353–362.
20 J. J. Kehoe, A. Brodkorb, D. Molle, E. Yokoyama, M. H. Famelart, S. Bouhallab, E. R. Morris and T. Croguennec, *J. Agric. Food Chem.*, 2007, **55**, 7107–7113.
21 L. K. Creamer, H. Bienvenue, H. Nilsson, M. Paulsson, M. van Wanroij, E. K. Lowe, S. G. Anema, M. J. Boland and R. Jimenez Flores, *J. Agric. Food Chem.*, 2004, **52**, 7660–7668.
22 Y. D. Livney, E. Verespej and D. G. Dalgleish, *J. Agric. Food Chem.*, 2003, **51**, 8098–8106.
23 Y. Surroca, J. Haverkamp and A. J. R. Heck, *J. Chromatogr., A*, 2002, **970**, 275–285.
24 M. A. M. Hoffmann and P. J. J. M. van Mil, *J. Agric. Food Chem.*, 1999, **47**, 1898–1905.
25 A. Bhattacharyay and A. Troisi, *Chem. Phys. Lett.*, 2008, **458**, 210–213.
26 D. Frenkel and B. Smit, *Understanding molecular simulation: from algorithms to applications*, Academic Pr, 2002, vol. 1.
27 R. Pool, PhD thesis, University of Amsterdam, 2006.
28 L. A. Pugnaloni, R. Ettelaie and E. Dickinson, *Langmuir*, 2003, **19**, 1923–1926.
29 M. Verheul, S. P. F. M. Roefs and K. G. de Kruif, *J. Agric. Food Chem.*, 1998, **46**, 896–903.
30 C. Schmitt, C. Moitzi, C. Bovay, M. Rouvet, L. Bovetto, L. Donato, M. E. Leser, P. Schurtenberger and A. Stradner, *Soft Matter*, 2010, **6**, 4876–4884.
31 A. C. Alting, R. J. Hamer, G. G. de Kruif and R. W. Visschers, *J. Agric. Food Chem.*, 2000, **48**, 5001–5007.

A multiscale approach to triglycerides simulations: from atomistic to coarse-grained models and back†

Antonio Brasiello,[a] Silvestro Crescitelli[b] and Giuseppe Milano[*c]

Received 24th February 2012, Accepted 3rd May 2012
DOI: 10.1039/c2fd20037f

The aim of this paper is to provide a simulation strategy to study the liquid–solid transition of triglycerides. The strategy is based on a multiscale approach. A coarse-grained model, parameterized on the basis of reference atomistic simulations, has been used to model the liquid–solid transition. A reverse mapping procedure has been proposed to reconstruct atomistic models from coarse-grained configurations and validated against experimental structural properties. The nucleation and growth of the crystalline order have been analysed in terms of several properties.

1 Introduction

Triglycerides are triesters of glycerol with fatty acids. The interest in triglycerides covers several fields ranging from combustion[1] to food technology. In food science, the occurrence of their crystal structures dictates end product quality loss.[2]

Triglycerides show several crystal modifications[3] and a liquid-crystalline phase.[4] The three major polymorphic forms are characterized by different packing and conformations of molecules. The main crystal forms are the α and β polymorphs. The α, β and β′ polymorphs are metastable, while β is stable. Polymorphs are characterized by precise molecular planar conformations named, according to their shape, tuning-fork, chair and trident conformation. FT-IR studies show that crystal forms are characterized by tuning-fork and chair conformations.[5] Trident conformations are observed only in monolayers at the hydrophobic–hydrophilic interface.[6]

The crystallization process of triglycerides is important from an industrial point of view and it is necessary for a rational design and engineering of food products.[7] For a very recent review covering several aspects of triglycerides in food material the reader can refer to ref. 7.

From a fundamental point of view, the mechanism of the crystallization in its earliest stages is an important problem.[8] The characterization of crystallization at an atomistic level of detail is a challenge to both experiment and theory. The processes of crystal nucleation and growth are slow if compared with the timescales reachable by atomistic simulations. The nucleation is a rare event and the crystal growth is characterized by a timescale longer than microseconds. The computational efforts to observe in an atomistic simulation a spontaneous crystallization process

[a]University of Salerno, Department of Industrial Engineering, via Ponte Don Melillo, 84084 Fisciano (SA), Italy
[b]University of Naples "Federico II", Department of Chemical Engineering, Piazzale Tecchio 80, 80125 Napoli, Italy
[c]University of Salerno, Department of Chemistry, via Ponte Don Melillo, 84084 Fisciano (SA), Italy. E-mail: gmilano@unisa.it

† Electronic Supplementary Information (ESI) available. See DOI: 10.1039/c2fd20037f

are typically too large and often special techniques (directed simulations) aimed to induce crystallization are needed to make the simulations possible.[9] The accuracy of these approaches depends on the choice of a suitable reaction coordinate. A very important point is the formulation of methods and procedures able to generate unbiased results. For a good and up to date overview of computational studies of crystallization processes the reader can refer to the review of Anwar et al.[9]

Despite the large difficulty of observing nucleation in MD simulations, the spontaneous nucleation of ice has been obtained using a large amount (years) of super-computing resources.[10]

On the other hand, simple generic models, coarse-grained (CG) models have been used to understand the basic features of crystallization processes. Auer and Frenkel pioneered the study of nucleation using hard spheres and Lennard-Jones particles.[11–14]

These simple models are very useful but they lack any molecular features and cannot reproduce more complex phenomena involving specific interactions and molecular conformations. To possibly include these aspects, more specific GC models can be developed.

In the last few years, several such coarse-grained models for biomolecules and synthetic polymers have been proposed.[15,16] Coarse graining strategies have been successfully proposed for several soft matter systems.[17]

In the study of crystallization processes several results have been obtained using coarse-grained models mapped on the basis of reference atomistic simulations.[16] Meyer reported the first computer experiments with polymer melts of chains long enough to form chain-folded structures.[18] In particular, using a model for poly(vinyl alcohol) (PVA) derived by a systematic coarse-graining procedure[19] from fully atomistic simulations of a melt, structures resembling the lamellae of polymer crystals have been obtained.[18] Furthermore, Vettorel and Meyer, using a similar approach, derived coarse-grained models of polyethylene in the melt state with the aim to study polymer crystallization.[20]

With these precedents, due to the relevance of triglycerides to food processing, very recently, three different coarse-grained models for tridecanoin melts, developed on the basis of reference atomistic simulations, were compared.[21] Interestingly, one of these models was found to be suitable to predict liquid–solid phase transitions.

In this paper, we analyse in detail, using the developed coarse-grained model, the crystallization process of tridecanoin. Furthermore, we propose a reverse mapping scheme able, starting from the coarse-grained systems, to rebuild the atomistic degrees of freedom. Following this approach, several models (at different stages of crystallization) can be obtained and analysed at an atomistic level of detail. According to this scheme, the main purpose of this contribution is the development and validation of a multiscale strategy suitable to study at an atomistic level the process of crystallization of triglycerides. The procedure of atomistic simulations → derivation of a coarse-grained model; coarse-grained simulations of mesoscale phenomena → reverse-mapping and local relaxation of the atomistic model is proposed as an efficient way to obtain detailed information on the crystallization process of this relevant class of molecules.

This paper is organized as follows: section 2 will guide the reader to the earlier work that is essential for understanding the present investigation. In section 3, technical details of both coarse-grained and atomistic simulations are reported. The proposed reverse-mapping strategy and the approach for the reinsertion of atomistic degrees of freedom are reported in sections 4.1 and 4.2. The application of the method and the calculated structural properties of relaxed atomistic models are described in section 5.

2 Mesocale model for triglycerides

The development of a coarse-grained model is based on the choice of a mapping scheme. According to this choice, molecules can be represented by effective beads grouping several atoms, which interact with each other through non-bonded and bonded potentials. A common practice is to derive coarse-grained models from reference atomistic simulations.[16] In this framework, the expression n to 1 mapping scheme refers to the number of atoms grouped in one effective particle. The potentials between these effective particles are adjusted to reproduce structural properties. The structure of an ensemble of molecules in an environment is described by the distributions of geometrical quantities. These distributions are extracted from atomistic simulations and are used as targets to be reproduced by the coarse-grained model.

The mapping scheme for the model adopted in the present work is schematised in Fig. 1. In this model, non-bonded interactions are calculated considering four effective particles of type C and N belonging to two different subtypes (C1, C2, N1, and N2). Bond and angle potentials are calculated considering only the particle types C and N. Bonded interactions consist of pairwise (two-body, see eqn (1)) and angles (three-body, see eqn (2)) harmonic potentials. Non-bonded interactions are modelled using 12–9 Lennard-Jones potential (eqn (3)).

$$V_{bond}\left(r_{ij}\right) = \frac{1}{2}k_{ij}\left(r_{ij} - r_{ij}^{o}\right)^2 \tag{1}$$

$$V_{angle}\left(\theta_{ijk}\right) = \frac{1}{2}k_{ijk}\left\{\cos\left(\theta_{ijk}\right) - \cos\left(\theta_{ijk}^{o}\right)\right\}^2 \tag{2}$$

$$V_{nonbonded}\left(r_{ij}\right) = 4\varepsilon_{ij}\left[\left(\frac{\sigma_{ij}}{r_{ij}}\right)^{12} - \left(\frac{\sigma_{ij}}{r_{ij}}\right)^{6}\right] \tag{3}$$

A complete list of bond, angle and non-bonded parameters is given in Tables 1 and 2. A full description of the model, its parameterization based on the matching between reference structural properties of the underlying atomistic model, is out of the scope of the present paper and can be found in ref. 21.

Technical details of coarse-grained simulations are reported in the next section.

3 Computational methods

All MD simulations, both atomistic and coarse-grained, have been performed using a leap-frog algorithm *via* GROMACS 4.5.4 software package. In the following, technical details of the simulations will be given.

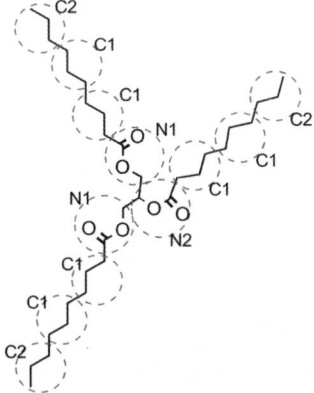

Fig. 1 Mapping scheme of a tridecanoin molecule.

Table 1 Bond and angle parameters

Bond	r^o [nm]	k_{ij} [kJ mol^{-1} nm^{-2}]
N–N	0.269	5000.0
C–N	0.337	2800.0
C–C	0.354	3000.0

Angle	k_{ijk} [kJ mol^{-1}]	θ^o_{ijk} [°]
N–N–N	120.0	114.5
C–N–N	110.0	127.0
C–C–N	19.0	142.7
C–C–C	34.0	143.0

Table 2 Lennard-Jones non-bonded parameters

Pair	ε_{ij} [kJ mol^{-1}]	σ_{ij} [nm]
N1 N1	0.4197	1.800
N2 N2	0.4490	5.700
C1 C1	0.4392	1.700
C2 C2	0.4257	1.500
N1 N2	0.4490	1.750
N1 C1	0.4685	1.350
N1 C2	0.4197	1.100
N2 C1	0.4392	1.190
N2 C2	0.4294	2.900
C1 C2	0.4392	1.667

3.1. Technical details of atomistic simulations

A suitable modified version of GROMOS96 united atoms force field recently applied to simulate the α phase of trioctanoin has been adopted for all atomistic simulations.[22] Molecular dynamics simulations have been performed in NPT ensemble. The temperature is kept constant using Berendsen thermostat (using a coupling constant $\tau_T = 0.1$ ps) and the pressure using Berendsen manostat ($\tau_P = 1.0$ ps) the time step used is 0.002 ps.

3.2. Technical details of coarse-grained simulations

For coarse-grained simulations, a time step of 0.03 ps has been used. Due to the smoothness of the potentials, time steps larger than atomistic simulations are common for coarse-grained models. This choice, as reported also in previous literature studies,[23] allows a good reproduction of both structural and thermodynamical properties. In particular, structural properties are very robust with respect to the time step; for example, Winger et al.[24] have shown that, even using larger time steps (0.050 ps), no noticeable effects on structural properties have been found.

4 Reverse mapping strategy

The quick generation of well-equilibrated atomistic structures is an important task of coarse-grained models. In the case of polymeric materials backmapping procedures have been extensively applied. The first polymers to be equilibrated in this

This journal is © The Royal Society of Chemistry 2012

way are polycarbonates.[25] More recently, backmapping procedures have also been applied to polystyrene[26] and to study gold nanoparticles polymer interfaces.[27] The extension of backmapping to non-equilibrium situations, to the determination of the rheology, has been recently demonstrated for atactic polystyrene under steady shear flow.[28] In the following the proposed strategy for the reverse mapping of triglycerides is explained in detail.

4.1. Rebuilding the atomistic models

The strategy chosen for the reverse-mapping is based on rigid superposition (rotation) of triglycerides atomistic models on the coarse-grained ones obtained from the mesoscale simulations. In particular, as schematized in Fig. 2, for a given coarse-grained molecule, several trial atomistic structures, belonging to a structure library, can be superimposed in order to minimize the root mean square deviation (RMSD) between the centre of the coarse-grained beads and the corresponding atomic sites. In order to give enough flexibility to this procedure, the structure library is made of configurations taken from an atomistic MD simulation of triglycerides in the liquid state. In particular, for this study, an atomistic simulation of 216 tridecanoin molecules, simulated at 446 K for 33 ns, has been utilized. From this simulation, 3300 configurations have been used. In fact, differently from the crystal structures, triglycerides in the liquid state are characterized by a larger flexibility of molecular conformations.

For a molecule, after a rotation able to maximize the superposition between the centre of the coarse-grained beads and their corresponding atomic sites

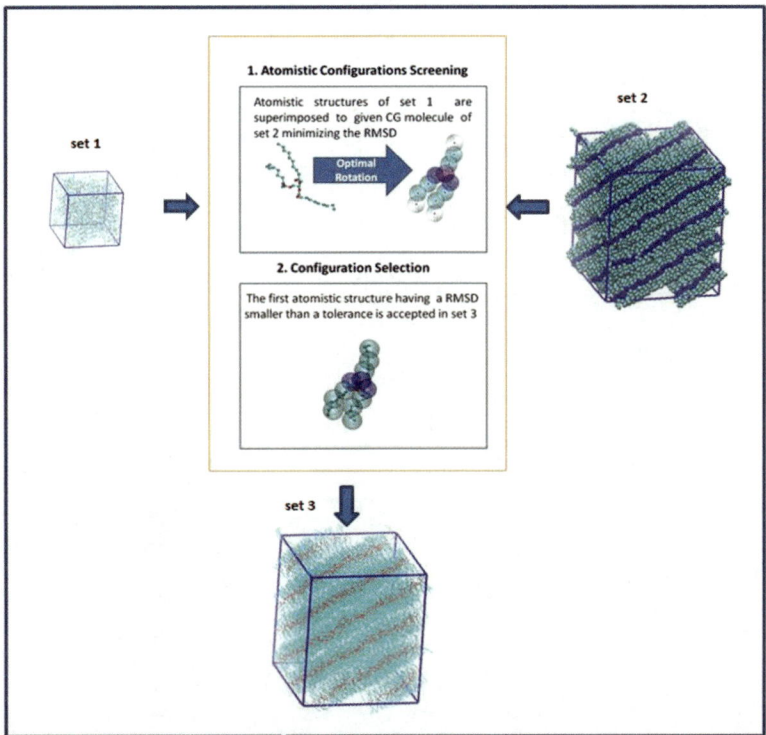

Fig. 2 Scheme representing the adopted strategy for reverse-mapping **Set 1** corresponds to atomistic configurations obtained from melt trajectories at 446 K, **set 2** is the single coarse-grained configuration to be back-mapped and **set 3** is the back-mapped atomistic configuration.

(see Fig. 3), a given trial atomistic structure is accepted if the maximum value obtained for the RMSD between two sites is smaller than a fixed tolerance. For the systems under investigation a value of tolerance of 10^{-2} nm has been found to be a good choice.

4.2. Structure superposition

As described in the preceding subsection, the reverse-mapping strategy involves a superposition obtained by rigid rotation of atomistic models on the coarse-grained coordinates. The structures obtained have to minimize the RMSD between some atomic sites and coarse-grained beads centres.

Structure superposition methods are useful to compare molecular structures. They facilitate visual comparisons and give a quantitative measure of shape similarity as the RMSD of distances between corresponding atoms. For these reasons different numerical and analytical approaches to the rigid fit of two structures have been proposed.[29,30]

In general, the problem can be solved by finding the optimal orthogonal transformation and requires determination of a rotation matrix R and a translation vector that will superimpose two sets of coordinates. Our choice is a method based on quaternions introduced by Kearsley.[31] This method is analytical (*i.e.* fast) and has been already successfully applied to generate atomistic configurations of polymer melts from coarse-grained models in equilibrium conditions,[26] under shear[27] and in the presence of nanoparticles.[28]

Quaternions are a non-commutative extension of complex numbers and have been used extensively to describe rotations in classical mechanics as well as quantum and relativistic physics. It can be shown that a quaternion can be used as a rotation operator for a vector. The vector can be considered as a quaternion with zero scalar components:

$$(0, r') = \hat{q}^{-1}(0, r)\hat{q} = (0, q_1^2 r + (r \cdot q)q + 2q_1(r \wedge q))$$

$$r' = Rr$$

$$= \begin{pmatrix} q_1^2 + q_2^2 - q_3^2 - q_4^2 & 2(q_2 q_3 + q_1 q_4) & 2(q_2 q_4 - q_1 q_3) \\ 2(q_2 q_3 - q_1 q_4) & q_1^2 + q_3^2 - q_2^2 - q_4^2 & 2(q_3 q_4 + q_1 q_2) \\ 2(q_2 q_4 + q_1 q_3) & 2(q_3 q_4 - q_1 q_2) & q_1^2 + q_4^2 - q_2^2 - q_3^2 \end{pmatrix} \begin{pmatrix} x \\ y \\ z \end{pmatrix}. \quad (4)$$

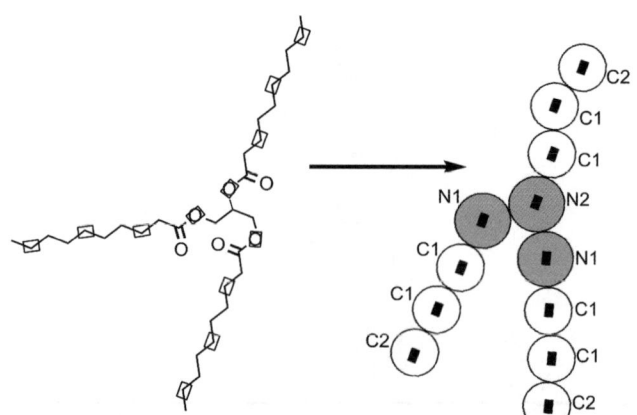

Fig. 3 Superposition between the centres of the coarse-grained beads and their corresponding atomic sites.

Kearsley has shown[31] that rotation matrices that minimize the sum of the squared distances between corresponding atoms for two structures can be calculated by posing a constrained least-squares problem in quaternion parameters. Posing: $x_- = x' - x$, $x_+ = x' + x$, with similar definitions for y_-, y_+, z_- and z_+, the resulting equations can be organized in the following eigenvalue problem:

$$
\begin{pmatrix}
\sum(x_-^2 + y_-^2 + z_-^2) & \sum(y_+z_- - y_-z_+) & \sum(x_-z_+ - x_+z_-) & \sum(x_+y_- - x_-y_+) \\
\sum(y_+z_- - y_-z_+) & \sum(x_-^2 + z_+^2 + y_+^2) & \sum(x_-y_- - x_+y_+) & \sum(x_-z_- - x_+z_+) \\
\sum(x_-z_+ - x_+z_-) & \sum(x_-y_- - x_+y_+) & \sum(x_+^2 + z_+^2 + y_-^2) & \sum(y_-z_- - y_+z_+) \\
\sum(x_+y_- - x_-y_+) & \sum(x_-z_- - x_+z_+) & \sum(y_-z_- - y_+z_+) & \sum(x_+^2 + y_+^2 + z_-^2)
\end{pmatrix}
$$

$$
\times
\begin{pmatrix} q_1 \\ q_2 \\ q_3 \\ q_4 \end{pmatrix}
= \lambda
\begin{pmatrix} q_1 \\ q_2 \\ q_3 \\ q_4 \end{pmatrix}
\tag{5}
$$

In the elements of the 4×4 matrix of eqn (5), the summation is made over the centres to superimpose. Diagonalizing this symmetric matrix will give four orthogonal unit quaternions. The eigenvalues give the value of the residual for the rotation produced by application of the corresponding eigenvector. The RMSD is given by $(\lambda/n)^{1/2}$ where n is the number of atoms compared. The smallest eigenvalue gives the rotations that minimize the sum of the distances between all corresponding atoms.

For the molecules under investigation, according to the chosen mapping scheme, the centres of coarse-grained and atomistic models to be superimposed are depicted in Fig. 3.

4.2. Structure finalization

Before atomistic equilibration and production runs, the configurations obtained from procedures described in the sections above have been energy minimized. Starting from scaled Lennard-Jones potentials (low values of ε and σ) in a few hundreds of steepest descent minimization steps (using 2 fs time step) followed by short MD NVT relaxations (200 steps using 1 fs time step) the nonbonded potentials were brought linearly to their full values.

5 Multiscale simulations of liquid–solid transition

In Fig. 4, the behavior of density during time, for a system made of 216 tridecanoin molecules (CG1), for a coarse-grained MD simulation 1.8 µs long is reported. The initial set up corresponds to the equilibrium conditions of the liquid state (446 K) obtained equilibrating a system for 500 ns with a NPT MD simulation. Starting from the equilibrium configuration the system was cooled from 446 to 200 K in 10 ns, and then the temperature was controlled keeping as a target value 200 K. In Fig. 4, the density is scaled by the equilibrium density of the liquid. It is clear there is a first sharp increase of the density starting from 250 ns. Further smaller increases can be observed up to 1 µs where the density reaches an equilibrium value. It is worth comparing these results with the preliminary ones reported in our previous study.[21] In that case, the first sharp transition in the density time behaviour can be obtained earlier (starting from 100 ns), this different behaviour can be ascribed to the different cooling rate, instantaneous for the previous study and linear (from 446 to 200 K in 10 ns) in the present case.

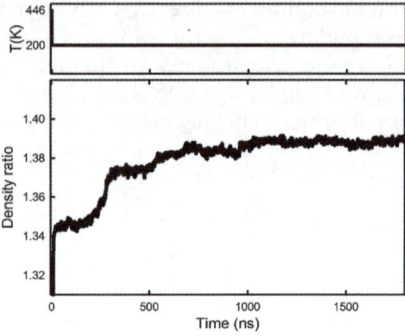

Fig. 4 Density evolution during time for system **CG1**. The density is scaled to the equilibrium density at 446 K.

In Fig. 5 a conformational analysis of the simulation of the system **CG1** in terms of tuning-fork, chair and trident conformations is reported. These conformations can be discriminated defining suitable order parameters.[21] In particular, according to the values of two scalars P and C, assessing planarity and conformation, the type of conformation can be calculated from molecules coordinates. Further details about the definition of P and C and their limiting values can be found in the ESI.†

From Fig. 5, it is clear from the conformational analysis results that a significant increase in the percentage of tuning-fork conformation is observed at the time instant corresponding to the first sharp increase of the density.

This behavior of the simulated system is consistent with the general behavior of triglycerides. As reported in the literature,[7] triglycerides during nucleation adopt tuning-fork conformation. Furthermore, as reported by Jensen, the crystal structure of tridecanoin in β form is characterized by tuning fork-conformation.[32]

In Fig. 6, some of the configurations obtained from the reverse mapping procedure applied on the coordinates of system **CG1** at 0, 94.5, and 730 ns together with their corresponding CG models are depicted.

From Fig. 6 it is clear that during the liquid–solid transition ordered layers made of effective beads of type N1 and N2 are formed. Correspondingly, in the atomistic configurations molecular regions close to carbonyl and ester oxygen atoms form ordered layers.

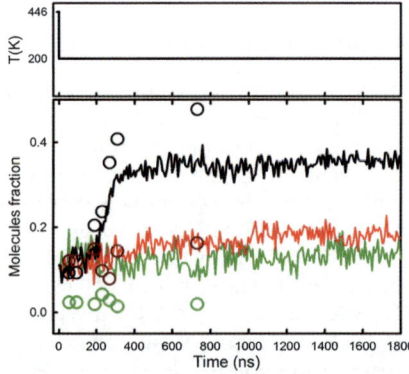

Fig. 5 Time evolution of the tuning-fork (black line CG model, black empty circles reverse mapped model), chair (red line CG model, red empty circles reverse mapped model) and trident (green line CG model, green empty circles reverse mapped model) during time for system **CG1**.

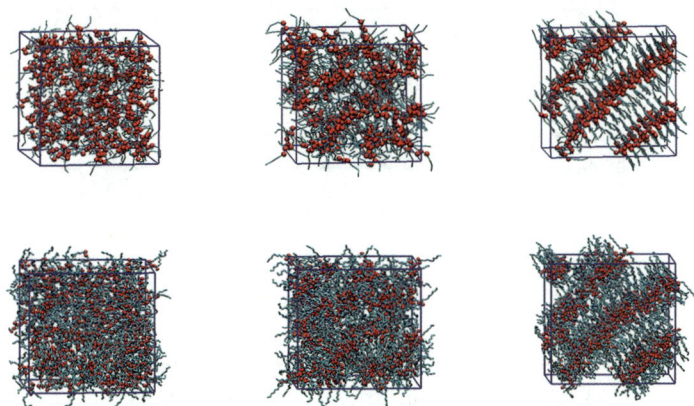

Fig. 6 Coarse-grained (top) and atomistic configurations (bottom) obtained from reverse mapping procedure. Effective beads of type N1 and N2 in coarse-grained models and oxygen atoms of the atomistic models are reported in red.

The reverse mapping procedure has been applied to several configurations of system **CG1** ranging from the first one up to 730 ns.

In Fig. 5, empty circles indicate data coming from the reverse mapped atomistic simulations. All the configurations obtained by reverse mapping of CG configurations are characterized by a slightly lower value of trident conformations. The population of tuning-fork conformations just after the first density increase is higher. This is probably due to rearrangements of molecular conformations connected to the more detailed model and more evident at higher densities.

Starting from the three reverse mapped configurations of Fig. 6, atomistic simulations 2 ns long have been employed to analyse atomistic features of the liquid–solid transition.

The total intermolecular radial distribution functions reported in Fig. 7 show an increase in order going from the liquid (0 ns) to the ordered phase (730.5 ns). This result is quite expected.

More interesting is the behaviour of partial intermolecular radial distribution functions. In particular, in Fig. 8 radial distribution functions between oxygen (carbonyl indicated as O= and ester indicated as –O–) and carbon atoms (carbonyl indicated as =C– and alkyl indicated as CH_n) are shown. From Fig. 8 it is clear that going from the liquid to the solid state the radial distributions functions between oxygen (both carbonyl and ester types) and between oxygen and carbonyl carbon show a dramatic change. In contrast, distributions involving alkyl carbons are almost unchanged during the transition.

From these results it can be understood that the polar groups of the TGA molecules are responsible for the aggregation preceding the nucleation and prepare the formation of ordered layers observed in the coarse-grained simulations discussed above.

In Fig. 9, the evolution of static structure factors calculated from the atomistic configurations at different stages of liquid–solid transition are reported. The structure factors were averaged over 100 configurations of the last nanosecond of the atomistic simulations obtained from the reverse mapping procedure.

The main pattern obtained for the ordered structure is in agreement with the experimental features of triglycerides X-ray diffraction spectra. Two main peaks can be observed corresponding to a short and a long spacing. The steep increase at short values of q can be ascribed to finite size effects caused by periodic boundary conditions (pbc). This feature is present also in the structure factor calculated for the liquid state (black curve). Furthermore, in these systems the shorter length of the pbc

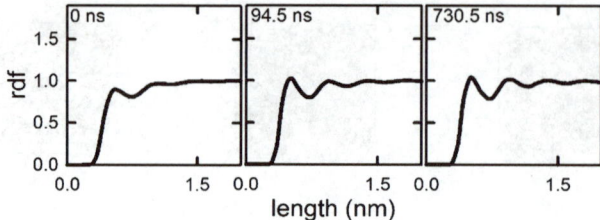

Fig. 7 Total radial distribution functions calculated from simulations starting from the reverse mapped configuration at 0, 94.5 and 730.5 ns.

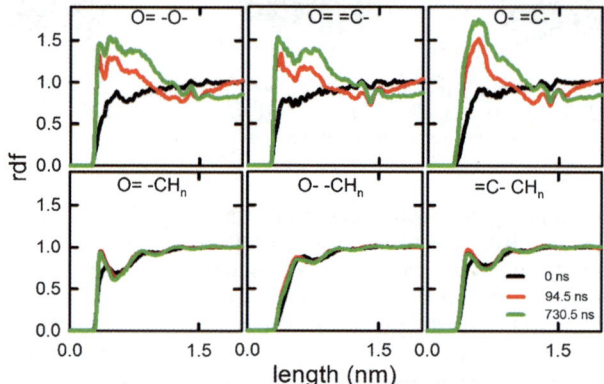

Fig. 8 Some partial radial distribution functions calculated from simulations starting from the reverse mapped configuration at 0, 94.5 and 730.5 ns.

simulation box is about 4.7, *i.e.* less than two times the expected length of the long spacing and then these effects interfere with a correct evaluation of the long spacing.

The developed coarse-grained models allow, with modest computational efforts, larger systems to be afforded, suitable to investigate at a reasonable length and time-scale the complex behaviours of the systems under study. Just to give an idea of the real time needed for these simulations, for the **CG1** system on 8 processors (Intel Xeon E7330 2.33 GHz) it is possible to simulate 3 μs per day. For the corresponding atomistic simulation it is possible to simulate, on the same number of processors, about 30 ns per day. For the larger CG systems described in the following, made of 1728 TGA molecules, it is possible to simulate on 24 processors about 1 μs per day, and for the corresponding atomistic simulations 8 ns per day on the same number of processors.

In Fig. 10, the behavior of density during time, for a system made of 1728 tridecanoin molecules (**CG2**), for a coarse-grained MD simulation 1.8 μs long is reported. The initial set up corresponding to the equilibrium conditions of the liquid state and the system cooling have been set in the same way as for system **CG1**.

Similarly to system **CG1**, from Fig. 10 a first sharp increase of the density is clear, but differently from the smaller system, for the **CG2** this increase starts later at 500 ns. Furthermore, no smaller steps are observed and the density grows almost linearly up to the equilibrium value at about 1.5 μs. The faster transition in the smaller system can be ascribed to finite size effects. In particular, the formation of ordered layers of coarse-grained beads of type N is helped by periodic boundary conditions.

According to the picture coming from the classical nucleation theory, liquids are characterized by continual density fluctuations. These fluctuations can give rise to transient aggregates that may either disappear or become stable and grow.

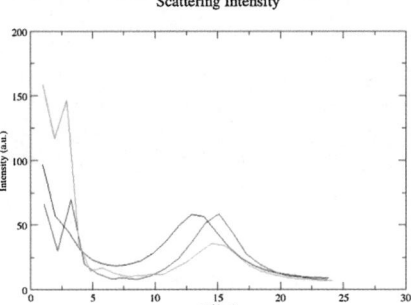

Fig. 9 Static structure factors for the reverse mapped configuration at 0 (black curve), 94.5 and 730 ns.

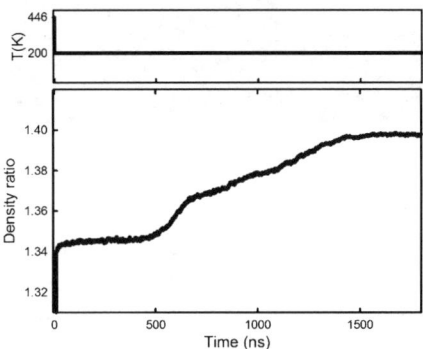

Fig. 10 Density evolution during time for the system **CG2**. The density is scaled to the equilibrium density at 446 K.

According to this picture, in addition to the behavior of average density of the NPT simulation it is interesting to investigate the behavior of the density field, *i.e.* the density as a spatial three dimensional function. For a given configuration, the space can be divided into cells and particles can be counted. From the 3D grids obtained in this way, isosurfaces plotted at a given value of the density can be obtained. The isosurfaces have been calculated for several configurations ranging from 0 to the end of simulation 1.8 μs. In particular, isosurfaces reported in Fig. 11 have been obtained at a value of density 1.20 times the average density of the ordered phase. For this value, no isosurfaces can be obtained for the first configuration at 0 ns corresponding to the equilibrium liquid state. After the cooling process, some small and few aggregates can be formed as shown in Fig. 11 for the configuration at 60 ns. From 100 and 500 ns aggregates become stable and are distributed in all the space (some of them look larger). After 500 ns the aggregates start to grow and become irregular. Finally, during the formation of crystalline order the regions become more regular and regularly spaced. The crystalline order formation passes through the breaking of irregular aggregates previously formed (the isosurface at time 0.558 μs in Fig. 11), and through the rearrangement of the molecules in a more ordered structure (the isosurface at the time 1.8 μs in Fig. 11).

The breakdown of aggregates leads to the formation of nuclei with a high density (see isosurface at time 1 μs in Fig. 11) that we hypothesize to be the real precursors to the formation of layers.

Results of conformational analysis (tuning-fork dominance *etc.*) are similar to the ones of system **CG1** and are reported in electronic supplementary information section (Fig. S2).†

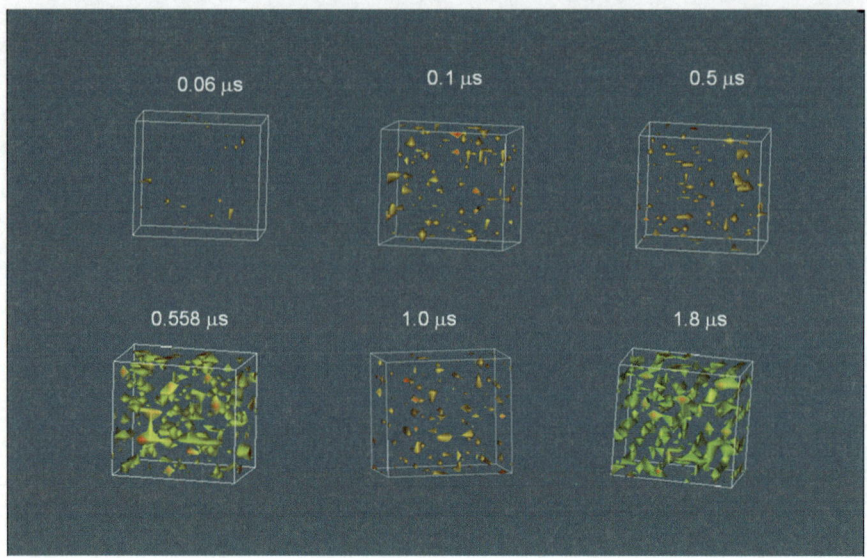

Fig. 11 Time evolution of density isosurface calculated for a value of 1.2 times the average density of the ordered phase obtained at 1.8 μs.

In Fig. 12, snapshots of configurations obtained from the simulation of system **CG2** are depicted. From this figure it is possible to see how the transition from the disordered structure of the liquid state and the ordered phase obtained starting from 1.2 ms is more smooth than in system **CG1** and is characterized by intermediate structures of increasing order. In particular, at 0.6 μs the formation of undulated layers is apparent from Fig. 12. Later (1.2 μs), the layers become regular and

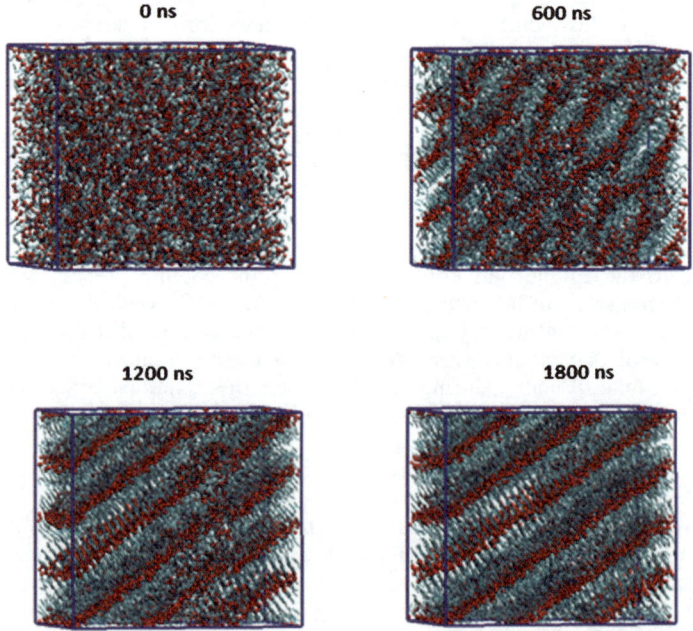

Fig. 12 Snapshots of configurations obtained from simulation of system **CG2**.

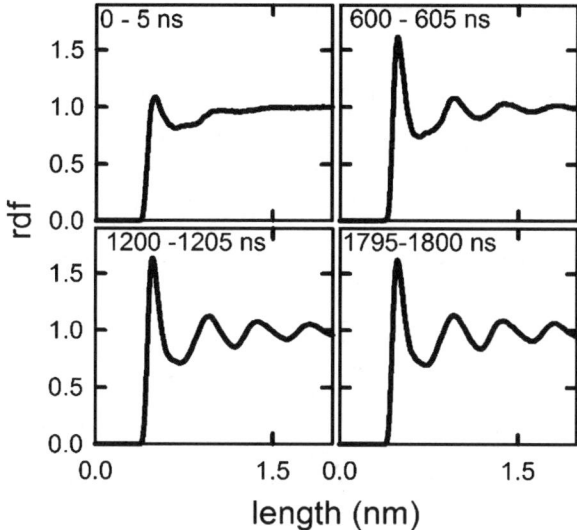

Fig. 13 Total radial distribution functions calculated from simulations of system **CG2**.

some defects in the distribution of beads of type N are still present. In the final structure obtained at 1.8 μs the defects are absent and an ordered structure is formed.

In order to follow in a more quantitative way the formation of long range order, the total radial distributions of the coarse-grained beads, shown in Fig. 13, have been calculated averaging configurations over several time intervals. In particular, starting from 1.2 μs, the radial distribution functions show long range structure and are unchanged for the further 600 ns.

As further validation of the proposed reverse mapping strategy, in Fig. 14 the static structure factor calculated from the atomistic model back-mapped from the configuration at 1.8 μs is reported. The structure factor was averaged over 100 configurations of the last nanosecond of the atomistic simulations obtained from the reverse mapping procedure. For comparison, the structure factor for the analogous system obtained by reverse mapping from the configuration of system **CG1** at 1.8 μs is also reported. The comparison between the two calculated spectra of Fig. 14 confirms that finite size effects affect mainly the first peak corresponding to the long spacing. From the peak positions, it is possible to calculate the two spacings characterizing the order of the atomistic model obtained with the proposed procedure.

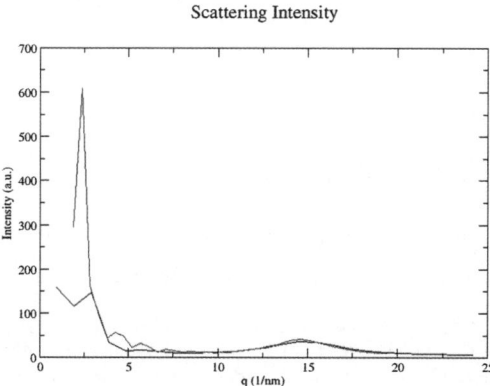

Fig. 14 Static structure factors for the reverse mapped configuration at 1.8 μs.

From the first peak ($q = 2.37$ nm^{-1}) a spacing of 2.65 nm and from the second peak ($q = 14.69$ nm^{-1}) a spacing of 0.43 nm can be calculated. These values compare very well with the ones reported by Clarkson and Malkin[33] for the long spacing of tridecanoin in β form (2.68 nm) and short spacing (approximately 0.46 nm).

Conclusions

A multiscale simulation strategy to study the crystallization of triglycerides is proposed. Coarse-grained models parameterized on the basis of reference atomistic simulations have been used to model liquid–solid transition. The nucleation and growth of the crystalline order of the coarse-grained models has been analysed in terms of several properties. A reverse mapping procedure, able to reconstruct atomistic models from coarse-grained configurations, has been proposed and validated against experimental structural properties. In particular, for tridecanoin the comparison between experimental and calculated structure factors suggests an ordered structure resembling the β crystalline phase. The proposed coarse-grained models and the reverse mapping procedure open the way to further studies aimed to understand at a molecular level the complex behaviour of triglycerides.

References

1 J. W. Goodrum and M. A. Eiteman, *Bioresour. Technol.*, 1996, **56**, 55–60.
2 P. Lonchampt and R. W. Hartel, *Eur. J. Lipid Sci. Technol.*, 2004, **106**, 241–274.
3 R. Wille and E. Lutton, *J. Am. Oil Chem. Soc.*, 1966, **43**, 491–496.
4 I. Heertje, E. C. Roijers and H. Hendrickx, *Food Sci. Technol.-Lebensm.-Wiss. Technol.*, 1998, **31**, 387–396.
5 J. Yano and K. Sato, *Food Res. Int.*, 1999, **32**, 249–259.
6 A. Hall, J. Repakova and I. Vattulainen, *J. Phys. Chem. B*, 2008, **112**, 13772–13782.
7 A. G. Marangoni, N. Acevedo, F. Maleky, E. Co, F. Peyronel, G. Mazzanti, B. Quinn and D. Pink, *Soft Matter*, 2012, **8**, 1275–1300.
8 *Faraday Discuss.*, 2007, 136, 1–426.
9 J. Anwar and D. Zahn, *Angew. Chem., Int. Ed.*, 2011, **50**, 1996–2013.
10 M. Matsumoto, S. Saito and I. Ohmine, *Nature*, 2002, **416**, 409–413.
11 S. Auer and D. Frenkel, *J. Chem. Phys.*, 2004, **120**, 3015–3029.
12 S. Auer and D. Frenkel, *Nature*, 2001, **413**, 711–713.
13 S. Auer and D. Frenkel, *Nature*, 2001, **409**, 1020–1023.
14 S. Auer and D. Frenkel, *Annu. Rev. Phys. Chem.*, 2004, **55**, 333–361.
15 S. J. Marrink, A. H. de Vries and D. P. Tieleman, *Biochim. Biophys. Acta, Biomembr.*, 2009, **1788**, 149–168.
16 F. Muller-Plathe, *ChemPhysChem*, 2002, **3**, 754–769.
17 C. Peter and K. Kremer, *Soft Matter*, 2009, **5**, 4357–4366.
18 H. Meyer and F. Müller-Plathe, *Macromolecules*, 2002, **35**, 1241–1252.
19 D. Reith, H. Meyer and F. Müller-Plathe, *Macromolecules*, 2001, **34**, 2335–2345.
20 T. Vettorel and H. Meyer, *J. Chem. Theory Comput.*, 2006, **2**, 616–629.
21 A. Brasiello, S. Crescitelli and G. Milano, *Phys. Chem. Chem. Phys.*, 2011, 13.
22 I. Chandrasekhar and W. F. van Gunsteren, *Eur. Biophys. J.*, 2002, **31**, 89–101.
23 S. J. Marrink, X. Periole, D. P. Tieleman and A. H. de Vries, *Phys. Chem. Chem. Phys.*, 2010, **12**, 2254–2256.
24 M. Winger, D. Trzesniak, R. Baron and W. F. van Gunsteren, *Phys. Chem. Chem. Phys.*, 2009, **11**, 1934–1941.
25 W. Tschöp, K. Kremer, O. Hahn, J. Batoulis and T. Bürger, *Acta Polym.*, 1998, **49**, 75–79.
26 G. Santangelo, A. Di Matteo, F. Müller-Plathe and G. Milano, *J. Phys. Chem. B*, 2007, **111**, 2765–2773.
27 G. Milano, G. Santangelo, F. Ragone, L. Cavallo and A. Di Matteo, *J. Phys. Chem. C*, 2011, **115**, 15154–15163.
28 X. Chen, P. Carbone, G. Santangelo, A. Di Matteo, G. Milano and F. Muller-Plathe, *Phys. Chem. Chem. Phys.*, 2009, 11.
29 R. B. Honzatko, *Acta Crystallogr., Sect. A: Found. Crystallogr.*, 1986, **42**, 172–178.
30 A. M. Lesk, *Acta Crystallogr., Sect. A: Found. Crystallogr.*, 1986, **42**, 110–113.
31 S. K. Kearsley, *Acta Crystallogr., Sect. A: Found. Crystallogr.*, 1989, **45**, 208–210.
32 M. A. J. Jensen and L.H, *Acta Crystallogr.*, 1966, **21**, 770.
33 E. Clarke and T. Malkin, *J. Chem. Soc.*, 1934, 666–671.

General discussion

Dr van der Sman opened the discussion of the paper by Professor Tanaka: For visco-elastic phase separation you state that hydrodynamic interaction is important, and should be included in the model. Frequently, in your papers, you do not apply an external driving force inducing fluid flow. In food processing, one usually does apply external force to induce new structures in the food, shear for example.[1] Would this external forcing still fit within the picture or phase diagram of viscoelastic phase separation? What other extra kinds of physical phenomena come into play?

1 S. H. Peighambardoust, R. J. Hamer, R. M. Boom and A. J. van der Goot, Migration of gluten under shear flow as a novel mechanism for separating wheat flour into gluten and starch, *J. Cereal Sci.*, 2008, **48**(2), 327–338.

Professor Tanaka replied: Yes, I think so. As discussed in Section 10 of the paper, the two-fluid model can describe an external driving inducing flow in a very natural manner. In viscoelastic phase separation under no external drive, self-generated mechanical stress induced by phase separation itself plays a major role. In this sense, there is no essential difference in the theoretical framework between a situation with and without an external drive. There are very nice works by Onuki and his coworkers on shear effects on viscoelastic phase separation (see, *e.g.*, Ref. 2 and 96 in the paper).

Dr van der Sman asked: Which kinds of changes do you expect in the phase diagram of VPS if shear flow is induced *via* external fields. In foods we have observed the creation of anisotropic structures in complex protein mixtures, which are cross-linked *via* enzymes as a complication.[1] Which kinds of structures do you expect if shear modulates VPS?

1 J. M. Manski, A. J. van der Goot and R. M. Boom, Formation of fibrous materials from dense calcium caseinate dispersions, *Biomacromolecules*, 2007, **8**(4), 1271–1279.

Professor Tanaka responded: As discussed in Section 10 of my paper, strictly speaking, viscoelastic effects cannot be included in the free energy and cannot be linked to the thermodynamic phase diagram. Historically, such efforts were made (see, *e.g.*, Ref. 152 and 153 in the paper). However, it was shown that such an approach is not appropriate (see, *e.g.*, Ref. 2 in the paper). We stress that the visco-elastic effects are dynamical and cannot be included in the static Hamiltonian. This is also clear from the fact that viscoelastic effects are anisotropic; for shear deformation, for example, the effects are very different between along the compression axis and along the extensional axis. This example clearly shows that shear effects cannot be expressed by a scalar quantity. Depending upon the relation between shear rate and the rheological relaxation time, various morphologies may be formed under shear. For weak shear we expect string-like structures, whereas for strong shear even chaotic behaviour may be observed. There are interesting experimental and theoretical studies on this issue (see Ref. 2 in the paper, and the references therein).

Dr Nicolai commented: Mixtures of protein particles and κ-carragheenan phase separation can phase separate. It has been observed that close to the binodal the phase separation is reversed when the carragheenan chains form an elastic network.[1,2] Very recently, it was found that the degree at which the phase separation is reversed is directly correlated to the elastic modulus.[3] Can you explain this behaviour by your theory?

1 K. Baussay *et al.*, *J. Coll. Int. Sci.*, 2006, **304**, 335–341.
2 K. Ako, *Soft Matter*, 2011, **7**, 2507–2516.
3 Nguyen *et al.*, *Food Hydrocolloids, submitted.*

Professor Tanaka answered: I speculate that the observed behaviour may be interpreted as a transformation from a pattern dominated by a thermodynamic force (interfacial tension) to that by a mechanical force. The change in the pattern selection rule can be induced by the temporal evolution of viscoelastic or elastic properties of the matrix phase (namely, the formation of an elastic network of the carragheenan chains), although the relevance of such a scenario needs to be checked carefully.

Dr van der Sman addressed Dr Nicolai and Professor Tanaka: Again, would the proteins denature or become more hydrophobic during heating in the system Dr Nicolai has shown in the video?

Dr Nicolai replied: The system I showed in the video was a mixture of β-lactoglobuline particles and κ-carrageenan. Phase separation is induced by cooling close to the binodal due to the reduction of the entropy of mixing. This is not surprising. The surprising feature is that phase seapration can be inversed by gelation of the κ-carrageenan.

Dr van der Sman remarked: I gathered from an earlier discussion of a problem raised by Professor Ettelaie that it is important to notice that the opening of a single macroscopic crack is induced by an external mechanical force – while in many of the simulations of Professor Tanaka periodic boundary conditions are applied and, thus, no external forcing. Thus, is it safe to say that without external forces one obtains more homogeneous fracturing on the microscale? and with external loading more inhomogeneous fracturing, with macroscopic features?

Professor Tanaka responded: There is a fundamental commonality between viscoelastic (or fracture) phase separation and mechanical fracture of glassy materials, as we discussed in Section 10 in our paper (see also Ref. 16 and 17 in our paper). The difference in the number density of cracks formed between the two cases mainly originates from the fact that in the former a system is thermodynamically unstable against phase separation whereas in the latter, a system is in a thermodynamically stable state and becomes unstable only by external fields. More precisely, in the former, concentration fluctuations start to grow just after a quench, which results in the spatial inhomogeneity in the viscoeastic properties of a system. Thus, cracks are formed in elastically weak parts of low polymer concentrations as a consequence of the concentration of the stress that is self-generated by phase separation. This leads to a high number density of cracks for fracture phase separation. Such a thermodynamic driving force for the growth of fluctuations is absent in mechanical fracture and there is only a mechanical driving force. I think that periodic boundary conditions are not relevant to the difference. For example, mechanical fracture behaviour under shear deformation, which we reported in the paper cited as Ref. 17 in our current paper, is also observed under a periodic boundary condition.

Dr van der Sman opened the discussion of the paper by Dr Hütter by asking: How do you obtain constitutive relations for the osmotic pressure? At my first reading of the paper I have the impression that you are missing the interactions between solids and solvents. If external forces compress the particle, the particle shrinks and the volume fraction of polymers increases, which changes the osmotic pressure inside the particle and there is a driving force to regain the expelled water – if the external forcing (compression by other particles) is lifted. In my opinion, a good candidate for the constitutive equation is the Flory–Rehner theory, which we have previously used for meat,[1,2,3] and are currently applying for the water holding capacity of (cell wall material) vegetables and mushrooms.[4] The Flory–Rehner theory is an extension of the well-known Flory–Huggins theory, and is extended with an elastic contribution. Of course, shrinkage or swelling can also be induced by changes in temperature, pH or ionic strength.

Does your model already account for the interaction between water and protein, which can make it swell and shrink *via* thermodynamic forces?

1 R. G. M. van der Sman, Moisture transport during cooking of meat: An analysis based on Flory–Rehner theory, *Meat science*, 2007, **76**(4), 730–738.
2 R. G. M. van der Sman, Soft condensed matter perspective on moisture transport in cooking meat, *AIChE Journal*, 2007, **53**(11), 2986–2995.
3 R. G. M. van der Sman, Thermodynamics of meat proteins, *Food Hydrocolloids*, 2012, **27**(2), 529–535.
4 A. K. Datta, R. G. M. van der Sman, T. Gulati and A. Warning, Soft matter approaches as enablers for food macroscale simulation, *Faraday Discussions*, 2012, **158**.

Dr Hütter replied: Each particle is seen as a porous structure that is permeated by the solvent. For example, due to the contact interaction with other particles, the porous structure gets compressed, *i.e.* the solvent is being expelled from the interior of the particle. Upon releasing the load due to the other contacting particles, the porous structure will expand again. In our work, this restoring force can have two contributions. On the one hand, there is the purely elastic response of the porous structure as such. On the other hand, there is the osmotic contribution that you mention. Both of these ways of restoring a preferred particle size can be captured in the potential energy, see eqn (31) in the paper, since these are purely thermodynamic effects. If particles are in contact, the elastic and osmotic effects enter into an interaction term, such as the ones described in eqn (33) in our paper. If the particles are not in contact with other particles, single particle contributions can be added to the potential energy, see eqn (31) in the paper. For an explicit procedure to account for these effects, we refer to the Flory–Rehner theory.[1] The volume fraction of the polymer used in the Flory-Rehner free energy relates, in our approach, to the inverse of the particle volume. As mentioned in our paper, the interaction potential is in principle a Helmholtz free energy density. As such, it does not only depend on the particle positions, but also on other parameters such as the temperature, concentration, or possibly pH and ionic strength, as done in the DLVO theory.

1 S. Wu, H. Li, J. P. Chen and K. Y. Lam, *Macromol. Theory Simulat.*, 2004, **13**, 13–29.

Dr Bot then opened the discussion of the paper by Professor Pink: The solid Crystalline NanoPlatelet (CNP) in your study is modeled as a solid continuum. In reality, these crystallites are ordered as lamellar phases and adsorbing triglycerides follow this template during the growth of the crystallites. It would be expected, therefore, that adsorption on a side face proceeds in a very different fashion than on a top or a bottom face. Your schematic diagram in Fig. 4 of the paper reflects the adsorption process I would expect to see on a side face (the face that exposes full triglyceride molecules, not just the tip of the fatty acid chain). Do your simulations predict exactly the same adsorption behaviour on side faces and on top or bottom faces of a CNP?

Also, the nano-phase separated layers at the edge of the crystal effectively soften the interface of the CNP – by replacing the "hard" boundary of the CNP by a "soft" boundary of the nano-phase separated layer. However, the layer is thin. Could you give an estimate of how much additional effective dispersed phase volume this adds in your system? Furthermore, would it be possible to calculate an effective viscosity of the nano-phase separated layer as function of the distance to the CNP interface?

Professor Pink replied: Since no distinction is made between the orientation of the molecules comprising the surface, it must predict that the same adsorption behaviour will be seen at all faces. Note that the faces are modelled as having no edges.

The simulation studied "nano phase separation" between two specific molecules, OOO + EEO or OOO + OOE (O = oleic, E = elaidic). We found that such phase separation would take place for the first case. Fig. 4 in the paper is a proposal, based on these simulation results, as to how oil might be retained in spaces between CNPs. The transition from the CNP surface to the bulk OOO oil is proposed to be mediated by molecules such as EEO and OOE. We have not carried out a simulation to test this proposal.

It should be possible to calculate an effective viscosity of the nano-phase separated layer as function of the distance to the CNP interface.

Dr van der Sman† then opened a general discussion on the universality in water holding capacity and oil binding capacity: In the papers presented by Professor Marangoni and Professor Pink, references are made to oil binding capacity (OBC). As Professor Marangoni has initially linked OBC to permeability (diffusivity) of the fat crystal network, it appears that there is not a clear understanding or definition of the concept of oil binding capacity. A similar misunderstanding is found concerning the concept of water holding capacity (WHC) of food materials.[9,11,15,8,6] We view that OBC and WHC are very similar concepts and can only be understood and defined within the framework of thermodynamics. Such an approach we have already undertaken for the WHC of meat,[17,18,19] where we put it in the framework of Flory–Rehner theory. We view the WHC as the amount of water that food can retain under specified con- ditions. In food science, there is no common understanding of what these specified conditions should be. By analysing the food using the Flory–Rehner theory, one can compute the WHC under any specified condition, provided all model parameters are known. The WHC definitely relates only to the driving force for moisture migration.[17,18] The permeability or moisture diffusivity determines the migration rate.

In our poster (displayed during the discussion meeting) and the paper presented by Professor Datta, we have shown that this framework can be extended to vegetables and mushrooms, which we will report in more detail in a forthcoming paper. We view that oil binding capacity can be put on a similar footing as WHC, if the thermodynamic approach is followed. This will enhance the engineering structuring of oil/fat crystals systems, as well as organo-gels,[12] which are discussed in the paper presented by Dr Bot.

In the studies on the WHC of meat we have considered the food to be a homogenous gel. Often, foods contain a pore-space next to a gel-phase. In mushrooms, the hyphae constitute the gel phase, while the interfibrilar space constitutes the pore space. In freeze-dried vegetables there is a pore space created by the sublimation of the ice-crystals, and the collapsed cells constitute the gel phase. These food systems we can generalize as hygroscopic-capillary media. In the fresh mushrooms and dry vegetables the pore space is filled with air. The pore space of mushrooms and freeze-dried vegetables can be filled with water *via* either vacuum impregnation or simple immersion in water. The water contained in the pore space adds to the water holding capacity of these hygroscopic-capillary media.

The water in both the gel phase and the pore space is characterized by its own thermodynamic potential. Water in the pore spaces is commonly characterized by the capillary pressure, which follows from the Young–Laplace law. Water in gels is commonly characterized by a swelling pressure, which follows from the Flory–Rehner theory. We can only define the WHC of hygroscopic-capillary media in a sensible way if the gel phase and pore space are in equilibrium, *i.e.* the thermodynamic potential should be equal, as is stated in the paper presented by Professor Datta.

Hence, with respect to moisture migration many foods can be regarded as hygroscopic porous media.[2,3] Liquid retention has been the subject of investigation in

---

† The hypotheses put forward by Dr van der Sman were contributed in collaboration with E. Paudel.

another class of porous media, namely those with an inert, solid phase, which is highly relevant to water and oil retention in the earth subsurface.[16] Here, also, the contribution of water absorbed to the solid surface *via* van der Waals forces is considered in the liquid retention of porous media. The characterizing thermodynamic potential is named the disjoining pressure, which scales with the film thickness h, as $1/h^3$ and is linear with the Hamaker constant. The disjoining pressure can be generalized if electrostatic interactions contribute to the liquid retention. For pores at the nanoscale the capillary pressure and disjoining pressure can be of equal importance. The sorption of water starts with the formation of a flat uniform film on the surface with thickness up to several nanometres. Subsequently, the pores will be filled *via* capillary condensation or *via* snap-off of the liquid films.[16] A similar theory holds for oil retention in this class of porous media.[14]

Summarizing, we have discussed three different systems capable of retaining liquid, which are sketched in Fig. 1. One class are hygroscopic-capillary foods, like the spongy protein particles discussed in the paper by presented by Dr Hütter, inert porous media, like soil retaining water and oil-based foods, as mixtures of oil and fat crystals, as discussed in the papers presented by Professor Pink and Professor Marangoni, and organogels as discussed by Dr Bot.

Comparison of the WHC of hygroscopic-capillary foods and the above class of porous media shows that: (1) it is probable that there exists a universal description for WHC and OBC in food materials, and (2) our previous framework has to be extended with contributions due to van der Waals and electrostatic interactions, which can be captured in the disjoining pressure, for example.

Electrostatic interactions can also play a role in the gel phase, as is evident in the water-binding of polyelectrolyte gels such as cartilage.[10] The Flory–Rehner theory is naturally extendable with electrostatic interactions,[5,17,18,19] which can be described by the Donnan-potential, largely accounting for the water binding of the free counterions. Due to its various contributions, one commonly divides the swelling pressure into partial pressures, each related to a specific contribution. For foods, three contributions are relevant, in general: (1) the osmotic pressure, (2) the elastic stress, and (3) the ionic (Donnan) potential. The osmotic pressure follows from the mixing of biopolymers and the water, and can be described by the Flory–Huggins theory.[21,19] Often, the ionic contribution can be absorbed in the osmotic contribution.

The interplay between elastic stresses and capillary pressure can lead to very intriguing physics[13] and is responsible for the collapse of the pore space in hygroscopic-capillary foods during air drying, for example.[7]

In hygroscopic-capillary foods one should also mind that small solutes can migrate from the gel phase to the pore space, if (partially) filled with water. Hence,

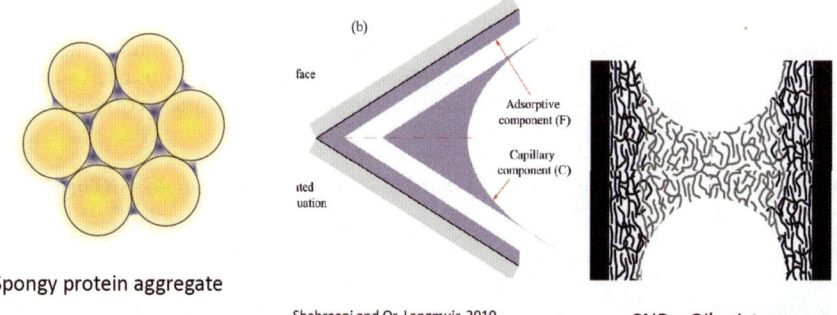

Fig. 1 Cartoon of different systems able to retain liquid: (a) spongy protein particles (as discussed in the paper presented by Dr Hütter), (b) inert porous media,[16] and (c) oil based food as discussed by Professor Pink, Professor Marangoni and Dr Bot.

the contributions of dissolved sugars and salts must be accounted for in the thermodynamic potential of both the gel phase and the pore space. For solutions of salts and sugars in water, one can compute the water activity *via* the Pitzer equation, or the Free-Volume-Flory–Huggins (FVFH) theory.[20,21,19] Our recent research has shown that FVFH theory even applies to polyols, which are commonly present in foods like mushrooms – this is to be reported in a forthcoming paper. The water activity can easily be reformulated in terms of an osmotic pressure. The thermodynamic potential of capillary water with dissolved solutes is obtained by adding the capillary pressure and the osmotic pressure. The contribution of dissolved solutes to the total thermodynamic potential can be quite significant, and it is said to play a vital role in trees to remove embolies in their woody vessels (this is performed by active dissolution of the sugars in the capillary water).[22]

In oil-based foods there will be no contribution of dissolved solutes, of course. Furthermore, it is likely that electrostatic interactions contribute little to the oil binding. Oils are typically mixtures of different oils, and can give rise to an osmotic pressure. Professor Pink has remarked that the osmotic pressure of mixtures of oils could be described by a Flory–Huggins type of model. His simulations also show that oils can be absorbed on solid interfaces (fat crystals) *via* van der Waals forces, which can probably also be expressed in terms of a disjoining pressure. Capillary forces can play a role if two nano-platelets (fat crystals) are oriented as a slit with a small interspacing. The absorbed oil films can coalesce *via* a snap-off mechanism. In the pore space of spherulites similar phenomena can play a role. Capillary pressures also arise in the presence of an oil–air interface, as in the case of oil-based food placed on a filter paper, *via* which one tries to measure the oil binding capacity. For computing the capillary pressure one requires knowledge of the wettability (contact angle) of the solid phase, and the interfacial tension between liquid and air.

There exists no systematic experimental method to determine the WHC of foods. We are developing two such methods, which are either based on centrifugation[1] or on confined compression.[4] In both methods we will apply defined external pressures on the food material. Due to permeability of the bottom of the food containers, moisture can be expelled from the compressed food. By varying the external pressure, we can determine the amount of retained moisture at various conditions. By analysis of the results *via* the Flory–Rehner theory we can obtain the relevant model parameters. By independent determination of the sorption isotherm and its analysis, *via* the FVFH theory, we can determine both the osmotic and elastic contribution to the swelling pressure. (Assuming that the ionic contribution can be absorbed in the osmotic pressure, and that the pore space is collapsed under sufficient load).

We summarize our discussion on the universality of WHC and OBC by listing the various contributions to the thermodynamic potential of water-based and oil-based food systems in Table 1. For water-based foods, we explicitly distinguish the gel phase and pore space, while for oil-based foods we assume that oil and fat crystals do not mix – these are thus missing contributions of the solid phase. In water-based foods the contribution of the disjoining pressure is hard to separate from the contribution of the osmotic pressure of the gel phase, and is thus thought to be absorbed in the osmotic pressure of the gel phase. The contribution of the dissolved solutes

Table 1 Partial pressures contribution to WHC or OBC of foods. The crosses (X) indicate the existence of this particle pressure in the food.

Potential	gel phase – WHC	pore space – WHC	OBC
Osmotic pressure	X	X	X
Disjoining pressure			X
Elastic stress	X		X
Capillary pressure		X	X

enters the osmotic pressure of the gel phase and pore space. A fat crystal network can give rise to an elastic contribution, as shown in the paper by Spicer and Hartel, and would contribute to the OBC when under an external load. Also, in organogels the fibre network can support elastic stresses, as indicated by Dr Bot. Osmotic pressure effects in the OBC arise if the liquid phase is a mixture of different oils.

We invite delegates to share their views on our hypotheses concerning WHC and OBC.

1 D. Curvers, H. Saveyn, P.J. Scales and P. van der Meeren, A centrifugation method for the assessment of low pressure compressibility of particulate suspensions, *Chem. Eng. J.*, 2009, **148**(2–3), 405–413.
2 A. K. Datta, Porous media approaches to studying simultaneous heat and mass transfer in food processes. i: Problem formulations, *J. Food Eng.*, 2007, **80**(1), 80–95.
3 A. Dhall and A. K. Datta, Transport in deformable food materials: A poromechanics approach, *Chem. Eng. Sci.*, 2011, **66**(24), 6482–6497.
4 M. R. Drost, P. Willems, H. Snijders, J. M. Huyghe, J. D. Janssen, and A. Huson, Confined compression of canine annulus fibrosus under chemical and mechanical loading, *J. biomech. eng.*, 1995, 117–390.
5 F. Horkay and G. McKenna, Polymer networks and gels, *Physical Properties of Polymers Handbook*, 2007, pp. 497–523.
6 E. Hu-Lonergan and S. M. Lonergan, Mechanisms of water-holding capacity of meat: The role of postmortem biochemical and structural changes, *Meat Sci.*, 2005, **71**(1), 194–204.
7 V. Karathanos, Collapse of structure during drying of celery, *Drying Technol.*, 1993, **11**(5), 1005–1023.
8 P. N. Kocher and E. A. Foegeding, Microcentrifuge-based method for measuring water-holding of protein gels, *J. Food Sci.*, **58**(5), 1040–1046, 1993.
9 T. P. Labuza and G. C. Busk, An analysis of the water binding in gels, *J. Food Sci.*, 1979, **44**(5), 1379–1385.
10 C. W. J. Oomens, H. J. De Heus, J. M. Huyghe, L. Nelissen and J. Janssen, Validation of the triphasic mixture theory for a mimic of intervertebral disk tissue, *Biomimetics*, 1995, **3**, 171–186.
11 J. A. Robertson and M. A. Eastwood, A method to measure the water-holding properties of dietary fibre using suction pressure, *Br. J. Nutr.*, 1981, **46**, 247–255.
12 M. A. Rogers, A. J. Wright and A.G. Marangoni, Engineering the oil binding capacity and crystallinity of self-assembled brillar networks of 12-hydroxystearic acid in edible oils, *Soft Matter*, 2008, **4**(7), 1483–1490.
13 B. Roman and J. Bico, Elasto-capillarity: deforming an elastic structure with a liquid droplet, *J. Phys.: Condens. Matter*, 2010, **22**, 493101.
14 N. Shahidzadeh, E. Bertrand, J. P. Dauplait, J. C. Borgotti, P. Vie and D. Bonn, Effect of wetting on gravity drainage in porous media, *Transp. Porous Media*, 2003, **52**(2), 213–227.
15 G.R. Trout, Techniques for measuring water-binding capacity in muscle foods – a review of methodology, *Meat Sci.*, 1988, **23**(4), 235–252.
16 M. Tuller, D. Or and L. M. Dudley, Adsorption and capillary condensation in porous media: Liquid retention and interfacial configurations in angular pores, *Water Resour. Res.*, 1999, **35**(7), 1949–1964.
17 R. G. M. van der Sman, Moisture transport during cooking of meat: An analysis based on Flory–Rehner theory, *Meat Sci.*, 2007, **76**(4), 730–738.
18 R. G. M. van der Sman, Soft condensed matter perspective on moisture transport in cooking meat, *AIChE J.*, 2007, **53**(11), 2986–2995.
19 R. G. M. van der Sman, Thermodynamics of meat proteins, *Food Hydrocolloids*, 2012, **27**(2), 529–535.
20 R. G. M. van der Sman and E. Boer, Predicting the initial freezing point and water activity of meat products from composition data, *J. Food Eng.*, 2005, **66**(4), 469–475.
21 R. G. M van der Sman and M. B. J. Meinders, Prediction of the state diagram of starch water mixtures using the Flory–Huggins free volume theory, *Soft Matter*, 2011, **7**(2), 429–442.
22 M. A. Zwieniecki and N. M. Holbrook, Confronting maxwell's demon: bio-physics of xylem embolism repair, *Trends in plant science*, 2009, **14**(10), 530–534.

Professor Pink, in response to the comment "it appears that there is not a clear understanding or definition of the concept of oil binding capacity. A similar misunderstanding...", answered: We have proposed a definition of OBC in eqn (4) and (5) in our paper. The question is perhaps not whether there is a misunderstanding, but whether one can relate an experimental measurement to a definition based upon the difference in free energies between the oil in, for example, a fats crystal network, and the oil in the bulk. It is possible that one can formulate the diffusivity of the oil constrained in a fat crystal network in terms of the free energy difference between oil in that environment and oil in the bulk outside the network.

With regards to the comments "we view that OBC and WHC are very similar concepts and can only be understood and defined within the framework of thermodynamics", this is exactly what eqn (4) and (5) in our paper do: defining OBC in terms of free energy differences.

Professor Pink further remarked, with respect to the comments "furthermore, it is likely that electrostatic interactions contribute little to the oil binding. Oils are typically mixtures of different oils, and can give rise to an osmotic pressure. Professor Pink has remarked that the osmotic pressure of mixtures of oils could be described by a Flory–Huggins type of model. His simulations also show that oils can be absorbed on solid interfaces (fat crystals) via van der Waals forces, which can probably also be expressed in terms of a disjoining pressure. Capillary forces can play a role if two nano-platelets (fat crystals) are oriented as a slit with a small interspacing": Although one might try to describe oil mixtures in terms of mean field theories, such as Flory-Huggins theory, one would likely encounter problems when dealing with effects on the nanoscale. In such cases, mean field theories will require the self-consistent definition of a number of mean fields, each one defined by, for example, the distance of a molecule from a solid nanoplatelet surface. It is possible, also, that correlations assume greater importance on the nanoscale than on the mesoscale – the existence of number density oscillations across the entire oil nano-space between nanoplatelets, shows this.

Professor Marangoni communicated: I think it is important to distinguish between the different ways that oil or water can be immobilized in a network. On one side, we have intermolecular interactions determining an adsorption or oil or water onto a macromolecular structure and, on the other side, we have capillary effects. I guess these could be related in the end. Ultimately, the treatment of WHC and OBC should be the same (except for osmotic effects) – bound water and oil, which would form the monolayer and condensation layers, and then the majority of the oil and water molecules, which are trapped within pores in the microstructure of the material. This is particularly obvious in the case of water, where high water activities are achieved at very low water contents...and it is similar with oil. Most of the water and oil molecules are trapped, in my view, in capillaries (pores) within a network. One has to add very high amounts of solutes to water to decrease the water activity.

Professor Datta commented: This is outside our area of expertise. The authors of the contribution make a reasonable case that there are enough parallels between water and oil so that OBC can be developed in an analogous way. Dr van der Sman's group has already published a soft matter based approach to predicting WHC. This being quite novel and not with enough precedence, only time will tell the extent to which this (prediction of WHC for a large number of food products) will be successful. The authors cite interesting examples to support their hypothesis that WHC and OBC have parallels and show how similar interactions could arise with oil as with water – thus attempting to extend their WHC work to OBC. In addition to the free volume Flory–Huggins interactions, the authors talk about disjoining pressure, which is related to the van der Waals forces between the oil and the solid surface, and that it may also contribute in determining the OBC. The discussions in the article, backed by appropriate examples, do indicate that universality in WHC and OBC is likely. We do see great promise in approaching OBC in an analogous way and encourage the authors to develop the concepts further than is already outlined. Surely, such a prediction of OBC can add to the list of auxiliary conditions that enable food process simulation – the title of this article. In particular, the modeling of oil transport during deep frying has been elusive and often has not appropriately included the surface interactions between oil and solid that are critical. A soft matter-based approach to OBC will help in both qualitative and quantitative understanding of the oil pick-up during a frying process.

Professor Bolhuis then opened the discussion of the paper by Dr Hütter by saying: In your approach you use effective potentials such as the DLVO potential. These potentials were developed for equilibrium conditions. Yet, your approach is in particular aimed at describing non-equilibrium dynamics such as flow, shear, gradients *etc.* Why is it allowed to still use these equilibrium potentials? Do you invoke a local equilibrium assumption? Is this based on an underlying separation of time scales?

Dr Hütter replied: We mentioned DLVO (as well as Hertzian) interactions only as examples, to indicate that the effective potentials (see eqn (31) in our paper) in our approach may, first, depend on temperature and, second, account for the elasticity of the particles. If we indeed used DLVO for practical applications, this would clearly imply that the electrostatic double layer in the DLVO-potential is always instantaneously relaxed, *i.e.*, that we look at comparably slow time scales. It should be pointed out that, at this stage of modeling, we are concerned with including many different features, *e.g.* particle elasticity, van der Waals and electrostatic double-layer interaction, effect of temperature, and particle concentration. Rather than being quantitative in a few aspects, we aim, at this point, to account for the large number of qualitative features, and to study the sensitivity of the system response to these features.

Dr van der Sman asked: You intend to apply Brownian dynamics. In this simulation model, can interacting particles exert mechanical/compressive forces onto one another if no hydrodynamics is included in the model?

Dr Hütter answered: The term Brownian dynamics refers to the fact that we are interested in time scales on which the relaxation of momentum is not relevant. Therefore, the acceleration term in the evolution equation for the positions can be dropped. However, all other force contributions are still retained in the model, particularly the forces due to potential interactions and hydrodynamic interactions. In the absence of many-particle hydrodynamic interactions, the particles still interact *via* potential forces. Particularly, if a particle is in contact with other surrounding particles, compressive forces arise. This is contained in our approach.

Dr van der Sman addressed Dr Hütter and Professor Tanaka: If the spongy particles are displaying some stickiness, would the system become more similar to viscoelastic phase separation, as discussed in the paper presented by Professor Tanaka? Should you not include hydrodynamic interactions for these systems of sticky, spongy particles? In any case, can you include hydrodynamic interaction into the model (i.e. the lubrication force approximation, which scales with $1/h$; with h being the gap between particles). In the paper by Vollebregt *et al.*, we state that hydrodynamic interactions are important in sheared particle suspensions, with sizes in the range of 0.1–10 micron. *Via* the Kirkwood–Irving equation, we can show that hydrodynamic interactions lead to the expression of particle stress, as shown in our paper (Vollebregt *et al.*). This can be done as shown in the following.

The particle stress tensor σ has a contribution due to contact forces,[4] and follows from the Kirkwood–Irving virial theorem:

$$\sigma = \frac{1}{V} \sum_{i<j} F_{ij} \times r_{ij} \tag{1}$$

Here, V is the volume of the system, F^{ij} is the contact force, and r^{ij} is the distance between particles. Lemaitre distinguishes three types of contact forces: (1) friction forces, (2) lubrication forces, and (3) collisional forces.[4] In the sheared suspensions lubrication forces (hydrodynamic interactions) are dominant. Now, we will derive an expression for σ showing the proper scaling with the particle pressure, as found in the state-of-the-art models of Brady and Morris on migration in sheared suspensions.[9,7,6] The magnitude of the lubrication force for monodisperse suspensions scales as:

$$F_{ij} = \eta_{\text{eff}} \dot{\gamma} \frac{a^2}{h_{ij}} \tag{2}$$

where η_{eff} is the suspension viscosity, $\dot{\gamma}$ is the shear rate, a is the particle radius and h_{ij} the gap between particles, indexed with i and j. Here, we have followed the mean field argument of Snabre and Mills,[5] that in a concentrated suspension the lubrication force scales with the suspension viscosity, rather than the fluid viscosity. Snabre and Mills have also argued that the particle pressure (the trace of the particle stress) is linear with the average distance between particles, a result we will use below:

$$\sigma = n \eta_{\text{eff}} \dot{\gamma} a^2 < h/a > \tag{3}$$

with n being the number density of the suspended particles, and $< h/a >$ is the average gap between particles, normalised against the radius a. Hence, from the above, it follows that the particle stress is linear with the viscous stress, which can be equated to an effective temperature.[8,12,2] Van den Brule and Jongschaap[11] and others[10] have argued that the average distance between particles is a function of the volume fraction, which is very similar to the free volume expression for the compressibility factor $Z(\phi)$. Hence, we can rewrite the particle pressure as:

$$\sigma = n(kT)_{\text{eff}} Z(\phi) \tag{4}$$

with the effective temperature defined by $(kT)_{\text{eff}} \sim \eta_{\text{eff}} \dot{\gamma} V_p$, with $V_p = 4/3\pi a^3$. Hence, the particle pressure is a generalisation of the *ideal gas law*, in the spirit of van der Waals. In our generalization we have introduced a generalized (effective) pressure (= particle pressure), the effective volume/free volume, and the effective temperature. This generalized pressure acts as an osmotic pressure, and has actually been measured in an experiment, where a suspension is sheared in a Couette cel, with a porous diafragm in the wall. Over the wall one can measure the difference in osmotic pressure between suspension and a non-flowing fluid.[1]

1 A. Deboeuf, G. Gauthier, J. Martin, Y. Yurkovetsky and J.F. Morris, Particle pressure in a sheared suspension: A bridge from osmosis to granular dilatancy, *Phys. Rev. Lett.*, 2009, **102**(10), 108301.
2 C. Eisenmann, C. Kim, J. Mattsson and D.A. Weitz, Shear melting of a colloidal glass, *Phys. Rev. Lett.*, 2010, **104**(3), 35502.
3 R.D. Kamien and A.J. Liu, Why is random close packing reproducible? *Phys. Rev. Lett.*, 2007, **99**(15), 155501.
4 A. Lemaitre, J. N. Roux and F. Chevoir, What do dry granular ows tell us about dense non-brownian suspension rheology? *Rheol. Acta*, 2009, **48**(8), 925–942.
5 P. Mills and P. Snabre, Apparent viscosity and particle pressure of a concentrated suspension of non-brownian hard spheres near the jamming transition, *Eur. Phys. J. E: Soft Matter and Biological Physics*, 2009, **30**(3), 309–316.

This journal is © The Royal Society of Chemistry 2012

6 J. F. Morris and F. Boulay, Curvilinear flows of noncolloidal suspensions: the role of normal stresses, *J. Rheol.*, 1999, **43**, 1213.

7 J. F. Morris and J. F. Brady, Self-diffusion in sheared suspensions, *J. Fluid Mech.*, 1996, **312**, 223–252.

8 A. Negi and C. Osuji, Dynamics of a colloidal glass during stress-mediated structural arrest, *EPL (Europhys. Lett.)*, 2010, **90**, 28003.

9 P. R. Nott and J. F. Brady, Pressure-driven flow of suspensions: simulation and theory, *J. Fluid Mech.*, 1994, **275**, 157–200.

10 R. F. Probstein, M. Z. Sengun and T.C. Tseng, Bimodal model of concentrated suspension viscosity for distributed particle sizes, *J. Rheol.*, 1994, **38**, 811.

11 B. Van den Brule and R. J. J. Jongschaap, Modeling of concentrated suspensions, *J. Stat. Phys.*, 1991, **62**(5), 1225–1237.

12 H. M. Vollebregt, R. G. M. van der Sman and R. M. Boom, Suspension flow modelling in particle migration and microfiltration, *Soft Matter*, 2010, **6**(24), 6052–6064.

Dr Hütter communicated in reply to Dr van der Sman: Depending on the suspension behaviour of interest, hydrodynamic interactions must be accounted for. In our approach, hydrodynamic interactions are incorporated *via* an appropriate choice of the mobility matrix, as stated after eqn (30) in the paper. As a result, the stress tensor shown in eqn (27a) in the paper is informed indirectly about the presence of hydrodynamic interactions through the evolution of the particle positions, which in turn include the hydrodynamic interactions. However, as shown in Chapter 7 of ref. 1, in general the stress tensor also contains contributions that depend explicitly on the hydrodynamic interactions, *via* the stresslets. Furthermore, it is demonstrated that not only relative velocities give rise to particle forces, but also there are force contributions that originate from the rate of strain tensor.[1] All this indicates that the hydrodynamics couples the macroscopic scale (stresslets contributions to the stress, rate of strain tensor) with the microscopic scale (particle forces and particle velocities). In our approach, when formulating the irreversible dynamics in Section 3.5 of our paper, we have assumed that the irreversible dynamics of the particles does not affect the momentum balance. Therefore, our approach falls short in this respect. To improve on this point in future work, we will follow the procedure developed in the context of Stokesian dynamics.[2]

1 S. Kim and S. J. Karrila, *Microhydrodynamics: Principles and Selected Applications*, Butterworth-Heinemann, Stoneham MA, 1991.
2 T. N. Phung, J. F. Brady and G. Bossis, *J. Fluid Mech.*, 1996, **313**, 181–207.

Professor Tanaka communicated in reply to Dr van der Sman: Spongy particles attracting with each other should show viscoelastic phase separation, whose principle is the same as that in colloidal suspensions and protein solutions. Hydrodynamic interactions including lubrication effects are naturally included in the two-fluid model of viscoelastic phase separation. If we simulate viscoelastic phase separation with an appropriately low polymer concentration, we expect that spongy viscoelastic polymer-rich balls are spontaneously formed as a consequence of phase separation and then they aggregate to form a transient gel network. Inclusion of thermal force noises in the momentum conservation equation should also allow us to incorporate thermal Brownian motion of particles. This model can also express characteristic features of spongy particles, such as the stress-induced solvent expelling behaviour. Furthermore, by introducing yield stress behaviour into our model, we can also prevent coalescence of particles. This may be a good way to simulate the behaviour of a suspension of spongy particles with full hydrodynamic interactions.

Professor Cates returned the discussion to the paper presented by Dr Hütter: I admit to being somewhat confused by eqn (11b) in your paper. If the velocity in this equation is the centre-of-mass velocity of the sponge and the solvent together, then for an incompressible system its divergence should be zero. Is there some reason not to set it to zero at this point rather than carry it through the calculation? An alternative approach to swelling and deswelling problems is to use a two-fluid model, in which there is a relative velocity between the two components as well as the mean velocity; clearly this would affect the form of eqn (11b). Is there a good reason why you do not need a two fluid approach for the class of problems you are considering? It seems that the hydrodynamics of fluid flowing into or out of the sponge particles is handled purely by setting the coefficients of certain terms in your generalized mobility tensor controlling the dissipative response, but I doubt that is adequate in all cases.

Dr Hütter replied: It is correct that the velocity used in the entire model is the centre-of-mass velocity of the entire suspension. If one considers a system with a very high bulk modulus, small volume changes are equivalent to large pressure variations. A hydrodynamic description can address this aspect in a dynamic way: The pressure with a high bulk modulus gives rise to changes in the velocity field (via the momentum balance) in such a way that the divergence of said velocity field is nearly zero. This argument also applies to the procedure adopted in our paper. From a practical perspective, though, one could indeed argue for enforcing the condition of diver-gence-free flow as a hard, mathematical constraint. The reason for not imposing such a rigid constraint in our case is twofold. First, having such a constraint on a dynamic variable results in constrained functional derivatives in eqn (2) in the paper, and, hence, in (unnecessarily confusing) technical complications. Second, the bulk pressure would at any moment be such as to make the flow field incompressible. In turn, all isotropic contributions in the expression for the stress tensor in eqn (35) of the paper would be meaningless. However, we believe that from a conceptual view-point, even the isotropic contributions in eqn (35) are interesting.

Clearly, a two-fluid approach is a possible alternative to our procedure. We expect a two-fluid approach to be particularly useful if the overall (average) motion of particles deviates significantly from the solvent velocity. Hence, our approach is not applicable to situations with *e.g.* a significant influence of gravitational or elec-tromagnetic effects, in the presence of strong thermophoresis, or in jammed systems where the solvent can still move through the particle pack.

Dr van der Sman queried: In the papers by Vollebregt *et al.* and Spicer *et al.* we have seen examples of (1) how non-equilibrium/irreversible phenomena (HI) at a lower scale give rise to reversible phenomena (thermodynamic potential = particle stress) at a higher length scale, or (2) the reverse: reversible (elastic) interactions at microscale lead to irreversible phenomena (= arrest). Is this admissible according to the GENERIC framework?

Similar coupling is observed in sheared polymers, as discussed by Jou and coworkers,[1] who developed a non-equilibrium chemical potential concept within the framework of (extended) non-equilibrium thermodynamics. Also, one can think of shear-induced/modulated phase transitions.[2,3]

1 M. Criado-Sancho, D. Jou, and J. Casas-Vázquez, Definition of nonequilibrium chemical potential: phase separation of polymers in shear flow, *Macromolecules*, 1991, **24**(10), 2834–2840.
2 P.Butler, Shear induced structures and transformations in complex fluids, *Curr. Opin. Colloid Interface Sci.*, 1999, **4**(3), 214–221.
3 A. Onuki, Phase transitions of fluids in shear flow, *J. Phys.: Condens. Matter*, 1997, **9**, 6119.

Dr Hütter responded: Clearly, reversible effects on the microscopic scale can give rise to irreversible behaviour on a larger scale: degrees of freedom are eliminated, and only lumped, collective variables are retained, leading to a loss of information because not all details are under mechanistic control. The emergence of irrevers-ibility is a major result of systematic coarse-graining procedures.[2,3,4] For example, water is purely reversible when described by molecular dynamics simulations, however, the very same material considered on macroscopic scale shows irreversible (viscosity and thermal conductivity) behaviour. All this is in agreement with the GENERIC framework.

However, I have serious reservations against the statement that irreversible phenomena at a lower scale can give rise to reversible phenomena on a higher scale. I am unaware of arguments stating that the amount of irreversible effects decreases when going to larger scales.

The work of Jou and coworkers discusses in detail the thermodynamic driving force for phase separation of polymers in shear flow.[1] In their paper, an extension of the Gibbs free energy is introduced that depends on the time-derivative of the viscous pressure tensor. First, it should be said that it is not clear how a thermodynamic potential in terms of such rate-type variables should be derived by coarse graining, e.g. within the framework of generalized ensemble theory.[2,3] Second, if such a thermodynamic potential is postulated on phenomenological grounds, Jou et al. just state that one of the effects of the viscous pressure tensor is that it occurs in the Gibbs free energy. So, it does not give rise to a truly reversible phenomenon, but rather enters in the thermodynamic driving force; there is still a kinetic prefactor that renders the entire phase separation irreversible. As an aside, I note that the extension of the Gibbs free energy by a viscous-pressure term may be related to the use of dissipation potentials[5] to model force-flux relations in irreversible processes; dissipation potentials are not thermodynamic potentials in the original sense.

1 M. Criado-Sancho, D. Jou, and J. Casas-Vázquez, Definition of nonequilibrium chemical potential: phase separation of polymers in shear flow, Macromolecules, 1991, 24(10), 2834–2840.
2 H. C. Öttinger, Phys. Rev. E, 1998, 57, 1416-1420.
3 H. C. Öttinger, Beyond Equilibrium Thermodynamics, Wiley, Hobroken, 2005.
4 M. Hütter and T. A. Tervoort, Adv. Appl. Mech., 2008, 42, 253–317.
5 M. Silhavy, The Mechanics and Thermodynamics of Continuous Media, Springer, 1997.

Dr van Gruijthuijsen continued the discussion of the paper by Professor Pink by saying: You showed us simulation data on the stacking of fat-crystal platelets, comparable to those discussed in the paper presented by Professor Marangoni. You attributed the strong tendency to stacking to the presence of van der Waals attractions.

To what extent do you think the excluded volume interactions play a role in the ordering of the platelets as well? Could you estimate an effective volume fraction, by comparing the real volume of the platelets to an effective volume of $\pi L^3/6$, if L is the longest dimension of the platelets? I can imagine that too strong van der Waals attractions could prevent stacking, since any encounter would lead to a permanent bond between two platelets, no matter their mutual orientation. Could you give some idea of the interaction strengths involved here?

Professor Pink replied: In order to model CNPs we represented the flat nanocrystals by a close-packed lattice of 3-dimensional solid unit structures composed of crystalline TAGs. It is convenient to represent each CNP as a close-packed structure of spheres, since the attractive part of the dispersion interaction between spheres has been established.[1] Each sphere represents a continuum of TAG molecules with a density characteristic of the crystalline phase of interest and Fig. 2 (shown below) shows representations of a CNP.

Although we were unaware of it, other works have represented solid objects as rigid aggregates of spheres.[2]

The interaction between two identical homogeneous spheres, each of radius R, a centre-to-centre distance, r, apart is the Hamaker hybrid form:[1]

$$V_d(r) = -\frac{A_H}{6}\left[2R^2\left(\frac{1}{s^2+4Rs}+\frac{1}{(s+2R)^2}\right)+\ell n\left(\frac{s^2+4Rs}{(s+2R)^2}\right)\right] \quad r \geq 2R \quad (1)$$

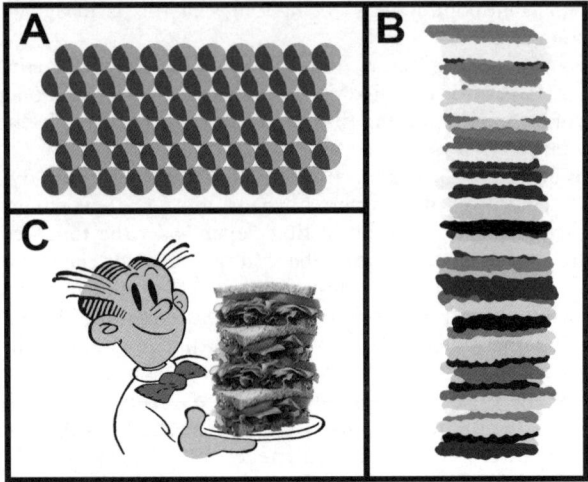

Fig. 2 A: A model CNP of dimensions $10 \times 6 \times 1$. B: A Dagwood formed from interacting CNPs. C: Dagwood and his sandwich.

where $s = r - 2R$ is the surface-to-surface separation of the two spheres and A_H is the Hamaker coefficient.

Aggregation of CNPs were modeled in an $L \times L \times L$ simulation box with periodic boundary conditions. One MC step involved translating and rotating CNPs and translating and rotating clusters of CNPs with respect to their centres of mass. We permitted movement of all clusters with the translational step size as well as the angle of rotation around a randomly-chosen axis through the centre of mass, proportional to $M^{-1/2}$, where M is the mass of the cluster. All CNPs were of size $m \times m \times 1$, We chose the radius of the spheres making up the CNPs to be $R = 0.5$ in arbitrary units and we chose $L = 100$.

If the concentration of nanoplatelets is not too high, we found essentially linear self-assembling aggregates composed of layers of model nanoplatelets.[3] These structures resemble dagwoods – the iconic multilayered sandwiches named after Dagwood Bumstead,[4] a central character in the comic strip Blondie. If these structures are shown to actually form in oils then we will have taken the next step in modelling and understanding the hierarchy of self-assembled structures in edible oils.

Naturally, the relative orientation of two adjacent nanoplatelets is constrained by the requirement that they cannot overlap and, in this sense, excluded volume plays a role. Does excluded volume play a role in the creation of the stacks? The platelets lie one on top the other because that configuration minimizes the van der Waals interaction energy (maximizes its absolute value), which likely dominates the free energy at room temperatures. In the absence of such an attractive interaction, no stacking was seen.

We also thought that a very strong van der Waals interaction might inhibit stacking. We found this not to be so at lower concentrations (~6% by volume). The reason is that thermal effects eventually maximize the overlap of pairs of nanoplatelets. At higher nanoplatelet concentrations, we did find that the stacks were shorter than at lower concentrations. This, however, might be due only to the creation of a greater number of nuclei for stack formation than at lower concentrations.

Essentially, what we have is analogous to diffusion limited aggregation (DLA) of highly anisotropic nanoplatelets, but one in which relaxation is permitted after they are in contact. When we replaced the van der Waals interaction by an infinitely strong interaction that forbids relaxation (thus forming a "permanent bond"), then we found the results of DLA. For all finite interactions that we used, we always found stacks unless the interactions were so weak as to be dominated by entropic effects.

We used interaction strengths varying from ~5 kT to ~1000 kT.

1 V. A. Parsegian, *Van der Waals Forces*, Cambridge University Press, 2005.
2 S. C. Glotzer, M. A. Horsc, C. R. Iacovella, Z. Zhang, E. R. Chan, and X. Zhang, *Curr. Opin. Colloid Interface Sci.*, 2005, **10**, 287.
3 D. A. Pink, *Intercrystalline Interactions in Structure-Function Analysis of Edible Fats*, ed. A. G. Marangoni, AOCS Press, 2012, Ch.6.
4 http://en.wikipedia.org/wiki/Blondie

Dr Smith asked: If your model included liquid molecules with S chains then one could expect a greater degree of interaction with the surface. Would this have a significant effect on the behaviour of the model?

Professor Pink answered: Not substantially, I think. Probably the free energy differences would be greater than that shown in Fig. 3 in our paper.

Dr Bot opened discussion on the paper presented by Professor Tanaka: Your paper links a wide range of phase separation phenomena to food systems. There is one particular basic food system that I consider excellently suited for a soft matter approach: dough. It has attracted interest at the present meeting too, as illustrated by two posters.[1,2] A few years ago, we attempted to formulate a qualitative model for dough, featuring a phase separated gluten + starch paste mixture.[3] The essential feature is that upon kneading, water is exchanged between both phases, because the gluten protein filaments take up water when stretched. Upon resting of the dough, the gluten relaxes again and the water is expelled again. The resulting change in consistency of the starch phase, which is close to close packing of the starch granules, explains the changes in consistency that are observed in practice. Could your models accommodate such a complicated exchange of water between both phase-separated phases?

1 A. Boire, P. Menut, M.H. Moret, C. Sanchez, Phase separation in wheat gluten protein solutions, Poster 1, *Faraday Discussion 158*, 2012.
2 J.H.J. van Opheusden, Brownian dynamics simulation of segregation of gluten and starch in shear flow, poster 18, *Faraday Discussion 158*, 2012.
3 A. Bot and D.W. de Bruijne, *Cereal Chem.*, 2003, **80**, 404–408.

Professor Tanaka answered: Yes, I think so. Our model includes the so-called stress-diffusion coupling, which is basically described by eqn (2) in the paper. This equation tells us that it is not only the osmotic stress but also the mechanical stress that control the water motion relative to a polymer matrix. Combined with an appropriate constitutive equation describing the temporal evolution of the stress tensor, we should be able to describe such complicated transport processes at least on a phenomenological level.

Dr van der Sman communicated: In your paper you refer to the idea that VPS might occur in sheared/kneaded dough, which contains gluten. It has been suggested that Flory-Rehner also applies to gluten,[1] similarly to meat proteins.[2] Does your model allow for the fact that shear stretches the protein phase, and that it attracts water? Dr Bot has posed earlier a similar picture for the sheared gluten.[3]

1 S. Domenek, L. Brendel, M. H. Morel and S. Guilbert, Swelling behaviour and structural characteristics of wheat gluten polypeptide films, *Biomacromolecules*, 2004, **5**(3), 1002–1008.
2 R. G. M. van der Sman, Thermodynamics of meat proteins, *Food Hydrocolloids*, 2012, **27**(2), 529–535.
3 A. Bot, and D. W. de Bruijne, Osmotic properties of gluten, *Cereal Chem.*, 2003, **80**(4), 404–408.

Professor Tanaka replied: As I answered to the question of Dr Bot, I think that it is possible to describe the behaviour of gluten including the transport of water under mechanical stress.

Dr Ubbink commented: In your paper, you describe a wide range of situations, of relevance to food, in which structure building can be analyzed in terms of phenomenological models of viscoelastic phase separation. To which extent would such models be helpful to not only develop a common basis for the understanding of structure formation in apparently disparate foods, but also to allow the prediction of structures, given quantitative knowledge of the relevant physical parameters? To which extent could, for instance, the structure of a baked product (such as bread) be predicted based on knowledge of the rheological, interfacial and colloidal properties of dough?

Professor Tanaka responded: This is a good question. At this moment, the weakness of our model comes from the phenomenological nature of the constitutive equation. This is because there is no firm physical basis for the constitutive relation of a thermodynamically 'not' stable state. This is an intrinsic problem since phase separation occurs only in a thermodynamically metastable or unstable state. Once we can establish a proper constitutive relation, there will be a good chance to make a quantitative prediction on the structural evolution on the basis of relevant material parameters. However, this is not an easy task. Nevertheless, our model would provide a physical basis for the mechanism of structural formation and key information on what physical factors control phase-separation morphologies.

Dr van der Sman enquired: Eqn (8) in your paper resembles very much the elastic contribution to the swelling pressure in the Flory-Rehner theory, which we think is quite valid for food materials.[1] In your paper you state that the bulk modulus should have sufficient steepness with respect to the volume fraction. Does the Flory–Rehner theory have sufficient steepness to show VPS? Are there constraints with respect to the model parameters? ϕ_0 for example? By the way, would ϕ_0 have any specific meaning for a food material? For a synthetic gel, ϕ_0 is the volume fraction of the gel at crosslinking. Food is not a synthetic gel, but it is constructed by nature *via* self-assembly like cell wall material.[2]

1 R. G. M. van der Sman, Thermodynamics of meat proteins, *Food Hydrocolloids*, 2012, **27**(2), 529–535.
2 D. J. Cosgrove, Growth of the plant cell wall, *Nat. Rev. Mol. Cell Biol.*, 2005, **6**(11), 850–861.

Professor Tanaka replied: We used eqn (8) in the paper to estimate the elastic modulus of a transient gel. We need a steep composition dependence of the structural relaxation time or the modulus to represent the transient nature of a gel network, which is not considered in usual theories of gel elasticity. ϕ_0 represents a reference state for a gel. For a transient gel this corresponds to the local ϕ when it is formed.

Dr van der Sman asked: In the paper by Vollebregt *et al.*, we discussed binary suspensions in shear flow. We are still wondering whether we have divided the stresses correctly over the two phases. In your paper, you discuss the point of stress division in VPS systems. We wonder whether your rules for division also apply to ternary systems (fluid + 2 particle phases having different sizes). We based the particle division on the Gibbs–Duhem relation, as given by Doi and Onuki,[1] where $\partial\sigma = \phi_1\partial\mu_1 + \phi_2\partial\mu_2$ (with σ being the particle pressure, and μ_i the chemical potential, ∂ represents the gradient operator). Hence, we have thought the gradient in the particle pressure is $\partial\sigma = \phi_i\partial\mu_i$. Furthermore, we thought that each particle phase has equal effective temperature. In your paper, you have proposed a stress division based on friction factors, and not based on thermodynamic arguments. Which do you think is more appropriate? (For polymer mixtures and for particle mixtures.) Your approach might be valid, and even give rise to non-equipartitioning of the effective temperature over the two particle phases (which does occur in granular media).

Professor Tanaka answered: In principle, we can generalize our expression for the stress division to a multi-component system. The stress division is the division of mechanical stress and thus it seems physically natural for me that the stress division is expressed by friction factors rather than thermodynamic factors.

Dr van der Sman then continued the discussion of the paper by Dr Hütter: In the paper by Vollebregt *et al.* we show that in sheared suspensions particles interact *via* hydrodynamic interactions, which give rise to a particle stress. Does this type of interaction fall also within the GENERIC framework? Although we have not modelled this, models by Jeff Morris show that particles can support normal stresses.[1] Does this support of normal stresses make the sheared suspensions also fit within the framework of VPS? The governing equations look very similar, especially if you look at the original two-fluid formulation.[2]

1 R. M. Miller and J. F. Morris, Normal stress-driven migration and axial development in pressure-driven flow of concentrated suspensions, *J. Non-Newtonian Fluid Mech.*, 2006, **135**(2), 149–165.
2 Suspension flow modelling in particle migration and microfiltration, H. M. Vollebregt, R. G. M. van der Sman and R. M. Boom, *Soft Matter*, 2010, **6**(24), 6052–6064.

Dr Hütter replied: Whether a dynamic model with its constitutive relations can be formulated also within the GENERIC framework must be worked out in detail. To my knowledge, this has not been shown for two-fluid models yet. It should be pointed out, however, that the GENERIC procedure does not make a statement about, *e.g.*, how the viscosity depends on structural details, as long as the viscosity is positive, see *e.g.* Sec. 3.1 in the paper by Vollebregt *et al.* within this volume.

To describe the flow of particle-laden fluids, classical hydrodynamics needs to be extended. A first step consists in using an effective medium approach, in which material properties are expressed in terms of, *e.g.*, the volume fraction of particles. A second level of sophistication is to consider a two-fluid model, or, more generally, a two-phase model. Such approaches are useful if one aims to study phenomena like phase separation. However, if one is interested in the microscopic origin of the macroscopic response in terms of particle properties, a different approach is needed. This was our motivation to formulate a model that supplements classical hydrodynamics with the evolution of the microstructure, where the latter is given by the particle positions and sizes. We have not yet studied the relevance of our approach to viscoelastic phase separation (VPS). Since our model was designed to serve a different purpose, it may well be that it must be extended to describe VPS. It seems plausible to incorporate an additional dependence of the energy and entropy on the macroscopic concentration of particles as in eqn (4b) in our article, akin to the Flory–Huggins–de Gennes form discussed in Sec. 3.1 of the paper presented by Professor Tanaka.

Professor Tanaka commented: Although our model of viscoelastic phase separation is not designed to describe particle systems, at a low polymer volume fraction the polymer-rich phase can form droplets. Then, hydrodynamic interactions between these droplets are properly taken into account in our scheme. If we introduce a feature having yield stress into the constitutive relation, we can simulate a particle system in that way. However, if we are interested in simulating the behaviour of a colloidal system, it is more suitable to use the fluid particle dynamics method,[1] which we developed to incorporate hydrodynamic interactions without suffering from solid-fluid boundary conditions. We can also apply a shear field rather easily in this method. It is quite difficult to describe the dynamics of such a system in a coarse-grained level, since we do not have a coarse-grained theory properly dealing with many-body hydrodynamic interactions.[2]

1 H. Tanaka and T. Araki, Simulation Method of Colloidal Suspensions with Hydrodynamic Interactions: Fluid Particle Dynamics, *Phys. Rev. Lett.*, 2000, **85**, 1338–1341.
2 H. Tanaka, Viscoelastic model of phase separation in colloidal suspensions and emulsions, *Phys. Rev. E*, 1999, **59**, 6842–6852.

Professor Cates continued the discussion of the paper presented by Professor Tanaka: None of your models addresses the case where the viscoelastic phase has a yield stress. Do you think this could be important, particularly in food systems where there is often a glass transition involved? We do know that rheological models of the glass transition predict a Herschel–Bulkley like behaviour, in which there is a finite limiting stress at low strain rates (*i.e.*, a yield stress).[1]

1 See *e.g.* J. M. Brader *et al.*, *Proc. Nat. Acad. Sci. U.S.A.*, 2009, **106**, 15186.

Professor Tanaka replied: I agree that the yield stress plays a very important role in arresting viscoelastic phase separation and making a phase-separated structure stable. This is briefly discussed in section 8 of our paper, but without any explicit forms of such constitutive equations. It is very useful to employ a constitutive relation that can take this effect of the yield stress into account, such as the one suggested by Brader *et al.*,[1] when we consider a coupling between viscoelastic phase separation and dynamic arrest. This may indeed be very relevant to many problems in foods science.

1 See *e.g.* J. M. Brader *et al.*, *Proc. Nat. Acad. Sci. U.S.A.*, 2009, **106**, 15186.

Dr van der Sman asked: To continue the discussion on the structure formation in bread, the structure of the crumb (core) of the bread is arrested *via* gelling in contrast to the crust, which is arrested due to a glass transition. Does your model allow for both an arrest *via* gelation and *via* a glass transition? If so, which information is required for this decription?

Professor Tanaka responded: We can differentiate dynamic arrest due to glass transition and that due to gelation by employing proper constitutive equations on a phenomenological level. To describe the dynamic arrest of viscoeloastic phase separation, we need to include the yield stress effects to constitutive relations (as I discussed in my response to the comment of Professor Cates). Previously, we simulated the dynamical arrest of viscoelastic phase separation by gelation due to chemical crosslinking.[1]

1 H. Nakazawa, S. Fujinami, M. Motoyama, T. Ohta, T. Araki, H. Tanaka, T. Fujisawa, H. Nakada, M. Hayashi, and M. Aizawa, Phase separation and gelation of polymer-dispersed liquid crystals, *Comp. Theor. Polym. Sci.*, 2001, **11**, 445–458.

Dr Royall opened the discussion of the paper by Professor Dr Bolhuis by communicating: What is the rationale for using the exponential decay for the electrostatic interactions instead of a Yukawa form?

Professor Bolhuis answered: We used the exponential form of the screened electrostatic interaction, as it seemed to make sense for macroions.[1] We agree that the Yukawa form is more commonly used for protein–protein interactions. However, for long distances, the difference between the Yukawa and the exponential form is small. We also performed simulations using the Yukawa potential and found qualitatively similar behavior. Since this is a very crude model anyway, we did not repeat all the simulations, also due to time restrictions.

1 See *e.g.* J. Israelachvili, *Intermolecular and Surface Forces: Revised Third Edition*, Academic press, 2011.

Dr Royall asked: How exactly are the covalent bonds implemented in your model?

Professor Bolhuis answered: The intermolecular covalent disulfide bonds between proteins are just a constraining potential that do not allow two proteins to move apart as long as they are bound together. By disulfide exchange reactions these bonds can be broken and formed. We assume the free energy difference for exchange is zero. Our simulations thus simulate an ideal situation where the disulfide exchange can take place between all neighboring protein pairs and reach equilibrium. This is not likely the case in reality.

Professor De Kruif commented: The initial formation of protein clusters is determined by a mechanism that involves a dissociation of the protein dimers, a partial unfolding and aggregation. After the formation of clusters there may be a formation of di-sulfide bonds, which "locks" the protein to the clusters and renders the process irreversible. It would be of interest to switch off the disulfide exchange. Then, results can be compared with the aggregation of porcine β-lactoglobulin, which lacks a free thiol group.[1]

1 See: T. V. Burova, N. V. Grinberg, R. W. Visschers, V. Y. Grinberg and C. G. de Kruif, Thermodynamic stability of porcine β-lactoglobulin: A structural relevance, *Eur. J. Biochem.* 2002, **269**, 3958–3968.

Professor Bolhuis responded: Our simulations indicate that switching off the disulfide bond formation has a tremendous effect on the location of the clustering transition, by shifting the coexistence concentrations with several orders of magnitude. This qualitatively corresponds to the finding that porcine β-lactoglobulin does not aggregate so easily. Our results suggest that at high concentration or at low pH even porcine β-lactoglobulin would aggregate, albeit possibly in a reversible way.

Professor Kulozik suggested: It should be considered that the rate limiting step in β-Lg aggregation at lower temperatures is the unfolding step. This means that the process depends on the temperature range used for the modelling. At low temperature (< 90 °C) monomers "wait" to be unfolded before they are available for aggregation, while at high temperatures there is an excess of unfolded molecules waiting to be incorporated in aggregates. These aspects are recommended to be considered in further developments of the model.

Professor Bolhuis commented: We did not include the unfolding transition in the model. Instead, we assumed that betalactoglobulin is largely denatured at the temperatures of the experiment. We therefore considered only a single (averaged) interaction potential between the denatured proteins. One can also take the viewpoint that the chemical potential imposed sets the (ideal gas) concentration of the unfolded protein rather than the total concentration. In that case the native protein is assumed to have no interaction. A more sophisticated model can also include the unfolding transition itself, and will thus be able to study temperature dependence (which we did not consider).

Dr Nicolai opened a general discussion based on the following: The simulation of the aggregation of β-lactoglobulin necessarily makes a number of simplifying assumptions that are, however, not true for the real system. Notably: (1) all proteins are reactive from the start (no denaturation process); (2) all sulfide groups are equally reactive; (3) disulfide bridges can be formed between any two sulfide groups between neighbours and within the same protein; (4) the proteins are spherical and the interaction potential is spherically symmetric. The question is whether the simulation can still predict the dramatic changes in the morphology of the aggregates as a function of the pH.[1] Above about pH 6.1, small elongated aggregates are formed, while below this pH larger spherical aggregates are formed.[2] The transition appears to be discontinuous and is induced by a small change of the charge density from about −6 to about −4. It cannot be induced by adding NaCl, *i.e.* by screening, but it can be induced by adding $CaCl_2$, which also reduces the charge density by specifically binding to the proteins.[3]

1 Nicolai *et al.*, *Food Hydrocolloids*, 2011, **25**, 1945–1962.
2 Phan-Xuan *et al.*, *Langmuir*, 2011, **27**, 15092.
3 Phan-Xuan *et al.*, *Food Hydrocolloids, accepted*.

Professor Bolhuis responded: It is true that our model makes strong assumptions, some of which are definitely severe oversimplifications of the real system. The model is meant to be a first step towards a more sophisticated approach.

However, we believe that the model captures some of the behavior of β-lactoglobulin, at least qualitatively. The morphology changes as a function of pH can be explained in our view as follows. The higher the pH, the higher is the charge on the protein. Thus, pH is proportional to our repulsion factor A. Assuming that the screening length and the interaction strength do not change, Fig. 7b of our article applies. For high pH above 7, the repulsion is high, say around 0.6. For the concentrations reported in the paper we would expect small clusters (as were also found in Ref. 1). Lowering the pH to 6 will lower the A to around 0.3–04, below the coexistence line with the ribbon phase in which we observe elongated aggregates. We predict that this transition is abrupt, as it is a (micro) phase separation. Lowering the pH even further reduces A to around 0.25–0.30 where we observe large aggregates (not shown in the paper). Note that due to the system sizes, we cannot really distinguish between large spherical aggregates and bulk aggregation. Adding NaCl will lower the screening length or increase κ. Our simulations predict that this will also induce a phase transition into the ribbon phase. That this is not found experimentally is indeed an indication that the model is too simplistic.

1 M. Verheul, S. P. F. M. Roefs and K. G. de Kruif, *J. Agric. Food Chem.*, 1998, **46**, 896–903.

Ms Zeiler communicated: There is a phenomenon that might be somewhat similar to the experimental results. In Fig. 9(d) of the paper presented by Professor Bolhuis you can see that at $k = 1.1$ the system goes very swiftly from the cluster phase into the ribbon phase. At $k = 1.2$, however, there no longer seems to be a ribbon phase, and the clusters grow in three dimensions. This, however, is an effect of screening and not a change in pH.

Dr Nicolai clarified: It is perhaps important to stress that the transition cannot be induced by adding monovalent salt. Adding monovalent salt simply leads to random association of the strands. This means that the actual charge density of the proteins is important and not just the strength of the electrostatic interactions.

Professor Bolhuis commented: The behaviour of your beta lactoglobulin system might be in qualitative agreement with our simulations. At high pH = 8 you see small aggregates that would be similar to the ones in our cluster phase. When the pH is lowered from 8, the protein charge is decreasing, thus, effectively, lowering the electrostatic repulsion term A. If you look at Fig. 9b or 9f in our paper, you can see that the effect of lowering A from the cluster phase induces a phase transition to a ribbon or wormlike micelle phase. This would be in agreement with your observation that wormlike structures form abruptly. Moreover, lowering the repulsion further would effectively lead to bulk aggregation (not shown in our phase diagrams), which would qualitatively explain the large aggregates at lower pH.

Dr Ettelaie asked: If the rate of production of unfolded β-lactoglobulin is the rate limiting step in the process, can a picture of dynamic evolution of the system be built up using your simulations? In this picture, the aggregates are in equilibrium with the monomers, and the chemical potential of the unfolded β-lactoglobulin is then varied in accord with that dictated by the rate limiting unfolding reaction/process.

Professor Bolhuis answered: We did not include the unfolding transition in the model. Instead, we assumed that the β-lactoglobulin is largely denatured at the temperatures of the experiment. We therefore considered only a single (averaged) interaction potential between the denatured proteins. Considering the fact that we use Monte Carlo simulations we cannot expect a dynamical picture of the evolution of the system, even if we would build in the unfolding transition. However, one could imagine a Brownian dynamics scheme or a kinetic Monte Carlo approach that does allow for time evolution. This might be a direction for future research.

Professor Kulozik opened the discussion of the paper presented by Ashim K. Datta by communicating: This comment includes a remark from Prof. Tanaka who commented on models that should be only used qualitatively, because of the many variables in complex food systems. Building on this, and referring to the earlier comment, I would like to reiterate that the chemical detail is not negligible, especially when one wants to go beyond the length scale of the dispersed phase, which leads to the chemical level of molecular particle properties.

Dr van der Sman commented: We are not scared by the complexity of food materials, and see this as a challenge. In the research project cited in the paper, the rehydration of freeze-dried vegetables, we have obtained most of the food material properties *via* predictive theories (which we have cited in our answer to question 712). Only elastic properties had been required to be measured, which are performed *via* centrifugation experiments where moisture is expelled from the food. Results have been analysed with Flory–Rehner theory and swelling properties plus the elastic modulus have been determined. The success of the application of the predictive theories (sorry that many results are awaiting publication) show that moisture migration at the meso/macroscale can be predicted using these theories. Many properties can be predicted using scaling rules based on coarse-grained molecular properties, like molecular weight. Also, these types of models sometimes require knowledge of the microstructure, which can be obtained nowadays *via* MRI or XRT. Of course, for some properties, like nutrition, taste, toxicity, enzyme kinetics, they depend on the specific molecular properties. These phenomena become relevant at the cellular scale, where food ingredients are absorbed into our body. This is definitely a lower scale than the meso/macroscale of food microstructure and food product. At the colloidal/mesoscale one often applies coarse-graining concepts, as sketched by Professor De Kruif for the casein micelles, which can be viewed as (adhesive) hard spheres – like many globular proteins. Subtleties in food materials are of course due to molecular specifics, like intramolecular hydrogen bonds, but often these can be regarded as second order effects. But taking universal behaviour as a first approximation can often be a fruitful strategy in tackling the complexity of food materials. This certainly has helped physicists in understanding other kinds of soft matter.

Professor Datta commented: Again, assuming we understand the question properly, continuum models in engineering do not become qualitative if the materials and processes are complex. Foods are biopolymers – while they are quite complex, they are not the only complex ones and there are many other classes of complex materials where continuum models are routinely applied. The macroscale continuum models will exist whether or not the soft matter approach succeeds in providing food properties data (if no predictive approach works, the macroscopic continuum level properties data can always be measured or empirically predicted). However, if models have to rely entirely on measured data, their predictive power is not as strong, since the predictions are valid over the range of measured data only. Also, measurements being complex enough, if properties can be predicted, it can enable modeling when it is impossible otherwise. Again, our focus is not that chemical details are negligible but when they are, as it would be in a soft matter approach, they would complement nicely the macroscale continuum-based approach. As we go down to lower scale (still continuum approach), properties appropriate to that scale need to be available from predictions (or measurements).

Professor Kulozik addressed Dr van der Sman and Professor Datta: Although generally agreeable, the paper appears to paint a somewhat overpolarised picture. The general aim to develop generic approaches based on physics and thermodynamics of food systems is not in question. However, whether or not this can be achieved independently of chemical molecular details to a satisfying degree remains to be seen for many food systems. Chemical properties should go hand in hand with the soft matter concept, depending on the system under consideration, when certain properties are required to feed models with facts and when some unavoidable assumptions are required to be validated.

Dr van der Sman communicated in reply: With the paper, we wanted to make readers realize that much of the food properties at the mesoscale to macroscale show universal behaviour, which depends much more on coarse-grained molecular properties than the actual chemical properties. Amongst others, our recent findings with respect to moisture sorption and moisture diffusion caused us to make the above statement. The moisture sorption of many food biopolymers (neutral polysaccharides and structural proteins), sugars and polyols are shown to follow the (FVFH) free volume extension of the Flory–Huggins (FH) theory.[1,2,3] Here, each compound is characterized by only two parameters: the glass transition of the dry compound $T_{g,s}$ and the Flory–Huggins interaction parameter χ_1. For the polysaccharides and animal, native, proteins we find very similar parameter values, $T_{g,s} = 475$ K, and $\chi_1 \approx 0.8$. For maltodextrins, we find that both $T_{g,s}$ and χ_1 depend only on the molecular weight – which is an example of a coarse-grained molecular property. Recently, we have found that the same scaling rule for maltodextrins applies to the FH-interaction parameter of polyols (glycerol, xylitol, mannitol *etc.*). The number of carbon atoms in the chain appears to be the explaining factor. In the future, we intend to explore the PEGs of different molecular weights.

The FVFH theory assumes a composition dependent interaction parameter, which makes it an approximation of a more fundamental theory, which accounts for hydrogen bonding of (bio)polymers with water, and self-association of biopolymers (again *via* hydrogen bonding with different energy levels). Of course, here one includes more molecular details (different hydrogen bonds: water–water, water–polymer, and polymer–polymer, each with a different energy level). But still, the systems behaviour is still largely independent of the exact chemical details. We think that the moisture sorption of (hydrophilic) biopolymers is more or less determined by the topology and density of the hydrogen bonded network.

The hydrogen bonded network appears also to determine the glass transition. Recent studies of glass transitions of polyols has shown that it largely determined by the number of OH-groups per molecule.[4,5] (The OH-group in polyols act as donor/acceptor of the hydrogen bond). It is slightly moderated by the number of carbon atoms per molecule – and the number of intramolecular bonds.[6] Now here, these intramolecular bonds of course depend on the actual topology of the molecule (at which atom the OH groups is attached, and their orientation). But, as said, it is a second order effect. The first order effect is determined by the number of OH-groups per molecule. For maltodextrins we have found that $T_{g,s}$ is (inverse) dependent on the molecular weight. For these compounds the number of OH-groups is equal to the number of carbon atoms per molecule, and, therefore, linear with the molar weight. Hence, our finding is in line with the findings for polyols. Differences in $T_{g,s}$ amongst sugars like sucrose, lactose, trelahose and maltose can likely be accounted for by intramolecular hydrogen bonds.[7] It is said that the glass transition behaviour is determined by the topology (p.e. coordination number of the nodes in this network). The coordination number ($n = 2$) explains the equal T_g of water and ethyl glycol.[8]

Proteins are special because they unfold upon heating, by which they expose more hydrophobic parts to the solvent, which is accounted for by adjustment of the FH interaction parameter.[9,10,3] Unfolding is often due to the disappearance of intramolecular bonds of the protein. A brief investigation of the denaturation temperature of several proteins show again very similar behaviour, which can be described by the Flory theory of melting point depression of polymer solutions. These results will be reported in the future. Hydrogen bonding also plays a role in the molecular mobility of biopolymers, sugars, polyols, and water in these systems.[11,12] These compounds have been shown to exhibit universality in water dynamics (beta-relaxations).[13] This universality is also reflected in the water self-diffusion coefficient in malto-oligomers, sugars, and hydrophilic biopolymers. This is shown by Job Ubbink and his co-worker Hanjo Limbach,[14] and in our upcoming paper.[15] In dry food systems the Fickian diffusion coefficient is equal to the water self diffusion coefficient, and our finding is thus very important for modelling drying at the macro/mesoscale. Other material properties like density, heat capacity and thermal conductivity can be computed using composition data, where the thermal conductivity also requires microstructural information.[16,17] Food ingredients are classified into water (ice), proteins, carbohydrates, and fats. This is again a sign of some universality of food material properties. In our current research we are investigating the prediction of sorption isotherms of fruits and vegetables based on composition using the above FVFH theory. We have succeeded already in the prediction for carrots and mushrooms.

1 X. He, A. Fowler and M. Toner, Water activity and mobility in solutions of glycerol and small molecular weight sugars: Implication for cryo-and lyopreservation, *J. Appl. Phys.*, 2006, 100:074702.
2 R. G. M. van der Sman and M. B. J. Meinders, Prediction of the state diagram of starch water mixtures using the Flory–Huggins free volume theory, *Soft Matter*, 2011, **7**(2), 429–442.
3 R. G. M. van der Sman, Thermodynamics of meat proteins, *Food Hydrocolloids*, 2012, **27**(2), 529–535.
4 L. Carpentier, M. Paluch and S. Pawlus, Dielectric studies of the mobility in pentitols, *J. Phys. Chem. B*, 2011.
5 M. Nakanishi and R. Nozaki, Systematic study of the glass transition in polyhydric alcohols, *Phys. Rev. E*, 2011, **83**(5), 051503.
6 S. Pawlus, A. Grzybowski, M. Paluch and P. Wlodarczyk, Role of hydrogen bonds and molecular structure in relaxation dynamics of pentiol isomers, *Phys. Rev. E*, **85**(5), 2012, 052501.
7 N. C. Ekdawi-Sever, P. B. Conrad and J. J. de Pablo, Molecular simulation of sucrose solutions near the glass transition temperature, *J. Phys. Chem. A*, 2001, **105**(4), 734–742.
8 M. Nakanishi and R. Nozaki, Model of the cooperative rearranging region for polyhydric alcohols, *Phys. Rev. E*, 2011, **84**(1), 011503.
9 R. G. M. van der Sman, Soft condensed matter perspective on moisture transport in cooking meat, *AIChE J.*, 2007, **53**(11), 2986–2995.

10 R. G. M. van der Sman, Moisture transport during cooking of meat: An analysis based on Flory–rehner theory, *Meat Sci.*, 2007, **76**(4), 730–738.

11 M. Naoki and S. Katahira, Contribution of hydrogen bonds to apparent molecular mobility in supercooled d-sorbitol and some polyols, *J. Phys. Chem.*, 1991, **95**(1), 431–437.

12 T. R. Noel, S. G. Ring and M. A. Whittam, Dielectric relaxations of small carbohydrate molecules in the liquid and glassy states, *J. Phys. Chem.*, 1992, **96**(13), 5662–5667.

13 S. Cerveny, A. Alegra and J. Colmenero, Universal features of water dynamics in solutions of hydrophilic polymers, biopolymers and small glass-forming materials, *Phys. Rev. E*, 2008, **77**(3), 031803.

14 H.J. Limbach and J. Ubbink, Structure and dynamics of maltooligomer–water solutions and glasses, *Soft Matter*, 2008, **4**(9), 1887–1898.

15 R. G. M. van der Sman and M. B. J. Meinders , Moisture diffusivity in food materials, *Food Chem.*, 2012, *accepted*.

16 R. G. M. van der Sman and E. Boer, Predicting the initial freezing point and water activity of meat products from composition data, *J. Food Eng.*, 2005, **66**(4), 469–475.

17 R. G. M. van der Sman, Prediction of enthalpy and thermal conductivity of frozen meat and fish products from composition data, *J. Food Eng.*, 2008, **84**(3), 400–412.

Professor Datta communicated in reply: I am unclear of what exactly is meant here. We do not do soft matter – we are not making the claim that chemical details can be ignored for all property predictions. Primarily, what we are saying is that if the soft matter approach provides the needed continuum properties, it will make it so much easier for food process simulations to become a part of the product, process and equipment design in addition to providing a quantum leap to understanding of food products and processes. This is because the rest of the continuum framework has been built (and it works), it is just that the properties remain one of the major stumbling blocks. Whether or not soft matter succeeds well for food in predicting many different properties remains to be seen but (1) the soft matter approach has already been proven to work for several properties (van der Sman group's work and those mentioned in the article); and (2) there are no alternatives that have been successful for the purpose of predicting properties (other than completely empirical ones).

Dr Bot opened the discussion of the paper by Dr Milano by communicating: Your paper addresses a single crystallising triglyceride model system. The size of the resulting crystals is of the order of the crystalline nanoplatelets.[1] In industrial application, it is always mixtures of triglycerides that are considered. These mixtures contain triglycerides with a rather wide range of physical properties, the most obvious being that some tend to remain in the liquid phase, whereas others end up in a crystalline phase. Would your simulations be able to handle such many-component mixtures? Would larger dimensions of the similation box also be in reach?

1 N.C. Acevedo, F. Peyronel, A.G. Marangoni, *Curr. Opin. Colloid Interface Sci.*, 2011, **16**, 374–383.

Dr Milano replied: Yes, the models can be straightforwardly extended to mixtures of triglycerides. I agree that this can be a natural extension of the work reported in our paper. About the possibility of larger dimensions of the simulation box, the answer is yes. The systems described in our paper (CG1 and CG2 having 2592 and 20 736 particles, respectively) are well below the size limit dictated by a reasonable computer power.

Dr Bot asked: The main result from your simulation is a triglyceride crystal structure. Would your simulation techniques allow the investigation of (re-)crystallisation phenomena? Or are the time scales involved much too long to simulate?

Dr Milano answered: The timescales reachable with this kind of model are upper limited between 0.1 and 1 ms, depending on the size of the simulated systems. Of course this limit can be pushed up using massive parallel hardware. The models described in our paper obtained by a 4 to 1 mapping scheme (every bead corresponds to four atoms) are still very close to atomistic models. The use of coarser models could lead to the observation of phenomena on larger time-scales.

Dr Bot continued the discussion: At the end of the Introduction of your paper, you asked what type of results would be of interest to a wider audience. Recently, Flöter wrote a nice overview concerning physical properties of fats and oils with industrial relevance.[1] His list includes solid fat content as a function of temperature, recrystallisation kinetics ('transition times'), miscibility of solid components, and interaction between emulsifiers and fat crystals. I think this list would be a good starting point.

1 E. Flöter, *Eur. J. Lipid Sci. Technol.*, 2009, **111**, 219-226.

Dr van der Sman addressed Dr Milano and Professor Pink: I work usually on the mesoscale. For that scale I particularly find the phase field method very attrac-tive.[1] It is a continuous method, which takes free energy functionals to derive the driving forces for physical transport phenomena and phase transitions. Furthermore, it is able to describe interfaces between phases in the same continuum framework using a squared gradient term in the free energy – much like in the spirit of van der Waals.[2,3] For solidification of crystals in alloys it has become the simulation method of choice.[1] I consider there is also much poten-tial for modelling fat crystallization with the phase field model. Of course, there is some lack of knowledge concerning model parameters for fat/oil mixtures. Espe-cially, the phase field method requires the interfacial tension (or surface tension) and the wettability (contact angle) of the fat crystal.[4] Do you have some idea how to extract this information from (coarse-grained) MD simulations? If you do the phase field method very properly, one also needs the free energy of mixing between molten fat and the oil phase. Could this be described by a Flory–Huggins type of expression? For water-based polymer systems this has been implemented earlier.[5]

1 A. A. Wheeler, W. J. Boettinger and G. B. McFadden, Phase-field model for isothermal phase transitions in binary alloys, *Phys. Rev. A*, 1992, **45**(10), 7424.
2 R. G. M. van der Sman and S. van der Graaf, Emulsion droplet deformation and breakup with lattice Boltzmann model, *Comput. Phys. Commun.*, 2008, **178**(7), 492–504.
3 R.G.M. van der Sman and S. van der Graaf, Diffuse interface model of surfactant adsorption onto flat and droplet interfaces, *Rheol. acta*, 2006, **46**(1), 3–11.
4 S. van der Graaf, T. Nisisako, C. Schroën, R. G. M. van der Sman, and R. M. Boom, Lattice Boltzmann simulations of droplet formation in a T-shaped microchannel, *Langmuir*, 2006, **22**(9), 4144–4152.
5 E.B. Nauman, and D.Q. He, Nonlinear diffusion and phase separation, *Chem. Eng. Sci.*, 2001, **56**(6), 1999–2018.

Dr Milano replied: Recently, I proposed, in the framework of molecular dynamics simulations techniques, a hybrid methodology combining particle and field models.[1] This hybrid strategy has been used to develop specific models of phospholipids.[2,3] For these models, the field parameters such as Flory Huggins chi parameters and field compressibility have been parametrized in order to reproduce structural properties of lipid bilayers. A similar approach could be extended to fat crystals. Probably, in order to describe the crystallization process it would be better to extend this approach from scalar to vector density fields.

1 G. Milano and T. J. Kawakatsu, *Chem. Phys.*, 2009, **130**, 214106.
2 A. De Nicola, Y. Zhao, T. Kawakatsu, D. Roccatano and G. J. Milano, *Chem. Theory Com-put.*, 2011, **7**, 2947.

3 A. De Nicola, Y. Zhao, T. Kawakatsu, D. Roccatano and G. J. Milano, *Theor. Chem. Acc.*, 2012, **131**, 1167.

Professor Marangoni commented: The simulated X-ray spectra for your tridecanoin agree very well with published data in terms of the d-spacing corresponding to the 001 plane reflection. However, you mention in the paper that your lattice constant for the short spacings is 0.43 nm, and you mention it agreed very well with published data by Willie and Lutton (1966). The original work on triacylglycerol polymorphism was that of Clarkson and Malkin (1934).[1] This reference should be used as the reference. Secondly, one reflection at 0.43 nm is not indicative of the beta polymorph! A strong reflection at 0.46 nm is. Obviously, your simulation does not give the correct chain packing conformation. What could be the cause for this?

1 Clarkson and Malkin, *J. Chem. Soc.*, 1934, 666-671.

Dr Milano responded: The value calculated for the short spacing of 0.43 nm overestimates slightly (0.03 nm) the experimental value. Usually, such a deviation obtained from simulations can be considered small. Probably, a more relevant difference is that the peak is broader in comparison with the usual behaviour found in the experiments. Probably, this behaviour is related to residual disorder in the reverse-mapped atomic structure. A smoother cooling in the CG simulations and a longer relaxation of the structure at atomistic level (the atomistic simulation for the short relaxation has been performed for 2 ns), probably would lead to a more ordered structure and a different value for the calculated short spacing.

Professor Bolhuis queried: Fig. 4, 5 and 10 in the paper show the time evolution of a coarse grained model. It is known that while the coarse grained model might represent structural features accurately, the dynamics is often not reproduced well. Can you comment on this issue?

Related to that, your coarse-grained potentials were obtained by Boltzmann inversion for a single density (or pressure) and temperature. How transferable are these potentials to other conditions?

Dr Milano replied: The transferability of the coarse-grained potentials should be always considered and tested carefully. From this point of view the behaviour of the CG model for triglycerides we developed is interesting. Using the same reference atomistic simulations, coarse-grained beads can be defined through the same 4 to 1 mapping but with a different multiplicity of bead types. In particular, three models having different bead types 2, 3 and 4 can be parameterised. All the models predict with sufficient accuracy the experimental density values in a wide range of temperatures, showing also similar results in terms of diffusivity and shear viscosity. The most significant difference concerns the possibility of predicting liquid–solid phase transitions: the 4T model has been the only one able to reproduce a phase transition from liquid to crystalline solid at temperature values close to the experimental one. As far as I know, this is the first case in which the behaviour of a coarse grained model has been tested as function of the number of bead types. Usually, the number of atoms grouped in one bead is expected to have influence the behaviour of the model. The possibility of having the same kind of mapping but just different bead types is interesting because the improvement of the model behaviour has no additional computational costs.

Professor Marangoni stated: You mentioned that the alpha form (hexagonal fatty acyl chain packing) in TAGs is not attainable by tridecanoin. This is not true. One can attain this polymorphic form by rapidly cooling the TAG. It is metastable, so it will not survive for long, but it is attainable. This brings up the question of what temperature was used for the simulation? At low degrees of undercooling, you are correct, the alpha form is not attainable, but at high degrees of undercooling, it is.

Dr Milano communicated in reply: Thanks for your suggestion. The reason of the misunderstanding is that I have been not able to find any resolved crystal structure of the alpha form of tridecanoin in the literature. Is it possible to obtain this crystal structure? Do you know of any crystallographic study about the α form of tridecanoin?

Professor Marangoni responded: Unfortunately there are absolutely no reports on the crystallographic structure of the alpha form or triglycerides. It definitely needs to be done.

Professor Marangoni asked: Can you estimate the size of the unit cell (length of crystallographic axes and crystallographic angles) from your simulations?

Dr Milano replied: Yes from simple geometrical considerations the simulation box lengths in x, y and z directions can be translated into unit cell lengths and angles.

Professor Marangoni commented: Your model would be particularly useful for the study of phase behaviour between specific TAGs and when trying to elucidate the molecular consequences of eutectic, monotectic, compound formation behaviour. It, however, would not be too useful, in its present state, to say anything about the mesoscale structure of TAGs. Many functional properties of fat crystal networks are a consequence of this mesoscale structure. Your model does not reach this length scale yet. Any thoughts of how you will go from molecules to crystal networks?

Dr Milano replied: The systems described in our paper are well below the size limit treatable using a reasonable amount of computer power. Just to give an idea about this, in the paper we reported the length in terms of computer time of our simulations. For example, 1 ns of a parallel simulation (on 24 processors) of the largest system reported in the paper (CG2) takes 24 hours. Then, the timescales reachable with this kind of model are between 0.1 and 1 ms depending on the size of the simulated systems. I believe that these models can reach the bottom of the mesoscale structure of TAGs. Answers related to phenomena occurring on the mesoscale, such as crystal networks, can be obtained extending the multiscale approach to coarser models. These coarser models can be still particle based, *i.e.* having beads representing more than four atoms or based on continuum models such as finite elements (FEM) or based on Self Consistent Field Theory (SCFT). Recently, we have proposed, in the context of Molecular Dynamics Simulations, a hybrid methodology combining particle and field models.[1] In this framework we developed also a formulation for the calculation of stress tensor suitable for these hybrid models.[3] Finally, this hybrid strategy has been used by us very recently to develop specific models of phospholipids.[3,4] This machinery can allow the connection, in a multiscale fashion, of TAGs models going from lower scales (atomistic, 4 to 1 mapped coarse-grained) to higher scales (hybrid particle-field, FEM). I believe that this strategy could be successful in the treatment of a typical multiscale problem such as TAGs behaviour.

1 G. Milano, T. Kawakatsu, *J. Chem. Phys.*, 2009, **130**, 214106.
2 G. Milano, T. Kawakatsu, *J. Chem. Phys.*, 2010, **133**, 21.
3 A. De Nicola, Y. Zhao, T. Kawakatsu, D. Roccatano and G. J. Milano, *Chem. Theory Comput.*, 2011, **7**, 2947.
4 A. De Nicola, Y. Zhao, T. Kawakatsu, D. Roccatano and G. J. Milano, *Theor. Chem. Acc.*, 2012, **131**, 1167.

Dr van der Sman continued the discussion of the paper by Professor Bolhuis by commenting: In dough it is known that disulfide-bridges are very important for the formation of the gluten network as occurs during kneading. Perhaps this plays a role during bread kneading.[1] In sourdough it is explicitly said that disulfide–thiol interchange plays a role in the structure formation.[2] This is mediated *via* enzymes and/or compounds with free thiol-groups (glutathione) produced by lactobacilli, which are added to the sourdough. Would there be other systems where it is known that disulfide–thiol interchange is important/applicable for structure formation? The importance of this interchange can be tested by the use of NEM, which blocks initially the free thiol groups.[3] Could this be used as a general mechanism for structure formation in proteins systems, as discussed in ref. 4? Does it play a role in soy?

1 I.K. Jones and P. R. Carnegie, Binding of oxidised glutathione to dough proteins and a new explanation, involving thiol disulfide exchange, of the physical properties of dough, *J. Sci. Food Agric.*, 1971, **22**(7), 358–364.
2 N. Vermeulen, J. Kretzer, H. Machalitza, R.F. Vogel and M.G. Gänzle, Influence of redox-reactions catalysed by homo-and hetero-fermentative lactobacilli on gluten in wheat sourdoughs, *J. Cereal Sci.*, 2006, **43**(2), 137–143.
3 K. J. R. Vallons, L.A.M. Ryan, P. Koehler and E. K. Arendt, High pressure–treated sorghum flour as a functional ingredient in the production of sorghum bread, *Eur. Food Res. Technol.*, 2010, **231**(5), 711–717.
4 A. Totosaus, J. G. Montejano, J. A. Salazar and I. Guerrero, A review of physical and chemical protein-gel induction, *Int. J. Food Sci. Technol.*, 2002, **37**(6), 589–601.

Dr van der Sman additionally asked: During the kneading of dough a gluten network is formed. It is said that the formation of intermolecular di-sulfide bridges are important in that.[1] Also, it is known that dough can be overkneaded. Is it possible that these intermolecular bonds are broken again during overkneading, and that disulfide-thiol interchange plays a significant role – perhaps enhanced by the shearing pulling on the intermolecular bonds? Due to the reversiblity of the disulfide–thiol interchange I thought this might be the mechanism why disulfide bonds between gluten molecules are continuously broken and made again. At the start of the kneading all disulfide bonds are intramolecular, and the overall action of the disulfide–thiol interchange would be the formation of intermolecular bonds, giving rise to the gluten network. After some time, all types of bonds (intramolecular *vs.* intermolecular) are approximately equally probable – where the shearing field might tip the balance towards breakage of the gluten network (intermolecular bonds).

1 A. Redl, S. Guilbert and M.H. Morel, Heat and shear mediated polymerisation of plasticized wheat gluten protein upon mixing, *J. Cereal Sci.*, 2003, **38**(1), 105–114.

Dr Bot replied: In the qualitative model introduced for dough involving stretching gluten filaments in a starch paste,[1] see also discussion of the paper presented by Professor Tanaka, overkneading happens if the filaments are streched so much that they break (and partly relax). So although disulfide bond exchange dynamics will be involved, and the observation on changes in disulfide bonding is not disputed, the relevant scale is actually bigger than the molecular scale. This seems to make more sense from a physicochemical point of view, and has the advantage that it removes the ambiguity that changes in the disulfide bonding pattern are considered positive during early stages of kneading whereas the same changes have an adverse effect late in the process.

1 A. Bot and D.W. de Bruijne, *Cereal Chem.*, 2003, **80**, 404–408 (2003).

Dr van der Sman added: It is said that gluten also contains Glutathione (having free SH groups), which compete with disulfide bridge formation between gluten/

gliadins[1,2] (via thiol/disulfide exchange), and are named as a cause for breaking the protein network.

Oxidation of free thiol-groups also plays a part,[2] which can be done also by enzymes.[3]

1 Lars Fischer, The chemistry of dough, *3rd Euchems Chem. Congress*, 2010.
2 Factors affecting dough breakdown during overmixing, K. Okada, Y. Negishi and S. Nagao, *Cereal Chem.*, 1987, **64**.
3 I. J. Joye, B. Lagrain and J.A. Delcour, Endogenous redox agents and enzymes that affect protein network formation during breadmaking – a review, *J. Cereal Sci.*, 2009, **50**(1), 1–10.

Dr van der Sman continued the discussion of the paper by Professor Datta: My question is with regards to the link between molecular scale and mesoscale, as detailed in the following.

The papers presented by Professor Pink, Professor Bolhuis and Dr Milano all deal with simulations at the molecular scale. I and my coworkers have been performing simulations of food (de)structuring processes using mesoscale simulations.[1,2,3] Thereby, we have employed Lattice Boltzmann (LB), and regularly the dispersed phases have been fully resolved on the computational grid. For suspensions, the particles have been described in a Lagrangian manner, and emulsions have been described in an Eulerian manner, using the phase field method. For a considerable time, I have been thinking about how to couple the results of simulations at the molecular scale to mesoscale simulations.[4,5,6] Here, I want restrict the discussion to so-called serial coupling, meaning that the molecular scale and meso-scale are coupled via "effective parameters", like a coarse-grained chemical potential, or friction (diffusion) coefficients. Coarse-grained force fields between colloids are easily incorporated in the LB simulations of particle suspensions. The coupling to the phase field is not that straightforward. In principle the LB phase field method can take any formulation of a free energy functional (or rather the derived chemical potentials and osmotic pressure), which seems a natural way of linking it to the molecular level. The phase field has proven to be very versatile, and can model polymer blends and solutions, (surfactant stabilized) emulsions, alloys with crystalline phases, liquid crystals *etc.*[7,8,2] Many of the models investigated by Professor Tanaka fall within the class of phase field models. Hence, there is a huge potential of simulating food systems with this method – for example fat crystal/oil mixtures as studied by Professor Pink. In MD simulations it is common practice to consider coarse-graining interactions *via* the potential-of-mean-force (pmf). Can this be linked to the chemical potential at the mesoscale? A complicating factor is the existence of a squared gradient term in the free energy functional of the phase field – representing the interfacial free energy. How can the surface free energy be extracted from MD or Monte-Carlo simulations? In the discussion, Dr Milano has stated that linking MD simulations to continuous methods, like self-consistent fields or density field theory, is a better way to approach the mesoscale. But, even here, I think the hurdle of translating the simulations results to the phase field – where the density fields are much smoother, and exhibit a diffuse interface between phases (due to the squared gradient terms). Are there hybrid schemes, where the solvent is represented as a continuum, and macromolecules are still represented as particles? Would these methods be helpful in the linking of molecular level to the mesoscale level? I have only touched upon the issue of thermodynamic potentials, but coarse-graining dynamics is also an important issue. By coarse-graining the molecular architecture, the friction between (macro)molecules and with solvent changes too, and a rescaling of time is required. Is this always a simple rescaling of time for all friction coefficients, or would this rescaling be different for friction between macromolecules, and for friction between macro-molecules and solvent?

1 J. Kromkamp, D. T. M Van Den Ende, D. Kandhai, R. G. M. van der Sman and R. M. Boom, Shear-induced self-diffusion and microstructure in non-Brownian suspensions at non-zero Reynolds numbers, *J. Fluid Mech.*, 2005, **529**, 253-278.
2 R. G. M. van der Sman and S. van der Graaf, Diffuse interface model of surfactant adsorption onto flat and droplet interfaces, *Rheol. Acta*, 2006, **46**(1), 3–11.
3 H. M. Vollebregt, R. G. M. van der Sman and R. M. Boom, Suspension flow modelling in particle migration and microfiltration, *Soft Matter*, 2010, **6**(24), 6052–6064.
4 R. G. M. van der Sman and A. J. van der Goot. The science of food structuring, *Soft Matter*, 2009, **5**(3), 501–510.
5 R. G. M. van der Sman, Simulations of conned suspension ow at multiple length scales, *Soft Matter*, 2009, **5**(22), 4376–4387.
6 R. G. M. van der Sman, Soft matter approaches to food structuring, *Adv. Colloid Int. Sci.*, 2012, **176**(177), 18–30.
7 V. E. Badalassi, H. D. Ceniceros and S. Banerjee, Computation of multi-phase systems with phase field models. *J. Comput. Phys.*, 2003, **190**(2), 371–397.
8 J. J. Feng, C. Liu, J. Shen, and P. Yue, An energetic variational formulation with phase field methods for interfacial dynamics of complex fluids: advantages and challenges, *Modeling of soft matter*, 2005, 1–26.

Dr Milano replied: I agree about the point that density fields are much smoother and exhibit diffuse interfaces. This behaviour can be ascribed mainly to the absence of short range pair interactions that dominate the excluded volume interactions in particle models. An extended discussion of this issue is present in the paper about phospholipids I mentioned previously, during the discussion about a possible connection between phase field models and coarse-grained particle based models.[1] As for the dynamics, I agree that a rescaling time is required. In principle, this rescaling can be different for friction between macromolecules and between macromolecules and solvent. A rigorous mapping of the dynamics is limited to simple model systems, while the complication in soft matter systems is the multitude of fluctuating energy barriers of comparable height. A recent and satisfactory discussion of this issue in the field of coarse-grained models has been done by Friz *et al.*[2]

1 A. De Nicola, Y. Zhao, T. Kawakatsu, D. Roccatano and G. J. Milano, *Chem. Theory Comput.*, 2011, **7**, 2947.
2 D. Fritz, K. Koschke, V. A. Harmandaris, N. F. A. van der Vegt and K. Kremer, Multiscale modeling of soft matter: scaling of dynamics, *Phys. Chem. Chem. Phys.*, 2011, **13**, 10412–10420.

Concluding remarks: the future of soft matter and food structure

C. G. (Kees) de Kruif[ab]

Received 2nd August 2012, Accepted 7th August 2012
DOI: 10.1039/c2fd20122d

Introduction

I was presented with the most honourable task to deliver the closing remarks of this Faraday Discussion. Of course I will be glad to do so and I thank the organizers for this privilege. I very much enjoyed reading all 23 papers and listening to the presentations and the lively and high quality scientific discussions. What I aim to do is to put this first Faraday Discussion on Soft Matter and Food Structures in a somewhat historical perspective in order to underline the relevance of *soft matter* and *food structure*. Let me start by defining *soft matter*.

Soft matter

Soft matter, la matière molle, Weiche Materie is considered to be a subfield of condensed matter physics comprising a variety of physical states that are easily deformed by thermal fluctuations and particularly by (external) stresses. Stresses may have a mechanical, electrical, magnetic, thermal or osmotic origin. Soft materials include liquids, colloids, polymers, foams, gels, granular materials. In addition, many biological and most, if not all, food systems fall within the definition. Simply put; the stress–deformation relation has a large proportionality factor. Note that no length scale is in this definition.

Length scales and stability

In the broad and extensive overview entitled 'Soft matter approaches to food structuring' by van der Sman, reference is made to the dispersed phase and length scales.[1] Ubbink *et al.* prefer softness as the defining property.[2]

In the review paper of van der Sman and van der Goot[3] in Soft Matter they say: "All soft matter is in principle thermodynamically unstable, and needs to be stabilized." with a reference to a review paper of Mezzenga in Food Hydrocolloids from 2007.[4] Well Mezzenga *et al.* argue that most processed foods are inherently unstable but do not generalize it to all foods and soft matter structures. I agree that most foods are in a dynamically arrested state as van der Sman points out.[3] But for instance the casein micelle in milk is thermodynamically stable.[5]

Literature reviews

There is a huge overlap with what are called nano-materials and/or colloids, but usually then reference is made to a length scale between 1 and 1000 nm. In recent years several papers appeared emphasizing the soft matter character of foods. For instance; Donald makes a point in saying that food and food science are not fashionable because everybody is familiar with food, and therefore familiarity breeds

[a]Van't Hoff Laboratory for Physical and Colloid Chemistry, Debye Institute, Utrecht University, Padualaan 8, 3584 CH Utrecht, the Netherlands
[b]NIZO Food Research, P.O. Box 20, 6710 BA Ede, The Netherlands

contempt.[6] Nevertheless designing and making microstructures is a pure material science problem not different from nano-structures or self-assembled systems. Publications in respected journals such as Soft Matter and Nature Materials[31] mentioned above will boost the image of food stuff structuring. I will come to this point later.

Food as soft material and soft matter modeling

Food systems can be considered to fall within the definition of both colloids and soft matter. I would not know a food that has no mesoscopic length scales in the colloidal region. Pure water would fall outside the definition but cannot be considered as a food although it is essential, just as oxygen for many organisms. This Faraday Discussion meeting focuses on the application of soft matter concepts, ideas, theories and computer simulations to food systems. I think that food science can and will benefit hugely from applying models from soft matter physics to food systems. So I agree wholeheartedly with Donald, Van der Sman, Mezzenga, Ubbink, Schurtenberger to name but a few. Interestingly they all come with a strong physics/engineering background.

In his Introductory lecture, Ubbink asked the question; can soft matter physics be used to improve food structures. He also said that 90% of the products in the food market are either naturally made or man-made without any explicit soft matter physics. Nevertheless I would answer Ubbinks question with: most certainly! and I will explain why I think this is the case.

The field of food science

Now let me point to an important driver for establishing a science field. The development of an adequate and rigorous theoretical framework is a catalyst for the development of a field of science. For instance the formulation of the three laws of thermodynamics has led to the beautiful and rigorous framework of chemical thermodynamics; almost forgotten nowadays but still applicable in all its strength. Without the theory on general relativity of Einstein we would not have a functioning global positioning system. The development, during the 1940s, of the Derjaguin and Landau, Verwey and Overbeek (DLVO) theory has made colloid science a recognized, respected and established scientific field and has been of tremendous importance for industry in producing *e.g.* polymers, paints, inks, cosmetics, pharmaceuticals, displays and much more.

The challenge to prove or disprove theoretical concepts and the subsequent application and implementation of these concepts drives the development of the science field. Whether or not a theory applies in all of its aspects to a practical system is not so important. What counts is the conceptual thinking, inspiration, ideas and challenges.

A further important development for colloid science came when it was recognized by Agienus Vrij[7,8] from Utrecht that the theoretical framework of statistical mechanics and in particular liquid state theory, developed by physicists for atomistic systems could be applied to super atoms *i.e.* the colloids as well, just as Perrin[9] did when studying Brownian motion of latex spheres in a continuum, already in 1914. Thus colloidal systems can be studied by considering them as supra molecules with an appropriate interaction potential. The solvent is considered as continuous background that plays a role in the transport and dynamic properties. This notion caught the attention of many physicists who then greatly aided in developing the (theoretical and practical) field of colloid science, which is now called soft matter physics. Among them were Hayter,[10] Ackerson and Clark,[11] and van Megen and Snook[12] in the 1980's (see Faraday Discussions, 1983). Parallel with that there was a great development in all kinds of computer simulations of colloidal systems which were appropriately called "designer atoms" *e.g.* by Frenkel.[13] The Utrecht group was among the first to exploit the supramolecular fluid ideas.[14]

These ideas actually initiated the development of the sterically stabilized silica colloids which in a good solvent behaved as perfect hard spheres, including crystallization, while in a poor solvent, an attraction could lead to a gas–liquid transition or a jammed or arrested system.

As the range of the interaction potential was (very) small compared to the size of the colloidal particles the application of the hard sphere model (and variations thereof) was justified much better than for atomic systems. Working at the van't Hoff laboratory in Utrecht during the 1980's, I showed that the structure of (adhesive) hard sphere colloids could be described using liquid state theory, while the rheology could be compared to the equations of Einstein and Batchelor describing hard sphere rheology.[15] These systems were amenable to both structural and kinetic experiments using small angle scattering[16,17] and microscopic techniques.

My first publications on food systems came in the early 1990's when I moved to NIZO food research. When I learned of renneting and acidification of milk, i.e. cheese and yoghurt making, it seemed to me very similar to the flocculation of colloidal particles with an adhesive hard sphere potential.[18,19] Although I never said that casein micelles *were* adhesive hard spheres it seems difficult to appreciate this conceptual approach by food scientists in order to describe the *initial* phases of milk coagulation. Casein micelles are association colloids and therefore changing conditions will change their integrity, often on a different (much longer) time scale than the initial flocculation. The difference in time scale allows a decoupling of the phenomena. See also the discussion of the paper by Gebhardt *et al.* in this Faraday Discussion.

Phase separation in food science

When I go to food science conferences I almost see a phase separation in the audience and papers at these meetings. On the one hand I see a very empirical approach and experiments on a host of different systems. Also in this conference one can see a watershed of papers with theoretical sections and calculations and papers which are predominantly descriptive in nature. A rough estimate would give a 50/50 distribution, which is really good, I think. A good example is, for instance, β-lactoglobulin, studied by almost any food science department in the world. It varies from heat, beat and eat or spray and pray, to sophisticated computer simulations as in the paper by Bolhuis *et al.*, in the paper by Milano on triglyceride crystallisation, or sophisticated SANS measurements on lysozyme as in the papers by Schurtenberger *et al.* and Krueger *et al.* in this Faraday Discussion.

On the other hand there is an increasing interest of theoretical chemists and physicists entering the field of food science. I think this is an excellent development the importance of which cannot be overestimated. I am convinced and there are several examples that future developments in food science will be based more and more on scientific insight and maybe serendipity. Take for instance the concepts of steric or polymeric stabilisation developed by Napper[20] but also here in Wageningen by Scheutjens and Fleer[21] and in Utrecht by van Helden and Vrij[22]

The concept of depletion flocculation dating back to the early 20th century[23] when polysaccharides were added to latex dispersions in order to concentrate them. Thus it was initially applied to colloids but now is of much more relevance for food systems.[24] Also because there is a sound theoretical framework: Asakura and Oosawa[25,26] and Vrij,[27] Lekkerkerker[28,29] and Fleer and Tuinier[30]. The application of computer simulation methods will contribute to further development and understanding of food systems. As is illustrated in this conference by several papers and posters. But I do recommend that the simulations are checked to represent reality as closely as possible.

Papers in Faraday Discussion 158

If I categorize the 23 papers in this Faraday Discussion using a course graining method I identify the following subjects.

4 papers on emulsions and in particular Pickering stabilisation
4 papers on structure of fat and protein gels
6 papers on phase separation and jamming
5 papers on computer simulations including jamming
8 papers address proteins

Thus papers on jamming and dynamical arrest are attracting the most attention and in particular when proteins are involved. Interestingly there is no paper on polysaccharides. Does it mean that there is a better connection between food science and colloids and less so with polymer physics? Food systems can be considered both as soft matter and colloidal systems and they often contain considerable amounts of polysaccharides. So I think we should involve polymer physicists as well! The paper of Tanaka in this Discussion on viscoelastic phase separation is a beautiful and important example of how polymer physics can and must be involved.

Recommendation

As said, I think that the contribution of (theoretical) physicists to the field is and will be enormous. In a major part of all the discussions words and concepts were used which come directly from liquid state physics, *e.g.* glass transition, percolation, attractive interaction, interaction potential, hard spheres, adhesive spheres, spinodal, structure factor *etc*. The Food Technologist will not be familiar with all the concepts, they usually have a much more applied hands on training. Until now the Food Technologists have done a great job because they developed the vast majority all food products one finds on the shelves of the supermarket. General De Gaulle once said "How can one reign a country with 250 varieties of cheese" (developed by food technologists and not by soft matter physics).

Having said this I think it is extremely important that the physicists translate their equations into practical advances and ideas and possibilities. In several (if not all) theoretical papers it says "this will be of great importance for further development…". Well this may be true, but it must be made more tangible. Not too long ago on my occasion of retirement from NIZO food research I gave a lecture called "the golden cheese" It defined the golden cheese as; it looks like Gouda it tastes like Gouda (it quacks like Gouda) but it has no fat. Reducing the fat content in traditional cheese making technology always ended up with a very rubbery product closely resembling shoe soles. As cheese can be considered as a concentrated polymer solution (see the work of Taco Nicolai) we looked at what was done in polymer physics and chemistry. Using their concepts we were able to make the "golden cheese". What I would suggest is a review paper on modern theoretical developments (no equations whatsoever) in relation to practical aspects and possibilities and opportunities; not in great detail but along the line of "if you want to achieve this, you should go in that direction using or varying the following conditions…"). Actually this Faraday Discussions could make a start with that if each contribution would translate their findings into very practical recommendations.

Maybe as part of this issue, I think that would be of value to this Faraday Discussions even more.

References

1 R. G. M. van der Sman, Soft matter approaches to food structuring, *Adv. Colloid Interface Sci.*, 2012, **176–177**, 18–30.
2 J. Ubbink, A. Burbidge and R. Mezzenga, Food structure and functionality: a soft matter perspective, *Soft Matter*, 2008, **4**, 1569.
3 R. G. M. van der Sman and A. J. van der Goot, The science of food structuring, *Soft Matter*, 2009, **5**, 501–510.
4 R. Mezzenga, *Food Hydrocolloids*, 2007, **21**, 674–682.

5 C. Holt, An equilibrium thermodynamic model of the sequestration of calcium phosphate by casein micelles and its application to the calculation of the partition of salts in milk, *Eur. Biophys. J.*, 2004, **33**, 421–434.

6 A. M. Donald, Food for thought, *Nat. Mater.*, 2004, **3**, 579–581; A. M. Donald, *Rep. Prog. Phys.*, 1994, **57**, 1081.

7 A. Vrij, *KNCV (Royal Dutch Chemical Society) Winter Meeting* (1976).

8 A. Vrij and H. M. Fijnaut, Colloidal fluids a new perspective?, *Yearbook, Slichting FOM*, 139 (1977).

9 J. Perrin, *Compt. Rend.*, 1914, **158**, 1168.

10 J. B. Hayter, *Faraday Discuss. Chem. Soc.*, 1983, **76**, 7.

11 B. J. Ackerson and N. A. Clark, *Phys. A*, 1983, **118**, 221.

12 W. van Megen and I. Snook, *Adv. Colloid Interface Sci.*, 1984, **21**, 119.

13 D. Frenkel, Playing Tricks with Designer "Atoms", *Science*, 2002, **296**, 65–66.

14 A. Vrij, E. A. Nicuwenhuis, H. M. Fijnaut and W. G. M. Agterof, *Faraday Discuss. Chem. Soc.*, 1978, **65**, 7.

15 C. G. de Kruif, E. M. F. van Iersel, A. Vrij and W. B. Russel, Hard sphere colloidal dispersions: viscosity as a function of shear rate and volume fraction., *J. Chem. Phys.*, 1985, **83**, 4717–4725.

16 C. G. de Kruif, W. J. Briels, R. P. May and A. Vrij, Hard-sphere colloidal silica dispersions. The structure factor determined with SANS, *Langmuir*, 1988, **4**, 668–676.

17 C. G. de Kruif, P. W. Rouw, W. J. Briels, M. H. G. Duits, A. Vrij and R. P. May, Adhesive hard-sphere colloidal dispersions. A small-angle neutron-scattering study of stickiness and the structure factor, *Langmuir*, 1989, **5**, 422–428.

18 C. G. de Kruif, T. J. M. Jeurnink and P. Zoon, The viscosity of milk during the initial stages of renneting, *Neth. Milk Dairy J*, 1992, **46**, 123–137.

19 C. G. de Kruif and E. B. Zhulina, κ-casein as a polyelectrolyte brush on the surface of casein micelles, *Colloids Surf., A*, 1996, **117**, 151–159.

20 D. H. Napper, *Polymer Stabilization of Colloidal Dispersions*, Academic Press, New York, 1983.

21 G. J. Fleer, J. M. H. M. Scheutjens and M. A. C. Stuart, Theoretical progress in polymer adsorption, steric stabilization and flocculation, *Colloids Surf.*, 1988, **31**, 1–29, http://dx.doi.org/10.1016/0166-6622(88)80178-1.

22 A. K. van Helden, J. W. Jansen and A. Vrij, *J. Colloid Interface Sci.*, 1981, **81**, 354.

23 J. Traube, *Gummi Z.*, 1925.

24 R. Tuinier, J. Rieger and C. G. de Kruif, Depletion-induced phase separation in colloid-polymer mixtures, *Adv. Colloid Interface Sci.*, 2003, **103**, 1–31.

25 S. Asakura and F. Oosawa, *J. Chem.Phys.*, 1954, **22**, 1255.

26 S. Asakura and F. Oosawa, *J. Polym. Sci.*, 1958, **33**, 183.

27 A. Vrij, *Pure Appl. Chem.*, 1976, **48**, 471.

28 H. N. W. Lekkerkerker, *ColloidsSurf.*, 1990, **51**, 419.

29 H. N. W. Lekkerkerker, W. C. K. Poon, P. N. Pusey, A. Stroobants and P. B. Warren, *Europhys. Lett.*, 1992, **20**, 559.

30 G. J. Fleer and R. Tuinier, Analytical phase diagrams for colloids and non-adsorbing polymer, *Advances in Colloid and Interface Science*, 4 November 2008, **143**(1–2), 1–47.

31 R. Mezzenga, P. Schurtenberger, A. Burbidge and M. Michel, Understanding foods as soft materials, *Nature Materials*, October 2005, **4**.

Poster titles

Phase diagrams: a tool to study the effect of concentration on wheat protein solutions, **A. Boire, P. Menut, P. Letting, M.-H. Morel and C. Sanchez** *UMR IATE – Montpellier SupAgro, France*

Effect of heat treatment on wheat dough rheology and wheat protein solubility, **J. Mann, B. Schiedt, A. Baumann, B. Conde-Petit, and T. Vilgis**, *Max-Planck Institute for Polymer Research*

Towards understanding gelation in complex model mixtures, **K. van Gruijthuijsen, P. Schurtenberger and A. Stradner**, *University of Fribourg, Switzerland.*

Micelle formation and gelation of amphiphilic multiblock copolymers in the bulk and at a surface: a Monte Carlo study, **V. Hugouvieux, M. A. V. Axelos and M. Kolb,** *Institut National de la Recherche Agronomique, France*

Change in water holding capacity (WHC) of mushroom during processing: an analysis based on Flory Rehner's approach, **E. Paudel, R. G. M van der Sman and R. Boom**, *Wageningen University, The Netherlands*

Thixotropic ethylcellulose oleogels, **T. Stortz and A. G. Marangoni**, *University of Guelph, Canada.*

β-Lactoglobulin nanofibrils by microwave heating, **C. Hettiarachchi, S. Loveday, J Gerrard and L. Melton**, *University of Auckland, New Zealand*

High resolution microscopy of casein micelles and casein monolayers: integration of complementary techniques, **M. Christensen, J. T. Rasmusen and A. C. Simonsen**, *University of Southern Denmark, Denmark*

Manipulation of casein interfacial activity using pH and ionic strength: effect on stability and rheology of oil in water concentrated emulsions, **M. Alayón and K. M. McGrath,** *University of Wellington, New Zealand*

Pore cell network modelling of gas exchange in fruit, **Q. T. Ho, P. Verboven, S. Fanta, M. K. Abera, M. A. Retta, E. Herremans, T. Defraeye and B. M. Nicolaï,** *Katholieke Universiteit Leuven, Belgium*

Microscale modelling of water transport in fruit tissue, **S. W. Fanta, M. K. Abera, Q. T. Ho, P. Verboven, B. M. Nicolaï and J. Carmeliet**, *Katholieke Universiteit Leuven, Belgium*

Understanding the rheology of concentrated carrot and tomato purées, **K. R. N. Moelants, R. Cardinaels, R. P. Jolie, T. A. J. Verrijssen, S. K. J. Palmers, S. Van Buggenhout, A. M. Van Loey, P. Moldenaers and M. E. Hendrickx**, *Katholieke Universiteit Leuven, Belgium*

Foaming emulsions and creaming under confinement, **Y. Yoshitake, F. Toquet, S. Heitkam, R. Lhermerout, E. Rio, A. Saint-Jalmes, D. Langevin and A. Salonen**, *Université Paris Sud, France*

Wetting dynamics on a model small intestine surface, **H. Mayama, S. Chida, T. Tanaka and Y. Nonomura**, *Hokkaido University, Japan*

Phase separation dynamics in colloid–polymer mixtures: the effect of interaction range, **I . Zhang, P. Barltett, C. P. Royall and M. Fears**, *University of Bristol, United Kingdom*

Molecular dynamics simulation of amphiphilic proteins, **D. L. Cheung**, *University of Warwick, United Kingdom*

Brownian dynamics simulation of segregation of gluten and starch in shear flow, **J. H. J. Van Opheusden**, *Wageningen University, the Netherlands*

Application of self-consistent field approach to rennetting of casein, stabilised by κ-casein, **R. Ettelaie, N. Khandelwal and R Wilkinson**, *University of Leeds, United Kingdom*

Manipulating interfacial rheology to rationally control emulsion viscoelasticity **P. Wilde, F. Husband, M. Ridout, N. Woodward, P. Gunning, R. Penfold and V. Morris**, *Institute of Food Research, United Kingdom*

Self assembly of natural and synthetic triblock copolymers: mixed micelles consisting of α_s-casein and PE6400, **A. Kessler, O. Menéndez-Aquirre, J. Hinrichs and J. Weiss**, *Universität Hohenheim, Germany*

The Skinner Prize for the best poster was awarded to Isla Zhang of the University of Bristol for her poster on phase separation dynamics in colloid–polymer mixtures.

The Soft Matter poster prize was awarded to Paul Menut of UMR IATE – Montpellier SupAgro for his poster on phase diagrams of protein wheat solutions.

List of participants

Mr M. Alayon Marichal, *Victoria University of Wellington, New Zealand*
Dr S. Bakalis, *University of Birmingham, United Kingdom*
Prof Dr P. Bolhuis, *Universiteit van Amsterdam, The Netherlands*
Professor R. Boom, *Wageningen University, The Netherlands*
Dr A. Bot, *Unilever R&D Vlaardingen, The Netherlands*
Dr R. Cardinaels, *Katholieke Universiteit Leuven, Belgium*
Prof M. Cates, *University of Edinburgh, United Kingdom*
Dr D. Cheung, *University of Warwick, United Kingdom*
Prof Dr M. Cohen Stuart, *Wageningen University, The Netherlands*
R. Cooper, *Royal Society of Chemistry, United Kingdom*
Professor A. Datta, *Cornell University, U.S.A.*
Professor K. De Kruif, *Utrecht University, The Netherlands*
Dr R. De Vries, *Wageningen University, The Netherlands*
Dr T. Defraeye, *Katholieke Universiteit Leuven, Belgium*
Mr S. Dhayal, *Wageningen University, The Netherlands*
Dr R. Ettelaie, *University of Leeds, United Kingdom*
Mr A. Finnemore, *University of Cambridge, United Kingdom*
Dr B. Frith, *Unilever Research Colworth, United Kingdom*
Professor P. Fryer, *University of Birmingham, United Kingdom*
Dr R. Gebhardt, *Technische Universität München, Germany*
Dr G. Gillies, *Fonterra Co-operative Group Limited, New Zealand*
Miss H. Gray, *Royal Society of Chemistry, United Kingdom*
Dr M. Hermes, *University of Edinburgh, United Kingdom*
Mr C. Hettiarachchi, *University of Auckland, New Zealand*
Dr S. Hill, *University Of Nottingham, United Kingdom*
Dr Q. Ho, *Katholieke Universiteit Leuven, Belgium*
Dr H. Hondoh, *University of Hiroshima, Japan*
Dr V. Hugouvieux, *Institut National de la Recherche Agronomique, France*
Associate Professor M. Hütter, *Eindhoven University of Technology, The Netherlands*
L. Jankowiak, *Wageningen University, The Netherlands*
Ms X. Jin, *Wageningen University, The Netherlands*
Dr E. John, *B.A.T. Industries Plc., United Kingdom*
Mrs A. Kessler, *University Hohenheim, Germany*
Mr M. Kolb, *Ecole Normale Supérieure de Lyon, France*
Dr S. Krueger, *National Institutes of Standards and Technology, U.S.A.*
Professor U. Kulozik, *Technische Universität München, Germany*
Professor A. Marangoni, *University of Guelph, Canada*
Ms S. Maurer, *Max Planck Institute for Polymer Research, Germany*
Dr H. Mayama, *Hokkaido University, Japan*
Dr P. Menut, *Montpellier Supagro, France*
Dr G. Milano, *University of Salerno, Italy*
Dr T. Nicolai, *University of Le Mans, France*
Professor I. Norton, *University of Birmingham, United Kingdom*
Mr C. Nwajagu, *University of Greenwich At Medway, United Kingdom*
Mr L. Oritte, *Wageningen University, The Netherlands*
E. Paudel, *Wageningen University, The Netherlands*
Miss Z. Peace, *Royal Society of Chemistry, United Kingdom*
Professor D. Pink, *St.Francis Xavier University, Canada*
Dr M. Rayner, *Lund University, Sweden*
Dr C. Royall, *University of Bristol, United Kingdom*

Ms M. Rutkowska, *Polish Academy of Sciences, Poland*
Dr A. Salonen, *Université Paris-Sud, France*
Dr B. Schiedt, *Max Planck Institute for Polymer Research, Germany*
Professor P. Schurtenberger, *University of Lund, Sweden*
Mr A. Siemens, *Leiden University, The Netherlands*
Professor A. Simonsen, *University of Southern Denmark, Denmark*
Dr P. Smith, *Cargill, Belgium*
Dr P. Spicer, *Procter and Gamble Co., U.S.A.*
Miss T. Stortz, *University of Guelph, Canada*
Prof Dr S. Stoyanov, *Unilever R&D, The Netherlands*
Prof Dr A. Stradner, *Lund University, Sweden*
Professor H. Tanaka, *University of Tokyo, Japan*
Mrs R. Thompson, *Royal Society of Chemistry, United Kingdom*
Dr V. Trappe, *University of Fribourg, Switzerland*
Dr J. Ubbink, *Nestle Research Centre, Switzerland*
Professor E. van der Linden, *Wageningen University, The Netherlands*
Assistant Professor R. van der Sman, *Wageningen University, The Netherlands*
Dr K. van Gruijthuijsen, *Adolphe Merkle Institute, University of Fribourg, Switzerland*
Dr J. van Opheusden, *Wageningen University, The Netherlands*
Dr K. Velikov, *Unilever R&D Vlaardingen, The Netherlands*
Dr P. Venema, *Wageningen University, The Netherlands*
Prof Dr T. Vilgis, *Max Planck Institute for Polymer Research, Germany*
ir. M. Vollebregt, *Wageningen University, The Netherlands*
Mr G. Waschatko, *Max Planck Institute for Polymer Research, Germany*
Prof Dr P. Wilde, *Institute of Food Research, United Kingdom*
Assistant Professor Hans Wyss, *Eindhoven University of Technology, The Netherlands*
Dr G. Zanchetta, *University of Fribourg, Switzerland*
Ms R. Zeiler, *University of Amsterdam, The Netherlands*
Ms Y. Zhang, *University of Bristol, United Kingdom*

Index of contributors*

Acevedo, N. C., **171**

Benyahia, L., **325**

Block, J. M., **171**

Bolhuis, P. G., 351, **461,** 493

Boom, R. M., **65, 89**

Bot, A., 105, **125, 223,** 239, 351, 493

Bouwman, W. G., **125, 223**

Brasiello, A., **479,**

Caggioni, M., **341**

Cardinaels, R., 105

Cates, M. E., 239, **313,** 351, 493

Crescitelli, S., **479**

Curtis, J. E., **285**

Danov, K. D., **195**

Datta, A. K., **435,** 493

De Kruif, C. G., 105, 39, 351, 493, **523**

De Vries, R., **51,** 105, 239

Defraeye, T., 105

Dejmek, P., **139**

den Adel, R., **125, 223**

Dhayal, S. K., **51**

Durand, D., **325**

Ettelaie, R., 105, 239, 351, 493

Faber, T. J., **407**

Flöter, E., **125, 223**

Frasch-Melnik, S., **37**

Frith, B., 351

Garamus, V. M., **223**

Garrec, D. A., **37**

Gebhardt, R., **77,** 105, 239

Gerkema, E., **65**

Gibaud, T., **267**

Gilbert, E. P., **223**

Golemanov, K, **195**

Gulati, T., **435**

Hanna, C. B., **425**

Hartel, R. W., **341**

Henry, J. V. L., **37**

Hettiarachchi, C., 239

Hütter, M., 105, **407,** 493

Jin, X., **65,** 105

Joshi, Y. M., **313**

Junghans, A., **157**

Kralchevsky, P. A., **195**

Kroes-Nijboer, A., **125**

Krueger, S., **285,** 351

Kulozik, U., **77,** 105, 239, 493

Lindner, L., **267**

MacDougall, C. J., **425**

Mahmoudi, N., **267**

Malins, A., **301**

Marangoni, A. G., 105, **171,** 239, **425,** 493

Mayama, H., **351**

McAuley, A., **285**

Menut, P., 105, 351

Meyer, P., **77**

Milano, G., **479,** 493

Nanda, H., **285**

Nicolai, T, 105, 239, **325,** 351, 493

Norton, I. T., **37,** 105, 239

Oberdisse, J., **267**

Oliveira, C. L. P., **267**

Papp-Szabo, E., **425**

Pawar, A. B., **341**

Pedersen, J. S., **267**

Perlich, J., **77**

Peyronel, F., **425**

Pink, D. A., **425,** 493

Radulova, G. M., **195**

Rayner, M., 105, 239

Razul, M. S., **425**

Royall, C. P., 105, **301,** 351, 493

Saricay, Y., **51**

Sawalha, H., **125, 223**

Schurtenberger, P., **267,** 351

Shahin, A., **313**

Sjöö, M., **139**

Smith, P., 239, 493

Spicer, P. T., **341,** 351

Spyropoulos, F., **37**

Steinhauer, T., **77**

Sterr, J., **77**

Stoyanov, S. D., 105, **195,** 239, 351

Stradner, A., **267**

Tanaka, H., 105, 351, **371,** 493

Thomar, P., **325**

Timgren, A., **139**

Trappe, V., 351

Ubbink, J., **9,** 105, 239, 493

Van As, H., **65**

van Boxtel, A. J. B., **65**

van der Linden, E., 105, **125, 223,** 239, 351

van der Sman, R. G. M., **65, 89,** 105, 239, 351, **435,** 493

van Gruijthuijsen, K., 105, 239, 351, 493

van Straten, G., **65**

Velikov, K., 239, 351

Venema, P., **125, 223**

Vergeldt, F. J., **65**
Vilgis, T. A., **157**
Vollebregt, H. M., **89**
Warning, A., **435**
Waschatko, G., **157,** 239

Wierenga, P. A., **51**
Wilde, P., 105, 239, 351
Wyss, H. M., 239, 351, **407**
Zanchetta, G., 105, 239
Zeiler, R. N. W., **461,** 493

*The page numbers in **bold** type indicate papers submitted for discussions.